OHM'S LAW COMBINED WITH JOULES LAW

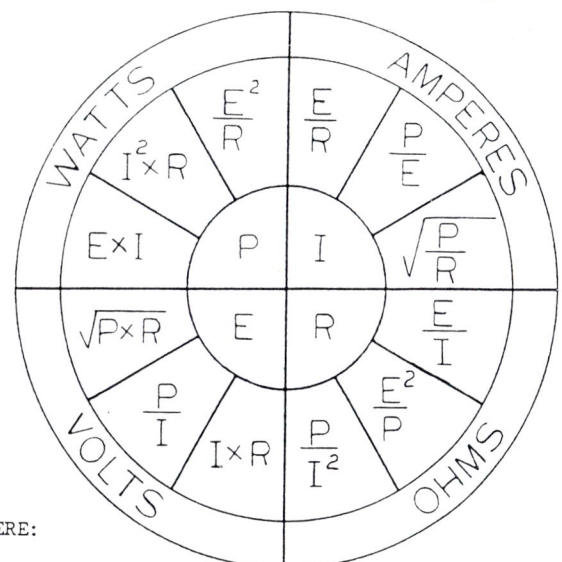

WHERE:

I = Current in Amperes
E = Potential in Volts
R = Resistance in Ohms
P = Power in Watts

RESISTANCE IN SERIES:

$$R_{total} = R_1 + R_2 + R_3 + \ldots$$

RESISTANCE IN PARALLEL:

$$\frac{1}{R_{total}} = \frac{1}{R_1} + \frac{1}{R_2} + \frac{1}{R_3} + \ldots$$

FOR TWO RESISTORS IN PARALLEL:

$$R = \frac{R_1 \times R_2}{R_1 + R_2}$$

POWER DISIPATED

$$P_1 = \frac{E_T^2}{R_1}, \text{ ect}$$

SECOND EDITION
DC/AC Fundamentals

Victor F. Veley, Ph.D., M.A. (Oxon.)
Dean and Professor of Electronics Technology
School of Science and High Technology
Los Angeles Trade-Technical College

John J. Dulin, Ph.D.
Professor Emeritus of Electronics Technology
Los Angeles Trade-Technical College

PRENTICE HALL, *Englewood Cliffs, New Jersey* 07632

Library of Congress Cataloging-in-Publication Data

VELEY, VICTOR F. C.
  DC/AC fundamentals / VICTOR F. VELEY, JOHN J. DULIN.—2nd ed.
    p.   cm.
  Rev. ed. of: Modern electronics. c1983.
  Includes index.
  ISBN 0-13-204587-7
  1. Electronics. I. Dulin, John J., (date). II. Veley, Victor
F. C. Modern electronics. III. Title.
TK7816.V42   1992
621.381—dc20
                                      91-19582

To Joyce and Margy

Design Director: JANET SCHMID
Text Designer: ANDREW ZUTIS
Cover Designer: ROSEMARIE PACCIONE
Prepress Buyer: MARY MCCARTNEY
Manufacturing Buyer: ED O'DOUGHERTY
Production Editor: PATRICK WALSH

 © 1992, 1983 by Prentice-Hall, Inc.
A Simon & Schuster Company
Englewood Cliffs, New Jersey 07632

First edition published under the title
MODERN ELECTRONICS: A First Course,
Veley & Dulin, Prentice Hall.

All rights reserved. No part of this book may be
reproduced, in any form or by any means,
without permission in writing from the publisher.

Printed in the United States of America
10 9 8 7 6 5 4 3 2 1

ISBN 0-13-204587-7

Prentice-Hall International (UK) Limited, *London*
Prentice-Hall of Australia Pty. Limited, *Sydney*
Prentice-Hall Canada Inc., *Toronto*
Prentice-Hall Hispanoamericana, S.A., *Mexico*
Prentice-Hall of India Private Limited, *New Delhi*
Prentice-Hall of Japan, Inc., *Tokyo*
Simon & Schuster Asia Pte. Ltd., *Singapore*
Editora Prentice-Hall do Brasil, Ltda., *Rio de Janeiro*

Contents

Preface   ix

Part 1 DC Principles

1   SI Units. Atomic Structure. Ohm's Law   1

- 1-0   Introduction   2
- 1-1   Mechanical SI Units. Scientific Notation. Prefixes   3
- 1-2   Atomic Structure. Conductors, Insulators, and Semiconductors   14
- 1-3   Units of Current and Charge   22
- 1-4   Unit of Electromotive Force   25
- 1-5   Ohm's Law. Resistance. Conductance   27
- 1-6   Resistance of a Cylindrical Conductor. Specific Resistance   35
- 1-7   Temperature Coefficient of Resistance   40
- 1-8   Practical Resistors. Color Code   44
  - Chapter Summary   53
  - Self-Review   56

2   Resistors In Series   60

- 2-0   Introduction   61
- 2-1   Current Flow in a Series Circuit   61
- 2-2   Individual Voltage Drops across the Resistors   64
- 2-3   Total Equivalent Resistance of a Series Circuit   67
- 2-4   Power Relationships in a Series Circuit   70
- 2-5   Step-by-Step Analysis of the Series Circuit   73
- 2-6   Voltage Division Rule   76
- 2-7   Voltage Divider Circuit   79
- 2-8   Ground. Voltage Reference Level   81
- 2-9   Potentiometer and Rheostat   85
- 2-10   Voltage Sources Connected in Series-Aiding and Series-Opposing   88
- 2-11   Effect of an Open Circuit   92

	2-12	Series Voltage-Dropping Resistor 94
	2-13	Troubleshooting a Series Circuit 97
		Chapter Summary 102
		Self-Review 103

3 Resistors In Parallel 111

- 3-0 Introduction 112
- 3-1 Relationship between Source Voltage and Voltage Drop across Each Parallel Resistor 112
- 3-2 Branch Currents Flowing through Resistors and Total Current Drawn from Voltage Source. Total Resistance 114
- 3-3 Total Equivalent Resistance of Parallel Circuit from Reciprocal and Product-over-Sum Formulas 117
- 3-4 Total Equivalent Conductance of Parallel Circuit 127
- 3-5 Power Relationships in Parallel Circuit 128
- 3-6 Step-by-Step Analysis of Parallel Circuit 132
- 3-7 Current Division Rule 134
- 3-8 Voltage Sources Connected in Parallel 138
- 3-9 Effect of Open Circuits on Resistors in Parallel 140
- 3-10 Effect of a Short Circuit on Resistors in Parallel 142
- 3-11 Troubleshooting a Parallel Circuit 143
- Chapter Summary 146
- Self-Review 149

4 Series–Parallel Arrangements of Resistors 156

- 4-0 Introduction 157
- 4-1 Circuit Containing Two Resistors in Series with a Third Resistor in Parallel 158
- 4-2 Circuit Containing Two Resistors in Parallel with a Third Resistor in Series 160
- 4-3 Step-by-Step Analysis of Complex Series–Parallel Networks 163
- 4-4 Ground Connections in Series–Parallel Networks 168
- 4-5 Wheatstone Bridge Circuit 171
- 4-6 Loaded Voltage Divider Circuit 174
- 4-7 Effect of Open and Short Circuits in Series–Parallel Networks 177
- 4-8 Troubleshooting a Series–Parallel Circuit 179
- Chapter Summary 183
- Self-Review 184

5 Voltage and Current Sources 193

- 5-0 Introduction 194
- 5-1 Dry Primary Cell 194
- 5-2 Secondary Cell 196
- 5-3 Constant Voltage Source 200
- 5-4 Maximum Power Transfer and Circuit Efficiency 205
- 5-5 Practical Sources in Series 211
- 5-6 Constant Current Source 212
- 5-7 Practical Sources in Parallel 213
- Chapter Summary 217
- Self-Review 219

6 Kirchhoff's Laws and the Network Theorems 228

- 6-0 Introduction 229
- 6-1 Kirchhoff's Voltage and Current Laws 229

- 6-2 Mesh Analysis 234
- 6-3 Nodal Analysis 237
- 6-4 Millman's Theorem 239
- 6-5 Superposition Theorem 242
- 6-6 Thévenin's Theorem 245
- 6-7 Norton's Theorem 252
- 6-8 Delta–Wye (Π–T) Transformations 256
  - Chapter Summary 263
  - Self-Review 264

7 Magnetism, Electromagnetism, and Electromagnetic Induction 278

- 7-0 Introduction 279
- 7-1 Bar Magnet, Magnetic Poles. Magnetic Field Patterns. Magnetic Induction 279
- 7-2 Magnetic Flux. Magnetic Flux Density 286
- 7-3 Electromagnetism. Motor Effect. Magnetomotive Force 288
- 7-4 Magnetic Field Intensity. Permeability of Free Space 294
- 7-5 Relative Permeability. Reluctance. Rowland's Law 296
- 7-6 Magnetization Curve. Hysteresis 300
- 7-7 Electromagnetic Induction. Faraday's Law. Lenz's Law 304
- 7-8 Principle of DC Electrical Generator 309
  - Chapter Summary 313
  - Self-Review 314

8 Inductance in DC Circuits 323

- 8-0 Introduction 324
- 8-1 Comparison between Electrical Properties of Resistance and Inductance 324
- 8-2 Physical Factors That Determine a Coil's Self-Inductance. Types of Inductors 329
- 8-3 Energy Stored in Magnetic Field of Inductor 333
- 8-4 Inductors in Series and Parallel 334
- 8-5 Time Constant of LR Circuit 340
- 8-6 Troubleshooting for Faults in Inductors 348
  - Chapter Summary 349
  - Self-Review 350

9 Capacitance and Electrostatics 356

- 9-0 Introduction 357
- 9-1 Electric Flux. Coulomb's Law 357
- 9-2 Charge Density. Electric Field Intensity. Permittivity of Free Space 360
- 9-3 Factors That Determine Capacitance. Relative Permittivity. Dielectric Strength. Types of Capacitors 363
- 9-4 Energy Stored in an Electric Field 370
- 9-5 Capacitors in Series and in Parallel 371
- 9-6 Time Constant of CR Circuit 378
- 9-7 Features of a Practical Capacitor 387
- 9-8 Troubleshooting for Faults in Capacitors 388
- 9-9 Differentiator Circuits 390
- 9-10 Integrator Circuits 401
  - Chapter Summary 407
  - Self-Review 408

10 DC Electrical Measurements — 416

- 10-0 Introduction 417
- 10-1 Moving Coil (D'Arsonval) Meter Movement 417
- 10-2 Ammeter 421
- 10-3 Voltmeter 425
- 10-4 Loading Effect of a Voltmeter 427
- 10-5 Ohmmeter 429
- 10-6 Moving-Iron Ammeter and Voltmeter 432
- 10-7 Thermocouple Meter 434
- 10-8 Electrodynamometer Movement, Wattmeter 435
- Chapter Summary 436
- Self-Review 437

Part 2 AC Principles

11 Alternating Currents — 443

- 11-0 Introduction 444
- 11-1 Alternating Current Fundamentals 444
- 11-2 Sine-Wave Alternating Current 452
- 11-3 Power and Effective Value 457
- 11-4 Alternating Current in an Inductor 463
- 11-5 Alternating Current in a Capacitor 469
- 11-6 Mathematical Representation of Sine Wave 472
- Chapter Summary 474
- Self-Review 476

12 Series and Parallel AC Circuits — 480

- 12-0 Introduction 481
- 12-1 Series Circuit with Resistance and Inductive Reactance 481
- 12-2 Series Circuit with Resistance and Capacitive Reactance 488
- 12-3 Parallel *RL* and *RC* Circuits 493
- 12-4 *RLC* Circuits 501
- Chapter Summary 523
- Self-Review 526

13 Power Circuits — 531

- 13-0 Introduction 532
- 13-1 Power, Reactive Power, and Apparent Power 532
- 13-2 Power Factor 535
- 13-3 Maximum Power Transfer with Resistive Load 536
- 13-4 Maximum Power Transfer with Reactive Circuit Components 538
- Chapter Summary 541
- Self-Review 542

14 Resonant Circuits and Filters — 547

- 14-0 Introduction 548
- 14-1 Series Resonance 548
- 14-2 Parallel Resonance 559
- 14-3 Electrical Filter Concepts: Low-Pass Filters 567
- 14-4 High-Pass Filters 572

- **14-5** Band-Pass Filters 574
- **14-6** Band-Elimination (Notch) Filters 577
- **14-7** Decibels 578
- **14-8** Bode Plot 581
- **14-9** *LC* Filter Sections 583
  Chapter Summary 588
  Self-Review 590

15 Mutually Coupled Circuits 595

- **15-0** Introduction 596
- **15-1** Transformers 596
- **15-2** Troubleshooting for Faults in Transformers 607
- **15-3** Mutual Inductance 608
  Chapter Summary 618
  Self-Review 620

16 Complex Algebra 625

- **16-0** Introduction 626
- **16-1** The *j* Operator 626
- **16-2** Rectangular and Polar Conversions 631
- **16-3** Addition and Subtraction of Phasors 639
- **16-4** Multiplication and Division of Phasors 642
- **16-5** Series–Parallel Circuits 650
  Chapter Summary 653
  Self-Review 654

17 Kirchhoff's Laws and the Network Theorems in AC Circuits 659

- **17-0** Introduction 660
- **17-1** Kirchhoff's Laws 660
- **17-2** Mesh Analysis 665
- **17-3** Millman's Theorem 667
- **17-4** Nodal Analysis 670
- **17-5** Superposition Theorem 673
- **17-6** Thévenin's Theorem 677
- **17-7** Norton's Theorem 681
- **17-8** Wye–Delta and Delta—Wye Transformations 685
- **17-9** Reciprocity Theorem 689
  Chapter Summary 691
  Self-Review 692

18 Polyphase Systems 702

- **18-0** Introduction 703
- **18-1** Revolving Field Single-Phase Alternator 704
- **18-2** Two-Phase Alternator 707
- **18-3** Three-Phase Alternator 711
- **18-4** Wye Connection 715
- **18-5** Delta Connection. Three-Phase Power 718
- **18-6** Three-Phase Transformers. Six-Phase and Twelve-Phase Voltages 721
  Chapter Summary 728
  Self-Review 728

19 Nonsinusoidal Waveforms   733

- 19-0 Introduction   734
- 19-1 Synthesis of Nonsinusoidal Waveforms   734
- 19-2 Fourier Series. Analysis of Nonsinusoidal Waveforms   740
- 19-3 Effective Value of a Nonsinusoidal Wave   743
- 19-4 Nonsinusoidal Current in a Series AC Circuit   744
- 19-5 Nonsinusoidal Currents in a Series–Parallel Circuit   746
  - Chapter Summary   749
  - Self-Review   750

20 AC Electrical Measurements   756

- 20-0 Introduction   757
- 20-1 Electromechanical Meters   757
- 20-2 Analog Electronic Meters   763
- 20-3 Digital Multimeters   765
- 20-4 Cathode Ray Oscilloscopes   769
  - Chapter Summary   776
  - Self-Review   777

Appendices   783

- A SI and CGS Units. Conversion Factors   784
- B Color Code for Low-Value Inductors   785
- C Capacitor Color Codes   786

Answers to End-of-Chapter Problems   788

Index   797

Preface

To the Instructor

The purpose of this book is the same as of the first edition: To TEACH the principles of electricity and electronics without saturating the student with mountains of information and masses of equations. In other words, we invite the student to THINK and COMMUNICATE in terms of electricity and electronics. Many instructors who used the first edition reported that their students were able to learn merely by reading the text; this second edition with its numerous improvements should be even more beneficial in this respect.

The book is the outcome of many years spent in the instruction of electricity and electronics as they relate to the training of technicians. To these students we have always emphasized that a mastery of basics is vital; only by achieving such a mastery can the technician hope to progress in his or her chosen field. We are still teaching at the community college level in the areas of systems, communications, computer science, and microwave technology. Therefore, this book deals thoroughly with the particular problems that face students as they grapple with the fundamentals of direct and alternating current.

The Format

In order to achieve our goals we have used a format that assumes nothing except an everyday knowledge of such items as the car battery and the light bulb. Virtually all equations are developed from first principles and are not "plucked out of thin air."

The important features of this format are as follows:

- Every chapter has an introduction that now lists the objectives for the various sections. Each section introduces a limited number of principles, which are thoroughly explored. At the end of each session there is a summary of the more important concepts; if the student is not fully conversant with the content of the summary, he or she has an opportunity to look back through the material for a second time. The principles contained in a particular section are illustrated by a number of worked examples. Wherever possible, we have used practical values that the student would normally encounter in electronics. It is therefore possible to build many of the circuits in the laboratory, and then compare calculated results with experimental values.

- Heavy emphasis is placed on the checking of results. It is then possible to look at answers obtained through a variety of approaches. Occasionally, the same problem is solved by a number of analytical methods so that the student can readily see the advantages and disadvantages of each method.

- The text is written using the International Metric System of Units. This avoids the confusion that arises when you attempt to compare the modern units with the cgs and British systems.

- The Chapter Summary lists the units, the equations, and the main principles as they were discussed. This enables the student to see at a glance whether he or she has covered the chapter adequately. The summary is also a valuable reference for solving the questions and problems at the end of each chapter.

- The questions and problems are numerous (80 per chapter) and are carefully presented in ascending order of difficulty. There are true-false and multiple-choice questions, followed by practice problems and advanced problems. We have avoided redundancy as far as possible, so the student should be strongly advised to attempt all questions and problems.

- The friendly writing style is aimed at involving the student in the discussion of basic principles. At the same time the material is presented in a logical sequence so that there is a minimum amount of cross-referencing. To this end there are only four appendices.

- Emphasis is laid on the most effective way to use the electronic calculator to solve problems.

- There is an accompanying Laboratory Manual that is referenced to the text. This manual contains over forty experiments that enable the student to verify the most important principles as outlined in the various chapters.

The level of the book is intended for students in community colleges and vocational trade schools. Moreover, it is intended to appeal to a broad spectrum of student abilities. We have attempted to tread the narrow path between a treatment that is too superficial to be of any real use and one in which you cannot separate the basic principles from vast quantities of relatively unimportant information. Once the student has learned to THINK in terms of electronics, he or she can readily progress from the passive DC and AC circuits of this book to the next level, which involves active solid-state devices.

Highlights of the Second Edition

We have increased the breadth of the material presented and created a balanced text with ten chapters each for the fundamentals of direct and alternating current. Many of the following additions are the result of detailed recommendations from the large number of instructors who used our First Edition.

- New Objectives for all chapters.
- New Chapter on AC Circuit Analysis involving Kirchhoff's Laws and the Network Theorems.
- New Chapter on Polyphase Voltages. Technicians are now frequently required to have some knowledge of high voltage equipment and, in particular, three-phase systems.
- New Chapter on AC Circuit Analysis that involves nonsinusoidal waveforms. The basic principles of this subject are a unique feature of our introductory text.
- New Chapter on AC Electrical Measurements; this complements the earlier chapter on DC Electrical Measurements.
- New Sections on Differentiator and Integrator circuits.
- New Sections on Troubleshooting have been integrated into a number of chapters.
- New Calculator procedures appear throughout the text.
- New Questions and Problems have been added at the end of each chapter. There is now a total of 1600 questions and problems in the text.
- New Section on the topic of Filters.
- New experiments on Filters, Differentiator and Integrator circuits have been included in the laboratory manual.
- New package of Supplements produced by the same authors.
- New two-color format to highlight the most important principles.

Complete Package of Supplements

We have authored a complete set of supplements to accompany our book.

- A *Laboratory Manual* containing more than 40 experiments that have been thoroughly tested by the authors, other instructors, and students. Each experiment is carefully referenced to the text and includes a list of all the components and equipment required. This is followed by a step-by-step procedure together with tables in which the experimental results are entered. Finally, the student is asked to answer ten questions that test his/her understanding of the experiment.
- New *Instructor's Resource Manual with Transparency Masters* (over 250 pages), which begins with a preface to the instructor and is divided into two parts. The first part includes 1200 complete solutions to the multiple-choice questions, practice problems, and advanced problems that occur at the end of each chapter. Every effort has been made to provide solutions in the form that can be presented directly to the class. For this reason each solution has a complete logical sequence and all "short cuts" have been avoided.
  The second part is an instructor's manual for the laboratory manual. Anticipated results are shown in the related tables and accompanying graphs, while the questions at the end of each experiment are fully answered.

In the Transparency Masters section of the *Instructor's Resource Manual* we have selected over 100 figures from the text to be made available as transparencies. Most of the transparencies illustrate one or more fundamental principles that are vital in developing the students' understanding of electricity and electronics. The remaining transparencies are involved with practical experiences such as troubleshooting and measurements.

- New instructor's *Test Item File*. We have written a basic test and an advanced test for each chapter. Every test contains 10 problems which require mathematical solutions so there are altogether 400 problems for the entire twenty chapters. When these are combined with the 1600 end-of-chapter questions in the text, the result is a grand total of 2000 problems.

  The purpose of including basic and advanced tests is to allow the instructor to set a test which corresponds to the ability of the students. If the students have a broad spectrum of abilities, the two tests may be suitably interleaved. A complete set of answers for all the tests is included at the end of the file.

Organization of the Text

Our book is logically divided into two parts, each comprising ten chapters.

- Part 1 is entitled "DC Principles" and begins by first looking at the modern system of SI units and then surveying the principles of atomic structure. This is followed by a thorough examination of DC resistive circuits and the techniques for their troubleshooting. The network theorems are introduced as methods of analyzing a variety of circuits. The properties of inductance and capacitance are examined in detail, and then Part 1 concludes with a chapter on DC measurements.

- Part 2 is entitled "AC Principles" and starts by looking at the square wave as an example of an AC quantity. This leads to a discussion of the sine wave and the application of such a voltage to various *RLC* arrangements including resonant circuits and filters. The next topic is transformers and their troubleshooting. Then, following the introduction of operator *j*, the network theorems are used in the analysis of various AC circuits. The final chapters are devoted to polyphase systems, nonsinusoidal waveforms, and AC measurements; the combination of these chapters is unique to our text.

In order to avoid extensive cross-referencing we include only four Appendices.

- Conversion Factors. SI and cgs units.
- Color Code for Low Value Inductors.
- Capacitor Color Codes.
- Answers to all True-False Questions and odd-numbered Multiple-Choice Questions, Practice Problems, and Advanced Problems at the end of each chapter.

Development of the Second Edition

The First Edition was exhaustively tested on hundreds of students at Los Angeles Trade-Technical College. We were then able to identify certain areas where the students experienced some degree of difficulty or frustration. Generally speaking, these areas reflected a lack of smoothness in the transition from one concept to the next. Where such a difficulty occurred, the material was revised to eliminate the problem.

Since the First Edition was adopted by more than 100 institutions throughout the world, literally hundreds of instructor critique cards were available to the authors. We analyzed the information and were able to identify a number of important topics which were not included in the First Edition. Virtually all of these topics appear in the Second Edition. Some instructor recommendations about presentation were heavily duplicated and the manuscript was revised accordingly. Finally the numerous reviewers made many valuable suggestions, most of which were incorporated into the final manuscript. All changes and additions were class tested at our institution.

Acknowledgments

A book like this is only produced by a fine team effort. First and foremost, we would like to thank the highly competent professionals at Prentice-Hall—our editors, Holly Hodder, Susan Willig, and Judith Casillo and production editor Patrick Walsh. Next we owe a very considerable debt to the following reviewers who made a number of important recommendations for improving the book.

Margaret M. Drake—McHenry County College, Crystal Lake IL

James Mumaw—Terra Technical College, Fremont, OH

David E. Tester—ITT Technical Institute, Evansville, IN

Edward Troyan—Lehigh County Community College, Schnecksville, PA

John J. Hatch—ITT Technical Institute, Carson, CA

Fred S. Kerr—DeVry Institute of Technology, Decatur, GA

Randall Epstein—Total Technical Institute, Norcross, GA

Arlyn I. Smith—Alfred State College, Wellsville, NY

Michael Merchant

Gregory S. Wood—Bessemer State Technical College, Bessemer, AL

James David Wilkes—Heart of Georgia Technical Institute, Dublin, GA

Leonard E. Laabs—Walla Walla College, College Place, WA

Howard Duhon—Lee College, Baytown, TX

John R. Paris—Madisonville State Vo-Ed, Madisonville, KY

Ronald K. Aust—Indiana Vocational Technical College, Evansville, IN

Finally we are extremely grateful to our wives, Joyce and Margy, for typing the manuscript; without their love and support, the writing of this second edition would not have been possible.

To the Student

The book has deliberately written with you in mind; without your need this book simply would not exist. To begin with, we have assumed nothing except the presence of everyday items such as the car battery and the light bulb. The writing style is extremely friendly so that it is possible for you to pick up the book, read a chapter, and LEARN by understanding the material as it is presented. This unique feature of our book has been verified by a number of instructors who used the First Edition.

Our purpose throughout the book is to give you the ability to THINK and COMMUNICATE in terms of electricity and electronics. Only then will you acquire a mastery of the basics. Such a mastery is essential for initial employment in the electronics industry and for subsequent advancement.

We hope that you will enjoy reading the book and that you will find the effort to have been worthwhile. The material as presented has launched literally thousands of students on their successful careers. It is your first step to a bright future and so, study hard and we wish you every success.

VICTOR F. VELEY   JOHN J. DULIN

SI Units.
Atomic Structure.
OHM's Law

In this chapter you will learn:

1. about the various mechanical SI units and how they are interrelated.
2. how to express large and small numbers in terms of the scientific notation.
3. about the various prefixes and how they are used in electronics.
4. about the structure of atoms and the subatomic particles.
5. the electrical units of current, charge, electromotive force, and power.
6. the meaning of Ohm's law as it applies to the concept of resistance.
7. about the property of conductance as the reciprocal of resistance.
8. the factors on which the resistance of a cylindrical conductor depends.
9. how the resistance of a conductor varies with its temperature.
10. about practical resistors, some of which are identified by a color code.

1-0 Introduction

In a field such as electronics it is necessary to establish a system of *units*. The units are the language of a subject. For example, it would be difficult to get far in describing an automobile without using feet for the car's dimensions, pounds or tons for the mass, and miles per hour for the speed. We could also talk about the gas consumption (miles per gallon), the acceleration (zero to 50 miles per hour in 6 seconds), and the engine capacity in liters. This is really a mixture of units: the foot, pound, and gallon belong to the old *British* system, whereas the liter is a modern *metric* unit, which is now used in the sale of soft drinks and gasoline.

As far as electrical units are concerned, we are fortunate to live in an age in which one unified system has been adopted. Prior to about 1960 there were really three systems of measurement units. To start with, we had the practical everyday units, some of which you probably already know: the ampere, the volt, and the watt. The second system of units was based on magnetism and the third on electrostatics. The last two were referred to as *cgs* systems because they used the *c*entimeter as the unit of length, the *g*ram for mass, and the *s*econd for time. Frankly, it was a complete mess! Not only were there three possible units in which an electrical quantity could be measured, but between the three systems there were horrible conversion factors which were virtually impossible to memorize!

In 1901, Professor Giorgi of Italy proposed a new system founded on the meter (100 centimeters, or slightly greater than 1 yard), the kilogram (1000 grams, or about 2.2 pounds), and the second as the units of length, mass, and time. For electricity it was necessary to define a fourth fundamental unit and then build on this foundation to establish other units. In 1948, the ampere, which measures electrical current, was internationally adopted as the fourth unit. We therefore have the MKSA (meter, kilogram, second, ampere) system, which is also referred to as the *international system* or *SI* (Système International d'Unités). The three previous systems have therefore been replaced by a single system, with the added attraction that the old practical units are part of the new system. To avoid confusing you, this text has been entirely written in terms of SI units; the only reference to the obsolete units is contained in Appendix A.

As well as knowing the SI units, we must have some concept of what electricity really is. For this we will briefly study the atomic structure of matter, in particular the electron particle, from which the subject of electronics gets its name. Once we know something about electricity and have established the various units, we can progress to a study of the laws that govern electronics and also look at some of the practical components that we use in electrical circuits.

1-1 Mechanical SI Units. Scientific Notation. Prefixes

In our study of electricity we often need to use mechanical units to measure such quantities as force and energy. We are therefore going to use this section to establish the mechanical SI units, and we shall also see that some of these units are directly transferable to the electrical system. For example, electrical energy and mechanical energy are each measured by the same unit.

UNIT OF FORCE Isaac Newton (1642–1727) stated that when a *force* is applied to an object or mass (Fig. 1-1), the object or mass accelerates so that its speed or velocity increases. In the SI system the velocity is measured in meters per second; we shall use the abbreviations m for meter and s for second, so that meters per second is written as m/s. Acceleration will be expressed as meters per second per second. For example, if a mass starts from rest with zero velocity and is given an acceleration of 3 meters per second per second, its velocity after 1 second is 3 meters per second, after 2 s is 6 m/s, after 3 s is 9 m/s and so on. Therefore,

$$v = a \times t, \qquad a = \frac{v}{t} \qquad (1\text{-}1\text{-}1)$$

where v = velocity (meters per second) of the mass, which starts from rest
a = acceleration (meters per second per second)

If the force applied to a particular mass is increased, the acceleration will be greater. However, if the force is kept the same but the mass is greater, the acceleration will be less. The unit of force in the SI system is the newton, which will give a mass of 1 kilogram an acceleration of 1 meter per second per second in the direction of the force. The letter symbols for force, mass, and acceleration are, respectively, F, M, and a; therefore,

$$a = \frac{F}{M} \qquad (1\text{-}1\text{-}2)$$

or

$$F = M \times a \qquad (1\text{-}1\text{-}3)$$

where F = force [newtons (N)]
M = mass [kilograms (kg)]
a = acceleration [meters per second per second (m/s^2)]

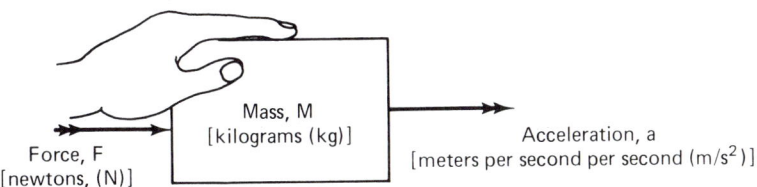

FIGURE 1-1 Relationships among force, mass, and acceleration.

For example, if a force of 10 N is applied to a mass of 5 kg, the acceleration will be 10 N/5 kg = 2 m/s².

When a mass is falling under the force of the earth's gravity, its acceleration is 9.81 m/s². Therefore, the gravitational force on the mass of 1 kg is 9.81 N; this force is sometimes referred to as 1-kilogram weight. Of course, on the moon the force of gravity exerted on 1 kg would be much less than 9.81 N.

UNIT OF ENERGY OR WORK *Energy* is the capacity for doing work, and therefore both these quantities are measured by the same unit. When a force is applied through a certain distance in the direction of the force, energy must be supplied so that work may be performed. A good example is lifting a mass at a constant speed against the force of gravity (Fig. 1-2); when the mass has been raised through a given height or distance, you have expended some energy in performing a certain amount of work. The larger the force and the longer the distance through which the force is applied, the greater is the amount of work that must be done. The value of the work performed is then found by multiplying the force in newtons by the distance in meters. The result of multiplying any two quantities together is called their *product*; we can therefore say that work is the product of force and distance. In the SI system the unit of mechanical energy or work is the joule (James Joule, 1818–1889) and the letter symbol for work is W. Therefore,

$$W = d \times F \qquad (1\text{-}1\text{-}4)$$

where W = work done [joules (J)]
F = force applied [newtons (N)]
d = distance [meters (m)]

For example, if a force of 5 N is applied through a distance of 2 m, the work done in joules is 2 m × 5 N = 10 J. One joule can therefore be thought of as 1 meter-newton, as it is the result of multiplying 1 meter by 1 newton.

Notice what we are doing. The newton was defined in terms of our fundamental units of mass (kilogram), length (meter), and time (second). The joule is derived from the newton and the meter; in other words, each new unit

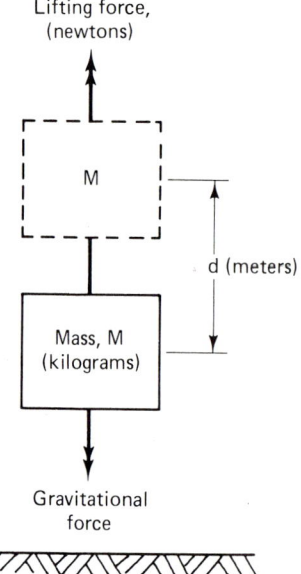

FIGURE 1-2 Relationships among work, force, and distance.

is defined in terms of its predecessors. This is the logical manner in which a system of units is established.

An electrical motor such as that used in a vacuum cleaner is a means whereby electrical energy is converted into mechanical energy. We plug the vacuum cleaner into the household outlet and use the electrical energy to drive the motor. The rotating shaft of the motor then provides mechanical energy to perform the task of cleaning. Since electrical energy can be converted into mechanical energy, both can be measured by the same unit; consequently, the SI unit of electrical energy is also the joule.

UNIT OF TORQUE A *torque* produces a twisting effect which we use every day when we open a door by either pushing or pulling its handle. The force that we apply is most effective when its direction is at right angles (90°) to the line joining the handle to the door hinge [Fig. 1-3(a)]; this is clearly so because if the force and the line were in the same direction, the door would not open at all. If the handle were positioned close to the hinge, it would be difficult to open the door; we are therefore led to the conclusion that the value of the torque must be equal to the result of multiplying the applied force, F, by the distance, d, which is at right angles (90°) to the force's direction. This distance is the length between the point, P, at which the force is applied and the pivot, O, about which the twisting effect or the rotation occurs [Fig. 1-3(b)]. Torque, like work, is therefore equal to the product of force and distance, but with one

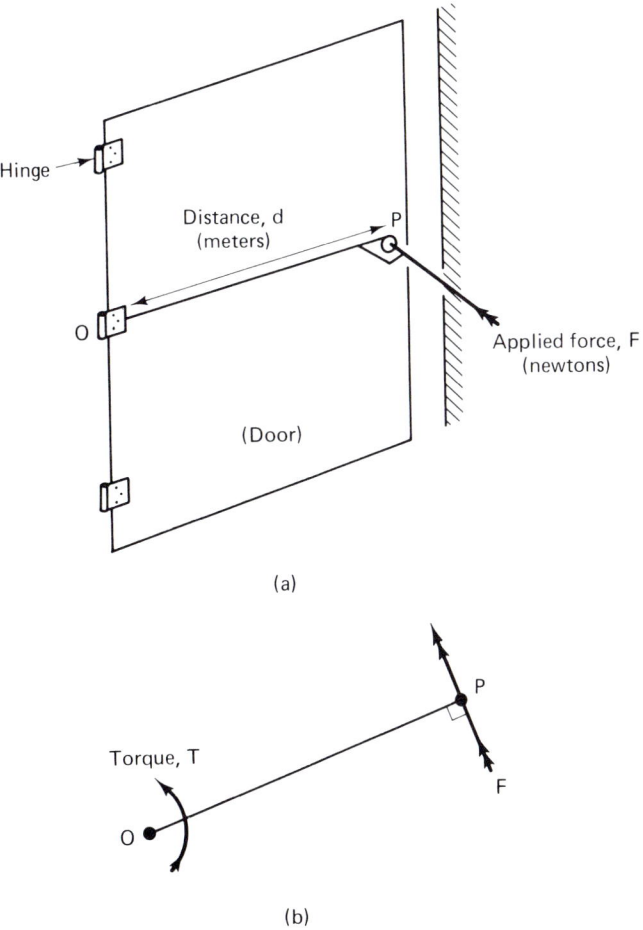

FIGURE 1-3 Relationships among torque, force, and distance.

Sec. 1-1 Mechanical SI Units. Scientific Notation. Prefixes

very important difference. In the case of work, the directions of the force and the distance are the same, but with torque, the directions are 90° apart. Consequently, torque is not measured in joules but in newton-meters. The letter symbol for torque is T; therefore,

$$T = F \times d \qquad (1\text{-}1\text{-}5)$$

where T = torque [newton-meters (N · m)]
F = applied force [newtons (N)]
d = distance [meters (m)]

UNIT OF POWER There is often much confusion over the distinction between "work" and "power." *Power* is the *rate* at which work is performed or energy is expended. As soon as you see the word "rate," you must realize that time is involved. Let us take a mechanical example; suppose that you have a heavy weight and you ask a powerful adult to lift it up a certain height against the force of gravity. The adult will be able to perform this task quickly (in a short time) because of his or her power. However, you could have a complex pulley system attached to the weight and at the end of the system there might be a wheel. A small child would be capable of turning the handle on the wheel and the weight would slowly rise to the same height achieved by the powerful adult. Neglecting the weight of the pulley system and its friction, the total work performed by the adult and the child is the same, but the child took a much longer time because he or she is much less powerful than the adult. An old British unit is the horsepower, which is equivalent to 550 foot-pounds per second. This means that a motor whose mechanical output is 1 horsepower (hp) would be capable of lifting a mass of 550 pounds through a distance of 1 foot against the force of gravity, and do it in a time of 1 second.

The SI unit of power is the watt (James Watt, 1736–1819), whose unit symbol is W. The power is 1 watt if 1 joule of energy is created or used every second. For example, when you switch on a 60-W electric light bulb, 60 J of energy is released from that bulb every second, mostly in the form of heat but a small amount as light. Therefore, watts are equivalent to joules per second, or joules are the same as watts × seconds, which may be written as watts-seconds. Since the horsepower and the watt both measure the same quantity, the two must be related and, in fact, 1 horsepower is equivalent to 746 watts.

As we have seen, power, whose letter symbol is P, is the result of dividing the work (W) by the time (t). Therefore,

$$P = \frac{W}{t}, \qquad W = P \times t \qquad (1\text{-}1\text{-}6)$$

From Eq. (1-1-4),

$$W = d \times F$$

Then

$$P = \frac{W}{t} = \frac{d}{t} \times F = v \times F \qquad (1\text{-}1\text{-}7)$$

so that the power, P, in watts is also the product of the force, F, in newtons and the velocity, v, of the moving mass in meters per second.

If the 60-W light bulb is left on for 1 h (3600 s), the energy consumed is 60 W × 3600 s = 216,000 J, and this consumption would appear on the electricity bill. It is clear that, for everyday purposes, the joule is too small a unit. In fact, it takes about 8 J of energy to lift a 2-lb book through a distance of 1 yard and about half a million joules to boil a kettle of water. A larger unit would be the watthour (Wh), which is the energy consumed when a power of 1 watt is operated for a time of 1 hour. Since 1 h is the same as 3600 s, 1 watthour = 1 W × 3600 s = 3600 J. The unit on the electricity bill is still larger; it is the kilowatthour (kWh), which will be equal to 1000 Wh or 3,600,000 J.

Scientific Notation

In electronics we are often required to use either very large or very small numbers in expressing the values of various quantities. It would be extremely laborious if all such numbers had to be written out in full. A shorthand system has therefore been developed through the use of *exponents*, sometimes referred to as "powers." The base 10 when raised to the exponent N is written as 10^N; if N is then a positive whole number, 10^N means 10 multiplied by itself N times. As an example, $10^6 = 10 \times 10 \times 10 \times 10 \times 10 \times 10 = 1,000,000$ or 1 million. Therefore, 2,714,000 could be written as 2.714×10^6; the number 2.714 is called the *coefficient* and the exponent "6" is found by counting the number of places the decimal point is shifted to the *left*. When numbers are written in scientific notation, there is a *single* figure before the decimal point followed by such decimal places as are required by the number's accuracy. This decimal coefficient, which is greater than or equal to 1 but is less than 10, must then be multiplied by 10, raised to the required exponent. As further examples:

$$8763943000 = 8.764 \times 10^9$$

This answer has been rounded off to four *significant figures*.

$$317.86 = 3.18 \times 10^2$$

rounded off to three significant figures.

$$41.423 = 4.142 \times 10^1$$

rounded off to four significant figures. Notice that 10^1 has the same value as 10.

When calculating a result from a formula, all numbers entered into the formula should be expressed in scientific notation. This makes the calculation easier and also allows the answer to appear in scientific notation for possible substitution into another formula. As a result, many electronic calculators display their results in scientific notation.

Let us look at the rules that we must observe when carrying out mathematical operations using exponents.

MULTIPLICATION OF NUMBERS We know that $200 \times 30{,}000 = 6{,}000{,}000$; these numbers are expressed in scientific notation as follows:

$$200 \times 30{,}000 = 6{,}000{,}000$$
$$(2 \times 10^2) \times (3 \times 10^4) = (6 \times 10^6)$$

Notice that the individual exponents "2" and "4" are *added* (*not* multiplied) to produce the exponent "6" in the answer. As a further example:

$$34.2 \times 7{,}531{,}000 \times 8740 = 3.42 \times 10^1 \times 7.531 \times 10^6 \times 8.74 \times 10^3$$
$$= 3.42 \times 7.531 \times 8.74 \times 10^1 \times 10^6 \times 10^3$$
$$= 225.1 \times 10^{10}$$
$$= 2.251 \times 10^2 \times 10^{10}$$
$$= 2.251 \times 10^{12}$$

The answer is rounded off to four significant figures because of the four significant figures in the number 7,531,000; the accuracy of the answer is therefore comparable with the accuracy of the problem.

As a step-by-step procedure, each number is expressed in scientific notation. The coefficients are grouped together and then multiplied. The exponents are added and the final answer is given in scientific notation.

When any number is raised to the exponent 2, we obtain the *square* of the number. The square of a number is therefore the result of multiplying a number by itself, so that $10^2 = 10 \times 10 = 100$ and $7^2 = 7 \times 7 = 49$. Remember that the square of a negative number is always positive. For example, $(-7)^2 = (-7) \times (-7) = +49$; consequently, the minimum value of a squared number is zero.

When calculating the area of a rectangle it is necessary to multiply the length by the width. If the length is 5 m and the width is 3 m, the area is 5 m $\times$ 3 m = 15 square meters; square meters are abbreviated as m². Since 100 centimeters = 1 meter, 1 square meter = 100 centimeters $\times$ 100 centimeters = 10,000 square centimeters; using abbreviations, 1 m² = 10^4 cm². $10^3 cm^2$

DIVISION OF NUMBERS If we divide 6,000,000 by 30,000, the answer is 200. Using scientific notation, we have

$$\frac{6{,}000{,}000}{30{,}000} = 200$$

$$\frac{6 \times 10^6}{3 \times 10^4} = 2 \times 10^2$$

In the operation of division, the exponent "4" is subtracted from the exponent "6" so that the exponent in the answer is 6 − 4 = "2." As an example to illustrate both multiplication and division:

$$\frac{4352 \times 17{,}400{,}000}{27.8 \times 135{,}200} = \frac{4.352 \times 10^3 \times 1.74 \times 10^7}{2.78 \times 10^1 \times 1.352 \times 10^5}$$
$$= \frac{4.352 \times 1.74 \times 10^{10}}{2.78 \times 1.352 \times 10^6}$$
$$= \frac{4.352 \times 1.74}{2.78 \times 1.352} \times 10^{10-6}$$
$$= 2.015 \times 10^4$$

One special case. If we divide 2000 by 2000, the answer must, of course, be 1. However, using scientific notation, we obtain

$$\frac{2000}{2000} = \frac{2 \times 10^3}{2 \times 10^3} = 1 \times 10^{3-3} = 1 \times 10^0$$

Consequently, $10^0 = 1$. This means that a number like 5.73 could be written in scientific notation as 5.73×10^0

MEANING OF THE NEGATIVE EXPONENT Suppose that we divide 400 by 2,000,000; the result is 0.0002. But if we use scientific notation and subtract the exponents,

$$\frac{400}{2,000,000} = \frac{4 \times 10^2}{2 \times 10^6} = 2 \times 10^{2-6} = 2 \times 10^{-4}$$

Since $0.0002 = 2 \times 10^{-4}$, we may use the negative exponent to express in scientific notation all numbers which lie between $+1$ and -1. For example, $0.00000358 = 3.58 \times 10^{-6}$, with the value of the negative exponent equal to the number of places that the decimal point is shifted to the *right*. If we again divide 400 by 2,000,000,

$$\frac{400}{2,000,000} = \frac{2}{10,000} = \frac{2}{10^4} = 2 \times \frac{1}{10^4}$$

Comparing the answers "2×10^{-4}" and "$2 \times 1/10^4$," it is clear that $1/10^4 = 10^{-4}$ and therefore $1/10^{-4} = 10^4$. We can say that 10^4 and 10^{-4} are *reciprocals*; the reciprocal of a number is the result of dividing 1 by that number. Therefore, the reciprocal of 2 is $1/2 = 0.5$; in the same way, the reciprocal of 0.5 is $1/0.5 = 2$.

Table 1-1 shows the application of positive and negative exponents to scientific notation. The combined use of positive and negative exponents is illustrated in the following example:

$$\frac{763 \times 0.000149}{0.000000527 \times 12.5} = \frac{7.63 \times 10^2 \times 1.49 \times 10^{-4}}{5.27 \times 10^{-7} \times 1.25 \times 10^1}$$

$$= \frac{7.63 \times 1.49 \times 10^{2+(-4)}}{5.27 \times 1.25 \times 10^{(-7)+1}}$$

$$= \frac{7.63 \times 1.49 \times 10^{-2}}{5.27 \times 1.25 \times 10^{-6}}$$

$$= 1.73 \times 10^{-2} \times 10^6$$

$$= 1.73 \times 10^{(-2)+6}$$

$$= 1.73 \times 10^4$$

RAISING A NUMBER TO A POWER Many of our elecronics formulas involve squares and we need to know how to use scientific notation in raising a number to a particular power. If we square 3000, the answer is $3000 \times 3000 = 9,000,000$. By using scientific notation,

$$(3000)^2 = 3000 \times 3000 = 9,000,000$$

Sec. 1-1 Mechanical SI Units. Scientific Notation. Prefixes

TABLE 1-1 Positive and Negative Exponents

Exponents of 10	Meaning
10^6	$\times$ 1,000,000
10^5	$\times$ 100,000
10^4	$\times$ 10,000
10^3	$\times$ 1000
10^2	$\times$ 100
10^1	$\times$ 10
10^0	$\times$ 1
$10^{-1} = \dfrac{1}{10^1}$	$\div$ 10 or $\times$ 0.1
$10^{-2} = \dfrac{1}{10^2}$	$\div$ 100 or $\times$ 0.01
$10^{-3} = \dfrac{1}{10^3}$	$\div$ 1000 or $\times$ 0.001
$10^{-4} = \dfrac{1}{10^4}$	$\div$ 10,000 or $\times$ 0.0001
$10^{-5} = \dfrac{1}{10^5}$	$\div$ 100,000 or $\times$ 0.00001
$10^{-6} = \dfrac{1}{10^6}$	$\div$ 1,000,000 or $\times$ 0.000001

or

$$(3000)^2 = (3 \times 10^3)^2 = 9 \times 10^6$$

Therefore, the coefficient is squared, but the exponent "3" is multiplied by 2 to equal the answer's exponent of "6." This rule of multiplication applies to both positive and negative exponents. As further examples:

$$(324)^4 = (3.24 \times 10^2)^4 = (3.24)^4 \times 10^8 = 110 \times 10^8$$
$$= 1.10 \times 10^2 \times 10^8 = 1.10 \times 10^{10}$$
$$(0.000743)^3 = (7.43 \times 10^{-4})^3 = (7.43)^3 \times 10^{-12}$$
$$= 410 \times 10^{-12}$$
$$= 4.10 \times 10^2 \times 10^{-12}$$
$$= 4.10 \times 10^{-10}$$

OBTAINING THE SQUARE ROOT OF A NUMBER As far as electronics formulas are concerned, square roots are every bit as important as squares. The *square root* of a number N, written as $\sqrt{N}$, is another number, M, such that $M^2 = N$. Therefore, the square root of 49 is $\sqrt{49}$ = either $+7$ or -7, since $7^2 = 7 \times 7 = 49$ and $(-7)^2 = (-7) \times (-7) = 49$; for most of our problems in electronics we will only need to consider the positive square root.

As a further example, $\sqrt{6,250,000} = 2500$; to obtain the same result by scientific notation:

$$\sqrt{6,250,000} = \sqrt{6.25 \times 10^6} = \sqrt{6.25} \times 10^3 = 2.5 \times 10^3$$

The rule is to take the square root of the coefficient, but the exponent "6" is divided by 2 to obtain the answer's exponent of "3."

One problem with taking square roots:

$$\sqrt{16{,}000{,}000} = \sqrt{1.6 \times 10^7} = \sqrt{1.6} \times 10^{7/2}$$
$$= 1.26 \times 10^{7/2}$$
$$= 1.26 \times 10^{3.5}$$

Although this is, in fact, a perfectly correct answer, it is difficult to interpret the value of "$10^{3.5}$." The problem is solved by rewriting 1.6×10^7 as 16.0×10^6; then

$$\sqrt{16.0 \times 10^6} = \sqrt{16.0} \times 10^{6/2} = 4.0 \times 10^3$$

Consequently, if when expressing the original number in scientific notation, you come up with an odd-number exponent, the number must be rewritten with two digits before the decimal point; the exponent will then be an even number so that after division by 2, the exponent in the answer is a whole number and not a fraction. This rule also applies to negative exponents, as illustrated in the following examples:

$$\sqrt{0.0327} = \sqrt{3.27 \times 10^{-2}} = \sqrt{3.27} \times 10^{-2/2} = 1.81 \times 10^{-1}$$
$$\sqrt{0.0000327} = \sqrt{3.27 \times 10^{-5}} = \sqrt{32.7 \times 10^{-6}} = \sqrt{32.7} \times 10^{-6/2}$$
$$= 5.72 \times 10^{-3}$$

Prefixes

As well as scientific notation, we commonly use word prefixes in electronics to represent both large and small quantities. We have already met the kilogram, which is equivalent to 1000 grams. The prefix "kilo" therefore means that we must multiply the gram unit by 1000. Such prefixes are based on the metric system, and the ones most commonly used in electronics are shown in Table 1-2.

In the following list we show the use of prefixes with the units that we have already derived. The values are those that we can encounter in practice.

1 megawatt (MW) = 1×10^6 watts = 1,000,000 W
1 kilometer (km) = 1×10^3 meters = 1000 m
1 kilowatt (kW) = 1×10^3 watts = 1000 W
1 kilogram (kg) = 1×10^3 grams = 1000 g

TABLE 1-2 Most Commonly Used Prefixes in Electronics

Prefix	Symbol	Scientific notation equivalent	Meaning
tera-	T	$\times 10^{12}$	$\times$ 1,000,000,000,000
giga-	G	$\times 10^9$	$\times$ 1,000,000,000
mega-	M	$\times 10^6$	$\times$ 1,000,000
kilo-	k	$\times 10^3$	$\times$ 1000
centi-	c	$\div 10^2$ or $\times 10^{-2}$	$\div$ 100 or $\times$ 0.01
milli-	m	$\div 10^3$ or $\times 10^{-3}$	$\div$ 1000 or $\times$ 0.001
micro-	μ	$\div 10^6$ or $\times 10^{-6}$	$\div$ 1,000,000 or $\times$ 0.000001
nano-	n or ν	$\div 10^9$ or $\times 10^{-9}$	$\div$ 1,000,000,000 or $\times$ 0.000000001
pico- or micromicro	p or μμ	$\div 10^{12}$ or $\times 10^{-12}$	$\div$ 1,000,000,000,000 or $\times$ 0.000000000001

Sec. 1-1 Mechanical SI Units. Scientific Notation. Prefixes

$1 \text{ centimeter (cm)} = 1 \times 10^{-2} \text{ or } \frac{1}{10^2} \text{ meter} = \frac{1}{100} \text{ or } 0.01 \text{ m}$

$1 \text{ millimeter (mm)} = 1 \times 10^{-3} \text{ or } \frac{1}{10^3} \text{ meter} = \frac{1}{1000} \text{ or } 0.001 \text{ m}$

$1 \text{ milliwatt (mW)} = 1 \times 10^{-3} \text{ or } \frac{1}{10^3} \text{ watt} = \frac{1}{1000} \text{ or } 0.001 \text{ W}$

$1 \text{ millisecond (ms)} = 1 \times 10^{-3} \text{ or } \frac{1}{10^3} \text{ second} = \frac{1}{1000} \text{ or } 0.001 \text{ s}$

$1 \text{ micrometer or micron } (\mu m) = 1 \times 10^{-6} \text{ or } \frac{1}{10^6} \text{ meter} = \frac{1}{1,000,000} \text{ or } 0.000001 \text{ m}$

$1 \text{ microwatt } (\mu W) = 1 \times 10^{-6} \text{ or } \frac{1}{10^6} \text{ watt} = \frac{1}{1,000,000} \text{ or } 0.000001 \text{ W}$

$1 \text{ microsecond } (\mu s) = 1 \times 10^{-6} \text{ or } \frac{1}{10^6} \text{ second} = \frac{1}{1,000,000} \text{ or } 0.000001 \text{ s}$

$1 \text{ microjoule } (\mu J) = 1 \times 10^{-6} \text{ or } \frac{1}{10^6} \text{ joule} = \frac{1}{1,000,000} \text{ or } 0.000001 \text{ J}$

$1 \text{ nanosecond (ns)} = 1 \times 10^{-9} \text{ or } \frac{1}{10^9} \text{ second} = \frac{1}{1,000,000,000} \text{ or } 0.000000001 \text{ s}$

$1 \text{ nanowatt (nW)} = 1 \times 10^{-9} \text{ or } \frac{1}{10^9} \text{ watt} = \frac{1}{1,000,000,000} \text{ or } 0.000000001 \text{ W}$

$1 \text{ picosecond (ps)} = 1 \times 10^{-12} \text{ or } \frac{1}{10^{12}} \text{ second} = \frac{1}{1,000,000,000,000} \text{ or } 0.000000000001 \text{ s}$

$1 \text{ picowatt (pW) or 1 micromicrowatt } (\mu\mu W) = 1 \times 10^{-12} \text{ or } \frac{1}{10^{12}} \text{ watt} = \frac{1}{1,000,000,000,000} \text{ or } 0.000000000001 \text{ W}$

When using the formulas that occur in electronics, all quantities must be entered in terms of their basic units. As an example, if the value of the power is 8 mW, it is entered into the formula as 8×10^{-3} W. This principle is illustrated throughout the book as we solve various examples at the end of each section.

To summarize: We introduced the concept of the SI or MKSA system of electrical units. We discussed the meanings of force, energy or work, torque, power, and derived their units. We described the use of scientific notation and word prefixes to express large or small values of these units.

EXAMPLE 1-1 A force of 200 N is continuously applied to a 40-kg mass which is initially at rest. Find the acceleration of the mass and its velocity after 10 s. When the mass has been moved through a distance of 2 km, what is the amount of the total work done?

Solution

$$F = 200 \text{ N}, \quad M = 40 \text{ kg}$$

$$\text{acceleration, } a = \frac{F}{M} \tag{1-1-2}$$

$$= \frac{200 \text{ N}}{40 \text{ kg}} = 5 \text{ meters per second per second}$$

$$= 5 \text{ m/s}^2$$

$$\text{velocity of mass after 10 s, } v = a \times t \tag{1-1-1}$$

$$= 5 \text{ m/s}^2 \times 10 \text{ s}$$

$$= 50 \text{ meters per second}$$

$$= 50 \text{ m/s}$$

$$F = 200 \text{ N}, \quad d = 2 \text{ km} = 2 \times 10^3 \text{ m}$$

$$\text{work done, } W = d \times F \tag{1-1-4}$$

$$= 2 \times 10^3 \text{ m} \times 200 \text{ N}$$

$$= 400 \times 10^3 \text{ J}$$

$$= 400{,}000 \text{ J}$$

$$= 4 \times 10^5 \text{ J}$$

EXAMPLE 1-2 An aluminum block whose mass is 150 g is given an acceleration of 20 cm/s². What is the value in newtons of the accelerating force?

Solution

$$M = 150 \text{ g} = \frac{150}{10^3} \text{ kg} = 150 \times 10^{-3} \text{ kg}$$

$$a = 20 \text{ cm/s}^2 = 20 \times 10^{-2} \text{ m/s}^2$$

$$\text{accelerating force, } F = M \times a \tag{1-1-3}$$

$$= 150 \times 10^{-3} \text{ kg} \times 20 \times 10^{-2} \text{ m/s}^2$$

$$= 3000 \times 10^{-5} \text{ N}$$

$$= 3.0 \times 10^{-2} \quad \text{or} \quad 0.03 \text{ N}$$

EXAMPLE 1-3 A mass of 2400 kg is lifted vertically with a velocity of 120 m/min. What is the value in kilowatts of the required power?

Solution The force, F, exerted on the mass due to gravity is 2400×9.81 N. The distance, d, through which the mass is lifted in 1 s is $120/60 = 2$ m. The work done in 1 s is

$$W = d \times F = 2 \text{ m} \times 2400 \times 9.81 \text{ N} \tag{1-1-4}$$

$$= 47{,}088 \text{ J}$$

$$W = 47{,}088 \text{ J}, \quad t = 1 \text{ s}$$

$$\text{power, } P = \frac{W}{t} \tag{1-1-6}$$

$$= \frac{47{,}088 \text{ J}}{1 \text{ s}} = 47{,}088 \text{ W} = \frac{47{,}088}{1000} \text{ kW}$$

$$= 47.1 \text{ kW (rounded off)}$$

Sec. 1-1 Mechanical SI Units. Scientific Notation. Prefixes

EXAMPLE 1-4 Figure 1-4 shows a copper loop that is capable of rotating about the axis AA'. What is the value of the torque exerted on the loop?

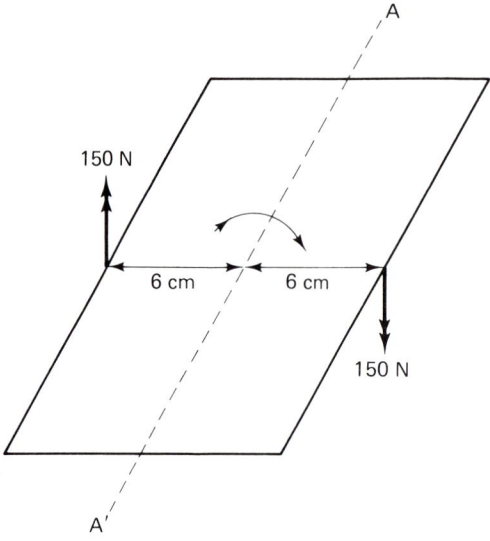

FIGURE 1-4 Diagram for Example 1-4.

Solution

$$F = 150 \text{ N}, \quad d = 6 \text{ cm} = 6 \times 10^{-2} \text{ m}$$

About the axis each of the forces will exert a torque which is equal to

$$\text{torque}, T = F \times d \quad (1\text{-}1\text{-}5)$$
$$= 150 \text{ N} \times 6 \times 10^{-2} \text{ m} = 9 \text{ N} \cdot \text{m}$$
$$\text{total torque} = 2 \times 9 = 18 \text{ N} \cdot \text{m}$$

1-2

Atomic Structure. Conductors, Insulators, and Semiconductors

All of matter is essentially electrical in nature. The page you are reading contains literally billions upon billions of electrically charged particles; the word *charge* means "a quantity of electricity." Half of these particles have a charge with a negative polarity while the other half carry an equal charge with a positive polarity. The result is that these charges with opposite polarities cancel out and we can say that the page is neutral in the sense that it is not electrified. Positive and negative polarities are fully explained later in this section; however, you are probably already familiar with the positive and negative terminals of a cell such as a flashlight battery.

As we shall learn, one substance is distinguished from another by the number and arrangement of its particles. We shall therefore make a start by examining a very common substance—water. However, we are not going to study tap water, in which a number of substances are dissolved; we are going to assume that all these substances have been removed so that we are left with pure distilled water, which has the properties of boiling at 100°C (Celsius or centigrade) and freezing at 0°C. Now take a drop of this water and divide it into smaller and smaller amounts (Fig. 1-5). Ultimately, you would come to the smallest quantity that still retains the property of water; such a quantity is called a *molecule*. Although water is, of course, a liquid, it is in fact a compound of two gases, hydrogen and oxygen, so that a molecule is the least amount of a compound that still retains the properties of the compound. The chemical formula for the water molecule is H_2O; so if we subdivide the molecule, we would obtain two parts of hydrogen gas to one part of oxygen. The gases would be separated and the water molecule would therefore no longer exist.

The particles of hydrogen and oxygen that combine to produce water are called *atoms*. Hydrogen and oxygen are themselves examples of *elements*, which are substances capable of a separate existence—they cannot be broken down further by chemical action. The atom of the element is therefore the smallest particle that still retains all the properties of the element. More than 100 different elements have been found; of these the first 92 occur naturally in the earth and the others have been human-made. Elements with similar properties can be grouped together into families, an arrangement called the *periodic table of the elements* (Table 1-3), developed by the Russian scientist, Dimitri Mendeleyev, in 1869.

We have said that compounds may be divided into molecules, which in turn are made from the atoms of elements. But what is the composition of the atom? Since hydrogen is the lightest element, it might be assumed to have the simplest structure. In 1913, the Norwegian physicist Niels Bohr proposed that the hydrogen atom contained a core called the *nucleus*, where most of the atom's mass was concentrated. This nucleus of the hydrogen (H) atom is the *proton* particle, which is positively charged. Revolving around the nucleus in a circle is a much lighter particle called the *electron*, which carries an equal negative charge (Fig. 1-6), so that hydrogen gas as a whole is electrically neutral. There is a law of electricity which states that positive and negative charges attract. As a result, there would be a tendency for the electron to merge with the proton. However, since the electron is revolving in a circular path, there is a *centrifugal force* acting outward to balance the inward force of attraction. An example of centrifugal force is produced by attaching a rock to a piece of string and then whirling the rock over your head. The faster the rock is whirled, the greater is the tension in the string; and if the string breaks, the rock will fly off as the result of the outward centrifugal force. A stable orbit for the electron requires that the radius of its circular path is approximately 5.3×10^{-11} m (now you can see why scientific notation is so necessary!). Some more figures: the mass of the proton is 1.6726×10^{-27} kg and the mass of the electron is 9.1096×10^{-31} kg. This means that the proton is more than 1800 times heavier than the electron. However, the radius of the electron's orbit

FIGURE 1-5 Concept of the molecule: (a) dividing a drop of water; (b) molecule of water H_2O.

(a)

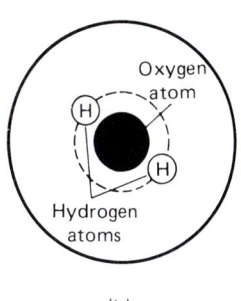

(b)

TABLE 1-3 Periodic Table of the Elements

Period	IA	IIA	IIIB	IVB	VB	VIB	VIIB	VIIIB			IB	IIB	IIIA	IVA	VA	VIA	VIIA	Inert Gases
1	1 H 1.00797																	2 He 4.0026
2	3 Li 6.939	4 Be 9.0122											5 B 10.811	6 C 12.01115	7 N 14.0067	8 O 15.9994	9 F 18.9984	10 Ne 20.183
3	11 Na 22.9898	12 Mg 24.312											13 Al 26.9815	14 Si 28.086	15 P 30.9738	16 S 32.064	17 Cl 35.453	18 Ar 39.948
4	19 K 39.102	20 Ca 40.08	21 Sc 44.956	22 Ti 47.90	23 V 50.942	24 Cr 51.996	25 Mn 54.9380	26 Fe 55.847	27 Co 58.9332	28 Ni 58.71	29 Cu 63.54	30 Zn 65.37	31 Ga 69.72	32 Ge 72.59	33 As 74.9216	34 Se 78.96	35 Br 79.909	36 Kr 83.80
5	37 Rb 85.47	38 Sr 87.62	39 Y 88.905	40 Zr 91.22	41 Nb 92.906	42 Mo 95.94	43 Tc (99)	44 Ru 101.07	45 Rh 102.905	46 Pd 106.4	47 Ag 107.870	48 Cd 112.40	49 In 114.82	50 Sn 118.69	51 Sb 121.75	52 Te 127.60	53 I 126.9044	54 Xe 131.30
6	55 Cs 132.905	56 Ba 137.34	57 La* 138.91	72 Hf 178.49	73 Ta 180.948	74 W 183.85	75 Re 186.2	76 Os 190.2	77 Ir 192.2	78 Pt 195.09	79 Au 196.967	80 Hg 200.59	81 Tl 204.37	82 Pb 207.19	83 Bi 208.980	84 Po (210)	85 At (210)	86 Rn (222)
7	87 Fr (223)	88 Ra (226)	89 Ac† (227)															

*Lanthanum Series

58 Ce 140.12	59 Pr 140.907	60 Nd 144.24	61 Pm (145)	62 Sm 150.35	63 Eu 151.96	64 Gd 157.25	65 Tb 158.924	66 Dy 162.50	67 Ho 164.930	68 Er 167.26	69 Tm 168.934	70 Yb 173.04	71 Lu 174.97

†Actinium Series

90 Th 232.038	91 Pa (231)	92 U 238.03	93 Np (237)	94 Pu (242)	95 Am (243)	96 Cm (247)	97 Bk (249)	98 Cf (251)	99 Es (254)	100 Fm (253)	101 Md (256)	102 No (253)	103 Lw (257)

The numbers in parentheses are the mass numbers of most stable or most common isotope.

Source: Clyde Herrick, *Unified Concepts of Electronics*, Prentice-Hall, Inc., Englewood Cliffs, N.J., 1970, p. 40.

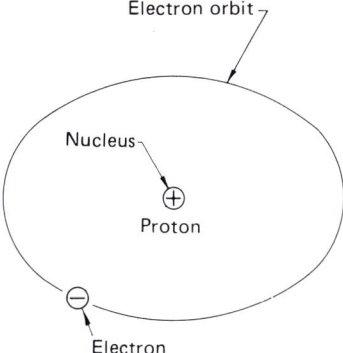

FIGURE 1-6 Hydrogen atom.

is about 10,000 times the diameter of the electron. Therefore, if you were asked "What does an atom mainly consist of?" the answer would have to be "Empty space!" The Bohr model of the atom is sometimes compared to our own planetary system: the earth (electron) revolves around the much heavier sun (proton), with a vast gulf of space in between.

Next to hydrogen, the lightest element is the gas helium (He), whose atom is about four times heavier than the atom of hydrogen. We would therefore assume that the helium nucleus contained four protons; however, there are only two orbiting electrons, and this would mean that all helium was positively charged, whereas, of course, helium is electrically neutral. This difficulty was solved in 1932 by the discovery of another particle, the *neutron*, which has about the same mass as the proton but is uncharged. Consequently, our picture of the helium atom is that of a nucleus with two neutrons and two protons, which are prevented from flying apart by powerful short-range nuclear forces. Around the nucleus revolve the two orbiting electrons (Fig. 1-7). In fact, the atoms of all the elements are composed of the same three subatomic particles: electrons, protons, and neutrons.

Each of the elements in the periodic table has two associated numbers: the atomic number and the atomic weight. The *atomic number* distinguishes one element from another and is equal to the number of protons in the nucleus; this is also the same as the number of the orbiting electrons. The atomic numbers start at 1 (hydrogen) and proceed in sequence to over 100. By contrast, *atomic weights* compare the masses of the atoms and are not in sequence, since they also account for the number of neutrons in the nucleus. To take a complex atom as an example, uranium (U) has an atomic number of 92

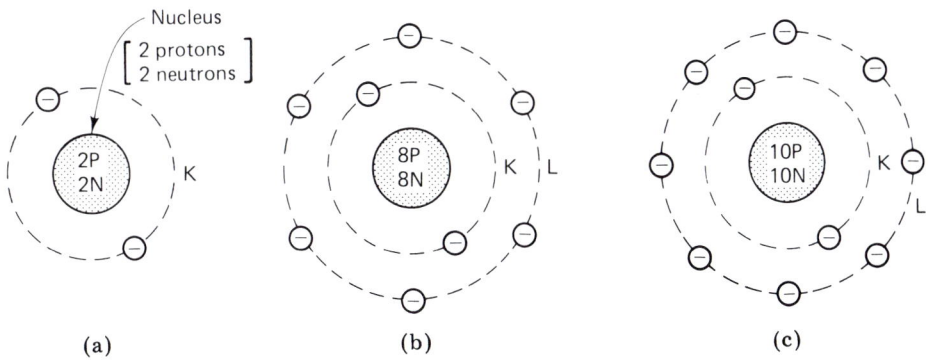

FIGURE 1-7 Atomic structure of various elements: (a) helium; (b) oxygen; (c) neon.

Sec. 1-2 Atomic Structure. Conductors, Insulators, and Semiconductors

and an atomic weight of 238. Consequently, the uranium atom will have 92 protons, 92 orbiting electrons, and 238 − 92 = 146 neutrons. This is the type of uranium that is commonly mined from the earth. However, the uranium used in an atom bomb has an atomic weight of only 235. There are still 92 protons and electrons but only 143 neutrons; because of the different number of neutrons, this other form of uranium is known as an *isotope*.

We have reached the point where we know that compounds are made up of molecules that can be broken down chemically into the atoms of the elements. These atoms consist of protons and neutrons within a nucleus around which revolves a number of electrons (Fig. 1-8). The question arises: "Do the electrons move in random orbits or is their movement structured in some manner?" In general, the electrons are distributed between a number of *shells*, which are imaginary surfaces on which the electrons are considered to revolve. The shells are arranged in steps which correspond to certain fixed energy levels. The major shells are labeled *KLMNOPQ* (Fig. 1-9), with the *K* shell representing the lowest energy level. Consequently, if sufficient energy is supplied to an electron in a particular shell, the electron will jump to a new orbit which is farther from the nucleus. Subsequently, the electron will return to its original orbit and energy will be emitted in the form of light (Fig. 1-10); this is the principle that enables a picture to be formed on the screen of a TV tube.

For each shell: (1) there is a maximum number of electrons that can be accommodated and (2) the shells must be filled in the sequence *K*, *L*, and so on. The shells are considered to be filled when they contain the numbers shown in Table 1-4.

The major shells may be divided into *subshells*, which also represent different energy levels. Within a major shell there can be up to four subshells, which are labeled *s*, *p*, *d*, and *f*. The maximum numbers of electrons that subshells can contain are also shown in Table 1-4.

The only elements with their major shells completely filled are the *inert gases*—helium (He), neon (Ne), argon (Ar), krypton (Kr), xenon (Xe), and radon (Rn)—which are grouped together on the right-hand side of the periodic table. For example, neon has an atomic number of 10; two of the ten orbiting electrons will reside in the *K* shell and the other eight in the *L* shell (Fig. 1-7).

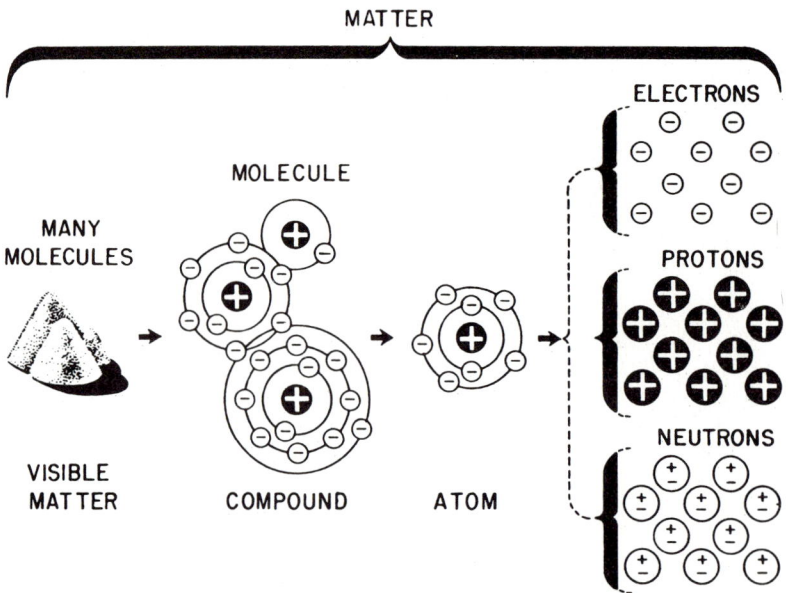

FIGURE 1-8 Breakdown of matter into subatomic particles.

FIGURE 1-9   Arrangement of shells in an atom.

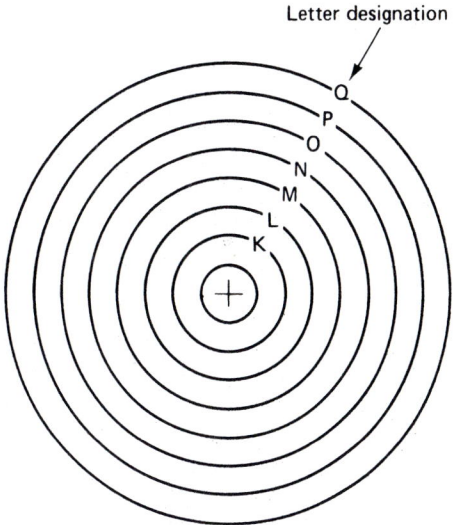

FIGURE 1-10   Movement of an electron between different energy levels.

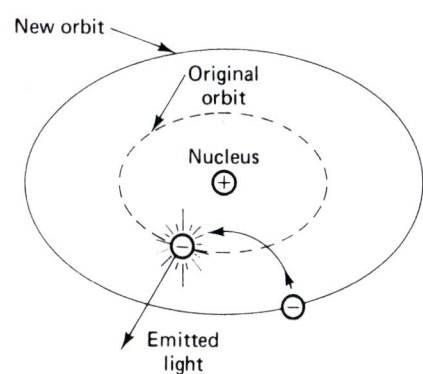

TABLE 1-4   Maximum Number of Electrons in Shells and Subshells

	Maximum number of electrons
Shell	
K	2
L	8
M	18
N	32
O	18
P	12
Q	2
Subshell	
s	2
p	6
d	10
f	14

Because their shells are completely filled, the inert gases are stable in the sense that they are reluctant to combine with the other elements.

By contrast with the inert gases, copper (Cu) has an atomic number of 29. Consequently, the K, L, and M shells will be completely filled, for a total of 28 electrons, and there will be only one electron in the N shell. The number of electrons in this last shell (or subshell) is called the *valence* of the atom, so that copper has a valence of 1. The outermost or valence shell therefore contains those electrons that are farthest away from the nucleus. A particular atom's valence governs its ability to either gain or lose an electron, which in turn governs the element's chemical and electrical properties. When the valence shell or subshell is stable, it will contain a total of eight electrons. If only one or two electrons are lacking, the atom will easily gain these, so that the shell is complete, but a large amount of energy will then be required to free any of the valence electrons from the shell. On the other hand, if there are only one or two electrons in the valence shell, these can be freed with only a small amount of energy.

When an atom gains one or more electrons, the total number of negative electrons will exceed the number of positive protons in the nucleus. The atom will then be negatively charged and is called a *negative ion*. However, if an atom loses one or more electrons, the number of protons will be greater than the number of the remaining electrons and the result will be a *positive ion*. To illustrate the formulation of ions we can refer back to the water molecule with its two hydrogen atoms and one oxygen atom. Hydrogen (H) has a valence of 1, while the valence of oxygen (O) is 6 (Fig. 1-7). The oxygen atom therefore captures the two electrons from the hydrogen atoms to fill its valence shell with eight electrons. Both the hydrogen atoms become positive ions, while the oxygen atom is a negative ion. The two positive ions and the single negative ion are then bonded together by their force of attraction to create a stable molecule of water (Fig. 1-11).

Conductors, Insulators, and Semiconductors

Conductors are those materials that allow electricity to flow easily with minimum opposition. Such materials will contain a large number of *free electrons*,

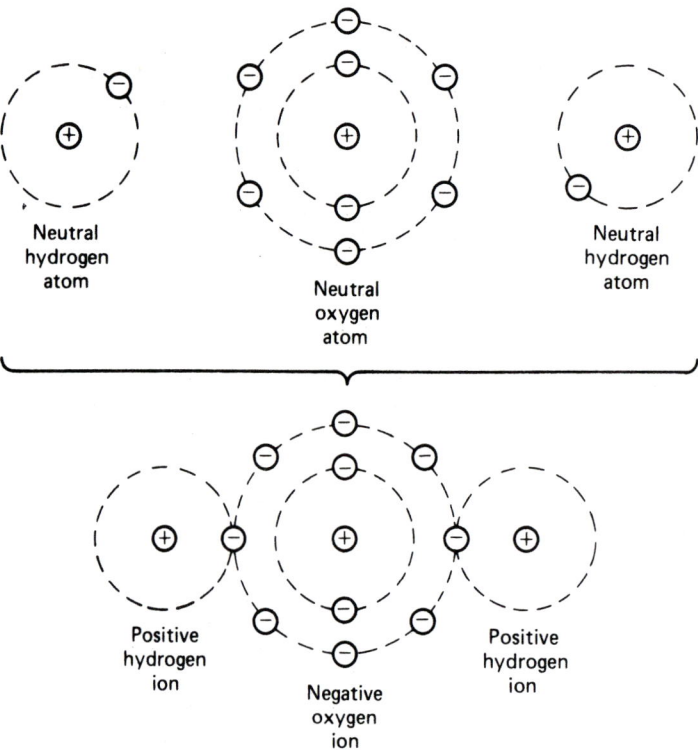

FIGURE 1-11 Formation of the water molecule.

which are loosely bound to their nuclei and are therefore able to move without difficulty from atom to atom. Generally speaking, all metals are good conductors and the three best in descending order are silver (Ag), copper (Cu), and gold (Au). It is no surprise to learn that each of these metals has a valence of 1. When an electric force is applied, this single valence electron is easily removed from its parent atom and becomes a free electron, capable of moving to another atom. Consequently, our concept of an electrical current is a flow of negative charges associated with the movement of the free electrons through a conductor. The next best conducting metal is aluminum, which has a valence of 3 and therefore offers more opposition to the flow of electricity than copper.

Insulators, such as rubber, paper, mica, and glass, are the opposite of conductors and are used to prevent the flow of electricity. These materials are normally compounds in which the valence electrons of certain atoms are used to fill the valence shells of other atoms. Consequently, no free electrons are naturally available and a large amount of energy must be added before any electrons can be freed from their orbits. Pure water, which is a compound, is an insulator and does not conduct electricity easily. It is the dissolved salts and traces of acid that cause tap water to be a good conductor.

Halfway between conductors and insulators are the *semiconductors*, such as germanium (Ge) and silicon (Si). Each of these elements has a valence of 4, so that the number of free electrons available is limited. The semiconductors are therefore neither good conductors nor good insulators; however, silicon and germanium are extremely important in the manufacture of solid-state devices such as diodes, transistors, and integrated circuits (ICs).

In this section we have looked at the atomic structure of the elements in terms of the electron, proton, and neutron particles. In particular we examined the arrangement of the electrons into shells and subshells, the development of ions, and the importance of valence in distinguishing among conductors, semiconductors, and insulators.

EXAMPLE 1-5 Referring to the periodic table, describe the structure of the atoms for the elements neon (Ne), gold (Au), silicon (Si), and sulfur (S).

Solution
Neon
Atomic number 10. Atomic weight 20.183. Nucleus contains 10 protons and $20 - 10 = 10$ neutrons. The 10 orbiting electrons are arranged: K shell, 2; L shell, 8. Valence = 8 (inert gas).
Gold
Atomic number 79. Atomic weight 196.967. Nucleus contains 79 protons and $196 - 79 = 117$ neutrons. The 79 orbiting electrons are arranged: K shell, 2; L shell, 8; M shell, 18; N shell, 32; O shell, 18; P shell, 1. Valence = 1 (conductor).
Silicon
Atomic number 14. Atomic weight 28.086. Nucleus contains 14 protons and $28 - 14 = 14$ neutrons. The 14 orbiting electrons are arranged: K shell, 2; L shell, 8; M shell, 4. Valence = 4 (semiconductor).
Sulfur
Atomic number 16. Atomic weight 32.064. Nucleus contains 16 protons and $32 - 16 = 16$ neutrons. The 16 orbiting electrons are arranged: K shell, 2; L shell, 8; M shell, 6. Valence = 6 (insulator).

1-3 Units of Current and Charge

In the 1830s, the English scientist Michael Faraday (1791–1867) performed a series of experiments on electrolytic conduction, which involved the flow of current through a solution. Figure 1-12 illustrates the type of experiment that Faraday carried out. A bar of silver and a nickel plate are immersed in a silver nitrate solution and are referred to as the *electrodes*. The silver bar is called the *anode* and is connected to one terminal of a battery, while the nickel plate or *cathode* is connected to the other terminal. Faraday observed that as time passed, silver was lost from the anode and an equal amount was deposited on the cathode; moreover, the rate of silver transfer was constant. This is the principle behind the electroplating of objects such as spoons and forks, which have the appearance of solid silver but are, of course, much cheaper.

At the time of these experiments the existence of the electron was unknown, and therefore Faraday logically concluded that the movement of the electricity through the circuit was carrying the silver from the anode to the cathode. The battery terminal connected to the anode was considered to be positive and the negative terminal was regarded as joined to the cathode (in the circuit symbol for the battery ─┤├─, the longer line is used to represent the positive terminal and the shorter line is the negative terminal). The movement of the electricity was then assumed to be in the direction from positive to negative. Unfortunately for Faraday, electrical flow is much more complex. Positive silver ions are moving from the anode to the cathode, while negative nitrate ions are traveling in the opposite direction through the electrolyte. In the external copper connecting wires, negative electrons are leaving the negative battery terminals and are entering the positive terminal. The *actual* or *physical electron flow of current* is therefore in the opposite direction to the Faraday assumption. However, extensive use is still made of the *conventional* or *mathematical current*, which flows from the positive battery terminal through the external circuit to the negative terminal. In this textbook we generally use the physical electron flow to represent the current.

Whether we choose electron flow or conventional current, the current we have met so far is in one direction only and never reverses. This type of flow, which is always associated with a battery, is called *direct current* (dc); by contrast, the normal household current regularly reverses its direction and is referred to as *alternating current* (ac).

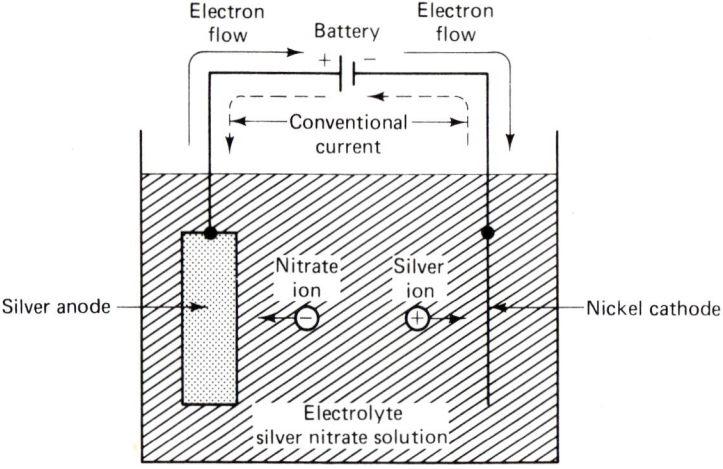

FIGURE 1-12 Electrolytic conduction.

In 1835, Faraday stated his *first law of electrolysis:* the mass of silver leaving the anode and deposited on the cathode is directly proportional to the quantity of electricity or charge passing through the electrolyte. At one time this law formed the basis for the international unit of charge, which was named the coulomb after the French scientist Charles A. Coulomb (1736–1806). The coulomb was that quantity of electricity which when passed through a silver nitrate solution, caused 1.118×10^{-6} kg of silver to be deposited on the cathode. The letter symbol for the charge is Q and the unit symbol for the coulomb is C. The coulomb is the practical unit of charge in the SI system. From our study of atomic structure it follows that the colomb must be equivalent to the negative charge associated with a certain number of electrons. In fact, the coulomb represents the electrical charge carried by 6.24×10^{18} electrons, so that the charge, e, possessed by a single electron is the incredibly small amount of $1/(6.24 \times 10^{18}) = 1.602 \times 10^{-19}$ C—further evidence of our need for scientific notation.

In the SI system we do not start with the coulomb, so we must establish another fundamental unit based on electricity. In 1948 this was selected to be the ampere, which is the unit of electrical current and is named after the French physicst André Marie Ampère (1775–1836). The letter symbol for current is I, which is derived from the French word *intensité*, and the unit symbol for the ampere (sometimes abbreviated "amp") is A.

The definition of the ampere is based on an effect that occurs in electromagnetism and is discussed fully in Section 7-3. For our present needs it will only be necessary to quote the definition. The international ampere is "that value of current which when flowing in each of two infinitely long parallel conductors whose centers are separated by 1 meter in a vacuum, causes a force of 2×10^{-7} newton per meter length to be exerted on each conductor." The current balance that is needed to standardize the ampere based on this definition is very expensive and complex, so that it is not normally available outside national physical laboratories.

The ampere, which measures the current, must represent the flow of coulombs around an electrical circuit. We can therefore derive the coulomb from the ampere. When a current of 1 ampere flows for a time of 1 second, a charge of 1 coulomb passes a particular point in the electrical circuit. Therefore, amperes are equivalent to coulombs per second, or coulombs are the same as amperes × seconds. The equations are

$$I = \frac{Q}{t}, \qquad Q = I \times t \qquad \text{(1-3-1)}$$

where I = current [amperes (A)]
Q = charge [coulombs (C)]
t = time [seconds (s)]

For example, if the current throughout an electrical circuit is 5 A, a charge of 5 C will pass any point in that circuit in a time of 1 s.

A larger unit of charge is the ampere-hour (Ah), which is equivalent to $1 \times 60 \times 60 = 3600$ C. For example, if a battery delivers a current of 4 A for a time of 3 h, the charge lost is $3 \times 4 = 12$ Ah $= 12 \times 3600 = 43{,}200$ C.

Referring back to the Faraday experiment with a silver nitrate solution, a current of 1 A flowing for a time of 1 s (equivalent to a charge of 1 coulomb) will deposit 1.118×10^{-6} kg of silver on the cathode. This quantity, 1.118×10^{-6} kg/C, is called silver's *electrochemical equivalent*, which has the letter symbol z. The total mass, M, liberated from the anode will therefore be given by

$$M = zQ = zIt \qquad (1\text{-}3\text{-}2)$$

where M = liberated mass of silver [kilograms (kg)]
z = electrochemical equivalent of silver [kilograms per coulomb (kg/C)]
Q = charge [coulombs (C)]
I = current [amperes (A)]
t = time [seconds (s)]

Other electrochemical equivalent values are: copper, 3.294×10^{-7} kg/C; nickel, 3.04×10^{-7} kg/C; and zinc, 3.38×10^{-7} kg/C.

The section defined the units of current and charge, and established their relationship. A current of 1 A is equivalent to a charge of 1 C passing a point in 1 s. The charge of 1 C is equivalent to the charge carried by 6.24×10^{18} electrons. The physical electron current flows from negative to positive, whereas the conventional current (a mathematical concept) is in the opposite direction. In the electrolytic action a charge of 1 coulomb results in a deposit of 1.118×10^{-6} kg of silver on the cathode.

EXAMPLE 1-6 If a steady current of 3 A is maintained at a point for a period of 20 min, calculate the amount of charge flowing past that point in coulombs and ampere-hours.

Solution

$$I = 3\ \text{A}, \qquad t = 20\ \text{min} = 20 \times 60\ \text{s} = 1200\ \text{s}$$

$$\text{charge},\ Q = I \times t \qquad (1\text{-}3\text{-}1)$$

$$= 3\ \text{A} \times 1200\ \text{s} = 3600\ \text{C}$$

Also

$$t = 20\ \text{min} = \frac{20}{60}\ \text{h} = \frac{1}{3}\ \text{h}$$

$$\text{charge},\ Q = I \times t \qquad (1\text{-}3\text{-}1)$$

$$= 3\ \text{A} \times \frac{1}{3}\ \text{h} = 1\ \text{Ah}$$

EXAMPLE 1-7 A charge of 900 C flows past a point in 3 min. Calculate the value of the current in amperes.

Solution

$$Q = 900\ \text{C}, \qquad t = 3\ \text{min} = 3 \times 60\ \text{s} = 180\ \text{s}$$

$$\text{current},\ I = \frac{Q}{t} \qquad (1\text{-}3\text{-}1)$$

$$= \frac{900\ \text{C}}{180\ \text{s}} = 5\ \text{A}$$

EXAMPLE 1-8 A constant current of 5 A flows through a copper sulfate solution for a period of 1 h. Calculate the mass liberated from the copper anode.

Solution

$$z = 3.294 \times 10^{-7} \text{ kg/C}, \quad I = 5\text{A}, \quad t = 1 \text{ h} = 60 \times 60 \text{ s} = 3600 \text{ s}$$

$$\text{mass liberated, } M = zIt \tag{1-3-2}$$

$$= 3.294 \times 10^{-7} \text{ kg/C} \times 5 \text{ A} \times 3600 \text{ s}$$

$$= 3.294 \times 10^{-7} \times 5 \times 3.6 \times 10^{3} \text{ kg}$$

$$= 59.29 \times 10^{-4}$$

$$= 5.929 \times 10^{-3} \text{ kg}$$

1-4 Unit of Electromotive Force

In establishing the electrical units, we started off by defining the ampere as the fourth fundamental unit of the MKSA or SI system. From the ampere and the second we derived the coulomb as the unit of charge or quantity of electricity. We also have the joule as the unit of electrical energy or work; this unit was borrowed from the mechanical system of units on the principle that the various forms of energy are interchangeable. But what drives the electricity around the circuit? In other words, what is the electrical equivalent of the mechanical force that accelerates a mass and therefore gives it a velocity? The answer is the *electromotive force* (the force that gives electricity its motion), whose letter symbol is E. The electromotive force is commonly abbreviated EMF, which in some textbooks is taken to mean "electron moving force." The unit of EMF is the volt, named after the inventor of the first chemical cell to generate electricity, the Italian Count Alessandro Volta (1745–1827); the unit symbol for the volt is the letter V.

We must already be familiar with the volt in our everyday lives. A car battery usually has an EMF of 12 V, while the EMF of the D cell for a flashlight is 1.5 V. In the case of a car battery this means that there are 12 V of EMF available to drive the electricity through such circuits as the car's ignition system, the lights, the radio, and so on. But, of course, you do not get something for nothing. Work must be done in driving the charge through the circuit and establishing the current; the battery must supply the energy in order to perform this work. We are therefore able to define the volt in terms of the joule and the coulomb. The EMF is 1 volt if when 1 coulomb of charge is driven around an electrical circuit, the work done is 1 joule. A much smaller unit of work is the electron-volt (eV), which is the work done when the charge carried by one electron is being driven by 1 volt. Since the electron charge is $1/(6.24 \times 10^{18})$ C, 1 J = 6.24×10^{18} eV. If we increase the EMF and the amount of charge driven through a circuit, the greater will be the amount of work done. Therefore,

$$W = Q \times E, \quad Q = \frac{W}{E}, \quad E = \frac{W}{Q} \tag{1-4-1}$$

where W = work done [joules (J)]
E = EMF [volts (V)]
Q = charge [coulombs (C)]

For example, if an EMF of 5 V drives 2 C around an electrical circuit, the work done is 5 V × 2 C = 10 J.

We already know that the power is a measure of the rate at which work is performed. From Eqs. (1-4-1), (1-1-6), and (1-3-1), the power, P, is given by

$$P = \frac{W}{t} = \frac{Q}{t} \times E = I \times E \qquad (1\text{-}4\text{-}2)$$

Therefore,

$$I = \frac{P}{E}, \quad E = \frac{P}{I} \qquad (1\text{-}4\text{-}3)$$

where P = power [watts (W)]
I = current [amperes (A)]
E = EMF [volts (V)]

The electrical power in watts is therefore the product of the current in amperes and the EMF in volts. This relationship is a direct result of the units' definitions and should be compared with Eq. (1-1-7), which shows that mechanical power is the product of the force and the velocity. Consequently, we can say that the force in newtons is the mechanical equivalent of the EMF in volts and that the velocity of the mass corresponds to the electrical current. You can remember the three equations connecting P, I, and E by using the "PIE" diagram of Fig. 1-13. For example, to find the expression for the current, cover up I with your hand; what remains is P/E, which is therefore the expression for I.

To take an everyday example, let us consider a 60-W light bulb which is connected across an electrical source whose EMF is 120 V. The current $I = P/E = $ 60 W/120 V = 0.5 A. However, if a 1.5-kW (1500-W) heater is connected across the same source, the heater current is 1500 W/120 V = 12.5 A. The heater takes far more current than the light bulb and is therefore more expensive to operate. In a time of 1 h the light bulb draws 60 watt-hours (Wh) of energy from the source while the heater requires 1.5 kilowatt-hour (kWh); we could in fact operate 25 such bulbs for the same cost as the single heater.

We have added the volt to our electrical system of units and therefore we

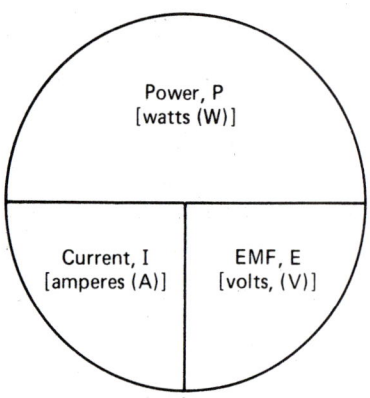

FIGURE 1-13 PIE diagram.

now have the ampere, coulomb, joule, volt, and watt. We know that the EMF, E, develops the current, I, but we are not aware of their exact relationship; this forms the main topic for the next section.

EXAMPLE 1-9 An EMF of 12 V drives a charge of 3 C through an electrical circuit. Calculate the amount of work done. If this work is performed in a time of 4 s, what is the power involved?

Solution

$$E = 12 \text{ V}, \quad Q = 3 \text{ C}$$

$$\text{work, } W = Q \times E \quad (1\text{-}4\text{-}1)$$

$$= 12 \text{ V} \times 3 \text{ C} = 36 \text{ J}$$

$$W = 36 \text{ J}, \quad t = 4 \text{ s}$$

$$\text{power, } P = \frac{W}{t} \quad (1\text{-}1\text{-}6)$$

$$= \frac{36 \text{ J}}{4 \text{ s}} = 9 \text{ W}$$

EXAMPLE 1-10 For a period of 4 h a current of 5 A is drawn from an electrical source whose EMF is 120 V. Calculate the power and the total energy consumed from the source.

Solution

$$E = 120 \text{ V}, \quad I = 5 \text{ A}$$

$$\text{power, } P = E \times I \quad (1\text{-}4\text{-}2)$$

$$= 120 \text{ V} \times 5 \text{ A} = 600 \text{ W}$$

$$P = 600 \text{ W}, \quad t = 4 \text{ h}$$

$$\text{energy consumed, } W = P \times t \quad (1\text{-}1\text{-}6)$$

$$= 600 \text{ W} \times 4 \text{ h}$$

$$= 2400 \text{ Wh}$$

$$= \frac{2400}{1000} \text{ kWh} = 2.4 \text{ kWh}$$

1-5
OHM'S Law. Resistance. Conductance

Ohm's law is probably the most quoted and most misunderstood relationship in electronics. It is only logical to suppose that if we increase the EMF applied to a circuit, the current will also increase. If that was the sum total of Ohm's law, it would not be saying very much. However, in 1827 Georg Simon Ohm

(1787–1854) stated that under constant physical conditions of temperature, humidity and pressure, the current flowing through a conductor is *directly proportional* to the EMF applied across a conductor. This means that if we triple the voltage, the current will be tripled, and if we halve the voltage, the current will be divided by 2; whatever we do to the voltage, the same will happen to the current. Do not imagine that this is a law which is obviously true. Although most components in electronics do obey Ohm's law, a number do not. For example, if we double the voltage applied across a transistor, the current increases but does not double, and therefore a transistor does not obey Ohm's law.

Figure 1-14 shows a battery with an EMF of E volts connected to a conductor which is in the form of a very long length of wire. We then have a closed circuit through which current can flow. A meter, A, is placed in the path of the current, I, and will record its value in amperes; such a meter is known as an *ammeter*. Let us assume that when E is 10 V, the recorded current is 2 A. If we double the EMF to 20 V, the current will also double, to 4 A. If we again double the voltage to 40 V, the new current will be 8 A. When the original voltage is multiplied by 10 so that the EMF becomes $10 \times 10 = 100$ V, the resulting current is $10 \times 2 = 20$ A. Finally, if we reduce the EMF to 10 V/2 = 5 V, the current will fall to 2 A/2 = 1 A. For this circuit the corresponding values of E and I are shown in Table 1-5.

For each corresponding voltage and current let us calculate the value of the ratio E to I. This is written as $E:I$ or E/I and is the result of dividing E by I. As Table 1-5 shows, the answer is always the same, "5," which is therefore a constant for the particular circuit of Fig. 1-14. Ohm's law can then be stated in another way: "under constant physical conditions the ratio of the voltage applied across a conductor to the current flowing through the conductor is a constant." What is the meaning of this constant? Suppose that we used more wire, so that the value of the constant in the new circuit was increased to "10." If the voltages remained the same, all the current values would be halved. The constant is therefore a measure of the conductor's opposition to current flow and is called its *resistance*. The letter symbol for resistance is R and its schematic symbol is ⌇⌇. The unit of resistance is the ohm, whose symbol is the Greek capital letter omega, Ω. In Fig. 1-14 the conductor's resistance is 5 Ω.

The equations for Ohm's law are

$$\frac{E}{I} = R, \quad E = I \times R, \quad R = \frac{E}{I} \qquad (1\text{-}5\text{-}1)$$

These relationships may be found with the aid of Fig. 1-15. The method is the same as we used for the equation relating P, E, and I.

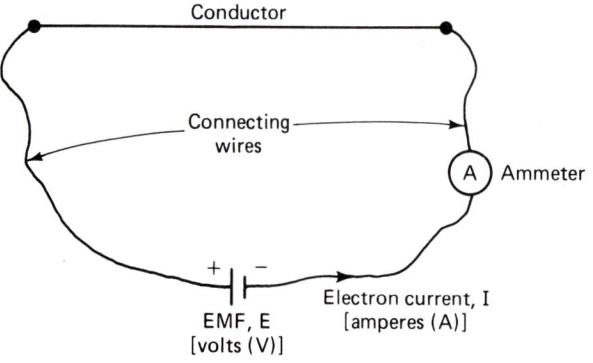

FIGURE 1-14 Circuit to illustrate Ohm's law.

TABLE 1-5 Values of E and I for the Circuit of Fig. 1-14

E(V)	I(A)	$\dfrac{E}{I}$
5	1	5
10	2	5
15	3	5
20	4	5
25	5	5
30	6	5
35	7	5
40	8	5
45	9	5
50	10	5
55	11	5
60	12	5
65	13	5
70	14	5
75	15	5
80	16	5
85	17	5
90	18	5
95	19	5
100	20	5

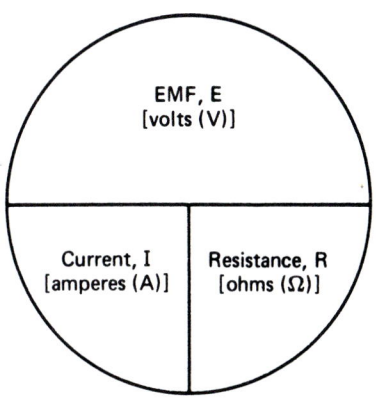

FIGURE 1-15 Diagram used to find the Ohm's law equations.

The statement that "$E/I = R$ is Ohm's law" is often made; but this is not so unless R has a constant value. For example, we could apply an EMF across a light bulb and measure the current in the circuit. Dividing the voltage by the current would give us the value of the resistance. If we then doubled the voltage, the current would increase but would be less than double its original value. When we divided the increased voltage by the greater current, the new value of resistance would be more than the old value. The light bulb therefore does not obey Ohm's law, although we have used the relationship $E/I = R$.

Resistance is one of the three properties which an electrical circuit can possess. The other two are inductance and capacitance, which are discussed in Chapters 8 and 9. Resistance is that property which opposes or limits current in an electrical circuit. The component that possesses the property of resistance is called a *resistor*, practical examples of which are described in Section 1-8.

If we plot a graph of E versus I using the values in Table 1-5, we obtain the straight line illustrated in Fig. 1-16. With a *straight-line graph* there is a linear relationship between the voltage and the current; this is the graphical way of saying that the current is directly proportional to the voltage. Any time there is a law of direct proportion between two quantities, their graphical relationship is a straight line which passes through the origin, O. The resistance

Sec. 1-5 OHM's Law. Resistance. Conductance

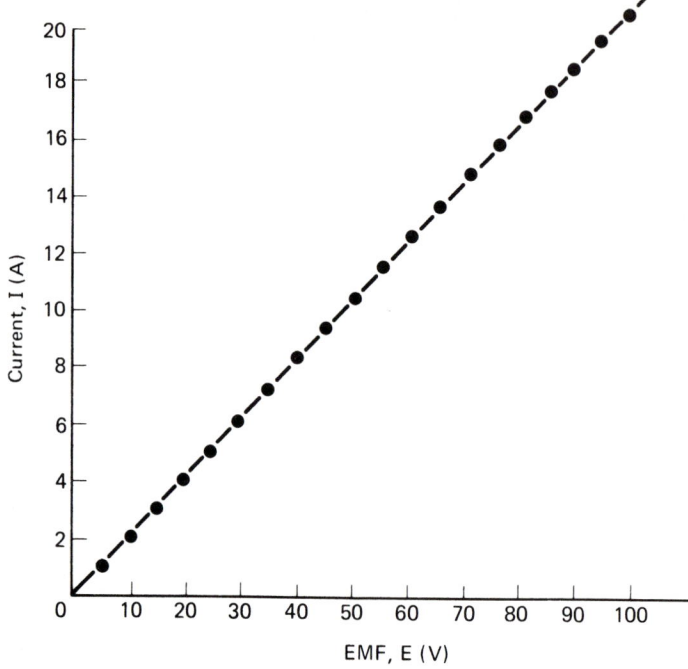

FIGURE 1-16 Voltage/current relationship for a linear resistor that obeys Ohm's law.

that obeys Ohm's law is therefore a linear resistance. By contrast, light bulbs and certain other electrical devices do not obey Ohm's law; their resistances are nonlinear and their voltage-current graphs are curves rather than straight lines. However, we must emphasize that most of the components used in electronics do obey Ohm's law over their operating range.

POTENTIAL DIFFERENCE ACROSS A RESISTOR In Fig. 1-17, a dc source whose EMF, E, is 10 V is connected across a resistor whose value is 5 kilohms (kΩ). The electron flow from A to B through the resistor develops a voltage, V_R, across the resistor. Because of the direction of the flow, point A is negative with respect to point B or, alternatively, we can say that point B is positive with respect to point A. This voltage across the resistor is sometimes called the *potential difference* (PD) or *difference of potential* (DP) between points A and B; the word "potential" has the same meaning as "voltage." Notice that a voltage exists between two points; for the time being we cannot refer to the voltage or potential at a single point.

The voltage across the resistor in this example must exactly balance the EMF of the electrical source, so that in our example V_R is also 10 V. As a

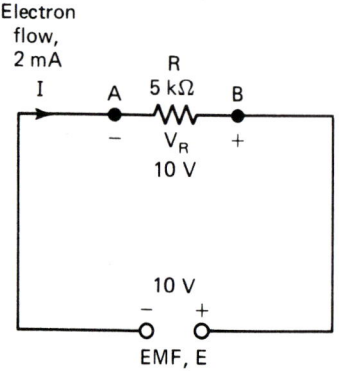

FIGURE 1-17 Voltage developed across a resistor.

mechanical analogy, think of using a pump to drive water through a thin pipe. The back pressure exerted by the pipe would balance the forward pressure of the pump. In the same way, a book resting on a table produces a downward force on the table due to gravity; this is balanced by an equal upward force exerted by the table on the book.

As well as being called the *potential difference*, V_R is also referred to as the *IR voltage* across the resistor. The meaning of "*IR*" is self-evident, since V_R can be calculated by multiplying the current, I, by the value of the resistor, R.

The three Ohm's law equations for the circuit of Fig. 1-17 are:

$$\text{current, } I = \frac{E}{R} = \frac{10 \text{ V}}{5 \text{ k}\Omega} = \frac{10 \text{ V}}{5000 \text{ }\Omega} = 0.002 \text{ A} = 2 \text{ mA}$$

$$\text{EMF, } E = I \times R = 2 \text{ mA} \times 5 \text{ k}\Omega = 0.002 \text{ A} \times 5000 \text{ }\Omega = 10 \text{ V}$$

$$\text{resistance, } R = \frac{E}{I} = \frac{10 \text{ V}}{2 \text{ mA}} = \frac{10 \text{ V}}{0.002 \text{ A}} = 5000 \text{ }\Omega = 5 \text{ k}\Omega$$

These equations show that when we divide volts (V) by kilohms (kΩ), we obtain a current in milliamperes (mA). Similarly, when we multiply a current in mA by a resistance in kΩ, the answer is an EMF in volts. Finally, if we divide an EMF in volts by a current in mA, the result is a resistance in kΩ. These are shortcuts that are important in electronics because most EMFs are given in volts, most resistances are measured in kilohms, and therefore most currents are in the order of milliamperes. Similar relationships hold true for volts, megohms (millions of ohms), and microamperes (millionths of an ampere). These are:

$$\frac{\text{volts (V)}}{\text{megohms (M}\Omega)} = \text{microamperes (}\mu\text{A)}$$

$$\text{volts (V)} = \text{microamperes (}\mu\text{A)} \times \text{megohms (M}\Omega)$$

$$\frac{\text{volts (V)}}{\text{microamperes (}\mu\text{A)}} = \text{megohms (M}\Omega)$$

POWER DISSIPATED IN A RESISTOR When a current, I, is drawn from a battery whose EMF is E, that source is supplying power to the resistor which is connected between the battery's terminals; the value of that power is IE watts [Eq. (1-4-2)]. Over a period of time, energy will be drawn from the source, but in what form does this energy appear? The answer is "heat!" The resistor gets hot as a result of friction between the free electrons and the atoms of the material from which the resistor is made. A good example is provided by a 60-W light bulb operated from a 120-V dc source. The current through the bulb's fine tungsten filament is $I = P/E = 60 \text{ W}/120 \text{ V} = 0.5 \text{ A}$, and the filament's resistance is $R = E/I = 120 \text{ V}/0.5 \text{ A} = 240 \text{ }\Omega$. When the current passes through the filament, 60 J of energy are drawn from the battery every second. This energy is dissipated as heat in the filament, whose temperature rises to over 2000°C. The filament is then white-hot, so that light is emitted. The word "dissipation" is used in the sense of "lost," since the heat energy cannot be returned to the battery. A fuse is a good example of heat dissipation; it contains a metal link which melts when the current through the fuse exceeds a certain value.

Household items such as toasters and light bulbs are called electrical *loads*, which perform a useful function by converting electrical energy into some other form of energy, such as heat. Each of these loads is rated at a certain power

Sec. 1-5 OHM's Law. Resistance. Conductance

for a given applied voltage, which is normally 120 V. For example, a heater might have a rating of 1.5 kW at 120 V; this means that the heater dissipates 1500 W when connected to a 120-V electrical source.

Notice that in our example with the light bulb, we were given the values of the power (wattage) rating and the voltage; we were then able to calculate the current and the resistance. In general, if we know any two of the four quantities P, E, I, and R, we can calculate the values of the other two. We already know six relationships: $P = IE$, $I = P/E$, $E = P/I$, $E = IR$, $I = E/R$, and $R = E/I$, but there are six more to help us with our calculations. From Eqs. (1-4-2) and (1-5-1), we obtain

$$\text{power, } P = I \times E$$

But

$$E = I \times R; \text{ therefore,}$$
$$P = I \times (I \times R)$$

or

$$\text{power, } P = I^2 R \quad \text{watts} \quad (1\text{-}5\text{-}2)$$

This equation may be arranged in two other forms:

$$\text{current, } I = \sqrt{\frac{P}{R}} \text{ amperes} \quad \text{and} \quad \text{resistance, } R = \frac{P}{I^2} \text{ ohms} \quad (1\text{-}5\text{-}3)$$

In addition, since $I = E/R$,

$$\text{power, } P = I \times E$$
$$= \frac{E}{R} \times E$$

or

$$\text{power, } P = \frac{E^2}{R} \quad \text{watts} \quad (1\text{-}5\text{-}4)$$

These equations may also be arranged in two other forms:

$$\text{voltage, } E = \sqrt{P \times R} \text{ volts} \quad \text{and} \quad \text{resistance, } R = \frac{E^2}{P} \text{ ohms} \quad (1\text{-}5\text{-}5)$$

We now have a total of 12 relationships connecting E, I, P, and R. These may be summarized as follows:

$$\text{electromotive force, } E = I \times R = \sqrt{P \times R} = \frac{P}{I} \quad \text{volts}$$

$$\text{current, } I = \frac{E}{R} = \sqrt{\frac{P}{R}} = \frac{P}{E} \quad \text{amperes}$$

$$\text{power, } P = \frac{E^2}{R} = I^2 R = I \times E \quad \text{watts}$$

$$\text{resistance, } R = \frac{E}{I} = \frac{P}{I^2} = \frac{E^2}{P} \quad \text{ohms}$$

These equations are illustrated in the memory aid of Fig. 1-18. To find a particular quantity such as the current, look at the I quadrant and select one of the three equations according to the information you have. For example, if you know the power and the resistance and you want to find the current, the equation to use is in the lower third of the quadrant, namely $I = \sqrt{P/R}$.

You will remember that we formed a group made up of values in volts, milliamperes, and kilohms. With these values the power would be measured in milliamperes × volts or milliwatts; the complete group is therefore V, mA, kΩ, and mW, so that in Fig. 1-17, the power would be $P = 2$ mA $\times 10$ V $= 20$ mW. The complete group involving voltage, microamperes, and megohms would be V, μA, MΩ, and μW. For example, if an EMF of 10 V is applied across a 2-MΩ resistor, the current is 10 V/2 MΩ $= 5$ μA and the power dissipated in the resistor is 5 μA $\times 10$ V $= 50$ μW.

Conductance

Our concept of resistance is one of "opposition" to current flow. However, another idea would be to introduce an electrical property which measures the "ease" with which current is allowed to flow in a circuit. Such a property is called the *conductance*, whose letter symbol is G. The SI unit for conductance is the siemens (S), but you may come across an older unit, the mho (ohm spelled backward), which is still widely used; 1 siemens has exactly the same value as 1 mho.

Thick copper wire offers little resistance and must therefore have a high conductance. Resistance and conductance are inversely related and are, in fact, reciprocals. Therefore,

$$\text{resistance, } R \, (\Omega) = \frac{1}{\text{conductance, } G \, (S)}$$

and

FIGURE 1-18 Ohm's law memory aid.

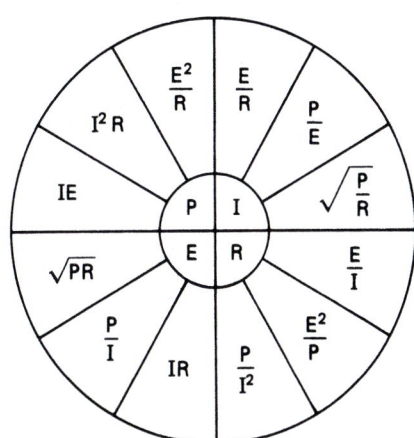

$$\text{conductance, } G \text{ (S)} = \frac{1}{\text{resistance, } R \text{ (}\Omega\text{)}} \qquad (1\text{-}5\text{-}6)$$

For example, a resistor with a value of 2 Ω would have a conductance of 1/2 Ω = 0.5 S; similarly, a conductor with a conductance of 500 S will have a resistance of 1/500 S = 0.002 Ω.

The equations relating G to E, I, and P are

$$\text{conductance, } G = \frac{I}{E} = \frac{I^2}{P} = \frac{P}{E^2} \quad \text{siemens} \qquad (1\text{-}5\text{-}7)$$

You may well ask: "Why do we need the idea of conductance when we already have the concept of resistance?" In Chapters 2 and 3 we join resistors together in two arrangements, which are called "series" and "parallel." With a series arrangement there is some advantage in using resistance rather than conductance but with parallel, the advantage lies with conductance.

You have seen that for a linear resistor which obeys Ohm's law, the voltage and the current are directly proportional and the resistance is constant. We derived the 12 relationships among E, I, P, and R, and introduced the concept of conductance as the reciprocal of resistance.

EXAMPLE 1-11   All of the following questions are related to Fig. 1-17.

(a) If $E = 20$ V and $I = 4$ A, find the values of P, R, and G.
(b) If $E = 100$ V and $P = 300$ W, find the values of I, R, and G.
(c) If $E = 36$ V and $R = 12$ kΩ, find the values of I and P.
(d) If $I = 6$ μA and $P = 54$ μW, find the values of E, R, and G.
(e) If $I = 7$ mA and $R = 2$ kΩ, find the values of E and P.
(f) If $P = 24$ mW and $R = 6$ kΩ, find the values of I and E.

Solution   In answering this question you should practice using Fig. 1-18.

(a) $E = 20$ V, $I = 4$ A

$$\text{Power, } P = I \times E = 4 \text{ A} \times 20 \text{ V} = 80 \text{ W}$$

$$\text{Resistance, } R = \frac{E}{I} = \frac{20 \text{ V}}{4 \text{ A}} = 5 \text{ }\Omega$$

$$\text{Conductance, } G = \frac{1}{R} = \frac{1}{5 \text{ }\Omega} = 0.2 \text{ S}$$

(b) $E = 100$ V, $P = 300$ W

$$\text{Current, } I = \frac{P}{E} = \frac{300 \text{ W}}{100 \text{ V}} = 3 \text{ A}$$

$$\text{Resistance, } R = \frac{E}{I} = \frac{100 \text{ V}}{3 \text{ A}} = 33.3 \text{ }\Omega$$

$$\text{Conductance, } G = \frac{I}{E} = \frac{3 \text{ A}}{100 \text{ V}} = 0.03 \text{ S}$$

(c) $E = 36$ V, $R = 12$ kΩ

$$\text{Current, } I = \frac{E}{R} = \frac{36 \text{ V}}{12 \text{ kΩ}} = 3 \text{ mA}$$

$$\text{Power, } P = I \times E = 3 \text{ mA} \times 36 \text{ V} = 108 \text{ mW}$$

(d) $I = 6$ μA, $P = 54$ μW

$$\text{EMF, } E = \frac{P}{I} = \frac{54 \text{ μW}}{6 \text{ μA}} = 9 \text{ V}$$

$$\text{Resistance, } R = \frac{E}{I} = \frac{9 \text{ V}}{6 \text{ μA}} = 1.5 \text{ MΩ}$$

$$\text{Conductance, } G = \frac{1}{R} = \frac{1}{1.5 \text{ MΩ}} = 0.67 \text{ μS}$$

(e) $I = 7$ mA, $R = 2$ kΩ

$$\text{EMF, } E = I \times R = 7 \text{ mA} \times 2 \text{ kΩ} = 14 \text{ V}$$

$$\text{Power, } P = I \times E = 7 \text{ mA} \times 14 \text{ V} = 98 \text{ mW}$$

(f) $P = 24$ mW, $R = 6$ kΩ

$$\text{Current, } I = \sqrt{\frac{P}{R}} = \sqrt{\frac{24 \text{ mW}}{6 \text{ kΩ}}} = \sqrt{4} = 2 \text{ mA}$$

$$\text{EMF, } E = I \times R = 2 \text{ mA} \times 6 \text{ kΩ} = 12 \text{ V}$$

1-6 Resistance of a Cylindrical Conductor. Specific Resistance

It is normal to use copper wiring to connect an electrical source to its load. Ideally, such wiring should have zero resistance, but in practice the resistance, although small, is not negligible. As the result, a small amount of the source's EMF will be lost across the connecting wires, and therefore a lower voltage will be applied to the load. In addition, some power will be dissipated as heat in the wiring. To reduce these effects to a minimum, it is necessary to use an adequate size for the connecting wire. If the current is of the order of milliamperes, American Wire Gage (AWG) No. 22 copper wire is sufficient, but, for household wiring, the wire must be thicker and AWG No. 14 is normally used. Notice that the higher the gage number, the thinner the wire and the less its *current-carrying capacity*, sometimes called its "ampacity."

A length of circular cross-section metal wire is an example of a cylindrical conductor (Fig. 1-19). If we join two identical lengths together, it is reasonable to assume that the resistance will be doubled; the resistance of the conductor is therefore directly proportional to its length, *l*. If the wire is made thicker by doubling its circular cross-sectional area, there are a greater number of free electrons available and the resistance will be halved; this means that the conductor's resistance is inversely proportional to the cross-sectional area, *A*; if

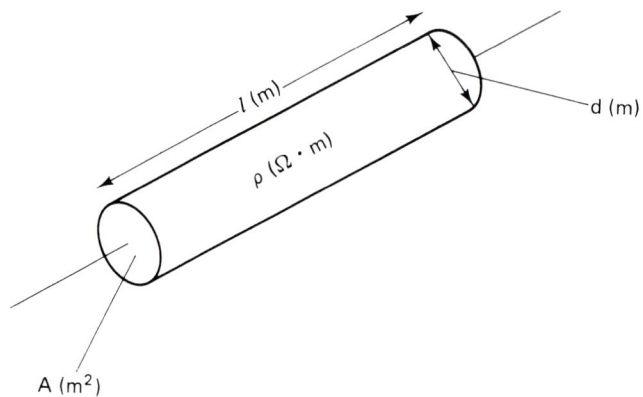

FIGURE 1-19 Resistance of a cylindrical conductor.

we know the conductor's diameter, d, the cross-sectional area, $A = \pi d^2/4$, where π is the circular constant, whose value is approximately 3.1416. Finally, the conductor's resistance must depend on the material used in its manufacture. The equation relating these factors is

$$R = \frac{\rho l}{A} = \frac{\rho l}{\pi d^2/4} = \frac{4\rho l}{\pi d^2} \qquad (1\text{-}6\text{-}1)$$

In terms of SI units:

R = conductor's resistance [ohms (Ω)]

l = length [meters (m)]

A = cross-sectional area [square meters (m²)]

d = diameter [meters (m)]

ρ = specific resistance or resistivity [ohm-meter ($\Omega \cdot$m)]

The specific resistance is the factor that allows us to compare the resistances of different materials. For a particular material a conductor that has a cross-sectional area of 1 m² and is 1 m long will have a resistance of ρ ohms. In another interpretation, the specific resistance is the resistance between opposite faces of a meter cube made from the material (a meter cube is a cube whose edges are all 1 m long). The SI unit of specific resistance is the ohm-meter ($\Omega \cdot$m).

Since the value of the specific resistance depends on the temperature, it is quoted for a particular value that is normally 20° or 25° Celsius (C), equivalent to 68° or 77° Fahrenheit (F); these temperatures are normally thought of as "room temperature." Table 1-6 shows the ρ values for various materials.

Copper is excellent for connecting wires but is far too good a conductor to be used for loads such as a light bulb or a toaster. For example, the element of a 600-W 120-V toaster has a resistance of $E^2/P = (120\text{ V})^2/600\text{ W} = 24\ \Omega$ and this could not be obtained with any practical length of copper wire with sufficient current-carrying capacity. For this reason a toaster element is made from nichrome wire, which has a much higher specific resistance than copper. As another example, the filament of a light bulb is made from tungsten, whose temperature can be raised to white heat without the filament melting or fracturing easily.

Unfortunately, it is still more common to quote ρ values which are based

TABLE 1-6 Specific Resistance (ρ) Values for Various Materials

Material	SI unit (specific resistance in $\Omega \cdot m$ at 20°C)	British unit (specific resistance in $\Omega \cdot cmil/ft$ at 20°C)
Silver	1.46×10^{-8}	9.86
Annealed copper	1.724×10^{-8}	10.37
Aluminum	2.83×10^{-8}	17.02
Tungsten	5.5×10^{-8}	33.08
Nickel	7.8×10^{-8}	46.9
Pure iron	1.2×10^{-7}	72.2
Constantan	4.9×10^{-7}	294.7
Nichrome	1.1×10^{-6}	660.0
Germanium (semiconductor)	0.55	3.3×10^{8}
Silicon (semiconductor)	550	3.3×10^{11}
Mica (insulator)	2×10^{10}	12.0×10^{18}

on the customary or British system of units. The length chosen is the foot and the diameter is given in linear mils, where 1 linear mil = 0.001 inch. For convenience the cross-sectional area is measured in circular mils, where 1 circular mil (cmil) is defined as the area of a circle whose diameter is 1 mil. The use of the circular mil avoids the complication of π since if the diameter is 3 mils, the area is $3^2 = 9$ circular mils. Notice that the circular mil is smaller than the square mil since

$$1 \text{ circular mil} = \frac{\pi \times 1^2}{4} = \frac{3.1416}{4} = 0.7854 \text{ square mil}$$

In the British system of units

$$R = \frac{\rho l}{A} = \frac{\rho l}{d^2} \qquad (1\text{-}6\text{-}2)$$

where R = conductor's resistance [ohms (Ω)]
l = length [feet (ft)]
d = diameter [linear mils (mil)]
A = cross-sectional area [circular mils (cmil)]
ρ = specific resistance [ohm · circular mil per foot ($\Omega \cdot cmil/ft$)]

The conversion factor between the British and SI systems is $1 \Omega \cdot m = 6.015 \times 10^8 \Omega \cdot cmil/ft$, so that the specific resistance of annealed copper is $1.724 \times 10^{-8} \Omega \cdot m$ (Table 1-6) = $1.724 \times 10^{-8} \times 6.015 \times 10^8 \Omega \cdot cmil/ft = 10.37 \Omega \cdot cmil/ft$. The values of specific resistance in British units are also shown in Table 1-6.

Table 1-7 shows the relationships between the American Wire Gage [also known as the Brown and Sharp (B&S) gage] number, the diameter in linear mils, the cross-sectional area in circular mils and the resistance per 1000 ft of standard annealed copper wire at 25°C. We have already seen that the diameters decrease as the gage numbers increase; in fact, as we go from one gage number to the next higher, the diameter is divided by a constant factor of 1.123. As an example, No. 5 wire has a diameter of 182.0 mil and therefore the diameter of No. 6 wire is 182.0/1.123 = 162.0 mil, rounded off. Since the cross-sectional area is proportional to the square of the diameter, the ratio of the cross-sectional area of one gage number to that of the next-higher gage number is $(1.123)^2 = 1.261$. Because $(1.261)^3 = 1.261 \times 1.261 \times 1.261 \approx$

TABLE 1-7 Standard Annealed Copper Wire

American Wire Gage number	Diameter (mil)	Cross-sectional area (cmil)	Ohms per 1000 ft at 25°C (= 77°F)
0	325.0	106,000.0	0.100
1	289.0	83,700.0	0.126
2	258.0	66,400.0	0.159
3	229.0	52,600.0	0.201
4	204.0	41,700.0	0.253
5	182.0	33,100.0	0.319
6	162.0	26,300.0	0.403
7	144.0	20,800.0	0.508
8	128.0	16,500.0	0.641
9	114.0	13,100.0	0.808
10	102.0	10,400.0	1.02
11	91.0	8,230.0	1.28
12	81.0	6,530.0	1.62
13	72.0	5,180.0	2.04
14	64.0	4,110.0	2.58
15	57.0	3,260.0	3.25
16	51.0	2,580.0	4.09
17	45.0	2,050.0	5.16
18	40.0	1,620.0	6.51
19	36.0	1,290.0	8.21
20	32.0	1,020.0	10.4
21	28.5	810.0	13.1
22	25.3	642.0	16.5
23	22.6	509.0	20.8
24	20.1	404.0	26.2
25	17.9	320.0	33.0
26	15.9	254.0	41.6
27	14.2	202.0	52.5
28	12.6	160.0	66.2
29	11.3	127.0	83.4
30	10.0	101.0	105.0
31	8.9	79.7	133.0
32	8.0	63.2	167.0
33	7.1	50.1	211.0
34	6.3	39.8	266.0
35	5.6	31.5	335.0
36	5.0	25.0	423.0

2 (the sign "≈" means "is approximately equal to"), the cross-sectional area is slightly more than halved as we increase by three gage numbers; as an illustration, gage 21 has an area of 810.0 cmil, while the area of gage 24 is 404.0 cmil. Finally, because $(1.261)^{10}$ is slightly greater than 10, the area will be approximately 10 times less for an increase of 10 gage numbers (for example, gage 20 area = 1020.0 cmil, gage 30 area = 101.0 cmil).

A wire gage is shown in Fig. 1-20. It will measure wire ranging in size from No. 0 to No. 36. The wire whose size is to be measured is inserted in the smallest slot that will accommodate the bare wire. The gage number corresponding to that slot indicates the wire size. The slot has parallel sides and should not be confused with the semicircular opening at the end of the slot. This opening simply permits the free movement of the wire through the slot.

Conductivity

Just as the conductance, G, is the reciprocal of the resistance, R, the *conductivity* is the reciprocal of the resistivity. The symbol for conductivity is the Greek

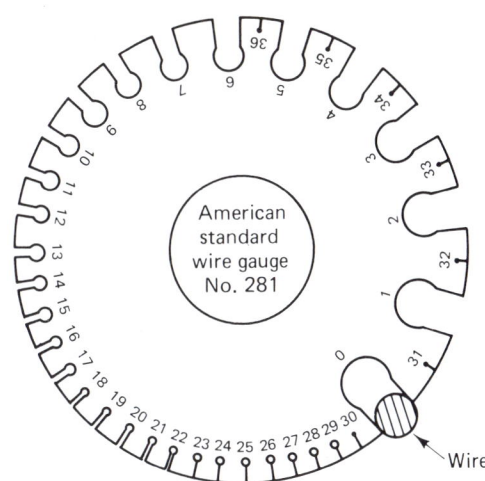

FIGURE 1-20 Wire gage.

lowercase letter sigma, σ, and its SI unit is siemens per meter (S/m). Then

$$\text{conductivity, } \sigma = \frac{1}{\rho} \quad \text{and} \quad \text{specific resistance, } \rho = \frac{1}{\sigma} \quad \text{(1-6-3)}$$

and

$$\text{conductor's resistance, } R = \frac{\rho l}{A} = \frac{l}{\sigma A} \quad \text{(1-6-4)}$$

$$\text{conductor's conductance, } G = \frac{A}{\rho l} = \frac{\sigma A}{l} \quad \text{(1-6-5)}$$

For example, the conductivity of annealed copper at 20°C is $1/(1.724 \times 10^{-8}\ \Omega \cdot m) = 5.8 \times 10^{7}$ S/m.

In this section we discussed the factors that influence the choice of wire size and material. We discovered that the resistance of a cylindrical conductor was directly proportional to its specific resistance and its length but was inversely proportional to its cross-sectional area. To compare different materials we introduced the specific resistance, which is measured in either the ohm-meter or the ohm-circular mil per foot.

EXAMPLE 1-12 Calculate the resistance of a cylindrical nickel conductor which is 4 m long and has a cross-sectional area of 5 mm².

Solution

$$l = 4\ m, \quad A = 5 \times 10^{-6}\ m^2, \quad \rho = 7.8 \times 10^{-8}\ \Omega \cdot m$$

$$\text{resistance, } R = \frac{\rho l}{A} \quad \text{(1-6-1)}$$

$$= \frac{7.8 \times 10^{-8}\ \Omega \cdot m \times 4\ m}{5 \times 10^{-6}\ m^2}$$

$$= 6.24 \times 10^{-2}\ \Omega$$

Sec. 1-6 Resistance of a Cylindrical Conductor. Specific Resistance

EXAMPLE 1-13 A 40-m length of annealed copper wire has a circular cross-sectional area whose diameter is 0.9 mm. What is the wire's resistance at 20°C?

Solution

$$d = 0.9 \times 10^{-3} \text{ m}, \quad \pi = 3.1416, \quad l = 40 \text{ m}$$
$$\rho = 1.724 \times 10^{-8} \, \Omega \cdot \text{m}$$

$$\text{resistance}, R = \frac{\rho l}{\pi d^2 / 4} \tag{1-6-1}$$

$$= \frac{4 \times 1.724 \times 10^{-8} \, \Omega \cdot \text{m} \times 40 \text{ m}}{3.1416 \times (0.9 \times 10^{-3})^2 \text{ m}^2}$$

$$= \frac{4 \times 1.724 \times 4}{3.1416 \times 8.1} = 1.08 \, \Omega$$

EXAMPLE 1-14 A tungsten filament has a length of 4 in. and a cross-sectional area of 2.5 cmil. What is the filament's resistance at 20°C?

Solution

$$\rho = 33.08 \, \Omega \cdot \text{cmil/ft}, \quad l = 4 \text{ in.} = \tfrac{1}{3} \text{ ft}, \quad A = 2.5 \text{ cmil}$$

$$\text{resistance}, R = \frac{\rho l}{A} \tag{1-6-2}$$

$$= \frac{33.08 \, \Omega \cdot \text{cmil/ft} \times \tfrac{1}{3} \text{ ft}}{2.5 \text{ cmil}} = 4.4 \, \Omega$$

1-7 Temperature Coefficient of Resistance

If we raise the temperature of a length of copper wire, the additional heat energy will cause the random motion of the free electrons to increase. If a voltage is then applied to the wire, it will be more difficult to move the electrons along the wire to produce the flow of current; the wire's resistance has therefore risen due to the increase in the temperature. The resistances of most of the good conductors, such as silver, copper, and aluminum, rise in a linear manner with an increase of temperature over normal temperature ranges; their graphs of resistance plotted against temperature are therefore straight lines.

It is possible to produce an alloy by fusing together two or more elements without any chemical action occurring. There are some alloys, for example, eureka (40% nickel, 60% copper), whose resistance is virtually independent of temperature; other examples are manganin and constantan, which are used to produce precision wirewound resistors.

As we already know, the semiconductors, such as carbon, germanium, and silicon, contain very few free electrons at room temperature. However, if the temperature is raised, more electrons are able to move out of the valence band and are then available as the charge carriers in a flow of current. This means that the resistance of semiconductors will fall as the temperature is raised.

The change in the resistance with a change in the temperature is measured by the temperature coefficient of resistance, whose symbol is the Greek lowercase letter alpha, α. If the changes in resistance and temperature are in the same direction, for example, the resistance increases as the temperature rises, the coefficient is said to be positive; this is the case with the pure metals, such as silver, copper, and aluminum. By contrast, the semiconductors will have a negative temperature coefficient, since their resistances fall as the temperature rises.

The value of α is normally referred to a temperature of 20°C. The temperature coefficient is defined as the change in ohms for every ohm of resistance at 20°C for every 1 degree Celsius rise above 20°C. For example, the value of α for copper is $+0.00393$ $\Omega/\Omega/°C$; this means that if a length of copper wire has a resistance of 1 Ω at 20°C, its resistance will *increase* by 0.00393 Ω for every degree Celsius *rise* in temperature above 20°C. If the temperature is increased to $T°C$,

$$\text{increase of resistance} = \alpha \times R_{20°C} \times (T°C - 20°C) \quad (1\text{-}7\text{-}1)$$

so that the new resistance, $R_{T°C}$, is given by

$$R_{T°C} = R_{20°C} + \alpha \times R_{20°C} \times (T°C - 20°C) \quad (1\text{-}7\text{-}2)$$
$$= R_{20°C}[1 + \alpha(T°C - 20°C)]$$

The values of the temperature coefficient for various materials are shown in Table 1-8. Notice that the temperature coefficient is extremely low for the alloy, constantan, and is negative for the semiconductor, carbon. From Eq. (1-6-1),

$$R_{20°C} = \frac{\rho_{20°C} \times l}{A}$$

and therefore

$$R_{T°C} = \frac{\rho_{20°C} \times l}{A}[1 + \alpha(T°C - 20°C)] \quad (1\text{-}7\text{-}3)$$

TABLE 1-8 Temperature Coefficient of Resistance for Various Materials

Material	Temperature coefficient of resistance, α, at 20°C ($\Omega/\Omega/°C$)
Silver	+0.0038
Copper	+0.00393
Aluminum	+0.0039
Tungsten	+0.0045
Iron	+0.0055
Nickel	+0.006
Constantan	+0.0000008
Carbon	−0.0005

and

$$\rho_{T°C} = \rho_{20°C} [1 + \alpha(T°C - 20°C)] \qquad (1\text{-}7\text{-}4)$$

Consider a copper conductor which has a resistance of 1 Ω at 20°C. Now if we cool the conductor, its resistance will decrease by 0.00393 Ω for every 1°C fall in temperature. It follows that if the temperature drops by

$$\frac{1\ \Omega \times 1°C}{0.00393\ \Omega} = 254°C$$

the resistance of the conductor should be zero; that would occur at a temperature of +20°C − 254°C = −234°C. In practice, the resistance/temperature graph becomes nonlinear at low temperatures and the resistance does not vanish until we reach a temperature that is close to −273.2°C (Fig. 1-21). This value is the ultimate theoretical limit to which temperature can fall, and is extremely important for many aspects of physics, particularly thermodynamics. The Kelvin (K) temperature scale is based on this limit, so that zero kelvins (0 K) is equal to −273.2°C. Since a difference in temperature of 1 K is the same as a difference of 1°C, 0°C will equal +273.2 K and 20°C is the same as +293.2 K.

Producing extremely low temperatures near 0 K is called *cryogenics* and will cause the resistance of conductors practically to vanish. We have then achieved the *superconducting* state, which can allow very high currents to flow without any appreciable voltage drop or power dissipation. Perhaps this will be the power distribution system of the future. The low-temperature power lines might be buried deep in the ground, and this would eliminate the need for the present (unsightly) system of high-voltage overhead lines. As a step in the right direction, recent research has shown that certain Lanthanum compounds exhibit the property of superconductivity at the temperature of boiling liquid nitrogen (−195.8°C).

To summarize, the resistance of the good conductor elements increases as the temperature rises, whereas the resistance of the semiconductors decreases. The increase or decrease of resistance is determined by the temperature coefficient, α, which is measured in ohms per ohm per degree Celsius (Ω/Ω/°C). For certain alloys, the values of α are extremely low.

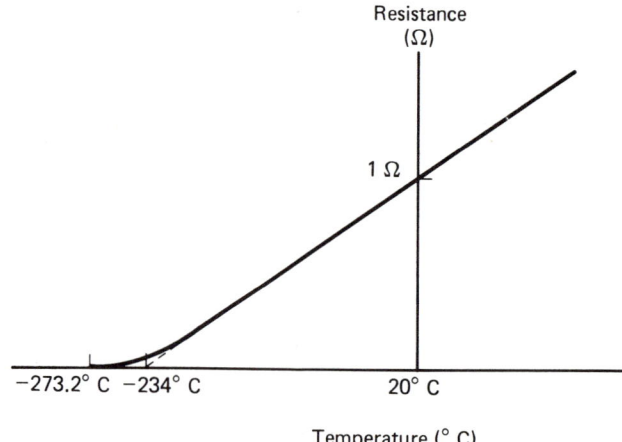

FIGURE 1-21 Temperature coefficient of resistance.

EXAMPLE 1-15 At 20°C a length of copper wire has a resistance of 15 Ω. What is the resistance at (a) 0°C and (b) 100°C?

Solution

(a) $R_{20°C} = 15\ \Omega$, $\quad \alpha = +0.00393\ \Omega/\Omega/°C$, $\quad T°C = 0°C$

$$R_{T°C} = R_{20°C}[1 + \alpha(T°C - 20°C)]$$

$$R_{0°C} = 15\ \Omega[1 + 0.00393(0°C - 20°C)] \quad\quad (1\text{-}7\text{-}2)$$

$$= 15(1 - 0.0786)$$

$$= 15 \times 0.9214 = 13.8\ \Omega$$

(b) $T°C = 100°C$

$$R_{100°C} = 15\ \Omega[1 + 0.00393(100°C - 20°C)]$$

$$= 15(1 + 0.3144)$$

$$= 15 \times 1.3144 = 19.7\ \Omega$$

EXAMPLE 1-16 What is the specific resistance of nickel at 80°C?

Solution

$\rho_{20°C} = 7.8 \times 10^{-8}\ \Omega \cdot m$, $\quad \alpha = +0.006\ \Omega/\Omega/°C$, $\quad T°C = 80°C$

$$\rho_{T°C} = \rho_{20°C}[1 + \alpha(T°C - 20°C)] \quad\quad (1\text{-}7\text{-}4)$$

$$\rho_{80°C} = \rho_{20°C}[1 + \alpha(80°C - 20°C)]$$

$$= 7.8 \times 10^{-8}(1 + 0.006 \times 60)$$

$$= 7.8 \times 10^{-8} \times 1.36 = 10.6 \times 10^{-8}\ \Omega \cdot m$$

EXAMPLE 1-17 The tungsten filament of a 60-W light bulb has a "hot" resistance of 240 Ω at 2500°C (white heat). What is its resistance at room temperature (20°C)?

Solution

$R_{T°C} = R_{2500°C} = 240\ \Omega$, $\quad \alpha = +0.0045\ \Omega/\Omega/°C$, $\quad T°C = 2500°C$

$$R_{T°C} = R_{20°C}[1 + \alpha(T°C - 20°C)] \quad\quad (1\text{-}7\text{-}2)$$

$$R_{20°C} = \frac{R_{T°C}}{1 + \alpha(T°C - 20°C)}$$

$$= \frac{240}{1 + 0.0045(2500°C - 20°C)}$$

$$= \frac{240}{1 + 0.0045 \times 2480} = \frac{240}{12.16} = 19.7\ \Omega$$

When taking readings with an ohmmeter, all power must be removed from the component whose resistance is being measured. Consequently, the ohmmeter can only measure the filament's resistance at room temperature. The "hot" resistance must be determined indirectly by measuring the voltage and current values and then using the equation $R = E/I$.

1-8 Practical Resistors. Color Code

The factors that influence the choice of a practical resistor, are its value in ohms, its tolerance, and its wattage rating. The tolerance is the accuracy of the resistance value as guaranteed by the manufacturer; the wattage rating is the maximum power that the resistor can safely dissipate without an excessive rise in temperature. Physically large resistors have sufficient surface area to radiate considerable heat and therefore have high wattage ratings.

To satisfy the requirements of electronics, resistors are commercially manufactured with a wide range of resistance values from a fraction of an ohm to several megohms; their normal power ratings extend from $\frac{1}{8}$ W upward. Depending on the materials from which the resistors are made, they fall into three categories: composition, wirewound, and film. All of these are illustrated in Fig. 1-22.

Composition Resistors

These are the most common resistors used in electronics. They cost only a few cents each and are manufactured in standard resistance values that range from 2.7 Ω to 22 MΩ. Over a limited range of temperature the resistance value remains reasonably constant; we can therefore say that composition resistors are linear and obey Ohm's law. However, these resistors can dissipate only a small amount of power and their normal wattage ratings are $\frac{1}{8}$ W, $\frac{1}{4}$ W, $\frac{1}{2}$ W, 1 W, and 2 W; these power ratings are not marked on the resistors but can be judged from their physical size [Fig. 1-22(a)]. Once you know the values of the ohms and the watts, the formulas $E = \sqrt{P \times R}$ and $I = \sqrt{P/R}$ will give you the maximum values of voltage and current you can use without exceeding the power rating.

Composition resistors are manufactured from powdered carbon which has a specific resistance in excess of 20,000 Ω · cmil/ft. The carbon is mixed with an insulating substance (for example, talc) and a binding material such as resin. By varying the proportions of carbon and talc, a wide range of resistance values can be manufactured. To protect the component from the effects of moisture and mechanical damage, the resistor is coated with an insulating material that is baked to a hard finish. At each end of the resistor the tinned copper "pigtail" wires provide a considerable contact area for making connections and are deeply embedded so that they cannot easily be pulled out. The construction of the resistor is fully exposed in the cutaway view of Fig. 1-23.

Apart from the low wattage ratings, the main disadvantage of composition resistors is that they cannot be produced in exact values. After manufacture the resistors are normally sorted into three groups. The first group contains those resistors whose resistances are within ±5% of their rated value; this is referred to as a 5% tolerance. The second group has a tolerance from ±5 to ±10%, and the tolerance of the third group is from ±10 to ±20%. Any resistor whose value lies outside the ±20% tolerance is thrown away. As an example let us assume that the rated or nominal value is 1 kΩ (1000 Ω). With a ±5% tolerance, the resistance value must be between 1000 ± [(5 × 1000)/

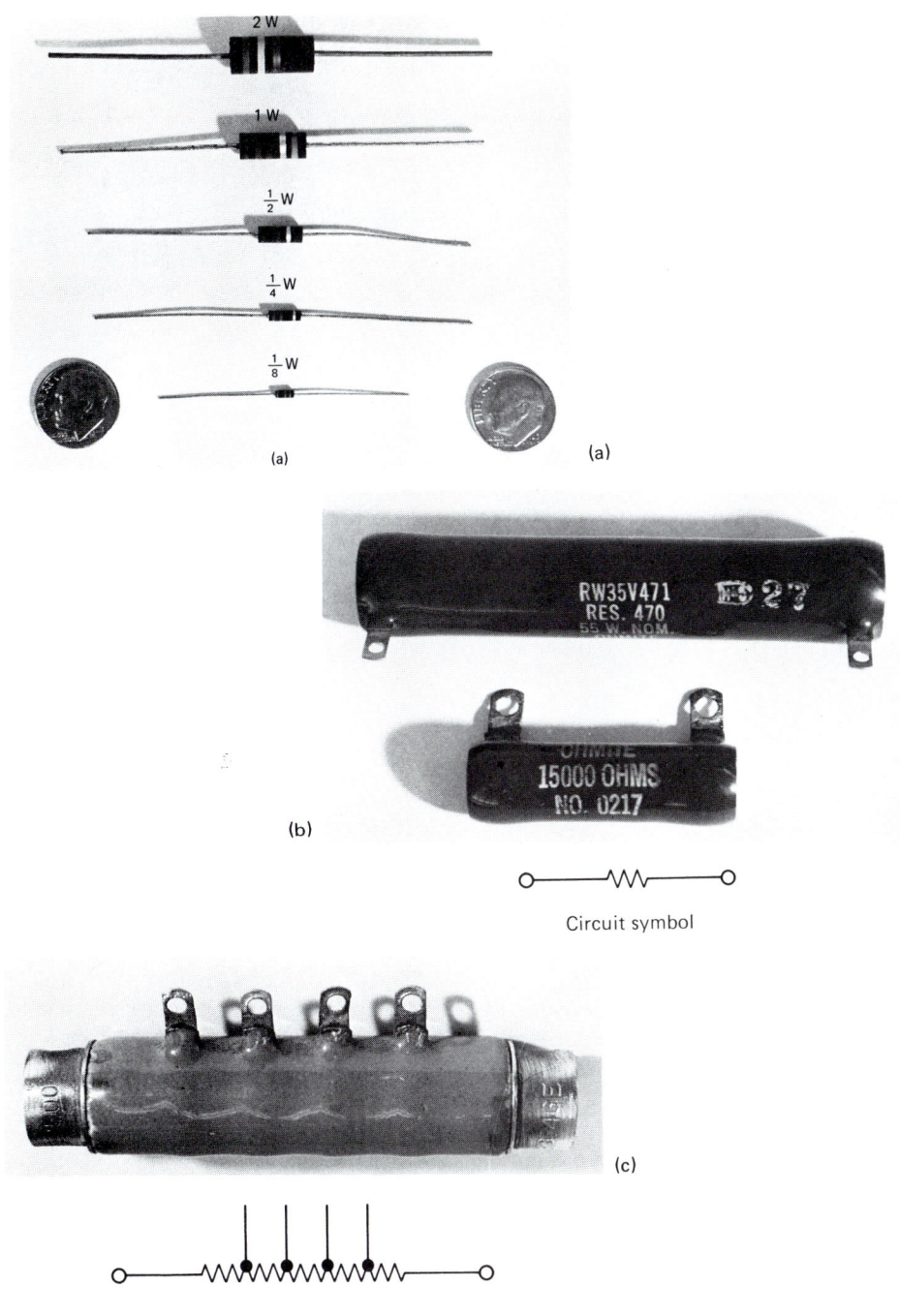

FIGURE 1-22 (a) Composition color-coded resistors; (b) wirewound fixed resistors; (c) wirewound tapped resistor; (d) adjustable or semivariable tapped resistor; (e) wirewound and composition-type potentiometers; (f) precision wirewound resistors; (g) carbon and metal film resistors.

100], or 950 Ω, and 1050 Ω. If the tolerance is ±10%, the upper and lower limits are 900 Ω and 1100 Ω, so that the value of the resistance should lie between either 900 and 950 Ω, or 1050 and 1100 Ω. Finally, the upper and lower limits for a ±20% tolerance are 800 Ω and 1200 Ω; consequently, the resistance value should lie between either 800 and 900 Ω, or 1100 and 1200 Ω.

You may well ask: "Are these wide tolerances acceptable for the resistors used in electronics?" The answer, perhaps surprisingly, is "yes." Electronics

Sec. 1-8 Practical Resistors. Color Code

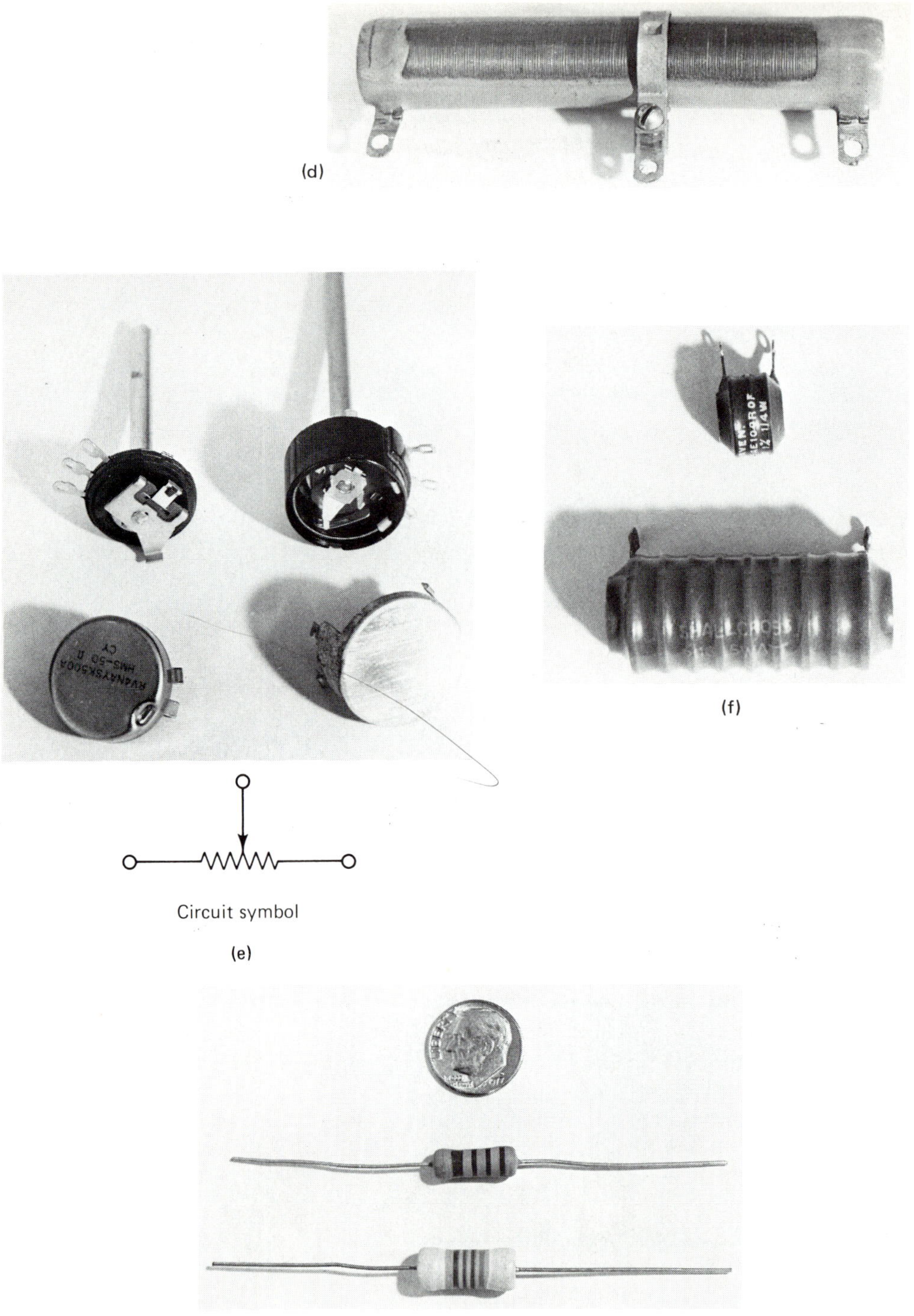

(d)

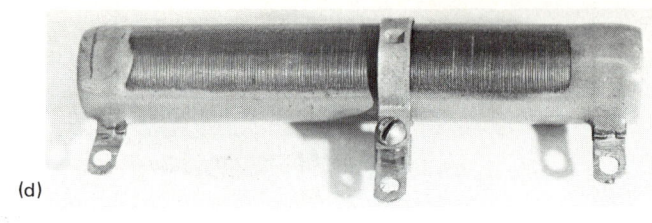
(f)

Circuit symbol
(e)

(g)

FIGURE 1-22 Continued

is sometimes referred to as a "20% tolerance science"; for example, in *some* circuits you can change a resistance value by a factor of 10 and it will make no appreciable difference to the circuit's performance.

Because of the tolerances, composition resistors are manufactured in only certain standard values. Starting with the ±20% range, we would manufacture a 10-Ω resistor because we use the decimal system for our values. The permitted upper and lower limits of the 10-Ω resistor are 8 Ω and 12 Ω; therefore,

FIGURE 1-23 Cutaway view of composition resistor.

we would not make an 11-Ω resistor in the ±20% range because, within the limits of their manufacture, the nominal 10-Ω resistor could have an actual value of 11 Ω, while the nominal 11-Ω resistor might have a value of 10 Ω. The next standard value is, in fact, 15 Ω, with upper and lower limits of 18 Ω and 12 Ω, so that the upper limit of the 10-Ω resistor is equal to the lower limit of the 15-Ω resistor. After the standard 15-Ω resistor comes the 22-Ω resistor, with limits of 26.4 Ω and 17.6 Ω; this shows that there is a small overlap between the 18-Ω upper limit of the 15-Ω resistor and the 17.6-Ω lower limit of the 22-Ω resistor. However, the next standard value is 33 Ω, with a lower limit of 33 − [(20 × 33)/100] = 26.4 Ω, which exactly equals the upper limit of the 22-Ω resistor. There are two more standard values at 47 Ω and 68 Ω, and then we arrive at 100 Ω, after which the two significant figures are repeated with values of 150 Ω, 220 Ω, 330 Ω, and so on. The procedure is then to add further zeros until the resistance values extend into megohms. The ±10% and ±5% ranges are formed by including additional standard values in the gaps of the 20% range (Table 1-9). For example, a nominal 47-Ω resistor may have a tolerance of either ±20% or ±10% or ±5%, but a nominal 51-Ω resistor can only have a tolerance of ±5%.

TABLE 1-9 Standard Resistor Values

5% tolerance	10% tolerance	20% tolerance
10	10	10
11		
12	12	
13		
15	15	15
16		
18	18	
20		
22	22	22
24		
27	27	
30		
33	33	33
36		
39	39	
43		
47	47	47
51		
56	56	
62		
68	68	68
75		
82	82	
91		

Sec. 1-8 Practical Resistors. Color Code

The least costly method of indicating the resistance value and the tolerance is to use a four-band color code (Fig. 1-24). The first two bands represent the significant figures of the resistance value in ohms; the third band is the multiplier, which gives the number of zeros; and the fourth band indicates the tolerance; if no fourth band exists, the tolerance is $\pm 20\%$. The following are examples of using the color code:

47 Ω $\pm$ 10%	yellow (4), violet (7), black (no zeros), silver ($\pm 10\%$)
680 Ω $\pm$ 20%	blue (6), gray (8), brown (1 zero), no fourth band
2000 Ω (2 kΩ) $\pm$ 5%	red (2), black (0), red (2 zeros), gold ($\pm 5\%$)
15,000 Ω (15 kΩ) $\pm$20%	brown (1), green (5), orange (3 zeros), no fourth band
390,000 Ω (390 kΩ) $\pm$10%	orange (3), white (9), yellow (4 zeros), silver ($\pm 10\%$)
2,700,000 Ω (2.7 MΩ) $\pm$5%	red (2), violet (7), green (5 zeros), gold ($\pm 5\%$)
22,000,000 Ω (22 MΩ) $\pm$20%	red (2), red (2), blue (6 zeros), no fourth band

Black is a forbidden color in the first band, so that for resistance values of less than 10 Ω, the four-band system cannot be used. Gold and silver then appear as multipliers of 0.1 and 0.01, respectively, in the third band. Examples are:

4.7 Ω = 47 $\times$ 0.1   yellow, violet, gold
0.68 Ω = 68 $\times$ 0.01   blue, gray, silver

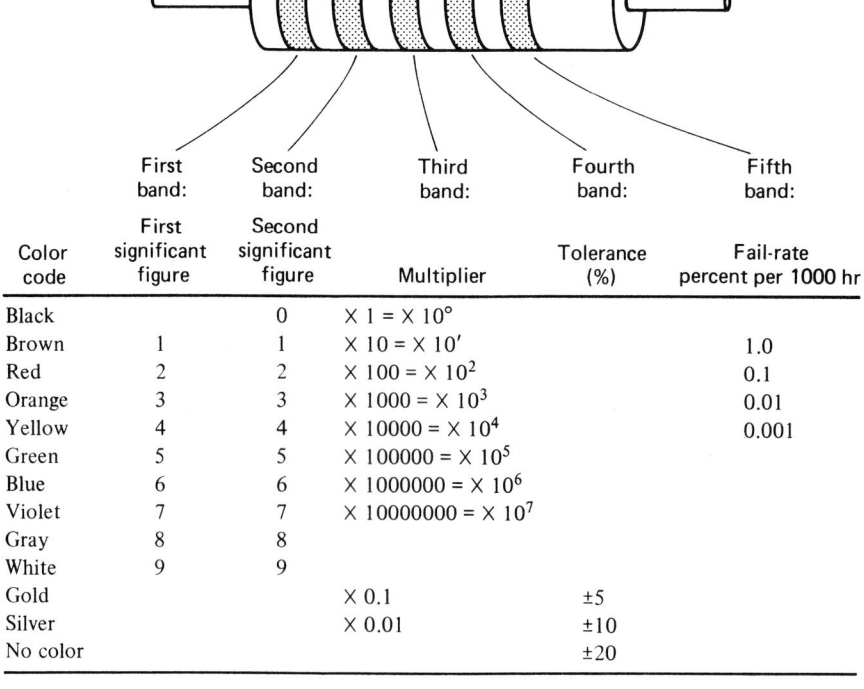

Color code	First band: First significant figure	Second band: Second significant figure	Third band: Multiplier	Fourth band: Tolerance (%)	Fifth band: Fail-rate percent per 1000 hr
Black		0	$\times 1 = \times 10^0$		
Brown	1	1	$\times 10 = \times 10^1$		1.0
Red	2	2	$\times 100 = \times 10^2$		0.1
Orange	3	3	$\times 1000 = \times 10^3$		0.01
Yellow	4	4	$\times 10000 = \times 10^4$		0.001
Green	5	5	$\times 100000 = \times 10^5$		
Blue	6	6	$\times 1000000 = \times 10^6$		
Violet	7	7	$\times 10000000 = \times 10^7$		
Gray	8	8			
White	9	9			
Gold			$\times 0.1$	± 5	
Silver			$\times 0.01$	± 10	
No color				± 20	

FIGURE 1-24 Resistor color code.

Some resistors, especially those in military use, have a fifth band which indicates the failure rate. As shown in Fig. 1-24, a fifth yellow band means a fail rate of 0.001% (1 in 100,000) per 1000 h of operation. Of course, with these resistors the fourth band would be either gold or silver.

In addition to their tolerance factors and low wattage ratings, composition resistors have the disadvantage of poor stability. If the resistor is over-heated, it can permanently change its value, which will no longer lie within its tolerance.

Wirewound Resistors

The construction of these resistors depends basically on Eq. (1-6-1), $R = \rho l/A$. The cross-sectional area of the wire is determined by the maximum current the resistor has to carry. The wire alloy then depends on whether we need a high or a low resistance. Constantan, with its negligible temperature coefficient, is suitable for the lower resistances; nichrome is used for the higher values. Once we know the ρ and A factors, we can calculate the length of the wire required to provide the necessary resistance; the wire-wound resistors can therefore be manufactured to a more precise value than the composition type, but at a greater cost. It is common practice to wind the wire on a hollow porcelain tube and then seal the wire in place by a further coating of procelain. Finally, there is an exterior molding on which is shown the resistor's data, such as its resistance value, power rating, tolerance, and stock number.

Wirewound resistors can dissipate considerable heat and therefore can have high power ratings, from 2 W upward. Over a wide range of temperature the resistance value is constant, so that the resistor is linear and obeys Ohm's law. These power resistors can be classified as follows:

FIXED RESISTORS The two ends of the winding are connected to two terminals as shown in Fig. 1-22(b). The larger the physical size of the resistor, the greater is its power rating.

TAPPED RESISTORS An example of this type with its schematic symbol is shown in Fig. 1-22(c). The tapped resistor is used where a number of different resistances are required (see Section 2-7) and is an alternative to using separate resistors, which may not be readily available in the required values. A resistor with fixed taps would have a limited application, but if the tap could be moved, the same resistor might be used in a variety of circuits. An example of this adjustable or semivariable resistor is shown in Fig. 1-22(d). To make the adjustment the metal band is moved to the required position and then locked into place. Once this has been done, there is no further adjustment, so this type of resistor is not suitable for circuits in which the resistance must be frequently varied.

VARIABLE RESISTORS These resistors, known as *potentiometers* ("pots"), are shown in Fig. 1-22(e) together with their circuit symbol. They are used when it is required to vary the resistance continuously from zero up to the full rated value, which is stamped, together with the power rating, on the component's metal casing. Bare resistance wire is wound on a circular form with sufficient spacing so that adjacent turns do not touch. A metallic slider is mounted on a rotating shaft and makes contact with the resistance wire. For potentiometers with higher resistance values, the circular track is partially coated with a powdered carbon composition on which the slider makes contact. The two ends of the wire (or track) and the slider are brought out to three external terminals. The slider is connected to the center terminal so that when the shaft is rotated, it varies the amount of resistance that appears between the center terminal and one of the end terminals. An alternative construction is to use a flat, straight form with the slider moving along the resistor. The potentiometer, with its three terminals, is a common form of voltage control (Section 2-9); if one of the end terminals is not used, the device becomes a *rheostat*, which is used to control current.

PRECISION RESISTORS These resistors are used with some of the meter circuits described in Chapter 10. They must normally be accurate to within 1% (or less) of their nominal resistance values; otherwise, the meter readings will be incorrect. To achieve this degree of accuracy, we use a length of resistance wire made from an alloy with a very low temperature coefficient. By ensuring that the cross-sectional area is constant all along the wire, the resistance will be directly proportional to the length and we can therefore cut off a calculated length to obtain the required resistance. The wire is then wound on some type of form such as a bobbin and connected to two terminals or leads as shown in Fig. 1-22(f).

Thin-Film Resistors

We now know that composition resistors are cheap, small in size and cover a wide range of resistance values; however they are unstable, inaccurate, noisy, and have low wattage ratings. By contrast, wirewound resistors can be manufactured to produce accurate values of resistance with high wattage ratings. Their disadvantages are bulk, cost, and a tendency for the wire to break and open-circuit. They are also unsuitable for high resistance values because of the large amount of wire that would be required and their inductive effect, which is explained in Chapter 8.

Film resistors are an attempt to have the best of both worlds. A thin film of carbon or metal alloy is deposited on a cylindrical ceramic core and the component is then sealed with an epoxy or glass coating so that it is not affected by moisture [Fig. 1-22(g)]. To achieve a high resistance, a spiral is cut into the film so that the current is forced to flow through the helical path of the resistive material; the closer the pitch of the spiral, the higher is the value of the resistance. Carbon film resistors have resistance values that range from 10 Ω to 2.5 MΩ, with a $\pm 1\%$ tolerance; they are normally manufactured with a $\frac{1}{2}$-W power rating. Metal film resistors are more costly and are available in power ratings of $\frac{1}{8}$, $\frac{1}{4}$, $\frac{1}{2}$, and 1 W; their resistance range is from 10 Ω to 1.5 MΩ, with a standard tolerance of $\pm 1\%$, although tolerances as low as 0.1% can be obtained.

A third type of film resistor is used in high-power applications. Manufactured at red heat, a film of tin oxide is fused into the surface of a glass substrate [Fig. 1-25(a)]. Tin oxide has a high specific resistance, so that we can use a thick film to provide power ratings that extend from $\frac{1}{8}$ W to over 1 kW. The resistance values are from 10 Ω to several MΩ, with a tolerance of $\pm 0.5\%$ for lower power resistors up to $\pm 10\%$ for those with high wattage ratings. It is common practice to use a substrate for a number of tin oxide resistors, which are then packaged into a complete unit [Fig. 1-25(b)].

The Fuse

When current passes through a resistor, electrical energy is dissipated as heat, which raises the resistor's temperature. If the current is excessive, the temperature will become too high, so that the metal of the wirewound resistor will melt and interrupt the flow of current. This effect is used to advantage in

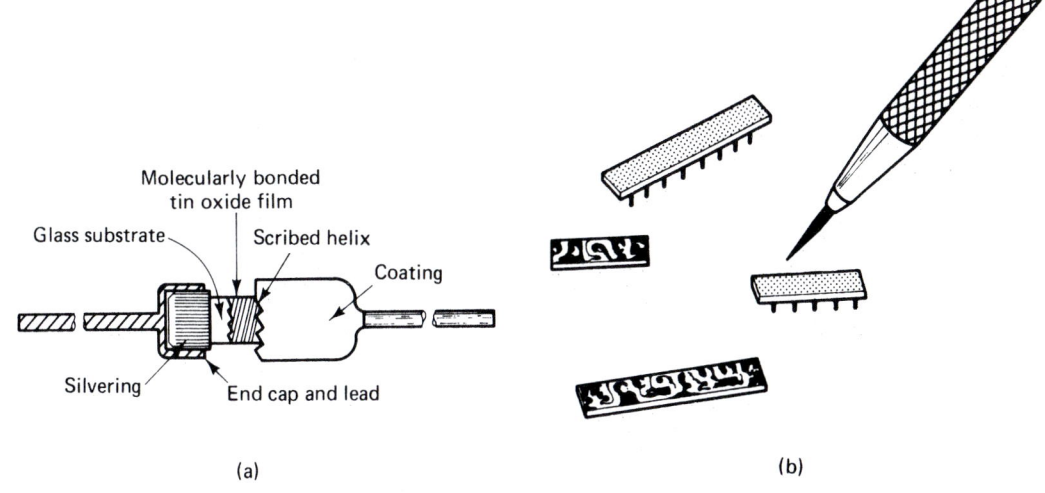

FIGURE 1-25 Tin-oxide film resistor. (From J. J. DeFrance, *Electrical Fundamentals*, Prentice-Hall, Inc., Englewood Cliffs, N.J., 1969, p. 67.)

fuses, which are actually metal resistors with very low resistance values. The metal used is either zinc or an alloy of lead and tin. This has a higher resistance than copper, so that the fuse heats up more rapidly than the connecting wire. However, because the fuse has a low melting point, it will melt ("blow") before the wires overheat and create a fire hazard. Fuses have a rating in amperes (or fractions of an ampere) at a specified voltage, and come in a variety of sizes and types (Fig. 1-26).

The purpose of a fuse is to protect the load, the wiring, and the source of an electrical circuit. The current through the circuit also flows through the fuse, so that if a fault develops, the current will rise above a certain level and the fuse will blow. The circuit is then broken and no further damage is possible. The blown fuse must, of course, be discarded and replaced by one with the same voltage, current, and current–time ratings. The current versus time characteristic is extremely important. We already know that the "room-temperature" resistance of a light bulb's tungsten filament is much lower than

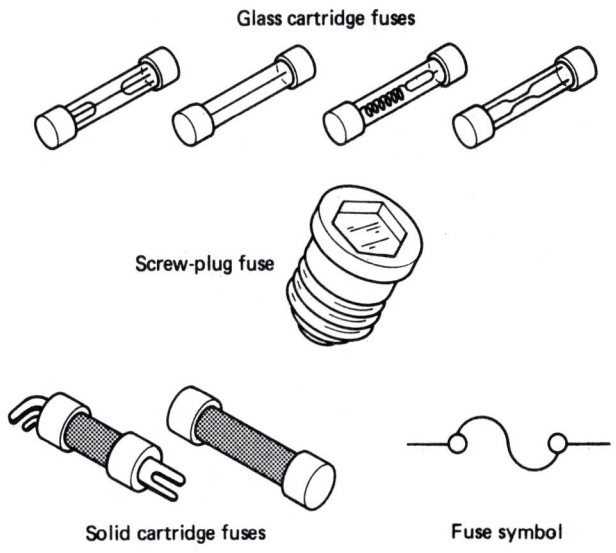

FIGURE 1-26 Various types of fuses.

Sec. 1-8 Practical Resistors. Color Code

its "hot" resistance. Consequently, when a light is switched on, there will be an initial surge current which is many times greater than the steady operating current. Another example is the electric motor, which takes far more current during starting than for normal running. Because of these short-duration current surges, fuses are divided into three time ranges, which indicate their overload "blowing" ability: fast (5 μs to 0.5 s), medium (0.5 to 5 s), and delayed (5 to 25 s).

A more costly alternative to the fuse is the *circuit breaker*, which trips and breaks the circuit when the current rises above a predetermined level. The circuit breaker differs from the fuse in that it can be reset, whereas the fuse melts and must be replaced.

To summarize, a resistor is characterized mainly by its type (composition, wirewound, or film), ohmic value, power rating, tolerance, and possibly, failure rate. Practical resistors may also be classified as fixed, tapped, adjustable, or variable. The main variable resistors are the potentiometer, to control voltage, and the rheostat, to control current. Circuit protection is provided by the fuse, which is metal resistor with a low melting point and a low resistance value.

EXAMPLE 1-18 What are the color codings for composition resistors whose nominal values are 3.3 Ω, 33 Ω, and 330 Ω?

Solution

$$3.3 \ \Omega = 33 \times 0.1 \ \Omega = \text{orange, orange, gold}$$
$$33 \ \Omega = 33 \times 10^0 \ \Omega = \text{orange, orange, black}$$
$$330 \ \Omega = 33 \times 10^1 \ \Omega = \text{orange, orange, brown}$$

EXAMPLE 1-19 A 1-W composition resistor is color-coded green, brown, red, and gold. What are the permitted upper and lower limits of its resistance value?

Solution The resistor's nominal value is 5100 Ω, with a ±5% tolerance. The permitted variation is therefore between

$$5100 \pm \frac{5}{100} \times 5100 \ \Omega = 5100 \pm 255 \ \Omega = 5355 \ \Omega \quad \text{and} \quad 4845 \ \Omega$$

EXAMPLE 1-20 The first three bands of a 1-W composition resistor are blue, gray, and brown. What is the maximum value of the current that can flow through the resistor without exceeding its power rating?

Solution

$$P = 1 \ \text{W}, \quad R = 680 \ \Omega$$

$$\text{maximum current, } I = \sqrt{\frac{P}{R}} \qquad (1\text{-}5\text{-}3)$$

$$= \sqrt{\frac{1 \text{ W}}{680 \text{ }\Omega}}$$

$$= 0.0383 \text{ A} = 38 \text{ mA} \quad \text{(rounded off)}$$

EXAMPLE 1-21 A 1-kΩ potentiometer has a power rating of 5 W. What is the maximum voltage that can be applied across the potentiometer without exceeding its power rating?

Solution

$$P = 5 \text{ W}, \quad R = 1 \text{ k}\Omega = 1000 \text{ }\Omega$$

$$\text{maximum voltage, } E = \sqrt{P \times R} \qquad (1\text{-}5\text{-}5)$$

$$= \sqrt{5 \text{ W} \times 1000 \text{ }\Omega} = 70.7 \text{ V}$$

EXAMPLE 1-22 When a current of 707 mA flows through a wirewound resistor, the power dissipated is 50 W. What is the value of the resistance?

Solution

$$I = 707 \text{ mA} = 0.707 \text{ A}, \quad P = 50 \text{ W}$$

$$\text{resistance, } R = \frac{P}{I^2} \qquad (1\text{-}5\text{-}3)$$

$$= \frac{50 \text{ W}}{(0.707 \text{ A})^2} = 100 \text{ }\Omega$$

Chapter Summary

1.1. Mechanical SI units:

$$\text{force (N)} = \text{mass (kg)} \times \text{acceleration (m/s}^2)$$

$$\text{or} \quad F = M \times a$$

$$\text{work (J)} = \text{distance (m)} \times \text{force (N)} = \text{power (W)} \times \text{time (s)}$$

$$\text{or} \quad W = d \times F = P \times t$$

$$\text{power (W)} = \text{work (J)/time (s)} = \text{force (N)} \times \text{velocity (m/s)}$$

$$\text{or} \quad P = W/t = F \times v$$

$$\text{torque (N} \cdot \text{m)} = \text{force (N)} \times \text{distance (m)}$$

$$\text{or} \quad T = F \times d$$

1.2. Prefixes:

tera (T) = $\times 10^{12}$	pico (p) [micromicro ($\mu\mu$)] = $\times 10^{-12}$
giga (G) = $\times 10^{9}$	nano (n or ν) = $\times 10^{-9}$
mega (M) = $\times 10^{6}$	micro (μ) = $\times 10^{-6}$
kilo (k) = $\times 10^{3}$	milli (m) = $\times 10^{-3}$

1.3. Atomic structure: A compound is made up of molecules; an element is composed of atoms. Atoms contain subatomic particles: positive protons, negative electrons, and neutrons. In a neutral atom the atomic number of the positive protons is equal to the number of negative electrons. An ion is a charged atom that has either lost or gained one or more electrons. The best conductors have a single-valence electron; semiconductors have four electrons in the valence band.

1.4. Electrolysis—Faraday's law:

mass liberated (kg) = electrochemical equivalent (kg/C) $\times$ charge (C)

= electrochemical equivalent (kg/C) $\times$ current (A) $\times$ time (s)

or $M = zIt$

1.5. Electrical SI units:

current (A) = charge (C)/time (s)

or $I = Q/t$

charge (C) = ampere (A) $\times$ time (s)

or $Q = I \times t$

electromotive force (V) = work (J)/charge (C)

or $E = W/Q$

work (J) = electromotive force (V) $\times$ charge (C)

= power (W) $\times$ time (s)

or $W = E \times Q = P \times t$

power (W) = work (J)/time (s)

= current (A) $\times$ electromotive force (V)

or $P = W/t = I \times E$

1.6. Ohm's law relationships:

$$\text{EMF, } E = I \times R = \sqrt{P \times R} = \frac{P}{I} \text{ [volts (V)]}$$

$$\text{current, } I = \frac{E}{R} = \sqrt{\frac{P}{R}} = \frac{P}{E} \text{ [amperes (A)]}$$

$$\text{power, } P = \frac{E^2}{R} = I^2 R = I \times E \text{ [watts (W)]}$$

$$\text{resistance, } R = \frac{E}{I} = \frac{P}{I^2} = \frac{E^2}{P} = \frac{1}{G} \text{ [ohms (}\Omega\text{)]}$$

$$\text{conductance, } G = \frac{I}{E} = \frac{I^2}{P} = \frac{P}{E^2} = \frac{1}{R} \text{ [siemens (S)]}$$

1.7. Resistance of cylindrical conductor:
SI units:

$$\text{resistance } (\Omega) = \frac{\text{specific resistance } (\Omega \cdot m) \times \text{length } (m)}{\text{cross-sectional area } (m^2)}$$

$$\text{or} \quad R = \frac{\rho l}{A}$$

$$= \frac{4 \times \text{specific resistance } (\Omega \cdot m) \times \text{length } (m)}{\pi \times \text{square of the diameter } (m^2)}$$

$$\text{or} \quad R = \frac{4\rho l}{\pi d^2}$$

British units:

$$\text{resistance } (\Omega) = \frac{\text{specific resistance } (\Omega \cdot \text{cmil/ft}) \times \text{length } (ft)}{\text{cross-sectional area (cmil)}}$$

$$= \frac{\text{specific resistance } (\Omega \cdot \text{cmil/ft}) \times \text{length } (ft)}{\text{square of the diameter (cmil)}}$$

$$\text{or} \quad R = \frac{\rho l}{d^2} = \frac{\rho l}{A}$$

1.8. Temperature coefficient of resistance:

$$R_{T°C} = R_{20°C}[1 + \alpha(T°C - 20°C)]$$

where the temperature coefficient α is measured in $\Omega/\Omega/°C$,

$$\rho_{T°C} = \rho_{20°C}[1 + \alpha(T°C - 20°C)]$$

1.9. Practical resistors:
Color code:
    First band   first significant figure
    Second band   second significant figure
    Third band   multiplier value;
                           tolerance (for resistors of less than 10 Ω)
    Fourth band   tolerance
    Fifth band   failure rate
    (if present)

black	0	blue	6	
brown	1	violet	7	
red	2	gray	8	
orange	3	white	9	
yellow	4	silver	10% tolerance or 0.01 multiplier	
green	5	gold	5% tolerance or 0.1 multiplier	

1.10. Summary of electrical units:

Quantity	Unit	Unit Symbol	Letter Symbol
Current	ampere	A	I
Charge	coulomb	C	Q
	ampere·hour	Ah	
EMF or PD	volt	V	E, V

Table continued on next page.

Chapter Summary

Quantity	Unit	Unit Symbol	Letter Symbol
Work, energy	joule	J	W
	watt hour	Wh	
	kilowatt hour	kWh	
Power	watt	W	P
Resistance	ohm	Ω	R
Conductance	siemens (mho)	S	G
Specific resistance, resistivity	ohm·meter	Ω·m	ρ
Conductivity	siemens per meter	S/m	σ
Temperature coefficient of resistance	ohms per ohm per degree Celsius	Ω/Ω/°C	α

Self-Review

True–False Questions

1. T. **F.** The SI unit of torque is the newton-meter or joule.
2. **T.** F. A copper atom whose atomic number is 29 has 28 orbiting electrons; that atom is a positive ion.
3. **T.** F. In electrolysis Faraday's law states that the mass liberated is directly proportional to the current.
4. **T.** F. The power in watts is equal to the product of the current in amperes and the electromotive force in volts.
5. **T.** F. Ohm's law states that under constant physical conditions the current flowing through a conductor is directly proportional to the voltage applied across the conductor.
6. **T.** F. When a power of 1 kW is operated for a time of 1 h, the energy consumed is 1 kWh.
7. **T.** F. Conductance in siemens is the reciprocal of resistance in ohms.
8. T. **F.** The resistance of a cylindrical conductor is directly proportional to the specific resistance and the cross-sectional area but is inversely proportional to its length.
9. **T.** F. Silver and copper have positive temperature coefficients of resistance; germanium and silicon have negative coefficients.
10. T. **F.** A 2-Ω composition resistor would be color coded black–red–black in its first three bands.
11. **T.** F. The power dissipated in a resistor is directly proportional to the square of the value of the current that flows through the resistor.
12. T. **F.** A resistor is color coded yellow–violet–black. Its rated resistance is 680 Ω.
13. T. **F.** Copper is a better conductor than silver.
14. **T.** F. A silver atom has 47 orbiting electrons; that atom is a negative ion.
15. **T.** F. The SI unit of work is the joule.
16. T. **F.** The higher the gage number of a particular length of wire, the lower its resistance value.
17. T. **F.** If gold is the third band of a color-coded resistor, its tolerance is 5%.
18. T. F. The charge carried by a single electron is 1.602×10^{19} C.
19. T. **F.** The prefix "milli" means "÷ 1,000,000."
20. **T.** F. The product of the force and the velocity is equal to the work done.

Multiple-Choice Questions

21. A current of 10 A flows through a circuit for a time of 5 min. The amount of charge which has passed a particular point in that circuit is
    (a) 50 C (b) 2 C (c) 3000 C (d) 500 C (e) 0.033 C

22. An electrical source has an EMF of 12 V. If a charge of 4 C is passed between the source terminals, the work done is
    (a) 48 J (b) 3 J (c) 0.33 J (d) 8 J (e) 16 J

23. A current of 6 A is drawn from a battery whose EMF is 24 V. The power delivered from the battery is
    (a) 4 W (b) 144 J (c) 0.25 W (d) 4 J (e) 144 W

24. An EMF of 36 V is applied across a 4-Ω resistor. The current flowing through the resistor is
    (a) 9 A (b) 144 A (c) 144 C (d) 40 A (e) 32 C

25. In Question 24, the values of the EMF and the resistance are both doubled. The value of the new current is
    (a) doubled (b) halved (c) unchanged (d) multiplied by 4
    (e) impossible to determine unless the resistor obeys Ohm's law

26. In Question 24, the power dissipated in the resistor is
    (a) 144 W (b) 324 W (c) 576 W (d) 5184 W (e) 2.25 W

27. In Question 24, the current flows for a time of 10 min. The total energy consumed is
    (a) 3240 J (b) 54 Wh (c) 3.24 kWh (d) 5.184 kWh (e) 32.4 Wh

28. In Question 24, the conductance of the resistor is
    (a) 4 S (b) 0.25 Ω (c) 0.4 S (d) 250 mS (e) 25 mS

29. A current of 20 mA flows through a composition resistor whose first three bands are blue, gray, and brown. Assuming that the actual resistance is equal to its nominal value, the voltage applied across the resistor is
    (a) 1.36 V (b) 34 V (c) 3.4 V (d) 13.6 V (e) 0.34 V

30. The minimum power rating required for the resistor of Question 29 is
    (a) $\frac{1}{8}$ W (b) $\frac{1}{4}$ W (c) $\frac{1}{2}$ W (d) 1 W (e) 2 W

31. What is the force required to give a mass of 4 kg an acceleration of 8 m/s² in the direction of the force?
    (a) 32 J (b) 2 N (c) 0.5 J (d) 2 J (e) 32 N

32. A 60-W light bulb is switched on for a period of 3 h. The total energy consumed is
    (a) 180 J (b) 20 Wh (c) 180 Wh (d) 0.018 kWh (e) 64,800 J

33. A resistor is color coded red–red–red–silver. The permitted upper limit of its resistance value is
    (a) 2420 Ω (b) 1980 Ω (c) 2640 Ω (d) 2310 Ω (e) 2222 Ω

34. A voltage of 8 V is applied across a resistor that is color coded red–black–yellow–gold. Assuming that the actual resistance is equal to its nominal value, the current flowing through the resistor is
    (a) 4 A (b) 4 mA (c) 4 μA (d) 40 mA (e) 40 μA

35. A mass that is initially at rest is subjected to an acceleration of 4 m/s² for a period of 3 s. Its velocity at the end of the time interval is
    (a) 36 m/s (b) 48 m/s (c) 12 m/s (d) 6 m/s (e) 18 m/s

36. A section of cylindrical wire is 1 m in length and has a resistance of 100 Ω. What is the resistance of a new section of wire made from the same material if both the length and the cross-sectional area are doubled?
    (a) 100 Ω (b) 200 Ω (c) 400 Ω (d) 50 Ω (e) 25 Ω

37. A section of cylindrical wire has a resistance of 2 Ω at room temperature. If the temperature coefficient of resistance for the material used is +0.004 Ω/Ω/°C and the temperature is raised by 20°C, the new resistance is
    (a) 2.04 Ω (b) 2.4 Ω (c) 2.08 Ω (d) 2.16 Ω (e) 2.32 Ω

38. When 6 V is applied across a resistor, the power dissipated is 15 mW. The value of the resistance is
    (a) 400 Ω (b) 240 Ω (c) 4000 Ω (d) 90 Ω (e) 2400 Ω

39. In an electrical circuit 240 C of charge passes a particular point in a period of 4 min. The current flowing at that point is
    (a) 60 A (b) 1 A (c) 0.96 A (d) 4 A (e) 2.4 A

40. A motor has a mechanical output of 2.8 hp. In the SI system this output is
(a) 209 W (b) 0.209 kW (c) 266 W (d) 2.09 kW
(e) both (a) and (b) are correct

Practice Problems

41. A mass of 5 kg is lifted through a height of 30 m against gravity. Calculate the work done in joules.

42. A current of 2.5 mA flows through a 4700-Ω resistor. Calculate the voltage applied across the resistor and the amount of power dissipated as heat.

43. A 100-W light bulb is operated from a 120-V dc source. What is the current flowing through the bulb's tungsten filament, and what is its resistance? If the bulb is switched on for 3 h, calculate the amount of energy consumed in kWh.

44. A thin-film resistor has a nominal value of 1500 Ω with a $\pm 1\%$ tolerance. What are the permited upper and lower limits of its resistance value?

45. What is the maximum voltage that can be applied across a 470-Ω 1-W resistor without exceeding its power rating?

46. What is the maximum current that can flow through a 560-kΩ $\frac{1}{2}$-W resistor without exceeding its power rating?

47. Calculate the resistance at 20°C of 200 m of nickel wire with a cross-sectional area of 4×10^{-6} m².

48. In Problem 47, calculate the resistance of the same wire at 100°C.

49. A current of 5 A flows for a time of 3 h through a copper sulfate solution. Calculate the mass liberated from the copper anode.

50. An EMF of 12 V is connected through a 5-A fuse to a load rated at 75 W. Draw the circuit and determine whether the rating of the fuse is adequate.

51. A 3-kg mass initially at rest falls under gravity. Calculate the amount of energy acquired by the mass after a period of 4 s has elapsed.

52. If a current flowing through a resistor is doubled but the resistance remains unchanged, by what factor is the dissipated power multiplied?

53. A voltage of 12 V is applied across a 2.7-kΩ resistor. Calculate the values of the current and the power dissipated.

54. If a voltage applied across a resistor is doubled but the resistance remains unchanged, by what factor is the dissipated power multiplied?

55. What is the resistance of 375 ft of AWG No. 24 standard annealed copper wire at a temperature of 25°C?

56. A resistor is color coded blue–gray–gold. Assuming that the resistance is the same as its nominal value, calculate the resistor's conductance. Express your answer in millisiemens.

57. If the power dissipated in a resistor is doubled but the resistance remains unchnged, by what factor must we multiply the voltage drop across the resistor?

58. A certain length of wire has a measured resistance of 240 Ω at 20°C and 319.2 Ω at 80°C. Calculate the value of the wire's temperature coefficient of resistance.

59. The total voltage drop across a cylindrical annealed copper conductor must not exceed 12 V when the current flowing through the conductor is 16 A. If the length of the wire is 50 m, calculate the value of its cross-sectional area.

60. What is the resistance per meter length at 20°C of nickel wire, which has a diameter of 2 mm?

Advanced Problems

61. In 1 min a motor is capable of lifting a mass of 250 kg at constant speed through a height of 50 m against gravity. What is the motor's output power in horsepower?

62. A current of 25 µA is flowing through a resistor whose value is 910 kΩ. Find the *IR* voltage across the resistor and the power dissipated as heat.

63. An EMF of 12 V is applied across a resistor of 1.5 MΩ. Find the energy consumed over a period of 3 h.
64. The first three bands of a ½-W composition resistor are orange, white, and yellow. Assuming that the actual resistance is the same as the nominal value, what is the maximum current in mA which can flow through the resistor without exceeding its power rating?
65. The first three bands of a 2-W composition resistor are green, brown, and red. Assuming that the actual resistance is the same as the nominal value, what is the highest voltage that can be applied across the resistor without exceeding its power rating?
66. Calculate the resistance at 0°C of 800 ft of No. 22 aluminum wire.
67. An electrical conductor, 3.5 m long, has a cross-sectional area of 0.85 mm^2 and a resistance of 0.017 Ω. What is the resistance of 60 m of wire that is made from the same material and has a cross-sectional area of 0.55 mm^2?
68. When a current of 450 mA flows through a wirewound resistor, the power dissipated is 65 W. What is the value of the resistor's conductance?
69. A charge of 30 C is moved through a potential difference of 18 V in a time of 5 s. Calculte the values of the power and the work done.
70. When a 75-W bulb is operated from a 120-V source, the temperature of the tungsten filament is 2600°C. What is the filament's resistance at room temperature (20°C)?
71. When a current of 22.3 mA flows through a color-coded resistor with a 5% tolerance, its power dissipation is 1 W. What is the value of the resistance, and what is the color coding of the first three bands?
72. A roll of nichrome wire is manufactured so that the diameter of the circular wire is 1 mm. What length of wire must be cut from the roll to provide a resistance of 100 Ω? Assume that the temperature is 20°C.
73. A mass is projected vertically upward with a velocity of 37 m/s. To what height will it rise before it starts to fall? Neglect any effect of air resistance.
74. A load has consumed 3 kWh in a period of 7 h. If the voltage applied across the load is 120 V, what is the value of the load current?
75. At a temperature of 0°C, a coil of wire is found to have a resistance of 10 Ω. When the temperature is raised to 100°C, the new resistance measurement is 13.9 Ω. Calculate the value of the temperature coefficient and state whether it is positive or negative.
76. If the power dissipated in a resistor is halved but the resistance remains unchanged, by what factor must we multiply the voltage drop across the resistor?
77. The resistance of a coil of aluminum wire is 800 Ω at 20°C. When the coil is submerged in a liquid for an appreciable time, its resistance drops to 780 Ω. What is the temperature of the liquid?
78. If the current flowing through a resistor is quartered but the resistance remains unchanged, by what factor is the dissipated power multiplied?
79. The rectangular copper conductors on a printed circuit board are 0.05 mm thick and 2 mm across. If a current of 0.5 A flows through one such copper conductor which is 1.5 cm in length, calculate the amount of voltage drop.
80. The temperature of the conductors described in Problem 79 is raised from 20°C to 55°C. If the current remains at 0.5 A, calculate the new voltage drop across the conductor whose length is 1.5 cm.

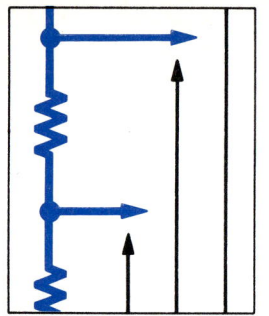

2

MUST KNOW ENTIRE CHAPTER

Resistors in Series

> ### In this chapter you will learn:
> 1. how to distinguish a series resistive circuit from other resistor arrangements.
> 2. about the manner in which a current flows through a series circuit.
> 3. how to determine the value of the individual voltages across resistors in series.
> 4. how to calculate the value of the total equivalent resistance.
> 5. how to determine the powers dissipated in the individual resistors and the total power in the series circuit.
> 6. how to analyze a series circuit by using a step-by-step procedure.
> 7. about using the voltage division rule to obtain the value of a particular voltage drop.
> 8. about the use of voltage dividers and how they are formed by connecting resistors in series.
> 9. about the use of ground as a common connection and a voltage reference level.
> 10. how to use a potentiometer to obtain a variable output voltage and a rheostat to control the current in a series circuit.
> 11. how to connect cells in series to increase the voltage available.
> 12. how to identify the presence of an open circuit with the aid of an ohmmeter or a voltmeter.
> 13. how to obtain the correct voltage for a load when only a higher voltage is available.
> 14. how to troubleshoot a series circuit.

2-0

Introduction

There are two basic configurations in which components such as resistors may be connected. These are called *series* and *parallel* and it is vital for you to be able to distinguish between these arrangements. Frequently, instructors draw on a chalkboard a circuit such as that shown in Fig. 2-1(a) and ask students whether the resistors R_1 and R_2 are in series or parallel. Some will say "series" and others will say "parallel," but the correct answer is: "I cannot tell." If the voltage source, which is the battery, E, is connected within the ring or loop containing the resistors [Fig. 2-1(b)], the result is a series arrangement; however, if the source is positioned external to the ring [Fig. 2-1(c)], the resistors are then in parallel (discussed in Chapter 3). In other words, the terms "series" and "parallel" refer to the *position* of the voltage source in relation to the components.

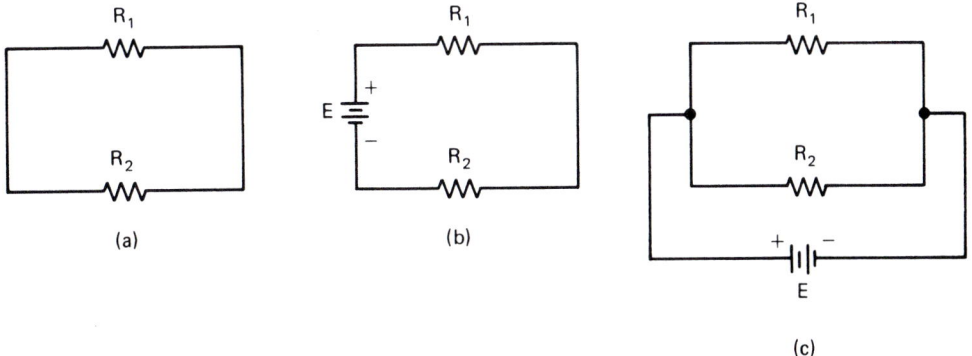

FIGURE 2-1 Comparison between series and parallel circuits.

2-1

Current Flow in a Series Circuit

Referring to Fig. 2-2(a), the resistors R_1, R_2, and R_3 are connected in series because they are joined end to end in succession. To cause a current to flow, the two ends of the series combination are attached to the positive and negative terminals of the battery, E_T. The equivalent wiring diagram for this circuit is shown in Fig. 2-2(b). Remember that a resistor has no polarity, so that if any or all of the resistors are turned around, the circuit is not affected. Moreover, as far as the current flow is concerned, the order in which two or more resistors are series connected does not matter, so that the resistors may be interchanged.

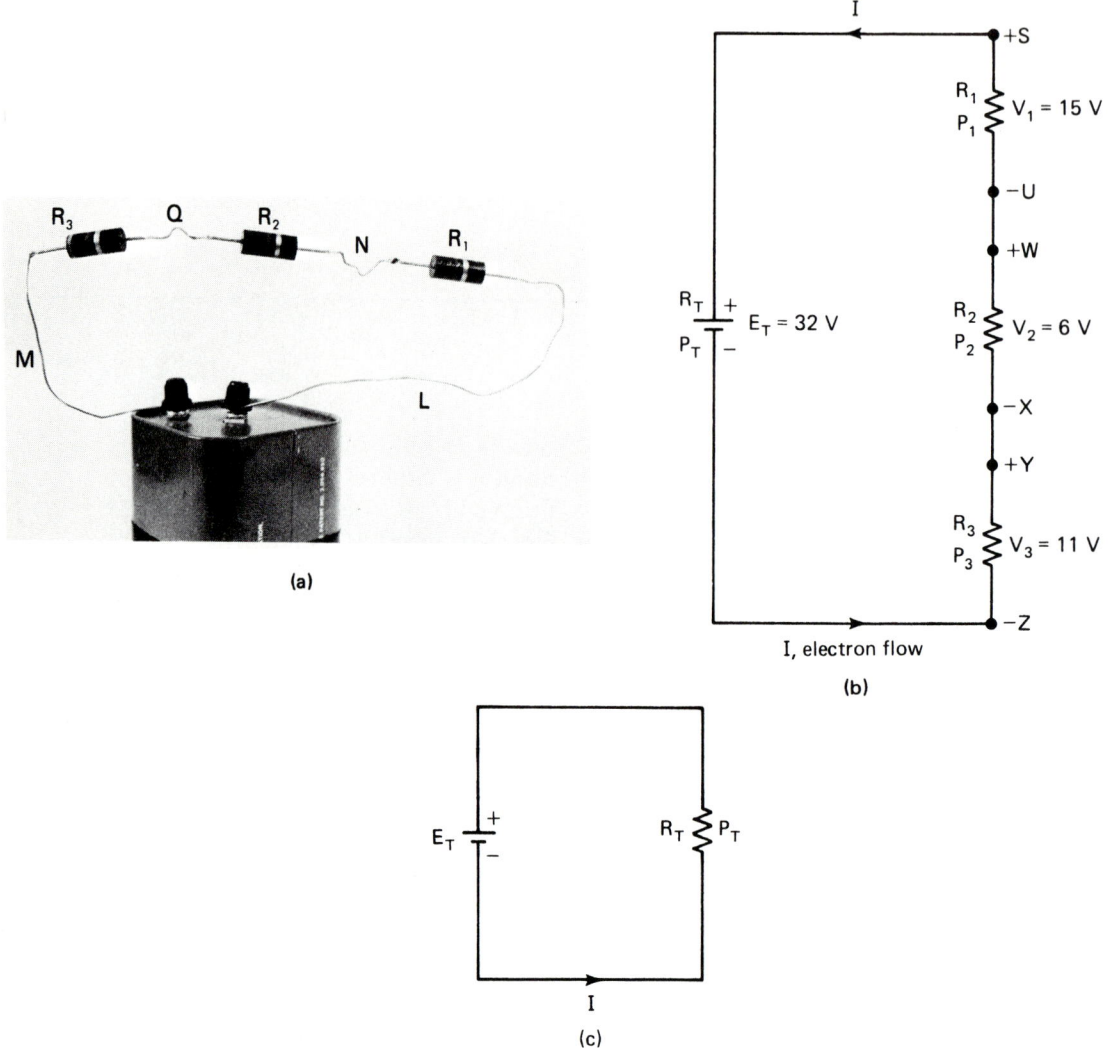

FIGURE 2-2 Series circuit.

Consequently, there are a wide variety of series configurations that can be mounted on a terminal strip, and some of these are shown in Fig. 2-3.

Obviously, in the series circuit of Fig. 2-2(a) there is only a single path for current flow. The negative terminal of the battery has a surplus of electrons and these will move into the copper connecting wire, M. Free electrons in this wire then travel from atom to atom and there will be an electron drift or current that will reach one end of the resistor R_3.

Under the repelling effect of the battery's negative terminal and the attractive effect of the positive terminal, there will be a negative electron drift through R_3, the connecting wire Q, the resistor R_2, the connecting wire N, the resistor R_1, and the connecting wire L. Eventually, electrons will enter the positive terminal of the battery but that is not the end of the story. Because the battery possesses energy as the result of its chemical action, it is capable of performing work, so that inside the battery there will be an electron movement away from the positive terminal and toward the negative terminal. This action will once more establish the deficit of electrons at the positive terminal and the surplus at the negative terminal. The single path is now complete and although we have described a sequence of events, the electron motion should be regarded as starting simultaneously in all parts of the circuit.

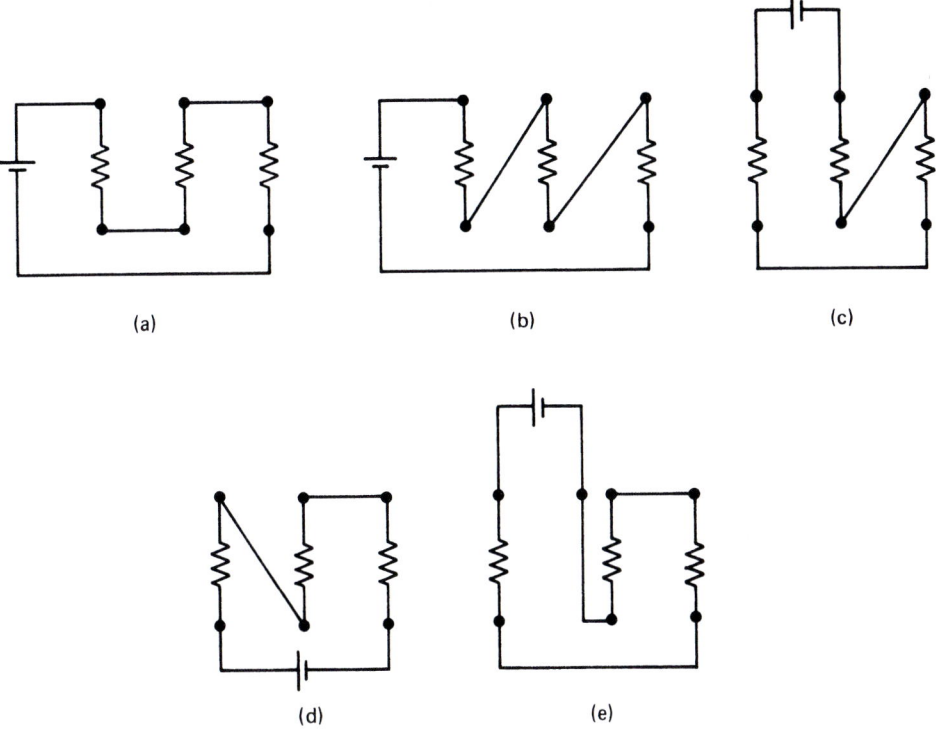

FIGURE 2-3 Series arrangements of resistors.

To delve a little deeper, the movement of electrons from *positive* to *negative* in the battery must be compared with the *negative-to-positive* electron drift through the resistors. Here it is useful to think in terms of a water analogy. The battery is equivalent to a pump that performs the work necessary to raise water to a certain height above ground level. The water is then allowed to flow back to ground through a series of pipes of differing lengths and diameters; this flow will naturally occur as the result of gravity. In the same way there will be a natural motion of the electrons from negative to positive through the resistors, but work has to be done on the electrons to cause the movement from positive to negative through the battery.

Although it is obvious that there is only a single path for current flow in a series circuit, it is not immediately clear that the value of the current is the same at every point. This can easily be verified by inserting current meters (ammeters) at various positions and showing that all the meters read the same. However, the question is frequently asked: If resistance is the electrical property that opposes current, why isn't the value of the electron flow leaving a resistor less than the value of the current entering? Here you must realize that the free electrons exiting a point in the conductor or resistor every second are balanced by an equal number of electrons entering at the same point. In the same way the number of electrons leaving the battery's negative terminal per second is equal to the number arriving at the positive terminal. Using the water analogy again, this is equivalent to saying that if the flow recorded at the entrance to a pipe is 2 gal/min, the flow at the pipe's exit must also be 2 gal/min. If this was not so and the exit flow was, for example, only 1 gal/min, the pipe would be retaining 1 gal/min and must ultimately burst! In the same way, if the electron flow entering one end of a resistor is 2 C/s (or 2 A), the current leaving the other end must also be 2 A. If the exit current was only 1 A, the resistor would be retaining 1 C every second and would become more and more negatively charged.

Sec. 2-1 Current Flow in a Series Circuit

To summarize, we have learned that in a series circuit containing resistors and a voltage source:

1. There is only a single path for current flow. As a result, a series circuit cannot contain any point where it is possible for the current to divide and take another path.
2. The current has the same value at every point in the circuit.

2-2 Individual Voltage Drops Across the Resistors

Referring to Fig. 2-2(b), we can regard the source voltage, E_T, as the total electromotive force (EMF) available to drive the current, I, around the series circuit. Part of this EMF will be used in overcoming the resistance of R_1, another part will overcome the resistance of R_2, and the remainder will be available for driving the current through R_3. Consequently, across each resistor there is developed a voltage which may be referred to as the *IR* drop or voltage drop; this same voltage may also be called the difference of potential or potential difference between the ends of the resistor (Section 1-5). The polarity of this voltage is determined by considering the direction of the electron drift or flow through the resistor. The motion of the negative electrons must naturally be from the negative end to the positive end of each resistor. These polarities are indicated in Fig. 2-2(b); also included are some possible values for the source voltage and the individual voltage drops across the resistors. The symbols V_1, V_2, and V_3 represent these voltage drops, but the values have only been chosen as an example; the actual voltages in a particular series circuit depend, as we shall see, on the values of the source voltage and the resistors.

Now let us discuss in detail an individual voltage drop in Fig. 2-2(b). V_1 has a value of 15 V, which means that point S is 15 V positive with respect to point U. Just as correctly, we could say that point U is 15 V negative with respect to point S. Alternatively, we could refer to a potential difference or difference of potential of 15 V between the points S and U; such a potential difference could be designated as V_{SU}. In every case we must consider a voltage as existing between *two* points; the statement "S is 15 V positive" would be meaningless at this stage. The value of V_1 is related by Ohm's law to the current, I, and the resistance, R_1 so that V_1 (volts) = I (amperes) $\times R_1$ (ohms). It is therefore logical to refer to V_1 as the *IR* drop across R_1, although we have not yet discussed the significance of the word "drop." Notice that since $V_1 = I \times R_1$ and I has the same value throughout the circuit, the highest-value resistor will have the greatest voltage drop and the lowest-value resistor the smallest voltage drop.

V_2 is equal to 6 V, with the point W positive with respect to the point X. You may be confused by the fact that there is a positive polarity at W and a negative polarity at U, since this appears to indicate that a voltage exists between these points. However, the positive polarity at W is related to the negative polarity at X, and does not in any way refer to point U. In fact, the points U and W are joined only by a connecting wire or clip lead, and ideally such connections are assumed to have zero resistance. With $R = 0\ \Omega$, there will be zero *IR* voltage or drop from U to W.

The fact that the negative end of V_1 is connected to the positive end of V_2 means that these two voltages are additive in the circuit, since they are the result of the current flowing in the same direction through the resistors R_1 and R_2. As an everyday example of voltage addition, think of inserting two $1\frac{1}{2}$-V D cells in a flashlight. To obtain the required 3 V you must arrange that the positive stud of one cell is in contact with the negative casing of the other cell.

In our example V_3 is 11 V and this voltage is also additive with V_1 and V_2. The total of the voltage drops across the three resistors is $V_1 + V_2 + V_3 = 15 + 6 + 11 = 32$ V, which exactly balances the source voltage, E_T. Such a voltage balance must always exist around any closed loop of an electrical circuit; this concept will be referred to later as Kirchhoff's voltage law. Looking at it another way, the source voltage is being divided between the resistors in a manner that is determined by their values. In equation form,

$$E_T = V_1 + V_2 + V_3 \qquad (2\text{-}2\text{-}1)$$

If there are N resistors connected in series,

$$E_T = V_1 + V_2 + V_3 + \cdots + V_N \qquad (2\text{-}2\text{-}2)$$

If the N resistors all have the same value,

$$E_T = NV \qquad (2\text{-}2\text{-}3)$$

where V is the voltage drop across one of the equal-value resistors.

Finally in this section we examine the significance of the word "drop." To avoid confusion the circuit diagram of Fig. 2-2(b) has been reproduced in Fig. 2-4. We now take a voltmeter and assuming that this meter does not affect the circuit in any way, we attach the negative or black probe permanently to the negative terminal of E_T. If we now touch the positive terminal with the red probe, the meter will read the source voltage of 32 V. When we move the red probe to point S, the reading will remain at 32 V since we are assuming a zero voltage drop across the connecting lead between S and the positive terminal. However, if the red probe is shifted to points U or W, the meter will be measuring the total voltage across R_2 and R_3, so that the reading is $V_2 + V_3 = 6 + 11 = 17$ V. This means that there is a voltage drop of $32 - 17 = 15$ V across the resistor R_1. With the red probe moved to points X and Y,

FIGURE 2-4 Voltage drops across resistors in series.

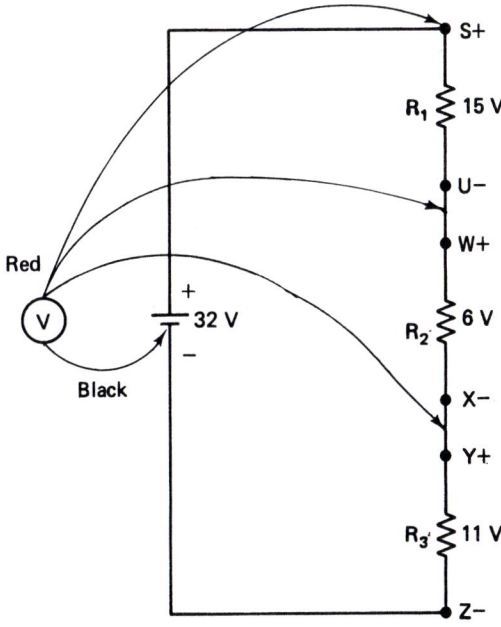

the reading will fall to the 11 V across R_3, so that there has been a further drop of $17 - 11 = 6$ V across R_2. Finally, with the red probe at point Z, both probes are joined together by a connecting lead and the reading is zero volts; the final 11 V has therefore been dropped across R_3.

To summarize the results of this section:

1. Across each resistor in a series circuit there is a voltage drop whose value is found by multiplying the current through the resistor by the value of the resistor. Since the current is the same throughout the circuit, the highest value resistor will carry the greatest voltage drop.
2. The voltage drops across the resistors are additive and their sum exactly balances the source voltage, which is divided between the resistors according to their values.

EXAMPLE 2-1 In Fig. 2-5, what are the values of the voltages V_{AB}, V_{BC}, V_{AF}, V_{EG}, V_{GN}, V_{JP}, V_{AQ}, V_{NQ}, V_{PQ}, and V_{KQ}?

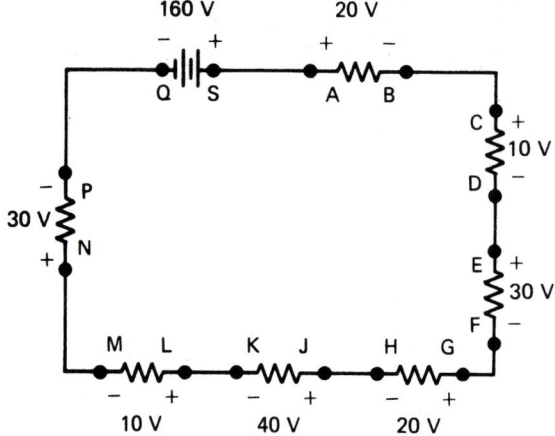

FIGURE 2-5 Circuit for Example 2-1.

Solution

$$V_{AB} = 20 \text{ V}$$

$$V_{BC} = 0 \text{ V}$$

Between B and C there is a connecting wire across which there is assumed to be a zero voltage drop.

$$V_{AF} = V_{AB} + V_{BC} + V_{CD} + V_{DE} + V_{EF}$$
$$= 20 \text{ V} + 0 \text{ V} + 10 \text{ V} + 0 \text{ V} + 30 \text{ V} = 60 \text{ V}$$
$$V_{EG} = V_{EF} + V_{FG} = 30 \text{ V} + 0 \text{ V} = 30 \text{ V}$$
$$V_{GN} = V_{GH} + V_{HJ} + V_{JK} + V_{KL} + V_{LM} + V_{MN}$$
$$= 20 \text{ V} + 0 \text{ V} + 40 \text{ V} + 0 \text{ V} + 10 \text{ V} + 0 \text{ V} = 70 \text{ V}$$
$$V_{JP} = V_{JK} + V_{KL} + V_{LM} + V_{MN} + V_{NP}$$
$$= 40 \text{ V} + 0 \text{ V} + 10 \text{ V} + 0 \text{ V} + 30 \text{ V} = 80 \text{ V}$$
$$V_{AQ} = 160 \text{ V}$$

The 160-V battery is positioned between points A and Q.

$$V_{NQ} = V_{NP} + V_{PQ} = 30 \text{ V} + 0 \text{ V} = 30 \text{ V}$$

$$V_{PQ} = 0 \text{ V}$$

There is only a connecting wire between points P and Q.

$$V_{KQ} = V_{KL} + V_{LM} + V_{MN} + V_{NP} + V_{PQ}$$
$$= 0 \text{ V} + 10 \text{ V} + 0 \text{ V} + 30 \text{ V} + 0 \text{ V}$$
$$= 40 \text{ V}$$

2-3 Total Equivalent Resistance of a Series Circuit

Referring to Fig. 2-2(b), we have already established that

$$E_T = V_1 + V_2 + V_3 \qquad (2\text{-}3\text{-}1)$$

and

$$V_1 = IR_1, \quad V_2 = IR_2, \quad V_3 = IR_3$$

Therefore,

$$E_T = V_1 + V_2 + V_3$$
$$= IR_1 + IR_2 + IR_3$$

or

$$E_T = I \times (R_1 + R_2 + R_3) \qquad (2\text{-}3\text{-}2)$$

We now want to consider the total resistance, R_T, which is equivalent to the series combination. By the total resistance we mean that value of resistance which offers the same total opposition to current flow and dissipates the same total power as R_1, R_2, and R_3; this is represented by the circuit of Fig. 2-2(c). In this circuit we have, by Ohm's law,

$$E_T = I \times R_T \qquad (2\text{-}3\text{-}3)$$

Comparing Eqs. (2-3-2) and (2-3-3), it is evident that

$$R_T = R_1 + R_2 + R_3 \qquad (2\text{-}3\text{-}4)$$

The total equivalent resistance is therefore equal to the sum of the individual resistances. Although this appears to be obvious, we have in fact derived the result of Eq. (2-3-4) by using Ohm's law and the concept of the voltage balance which exists in a series electrical loop.

The total equivalent resistance of any number of series resistors may be found in the same way. For example, with N resistors in series,

$$R_T = R_1 + R_2 + R_3 + \cdots + R_N \qquad (2\text{-}3\text{-}5)$$

If the N resistors are all *equal* in value,

$$R_T = NR \qquad (2\text{-}3\text{-}6)$$

where R is the value of one of the equal resistors.

Equation (2-3-5) tells us that the total resistance, R_T, is always greater than the ohms of the highest-value resistor connected in series. It also follows that we may increase the total resistance by adding an additional resistor in series; in practice this could be used to limit a circuit current to a safe value.

The circuit we have discussed may also be referred to as a series string of resistors. Summarizing this section, the total equivalent resistance of a series string of resistors is equal to the sum of the individual resistances.

EXAMPLE 2-2 In Fig. 2-6, what are the values of the voltage drops V_{AB}, V_{BD}, V_{CE}, V_{BE}, and V_{DE}?

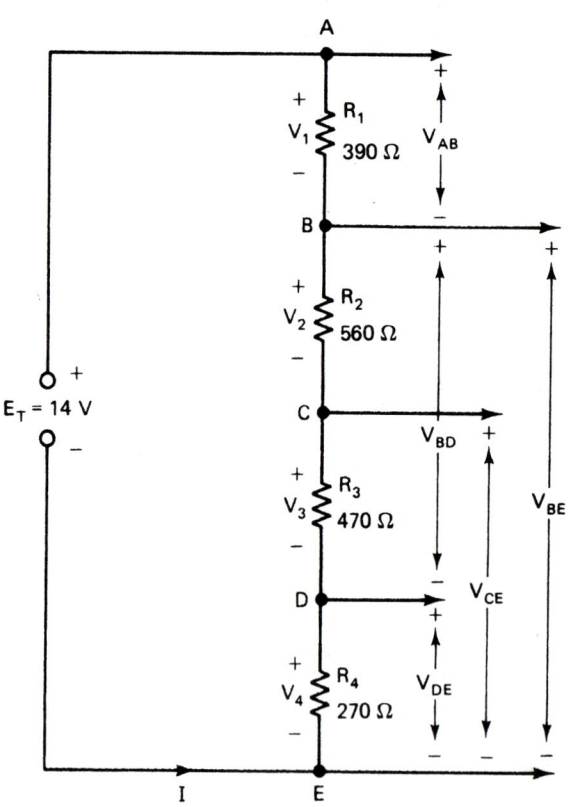

FIGURE 2-6 Circuit for Example 2-2.

Solution

$$\text{total resistance, } R_T = 390 + 560 + 470 + 270 = 1690 \ \Omega$$

$$\text{current, } I = \frac{E_T}{R_T} = \frac{14 \text{ V}}{1690 \ \Omega} = 0.008284 \text{ A}$$

$$\text{voltage drop, } V_1 = IR_1 = 0.008284 \text{ A} \times 390 \text{ }\Omega = 3.23077 \text{ V}$$

$$\text{voltage drop, } V_2 = IR_2 = 0.008284 \text{ A} \times 560 \text{ }\Omega = 4.63905 \text{ V}$$

$$\text{voltage drop, } V_3 = IR_3 = 0.008284 \text{ A} \times 470 \text{ }\Omega = 3.89349 \text{ V}$$

$$\text{voltage drop, } V_4 = IR_4 = 0.008284 \text{ A} \times 270 \text{ }\Omega = 2.23669 \text{ V}$$

This type of problem is very easily solved on an electronic calculator. The current is

$$I = \frac{14 \text{ V}}{390 \text{ }\Omega + 560 \text{ }\Omega + 470 \text{ }\Omega + 270 \text{ }\Omega}$$

The procedure involves adding the individual resistances so that the display shows 1690 Ω as the value of R_T. With a scientific calculator there can be several different ways of obtaining the value of I without having to manually record and reenter the value of R_T. Examples of such methods are:

1. Divide R_T by E_T and then use the reciprocal key to obtain $I = E_T/R_T$.

Operation		Display	
1690	$\div$	1690	
14	$=$	120.7	
	$1/x$	0.008284	(I)

2. Obtain the reciprocal of R_T and multiply this reciprocal by E_T.

Operation			Display	
1690	$1/x$	$\times$	0.0005917	
14	$=$		0.008284	(I)

A word of warning! If the value of R_T is entered as a very large number, the final answer can have a low degree of accuracy unless the calculator has internal scientific notation.

3. Store R_T in the calculator's memory. Then enter the value of E_T and divide it by R_T when retrieved from the memory.

Since each of the resistors carries the same current, the individual voltage drops may be found on the calculator by retaining the current as a "constant multiplying factor." You should consult your manual for the necessary series of operations, but a common procedure is:

Operation		Display	
0.008284	×	0.008284	
390	=	3.23077	(V_1)
560	=	4.63905	(V_2)
470	=	3.89349	(V_3)
270	=	2.23669	(V_4)

Note that all display figures have been rounded off.

Voltage Check

$$\text{sum of the voltage drops} = V_1 + V_2 + V_3 + V_4$$
$$= 3.23077 + 4.63905 + 3.89349 + 2.23669$$
$$= 14.00000 \text{ V} = E_T$$

$$V_{AB} = V_1 = 3.23077 = 3.23 \text{ V} \quad \text{(rounded off)}$$
$$V_{BD} = V_2 + V_3 = 8.53254 = 8.53 \text{ V} \quad \text{(rounded off)}$$
$$V_{CE} = V_3 + V_4 = 6.13018 = 6.13 \text{ V} \quad \text{(rounded off)}$$
$$V_{BE} = V_2 + V_3 + V_4 = 10.76922 = 10.8 \text{ V} \quad \text{(rounded off)}$$
$$V_{DE} = V_4 = 2.23669 = 2.24 \text{ V} \quad \text{(rounded off)}$$

The rounding off of the answers is dictated by the accuracy of the resistor values given in the problem.

2-4 Power Relationships in a Series Circuit

Referring to Fig. 2-2(b), each one of the three resistors in the series circuit dissipates power in the form of heat. It is of course important that the power dissipated in a particular resistor does not exceed its power rating; otherwise, the resistor may burn out. As discussed in Section 1-5, the power dissipated in the resistor R_1 is given by

$$P_1 = IV_1 = I^2 R_1 = \frac{V_1^2}{R_1} \quad \text{watts}$$

Similarly,

$$P_2 = IV_2 = I^2 R_2 = \frac{V_2^2}{R_2} \quad \text{and} \quad P_3 = IV_3 = I^2 R_3 = \frac{V_3^2}{R_3} \quad (2\text{-}4\text{-}1)$$

Since the powers dissipated in the resistors can only be obtained from the voltage source and the heating effects are additive, it follows that the total power, P_T, taken from the source must equal the sum of the individual powers dissipated in the resistors. Since

$$E_T = V_1 + V_2 + V_3$$

then

$$IE_T = IV_1 + IV_2 + IV_3$$

and

$$P_T = IE_T = P_1 + P_2 + P_3 \qquad (2\text{-}4\text{-}2)$$

In terms of the total resistance,

$$P_T = I^2 R_T = \frac{E_T^2}{R_T} \qquad (2\text{-}4\text{-}3)$$

For N resistors in series,

$$P_T = P_1 + P_2 + P_3 + \cdots + P_N \qquad (2\text{-}4\text{-}4)$$

For N equal value resistors in series,

$$P_T = NP \qquad (2\text{-}4\text{-}5)$$

where P is the power dissipated in one of the equal-value resistors.

A word of warning! Unless the resistors are of equal value, the total wattage rating of a series string is not equal to the sum of the individual wattage ratings. In other words, if 180 Ω, 220 Ω, 330 Ω, and 470 Ω, 1-W resistors are connected in series, they are not equivalent to a 1200-Ω resistor with a 4-W rating. The curent in each resistor to produce its 1-W rating is different; using the relationship $I = \sqrt{P/R}$, these currents are

$$I_{180\Omega} = \sqrt{\frac{1\text{ W}}{180\text{ }\Omega}} = 0.0745\text{ A}$$

$$I_{220\Omega} = \sqrt{\frac{1\text{ W}}{220\text{ }\Omega}} = 0.0674\text{ A}$$

$$I_{330\Omega} = \sqrt{\frac{1\text{ W}}{330\text{ }\Omega}} = 0.0550\text{ A}$$

$$I_{470\Omega} = \sqrt{\frac{1\text{ W}}{470\text{ }\Omega}} = 0.0461\text{ A}$$

We must select the smallest of these currents, 0.0461 A, since if the current were allowed to exceed this value, the power dissipation in the 470-Ω resistor would exceed its wattage rating. Using the relationship $P = I^2 R$, the wattage rating of the series string is only

$$(0.0461\text{ A})^2 \times 1200\text{ }\Omega = 2.55\text{ W}$$

Sec. 2-4 Power Relationships in a Series Circuit

Summarizing, the total power delivered from a voltage source to a series string of resistors is equal to the sum of the individual powers dissipated in the resistors. Since the highest-value resistor carries the greatest voltage drop, it will dissipate the most power; similarly, the lowest-value resistor will dissipate the least power.

EXAMPLE 2-3 Calculate the values of V_1, V_2, V_3, P_1, P_2, P_3, and the total power, P_T, dissipated in the circuit of Fig. 2-7.

Solution

$$\text{total resistance, } R_T = 3.3 \text{ k}\Omega + 4.7 \text{ k}\Omega + 1 \text{ k}\Omega = 9 \text{ k}\Omega$$

$$\text{circuit current, } I = \frac{E_T}{R_T} = \frac{27 \text{ V}}{9 \text{ k}\Omega} = 3 \text{ mA}$$

Remember that when a voltage in volts is divided by a resistance in kilohms, the result is a current in milliamperes. Similarly, the result of dividing a voltage in volts by a current in milliamperes is a resistance in kilohms; it follows that multiplying a resistance in kilohms by a current in milliamperes produces a voltage in volts.

The individual voltage drops are

$$V_1 = IR_1 = 3 \text{ mA} \times 3.3 \text{ k}\Omega = 9.9 \text{ V}$$

$$V_2 = IR_2 = 3 \text{ mA} \times 4.7 \text{ k}\Omega = 14.1 \text{ V}$$

$$V_3 = IR_3 = 3 \text{ mA} \times 1 \text{ k}\Omega = 3 \text{ V}$$

The sum of the voltage drops is

$$V_1 + V_2 + V_3 = 9.9 \text{ V} + 14.1 \text{ V} + 3 \text{ V} = 27 \text{ V}$$

which exactly balances the source voltage, E_T.

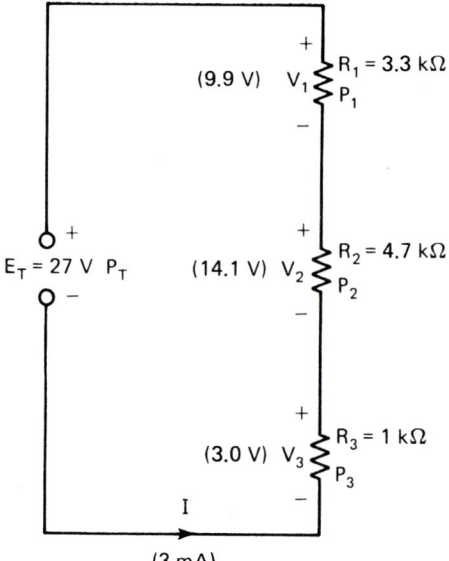

FIGURE 2-7 Circuit for Example 2-3.

The individual powers dissipated in the resistors are

$$P_1 = IV_1 = 3\text{mA} \times 9.9 \text{ V} = 29.7 \text{ mW}$$

$$P_2 = IV_2 = 3 \text{ mA} \times 14.1 \text{ V} = 42.3 \text{ mW}$$

$$P_3 = IV_3 = 3 \text{ mA} \times 3 \text{ V} = 9.0 \text{ mW}$$

Notice that the result of multiplying a current in milliamperes by a voltage in volts is a power expressed in milliwatts.

On the calculator the individual powers are found by entering the current as the "constant factor," which is then multiplied in turn by each of the voltage drops.

The total power dissipated by the resistors is

$$P_T = P_1 + P_2 + P_3 = 29.7 \text{ mW} + 42.3 \text{ mW} + 9.0 \text{ mW} = 81 \text{ mW}$$

The total power delivered by the source is

$$E_T \times I = 27 \text{ V} \times 3 \text{ mA} = 81 \text{ mW}$$

which exactly balances the value of P_T.

2-5 Step-by-Step Analysis of the Series Circuit

By now we have derived all the basic concepts and equations required for the solution of a series circuit if you are given only the values of the resistors and the source voltage. We therefore take an example of a series string with four resistors and use a step-by-step method for its solution. Solving the circuit shown in Fig. 2-8 would require finding the total resistance, the current, the individual voltage drops across the resistors, the individual powers dissipated in the resistors, and the total power in the circuit.

STEP 1   Find the total resistance of the circuit.

$$\begin{aligned}R_T &= R_1 + R_2 + R_3 + R_4 \\ &= 8.2 \text{ k}\Omega + 1800 \text{ }\Omega + 3.3 \text{ k}\Omega + 4.7 \text{ k}\Omega \\ &= 8.2 \text{ k}\Omega + 1.8 \text{ k}\Omega + 3.3 \text{ k}\Omega + 4.7 \text{ k}\Omega \\ &= 18 \text{ k}\Omega\end{aligned}$$

When adding the resistances together you must make certain that the values are in the same units. In our example the value of 1800 Ω had to be converted to 1.8 kΩ before the addition was carried out.

STEP 2   Calculate the circuit current.

$$I = \frac{E_T}{R_T} = \frac{72 \text{ V}}{18 \text{ k}\Omega} = 4 \text{ mA}$$

Remember that this current is the same at every point in the circuit.

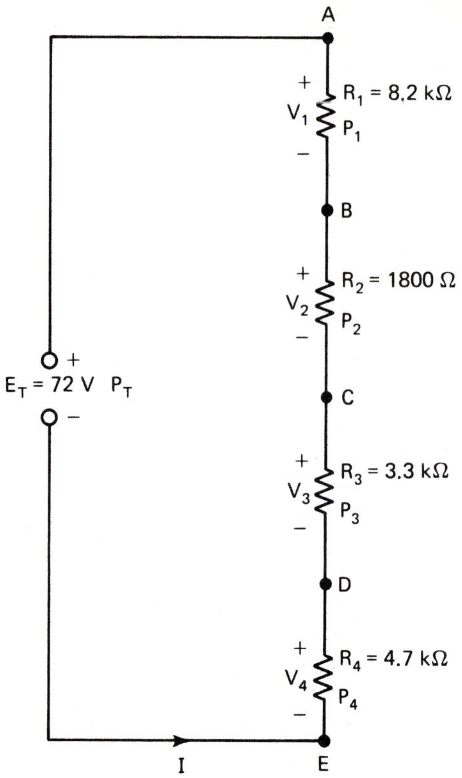

FIGURE 2-8 Step-by-step analysis of the series circuit.

When using your calculator to find the current, you must first add up the resistances to obtain R_T, then divide by E_T and take the reciprocal of the result. This enables you to find I by continuous chain operations without recording any intermediate results.

STEP 3 Determine the individual voltage drops.

$$V_1 = IR_1 = 4 \text{ mA} \times 8.2 \text{ k}\Omega = 32.8 \text{ V}$$
$$V_2 = IR_2 = 4 \text{ mA} \times 1.8 \text{ k}\Omega = 7.2 \text{ V}$$
$$V_3 = IR_3 = 4 \text{ mA} \times 3.3 \text{ k}\Omega = 13.2 \text{ V}$$
$$V_4 = IR_4 = 4 \text{ mA} \times 4.7 \text{ k}\Omega = 18.8 \text{ V}$$

Remember to enter the current in the calculator as the "constant factor."

VOLTAGE CHECK Make certain that the sum of the individual voltage drops exactly balances the supply voltage. Therefore,

$$V_1 + V_2 + V_3 + V_4 = 32.8 \text{ V} + 7.2 \text{ V} + 13.2 \text{ V} + 18.8 \text{ V}$$
$$= 72 \text{ V} = E_T$$

Notice that other potential differences exist in the circuit. For example,

$$V_{AC} = \text{potential difference between points } A \text{ and } C$$
$$= V_{AB} + V_{BC} = V_1 + V_2 = 32.8 \text{ V} + 7.2 \text{ V} = 40 \text{ V}$$
$$V_{AD} = V_{AB} + V_{BC} + V_{CD}$$
$$= V_1 + V_2 + V_3 = 32.8 \text{ V} + 7.2 \text{ V} + 13.2 \text{ V} = 53.2 \text{ V}$$

$$V_{BD} = V_{BC} + V_{CD}$$
$$= V_2 + V_3 = 7.2 \text{ V} + 13.2 \text{ V} = 20.4 \text{ V}$$
$$V_{BE} = V_{BC} + V_{CD} + V_{DE}$$
$$= V_2 + V_3 + V_4 = 7.2 \text{ V} + 13.2 \text{ V} + 18.8 \text{ V} = 39.2 \text{ V}$$
$$V_{CE} = V_{CD} + V_{DE}$$
$$= V_3 + V_4 = 13.2 \text{ V} + 18.8 \text{ V} = 32 \text{V}$$

STEP 4 Calculate the power dissipated in the individual resistors.

$$P_1 = IV_1 = 4 \text{ mA} \times 32.8 \text{ V} = 131.2 \text{ mW}$$

As discussed in Section 1-5, when a current in milliamperes is multiplied by a potential difference in volts, the result is the power dissipated in milliwatts.

P_1 may also be found from the formulas

$$P_1 = I^2 R_1 = (4 \text{ mA})^2 \times 8.2 \text{ k}\Omega$$
$$= 4 \text{ mA} \times 4 \text{ mA} \times 8.2 \text{ k}\Omega = 131.2 \text{ mW}$$

and

$$P_1 = \frac{V_1^2}{R_1} = \frac{(32.8 \text{ V})^2}{8.2 \text{ k}\Omega} = \frac{32.8 \text{ V} \times 32.8 \text{ V}}{8.2 \text{ k}\Omega} = 131.2 \text{ mW}$$

Then

$$P_2 = IV_2 = 4 \text{ mA} \times 7.2 \text{ V} = 28.8 \text{ mW}$$
$$P_3 = IV_3 = 4 \text{ mA} \times 13.2 \text{ V} = 52.8 \text{ mW}$$
$$P_4 = IV_4 = 4 \text{ mA} \times 18.8 \text{ V} = 75.2 \text{ mW}$$

The total power delivered from the voltage source is given by

$$P_T = IE_T = 4 \text{ mA} \times 72 \text{ V} = 288 \text{ mW}$$

The current should be entered as the "constant factor" when obtaining the power values on a calculator.

POWER CHECK Make certain that the sum of the individual powers dissipated exactly balances the total power delivered from the source. Therefore,

$$P_1 + P_2 + P_3 + P_4 = 131.2 \text{ mW} + 28.8 \text{ mW} + 52.8 \text{ mW} + 75.2 \text{ mW}$$
$$= 288 \text{ mW} = P_T$$

Notice that the highest value resistor, R_1, develops the greatest voltage drop and dissipates the most power (however, P_1 is only 131.2 mW and could therefore be safely dissipated by a ½-W 8.2-kΩ 10% color-coded resistor). By contrast, the lowest-value resistor, R_2, develops the smallest voltage drop and dissipates the least power.

When the value of the power dissipated has been determined, this power level is doubled to arrive at the required wattage rating for the resistor; this procedure results in an adequate safety factor.

2-6 Voltage Division Rule

The step-by-step analysis method can be used only when the resistor values and the source voltage are known. However, we may, for example, be given the voltage drop across the resistor and then be asked to calculate the voltage drop or the power dissipated for another one of the resistors in the series string. In this section we are therefore going to emphasize some of the other relationships in a series circuit and how they can be used to solve a variety of problems.

Referring back to Fig. 2-2(b), we have already seen that

$$V_1 = IR_1 \text{ and } V_2 = IR_2$$

Then

$$I = \frac{V_1}{R_1} = \frac{V_2}{R_2} \tag{2-6-1}$$

or

$$\frac{V_1}{V_2} = \frac{R_1}{R_2} \tag{2-6-2}$$

Equation (2-6-2) shows that the voltage drops are in proportion to the values of the resistors. We may therefore use the method of proportion to solve a problem in which we are given R_1, R_2, and V_2 and asked to find the value of V_1. We would do this by using the formula

$$V_1 = V_2 \times \frac{R_1}{R_2} \tag{2-6-3}$$

Similarly,

$$V_2 = V_1 \times \frac{R_2}{R_1} \tag{2-6-4}$$

The same rule of proportion can be applied to the powers dissipated in the individual resistors. Since

$$P_1 = I^2 R_1 \text{ and } P_2 = I^2 R_2$$

$$I^2 = \frac{P_1}{R_1} = \frac{P_2}{R_2} \text{ and } \frac{P_1}{P_2} = \frac{R_1}{R_2} \tag{2-6-5}$$

Then

$$P_1 = P_2 \times \frac{R_1}{R_2} \text{ and } P_2 = P_1 \times \frac{R_2}{R_1} \qquad (2\text{-}6\text{-}6)$$

These equations involving voltage drops and the powers dissipated apply to any number of resistors in series.

Section 2-2 showed that the source voltage was divided between the series resistors in a manner that was determined by the resistor values. Since

$$I = \frac{E_T}{R_T} \text{ and } V_1 = IR_1$$

$$V_1 = \frac{E_T}{R_T} \times R_1 = E_T \times \frac{R_1}{R_T} \qquad (2\text{-}6\text{-}7)$$

Then

$$\frac{V_1}{E_T} = \frac{R_1}{R_T} \text{ and } E_T = V_1 \times \frac{R_T}{R_1} \qquad (2\text{-}6\text{-}8)$$

Equations (2-6-7) and (2-6-8) represent the *voltage division rule* and show that, in a series circuit, the fraction of the source voltage developed across a particular resistor is equal to the ratio of the resistor's value to the circuit's total resistance. Since $V_1 = (E_T/R_T) \times R_1$, $V_2 = (E_T/R_T) \times R_2, \ldots$, the value of E_T/R_T should first be obtained on the calculator and then used as a "constant multiplier" when finding the individual voltage drops.

In the case of two resistors, R_1, and R_2, in series,

$$V_1 = E_T \times \frac{R_1}{R_T} = E_T \times \frac{R_1}{R_1 + R_2} \qquad (2\text{-}6\text{-}9)$$

and

$$V_2 = E_T \times \frac{R_2}{R_1 + R_2} \qquad (2\text{-}6\text{-}10)$$

The total power delivered from the source will divide between the resistors in the same manner as the total voltage. Since

$$P_T = I^2 R_T \text{ and } P_1 = I^2 R_1$$

$$I^2 = \frac{P_T}{R_T} = \frac{P_1}{R_1} \qquad (2\text{-}6\text{-}11)$$

This leads to

$$P_1 = P_T \times \frac{R_1}{R_T} \text{ and } P_2 = P_T \times \frac{R_2}{R_T} \qquad (2\text{-}6\text{-}12)$$

EXAMPLE 2-4 In Fig. 2-9, calculate the value of V_3, E_T, and P_2.

Solution

$$V_3 = V_1 \times \frac{R_3}{R_1} \qquad (2\text{-}6\text{-}4)$$

$$= 6.2 \text{ V} \times \frac{3.9 \text{ k}\Omega}{2.7 \text{ k}\Omega} = 8.96 \text{ V}$$

$$E_T = V_1 \times \frac{R_T}{R_1} \qquad (2\text{-}6\text{-}8)$$

$$= 6.2 \text{ V} \times \frac{2.7 \text{ k}\Omega + 6.8 \text{ k}\Omega + 3.9 \text{ k}\Omega}{2.7 \text{ k}\Omega}$$

$$= 6.2 \times \frac{13.4}{2.7} = 30.8 \text{ V}$$

$$P_1 = \frac{V_1^2}{R_1} \qquad (1\text{-}5\text{-}4)$$

$$= \frac{(6.2 \text{ V})^2}{2.7 \text{ k}\Omega}$$

$$= 14.2 \text{ mW}$$

$$P_2 = P_1 \times \frac{R_2}{R_1} \qquad (2\text{-}6\text{-}6)$$

$$= 14.2 \text{ mW} \times \frac{6.8 \text{ k}\Omega}{2.7 \text{ k}\Omega} = 35.9 \text{ mW}$$

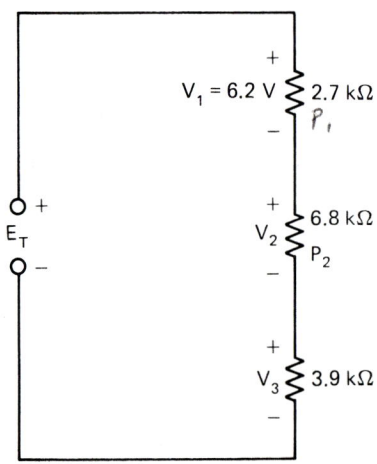

FIGURE 2-9 Circuit for Example 2-4.

EXAMPLE 2-5 In Fig. 2-10, find the values of V_1 and V_2.

Solution

$$V_1 = E_T \times \frac{R_1}{R_T}$$

$$= 31 \text{ V} \times \frac{2.7 \text{ k}\Omega}{2.7 \text{ k}\Omega + 3.3 \text{ k}\Omega + 1.8 \text{ k}\Omega + 5.6 \text{ k}\Omega} \qquad (2\text{-}6\text{-}7)$$

$$= 6.25 \text{ V}$$

FIGURE 2-10 Circuit for Example 2-5.

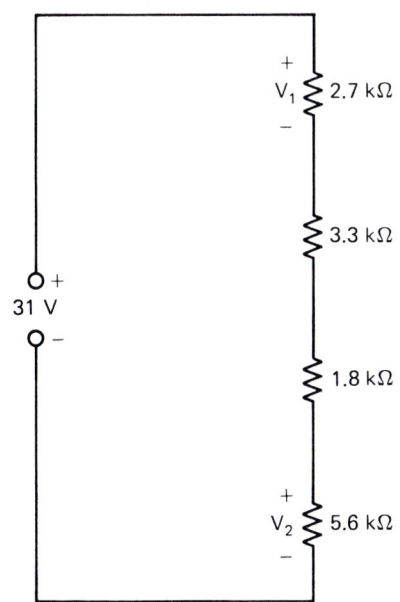

$$V_2 = V_1 \times \frac{R_2}{R_1} \qquad (2\text{-}6\text{-}4)$$

$$= 6.25 \text{ V} \times \frac{5.6 \text{ k}\Omega}{2.7 \text{ k}\Omega} = 12.96 \text{ V}$$

2-7
Voltage Divider Circuit

This circuit is a practical outcome of the voltage division rule. By its use we can obtain a number of different voltages from a single voltage source. These voltages will then be available as the supplies for various loads.

Referring to Fig. 2-11, V_1 is the voltage drop across the series combination of R_1, R_2, R_3, and R_4 and is therefore equal to the source voltage, E_T. V_2 is dropped across R_2, R_3, and R_4 in series; therefore, using the voltage division rule, we obtain

$$V_2 = E_T \times \frac{R_2 + R_3 + R_4}{R_T} = E_T \times \frac{R_2 + R_3 + R_4}{R_1 + R_2 + R_3 + R_4} \qquad (2\text{-}7\text{-}1)$$

Similarly,

$$V_3 = E_T \times \frac{R_3 + R_4}{R_T} = E_T \times \frac{R_3 + R_4}{R_1 + R_2 + R_3 + R_4} \qquad (2\text{-}7\text{-}2)$$

and

$$V_4 = E_T \times \frac{R_4}{R_T} = E_T \times \frac{R_4}{R_1 + R_2 + R_3 + R_4} \qquad (2\text{-}7\text{-}3)$$

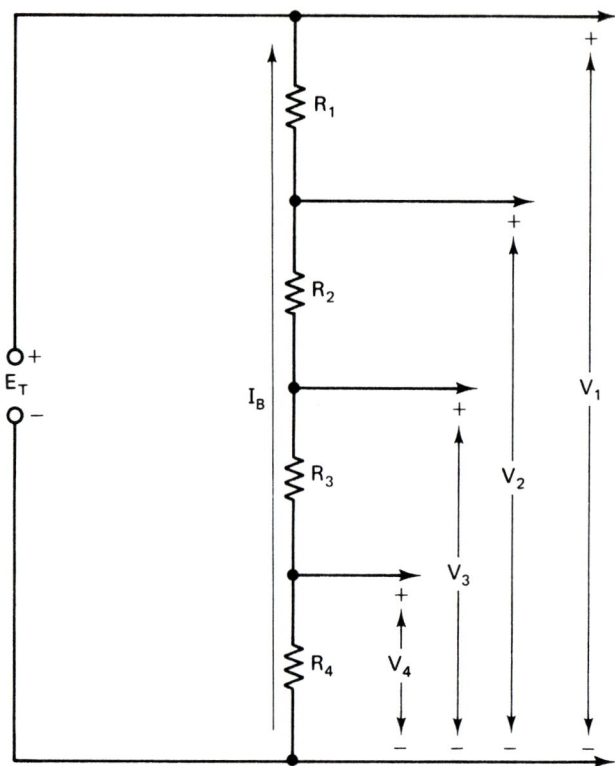

FIGURE 2-11 Voltage divider circuit.

To develop the voltages V_1, V_2, V_3, V_4, it is necessary for a so-called "bleeder current," I_B, to flow through the resistors. The provision of the different output voltages is therefore achieved at the expense of the power dissipated in the series string.

The voltage divider described so far would be regarded as operating under no-load conditions. As soon as loads are connected across the voltages V_1, V_2, V_3, and V_4, the circuit becomes more complex (Section 4-6) and Eqs. (2-7-1), (2-7-2), and (2-7-3) no longer apply.

EXAMPLE 2-6 In Fig. 2-12, find the values of V_1, V_2, V_3, and V_4.

Solution

$$V_1 = 65 \text{ V}$$

Using Eqs. (2-7-1), (2-7-2), and (2-7-3), we obtain

$$V_2 = 65 \text{ V} \times \frac{90 \text{ k}\Omega + 9 \text{ k}\Omega + 1 \text{ k}\Omega}{900 \text{ k}\Omega + 90 \text{ k}\Omega + 9 \text{ k}\Omega + 1 \text{ k}\Omega} = 65 \times \frac{100}{1000}$$
$$= 6.5 \text{ V}$$

$$V_3 = 65 \text{ V} \times \frac{9 \text{ k}\Omega + 1 \text{ k}\Omega}{900 \text{ k}\Omega + 90 \text{ k}\Omega + 9 \text{ k}\Omega + 1 \text{ k}\Omega} = 65 \times \frac{10}{1000}$$
$$= 0.65 \text{ V}$$

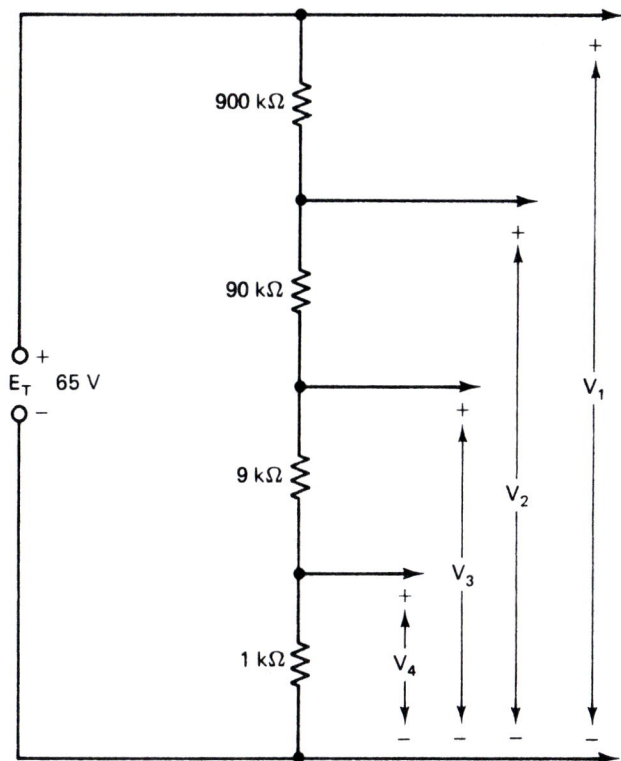

FIGURE 2-12 Circuit for Example 2-6.

$$V_4 = 65 \text{ V} \times \frac{1 \text{ k}\Omega}{900 \text{ k}\Omega + 90 \text{ k}\Omega + 9 \text{ k}\Omega + 1 \text{ k}\Omega} = 65 \times \frac{1}{1000}$$
$$= 0.065 \text{ V}$$

Figure 2-12 could be referred to as a "divide-by-10" circuit since V_2 is $\frac{1}{10}$ of V_1, V_3 is $\frac{1}{10}$ of V_2 and so on.

2-8 Ground. Voltage Reference Level

Ground may be regarded as any large mass of conducting material, so that there is essentially zero resistance between any two ground points. Examples of ground are the metal chassis of a transmitter, the aluminum chassis of a receiver, and a wide strip of copper plating on a printed-circuit board. Sometimes the damp earth itself is used as a ground, and at this point a student might ask: "Damp earth is not a particularly good conducting material, so how do we obtain the virtually zero resistance that is necessary for a good ground?" The answer is that the earth is part of a conductor that has an enormous cross-sectional area with the current able to penetrate the soil to a considerable depth. As we learned in Section 1-6, the resistance of a conductor is inversely proportional to its cross-sectional area, and therefore the requirement for a zero resistance ground is fulfilled.

The next obvious question: "Why do we use a ground system?" The main reason is to simplify our circuitry by saving on the amount of wiring. Ground may be used as the return path for many circuits, so that at any time there are a number of currents flowing through the ground system. Here we should mention another popular misconception; ground is not an enormous trashcan for collecting electrons but is merely part of a complete current path, so that at any instant the numbers of electrons per second entering and leaving ground are the same.

Figure 2-13(a) and (b) show two circuits which contain four connecting wires A, B, C, D. However, if a common ground (alternative symbols ⏚, ⏛, ⏜) is used for the return path as in Fig. 2-13(c), only two connecting wires A, D are required. Here is where the concept of zero resistance is important. If the resistance were not zero, the current flow in one circuit could develop a voltage between the ground points X and Y, and this would then cause a current to flow in the other circuit; in other words, the circuits would interfere with each other. The zero-resistance property of ground is therefore essential in order to achieve isolation between all those circuits that use the same ground.

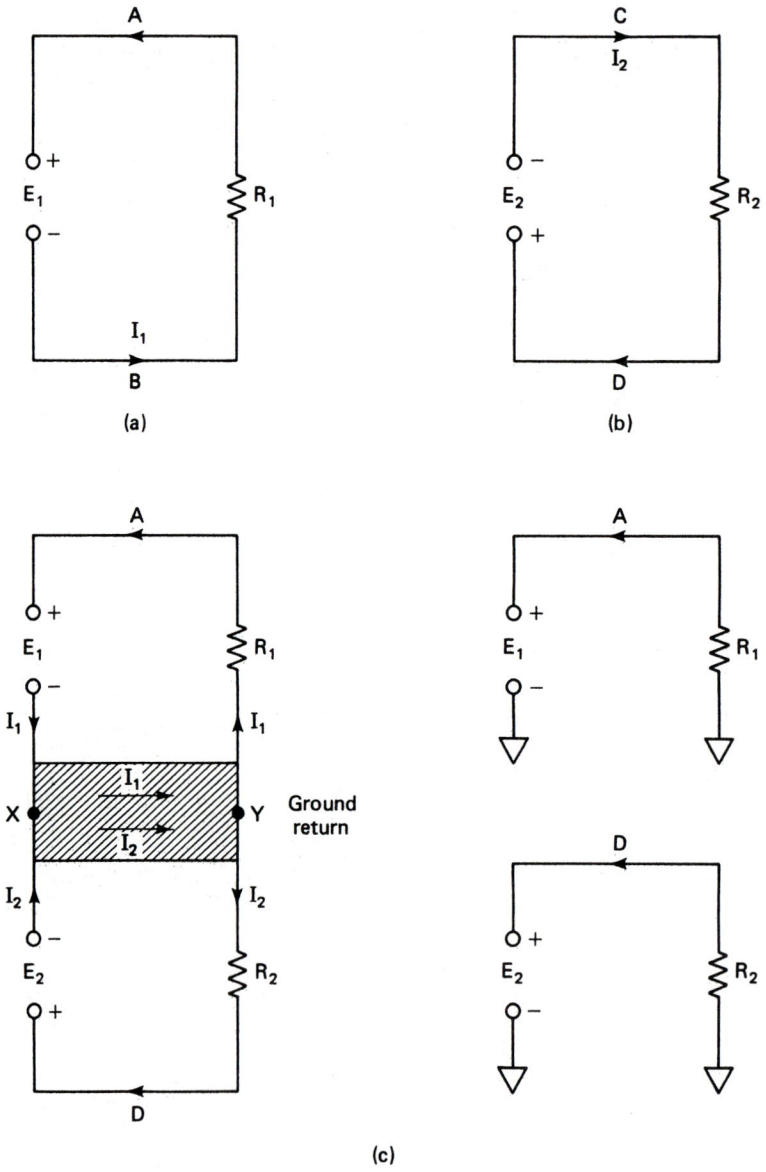

FIGURE 2-13 Use of a ground return.

Up to this point we have considered a voltage drop, potential difference, or difference of potential as existing between two points; to refer to the voltage at a single point would have been meaningless. However, if we create a common reference level to represent one point, we can quote the voltages at other points in relation to the common level. Ground is therefore chosen as a *reference level* of zero volts and the voltage or potential at any point may then be measured relative to ground. In Fig. 2-14(a) we would either say that point X is 12 V positive with respect to point Y or that point Y is 12 V negative with respect to point X. However, if point Y is grounded at zero volts [Fig. 2-14(b)], point X carries a potential of positive 12 V ($+12$ V) with respect to ground, whereas if X is grounded [Fig. 2-14(c)], the potential at Y is negative 12 V (-12 V) with respect to ground. Of course, both X and Y cannot be grounded simultaneously since this would mean zero resistance between the two points, and the load, R, would effectively be eliminated from the circuit. When troubleshooting a circuit it is normal practice to connect the black or common probe of a voltmeter permanently to ground and to measure the potentials at various points using the red probe. The voltmeter's function switch has two positions, marked "$+$DC" and "$-$DC"; these enable both positive and negative potentials to be measured without removing the black probe from ground.

As indicated, points may either carry a positive or a negative potential with respect to ground. This may be thought of in terms of electron flow; if the flow is from ground to a point, that point's potential will be positive, but if the flow's direction is reversed, the potential is negative. As an example, think of using an automobile chassis as ground. With some cars the negative terminal of the battery is strapped to the chassis and the positive lead is "live" or "hot," but with other cars the positive terminal is grounded and the negative lead is "live."

In any particular circuit the potentials at the various points will depend on which point is grounded; however, the voltage drops between any two points must always remain the same.

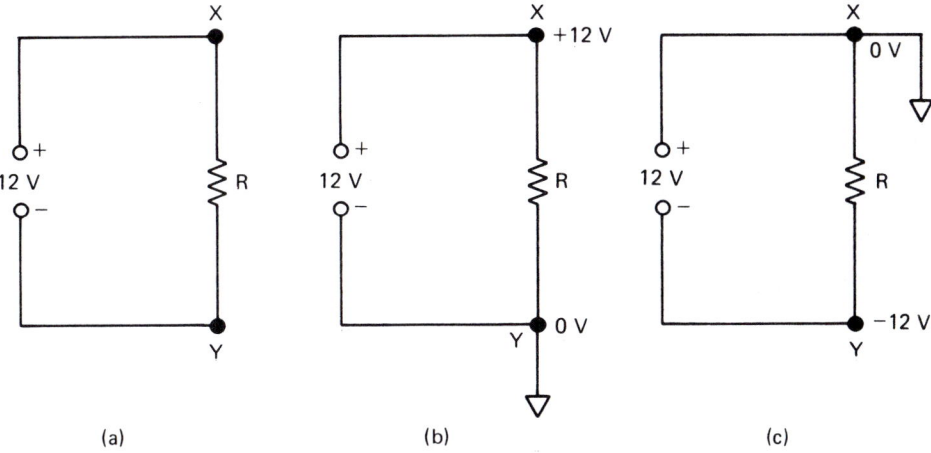

FIGURE 2-14 Ground as a voltage reference level.

EXAMPLE 2-7 In the circuit of Fig. 2-15, ground each of the points in turn. What potentials then exist at all the points in the circuit?

Sec. 2-8 Ground. Voltage Reference Level

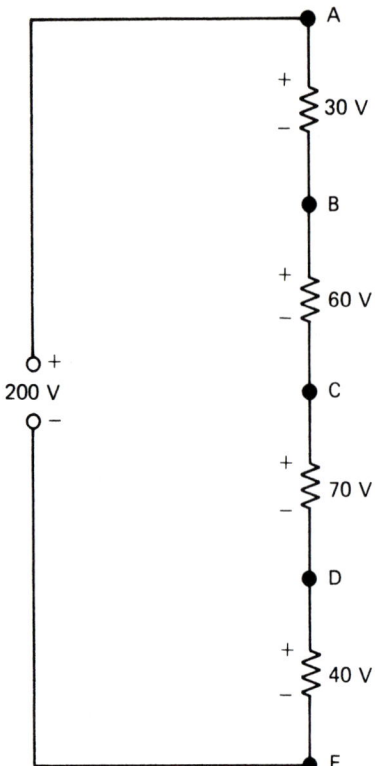

FIGURE 2-15 Circuit for Example 2-7.

Solution   *A Grounded*

$$\text{potential at } A = 0 \text{ V}$$
$$\text{potential at } B = -30 \text{ V}$$
$$\text{potential at } C = -(30 + 60) = -90 \text{ V}$$
$$\text{potential at } D = -(30 + 60 + 70) = -160 \text{ V}$$
$$\text{potential at } E = -(30 + 60 + 70 + 40) = -200 \text{ V}$$

B Grounded

$$\text{potential at } A = +30 \text{ V}$$
$$\text{potential at } B = 0 \text{ V}$$
$$\text{potential at } C = -60 \text{ V}$$
$$\text{potential at } D = -(60 + 70) = -130 \text{ V}$$
$$\text{potential at } E = -(60 + 70 + 40) = -170 \text{ V}$$

Grounding a point such as *B* would be referred to as an *intermediate ground*.

C Grounded

$$\text{potential at } A = +(30 + 60) = +90 \text{ V}$$
$$\text{potential at } B = +60 \text{ V}$$
$$\text{potential at } C = 0 \text{ V}$$
$$\text{potential at } D = -70 \text{ V}$$
$$\text{potential at } E = -(70 + 40) = -110 \text{ V}$$

D Grounded

$$\text{potential at } A = +(70 + 60 + 30) = +160 \text{ V}$$
$$\text{potential at } B = +(70 + 60) = +130 \text{ V}$$
$$\text{potential at } C = +70 \text{ V}$$
$$\text{potential at } D = 0 \text{ V}$$
$$\text{potential at } E = -40 \text{ V}$$

E Grounded

$$\text{potential at } A = +(40 + 70 + 60 + 30) = +200 \text{ V}$$
$$\text{potential at } B = +(70 + 60 + 40) = +170 \text{ V}$$
$$\text{potential at } C = +(70 + 40) = +110 \text{ V}$$
$$\text{potential at } D = +40 \text{ V}$$
$$\text{potential at } E = 0 \text{ V}$$

Now let us work out the potential difference between B and D as the various points are grounded in turn.

$$A \text{ grounded: } V_{BD} = (-30 \text{ V}) - (-160 \text{ V}) = 130 \text{ V}$$

This means that point B is 130 V positive with respect to point D, or point D is 130 V negative with respect to point B.

$$B \text{ grounded: } V_{BD} = 0 \text{ V} - (-130 \text{ V}) = 130 \text{ V}$$
$$C \text{ grounded: } V_{BD} = +60 \text{ V} - (-70 \text{ V}) = 130 \text{ V}$$
$$D \text{ grounded: } V_{BD} = +130 \text{ V} - 0 \text{ V} = 130 \text{ V}$$
$$E \text{ grounded: } V_{BD} = +170 \text{ V} - (+40 \text{ V}) = 130 \text{ V}$$

We have seen that the voltage drop between points B and D remains at 130 V irrespective of which point is grounded.

2-9
Potentiometer and Rheostat

The potentiometer (normally spoken of as a "pot") is a type of variable resistor whose construction has been described in Section 1-8. Basically, it consists of a length of resistance wire (wire possessing appreciable resistance) or a thin carbon track along which a moving contact or slider may be set to any point (Fig. 2-16). Its purpose is to obtain an output *voltage*, V_o, which may be varied from zero up to the full value of the input voltage, V_i. There are two terminals at the ends of the potentiometer and the slider is connected to a third terminal. The input voltage is then applied to the end terminals while the output voltage appears between the slider terminal and one end terminal, which is usually grounded. A good example of a potentiometer is a radio or TV receiver's volume control.

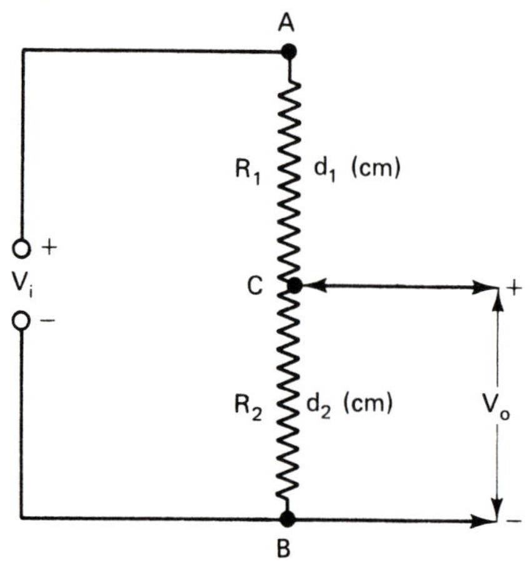

FIGURE 2-16 Potentiometer as a voltage control.

If the slider is set to position C, the total resistance of the potentiometer is $R_1 + R_2$, and this value would be stamped on the casing of the potentiometer. The slider splits the total resistance between R_1 and R_2, so that we can use the voltage division rule of Section 2-6 and

$$V_o = V_i \times \frac{R_2}{R_1 + R_2} \qquad (2\text{-}9\text{-}1)$$

When a slider is moved to position B, $R_2 = 0\ \Omega$ and $V_o = 0$ V.

If C is at the center position, $R_1 = R_2$ and $V_o = V_i/2$. With the slider at A, $R_1 = 0\ \Omega$ and $V_o = V_i$. These results will only be true provided that no load is connected across the output voltage. If a load is connected, the load current would flow through R_1, increasing its voltage drop so that the output voltage would fall.

The use of the potentiometer is, of course, not confined to dc, so that the input voltage may take any form provided that the power rating (also stamped on the potentiometer's casing) is not exceeded.

If the ratio of R_2 to the total resistance of the potentiometer is equal to the ratio of the slider's travel, d_2 (from its starting position), to the total travel available, $d_1 + d_2$, the potentiometer is said to be linear. Whether or not a potentiometer is linear depends on the winding of the resistance wire or the construction of the carbon track. For example, if the resistance and travel ratios were connected by a logarithmic relationship, we would have a "log.pot." However, with a linear pot

$$V_o = V_i \times \frac{d_2}{d_1 + d_2} \qquad (2\text{-}9\text{-}2)$$

where d_1 and d_2 may conveniently be measured in centimeters.

The rheostat is used for controlling the *current* in a series circuit (as opposed to the potentiometer, which controls an output *voltage*); its construction is similar to that of the potentiometer but as shown in Fig. 2-17, the rheostat only needs one end terminal and the terminal connected to the moving contact. A suitable potentiometer may therefore be adapted for use as a rheostat either by leaving the third terminal free [Fig. 2-17(a)] or by connecting it to the moving contact [Fig. 2-17(b)]. An everyday example of a rheostat is the dimmer control for the lights on a car's dashboard.

Referring again to Fig. 2-17(a) let us move the sliding contact to position

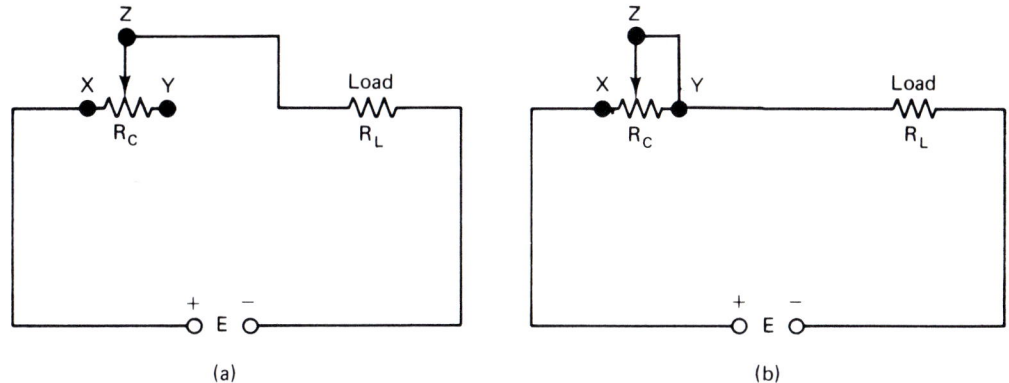

FIGURE 2-17 Rheostat as a current control.

X. None of the rheostat's resistance is included in the circuit, so that the current will have its maximum value,

$$I_{max} = \frac{E}{R_L} \qquad (2\text{-}9\text{-}3)$$

If the slider is now moved to position Y [Fig. 2-17(b)], the whole of the rheostat's resistance, R_C, is in series with the load, R_L, and minimum current will flow.

$$I_{min} = \frac{E}{R_L + R_C} \qquad (2\text{-}9\text{-}4)$$

The rheostat can therefore set any level of current between the maximum and minimum values.

EXAMPLE 2-8 Referring to Fig. 2-16, a 2-kΩ linear potentiometer is connected across a 50-V source. Find the minimum wattage rating for the potentiometer. If $d_1 = 2$ cm and $d_2 = 2.5$ cm, calculate the output voltage.

Solution

$$\text{minimum wattage rating, } \frac{(50 \text{ V})^2}{2 \text{ k}\Omega} = \frac{2500}{2000} = 1.25 \text{ W} \qquad \frac{E^2}{R} = W$$

$$\text{output voltage, } V_o = V_i \times \frac{d_2}{d_1 + d_2}$$

$$= 50 \times \frac{2.5}{2.0 + 2.5} \qquad (2\text{-}9\text{-}2)$$

$$= \frac{50 \times 2.5}{4.5} = 27.78 \text{ V}$$

EXAMPLE 2-9 In the circuit of Fig. 2-17(a), $E = 12$ V, $R_L = 3$ kΩ, and $R_C = 5$ kΩ. Calculate the maximum and minimum values of the circuit current.

Sec. 2-9 Potentiometer and Rheostat

Assuming that the rheostat is linear, what is the value of the current with the moving contact in the center position?

Solution

$$\text{maximum current, } I_{max} = \frac{E}{R_L} \quad (2\text{-}9\text{-}3)$$

$$= \frac{12 \text{ V}}{3 \text{ k}\Omega}$$

$$= 4 \text{ mA}$$

$$\text{minimum current, } I_{min} = \frac{E}{R_L + R_C} \quad (2\text{-}9\text{-}4)$$

$$= \frac{12 \text{ V}}{3 \text{ k}\Omega + 5 \text{ k}\Omega} = \frac{12 \text{ V}}{8 \text{ k}\Omega}$$

$$= 1.5 \text{ mA}$$

With the moving contact in the center position,

$$R_C = \frac{5 \text{ k}\Omega}{2} = 2.5 \text{ k}\Omega$$

The current is then

$$I = \frac{E}{R_L + R_C} = \frac{12 \text{ V}}{3 \text{ k}\Omega + 2.5 \text{ k}\Omega} = \frac{12}{5.5} = 2.18 \text{ mA}$$

2-10

Voltage Sources Connected in Series-Aiding and Series-Opposing

The form of series circuit so far discussed consists of resistors joined end to end with a single voltage source. However, a series circuit may contain any number of resistors together with any number of voltage sources which, may be connected either in series-aiding or series-opposing. Two sources are said to be *series-aiding* when their polarities are such as to drive current in the same direction around the circuit. This is illustrated in Fig. 2-18, where the positive terminal of E_1 is directly connected to the negative terminal of E_2; both voltages will then drive the electrons around the circuit in a clockwise direction. If both voltage sources are now reversed, they will still be connected in series-aiding but the electron flow will be in the counterclockwise direction. The normal purpose of connecting sources in series-aiding is to increase the voltage applied to a circuit. A good example is the insertion of two $1\frac{1}{2}$-V D cells into a flashlight to produce a total of 3 V; here the positive center terminal of one cell must be in contact with the negative casing of the other cell to produce the series-aiding connection.

Referring again to Fig. 2-18, the voltage sources may be added in the same way as the voltage drops across series resistors. The equivalent EMF is therefore the sum of the individual EMFs or $E_T = E_1 + E_2$, and since the

FIGURE 2-18 Two cells in series-aiding.

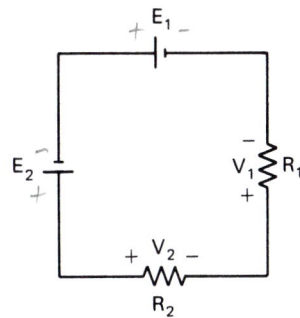

total resistance, R_T, is $R_1 + R_2$, the current is

$$I = \frac{E_T}{R_T} = \frac{E_1 + E_2}{R_1 + R_2} \qquad (2\text{-}10\text{-}1)$$

This current may also be regarded as the sum of the currents due to E_1 and E_2 separately. However, the same idea cannot be applied toward finding the total power in the circuit; P_T can only be found by

$$P_T = I \times E_T = I \times (E_1 + E_2) = I \times (V_1 + V_2)$$
$$= I^2 \times (R_1 + R_2) = \frac{(E_1 + E_2)^2}{R_1 + R_2} \qquad (2\text{-}10\text{-}2)$$

If one of the sources, E_2, is reversed as in Fig. 2-19, the polarities of the sources will be such as to drive currents in opposite directions around the circuit. The connection is therefore *series-opposing* and the total equivalent EMF is the difference of the individual EMFs; the greater of the two EMFs will then determine the direction of the current flow. In the particular case when two identical voltage sources are connected in series-opposing, the total equivalent EMF is zero; the current is also zero, as are the individual drops and the powers dissipated in the resistors.

FIGURE 2-19 Two cells in series-opposing.

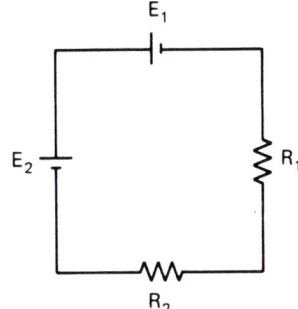

EXAMPLE 2-10 In the circuit of Fig. 2-20, calculate the values of I, V_1, V_2, P_1, P_2, and P_T.

Solution The connection of the two voltage sources is series-aiding.

total equivalent EMF, $E_T = 10\text{ V} + 6\text{ V} = 16\text{ V}$

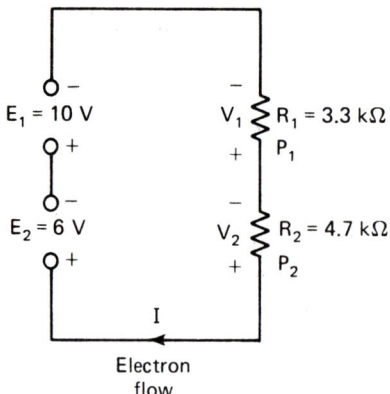

FIGURE 2-20 Circuit for Example 2-10.

$$\text{total equivalent resistance, } R_T = 3.3 \text{ k}\Omega + 4.7 \text{ k}\Omega = 8 \text{ k}\Omega$$

$$\text{current, } I = \frac{E_T}{R_T} = \frac{16 \text{ V}}{8 \text{ k}\Omega} = 2 \text{ mA}$$

Notice that the current I_1, solely due to the EMF E_1, is $I_1 = E_1/R_T = 10 \text{ V}/8 \text{ k}\Omega = 1.25 \text{ mA}$; the current I_2, solely due to the EMF E_2, is $I_2 = E_2/R_T = 6 \text{ V}/8 \text{ k}\Omega = 0.75 \text{ mA}$. Then $I_1 + I_2 = 1.25 \text{ mA} + 0.75 \text{ mA} = 2 \text{ mA} = I$.

$$\text{voltage drop, } V_1 = 2 \text{ mA} \times 3.3 \text{ k}\Omega = 6.6 \text{ V}$$

$$\text{voltage drop, } V_2 = 2 \text{ mA} \times 4.7 \text{ k}\Omega = 9.4 \text{ V}$$

$$\text{power dissipated, } P_1 = I \times V_1 = 2 \text{ mA} \times 6.6 \text{ V} = 13.2 \text{ mW}$$

$$\text{power dissipated, } P_2 = I \times V_2 = 2 \text{ mA} \times 9.4 \text{ V} = 18.8 \text{ mW}$$

$$\text{total power dissipated, } P_T = P_1 + P_2 = 13.2 \text{ mW} + 18.8 \text{ mW} = 32 \text{ mW}$$

Voltage Check

$$\text{sum of individual voltage drops} = V_1 + V_2 = 6.6 \text{ V} + 9.4 \text{ V}$$
$$= 16 \text{ V}$$
$$= \text{total equivalent EMF, } E_T$$

Power Check

$$\text{total power delivered from the two sources, } P_T = I \times E_T$$
$$= 2 \text{ mA} \times 16 \text{ V} = 32 \text{ mW}$$

Notice that, with E_1 in the circuit (E_2 removed), the power dissipated would be

$$\frac{E_1^2}{R_T} = \frac{(10 \text{ V})^2}{8 \text{ k}\Omega} = 12.5 \text{ mW}$$

If only E_2 were present, the power dissipated would be

$$\frac{E_2^2}{R_T} = \frac{(6 \text{ V})^2}{8 \text{ k}\Omega} = 4.5 \text{ mW}$$

The total of these two powers is only 12.5 + 4.5 = 17 mW, which is less than the actual total power dissipated, 32 mW. You are probably asking yourself: "Why doesn't the method work for power when it works so well for current?" The argument involves the principle of superposition (described in Section 6-5), but basically it boils down to the fact that whereas the voltage and the current are directly proportional (Ohm's law), the power dissipated is proportional to the *square* of the voltage or current. When a square is involved, the principle of superposition does not apply.

EXAMPLE 2-11 In the circuit of Fig. 2-21, calculate the values of I, V_1, V_2, P_1, P_2, and P_T.

FIGURE 2-21 Circuit for Example 2-11.

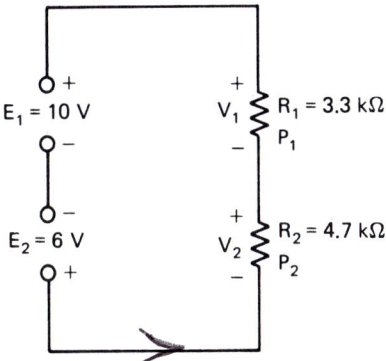

Solution For comparison purposes we have reversed the polarity of E_1 in the circuit of Example 2-10, so that the connection is now series-opposing. Since E_1 is greater than E_2, the direction of the current is reversed.

total equivalent EMF, $E_T = E_1 - E_2 = 10\text{ V} - 6\text{ V} = 4\text{ V}$

current, $I = \dfrac{E_T}{R_T} = \dfrac{4\text{ V}}{8\text{ k}\Omega} = 0.5\text{ mA}$

voltage drop, $V_1 = 0.5\text{ mA} \times 3.3\text{ k}\Omega = 1.65\text{ V}$

voltage drop, $V_2 = 0.5\text{ mA} \times 4.7\text{ k}\Omega = 2.35\text{ V}$

power dissipated, $P_1 = I \times V_1 = 0.5\text{ mA} \times 1.65\text{ V} = 0.825\text{ mW}$

power dissipated, $P_2 = I \times V_2 = 0.5\text{ mA} \times 2.35\text{ V} = 1.175\text{ mW}$

total power dissipated, $P_T = P_1 + P_2 = 0.825\text{ mW} + 1.175\text{ mW} = 2\text{ mW}$

Voltage Check

sum of voltage drops, $V_1 + V_2 = 1.65\text{ V} + 2.35\text{ V} = 4\text{ V} = E_T$

Power Check

total power delivered from the two sources $= I \times E_T$

$= 0.5\text{ mA} \times 4\text{ V} = 2\text{ mW}$

EXAMPLE 2-12 Calculate the total power dissipated in the circuit of Fig. 2-22.

Solution The connection of the two voltage sources is series-opposing. The total equivalent EMF is $E_T = 100\text{ V} - 100\text{ V} = 0\text{ V}$, and therefore the circuit current is zero. There will be no power dissipated in any of the resistors; consequently, the total power dissipated in the circuit is zero.

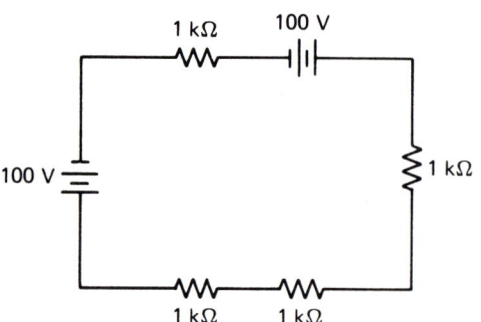

FIGURE 2-22 Circuit for Example 2-12.

2-11 Effect of an Open Circuit

An *open circuit* may be regarded as a break that occurs in the series string. This may be caused by a blown fuse; other causes can be a resistor which for some reason has burned out due to excessive power dissipation or a resistor becoming disconnected from its neighbor. In the last case the result would be an air gap, which could have a resistance of billions of ohms. This resistance may be regarded as part of the series circuit (Fig. 2-23) and since its value is so large, it will swamp out the effect of all the other resistors. The circuit current will then be practically zero and the voltage drops across the individual resistors will be negligible.

To a first approximation we can say that if an open circuit occurs in a series string the total resistance becomes infinite and the current is therefore zero. There will be no voltage drops across the resistors and no power dissipated in the circuit. Remember that zero voltage drop can occur either when a current is flowing but the resistance is zero, as in the case of a connecting wire, or when resistance is present but the current is zero, as with the open circuit.

Although the current in the circuit of Fig. 2-23 is theoretically zero, the voltage of the source will still be present and will in fact appear between points D and E where the open circuit exists. To the question "By what means does the voltage get there?" we can only mention at this stage that the source charges the capacitance between D and E (Section 9-6). However, here is a good point to remember: a source voltage exists irrespective of whether a load or a complete circuit is connected across the source. For example, there is 110 V present at a wall outlet even though the electrical generator is a considerable distance away and no load is connected to the outlet.

To summarize, the resistance of an open circuit is theoretically infinite; no current may flow but a voltage can be developed across the open circuit. A good example is a series string of Christmas tree lights; as you know, when one of the bulbs burns out to create the open circuit, all the other lights go

FIGURE 2-23 Effect of an open circuit.

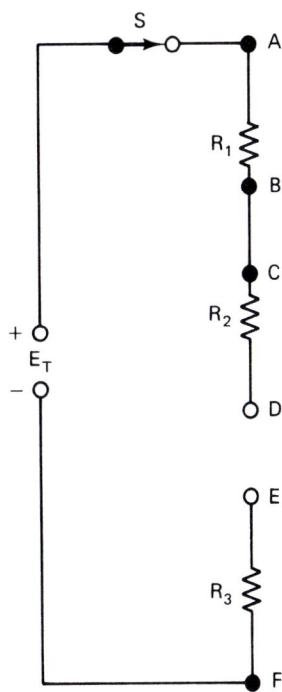

out as well, so that the current in the circuit is zero. You then try to find the defective bulb by substituting a good one for each of the lights in turn; this will solve the problem provided that not more than one bulb has burned out. However, let us see how we would find the open circuit by using our measuring instruments; this can be done with either an ohmmeter for measuring resistance or a voltmeter.

All resistance measurements must be made with the power removed, so in Fig. 2-23 we must first open the switch, S. Using a high-resistance range such as R × 1 MΩ, connect the black or common probe to point F. With the red probe at point A, the measured reading would be virtually infinite since the resistance contains the open circuit. The reading will show no change as the red probe is moved to points B, C, and D. However, as soon as the red probe is moved to point E, the resistance falls to the relatively low value of R_3. This sudden change in reading then indicates that the open circuit exists between points D and E. We have used an ohmmeter to conduct a *continuity* test that will locate a discontinuity—the open circuit.

To use the voltmeter, S is closed and the black probe is connected to point F. With the red probe at point A the voltmeter reads the value of the source voltage. The reading will not change when the red probe is moved in turn to points B, C, and D since with no current flowing, there will be zero voltage drops across the resistors R_1 and R_2. However, as soon as the red probe is moved to point E, the reading drops to zero since the voltmeter is connected across R_3 and there is no voltage drop for this resistor. Once again the abrupt change in the voltage readings shows that the open circuit is between points D and E.

It is appropriate at this stage to mention that the *short circuit* has theoretically zero resistance and therefore current may flow through the short circuit without developing any voltage drop. However, resistors themselves rarely become shorted and the short circuit normally occurs as the result of bare connecting wires coming into contact or being joined together by a stray drop of solder. The short circuit is therefore a zero-resistance path across or in parallel with the resistor and this will be fully discussed in Chapter 3.

EXAMPLE 2-13 In Fig. 2-24, what is the voltage drop between points B and C?

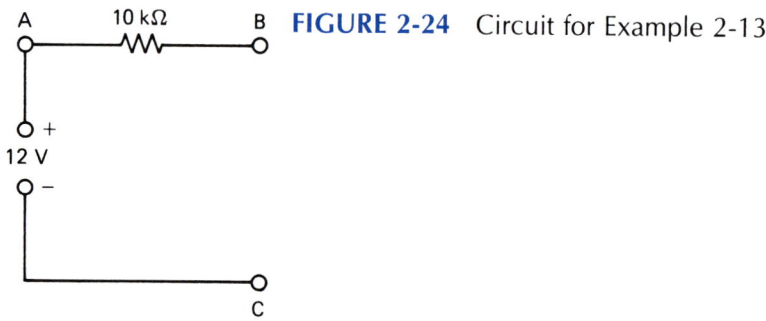

FIGURE 2-24 Circuit for Example 2-13.

Solution Since there is an open circuit between B and C, there is zero current flow through the 10-kΩ resistor and therefore no voltage drop between A and B. Consequently, the voltage between B and C is 12 V.

EXAMPLE 2-14 In Fig. 2-25, what are the values of the voltage potentials at points A, B, and C?

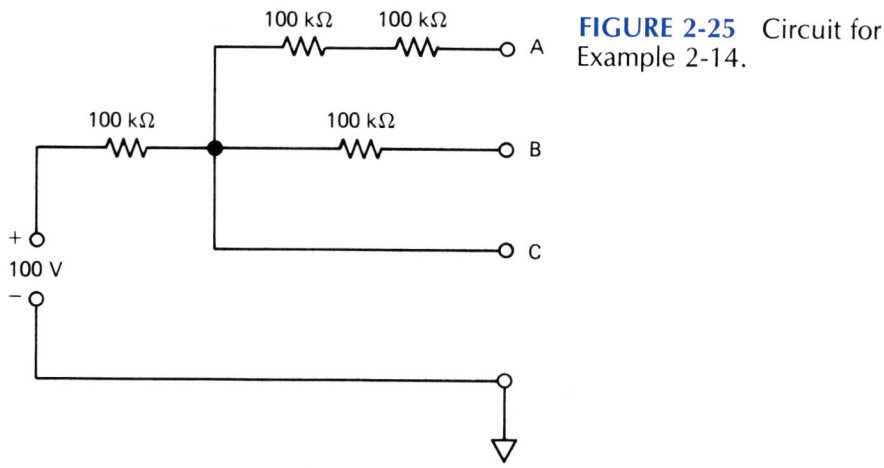

FIGURE 2-25 Circuit for Example 2-14.

Solution There are open circuits between ground and points A, B, and C. There is no current flowing in any part of the circuit and the voltage drops across the resistors are all zero. Consequently, the voltage potential at each of points A, B, and C is +100 V.

2-12

Series Voltage-Dropping Resistor

One practical example of a series circuit is the use of a resistor to drop the source voltage down to the lower level required by a load. Figure 2-26 shows a load requiring a voltage of 9 V for which a current of 10 mA is drawn (these values are typical for a transistor amplifier). Since the only supply voltage is

FIGURE 2-26 Use of a series voltage dropping resistor.

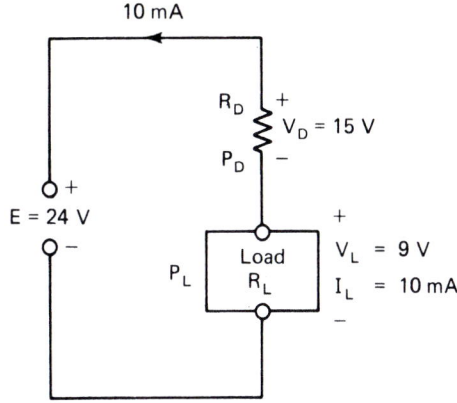

24 V, the series resistor, R_D, is inserted to drop the voltage down to the 9 V required by the load. The voltage across the series dropping resistor, R_D, is therefore 24 V − 9 V = 15 V. Since the resistor is in series with the load, the current through R_D must be the load current of 10 mA. The value of the dropping resistor is

$$R_D = \frac{15 \text{ V}}{10 \text{ mA}} = 1.5 \text{ k}\Omega$$

As well as knowing the ohmic value of the resistor, you must be careful to calculate its required power dissipation since too low a power rating would cause the resistor to burn out. The power dissipated in R_D is

$$P_D = 15 \text{ V} \times 10 \text{ mA} = 150 \text{ mW} = 0.15 \text{ W}$$

A suitable resistor would be 1.5 kΩ, $\frac{1}{4}$ W. *NEXT ZIZE UP.*

Another example involves the heater circuit for an electronics system using tubes. Such a system might be, for example, a radio receiver or a TV receiver. Each tube needs a certain rated heater voltage and heater current for its correct operation; the heater voltage is given by the first digit or first two digits of the tube number, so that the 35W4, a rectifier tube, requires 35 V for its heater circuit. The corresponding heater current is 0.3 A as shown in the tube manual, and therefore the power supply to the heater circuit is 35 V × 0.3 A = 10.5 W. The heater or filament resistance at its correct operating temperature is 35 V/0.3 A = 117 Ω. We should mention that the same numbering system does not apply to TV picture tubes, whose first two digits indicate the diagonal distance in inches across the screen.

With the 6BA6 amplifier tube, the first digit, which is the "six" before the letter "B," means that the necessary heater voltage is 6.3 V (for a 12BA6 tube the heater voltage is 12.6 V); the tube manual shows that the heater current is again 0.3 A, so that the heater circuit power is 6.3 V × 0.3 A = 1.89 W and the filament resistance is 6.3 V/0.3 A = 21 Ω. It is impossible to measure this resistance directly with an ohmmeter since all resistance measurements must be made with the power off and as soon as the power is removed from the heater circuit, the filament rapidly cools down and its resistance is then much less.

The heater circuits for the 35W4 and the 6BA6 have different voltage drops, power requirements, and filament resistances but the *same* heater current; they could therefore be operated in series and the required supply voltage would be 35 V + 6.3 V = 41.3 V. To summarize, heater circuits may be

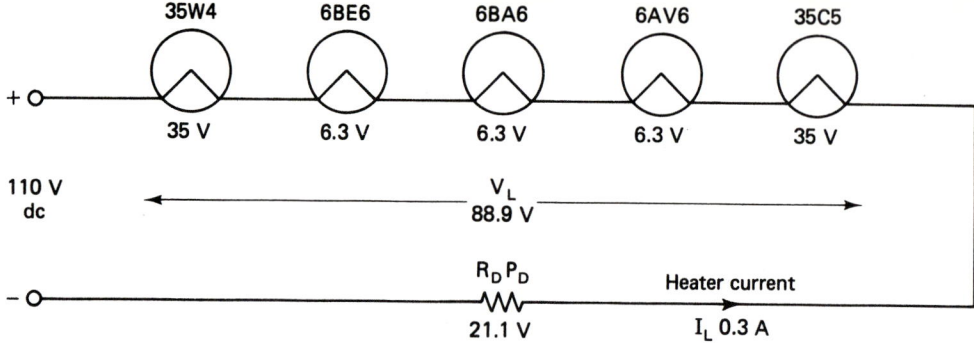

FIGURE 2-27 Application of a dropping resistor.

operated in series provided that the heater current required for each tube is the same; the total supply voltage necessary will be the sum of the heater voltages.

Now let us take the practical case of a radio receiver which uses the following five tubes: 35W4, 6BE6, 6BA6, 6AV6, and 35C5 (Fig. 2-27). All of these tubes need a heater current of 0.3 A and the total voltage required is 35 V + 6.3 V + 6.3 V + 6.3 V + 35 V = 88.9 V. However, the voltage available is a 110-V dc supply. To reduce the 110-V supply down to the 88.9-V level required by the heaters, we insert the series resistor, R_D, across which the voltage drop must be 110 V − 88.9 V = 21.1 V. Without this resistor the current would exceed 0.3 A, the filaments would over-heat, so that one might burn out and create an open circuit. The required value for R_D is 21.1 V/0.3 A = 70 Ω approximately and its power dissipation, P_D = 21.1 V × 0.3 A = 6.3 W; because of the high power dissipation necessary, R_D could not be a color-coded composition-type resistor but would have to be of the wire-wound variety.

Referring to Fig. 2-26, the following equations apply to problems that involve the series dropping resistor:

$$V_D = E - V_L = E - I_L R_L \qquad (2\text{-}12\text{-}1)$$

$$R_D = \frac{V_D}{I_L} \qquad (2\text{-}12\text{-}2)$$

$$P_D = I_L \times V_D \qquad (2\text{-}12\text{-}3)$$

EXAMPLE 2-15 A relay whose coil resistance is 500 Ω is designed to operate when 200 mA flows through the coil. If the relay is operated from 110 V dc, find the value of the series dropping resistor required and calculate its power dissipation.

Solution

$$R_L = 500\ \Omega, \quad I_L = 200\ \text{mA}, \quad E = 110\ \text{V}$$

$$V_D = E - I_L R_L \qquad (2\text{-}12\text{-}1)$$

$$= 110\ \text{V} - 200\ \text{mA} \times 500\ \Omega = 110 - 100\ \text{V} = 10\ \text{V}$$

$$R_D = \frac{V_D}{I_L} \tag{2-12-2}$$

$$= \frac{10 \text{ V}}{200 \text{ mA}} = 50 \text{ }\Omega$$

$$P_D = I_L \times V_D \tag{2-12-3}$$

$$= 200 \text{ mA} \times 10 \text{ V} = 2 \text{ W}$$

2-13
Troubleshooting a Series Circuit

The best way of testing your knowledge of series circuits is to troubleshoot an example of such a network. Troubleshooting itself is the use of logical procedures that enable you to locate the position of one or more faults in a circuit. During this process you take a number of meter readings, which involve resistance, voltage, and current. These readings are the clues in finding the faults. The process can be compared with mystery solving, especially if a circuit has multiple faults. The art of troubleshooting is one of the major reasons for learning the theory of electronic circuitry; anyone can take readings but it requires theoretical knowledge to interpret the readings correctly.

Locating an Open Circuit

As we mentioned in Section 2-11, one of the major faults is the *open*, which is due to a break in the circuit; this might be caused by a burned-out resistor or a disconnected wire. Incidentally, the movement of a resistor's value outside its tolerance is an example of a relatively minor fault. By contrast, remember that an open circuit has theoretically infinite resistance and therefore cannot carry any current; however, a voltage can exist across the open circuit.

As an example, let us refer to Fig. 2-28(a), which illustrates a series circuit in which resistor R_3 has burned out and a wire has become disconnected from point B. Frequently, such a resistor will appear to be charred, but sometimes the damage will not be visible to the naked eye. However, we would know that one or more open circuits has occurred because the ammeter shows zero and if the switch S were opened, an ohmmeter connected across the entire circuit would indicate a reading of infinite ohms.

One method of locating the open circuit is to use an ohmmeter and select a resistance range that is appropriate to the rated values of the resistors. After opening switch S, connect the ohmmeter's common (black) probe permanently to ground point H and use the movable red probe to touch point G. The ohmmeter will record the value of resistor R_4. The reading will not change as the red probe is shifted to point F. However, when the probe is connected to point E, the ohmmeter reading changes to infinite ohms. This shows that the open circuit exists between points E and F so that resistor R_3 must be replaced. Of course, it would also be possible to connect the ohmmeter across each of the resistors in turn. The readings across R_1, R_2, and R_3 would be close to the rated value of the resistors and the reading across R_3 would be infinite ohms. Although the movement of both leads is simple to do under laboratory conditions, it may not be so easy in a practical circuit, especially if there are long connecting wires.

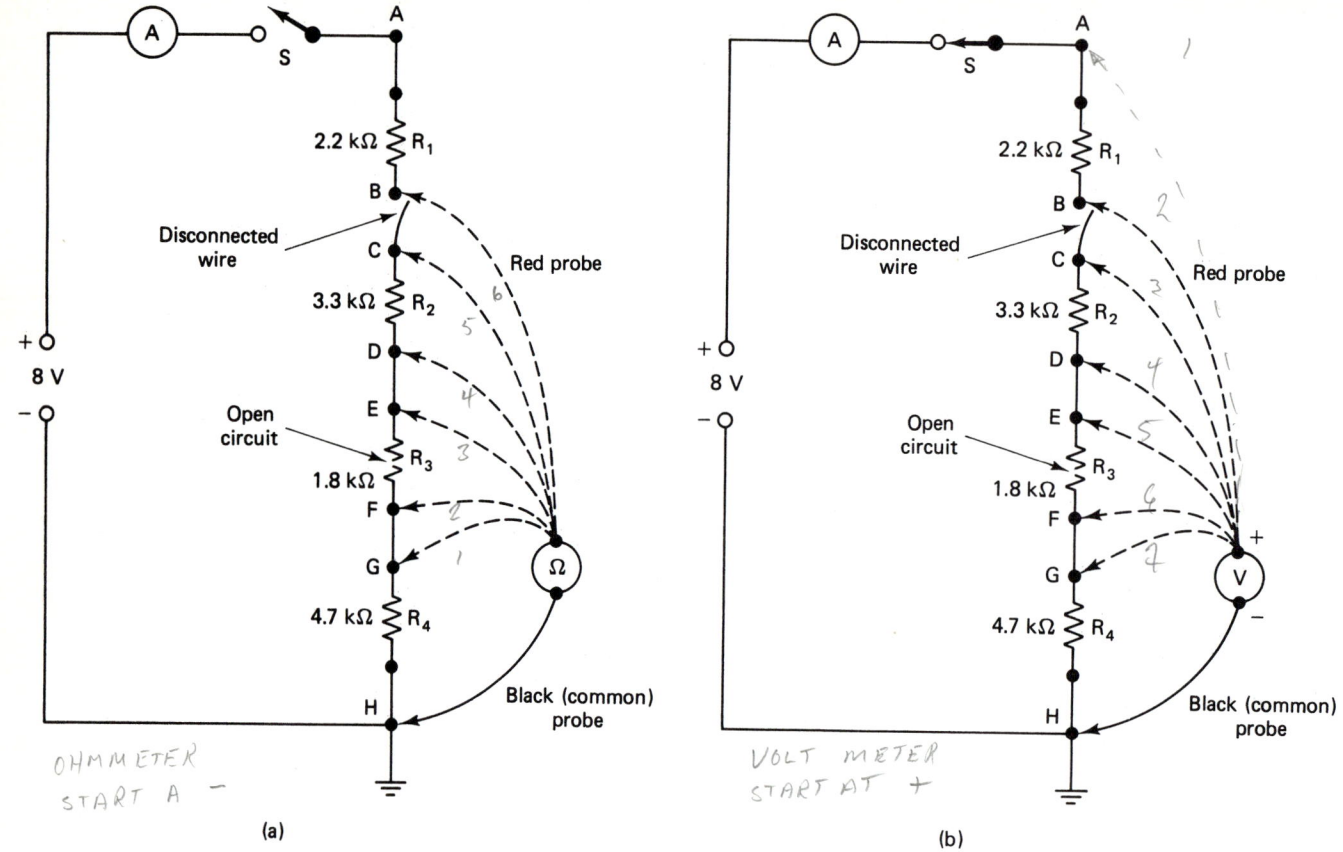

FIGURE 2-28 Use of the ohmmeter and the voltmeter in locating the position of an open in a series circuit: (a) ohmmeter method; (b) voltmeter method.

After R_3 has been replaced, the procedure is repeated until you discover the open circuit between points B and C. The disconnected wire is then soldered back to point B and the circuit has been repaired as far as this fault is concerned.

You may well ask: "Does it matter whether I begin the procedure by connecting the red probe to point A and work my way downward rather than starting at point G?" The answer is: "Yes, it does." If you start at point A the ohmmeter reads infinite ohms and the same reading would occur at point B. When you move to point C, the reading will still be infinite because of open-circuited resistor R_3, so you have no evidence that the connecting wire is not joined to point B.

The second method of finding the open circuits is to close switch S and to use a voltmeter on its 0 to 10 V range [Fig. 2-28(b)]. The (black) common probe is again attached permanently to ground point H and the movable red probe is used to touch point A; the reading is then 8 V, which represents the source voltage. When the probe is moved to point B, the reading stays at 8 V since without any current flow, there can be no voltage drop across resistor R_1. However, when the red probe touches point C, the reading will fall to zero. This indicates the presence of the open circuit between points B and C and allows the disconnected wire to be repaired. Subsequently, if the red probe touches point C again, the reading will be 8 V, and this indicates that there is at least one more open circuit remaining. The procedure is then repeated until the red probe touches points E and F, when the readings will be 8 V and zero, respectively. This indicates that R_3 has in some way open-circuited and must be replaced.

Locating a Short Circuit

A *short* is the second major fault that can occur in a circuit. As we have already mentioned, a short circuit has theoretically zero resistance and therefore a current can flow through the short without developing any voltage drop. Although resistors themselves rarely become shorted, short circuits can occur as the result of the bare connecting wires coming into contact or being joined together by a stray "splash" of solder. If such a short exists in a series circuit, the result must be to lower the circuit's resistance and raise the value of the current drawn from the source.

Let us now refer to the series circuit of Fig. 2-29(a) and consider that a short has occurred across resistor R_3. Before the short occurs:

$$\text{total circuit resistance} = 2.2 + 3.3 + 1.8 + 4.7$$
$$= 12.0 \text{ k}\Omega$$

$$\text{source current} = \frac{8 \text{ V}}{12 \text{ k}\Omega} = 0.67 \text{ mA}$$

After the short occurs:

$$\text{total current resistance} = 2.2 + 3.3 + 0 + 4.7 = 10.2 \text{ k}\Omega$$

$$\text{source current} = \frac{8 \text{ V}}{10.2 \text{ k}\Omega} = 0.78 \text{ mA}$$

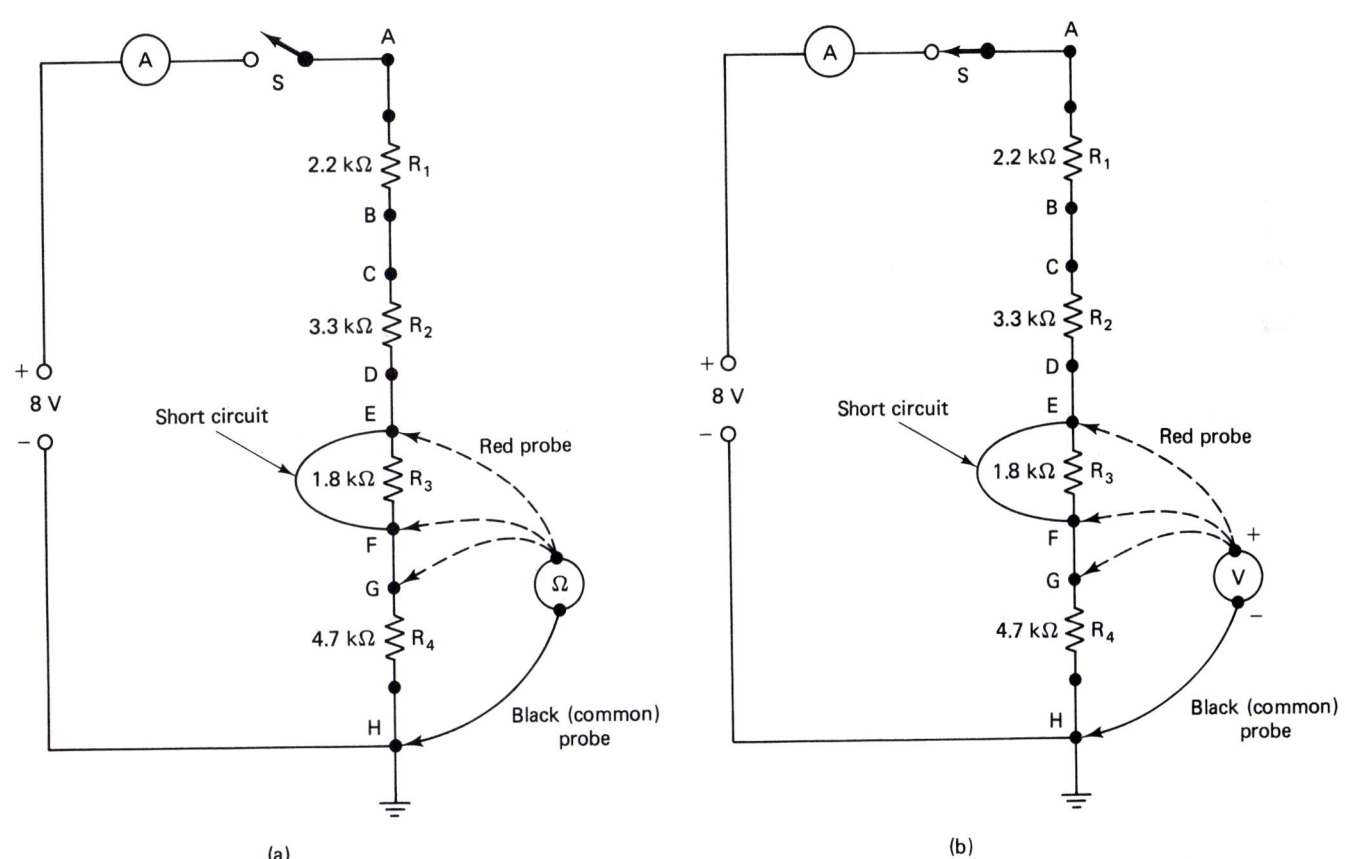

FIGURE 2-29 Use of the ohmmeter and the voltmeter in locating the position of a short in a series circuit: (a) ohmmeter method; (b) voltmeter method.

Sec. 2-13 Troubleshooting a Series Circuit

One method of locating the short circuit is to use an ohmmeter and select a resistance range that is appropriate to the rated values of the resistors. After opening switch S, connect the ohmmeter's common (black) probe to ground point H and use the movable red probe to touch point G. The ohmmeter will record the value of resistor R_4. The reading is not expected to change as the red probe is shifted to point F. However, since a short has zero resistance, the reading will remain the same as the movable probe is connected to point E. This means that a short must exist between points E and F. The reason for the fault is then determined and the necessary repair is made in order to remove the short.

The second method of locating the short circuit is to close switch S and to use a voltmeter on its 0 to 10 V range [Fig. 2-29(b)]. The (black) common probe is again attached permanently to ground point H and the movable red probe is used to touch point A; the reading is then $+8$ V. When the probe is shifted to point B, the reading drops to $+8$ V $\times$ (3.3 kΩ + 4.7 kΩ)/(2.2 kΩ + 3.3 kΩ + 4.7 kΩ) = $+6.27$ V (voltage division rule) and the same voltage appears at point C. At points D and E the reading falls to $+8$ V $\times$ 4.7 kΩ/(2.2 kΩ + 3.3 kΩ + 4.7 kΩ) = $+3.69$ V. However, the same reading will also occur at point F, and this is a clear indication that a short exists across resistor R_3 and that this short must be removed.

To summarize the procedures for locating major faults in a series circuit, there is an infinite change in the resistance reading between one side and the other side of the open circuit; if the voltmeter method is used, there are corresponding changes in the voltage readings between one side and the other side of the open circuit. By contrast, there are no changes in the resistance or voltage readings as we move the red probe across from one side to the other side of a short circuit. We should notice that all the troubleshooting procedures we have looked at so far, have involved only one test instrument.

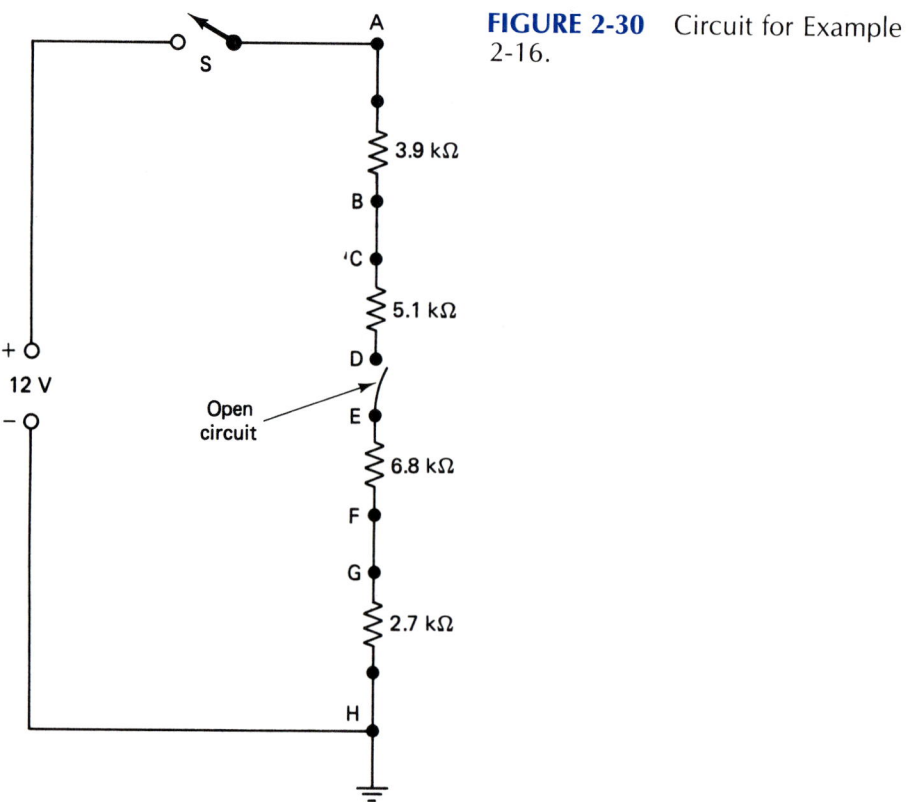

FIGURE 2-30 Circuit for Example 2-16.

EXAMPLE 2-16 In the series circuit of Fig. 2-30, a connecting wire has broken away from point D. What are the resistance readings between (a) points E and H and (b) points C and H? If S is then closed, what are the voltmeter readings between (c) points F and H and (d) points D and H?

Solution

(a) Resistance reading between points E and H is 6.8 + 2.7 = 9.5 kΩ.

(b) Resistance reading between points C and H is infinite ohms because an open circuit exists between these two points.

(c) Voltmeter reading between points F and H is zero because zero current flows through the 2.7-kΩ resistor.

(d) Voltmeter reading between points D and H is 12 V since there are zero voltage drops across the 3.9-kΩ and 5.1-kΩ resistors.

EXAMPLE 2-17 In the series circuit of Fig. 2-31 a short circuit occurs between points A and D. Determine (a) the total resistance of the circuit and the source current before and after the short circuit occurs and (b) the voltage between points C and H before and after the short circuit occurs.

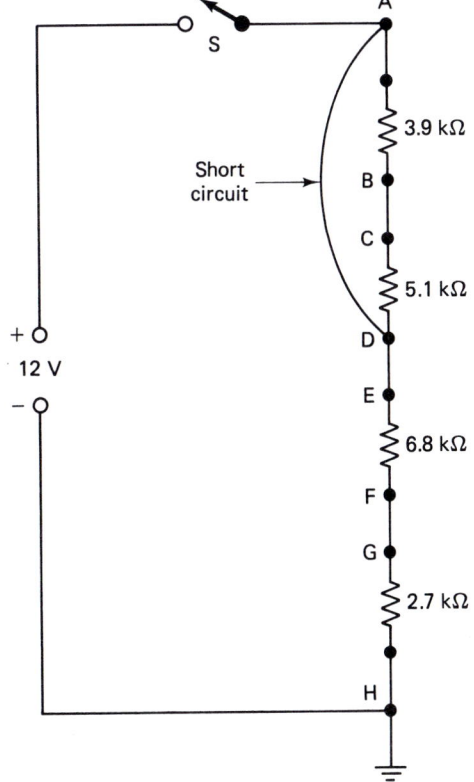

FIGURE 2-31 Circuit for Example 2-17.

Solution

(a) Total resistance before the short circuit occurs = 3.9 + 5.1 + 6.8 + 2.7 = 18.5 kΩ.
Source current before the short circuit occurs = 12 V/18.5 kΩ = 0.65 mA.

Total resistance after the short circuit occurs = 6.8 + 2.7 = 9.5 kΩ.
Source current after the short circuit occurs = 12 V/9.5 kΩ = 1.26 mA.

(b) Using the voltage division rule, the voltage between points C and H before the short circuit occurs is

$$\frac{12\text{ V} \times (5.1\text{ k}\Omega + 6.8\text{ k}\Omega + 2.7\text{ k}\Omega)}{3.9\text{ k}\Omega + 5.1\text{ k}\Omega + 6.8\text{ k}\Omega + 2.7\text{ k}\Omega} = \frac{12\text{ V} \times 14.6\text{ k}\Omega}{18.5\text{ k}\Omega} = 9.47\text{ V}$$

The voltage between points C and H after the short circuit occurs is 12 V since there are zero voltage drops across the 3.9-kΩ and 5.1-kΩ resistors.

Chapter Summary

2.1. In a series circuit there is only a single path flow and the current has the same value throughout the circuit.

2.2. The sum of the individual drops across the resistors is equal to the value of the source voltage (*Kirchhoff's voltage law*).

$$E_T = V_1 + V_2 + V_3 + \cdots + V_N$$

where

$$V_1 = IR_1, \quad V_2 = IR_2, \quad \cdots, \quad V_N = IR_N$$

2.3. The total equivalent resistance is equal to the sum of the individual resistances.

$$R_T = R_1 + R_2 + R_3 + \cdots + R_N$$

$$I = \frac{E_T}{R_T} = \frac{E_T}{R_1 + R_2 + R_3 + \cdots + R_N}$$

2.4. The total power delivered from the source is equal to the sum of the individual powers dissipated in the resistors.

$$P_T = P_1 + P_2 + P_3 + \cdots + P_T$$

$$= I^2 R_T = \frac{E_T^2}{R_T} = IE_T$$

where

$$P_1 = I^2 R_1 = \frac{V_1^2}{R_1} = IV_1, \quad \text{etc.}$$

2.5. The source voltage divides between the series resistors in direct proportion to the resistance values.

$$V_1 = V_2 \times \frac{R_1}{R_2} = E_T \times \frac{R_1}{R_T}, \quad \text{etc.}$$

For two resistors

$$V_1 = E_T \times \frac{R_1}{R_1 + R_2}, \quad V_2 = E_T \times \frac{R_2}{R_1 + R_2}$$

The power delivered from the source is also divided between the resistors in direct proportion to the resistance values.

$$P_1 = P_2 \times \frac{R_1}{R_2} = P_T \times \frac{R_1}{R_T}$$

2.6. A ground system has effectively zero resistance. Ground is also regarded as a reference level of zero volts. The positive or negative potentials at individual points in a circuit are measured relative to ground.

2.7. The potentiometer is a variable resistor that is used to obtain an adjustable output voltage.

$$V_o = V_i \times \frac{R_2}{R_1 + R_2} = V_i \times \frac{d_2}{d_1 + d_2}$$

The rheostat is a variable resistor used to control the current level in a series circuit.

2.8. When voltage sources are connected in series-aiding, the total equivalent EMF is the sum of the individual EMFs. If two voltage sources are connected in series-opposing, the total equivalent EMF is the difference of the individual EMFs.

2.9. An open circuit has theoretically infinite resistance. No current may flow through an open circuit, but in a series string of resistors the source voltage appears across the open circuit.

2.10. A series dropping resistor may be used to reduce the source voltage down to the level required by a load.

$$V_D = E - V_L = E - I_L R_L$$

$$R_D = \frac{V_D}{I_L}$$

$$P_D = I_L \times V_D$$

Self-Review

True–False Questions

1. T. F. In a circuit consisting of a series string of resistors, there are many paths available for current flow.
2. T. F. In a series circuit the highest-value resistor dissipates the least power and develops the smallest voltage drop.
3. T. F. A circuit consists of 470-Ω, 2.2-kΩ, 1500-Ω, and 4.7-kΩ resistors in series across a voltage source. The total equivalent resistance of the circuit is 8.87 kΩ.
4. T. F. A series dropping resistor is used to raise the source voltage up to the level required by the load.
5. T. F. A 1.2-kΩ resistor and a 2400-Ω resistor are connected in series across a voltage source. The current in the 2400-Ω resistor is twice that in the 1.2-kΩ resistor.

6. T. F. Electrons are flowing away from a point toward ground. The potential at that point is negative.
7. T. F. Two voltage sources are joined in series with their negative terminals connected together. The total equivalent EMF is the sum of the individual EMFs.
8. T. F. An open circuit has zero resistance and allows infinite current flow.
9. T. F. If the positive terminal of a 12-V car battery is grounded, the potential at that terminal is zero volts.
10. T. F. A rheostat is a device that produces a variable output voltage.
11. T. F. In a series arrangement of resistors the value of the voltage drop across one resistor is twice the value of the voltage drop across another resistor. It follows that the current through the first resistor is double the current through the second resistor.
12. T. F. In a series arrangement of resistors the power dissipated in one resistor is twice the power dissipated in another resistor. Consequently, the resistance of the first resistor is double the resistance of the second resistor.
13. T. F. For an EMF to exist in a circuit, there must be a complete path for the current to flow.
14. T. F. The sum of all the voltage drops in a series string of resistors must exactly balance the source voltage.
15. T. F. If a 15-V battery and a 5-V battery are series connected, the total voltage available is either 10 V or 20 V.
16. T. F. The reading of an ammeter in a series circuit is always the same and is independent of the position where the ammeter is connected.
17. T. F. If two batteries are joined in a series-opposing arrangement, it is essential to connect the positive terminal of one battery to the negative terminal of the other battery.
18. T. F. A resistive load is connected across a dc source whose negative terminal is grounded. The electrons will then flow through the load toward ground.
19. T. F. A potentiometer is a device that is used to control the level of the current flowing in a series circuit.
20. T. F. A short circuit occurs across one of the resistors in a series string that is connected across a dc voltage source. The current flowing through the short circuit is zero.

Multiple-Choice Questions

21. A series circuit consists of a 60-V battery and five resistors whose values are 6, 5, 4, 3 and 2 Ω. The total equivalent resistance of the circuit is
    (a) 6 Ω   (b) 2 Ω   (c) 80 Ω   (d) 20 Ω   (e) 4 Ω
22. In Question 21, the value of the circuit current is
    (a) 10 A   (b) 30 A   (c) 3 A
    (d) 30 A at the positive terminal of the battery and 10 A at the negative terminal
    (e) dependent on the position at which the current is measured
23. In Question 21, the value of the voltage drop across the 4-Ω resistor is
    (a) 12 V   (b) 16 V   (c) 8 V   (d) 24 V
    (e) dependent on the position of the ground
24. In Question 21, the power dissipated in the 5-Ω resistor is
    (a) 15 W   (b) 45 V   (c) 75 W   (d) 75 V   (e) 45 W
25. In Question 21, the total power dissipated in the circuit is
    (a) 100 W   (b) 60 W   (c) 180 W   (d) 240 W
    (e) no correct choice
26. In Question 21, the 5-Ω resistor open-circuits. The new voltage across this resistor is
    (a) 20 V   (b) 40 V   (c) zero   (d) 60 V   (e) infinite
27. In Question 26, the power dissipated in the 2-Ω resistor is
    (a) zero   (b) 32 W   (c) 8 W   (d) 240 W   (e) 120 W

28. The positive terminal of a 4-V battery is connected to the negative terminal of a 5-V battery. The positive terminal of the 5-V battery is grounded. The potential at the negative terminal of the 4-V battery is
    (a) +9 V   (b) −4 V   (c) −5 V   (d) −1 V   (e) −9 V
29. In Question 28, the potential at the positive terminal of the 4-V battery is
    (a) −5 V   (b) +4 V   (c) −4 V   (d) +5 V   (e) +9 V
30. The voltage rating of an electrical load is 110 V. If the source voltage is 200 V dc, the voltage across a series dropping resistor is
    (a) 220 V   (b) 90 V   (c) 200 V   (d) 310 V
    (e) no correct choice
31. A 10-kΩ linear potentiometer is connected across a 24-V supply. If the shaft is rotated through three-fourths of its entire movement, the output voltage is
    (a) 24 V   (b) 0 V   (c) 4 V   (d) 20 V   (e) 18 V
32. In Question 31 the power dissipated in the potentiometer is
    (a) 57.6 mW   (b) 43.2 mW   (c) 27.8 mW   (d) 21.6 mW   (e) 24.0 mW
33. Three resistors are connected in series across a 24-V dc source. A current of 3 mA flows through the first resistor, a 12-V drop exists across the second resistor, and the value of the third resistor is 2 kΩ. The value of the first resistor is
    (a) 2 kΩ (b) 4 kΩ (c) 1 kΩ (d) 3 kΩ (e) 5 kΩ
34. A 2.2-kΩ resistor is one of a series string whose total resistance is 6.6 kΩ. When the string is connected across a 12-V dc source, the voltage drop across the 2.2-kΩ resistor is
    (a) 4 V   (b) 3 V   (c) 8 V   (d) 9 V   (e) 4.4 V
35. Four equal-value resistors are connected across a 100-V dc source. The voltage drop across each resistor is
    (a) 25 V   (b) 50 V   (c) 75 V   (d) 100 V
    (e) insufficient information to solve
36. An 11.0-V battery is to be charged through a series resistor from a 12.5-V dc source. If the initial charging current is 5 A, the required value of the series resistor is
    (a) 1 Ω (b) 5 Ω (c) 3 Ω (d) 0.5 Ω (e) 0.3 Ω
37. A 100-W 100-V lamp and a 50-W 100-V lamp are connected in series across a 200-V dc source. If the resistances of the two lamps remain constant, the voltage across the 100-W lamp is
    (a) 66.7 V   (b) 133.3 V   (c) 50 V   (d) 150 V   (e) 100 V
38. A series string of resistors has a total resistance of 8 kΩ and includes two resistors whose values are 2 kΩ and 3 kΩ. When a voltage is applied across the complete circuit, the voltage drop across the 3-kΩ resistor is 60 V. The voltage drop across the 2-kΩ resistor is
    (a) 30 V   (b) 40 V   (c) 50 V   (d) 60 V   (e) 20 V
39. Certain identical cells, each of EMF 1.5 V, are connected in series aiding to produce a total output voltage of 12 V. If one of the cells is reversed, the new value of the total output voltage is
    (a) 12 V   (b) 10.5 V   (c) 9 V   (d) 7.5 V   (e) 6 V
40. Three resistors are connected in series across a dc source. The second resistor has twice the resistance of the first resistor, and the third resistor has three times the resistance of the second resistor. If the voltage across the first resistor is 4 V, the value of the source voltage is
    (a) 36 V   (b) 32 V   (c) 28 V   (d) 40 V   (e) 24 V

Practice Problems

41. Three resistors are connected in series across a voltage source. The color coding of these resistors is brown−black−red−silver, gray−red−brown−silver, and green−blue−black−silver. What is the total equivalent rated resistance of this series string?
42. Three resistors whose values are 1.5 kΩ, 3.3 kΩ, and 820 Ω are connected in series across a 120-V dc source. What is the value of the current that is flowing in the series circuit?

43. A 22-kΩ resistor and a 47-kΩ resistor are connected in series across a 50-V dc source. What is the voltage drop across the 47-kΩ resistor?
44. In the voltage divider circuit of Fig. 2-32, what is the value of the potential at point X?

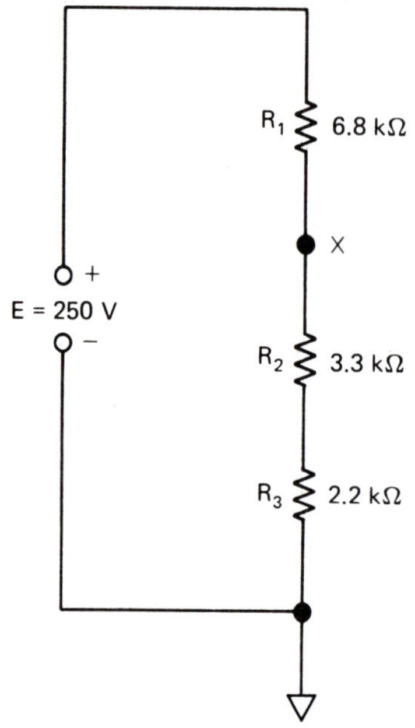

FIGURE 2-32   Circuit for Problem 44.

45. In the circuit of Problem 44, what is the amount of power dissipated in the 6.8-kΩ resistor?
46. In the circuit of Problem 44, what is the amount of the total power dissipated?
47. Five identical resistors are connected in series across a 150-V dc source. If the total power dissipated is 375 W, what is the value of each resistor?
48. Four 10-W 100-Ω resistors are connected in series. What is the total power dissipation capability of this combination?
49. A 20,000-Ω 200-W resistor, a 40,000-Ω 100-W resistor, and a 5000-Ω 20-W resistor are connected in series. What is the maximum value of the total applied voltage that will not cause the wattage rating of any of the resistors to be exceeded?
50. A relay coil has an operating current of 0.2 A and a resistance of 250 Ω. If it is connected to a 120-V dc source, what is the required value for the dropping resistor, and what is its power dissipated?
51. A 100-kΩ potentiometer is connected in series with an unknown resistor across a 20-V dc source. Calculate the value of the unknown resistor if the maximum output voltage from the potentiometer is 16 V. What is the minimum power rating required for the potentiometer?
52. Two conductors are used to join a 10-Ω load to a 120-V source. If the voltage across the load is 115 V, determine the total resistance of the conductors and their total power dissipation.
53. Ten 6-V 5-W lamps are connected in series with a voltage dropping resistor across a 100-V dc source. Calculate the values of the resistance of the dropping resistor and its power dissipation.
54. A 1000-Ω 5-W resistor and a 5000-Ω 20-W resistor are connected in series. What is the maximum voltage that can be applied across this combination without exceeding the power rating of either resistor? What is the total power dissipated under these conditions?
55. A 470-Ω resistor and a 330-Ω resistor are connected in series across a dc voltage source. If the 330-Ω resistor dissipates 1.5 W, calculate the value of the source voltage.
56. Four resistors are connected in series to a 24-V source that supplies 120 mW of

the total power. The first resistor has twice the resistance of the second resistor, and each of the third and fourth resistors has a voltage drop of 6 V. Determine the resistance value of each resistor.

57. There are a number of 150-Ω 1-W resistors available. How would you connect some of these resistors to produce a total resistance of 600 Ω with a power rating of 4 W?

58. Three resistors whose values are 220 Ω, 390 Ω, and 470 Ω are connected in series across a dc voltage source. If the voltage drop across the 390-Ω resistor is 6.7 V, what is the value of the source voltage?

59. Three resistors are connected in series across a 70-V dc source. Two of the resistors are equal in value, while the third resistor has a value of 2.2 kΩ. If the current drawn from the source is 6.7 mA, calculate the individual resistance of one of the two equal-value resistors.

60. Twenty-four 8-W lamps are connected in series across a 120-V dc supply. Determine the value of the total equivalent "hot" resistance of the circuit.

Advanced Problems

61. In Fig. 2-33, the 2000-Ω resistor is dissipating 1 W. What is the voltage drop across the 5000-Ω resistor?

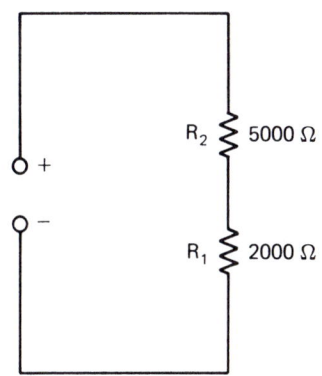

FIGURE 2-33 Circuit for Problem 61.

62. A 500-Ω resistive load has a power rating of 30 W and is connected in series with a dropping resistor across a 220-V dc source. What is the power dissipated in the dropping resistor?

63. It is required to operate a resistive load rated at 12 V, 100 W from a 120-V dc source. What is the value of the dropping resistor that must be connected in series with the load?

64. In Fig. 2-34, what is the value of the voltage between points X and Z?

FIGURE 2-34 Circuit for Problem 64.

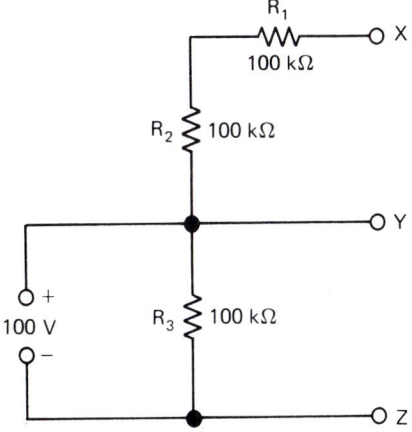

65. In the circuit of Fig. 2-35, what is the value of the potential at the point *X*?

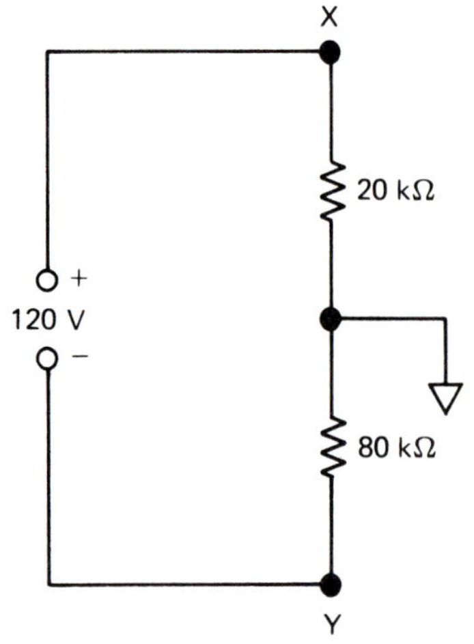

FIGURE 2-35 Circuit for Problem 65.

66. A 470-Ω resistor and a second resistor are connected in series across a 25-V supply. If the voltage drop across the unknown resistor is 8 V, calculate its power dissipation.

67. A voltage divider consists of two resistors connected across a 15-V supply. The output voltage, 6 V, is developed across a 2.2-kΩ resistor. Calculate the total power dissipated in the divider circuit.

68. In Fig. 2-36, calculate the values of the potentials at points *A*, *B*, *C*, and *D*.

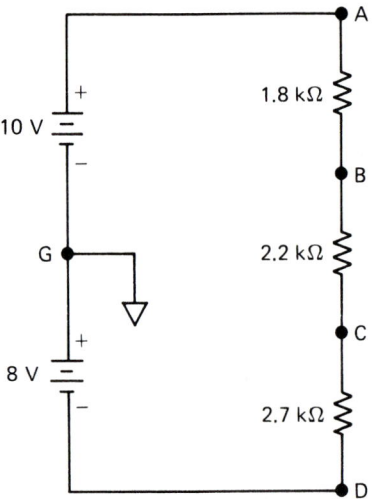

FIGURE 2-36 Circuit for Problems 68 and 69.

69. In Fig. 2-36, the polarity of the 8-V battery is reversed. Find the new values of the potentials at points *A*, *B*, *C*, and *D*.

70. Three resistors are connected in series across a 10-V supply. The first resistor has a value of 1 Ω, the second has a voltage drop of 1 V, and the third has a power dissipation of 1 W. Find the value of the circuit current. *Note:* This is a perennial problem that students often ask about. Its solution involves algebra and a knowledge of quadratic equations; there are two possible answers.

71. In the circuit of Fig. 2-37, the current flowing through lamp L_1 is 110 mA. Find the voltage drop across the lamp, its resistance, and its power dissipation.

FIGURE 2-37 Circuit for Problem 71.

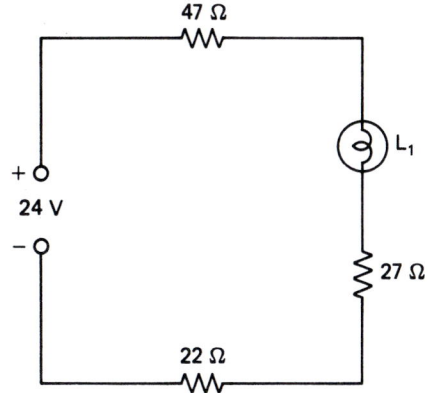

72. In the circuit of Fig. 2-38, calculate the maximum and minimum values of the output voltage, V_o.

FIGURE 2-38 Circuit for Problems 72 and 73.

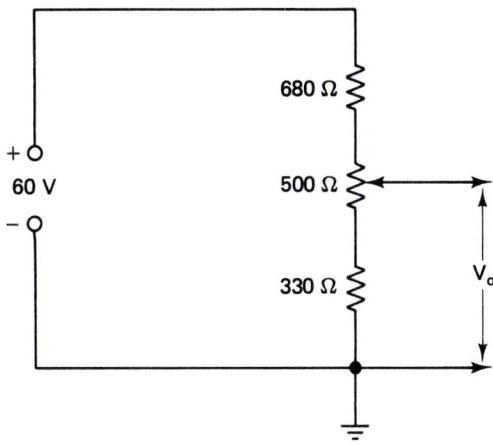

73. In Problem 72, recalculate the maximum and minimum values of V_o when a short circuit develops across the 330-Ω resistor.
74. In the circuit of Fig. 2-39, determine the values of the potentials at points X, Y, and Z.

FIGURE 2-39 Circuit for Problems 74 and 75.

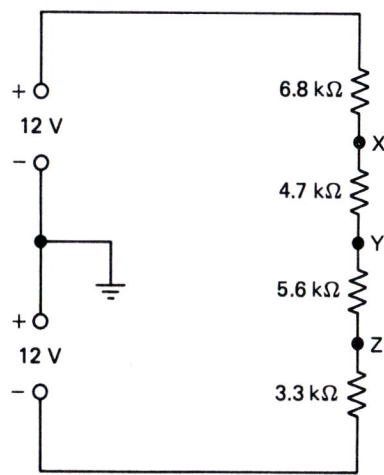

75. In Problem 74, recalculate the value of the potentials at points X, Y, and Z if (a) a short circuit develops across the 4.7-kΩ resistor and (b) the 5.6-kΩ resistor is open circuited.

76. For the circuit of Fig. 2-40, calculate the values of the maximum and minimum powers dissipated in the lamp L_1.

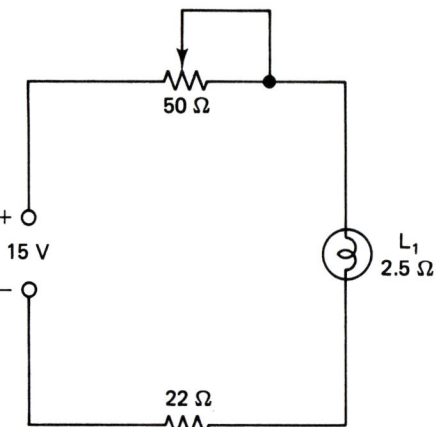

FIGURE 2-40 Circuit for Problem 76.

77. In the circuit of Fig. 2-41, calculate the value of the potential at point X (both voltages shown are measured with respect to ground).

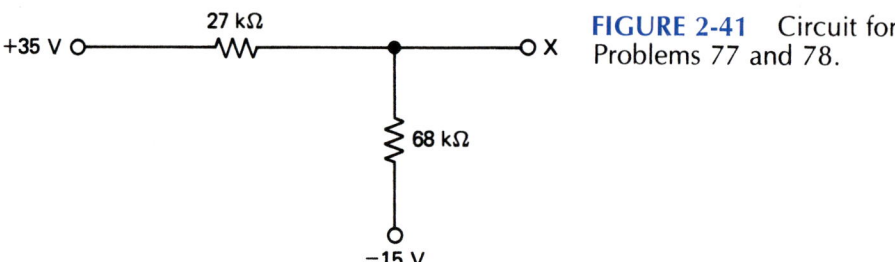

FIGURE 2-41 Circuit for Problems 77 and 78.

78. In Problem 77, the two resistors are interchanged. Recalculate the value of the potential at the point, X.

79. In Fig. 2-42, the points X and Y are joined by a path of zero resistance. Calculate the value of the potential at point X.

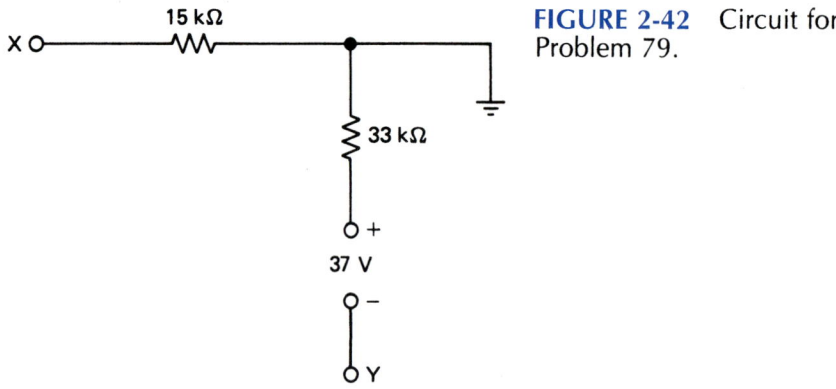

FIGURE 2-42 Circuit for Problem 79.

80. Two resistors whose values are 680 Ω and 1.2 kΩ are connected in series across a dc source. If the power dissipated in the 680-Ω resistor is 0.8 W, calculate the value of the source voltage.

3

Resistors in Parallel

By examining in detail the important properties of parallel circuits, you will learn:

1. about the equality between the source voltage and the voltage drop across each of the parallel resistors.
2. that the sum of the branch currents is equal to the total current drawn from the voltage source.
3. about the concept of total equivalent resistance and how its value may be determined from the reciprocal or product-over-sum formulas.
4. how to calculate the value of the total equivalent conductance in a parallel circuit.
5. to understand the relationship between the total power and the powers dissipated in the individual parallel resistors.
6. how to analyze a parallel circuit by using a step-by-step procedure.
7. to apply the current division rule in the analysis of a parallel circuit.
8. about the reasons for connecting voltage sources in parallel.
9. how open circuits affect a parallel arrangement of resistors.
10. how short circuits affect a parallel arrangement of resistors.
11. how to troubleshoot a parallel circuit.

3-0

Introduction

In Chapter 2 we learned that in a series string the resistors were joined end to end across a voltage source. If a break, such as a switch, occurred anywhere in the series circuit, all current ceased. Clearly, the lights in your home cannot be arranged in series since if the lights in the kitchen and bathroom are both switched on but you then switch off the kitchen light, the bathroom light remains lit. If you examine these light bulbs you will find that they are stamped "120 V," which means that the bulbs can only be correctly operated from a 120-V source. All bulbs must therefore be connected directly across the household supply and the same applies to other loads, such as vacuum cleaners, air conditioners, refrigerators, garbage disposals, hair dryers, and toasters. In every case the required voltage value for the device is the same, although each load will have its own power (wattage) rating. This type of arrangement is known as *parallel* and is illustrated in Fig. 3-1(a); the figure shows five color-coded resistors, each of which is directly connected to a $1\frac{1}{2}$-V battery. Since the connecting wires are assumed to have zero resistance, the parallel circuit can be changed to either of the configurations in Fig. 3-1(b) and (d), whose schematic diagrams are shown, respectively, in Fig. 3-1(c) and (e). A parallel arrangement of resistors may therefore be defined as one in which the resistors are connected between two common points X and Y, or two common lines; the points or lines are then directly connected to a voltage source.

3-1

Relationship Between Source Voltage and Voltage Drop Across each Parallel Resistor

Referring to Fig. 3-1(e), the junction point X is directly connected to the positive terminal, A, of the battery, while point Y is taken around to the negative terminal, B; the voltage between these points (X and Y) is therefore equal to the source EMF, E_T. At the same time one side of each resistor can be traced to junction X, while the other side is similarly connected to junction Y. It therefore follows that the potential difference across each of the parallel resistors must be the same and equal to the voltage between points X and Y. Since the voltage between these points is the source EMF, E_T,

$$E_T = V_1 = V_2 = V_3 = V_4 = V_5 \qquad (3\text{-}1\text{-}1)$$

For N resistors in parallel,

$$E_T = V_1 = V_2 = V_3 = \cdots = V_N \qquad (3\text{-}1\text{-}2)$$

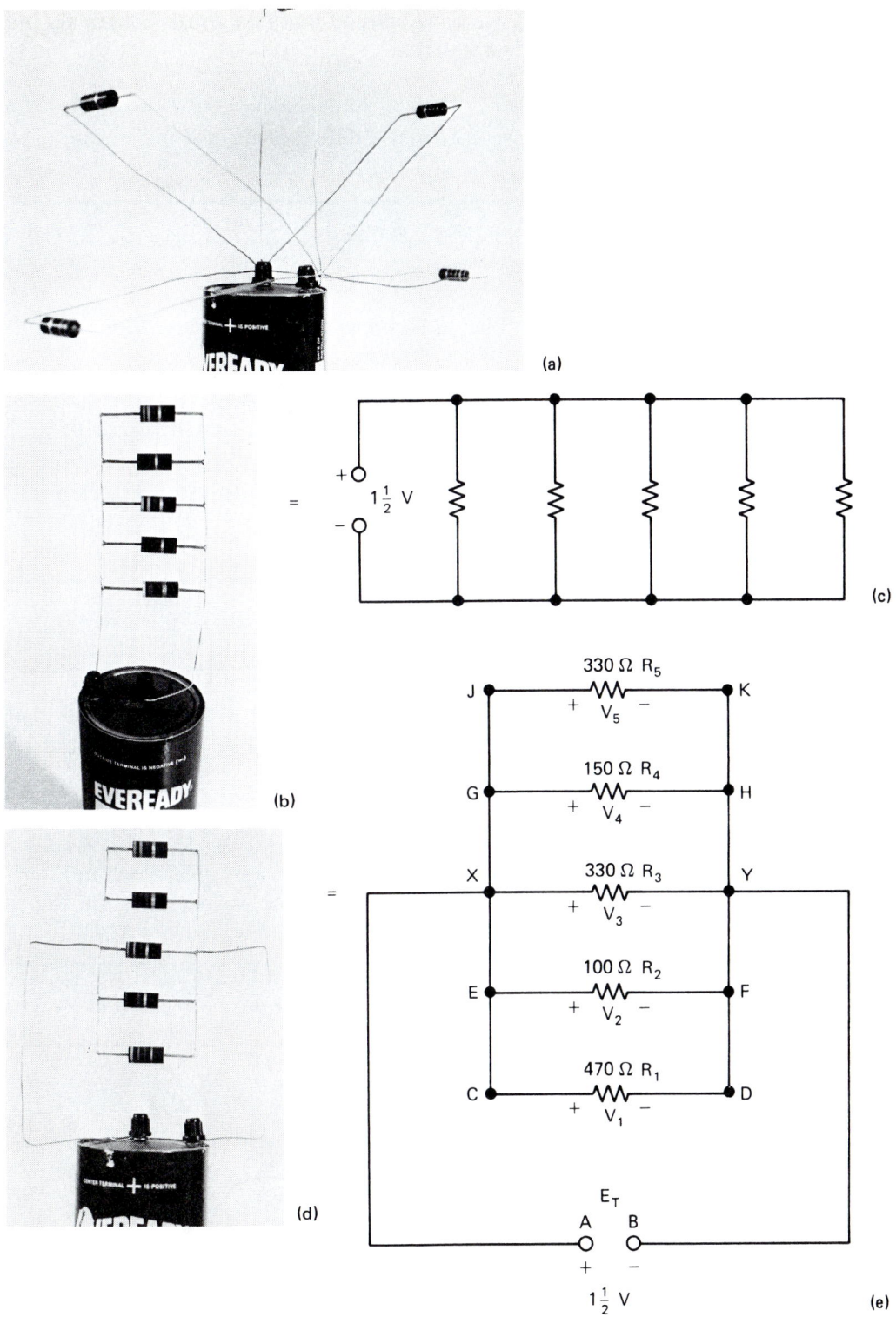

FIGURE 3-1 Parallel arrangements of resistors.

Just as there was only one current in a series circuit, there is only one voltage present in a parallel circuit. As we have already discussed, household loads are normally parallel-connected to the 120-V supply, and in the same way, the various electrical circuits of a car are joined in parallel across the terminals of the battery.

Sec. 3-1 Relationship Between Source Voltage and Voltage Drop Across each Parallel Resistor

To summarize, we have learned that in a circuit containing only parallel resistors and a voltage source:

1. The potential difference across each parallel resistor is the same.
2. Each of these potential differences is equal to the source voltage.

EXAMPLE 3-1 In Fig. 3-1(e), what are the values of the voltages V_{XB}, V_{YB}, V_{CJ}, V_{DG}, and V_{EH}?

Solution $V_{XB} = 1.5$ V. X is connected to the positive terminal of the battery; B is the battery's negative terminal.
$V_{YB} = 0$ V. B and Y are joined by a connecting wire whose resistance is assumed to be negligible.
$V_{CJ} = 0$ V. C and J are directly joined by connecting wires of zero resistance.
$V_{GD} = V_{EH} = 1.5$ V. G and E are effectively connected to X while D and H are joined to Y.

3-2 Branch Currents Flowing Through Resistors and Total Current Drawn from Voltage Source. Total Resistance

Each of the resistors in a parallel circuit is a path for current flow and each such path is called a *branch*. We can therefore refer to a *branch current* as the current flowing through a particular resistor. Referring to Fig. 3-1(e), the voltage drop across each branch is the same and equal to the source voltage. Using the Ohm's law relationship $I = E/R$, the individual branch currents are

$$I_1 = \frac{E_T}{R_1}, \quad I_2 = \frac{E_T}{R_2}, \quad I_3 = \frac{E_T}{R_3}, \quad I_4 = \frac{E_T}{R_4}, \quad I_5 = \frac{E_T}{R_5} \qquad (3\text{-}2\text{-}1)$$

These equations lead to

$$\frac{I_1}{I_2} = \frac{R_2}{R_1}, \quad \frac{I_2}{I_3} = \frac{R_3}{R_2}, \quad \text{etc.}$$

or

$$I_1 = I_2 \times \frac{R_2}{R_1}, \quad I_2 = I_3 \times \frac{R_3}{R_2}, \quad \text{etc.} \qquad (3\text{-}2\text{-}2)$$

The current in a particular branch is therefore inversely proportional to the value of the resistance in that branch. This means that the lowest-value resistor will carry the greatest current, while the smallest current will flow through the highest-value resistor. However, if the resistors are of equal value, the branch currents will all be the same. In terms of the electron flow, the currents I_1, I_2, I_3, I_4, and I_5 are all leaving the junction point, Y, while the total current,

I_T, drawn from the voltage source is entering the same junction point. Using the argument we developed in Section 2-1 yields

$$I_T = I_1 + I_2 + I_3 + I_4 + I_5 \qquad (3\text{-}2\text{-}3)$$

This equation is an expression of *Kirchhoff's current law* (Section 6-1).

The sum of the individual branch currents is equal to the total current drawn from the voltage source. This compares with the series circuit property in which the sum of the individual voltage drops across the resistors is equal to the source EMF (Kirchhoff's voltage law).

For N resistors in parallel,

$$I_T = I_1 + I_2 + I_3 + \cdots + I_N \qquad (3\text{-}2\text{-}4)$$

The total equivalent resistance, R_T, as presented to the source, will be given by

$$R_T = \frac{E_T}{I_T} = \frac{E_T}{I_1 + I_2 + I_3 + \cdots + I_N} \qquad (3\text{-}2\text{-}5)$$

If you choose to find R_T by this method and no value is given for E_T, any convenient value may be assumed (see Example 3-4).

If the N resistors are of equal value,

$$I_1 = I_2 = I_3 = \cdots = I_N = I$$

and

$$I_T = NI \qquad (3\text{-}2\text{-}6)$$

where I is one of the equal branch currents.

To summarize this section on resistors in parallel:

1. Each of the branch currents is equal to the result of dividing the source EMF by the branch resistance.
2. The sum of the individual branch currents is equal to the total current drawn from the source EMF.
3. The total equivalent resistance is the result of dividing the source EMF by the total source current.

EXAMPLE 3-2 In Fig. 3-1(e), what are the values of I_1, I_2, I_3, I_4, I_5, I_T, and R_T?

Solution The branch currents are:

$$I_1 = \frac{E_T}{R_1} = \frac{1.5 \text{ V}}{470 \text{ }\Omega} = 3.19 \text{ mA}$$

$$I_2 = \frac{E_T}{R_2} = \frac{1.5 \text{ V}}{100 \text{ }\Omega} = 15.00 \text{ mA}$$

$$I_3 = \frac{E_T}{R_3} = \frac{1.5 \text{ V}}{330 \text{ }\Omega} = 4.55 \text{ mA} \qquad (3\text{-}2\text{-}1)$$

$$I_4 = \frac{E_T}{R_4} = \frac{1.5 \text{ V}}{150 \text{ }\Omega} = 10.00 \text{ mA}$$

$$I_5 = \frac{E_T}{R_5} = \frac{1.5 \text{ V}}{330 \text{ }\Omega} = 4.55 \text{ mA}$$

With the electronic calculator the branch currents are found by using the Memory-in ($x{\to}M$) or STO (store) and Recall Memory (RM) or RCL (recall) keys. In our example the procedure is:

Operation			Display	
1.5	$x{\to}M$	$\div$	1.5	
470	=		0.00319149	(I_1)
	RM	$\div$	1.5	
100	=		0.015	(I_2)
	RM	$\div$	1.5	
330	=		0.00454545	(I_3, I_5)
	RM	$\div$	1.5	
150	=		0.01	(I_4)

Before starting the procedure, you must make certain that everything is cleared from the memory. This is normally done by first depressing the Clear key and then the Memory-in key.

Notice that the largest current (15 mA) is carried by the smallest-value resistor (100 Ω). By contrast, the smallest current (3.19 mA) is associated with the highest-value resistor (470 Ω). Equal currents (4.55 mA) flow through the two 330-Ω resistors.

The total current drawn from the voltage source is

$$I_T = I_1 + I_2 + I_3 + I_4 + I_5 \qquad (3\text{-}2\text{-}3)$$
$$= 3.19 + 15.00 + 4.55 + 10.00 + 4.55$$
$$= 37.29 \text{ mA}$$

total equivalent resistance, $R_T = \dfrac{E_T}{I_T}$ \qquad (3-2-5)

$$= \frac{1.5 \text{ V}}{37.29 \text{ mA}}$$

$$= 40.2 \text{ }\Omega \quad \text{(rounded off)}$$

EXAMPLE 3-3 In Fig. 3-2, find the values of I_3, I_4, I_T, R_1, and E_T.

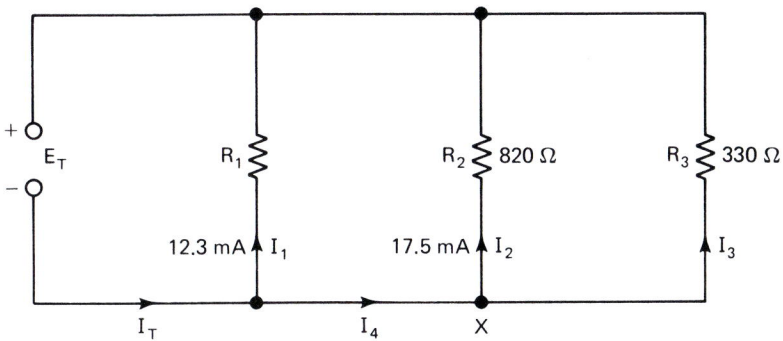

FIGURE 3-2 Circuit for Example 3-3.

Solution

$$I_3 = I_2 \times \frac{R_2}{R_3} \tag{3-2-2}$$

$$= 17.5 \text{ mA} \times \frac{820 \text{ }\Omega}{330 \text{ }\Omega} = 43.5 \text{ mA}$$

At point X, I_2 and I_3 are leaving this junction while I_4 is entering. Then

$$I_4 = I_2 + I_3 = 17.5 + 43.5 = 61.0 \text{ mA}$$

source current, $I_T = I_1 + I_2 + I_3 = I_1 + I_4$ \quad (3-2-4)

$$= 12.3 + 61.0 = 73.3 \text{ mA}$$

$$R_1 = R_2 \times \frac{I_2}{I_1} \tag{3-2-2}$$

$$= 820 \text{ }\Omega \times \frac{17.5 \text{ mA}}{12.3 \text{ mA}} = 1167 \text{ }\Omega$$

$$E_T = I_2 \times R_2 \tag{3-2-1}$$

$$= 17.5 \text{ mA} \times 820 \text{ }\Omega = 14.35 \text{ V}$$

.0175 A × 820 = 14.35

3-3
Total Equivalent Resistance of Parallel Circuit from Reciprocal and Product-Over-Sum Formulas

Figure 3-3(a) shows a bank of N resistors across a voltage source. The word "bank" is often used to denote parallel resistors just as the word "string" refers to resistors in series. We want to find the bank's total equivalent resistance R_T [Fig. 3-3(b)]; this is the single value of resistance that will draw the total current, I_T, from the voltage source. Using Eq. (3-2-4), we obtain

$$I_T = I_1 + I_2 + I_3 + \cdots + I_N$$

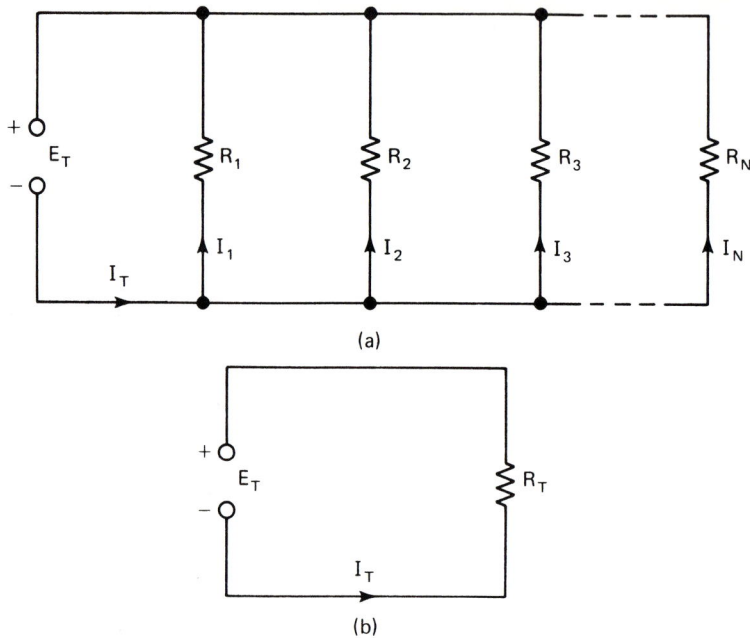

FIGURE 3-3 N resistors in parallel.

From Eq. (3-2-1),

$$I_1 = \frac{E_T}{R_1}, \quad I_2 = \frac{E_T}{R_2}, \quad \text{etc.}$$

and in Fig. 3-3(b),

$$I_T = \frac{E_T}{R_T}$$

Therefore,

$$\frac{E_T}{R_T} = \frac{E_T}{R_1} + \frac{E_T}{R_2} + \frac{E_T}{R_3} + \cdots + \frac{E_T}{R_N}$$

Dividing both sides of this equation by the common factor E_T yields

$$\frac{1}{R_T} = \frac{1}{R_1} + \frac{1}{R_2} + \frac{1}{R_3} + \cdots + \frac{1}{R_N} \qquad (3\text{-}3\text{-}1)$$

or

$$R_T = \frac{1}{\dfrac{1}{R_1} + \dfrac{1}{R_2} + \dfrac{1}{R_3} + \cdots + \dfrac{1}{R_N}} \qquad (3\text{-}3\text{-}2)$$

This relationship is called the *reciprocal formula*. The reciprocal of a quantity x is defined as $1/x$, which is how the reciprocal key is normally shown on an electronic calculator; as an example, the reciprocal of the number 2 is $\frac{1}{2}$ or 0.5.

Consider a bank of three parallel resistors whose values are 1.2 kΩ (R_1), 820 Ω (R_2), and 680 Ω (R_3). To calculate R_T we must use the same units throughout the reciprocal formula, and therefore the first step is to convert the 1.2 kΩ into 1.2 × 1000 = 1200 Ω. Then

$$\frac{1}{R_T} = \frac{1}{R_1} + \frac{1}{R_2} + \frac{1}{R_3} \qquad (3\text{-}3\text{-}1)$$

$$= \frac{1}{1200} + \frac{1}{820} + \frac{1}{680}$$

$$= 0.00083333 + 0.0012195 + 0.0014706$$

$$= 0.0035234$$

$$R_T = \frac{1}{0.0035234} = 284 \text{ Ω} \quad \text{(rounded off)}$$

Note that R_T was only obtained after inverting the value for $1/R_T$.

The calculator greatly simplifies the use of the reciprocal formula. The individual resistors are entered in turn and their reciprocals can then be directly added to obtain $1/R_T$. Depressing the $1/x$ key one more time will give the value of R_T. Using our example, the procedure would be as follows:

Operation			Display
1200	$1/x$	+	0.0008333
820	$1/x$	+	0.002053
680	$1/x$	=	0.003523
	$1/x$		283.8 (R_T)

This procedure can readily be applied to a large number of resistors in parallel and is much superior to any other method of finding R_T.

The answer of 280 Ω for R_T is at first surprising. For parallel resistors R_T is in fact always *less* than the lowest-value resistor in parallel. It follows that if another resistor is added in parallel, the value of R_T is further reduced. Here is where many students become baffled and ask: "How can it be that when we add more and more resistors in parallel, the value of R_T gets less and less?" To answer that question, let us look at the circuit of Fig. 3-4; this shows

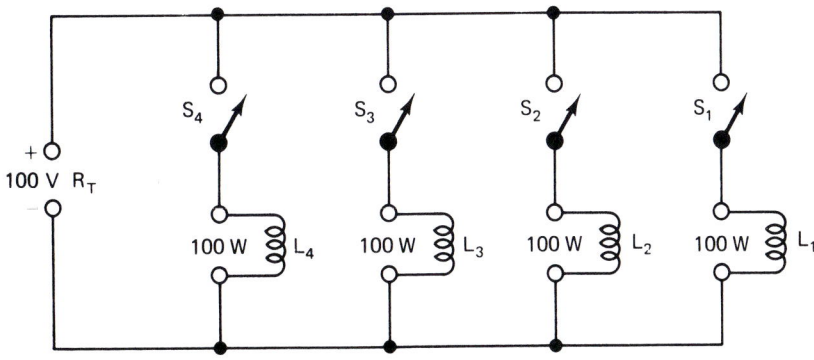

FIGURE 3-4 Four 100-V, 100-W electric light bulbs connected in parallel.

four 100-W 100-V light bulbs each of which is controlled by a single-pole, single-throw switch. When the switch is closed for a particular bulb, the bulb current is 100 W/100 V = 1 A and its "hot"-filament resistance is 100 V/1 A = 100 Ω.

When only S_1 is closed, the current through L_1 is 1 A, which will also be the current drawn from the 100-V source. Therefore, the effective resistance presented to the source is the 100-Ω associated with the filament of L_1. However, if S_2 is also closed, the current through L_2 will again be 1 A and the total source current will then be 1 A + 1 A = 2 A. This means that the R_T falls from 100 Ω to 100 V/2 A = 50 Ω. With S_1, S_2, and S_3 closed, the source current is 1 A + 1 A + 1 A = 3 A and R_T is down to 100 V/3 A = $33\frac{1}{3}$ Ω. Finally, if all four switches are thrown, I_T is 4 A and R_T is only 100 V/4 A = 25 Ω. If there were 10 such bulbs in parallel, I_T would be 10 A, with a corresponding R_T of 10 Ω and with a hundred bulbs, I_T and R_T would be 100 A and 1 Ω, respectively. Every additional bulb in parallel provides another path for current flow, so that the source current is increased and the total opposition, R_T, to current flow is reduced. To use an analogy, think of a pump that is providing a constant pressure and is driving water through a number of parallel pipes with different diameters. If another pipe is included, the additional flow of water through this pipe will mean that the total flow is increased and the effective opposition to the flow of water at the pump is less.

Two Resistors in Parallel

If there are only two resistors, R_1 and R_2, in parallel, the reciprocal formula for R_T will be

$$\frac{1}{R_T} = \frac{1}{R_1} + \frac{1}{R_2}$$

This leads to

$$R_T = \frac{R_1 \times R_2}{R_1 + R_2} \qquad (3\text{-}3\text{-}3)$$

Since $R_1 \times R_2$ represents the product of the resistances while $R_1 + R_2$ is their sum, Eq. (3-3-3) is referred to as the *product-over-sum formula*. It must be emphasized that this method can be applied directly only to *two* resistors; for example, with three parallel resistors, R_T is

$$\frac{R_1 \times R_2 \times R_3}{R_1 R_2 + R_2 R_3 + R_3 R_1} \quad \text{and not} \quad \frac{R_1 \times R_2 \times R_3}{R_1 + R_2 + R_3}$$

However, if there are more than two resistors, the resistances may be taken two at a time and the method can then be repeated until R_T is finally determined. Using this method, let us again solve the problem of finding the total equivalent resistance of 1200 Ω (R_1), 820 Ω (R_2), and 680 Ω (R_3) in parallel. As a first step, any two of the resistors may initially be chosen in the use of the product-over-sum formula. The choice of 820 Ω and 680 Ω gives us a slight advantage, since their sum is a convenient 1500 Ω; their equivalent resistance, R_{23} (we shall use R_{23} to mean the total equivalent resistance of R_2 and R_3 in parallel) will be

$$R_{23} = \frac{R_2 \times R_3}{R_2 + R_3} = \frac{820 \times 680}{820 + 680} = \frac{820 \times 680}{1500} = 371.73 \text{ Ω}$$

To find R_T, the formula will now be repeated for R_{23} and R_1 in parallel, so that

$$R_T = \frac{R_{23} \times R_1}{R_{23} + R_1} = \frac{371.73 \times 1200}{371.73 + 1200} = 284 \ \Omega \quad \text{(rounded off)}$$

Naturally, if there are a large number of parallel resistors, the repeated product-over-sum method is very laborious and the reciprocal formula is far superior.

Equation (3-3-3) may be rearranged as

$$R_1 = \frac{R_T \times R_2}{R_2 - R_T} \quad \text{and} \quad R_2 = \frac{R_T \times R_1}{R_1 - R_T} \qquad (3\text{-}3\text{-}4)$$

These equations enable one of the parallel resistances to be found provided that the values of R_T and the other resistance are known.

The scientific calculator easily allows us to obtain the value of R_T for two resistors from the "product-over-sum" formula. For example, if a 470-Ω resistor and a 680-Ω resistor are connected in parallel, the calculator procedure is as follows:

Operation		Display	
470	$\times$	470	(R_1)
680	$\div$	319600	($R_1 \times R_2$)
(470	+	470	
680	)	1150	($R_1 + R_2$)
	=	277.9	(R_T)

The value of R_T is 278 Ω, rounded off.

To obtain the value of R_1 when the values of R_T and R_2 are known, the formula is $R_2 \times R_T/(R_2 - R_T)$, and therefore the procedure is the same except that the $\boxed{-}$ key must be used instead of the $\boxed{+}$ key.

Equal-Value Resistors in Parallel

We have already considered one such example in which four 100-V bulbs were connected in parallel across a 100-V source. If the value of each identical resistor is R and there are N such resistors, then by Eq. (3-3-1),

$$\frac{1}{R_T} = \frac{1}{R} + \frac{1}{R} + \frac{1}{R} + \cdots + \frac{1}{R}$$

$$= \frac{N}{R}$$

This yields,

$$R_T = \frac{R}{N} \qquad (3\text{-}3\text{-}5)$$

R_T is therefore equal in value to one of the resistors divided by the number of the resistors. If you are faced with a number of resistors in parallel and some of these have the same value, it is a good idea to start combining the equal-value resistors into a single-value resistance. Suppose, for example, that you have a circuit of 10-kΩ, 5-kΩ, and 10-kΩ resistors in parallel. The total equivalent resistance of the two 10-kΩ resistors is 5 kΩ, and when this is considered in parallel with the 5-kΩ resistor, the final value of R_T is 5 kΩ/2 = 2.5 kΩ.

On some occasions you can also use the formula for equal resistors in parallel as a shortcut in finding R_T. As an illustration, think of a parallel circuit consisting of one 18-Ω and one 54-Ω resistor. The 18-Ω resistor may be regarded as equal to three 54-Ω resistors in parallel, so that the total circuit is equivalent to a bank of four 54-Ω resistors; therefore, the value of R_T is 54 Ω/4 = 13.5 Ω. You should use this form of shortcut when the value of one of the parallel resistors is a multiple of one of the other values.

This section may be summarized as follows:

1. For any number of resistors in parallel, the total equivalent resistance can be found by means of the reciprocal formula.

2. For *two* resistors in parallel, R_T can be calculated from the product-over-sum formula. For more than two resistors, the product-over-sum formula must be repeated.

3. For equal resistors in parallel, the total resistance is equal to the resistance of one resistor divided by the number of resistors.

EXAMPLE 3-4 In the circuit of Fig. 3-5(a), find the total equivalent resistance by (a) the assumed voltage method, (b) the reciprocal formula, and (c) the product-over-sum formula.

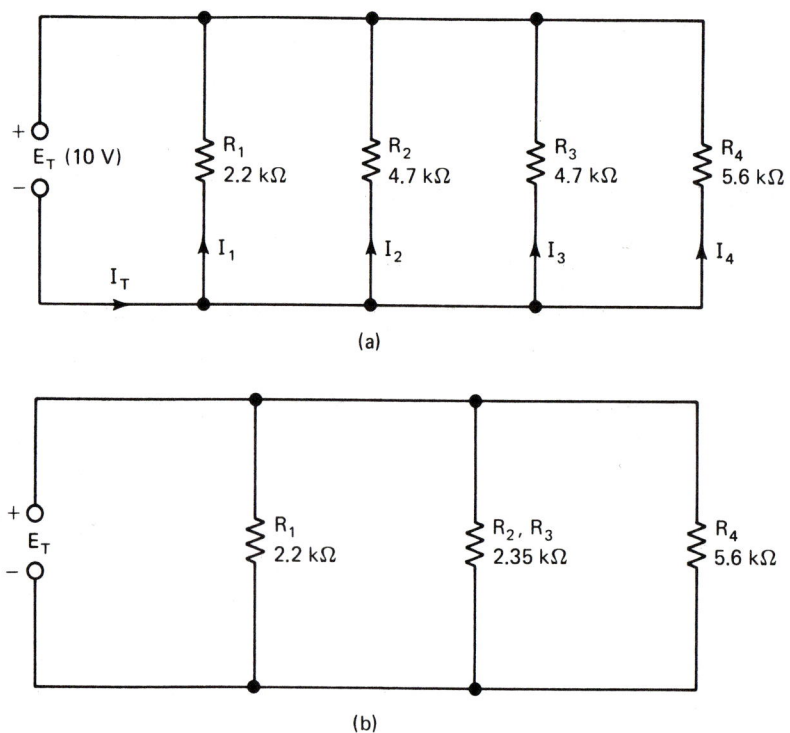

FIGURE 3-5 Circuit for Example 3-4.

Solution

(a) Since the values of the resistors are not simple numbers, there is no advantage in assuming any particular value for E_T. Therefore, let E_T be, say, 10 V.
The branch currents are

$$I_1 = \frac{10 \text{ V}}{2.2 \text{ k}\Omega} = 4.545 \text{ mA}$$

$$I_2 = I_3 = \frac{10 \text{ V}}{4.7 \text{ k}\Omega} = 2.128 \text{ mA} \qquad (3\text{-}2\text{-}1)$$

$$I_4 = \frac{10 \text{ V}}{5.6 \text{ k}\Omega} = 1.786 \text{ mA}$$

source current, $I_T = I_1 + I_2 + I_3 + I_4$ \hfill (3-2-3)

$$= 4.545 + 2.128 + 2.128 + 1.786$$

$$= 10.587 \text{ mA}$$

total equivalent resistance, $R_T = \dfrac{E_T}{I_T}$ \hfill (3-2-5)

$$= \frac{10 \text{ V}}{10.587 \text{ mA}}$$

$$= 0.945 \text{ k}\Omega$$

$$= 945 \text{ }\Omega \text{ (rounded off)}$$

(b) As a first step, combine the two parallel 4.7-kΩ resistors into a single equivalent resistance of 4.7 k$\Omega/2$ = 2.35 kΩ [Fig. 3-5(b)].

USE RECIPROCAL FORMULA.

$$\frac{1}{R_T} = \frac{1}{2.2} + \frac{1}{2.35} + \frac{1}{5.6} \qquad (3\text{-}3\text{-}1)$$

$$= 0.4545 + 0.4255 + 0.1786 = 1.0586$$

total equivalent resistance, $R_T = \dfrac{1}{1.0586} \text{ k}\Omega$

$$= 945 \text{ }\Omega \text{ (rounded off)}$$

(c) Again combine the two parallel 4.7-kΩ resistors into a single equivalent resistance of 2.35 kΩ. Next, by arbitrary choice, the equivalent resistance of $R_1 \| R_4$ is calculated (the notation "$R_1 \| R_4$" means "R_1 in parallel with R_4").

$$R_1 \| R_4 = \frac{R_1 \times R_4}{R_1 + R_4} \qquad (3\text{-}3\text{-}3)$$

$$= \frac{2.2 \times 5.6}{2.2 + 5.6} = 1.58 \text{ k}\Omega$$

Repeating the product-over-sum formula for $R_1 \| R_4$ and $R_2 \| R_3$ (= 2.35 kΩ) yields

$$\text{total equivalent resistance, } R_T = \frac{1.58 \times 2.35}{1.58 + 2.35}$$

$$= 0.945 \text{ k}\Omega$$

$$= 945 \text{ }\Omega \text{ (rounded off)}$$

EXAMPLE 3-5 In Fig. 3-6, calculate the value of the resistor, R_1.

Solution

$$\text{total equivalent resistance, } R_T = \frac{E_T}{I_T} \quad (3\text{-}2\text{-}5)$$

$$= \frac{24 \text{ V}}{4.7 \text{ mA}} = 5.106 \text{ k}\Omega$$

$$R_1 = \frac{R_2 \times R_T}{R_2 - R_T} \quad (3\text{-}3\text{-}4)$$

$$= \frac{8.2 \times 5.106}{8.2 - 5.106} = 13.5 \text{ k}\Omega \text{ (rounded off)}$$

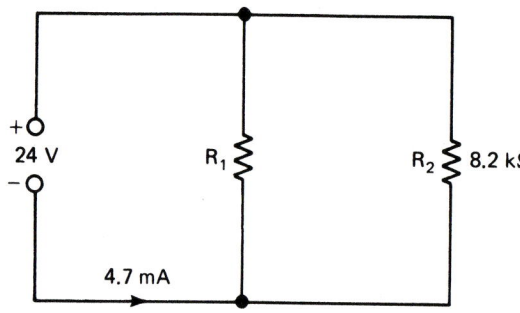

FIGURE 3-6 Circuit for Example 3-5.

EXAMPLE 3-6 In Fig. 3-7, calculate the value of R_3 if the total equivalent resistance presented to the source is 5.3 kΩ.

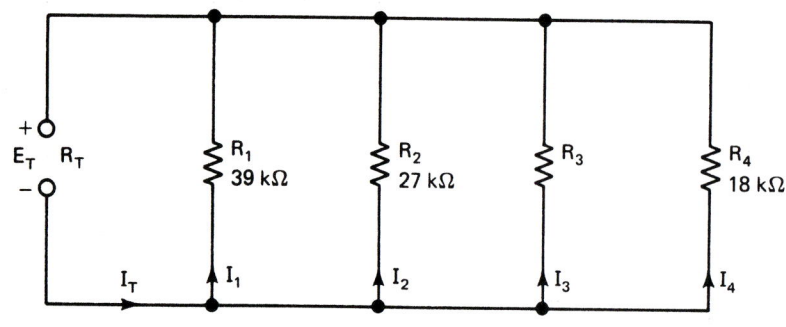

FIGURE 3-7 Circuit for Example 3-6.

Solution Using Eq. (3-3-1), we obtain

$$\frac{1}{R_T} = \frac{1}{5.3} = \frac{1}{39} + \frac{1}{27} + \frac{1}{R_3} + \frac{1}{18}$$

Therefore,

$$\frac{1}{R_3} = \frac{1}{5.3} - \frac{1}{39} - \frac{1}{27} - \frac{1}{18}$$

$$= 0.1887 - 0.02564 - 0.03704 - 0.05556$$

$$= 0.07046$$

$$R_3 = \frac{1}{0.07046} \text{ k}\Omega = 14.2 \text{ k}\Omega \text{ (rounded off)}$$

This problem may also be solved by the assumed voltage method. If $E_T = 100$ V, say,

$$I_T = \frac{E_T}{R_T} \qquad (3\text{-}2\text{-}5)$$

$$= \frac{100 \text{ V}}{5.3 \text{ k}\Omega} = 18.87 \text{ mA}$$

$$I_1 = \frac{100 \text{ V}}{39 \text{ k}\Omega} = 2.564 \text{ mA} \qquad (3\text{-}2\text{-}1)$$

$$I_2 = \frac{100 \text{ V}}{27 \text{ k}\Omega} = 3.704 \text{ mA}$$

$$I_4 = \frac{100 \text{ V}}{18 \text{ k}\Omega} = 5.556 \text{ mA}$$

$$I_3 = I_T - I_1 - I_2 - I_4 \qquad (3\text{-}2\text{-}4)$$

$$= 18.87 - 2.564 - 3.704 - 5.556$$

$$= 7.046 \text{ mA}$$

$$R_3 = \frac{E_T}{I_3} \qquad (3\text{-}2\text{-}1)$$

$$= \frac{100 \text{ V}}{7.046 \text{ mA}} = 14.2 \text{ k}\Omega \text{ (rounded off)}$$

This example has been used to indicate the similarity between the reciprocal formula and the assumed voltage method.

Another method of solution would involve finding the equivalent resistance, R_{124}, of $R_1 \| R_2 \| R_4$ and then using Eq. (3-3-4).

$$\frac{1}{R_{124}} = \frac{1}{R_1} + \frac{1}{R_2} + \frac{1}{R_4} \qquad (3\text{-}3\text{-}1)$$

$$\frac{1}{R_{124}} = \frac{1}{39} + \frac{1}{27} + \frac{1}{18}$$

$$= 0.02564 + 0.02704 + 0.05556$$

$$= 0.11824$$

$$R_{124} = \frac{1}{0.11824} = 8.46 \text{ k}\Omega$$

$$R_3 = \frac{R_{124} \times R_T}{R_{124} - R_T} \tag{3-3-4}$$

$$= \frac{8.46 \times 5.3}{8.46 - 5.3} = 14.2 \text{ k}\Omega \text{ (rounded off)}$$

EXAMPLE 3-7 Eight 100-W electric light bulbs are being correctly operated in parallel from a 120-V supply. Calculate the values of the total equivalent resistance and the total current drawn from the supply.

Solution

$$\text{"hot" filament resistance of each bulb, } R = \frac{E^2}{P} \tag{1-5-5}$$

$$= \frac{(120 \text{ V})^2}{100 \text{ W}} = 144 \text{ }\Omega$$

$$\text{total equivalent resistance, } R_T = \frac{R}{N} \tag{3-3-5}$$

$$= \frac{144}{8} = 18 \text{ }\Omega$$

$$\text{total current drawn from the supply, } I_T = \frac{E_T}{R_T} \tag{3-2-5}$$

$$= \frac{120 \text{ V}}{18 \text{ }\Omega}$$

$$= 6.67 \text{ A (rounded off)}$$

Alternatively,

$$\text{current flowing through each light bulb, } I = \frac{P}{E} \tag{1-4-3}$$

$$= \frac{100 \text{ W}}{120 \text{ V}}$$

$$= \frac{5}{6} \text{ A}$$

$$\text{total current drawn from the supply, } I_T = NI \tag{3-2-6}$$

$$= 8 \times \frac{5}{6}$$

$$= 6.67 \text{ A (rounded off)}$$

3-4 Total Equivalent Conductance of Parallel Circuit

As we have discussed in Section 1-5, the property of conductance is a measure of the ease with which current flows through a given material. Conductance, G, is therefore the inverse or reciprocal of resistance and is measured in siemens (formerly known as mhos). Substituting the relationship $G = 1/R$ in Eq. (3-3-1), the total equivalent conductance, G_T, of N resistors in parallel is

$$G_T = G_1 + G_2 + G_3 + \cdots + G_N \tag{3-4-1}$$

and

$$I_T = E_T \times G_T, \quad I_1 = E_T \times G_1, \quad I_2 = E_T \times G_2, \quad \text{etc.} \tag{3-4-2}$$

Each branch current is therefore proportional to the branch conductance. If the N resistors all have the same conductance, G,

$$G_T = NG \tag{3-4-3}$$

The additive formula for G_T is, of course, much simpler to use than the reciprocal formula for R_T. The question then arises: "Why not always use conductance (rather than resistance) in our formulas?" The answer lies in the R_T formula for resistors in series; in Eq. (2-3-5),

$$R_T = R_1 + R_2 + R_3 + \cdots + R_N$$

Then

$$\frac{1}{G_T} = \frac{1}{G_1} + \frac{1}{G_2} + \frac{1}{G_3} + \cdots + \frac{1}{G_N} \tag{3-4-4}$$

Therefore, the reciprocal formula applies to conductances in series. To have the best of both worlds, it would be logical to use the additive formulas of R_T for the series circuit and G_T for the parallel circuit—this is normally done when using sophisticated techniques to analyze electrical circuits.

To summarize, the total conductance, G_T, measured in siemens (S), of resistors in parallel is the sum of the individual conductances. For resistors in series, G_T is found by the reciprocal formula.

EXAMPLE 3-8 Three conductances, G_1, G_2, and G_3, are connected in parallel across a 24-V source. If $G_1 = 0.00370$ S, $G_2 = 0.00213$ S, and $G_3 = 0.00147$ S, find the values of I_1, I_2, I_3, G_T, I_T, and R_T.

Solution From Eq. (3-4-2),

$$\text{branch current, } I_1 = E_T G_1 = 24 \times 0.00370 \text{ A} = 88.8 \text{ mA}$$

$$I_2 = E_T G_2 = 24 \times 0.00213 \text{ A} = 51.1 \text{ mA}$$

$$I_3 = E_T G_3 = 24 \times 0.00147 \text{ A} = 35.3 \text{ mA}$$

With the electronic calculator the branch currrents are found by entering the value of E_T as the "constant multiplying factor."

$$\text{total conductance, } G_T = G_1 + G_2 + G_3 \qquad (3\text{-}4\text{-}1)$$

$$= 0.00370 + 0.00213 + 0.00147$$

$$= 0.00730 \text{ S}$$

$$\text{total resistance, } R_T = \frac{1}{G_T}$$

$$= \frac{1}{0.00730} = 137 \text{ }\Omega \text{ (rounded off)}$$

$$\text{total source current, } I_T = E_T G_T \qquad (3\text{-}4\text{-}2)$$

$$= 24 \times 0.00730 \text{ A} = 175.2 \text{ mA}$$

Current Check

$$I_T = I_1 + I_2 + I_3 \qquad (3\text{-}2\text{-}4)$$

$$= 88.8 + 51.1 + 35.3 = 175.2 \text{ mA}$$

3-5 Power Relationships in Parallel Circuit

Referring back to Fig. 3-1(e), all five resistors in the parallel circuit dissipate power in the form of heat. The source voltage, E_T, is dropped across each of the resistors, and therefore from the equations of Section 1-5, the power dissipated in the resistor R_1 is

$$P_1 = \frac{E_T^2}{R_1} = E_T I_1 = I_1^2 R_1 \quad \text{watts}$$

Similar formulas apply to P_2, P_3, P_4, and P_5. Then

$$\frac{P_1}{P_2} = \frac{E_T^2/R_1}{E_T^2/R_2} = \frac{R_2}{R_1} \quad \text{or} \quad P_1 = P_2 \times \frac{R_2}{R_1}, \quad \text{etc.} \qquad (3\text{-}5\text{-}1)$$

Using Eq. (3-2-3), we have

$$I_T = I_1 + I_2 + I_3 + I_4 + I_5$$

Therefore,

$$EI_T = EI_1 + EI_2 + EI_3 + EI_4 + EI_5$$

so that

$$P_T = P_1 + P_2 + P_3 + P_4 + P_5 \qquad (3\text{-}5\text{-}2)$$

and
$$P_T = \frac{E_T^2}{R_T} = I_T^2 R_T = E_T I_T$$

Then

$$\frac{P_1}{P_T} = \frac{E_T^2/R_1}{E_T^2/R_T} = \frac{R_T}{R_1} \quad \text{or} \quad P_1 = P_T \times \frac{R_T}{R_1}, \quad \text{etc.} \qquad (3\text{-}5\text{-}3)$$

For N resistors in parallel,

$$P_T = P_1 + P_2 + P_3 + \ldots + P_N \qquad (3\text{-}5\text{-}4)$$

For N equal-value resistors in parallel,

$$P_T = NP$$

where P is the power dissipated in one of the equal-value resistors.

To summarize, the P_T formula of Eq. (3-5-4), for resistors in parallel is exactly the same as the corresponding formula [Eq. (2-4-4)] for resistors in series. This must be true since the powers dissipated in the resistors, whether in series or parallel, can only be obtained from the source and the individual heating effects are naturally additive. However, you will remember that in the series circuit the highest-value resistor, which developed the greatest voltage drop, dissipated the most power. By contrast, in the parallel circuit the greatest power is dissipated by the lowest-value resistor, which carries the greatest branch current; for example, the resistance of a nickel-iron 1000-W heater element is only one-tenth that of the tungsten 100-W electric light bulb. It follows that the parallel branch with the highest resistance will dissipate the least power, while branches with equal resistances will dissipate the same powers.

EXAMPLE 3-9 In Fig. 3-8, calculate the values of the individual powers dissipated in the three resistors and the total power delivered by the source.

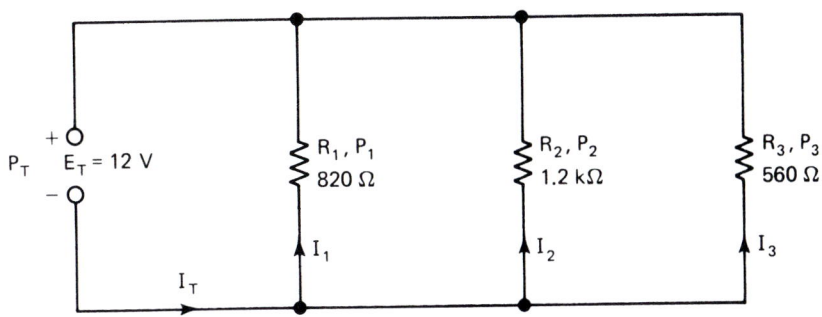

FIGURE 3-8 Circuit for Example 3-9.

Solution Using Eq. (3-5-1), we obtain

$$P_1 = \frac{E_T^2}{R_1} = \frac{(12\text{ V})^2}{820\text{ }\Omega} = 0.1756\text{ W}$$

Sec. 3-5 Power Relationships in Parallel Circuit

$$P_2 = \frac{(12\text{ V})^2}{1.2\text{ k}\Omega} = \frac{(12\text{ V})^2}{1200\ \Omega} = 0.1200\text{ W}$$

$$P_3 = \frac{(12\text{ V})^2}{560\ \Omega} = 0.2571\text{ W}$$

total power dissipated by the resistors, P_T

$$= P_1 + P_2 + P_3 \qquad (3\text{-}5\text{-}2)$$

$$= 0.1756 + 0.1200 + 0.2571 = 0.553\text{ W} \quad \text{(rounded off)}$$

Power Check

Total equivalent resistance, R_T, is given by

$$\frac{1}{R_T} = \frac{1}{820} + \frac{1}{1200} + \frac{1}{560} = 0.00122 + 0.000833 + 0.00179$$

$$= 0.003843\text{ S}$$

$$R_T = \frac{1}{0.003843\text{ S}} = 260.2\ \Omega$$

total power dissipated from the source, $P_T = \dfrac{E_T^2}{R_T}$

$$= \frac{(12\text{ V})^2}{260.2\ \Omega} = 0.553\text{ W} \quad \text{(rounded off)}$$

Notice the lowest-value resistor (560 Ω) dissipates the greatest power, while the highest-value resistor (1.2 kΩ) dissipates the least power.

A word of warning about connecting resistors in parallel; unless the resistors are identical, the total power rating of the parallel combination is not the same as the sum of the individual wattage ratings of the resistors. Each of the parallel resistors will have a certain voltage rating which must not be exceeded; otherwise, the resistor will overheat. It is therefore only safe to apply the lowest individual rating to the parallel combination.

For comparison purposes let us use exactly the same resistors as those we chose in Section 2-4, so that we are now considering 180 Ω, 220 Ω, 330 Ω, and 470 Ω, 1-W resistors in parallel. Using the Ohm's law relationship $E = \sqrt{P \times R}$, the voltage ratings for the four resistors are:

$$E_1 = \sqrt{1\text{ W} \times 180\ \Omega} = 13.4\text{ V}$$

$$E_2 = \sqrt{1\text{ W} \times 220\ \Omega} = 14.8\text{ V}$$

$$E_3 = \sqrt{1\text{ W} \times 330\ \Omega} = 18.2\text{ V}$$

$$E_4 = \sqrt{1\text{ W} \times 470\ \Omega} = 21.7\text{ V}$$

The maximum voltage that can safely be applied to the parallel combination is therefore 13.4 V, since if this is exceeded, the 180-Ω resistor would overheat. The total equivalent resistance, R_T, of the parallel combination is given by

$$\frac{1}{R_T} = \frac{1}{180} + \frac{1}{220} + \frac{1}{330} + \frac{1}{470}$$

$$= 0.005556 + 0.004545 + 0.003030 + 0.002128$$

$$= 0.015259 \text{ S}$$

$$R_T = \frac{1}{0.015259 \text{ S}} = 65.5 \text{ }\Omega$$

The total wattage rating of the parallel combination is

$$\frac{E_T^2}{R_T} = \frac{(13.4 \text{ V})^2}{65.5 \text{ }\Omega} = 2.74 \text{ W} \quad \text{(rounded off)}$$

That is, of course, totally different from the sum of the individual power ratings, namely 4 W.

EXAMPLE 3-10 In Fig. 3-9, find the values of P_1, R_3, P_T, and R_T.

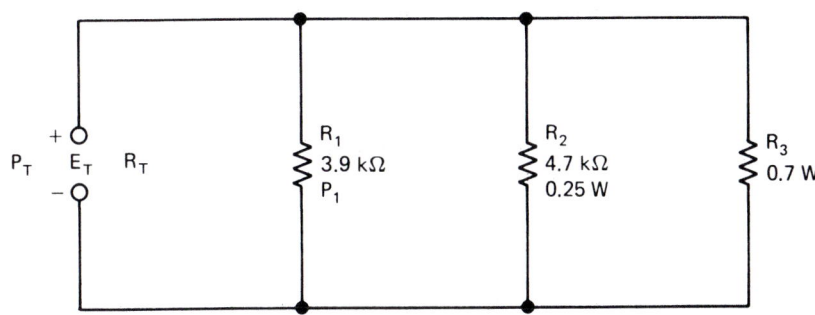

FIGURE 3-9 Circuit for Example 3-10.

Solution

$$\text{power dissipated, } P_1 = P_2 \times \frac{R_2}{R_1} \tag{3-5-1}$$

$$= 0.25 \text{ W} \times \frac{4.7 \text{ k}\Omega}{3.9 \text{ k}\Omega} = 0.30 \text{ W}$$

Also,

$$R_3 = R_2 \times \frac{P_2}{P_3} = 4.7 \text{ k}\Omega \times \frac{0.25 \text{ W}}{0.7 \text{ W}} = 1.68 \text{ k}\Omega$$

$$\text{total power, } P_T = P_1 + P_2 + P_3 \tag{3-5-2}$$

$$= 0.30 + 0.25 + 0.70 = 1.25 \text{ W}$$

$$\text{total equivalent resistance, } R_T = R_2 \times \frac{P_2}{P_T} \tag{3-5-3}$$

$$= 4.7 \text{ k}\Omega \times \frac{0.25 \text{ W}}{1.25 \text{ W}}$$

$$= 0.94 \text{ k}\Omega$$

Sec. 3-5 Power Relationships in Parallel Circuit

Resistance Check
Reciprocal formula:

$$\frac{1}{R_T} = \frac{1}{3.9} + \frac{1}{4.7} + \frac{1}{1.68} = 0.256 + 0.213 + 0.596 = 1.065 \text{ mS}$$

$$R_T = \frac{1}{1.065 \text{ mS}} = 0.94 \text{ k}\Omega$$

This example was included to illustrate the need for care in selecting the necessary equations, so that you can get the required answers with a minimum number of steps.

3-6 Step-by-step Analysis of Parallel Circuit

This analysis assumes that you know the value of the parallel resistors and the source voltage. Such would normally be the case when you are conducting an experiment on parallel resistors in the laboratory. For comparison we will use precisely the same resistors and the source voltage as we chose for the step-by-step analysis of the series circuit in Section 2-5; the parallel circuit we are therefore going to analyze is shown in Fig. 3-10.

STEP 1 Determine the branch currents (remember that if the value of E_T is not in fact given, any convenient value may be assumed).
  Using Eq. (3-2-1), we obtain

$$I_1 = \frac{E_T}{R_1} = \frac{72 \text{ V}}{8.2 \text{ k}\Omega} = 8.8 \text{ mA}$$

$$I_2 = \frac{E_T}{R_2} = \frac{72 \text{ V}}{1.8 \text{ k}\Omega} = 40.0 \text{ mA}$$

$$I_3 = \frac{E_T}{R_3} = \frac{72 \text{ V}}{3.3 \text{ k}\Omega} = 21.8 \text{ mA}$$

$$I_4 = \frac{E_T}{R_4} = \frac{72 \text{ V}}{4.7 \text{ k}\Omega} = 15.3 \text{ mA}$$

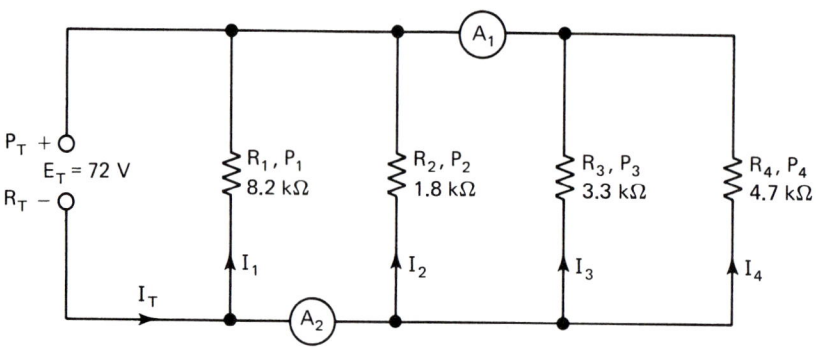

FIGURE 3-10 Step-by-step analysis of the parallel circuit.

Notice that the highest current, I_2 (40 mA), is carried by the lowest-value resistor, R_2 (1.8 kΩ).

STEP 2 Calculate the total circuit current, I_T, drawn from the voltage source, E_T.

$$I_T = I_1 + I_2 + I_3 + I_4 = 8.8 + 40.0 + 21.8 + 15.3$$
$$= 85.9 \text{ mA}$$

The individual branch currents and the total circuit current are not the only current values that can be measured in the circuit of Fig. 3-10. For example, the milliammeter, A_1, will read the sum of the currents, I_3 and I_4; this is 21.8 + 15.3 = 37.1 mA. Similarly, A_2 indicates the sum of the currents I_2, I_3, and I_4 or 40.0 + 21.8 + 15.3 = 77.1 mA.

STEP 3 Find the total equivalent resistance, R_T.
 (a) Assumed (if necessary) voltage method:

$$\text{equivalent resistance, } R_T = \frac{E_T}{I_T}$$

$$= \frac{72 \text{ V}}{85.9 \text{ mA}} = 0.838 \text{ kΩ} = 838 \text{ Ω}$$

 (b) Reciprocal formula:

$$\frac{1}{R_T} = \frac{1}{R_1} + \frac{1}{R_2} + \frac{1}{R_3} + \frac{1}{R_4}$$

$$= \frac{1}{8.2 \text{ kΩ}} + \frac{1}{1.8 \text{ kΩ}} + \frac{1}{3.3 \text{ kΩ}} + \frac{1}{4.7 \text{ kΩ}}$$

$$= 0.122 + 0.556 + 0.303 + 0.213$$

$$= 1.194 \text{ mS}$$

$$R_T = \frac{1}{1.194 \text{ mS}} = 0.838 \text{ kΩ} = 838 \text{ Ω}$$

Again notice that the value of R_T is less than the smallest-value resistor (1800 Ω) in parallel.
 (c) Repeated product-over-sum formula: The total resistance, R_{12}, of R_1 and R_2 in parallel is given by

$$R_{12} = \frac{R_1 \times R_2}{R_1 + R_2}$$

$$= \frac{8.2 \times 1.8}{8.2 + 1.8} = \frac{14.76}{10} = 1.476 \text{ kΩ}$$

Similarly,

$$R_{34} = \frac{R_3 \times R_4}{R_3 + R_4}$$

$$= \frac{3.3 \times 4.7}{3.3 + 4.7} = \frac{15.51}{8} = 1.939 \text{ kΩ}$$

Sec. 3-6 Step-by-step Analysis of Parallel Circuit

Then

$$R_T = \frac{R_{12} \times R_{34}}{R_{12} + R_{34}}$$

$$= \frac{1.476 \times 1.939}{1.476 + 1.939}$$

$$= \frac{2.862}{3.415} = 0.838 \text{ k}\Omega = 838 \text{ }\Omega$$

STEP 4 Determine the powers dissipated in the individual branches. Using Eq. (3-5-1), we obtain

$$\text{power dissipated, } P_1 = \frac{E_T^2}{R_1} = \frac{(72 \text{ V})^2}{8.2 \text{ k}\Omega} = 632.2 \text{ mW}$$

$$P_2 = \frac{E_T^2}{R_2} = \frac{(72 \text{ V})^2}{1.8 \text{ k}\Omega} = 2880 \text{ mW}$$

$$P_3 = \frac{E_T^2}{R_3} = \frac{(72 \text{ V})^2}{3.3 \text{ k}\Omega} = 1571 \text{ mW}$$

$$P_4 = \frac{E_T^2}{R_4} = \frac{(72 \text{ V})^2}{4.7 \text{ k}\Omega} = 1103 \text{ mW}$$

$$\text{total power dissipated, } P_T = P_1 + P_2 + P_3 + P_4$$

$$= 632.2 + 2880 + 1571 + 1103 \quad (3\text{-}5\text{-}2)$$

$$= 6186.2 \text{ mW}$$

$$= 6.19 \text{ W (rounded off)}$$

POWER CHECK Make certain that the total power dissipated in the parallel branches is equal to the total power delivered from the source voltage.

total power delivered from the source voltage, $P_T = E_T \times I_T$

$$= 72 \text{ V} \times 85.9 \text{ mA} = 6.19 \text{ W (rounded off)}$$

Again notice that the lowest-value (1.8 kΩ) resistor dissipates the most power, and the highest-value (8.2 kΩ) resistor dissipates the least power.

3-7

Current Division Rule

The step-by-step method of analysis can only be used when the source voltage and all the values of the parallel resistances are known. However, we might be given the values of the parallel resistances and the total current and then be asked to determine the individual branch currents. The purpose of this section is therefore to provide you with some of the other relationships that exist in a parallel circuit.

From Eqs. (3-2-1) and (3-2-5) we have already seen that

$$E_T = I_1 R_1 = I_2 R_2 = I_3 R_3 = \cdots = I_T R_T$$

Then

$$\frac{I_1}{I_2} = \frac{R_2}{R_1}, \quad \text{etc.} \qquad (3\text{-}7\text{-}1)$$

This equation shows that the branch currents are in *inverse* proportion to the values of the resistors. We can therefore use the method of proportion to solve a problem in which we are given R_1, R_2, and I_2 and asked to find the value of I_1. This would be done by using the formula

$$I_1 = I_2 \times \frac{R_2}{R_1} \qquad (3\text{-}7\text{-}2)$$

Similarly,

$$I_2 = I_1 \times \frac{R_1}{R_2}, \quad \text{etc.} \qquad (3\text{-}7\text{-}3)$$

The same rule of inverse proportion can be applied to the powers dissipated in the individual branches. Since

$$P_1 = \frac{E_T^2}{R_1} \quad \text{and} \quad P_2 = \frac{E_T^2}{R_2}$$

$$E_T^2 = P_1 R_1 = P_2 R_2 \quad \text{and} \quad \frac{P_1}{P_2} = \frac{R_2}{R_1} \qquad (3\text{-}7\text{-}4)$$

Therefore,

$$P_1 = P_2 \times \frac{R_2}{R_1} \quad \text{and} \quad P_2 = P_1 \times \frac{R_1}{R_2}, \quad \text{etc.} \qquad (3\text{-}7\text{-}5)$$

The total source current, I_T, is divided among the parallel branches in inverse proportion to the values of the resistances. Since

$$E_T = I_1 R_1 = I_T R_T$$

$$\frac{I_1}{I_T} = \frac{R_T}{R_1} \quad \text{and} \quad I_1 = I_T \times \frac{R_T}{R_1} = I_T \times \frac{G_1}{G_T} \qquad (3\text{-}7\text{-}6)$$

Similar equations apply to I_2, I_3, and so on.

Equation (3-7-6) represents the *current division rule* and shows that, in a parallel circuit, the fraction of the source current flowing through a particular branch is equal to the ratio of the total equivalent resistance to the branch resistance.

In the case of two resistors, R_1 and R_2, in parallel,

$$R_T = \frac{R_1 R_2}{R_1 + R_2}$$

and therefore,

$$I_1 = I_T \times \frac{R_T}{R_1} = I_T \times \frac{R_1 R_2}{R_1(R_1 + R_2)} = I_T \times \frac{R_2}{R_1 + R_2}$$

$$I_2 = I_T \times \frac{R_1}{R_1 + R_2}$$
(3-7-7)

To calculate one of the branch currents for two resistors in parallel, you first multiply the source current by the resistance *in the other branch* and then divide by the sum of the two resistances.

The total power delivered from the source also divides between the parallel resistors in the same way as the total current. Since

$$P_T = \frac{E_T^2}{R_T} \quad \text{and} \quad P_1 = \frac{E_T^2}{R_1}$$

$$E_T^2 = P_T R_T = P_1 R_1 \quad \text{and} \quad \frac{P_1}{R_T} = \frac{P_T}{R_1}$$
(3-7-8)

This leads to

$$P_1 = P_T \times \frac{R_T}{R_1} = P_T \times \frac{G_1}{G_T}, \quad P_2 = P_T \times \frac{R_T}{R_2} = P_T \times \frac{G_2}{G_T}$$
(3-7-9)

To summarize, the individual branch currents and their dissipated powers are in inverse proportion to the values of the branch resistances. The current division rule states that the fraction of the total source current flowing through a particular branch is equal to the ratio of the total equivalent resistance to the branch resistance (or the ratio of the branch conductance to the total equivalent conductance).

All these results should be carefully compared with those for the voltage division rule in Section 2-7.

EXAMPLE 3-11 In Fig. 3-11, find the values of I_1, I_2, E_T, and P_2.

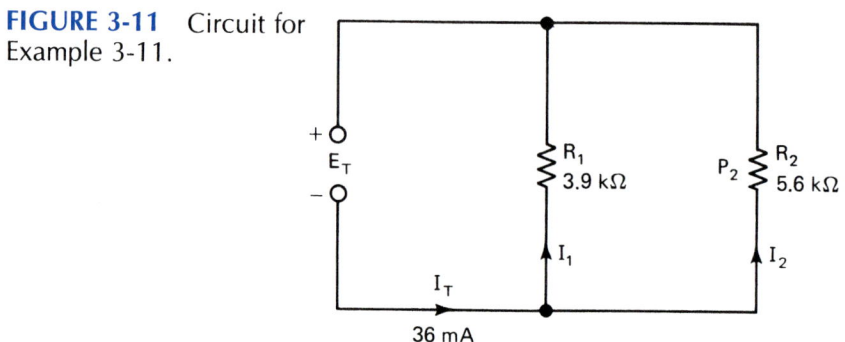

FIGURE 3-11 Circuit for Example 3-11.

Solution By Eq. (3-7-7),

$$\text{branch current, } I_1 = I_T \times \frac{R_2}{R_1 + R_2}$$

$$= 36 \text{ mA} \times \frac{5.6 \text{ k}\Omega}{3.9 \text{ k}\Omega + 5.6 \text{ k}\Omega}$$

$$= 36 \times \frac{5.6}{9.5} \text{ mA} = 21.22 \text{ mA}$$

$$I_2 = I_T \times \frac{R_1}{R_1 + R_2}$$

$$= 36 \text{ mA} \times \frac{3.9 \text{ k}\Omega}{3.9 \text{ k}\Omega + 5.6 \text{ k}\Omega}$$

$$= 36 \times \frac{3.9}{9.5} \text{ mA} = 14.78 \text{ mA}$$

Current Check

Make certain that the sum of the calculated branch currents is equal to the total source current.

$$I_1 + I_2 = 21.22 + 14.78 = 36 \text{ mA} = I_T$$

$$E_T = I_1 R_1 \qquad (3\text{-}2\text{-}1)$$

$$= 21.22 \text{ mA} \times 3.9 \text{ k}\Omega = 82.76 \text{ V}$$

Voltage Check

$$E_T = I_2 R_2 = 14.78 \text{ mA} \times 5.6 \text{ k}\Omega = 82.77 \text{ V}$$

$$P_2 = I_2^2 \times R_2 \qquad (3\text{-}5\text{-}1)$$

$$= (14.78 \text{ mA})^2 \times 5.6 \text{ k}\Omega = 1223 \text{ mW}$$

Power Check

$$P_2 = I_2 \times E_T = 14.78 \text{ mA} \times 82.77 \text{ V} = 1223 \text{ mW}$$

EXAMPLE 3-12 In Fig. 3-12, calculate the value of I_T.

Solution

$$\frac{1}{R_T} = \frac{1}{R_1} + \frac{1}{R_2} + \frac{1}{R_3} \qquad (3\text{-}3\text{-}1)$$

$$= \frac{1}{27} + \frac{1}{82} + \frac{1}{47}$$

$$= 0.03704 + 0.01220 + 0.02128 = 0.07052 \text{ mS}$$

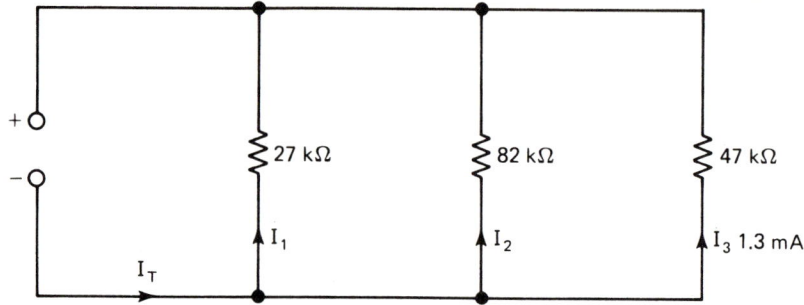

FIGURE 3-12 Circuit for Example 3-12.

$$R_T = \frac{1}{0.07052 \text{ mS}} = 14.2 \text{ k}\Omega$$

$$I_T = I_3 \times \frac{R_3}{R_T} \tag{3-7-6}$$

$$= 1.3 \text{ mA} \times \frac{47 \text{ k}\Omega}{14.2 \text{ k}\Omega} = 4.3 \text{ mA}$$

3-8 Voltage Sources Connected in Parallel

In this section we consider the connection of identical cells in parallel (non-identical cells in parallel are covered in Section 5-7). An everyday example occurs when you give your friend a "jump start" for his or her car. As you know, you join the positive terminals of the two batteries together and make the same type of connection with the negative terminals. For this parallel arrangement the total voltage available is only equal to that of one battery. However, the current capability has been increased since the total load current will be shared between the batteries.

Figure 3-13(a) and (b) shows N identical cells, each of EMF, E, which are parallel-connected across a number of resistive loads. All the positive terminals are joined together so that any one of these terminals may be used as the output positive terminal; the negative terminals are similarly connected. Then

$$\text{total voltage of the parallel combination, } E_T = E \tag{3-8-1}$$

$$\text{total load current, } I_T = \frac{E_T}{R_T} \tag{3-8-2}$$

where R_T is the total equivalent resistance of the parallel loads [Fig. 3-13(c)].

$$\text{current supplied by each cell} = \frac{I_T}{N} \tag{3-8-3}$$

Notice that in the loops containing the cells, the arrangement is series-opposing, so that there is no loading effect of one cell on another. With identical cells,

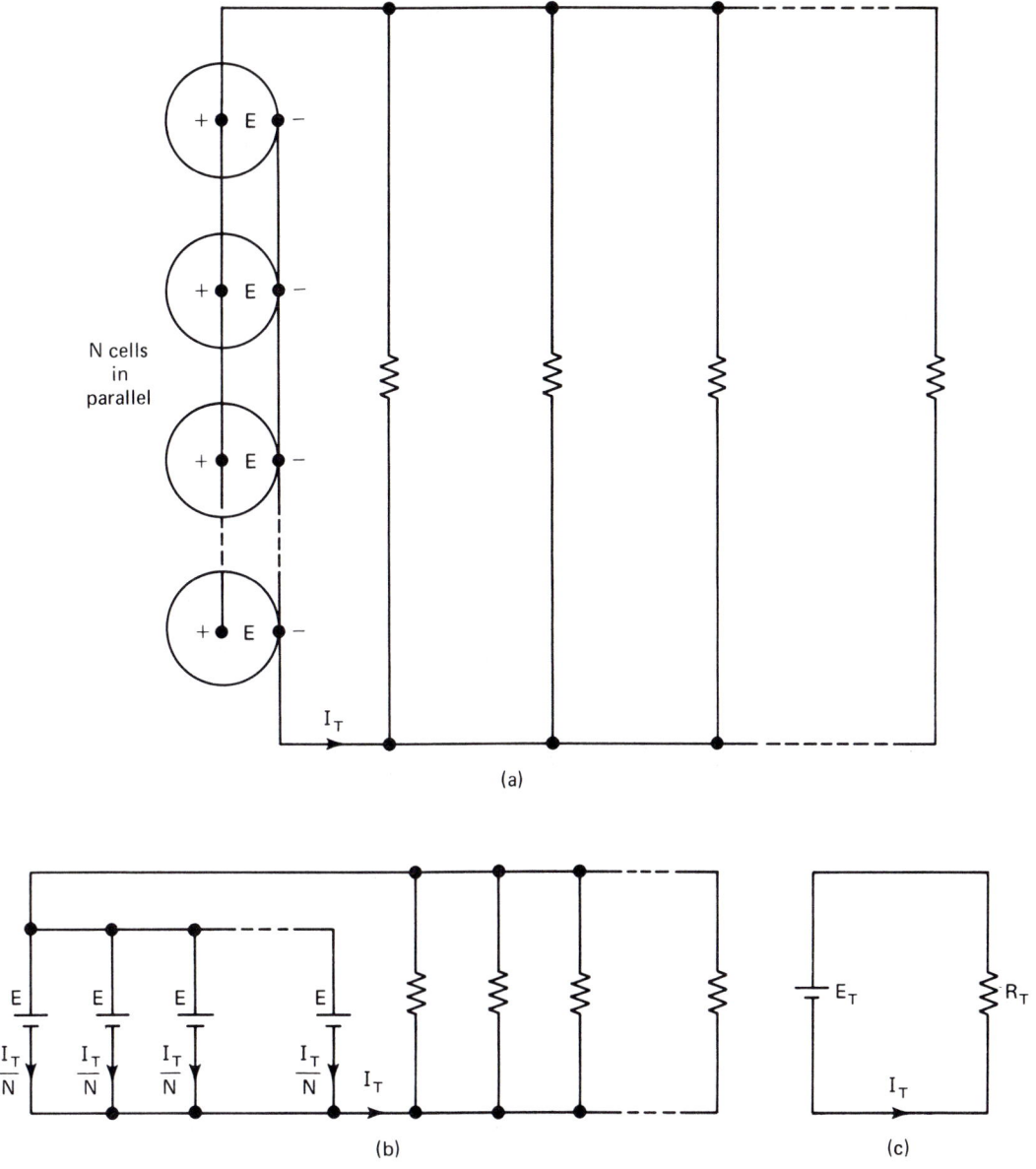

FIGURE 3-13 Identical cells in parallel.

no parallel-opposing arrangement is possible with respect to the load since the cells themselves would be connected in series-aiding around the loops. A very high circulating current would flow between the cells but no current would be supplied to the load.

To summarize this section, the total EMF of parallel-connected identical cells is equal to the EMF of only one cell. The purpose of the parallel arrangement is to increase the load current capability by allowing the total load current to be shared among the cells. This will be better understood after you have read Section 5-3, which discusses the property of a cell's internal resistance.

EXAMPLE 3-13 Five identical $1\frac{1}{2}$-V cells are connected in parallel across three parallel loads whose resistances are 600 Ω, 300 Ω, and 200 Ω. Calculate the value of the current supplied by each cell.

Solution

total EMF, E_T, applied to the parallel loads

$$= 1.5 \text{ V (EMF of one cell)} \quad (3\text{-}8\text{-}1)$$

$$\text{total load, } R_T = \frac{1}{\frac{1}{600} + \frac{1}{300} + \frac{1}{200}} = \frac{1}{1/100} = 100 \text{ } \Omega$$

$$\text{total load current, } I_T = \frac{E_T}{R_T} \quad (3\text{-}8\text{-}2)$$

$$= \frac{1.5 \text{ V}}{100} = 0.015 \text{ A} = 15 \text{ mA}$$

$$\text{current supplied by each cell} = \frac{15}{5} = 3 \text{ mA} \quad (3\text{-}8\text{-}3)$$

3-9 Effect of Open Circuits on Resistors in Parallel

As we saw in Section 2-11, an open circuit represents a break that will possess virtually infinite resistance, so that no current will flow. In a series string of resistors a break, wherever it occurred, caused all current to cease and the source voltage then appeared across the open circuit. However, with a parallel arrangement of loads, we must distinguish between breaks that occur in the individual branches [Fig. 3-14(a), where the resistor, R_2, has completely burned out, so that there is an open circuit between points A and B] and a break in the line which is directly connected to the voltage source [Fig. 3-14(b)].

Referring to Fig. 3-14(a), such an open circuit could be caused by a broken or disconnected wire, a burned-out resistor, or the fractured element of a tube's

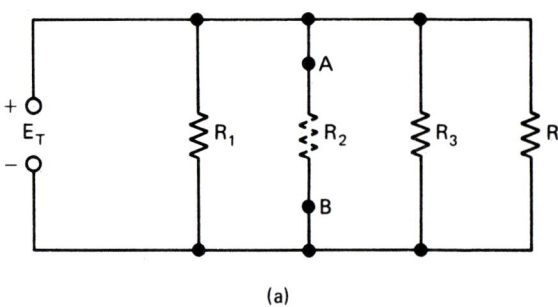

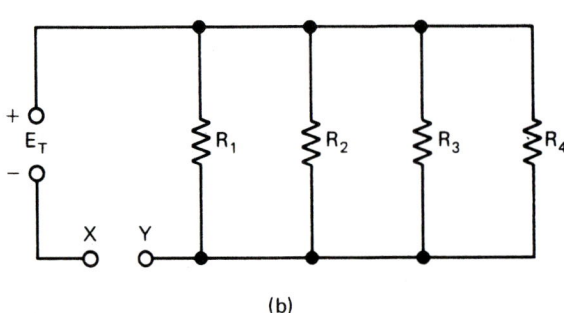

FIGURE 3-14 Effects of open circuits on parallel banks of resistors.

heater. In each of these examples the current in that particular branch would be zero, but the source voltage would still be across the break. The currents in the other branches would then continue to flow with the same values as before the open-circuit occurred. In Section 2-11 we discussed what would happen in a series heater string if one of the tube filaments burned out. The result was that all current ceased so that all the heaters were extinguished; the open circuit could then be found by using an ohmmeter or a voltmeter. For the heaters to be connected in parallel they must all have the same voltage rating. Then if one of the heaters open-circuits, only that tube is extinguished and the heaters of the other tubes remain red-hot; it is therefore a simple matter to find the defective tube.

Using an ohmmeter with the power off, you cannot easily locate the open circuit in a parallel branch such as R_2 in Fig. 3-14(a). Although the resistance of the open circuit is virtually infinite, an ohmmeter connected across the break will not read infinity because of the effect of the other resistors which are connected to A and B. Therefore, when making tests with an ohmmeter, you must always be careful to look out for all parallel paths that exist between the ohmmeter leads (see Example 3-14).

In Fig. 3-14(b) the open circuit is between points X and Y, so that the total current, I_T, drawn from the source is zero; the voltage drop across all the resistors is also zero, and therefore the source voltage E_T appears between X and Y. This in turn means that all the branch currents are zero; the total resistance presented to the source is infinite and no power is dissipated in the circuit.

The effect of open circuits on a parallel branch of resistors may be summarized as follows:

1. If the open circuit occurs in an individual branch, the current in that branch is zero but the currents in the other branches remain unchanged.

2. If the break is in the main line connected to the voltage source, all the branch currents and the total line current are zero.

EXAMPLE 3-14 In Fig. 3-15 the resistor R_3 is completely burned out, so that an open circuit exists between the points P and Q. What are the values of I_1, I_2, I_3, and V_{PQ}? The switch, S, is opened and an ohmmeter is connected between P and Q. What is the reading of the ohmmeter?

FIGURE 3-15 Circuit for Example 3-14.

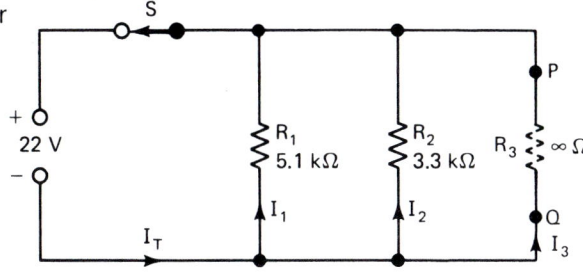

Solution Substituting in Eq. (3-2-1), we obtain

$$\text{branch current, } I_1 = \frac{22 \text{ V}}{5.1 \text{ k}\Omega} = 4.31 \text{ mA}$$

$$I_2 = \frac{22 \text{ V}}{3.3 \text{ k}\Omega} = 6.67 \text{ mA}$$

Sec. 3-9 Effect of Open Circuits on Resistors in Parallel

Since there is an open circuit between P and Q, I_3 is zero and $V_{PQ} = 22$ V.

Notice that $I_T = 4.31 + 6.67 = 10.98$ mA and $P_T = 22$ V $\times$ 10.98 mA = 242 mW (rounded off). When the switch, S, is opened and the ohmmeter is connected between P and Q, the resistors R_1 and R_2 are in parallel with respect to the ohmmeter's position. The ohmmeter reading is therefore equal to

$$\frac{R_1 R_2}{R_1 + R_2} = \frac{5.1 \times 3.3}{5.1 + 3.3} = 2.0 \text{ k}\Omega \quad \text{(rounded off)}$$

EXAMPLE 3-15 In Fig. 3-16 there is an open circuit between points X and Y. What are the values of I_1, I_2, I_3, I_T, and V_{XY}?

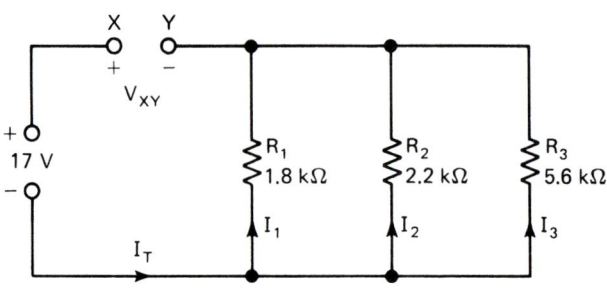

FIGURE 3-16 Circuit for Example 3-15.

Solution Since there is infinite resistance between X and Y, I_T is zero and $I_1 = I_2 = I_3 = 0$ A. There is zero voltage drop across all the resistors and therefore $V_{XY} = 17$ V.

3-10

Effect of a Short Circuit on Resistors in Parallel

In Section 2-11 we mentioned that a short circuit has theoretically zero resistance so that current may flow through a short circuit without developing any voltage drop. Although resistors rarely become completely shorted, a short circuit can occur as the result of bare connecting wires coming into contact or being joined together by a stray drop of solder. The short circuit is therefore a zero-resistance path across or in parallel with the resistor.

In Fig. 3-17 a short circuit has developed between the ends of the resistor load, R_3. Across the source there is now a very low resistance path which consists of the connecting lead (BC), the short circuit (CD), and the return lead (DA). The current drawn from the source may be extremely high, although it is limited by the resistance of the connecting wires and the short circuit as well as the source's internal resistance (Section 5-3). To prevent possible damage to the source and the connecting wires, the circuit may be protected by a fuse (Section 1-8) or a magnetic circuit breaker. Then, should any short circuit occur, the excessive current will cause the fuse to blow or the magnetic circuit breaker to trip.

Since the short circuit offers virtually zero resistance, all the current drawn from the source will be diverted through the short and no current will flow through any of the parallel loads; therefore, $I_1 = I_2 = I_3 = 0$ A. When this occurs we say that the loads R_1, R_2, and R_3 have been shorted out of the circuit; this result is the same no matter how many loads are involved. With no current

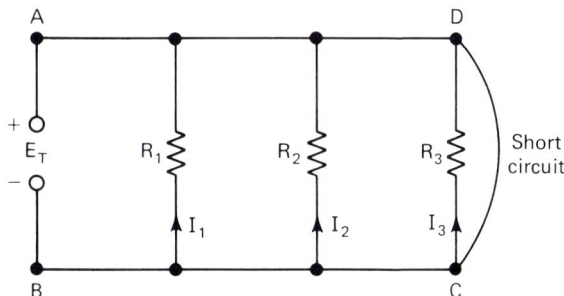

FIGURE 3-17 Effect of a short circuit on a parallel bank of resistors.

flowing in any of the branches, no power is being dissipated by the resistive loads, which therefore cannot function properly. However, you must realize that the presence of the short circuit is not causing any damage to these loads; once the short has been located and removed, the whole circuit can be restored to normal operation.

EXAMPLE 3-16 In Fig. 3-18 a short circuit occurs across R_2. What is then the resistance between points A and B and the values of the currents I_1, I_2, I_3, and I_4?

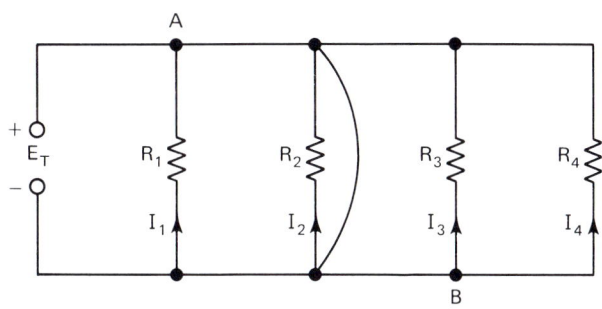

FIGURE 3-18 Circuit for Example 3-16.

Solution The resistance between A and B is equal to the resistance of the short circuit and is therefore virtually zero. The branch currents are also zero and therefore $I_1 = I_2 = I_3 = I_4 = 0$ A.

3-11
Troubleshooting a Parallel Circuit

It is more difficult to troubleshoot a parallel arrangement of resistors that are connected to a single source than it was to troubleshoot the simple series circuit of Section 2-13. For example, only the value of the source voltage exists in a parallel circuit, so that the presence of an open or a short cannot be found by taking voltmeter readings. Our troubleshooting procedures will therefore be confined to using the ohmmeter and the current meter.

Locating an Open Circuit

Let us refer to the parallel circuit of Fig. 3-19. Assuming that no opens exist, the total resistance of the network is

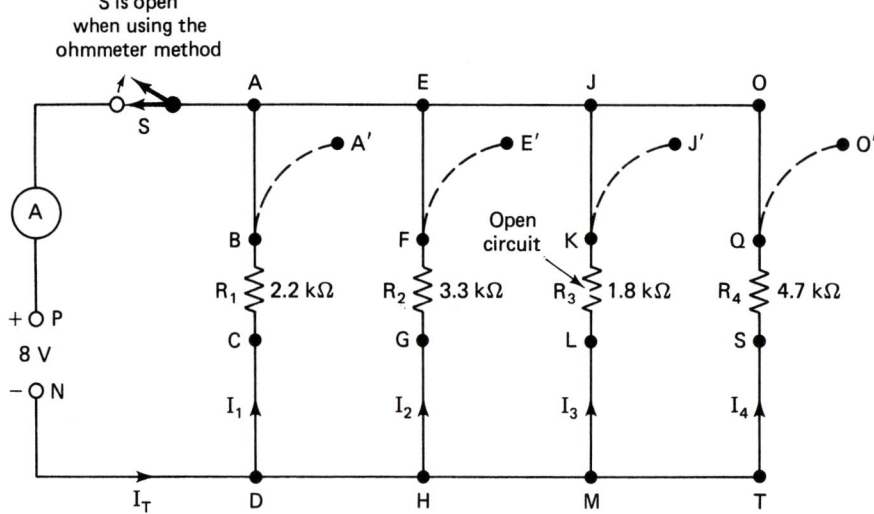

FIGURE 3-19 Ohmmeter method of locating the position of an open in a parallel circuit.

$$\frac{1}{\frac{1}{2.2} + \frac{1}{3.3} + \frac{1}{1.8} + \frac{1}{4.7}} \text{ k}\Omega = 655 \text{ }\Omega$$

and the reading of the current meter, A, is 8 V/655 Ω = 12.21 mA. If the reading of A drops to zero, an open circuit must exist in one or both of the connecting wires PA and ND; a less likely possibility is that all the parallel branches have open-circuited simultaneously.

If the reading of A does not drop to zero but merely decreases to a lower level, it is possible that an open has occurred in one (or more) of the parallel branches. Remember that the same currents continue to flow in the remaining branches, so that the decrease in the source current must equal the total amount of current that was flowing in the branch or branches that are now open.

In our circuit we will assume that the reading of A falls to 7.76 mA. The decrease of current is therefore 12.21 − 7.76 = 4.45 mA, so that the branch resistance (which is now open-circuited) had a value of 8 V/4.45 mA = 1.8 kΩ under normal conditions. Clearly, there is an open circuit in the branch that contains resistor R_3.

One method of locating the open is to use an ohmmeter. As a first step, we must open the switch, S, so that all power is removed from the circuit. It is then tempting to connect the ohmmeter between points J and M in the expectation that the reading will be infinite ohms. But such a procedure is worthless because the ohmmeter is also connected across $R_1 \| R_2 \| R_4$ and its reading would in fact be

$$\frac{1}{\frac{1}{2.2} + \frac{1}{3.3} + \frac{1}{4.7}} \text{ k}\Omega = 1031 \text{ }\Omega$$

The same reading would also exist between points A and D, points E and H, and points O and T.

Since when troubleshooting, we have no idea where the open circuit is, we will proceed to test each resistor in turn. Starting with the branch containing resistor R_1, disconnect lead BA from point A. Referring to the disconnected

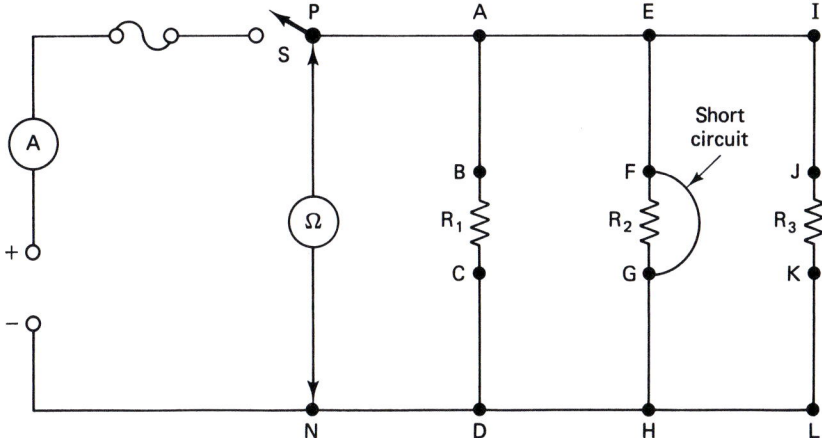

FIGURE 3-20 Ohmmeter method of locating the position of a short in a parallel circuit.

end as A', measure the resistance between point A' and point D. The ohmmeter's reading is close to the rated value of the resistor, so that the open does not exist in branch R_1 and A' is therefore reconnected to point A. The procedure is then repeated for branch R_2 and again no open circuit is found. The procedure is again repeated for the R_3 branch. However, when the ohmmeter is connected between J' and point M, the reading will be infinite ohms, so the open exists in branch R_3. After correcting the fault, branch R_4 should also be checked in case more than one open circuit exists.

A more cumbersome method of finding the open circuit is to use a current meter. After opening the switch, S, lead AB is removed and a milliammeter is inserted between these points. When the switch is closed, the milliammeter records the normal current and no open exists in branch R_1. The same results would be obtained for branch R_2 but when the milliammeter is inserted between points J and K, its reading is zero, which indicates the presence of the open circuit in branch R_3.

Locating a Short Circuit

As we learned in Section 3-10, a short circuit across any parallel branch provides an extremely high value of the current drawn from the source, which is, hopefully, protected by either a fuse or a magnetic circuit breaker. The problem is therefore to locate the short, remove it, then restore the circuit to normal operation.

Referring to Fig. 3-20, the short exists between points F and G. With switch S open, connect an ohmmeter between points P and N. Because of the presence of the short, the ohmmeter reading will be virtually zero. Start the troubleshooting procedure by disconnecting lead BA from point A. Since the short is still in parallel with the ohmmeter, the reading of the meter remains at zero; lead BA is therefore soldered back to point A. However, when lead FE is disconnected from point E, the short is no longer across the ohmmeter, so that the resistance reading rises to a value that is the result of R_1 in parallel with R_3. The short has been located between points E and H and the necessary repair can be carried out.

EXAMPLE 3-17 In Fig. 3-21 the reading of milliammeter A falls from its normal value to 3.23 mA. What fault has occurred in the circuit?

Sec. 3-11 Troubleshooting a Parallel Circuit

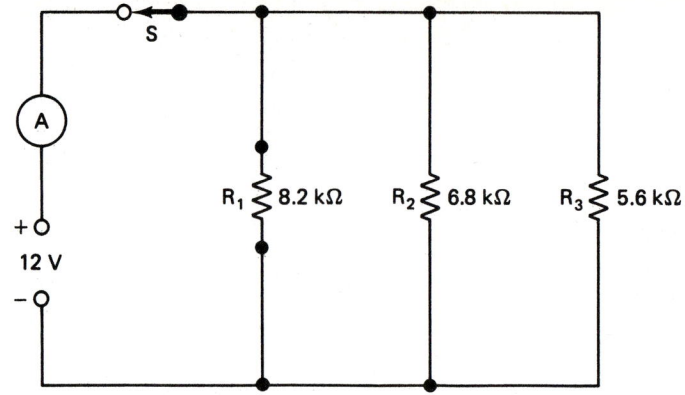

FIGURE 3-21 Circuit for Example 3-17.

Solution Before the fault, the total resistance of the circuit is

$$\frac{1}{\frac{1}{8.2} + \frac{1}{6.8} + \frac{1}{5.6}} \text{ k}\Omega = 2.23 \text{ k}\Omega$$

and the reading of A is 12 V/2.23 kΩ = 5.37 mA. The decrease in current is therefore 5.37 − 3.23 = 2.14 mA, which corresponds to a branch resistance of 12 V/2.14 mA = 5.6 kΩ. Consequently, the fault is an open circuit in resistor R_3.

Chapter Summary

3.1. In any parallel circuit the voltage drop across each branch is the same and equal to the source voltage.

3.2. The sum of the individual branch currents flowing through the resistors is equal to the value of the total current drawn from the voltage source (Kirchhoff's current law).

$$I_T = I_1 + I_2 + I_3 + \cdots + I_N$$

where

$$I_1 = \frac{E_T}{R_1}, \quad I_2 = \frac{E_T}{R_2}, \quad \ldots, \quad I_N = \frac{E_T}{R_N}$$

3.3. The reciprocal of the total equivalent resistance is equal to the sum of the reciprocals of the individual branch resistances

$$\frac{1}{R_T} = \frac{1}{R_1} + \frac{1}{R_2} + \frac{1}{R_3} + \cdots + \frac{1}{R_N}$$

or

$$R_T = \frac{E_T}{I_T} = \frac{1}{\frac{1}{R_1} + \frac{1}{R_2} + \frac{1}{R_3} + \cdots + \frac{1}{R_N}}$$

For two resistors in parallel,

$$R_T = \frac{R_1 R_2}{R_1 + R_2} \quad \text{(product-over-sum formula)}$$

For equal resistors in parallel,

$$R_T = \frac{R}{N}$$

3.4. For conductances in parallel the total equivalent conductance is the sum of the individual branch conductances.

$$G_T = G_1 + G_2 + G_3 + \cdots + G_N = \frac{I_T}{E_T}$$

3.5. The total power delivered from the source is equal to the sum of the powers dissipated in the individual branches.

$$P_T = P_1 + P_2 + P_3 + \cdots + P_N$$
$$= I_T^2 R_T = \frac{E_T^2}{R_T} = I_T E_T$$

where

$$P_1 = I_1^2 R_1 = \frac{E_T^2}{R_1} = I_1 E_T, \quad \text{etc.}$$

3.6. The total source current divides among the parallel resistors in inverse proportion to the resistance values (current division rule).

$$I_1 = I_2 \times \frac{R_2}{R_1} = I_T \times \frac{R_T}{R_1} = I_T \times \frac{G_1}{G_T}, \quad \text{etc.}$$

For two resistors in parallel,

$$I_1 = I_T \times \frac{R_2}{R_1 + R_2}, \quad I_2 = I_T \times \frac{R_1}{R_1 + R_2}$$

The total power delivered from the voltage source is also divided among the parallel resistors in inverse proportion to the resistance values.

$$P_1 = P_2 \times \frac{R_2}{R_1} = P_T \times \frac{R_T}{R_1}, \quad \text{etc.}$$

3.7. When identical cells are connected in parallel, the EMF available is equal to the EMF of one cell. The total load current is then shared equally among the cells.

3.8. When an open circuit occurs in a particular branch, the current in that branch is zero, but the currents in all the other branches remain the same. However, if the break occurs in the main line which is connected to the voltage source, all the currents in the circuit fall to zero.

3.9. A short circuit is a path of zero resistance so that a current can flow through the short circuit without developing any voltage drop. If a short occurs across any one of a number of loads in parallel, all the loads are shorted out of the circuit, so that their currents are then zero. A very high current will be drawn from the voltage source, which may be protected by either a fuse or a magnetic circuit breaker.

Chapter Summary

3.10. Comparison between series and parallel circuits:

	Series Circuit	Parallel Circuit
Source voltage, E_T (volts)	The sum of the individual voltage drops is equal to the source voltage: $$E_T = V_1 + V_2 + V_3 + \cdots + V_N$$	The voltage across each parallel branch is the same and equal to the source voltage: $$E_T = V_1 = V_2 = V_3 = \cdots = V_N$$
Source current, I_T (amperes)	The current is the same through each series component and is therefore equal to the source current.	The sum of the individual branch currents is equal to the source current: $$I_T = I_1 + I_2 + I_3 + \cdots + I_N$$
Total equivalent resistance, R_T (ohms)	The total equivalent resistance is the sum of the individual resistances: $$R_T = R_1 + R_2 + R_3 + \cdots + R_N$$ The value of R_T is greater than the highest-value resistor in series.	The total equivalent resistance is found by the use of the reciprocal formula: $$R_T = \frac{1}{\frac{1}{R_1} + \frac{1}{R_2} + \frac{1}{R_3} + \cdots + \frac{1}{R_N}}$$ The value of R_T is less than the lowest-value resistor in parallel. Two resistors in parallel: $$R_T = \frac{R_1 R_2}{R_1 + R_2}$$
Total equivalent conductance, G_T (siemens)	The total equivalent conductance is found by the use of the reciprocal formula: $$G_T = \frac{1}{\frac{1}{G_1} + \frac{1}{G_2} + \frac{1}{G_3} + \cdots + \frac{1}{G_N}}$$	The total equivalent conductance is the sum of the individual branch conductances: $$G_T = G_1 + G_2 + G_3 + \cdots + G_N$$
Total power, P_T (watts)	The total power delivered from the source is equal to the sum of the individual powers dissipated in the resistors: $$P_T = P_1 + P_2 + P_3 + \cdots + P_N$$	The total power delivered from the source is equal to the sum of the individual powers dissipated in the resistors: $$P_T = P_1 + P_2 + P_3 + \cdots + P_N$$
Division rules	*Voltage division:* $$V_1 = V_2 \times \frac{R_1}{R_2} = E_T \times \frac{R_1}{R_T}$$ $$= E_T \times \frac{G_T}{G_1}$$ *Power division:* $$P_1 = P_2 \times \frac{R_1}{R_2} = P_T \times \frac{R_1}{R_T}$$ $$= P_T \times \frac{G_T}{G_1}$$ The highest-value resistor carries the greatest voltage drop and dissipates the most power.	*Current division:* $$I_1 = I_2 \times \frac{R_2}{R_1} = I_T \times \frac{R_T}{R_1}$$ $$= I_T \times \frac{G_1}{G_T}$$ *Power division:* $$P_1 = P_2 \times \frac{R_2}{R_1} = P_T \times \frac{R_T}{R_1}$$ $$= P_T \times \frac{G_1}{G_T}$$ The lowest-value resistor carries the highest branch current and dissipates the most power.

	Series Circuit	Parallel Circuit
Effect of an open circuit	Current throughout the circuit is zero. Voltage drop across each resistor is zero. Power dissipated in each resistor and the total power delivered from the source are zero. The source voltage appears across the open circuit.	If the open circuit occurs in a particular branch, the branch current is zero, but all other branch currents remain the same. The source current and the total power delivered from the source are both reduced. If the open circuit occurs in the main line connected to the voltage source, all branch currents are zero and therefore the total source current is zero and the total power is zero.
Effect of a short circuit	The voltage drop across the shorted resistor(s) is zero. The total equivalent resistance is less and therefore the source current is greater. The total power delivered from the source is increased.	If a short circuit occurs across any parallel branch, all branch currents are reduced to zero. A very large source current will then flow through the short circuit. If the source is not protected by a fuse or circuit breaker, damage may occur.

Self-Review

True–False Questions

1. T. F. In a circuit containing a parallel bank of resistors there is only a single path available for current flow.
2. T. F. In a parallel circuit the lowest-value resistor dissipates the most power and carries the highest current.
3. T. F. A circuit consists of two parallel resistors whose values are 20 kΩ and 5 kΩ. The total equivalent resistance of the circuit is 4 kΩ.
4. T. F. The current division rule states that the total source current divides between the resistors in direct proportion to the resistance values.
5. T. F. A 1-kΩ resistor and a 2000-Ω resistor are connected in parallel across a voltage source. The voltage drop across the 2000-Ω resistor is equal to twice the voltage drop across the 1-kΩ resistor.
6. T. F. Five identical 10-V cells are connected in parallel. The total EMF available is 50 V.
7. T. F. Three resistors, whose values are 60 Ω, 30 Ω, and 20 Ω, are connected in parallel across a 120-V source. The total source current is 12 A.
8. T. F. In Question 7, a short circuit occurs across the 30-Ω resistor. As a result, the current in the 60-Ω resistor falls to zero.
9. T. F. In Question 7, the 20-Ω resistor open-circuits. The source current then drops to 4 A.
10. T. F. In Question 7, the power dissipated in the 60-Ω resistor is three times the power dissipated in the 20-Ω resistor.
11. T. F. The total resistance of a parallel network of resistors is always less than that of the lowest-value resistor.
12. T. F. Kirchhoff's voltage law does not apply to a parallel arrangement of the resistors that are conneced to a dc voltage source.
13. T. F. In a parallel network of resistors, equal currents flow in the individual branches.
14. T. F. If a number of resistors are connected in parallel across a dc voltage, the source voltage is divided between the resistors in proportion to their resistance values.
15. T. F. An open circuit in any branch of a parallel arrangement of resistors causes the total equivalent resistance to increase.

16. T. F. A short circuit across any branch of a parallel arrangement of resistors causes a high value of source current and reduces the other branch currents to zero.
17. T. F. The reciprocal formula must be used to determine the total equivalent conductance of a parallel arrangement of resistors, whose conductance values are known.
18. T. F. Two resistors are connected in parallel across a dc voltage source. If the resistance of one resistor is doubled, the current through the other resistor will be halved.
19. T. F. If three 12-Ω resistors are connected in parallel, the total equivalent resistance is 36 Ω.
20. T. F. In a parallel arrangement of resistors the total power dissipated is always equal to the sum of the individual power ratings.

Multiple-Choice Questions

21. A parallel circuit consists of a 60-V battery and five resistors whose values are 20 Ω, 12 Ω, 6 Ω, 5 Ω, and 2 Ω. The total equivalent resistance of the circuit is
    (a) 45 Ω (b) 20 Ω (c) 2 Ω (d) 1 Ω (e) 0.5 Ω
22. In Question 21, the value of the total circuit current drawn from the battery is
    (a) 1.25 A (b) 3 A (c) 60 A (d) 30 A (e) 120 A
23. In Question 21, the current through the 12-Ω resistor is
    (a) 5 A (b) 3 A (c) 15 A (d) 55 A (e) 12 A
24. In Question 21, the value of the voltage drop across the 2-Ω resistor is
    (a) 2.67 V (b) 30 V (c) 20 V (d) 45 V (e) 60 V
25. In Question 21, the power dissipated in the 5-Ω resistor is
    (a) 300 W (b) 720 W (c) 12 W (d) 1800 W (e) 150 W
26. In Question 21, the total power dissipated in the circuit is
    (a) 3600 W (b) 1800 W (c) 900 W (d) 7200 W (e) 2400 W
27. In Question 21, the 2-Ω resistor open-circuits. The new total conductance of the circuit is
    (a) 2 S (b) 1 S (c) 0.5 S (d) 0.25 S (e) 1.5 S
28. In Question 21, a short circuit develops across the 5-Ω resistor. The power dissipated in the 12-Ω resistor is then
    (a) theoretically infinite (b) 720 W (c) 1440 W (d) 360 W
    (e) zero
29. In Question 21, the 60-V battery is joined in parallel with another identical 60-V battery. The current supplied by the new battery is
    (a) 2.5 A (b) 3 A (c) 60 A (d) 30 A (e) 15 A
30. In Question 21, an open circuit occurs in the main line connected to the battery. The current flowing through the 6-Ω resistor is then
    (a) 10 A (b) 5 A (c) 20 A (d) zero
    (e) theoretically infinite
31. A 470-Ω resistor and a 330-Ω resistor are connected in parallel across a dc voltage source. Their total equivalent resistance is
    (a) 800 Ω (b) 274 Ω (c) 194 Ω (d) 174 Ω (e) 140 Ω
32. In Question 31, a current of 19 mA flows through the 470-Ω resistor. The current flowing through the 330-Ω resistor is
    (a) 19 mA (b) 27 mA (c) 13.3 mA (d) 46 mA (e) 30.3 mA
33. In Question 32, the total power dissipated in the parallel combination is
    (a) 372 mW (b) 170 mW (c) 241 mW (d) 411 mW (e) 438 mW
34. In Question 32, the 330-Ω resistor open-circuits. The new total current drawn from the source is
    (a) 27 mA (b) 46 mA (c) 19 mA (d) zero (e) 12 mA
35. In Question 32, a short circuit develops across the 330-Ω resistor. The new current flowing through the 470-Ω resistor is
    (a) zero (b) 19 mA (c) 12 mA (d) 46 mA
    (e) theoretically infinite
36. A parallel arrangement of 150-Ω resistors is connected across a 200-Ω resistor.

How many of the 150-Ω resistors are necessary in order to produce a total equivalent resistance of 40 Ω?

(a) 2   (b) 3   (c) 4   (d) 5   (e) 1

37. One thousand 1-kΩ 20-W resistors are connected in parallel across a 100-V dc supply. The total equivalent resistance is

(a) 1 kΩ   (b) 1 MΩ   (c) 100 Ω   (d) 1 Ω   (e) 10 Ω

38. In Question 37, the current flowing through each branch is

1 A   (b) 1 mA   (c) 10 mA   (d) 100 mA   (e) 10 A

39. The total power dissipated in the circuit of Question 37 is

(a) 20 kW   (b) 1 kW   (c) 20 mW   (d) 20 W   (e) 10 kW

40. In Question 37, one of the 1-kΩ resistors open-circuits. The new total current is

(a) 99.9 A   (b) 9.99 A   (c) 999 A   (d) 999 mA   (e) 99.9 mA

Practice Problems

41. Three resistors are connected in parallel across a voltage source. The color coding of these resistors is brown–black–red–silver, gray–red–brown–silver, and blue–gray–brown–silver. What are the total equivalent rated resistance and conductance values of this parallel bank?

42. Three resistors, whose values are 2.2 kΩ, 3.3 kΩ, and 4.7 kΩ, are connected in parallel across a 25-V source. What are the values of the total current drawn from the source and the total power dissipated?

43. A 33-kΩ resistor and a 47-kΩ resistor are connected in parallel across a voltage source. If the total current drawn from the source is 2.7 mA, what is the current flowing in the 47-kΩ resistor and the power dissipated in the 33-kΩ resistor?

44. Five identical resistors are connected in parallel across an 80-V source. If the total power dissipated is 70 W, what is the value of each resistor?

45. Four 10-W 100-Ω resistors are connected in parallel. What is the total power dissipation capability of this combination?

46. Three identical 12-V batteries are connected in parallel across a load which consists of 1.2-kΩ, 2.2-kΩ, and 3.3-kΩ resistors in parallel. What is the value of the current supplied by each battery?

47. A 20,000-Ω 200-W resistor, a 40,000-Ω 100-W resistor and a 5000-Ω 20-W resistor are connected in parallel. What is the maximum value of the source voltage that will not cause the wattage rating of any of the resistors to be exceeded?

48. The values of three parallel resistors are 330 Ω, 470 Ω, and 560 Ω. If the current through the 470-Ω resistor is 8 mA, what is the value of the total current drawn from the source?

49. In Problem 48, the 560-Ω resistor open-circuits. What is the new total power dissipated by the circuit?

50. Three resistors in parallel have a total equivalent resistance of 2.1 kΩ. If two of the resistors have values of 6.8 kΩ and 8.2 kΩ, what is the value of the third resistor?

51. Two resistors in parallel possess a total conductance of 55 μS. If the value of one resistor is 39 kΩ, what is the value of the other resistor?

52. A 120-V dc source is connected through a 15-A fuse to a number of parallel-connected 120-V 25-W lamps. What is the maximum possible number of lamps that can be safely connected?

53. Five equal-value resistors are connected in parallel with a 2-kΩ resistor to lower the overall resistance to 400 Ω. What is the value of each resistor?

54. Two resistors, whose conductances are 15 mS and 20 mS, are connected in parallel across a dc voltage source. If the source current is 600 mA, determine the current flowing through each resistor.

55. What is the total equivalent conductance of three parallel resistors whose values are 150 Ω, 270 Ω, and 390 Ω?

56. Four 120-V 100-W lamps are connected in parallel across a 120-V dc source. What is the total equivalent "hot" resistance of the parallel combination?

57. The total equivalent resistance of four resistors in parallel is 245 Ω. If the values of three of the resistors are 820 Ω, 1 kΩ, and 1.2 kΩ, what is the value of the fourth resistor? If a short circuit develops across the 1-kΩ resistor, what is the value of the new total equivalent resistance?

58. Three resistors, whose values are 15 kΩ, 22 kΩ, and 33 kΩ, are connected in parallel. If the current flowing through the 22-kΩ resistor is 14 mA, calculate the amount of the total power dissipated in the circuit.

59. In Problem 58, the 22-kΩ resistor open-circuits. What is the new value of the total power dissipated?

60. Three lamps, whose ratings are 120 V 40 W, 120 V 60 W, and 120 V 100 W, are connected in parallel across a 120-V dc supply. If the total current must not exceed 3 A, what is the maximum power rating of a fourth lamp?

Advanced Problems

61. In Fig. 3-22, $R_1 = 3.9$ kΩ, $I_1 = 5.7$ mA, and $R_2 = 4.7$ kΩ. Find the values of I_2 and I_3 and the total power dissipated.

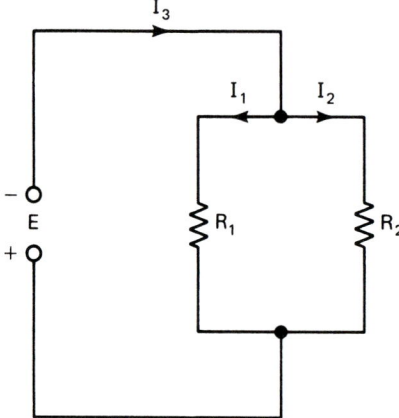

FIGURE 3-22 Circuit for Problems 61, 62, and 63.

62. In Fig. 3-22, $R_1 = 560$ kΩ, $E = 185$ V, and $I_3 = 3.7$ mA. What is the value of R_2?

63. In Fig. 3-22, $I_1 = 9.8$ mA, $R_2 = 510$ Ω, and $I_3 = 15.3$ mA. What are the values of R_1 and R_T?

64. In Fig. 3-23, what are the values of R_2 and the total power dissipated?

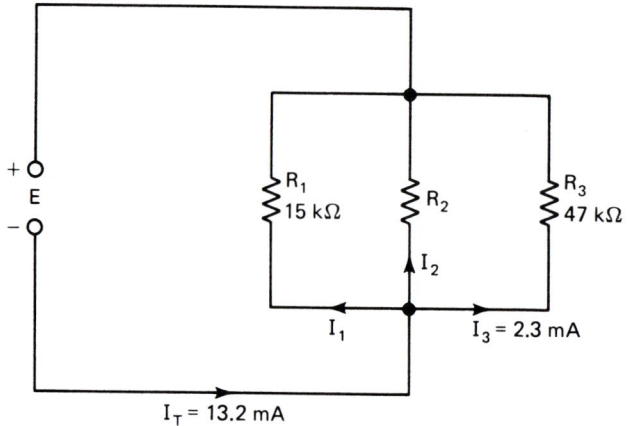

FIGURE 3-23 Circuit for Problems 64 and 65.

65. In Fig. 3-23, an additional 33-kΩ resistor is added in parallel with the circuit. What is the new value of I_T?

66. In Fig. 3-24, determine the values of R_1 and R_2.

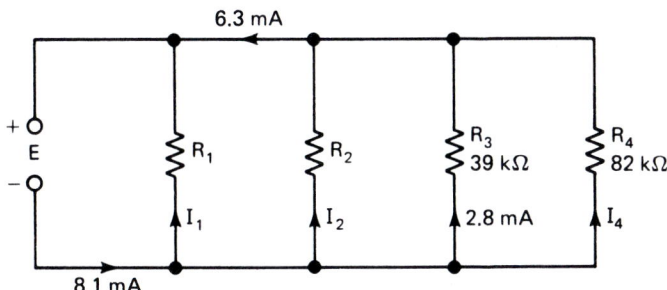

FIGURE 3-24 Circuit for Problem 66.

67. In Fig. 3-25, the power dissipated in the 3.3-kΩ resistor is 2 W. What is the total power dissipated in the circuit?

FIGURE 3-25 Circuit for Problem 67.

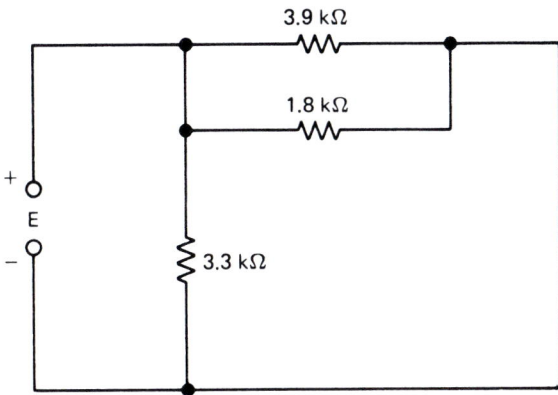

68. In Fig. 3-26, a short circuit develops between points C and D. Calculate the new values of the potentials at points A, B, C, D, and E.

FIGURE 3-26 Circuit for Problem 68.

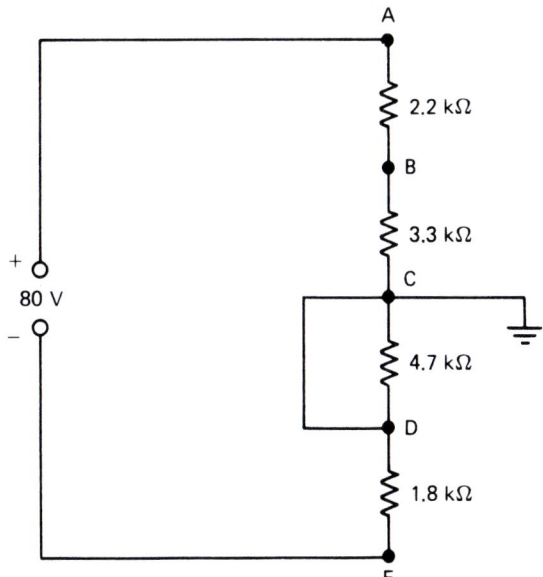

69. In Fig. 3-27, the total power dissipated is 2.5 W. Find the value of the current I.

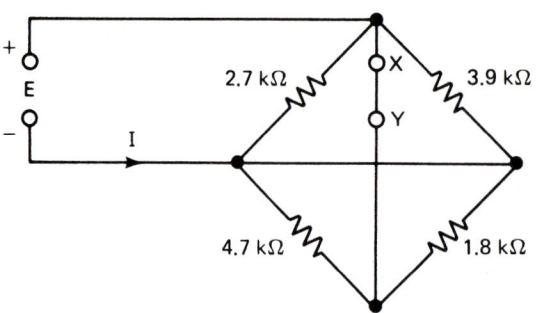

FIGURE 3-27 Circuit for Problems 69 and 70.

70. In Problem 69, an open circuit occurs between points X and Y. Calculate the new value of the total dissipated power.

71. A 100-Ω 10-W resistor and a 50-Ω 10-W resistor are connected in parallel across a dc voltage source. What is the maximum power that this circuit can safely dissipate without exceeding the power rating of either resistor?

72. A resistor whose conductance is 650 μS is connected in parallel with another resistor whose conductance is 1.4 mS. What is the total equivalent resistance of this combination?

73. Three 120-V light bulbs, whose power ratings are respectively 60 W, 100 W, and 150 W, are connected in parallel across a 120-V dc source. What is the total equivalent "hot" resistance of this combination?

74. The sum of the branch currents is 8 mA for three resistors in parallel. The voltage drop across the first resistor is 16 V, the second resistor has a value of 12 kΩ, and the third resistor dissipates 40 mW. What are the resistance values of the first and third resistors?

75. What is the power rating of the highest-value resistor that can be connected in parallel with a 12-kΩ 1-W resistor, so that the combination can accommodate a total current of 20 mA?

76. A dc voltage is applied across three parallel-connected resistors that dissipate a total power of 400 mW. The total current is 8 mA, the current flowing through the first resistor is 4 mA, and the second and third resistors are equal in value. Find the resistance values of all three resistors.

77. In Fig. 3-28, the battery voltage drops from 12 V to 11 V. Calculate the new value of R that will restore the battery current to its initial level.

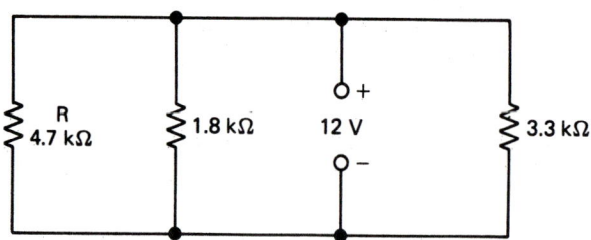

FIGURE 3-28 Circuit for Problem 77.

78. In Fig. 3-29, calculate the value of the resistor, R.

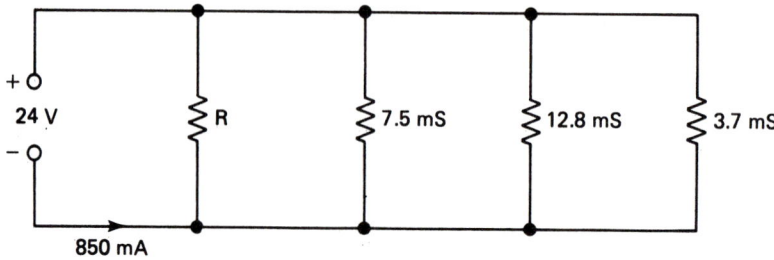

FIGURE 3-29 Circuit for Problem 78.

79. In Fig. 3-30, what is the value of the voltage, E?

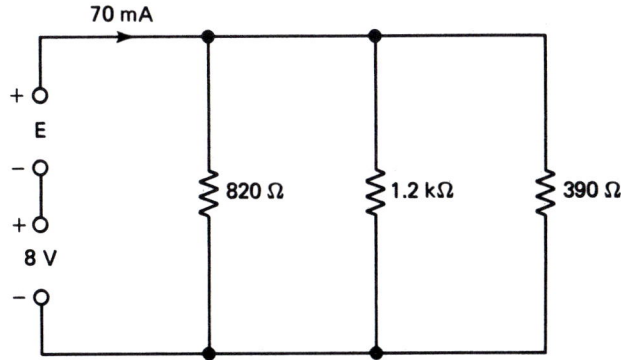

FIGURE 3-30 Circuit for Problems 79 and 80.

80. In Problem 79, the voltage E is replaced by a short circuit. Calculate the resistance value of the additional parallel resistor required to raise the level of the source curent to 100 mA.

4

Series-Parallel Arrangements of Resistors

As we discuss the properties of series–parallel circuits, you will learn:

1. how to analyze simple series–parallel networks that contain only three resistors.
2. how to calculate the total equivalent resistance of a series–parallel circuit.
3. to use a step-by-step method of analyzing complex series–parallel networks.
4. what happens when the position of a ground connection is changed in a series–parallel circuit.
5. what a Wheatstone bridge circuit is and how it is used to obtain an accurate measurement of resistance.
6. how a voltage divider circuit is affected by the resistive loads to which it is connected.
7. how a series–parallel circuit is affected by the presence of open circuits.
8. how a series–parallel circuit is affected by the presence of short circuits.
9. how to troubleshoot a series–parallel circuit.

4-0

Introduction

In Chapter 2 we learned that for a series string of resistors the current was the same throughout the circuit and that the source voltage was divided among the resistors. By contrast, Chapter 3 was concerned with parallel banks of resistors in which the source voltage was applied across each resistor and the source current was divided among the various branches. However, series strings and parallel banks may be combined to produce more complex networks, which are referred to as *series–parallel circuits*. There is an infinite variety of such circuits, but the two simplest are shown in Fig. 4-1.

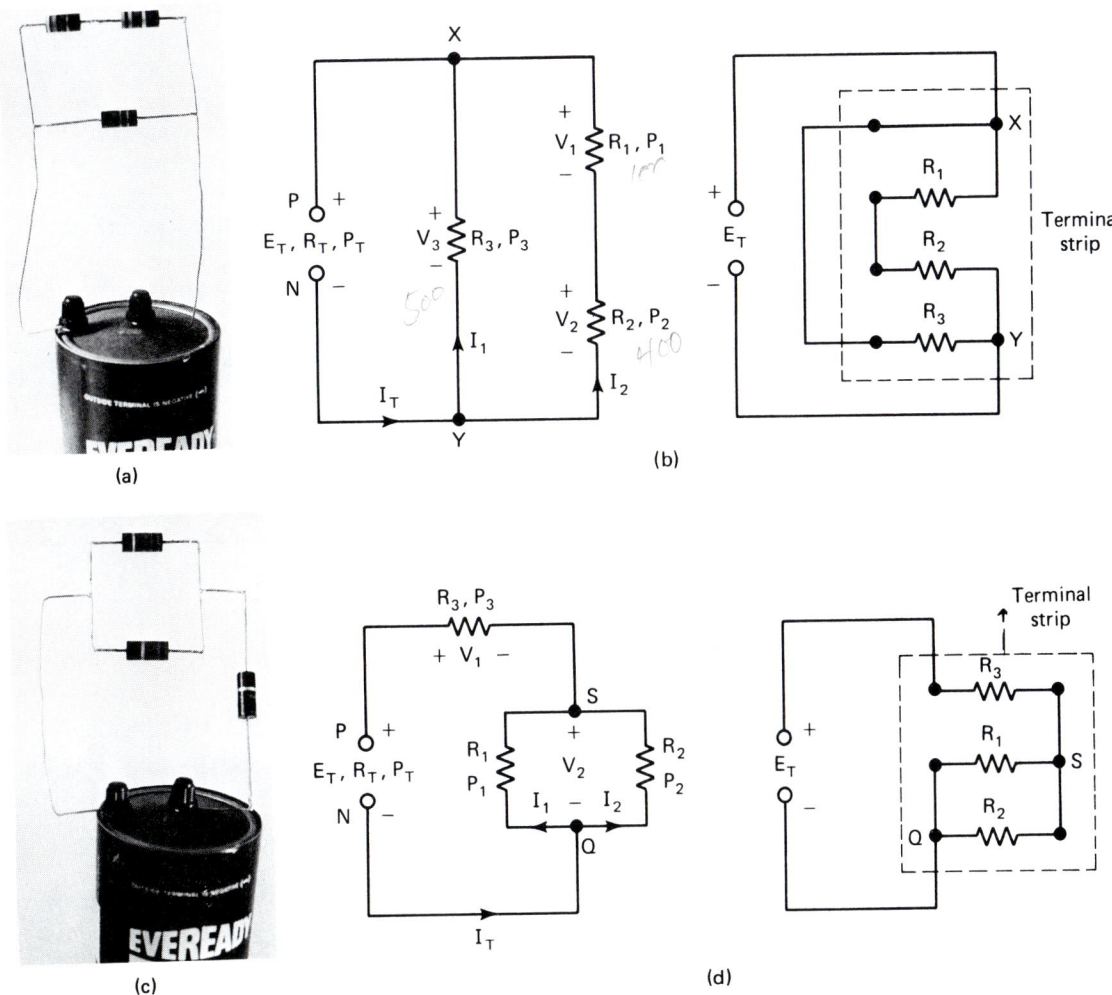

FIGURE 4-1 Examples of series–parallel circuits.

Sec. 4-0   Introduction

157

The resistors R_1 and R_2 of Fig. 4-1(a) and (b) are joined end to end, so that the same current flows through each of these resistors. However, the current splits at point Y and recombines at point X, so that R_3 is in parallel with the series combination of R_1 and R_2.

In Fig. 4-1(c) and (d) the current splits at point Q and recombines at point S, so that R_1 and R_2 are in parallel. However, if R_1 and R_2 are replaced by their equivalent resistance R_{12}, this resistance would be in series with R_3.

It is the purpose of this chapter to explore the properties of these circuits and to develop a step-by-step procedure for the analysis of the more complex series–parallel networks.

4-1 Circuit Containing Two Resistors in Series with a Third Resistor in Parallel

Referring again to Fig. 4-1(b), we can start at point N and consider that the current or electron flow is leaving the negative terminal of the voltage source. When the current, I_T, reaches the point, Y, it is faced with two possible paths and splits into I_1 and I_2 while obeying Kirchhoff's current law and the current division rule. At point X the two branch currents, I_1 and I_2, recombine to reestablish the total current, I_T, which then flows to the positive terminal, P, of the voltage source. As far as the currents are concerned, we have only used the properties of a parallel circuit.

The whole of the source voltage will be dropped across R_3 but will be divided between R_1 and R_2 in accordance with the voltage division rule, which is applicable to a series string. The complete analysis of the circuit will therefore not involve any new ideas but will require the most effective use of the series and parallel concepts introduced in Chapters 2 and 3.

Normally, the first step is to find the total equivalent resistance of the complete circuit. Since R_1 and R_2 are in series, their equivalent resistance, R_{12} is $R_1 + R_2$. Then R_{12} is in parallel with R_3, so that the total equivalent resistance, R_T, of the entire circuit is

$$R_T = \frac{R_{12} \times R_3}{R_{12} + R_3} \quad \text{(product-over-sum formula)}$$

$$= \frac{(R_1 + R_2) \times R_3}{R_1 + R_2 + R_3} \tag{4-1-1}$$

Then the total current, I_T, drawn from the source is

$$I_T = \frac{E_T}{R_T}$$

The whole of the supply voltage is dropped across R_3, so that $E_T = V_3$ and

$$I_1 = \frac{E_T}{R_3}$$

Since I_T has split into I_1 and I_2, then by Kirchhoff's current law,

$$I_T = I_1 + I_2 \text{ and } I_2 = I_T - I_1 \tag{4-1-2}$$

I_2 is flowing through the series combination of R_1 and R_2; in addition, E_T will be split between R_1 and R_2 in accordance with the voltage division rule. Therefore

$$V_1 = I_2 R_1 = E_T \times \frac{R_1}{R_1 + R_2}, \quad V_2 = I_2 R_2 = E_T \times \frac{R_2}{R_1 + R_2}$$

and

$$E_T = V_1 + V_2 \qquad (4\text{-}1\text{-}3)$$

The powers dissipated in the individual resistors may be found from the $P = IV$ relationship. The total power delivered from the source is the sum of the individual powers dissipated so that

$$P_T = E_T I_T = P_1 + P_2 + P_3 \qquad (4\text{-}1\text{-}4)$$

To summarize, the circuit analysis for Fig. 4-1(b) involved the following steps:

1. Combine the series resistors R_1 and R_2 into their equivalent resistance. Using the product-over-sum formula for this equivalent resistance and R_3, calculate the value of R_T.
2. Find $I_T (= E_T/R_T)$ and then calculate the values of the branch currents.
3. Determine the individual voltage drops, the powers dissipated in the individual resistors, and the total power delivered from the source.

EXAMPLE 4-1 In Fig. 4-1(b), $R_1 = 4.7 \text{ k}\Omega$, $R_2 = 3.3 \text{ k}\Omega$, and $R_3 = 10 \text{ k}\Omega$. If $E_T = 12$ V, find the values of R_T, I_T, I_1, I_2, V_1, V_2, V_3, P_1, P_2, P_3, and P_T.

Solution

$$\text{total equivalent resistance, } R_T = \frac{(R_1 + R_2) \times R_3}{R_1 + R_2 + R_3} \qquad (4\text{-}1\text{-}1)$$

$$= \frac{(4.7 + 3.3) \times 10}{4.7 + 3.3 + 10}$$

$$= \frac{80}{18} = 4.44 \text{ k}\Omega$$

$$\text{total current, } I_T = \frac{12 \text{ V}}{4.44 \text{ k}\Omega} = 2.7 \text{ mA}$$

$$\text{branch current, } I_1 = \frac{12 \text{ V}}{10 \text{ k}\Omega} = 1.2 \text{ mA}$$

$$\text{branch current, } I_2 = I_T - I_1 \qquad (4\text{-}1\text{-}2)$$

$$= 2.7 - 1.2 = 1.5 \text{ mA}$$

$$\text{voltage drop, } V_1 = I_2 \times R_1 = 1.5 \text{ mA} \times 4.7 \text{ k}\Omega = 7.05 \text{ V}$$

$$V_2 = I_2 \times R_2 = 1.5 \text{ mA} \times 3.3 \text{ k}\Omega = 4.95 \text{ V}$$

Voltage Check

Make certain that the sum of V_1 and V_2 exactly balances the source voltage, E_T.

$$V_1 + V_2 = 7.05 + 4.95 = 12.0 \text{ V} = E_T$$

power dissipated, $P_1 = I_2 V_1 = 1.5 \text{ mA} \times 7.05 \text{ V} = 10.575 \text{ mW}$

$$P_2 = I_2 V_2 = 1.5 \text{ mA} \times 4.95 \text{ V} = 7.425 \text{ mW}$$

$$P_3 = I_1 V_3 = I_1 E_T = 1.2 \text{ mA} \times 12 \text{ V} = 14.4 \text{ mW}$$

power, P_T, delivered from the source $= E_T I_T$

$$= 12 \text{ V} \times 2.7 \text{ mA} = 32.4 \text{ mW} \qquad (4\text{-}1\text{-}4)$$

Power Check

$$P_1 + P_2 + P_3 = 10.575 + 7.425 + 14.4 = 32.4 \text{ mW} = P_T$$

4-2

Circuit Containing Two Resistors in Parallel with a Third Resistor in Series

Referring back to Fig. 4-1(d), we can again start at point N, so that the electron current, I_T, leaves the negative terminal of the voltage source and reaches point Q. Since there are two possible paths available, the current splits into I_1 and I_2, which then recombine at point S; the same total current will therefore flow through R_3 to the positive terminal at the point, P.

From the description of the current flow you have seen that R_1 and R_2 are in parallel and that R_3 is in series with this parallel combination. Using the product-over-sum formula, the equivalent resistance of R_1 and R_2 is

$$R_{12} = \frac{R_1 \times R_2}{R_1 + R_2}$$

and the total resistance, R_T, of the complete circuit is

$$R_T = R_{12} + R_3 = \frac{R_1 R_2}{R_1 + R_2} + R_3$$

$$= \frac{R_1 R_2 + R_2 R_3 + R_3 R_1}{R_1 + R_2} \qquad (4\text{-}2\text{-}1)$$

Then the total current, I_T, drawn from the voltage source is

$$I_T = \frac{E_T}{R_T} = \frac{E_T \times (R_1 + R_2)}{R_1 R_2 + R_2 R_3 + R_3 R_1} \qquad (4\text{-}2\text{-}2)$$

Using the current division rule (Section 3-7), we obtain

$$I_1 = I_T \times \frac{R_2}{R_1 + R_2} = \frac{E_T \times R_2}{R_1 R_2 + R_2 R_3 + R_3 R_1} \qquad (4\text{-}2\text{-}3)$$

and

$$I_2 = I_T \times \frac{R_1}{R_1 + R_2} = \frac{E_T \times R_1}{R_1R_2 + R_2R_3 + R_3R_1} \qquad (4\text{-}2\text{-}4)$$

Also,

$$I_T = I_1 + I_2, \quad I_1 = \frac{V_2}{R_1} = I_T - I_2, \quad I_2 = \frac{V_2}{R_2} = I_T - I_1 \qquad (4\text{-}2\text{-}5)$$

Since the current, I_T, flows through R_3, the voltage drop across this resistor is

$$V_1 = I_T R_3 \qquad (4\text{-}2\text{-}6)$$

The sum of the voltages V_1 and V_2 must exactly balance the applied voltage, E_T. Therefore,

$$E_T = V_1 + V_2$$

Using the voltage division rule (Section 2-6), we obtain

$$V_1 = E_T \times \frac{R_3}{R_T} \quad \text{and} \quad V_2 = I_1 R_1 = I_2 R_2 = E_T \times \frac{R_{12}}{R_T}$$
$$= E_T \times \frac{R_1 R_2}{R_1 R_2 + R_2 R_3 + R_3 R_1} \qquad (4\text{-}2\text{-}7)$$

The powers dissipated in the individual resistors may be calculated from the basic Ohm's law relationships. The total power from the source is the sum of the individual powers dissipated, so that

$$P_T = E_T I_T = P_1 + P_2 + P_3 \qquad (4\text{-}2\text{-}8)$$

To summarize, the analysis for the circuit of Fig. 4-1(d) required the following procedure:

1. Using the product-over-sum formula, find the equivalent resistance of R_1 and R_2 in parallel. Add this equivalent resistance to the value of the series resistor R_3; the result is the total equivalent resistance, R_T, of the entire circuit.
2. Find $I_T \, (= E_T/R_T)$ and the values of the branch currents.
3. Determine the voltage drops, the powers dissipated in the individual resistors, and the total power delivered from the source.

EXAMPLE 4-2 In Fig. 4-1(d) $R_1 = 1.8 \text{ k}\Omega$, $R_2 = 2.2 \text{ k}\Omega$, and $R_3 = 1 \text{ k}\Omega$. If $E_T = 8 \text{ V}$, find the values of R_T, I_T, I_1, I_2, V_1, V_2, P_1, P_2, P_3, and P_T.

Solution The equivalent resistance of $R_1 \| R_2$ is

$$R_{12} = \frac{R_1 \times R_2}{R_1 + R_2} = \frac{1.8 \times 2.2}{1.8 + 2.2} = \frac{1.8 \times 2.2}{4} = 0.99 \text{ k}\Omega$$

$$\text{total equivalent resistance, } R_T = R_{12} + R_3 \qquad (4\text{-}2\text{-}1)$$
$$= 0.99 + 1.0 = 1.99 \text{ k}\Omega$$

$$\text{total source current, } I_T = \frac{E_T}{R_T} \qquad (4\text{-}2\text{-}2)$$
$$= \frac{8 \text{ V}}{1.99 \text{ k}\Omega} = 4.02 \text{ mA}$$

Using Eqs. (4-2-3) and (4-2-4), we obtain

$$I_1 = 4.02 \text{ mA} \times \frac{2.2}{1.8 + 2.2} = 2.21 \text{ mA}$$

$$I_2 = 4.02 \text{ mA} \times \frac{1.8}{1.8 + 2.2} = 1.81 \text{ mA}$$

Current Check

The sum of the currents I_1 and I_2 must equal the total current, I_T.

$$I_1 + I_2 = 2.21 + 1.81 = 4.02 \text{ mA} = I_T$$

$$V_1 = I_T R_3 \qquad (4\text{-}2\text{-}6)$$
$$= 4.02 \text{ mA} \times 1 \text{ k}\Omega = 4.02 \text{ V}$$

$$V_2 = I_1 R_1 \qquad (4\text{-}2\text{-}7)$$
$$= 2.21 \text{ mA} \times 1.8 \text{ k}\Omega = 3.98 \text{ V}$$

Voltage Checks

$$V_2' = I_2 R_2 = 1.81 \text{ mA} \times 2.2 \text{ k}\Omega = 3.98 \text{ V}$$
$$V_1 + V_2 = 4.02 + 3.98 = 8 \text{ V} = E_T$$

The individual powers dissipated are

$$P_1 = I_1 V_2 = 2.21 \text{ mA} \times 3.98 \text{ V} = 8.80 \text{ mW}$$
$$P_2 = I_2 V_2 = 1.81 \text{ mA} \times 3.98 \text{ V} = 7.20 \text{ mW}$$
$$P_3 = I_T V_1 = 4.02 \text{ mA} \times 4.02 \text{ V} = 16.16 \text{ mW}$$

The total power, P_T, delivered from the source is

$$P_T = E_T I_T \qquad (4\text{-}2\text{-}8)$$
$$= 8 \text{ V} \times 4.02 \text{ mA} = 32.16 \text{ mW}$$

Power Check

$$P_1 + P_2 + P_3 = 8.80 + 7.20 + 16.16 = 32.16 \text{ mW} = P_T$$

4-3

Step-by-step Analysis of Complex Series–Parallel Networks

As we have previously said, there is an infinite variety of complex series-parallel networks and we really have to look at each circuit on its own merits. However, there are certain general principles we can establish and these are best illustrated by the analysis of a particular circuit in which we will find the total equivalent resistance, the source current, the branch currents, the voltage drops across the various resistors, the powers dissipated in the individual resistors, and the total power delivered from the source. Such a circuit is shown in Fig. 4-2(a) and probably your first reaction is to recoil in horror! In fact, such a circuit would probably never occur in practice, but it will serve well to illustrate the principles of series-parallel analysis and the steps that must be taken in carrying out such an analysis.

STEP 1 Find the total equivalent resistance of the entire circuit as presented to the voltage source, E_T. You can start by identifying all single resistors that are directly in series with each other; such resistors must be joined end to end so that their voltages are additive and the same current flows through each.

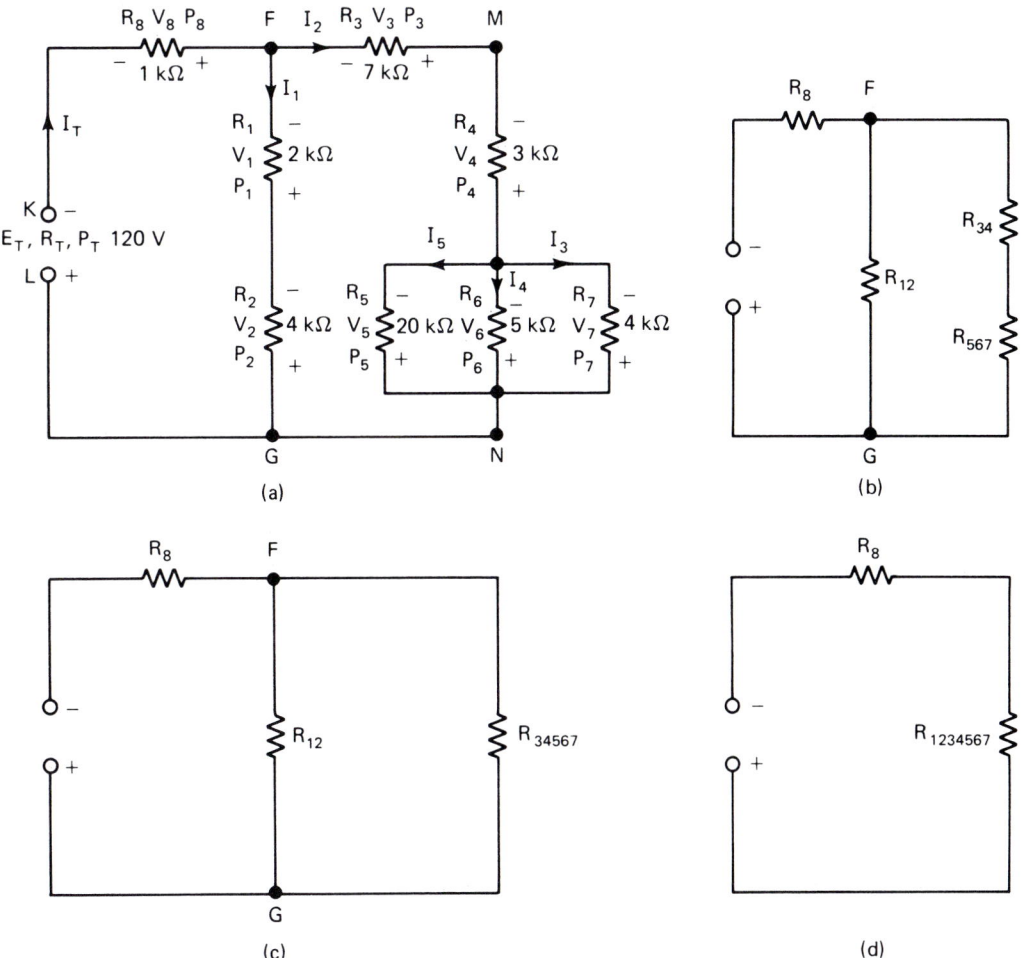

FIGURE 4-2 Step-by-step analysis of a complex series–parallel circuit.

Sec. 4-3 Step-by-step Analysis of Complex Series–Parallel Networks 163

Combine these resistors into their equivalent resistances; for example R_1 and R_2 in series and their equivalent resistance, R_{12}, is $2 + 4 = 6$ kΩ. In the same way R_3 and R_4 are in series with a total resistance, R_{34}, equal to $7 + 3 = 10$ kΩ [Eq. (2-3-5)].

There are no more series resistors to combine, so we now turn our attention to those single resistors that are connected directly in parallel with each other. Of course, if there are originally no single resistors in series, we would immediately proceed to the parallel resistors. Such parallel resistors must be considered between points or lines where a current division occurs; therefore, resistors R_5, R_6, and R_7 are in parallel and using the reciprocal formula [Eq. (3-3-1)], their equivalent resistance R_{567} is given by

$$\frac{1}{R_{567}} = \frac{1}{R_5} + \frac{1}{R_6} + \frac{1}{R_7} = \frac{1}{20} + \frac{1}{5} + \frac{1}{4}$$

$$= \frac{10}{20} = \frac{1}{2}$$

or

$$R_{567} = 2 \text{ k}\Omega$$

By replacing the series and parallel resistors (which we have now combined) with their equivalent resistances, the circuit can be redrawn as shown in Fig. 4-2(b).

In the new diagram we look for series arrangements of the equivalent resistances. In other words, we are alternating between series and parallel and breaking the circuit down a bit at a time. Normally, we start farthest away from the source and gradually work our way toward the source. At the same time our purpose is to reduce the circuit by a number of redrawings to a single resistance, R_T, which will represent the total equivalent resistance connected across the source. In Fig. 4-2(b) the resistances R_{34} and R_{567} are in series, so that their total equivalent resistance R_{34567} is $10 + 2 = 12$ kΩ; the circuit can now be redrawn again as in Fig. 4-2(c); this arrangement is now comparable with the series-parallel network of Fig. 4-1(d).

Since in Fig. 4-2(c) there is a current split at points F and G, R_{12} and R_{34567} are in parallel, and therefore the total resistance of these seven resistors is

$$\frac{6 \times 12}{6 + 12} = 4 \text{ k}\Omega$$

In the final redrawing of Fig. 4-2(d), R_8 is in series with $R_{1234567}$, so that the total equivalent resistance of the entire circuit is $R_T = 1 + 4 = 5$ kΩ, which is directly connected across the 120-V source. As you become more familiar with the analysis of series-parallel circuits, you will find that you can omit most, if not all, of the intermediate drawings.

STEP 2   We will now start to work away from the source by finding I_T, the branch currents, and the voltage drops across the resistors. When carrying out these calculations we must check that the voltages in any closed loop as well as the currents at any point must exactly balance (Kirchhoff's voltage and current laws).

$$\text{total current, } I_T = \frac{E_T}{R_T} = \frac{120 \text{ V}}{5 \text{ k}\Omega} = 24 \text{ mA}$$

voltage drop, $V_8 = I_T R_8 = 24$ mA $\times$ 1 kΩ = 24 V

Then

$$V_1 + V_2 = E_T - V_8 = 120 \text{ V} - 24 \text{ V} = 96 \text{ V}$$

$$\text{branch current, } I_1 = \frac{V_1 + V_2}{R_{23}} = \frac{96 \text{ V}}{6 \text{ k}\Omega} = 16 \text{ mA}$$

$$\text{voltage drop, } V_1 = I_1 R_1 = 16 \text{ mA} \times 2 \text{ k}\Omega = 32 \text{ V}$$

$$V_2 = I_1 R_2 = 16 \text{ mA} \times 4 \text{ k}\Omega = 64 \text{ V}$$

VOLTAGE CHECK For the closed loop *FGLKF*:

$$V_1 + V_2 + V_8 = 32 + 64 + 24 = 120 \text{ V} = E_T$$

Since the currents at point *F* must exactly balance,

$$\text{branch current, } I_2 = I_T - I_1 = 24 - 16 = 8 \text{ mA}$$

$$\text{voltage drop, } V_3 = I_2 R_3 = 8 \text{ mA} \times 7 \text{ k}\Omega = 56 \text{ V}$$

$$V_4 = I_2 R_4 = 8 \text{ mA} \times 3 \text{ k}\Omega = 24 \text{ V}$$

$$V_5 = V_6 = V_7 = I_2 \times R_{567} = 8 \text{ mA} \times 2 \text{ k}\Omega = 16 \text{ V}$$

VOLTAGE CHECK For the closed loop *KFMNGLK*,

$$V_8 + V_3 + V_4 + V_5 = 24 + 56 + 24 + 16 = 120 \text{ V} = E_T$$

$$\text{branch current, } I_5 = \frac{V_5}{R_5} = \frac{16 \text{ V}}{20 \text{ k}\Omega} = 0.8 \text{ mA}$$

$$I_4 = \frac{V_6}{R_6} = \frac{16 \text{ V}}{5 \text{ k}\Omega} = 3.2 \text{ mA}$$

$$I_3 = \frac{V_7}{R_7} = \frac{16 \text{ V}}{4 \text{ k}\Omega} = 4.0 \text{ mA}$$

CURRENT CHECK For the currents entering and leaving point *Q*,

$$I_3 + I_4 + I_5 = 4.0 + 3.2 + 0.8 = 8 \text{ mA} = I_2$$

STEP 3 We have now calculated all the currents and the voltage drops across the individual resistors. The simplest way of calculating the powers dissipated in the individual resistors is to multiply the voltage drop across each resistor by the current through the resistor. Then

$$P_1 = I_1 V_1 = 16 \text{ mA} \times 32 \text{ V} = 512 \text{ mW}$$

$$P_2 = I_1 V_2 = 16 \text{ mA} \times 64 \text{ V} = 1024 \text{ mW}$$

$$P_3 = I_2 V_3 = 8 \text{ mA} \times 56 \text{ V} = 448 \text{ mW}$$

$$P_4 = I_2 V_4 = 8 \text{ mA} \times 24 \text{ V} = 192 \text{ mW}$$

$$P_5 = I_5 V_5 = 0.8 \text{ mA} \times 16 \text{ V} = 12.8 \text{ mW}$$

$$P_6 = I_4 V_6 = 3.2 \text{ mA} \times 16 \text{ V} = 51.2 \text{ mW}$$

Sec. 4-3 Step-by-step Analysis of Complex Series–Parallel Networks

$$P_7 = I_3V_7 = 4 \text{ mA} \times 16 \text{ V} = 64 \text{ mW}$$

$$P_8 = I_TV_8 = 24 \text{ mA} \times 24 \text{ V} = 576 \text{ mW}$$

total power delivered by the source, $P_T = I_T E_T$

$$= 24 \text{ mA} \times 120 \text{ V} = 2880 \text{ mW}$$

POWER CHECK The total power delivered from the source must equal the sum of the powers dissipated in the individual resistors. Then

$$P_1 + P_2 + P_3 + P_4 + P_5 + P_6 + P_7 + P_8$$
$$= 512 + 1024 + 448 + 192 + 12.8 + 51.2 + 64 + 576$$
$$= 2880 \text{ mW} = P_T$$

In carrying out the analysis you should remember that:

1. Each branch current must be less than the total current drawn from the voltage source.
2. Each voltage drop must be less than the value of the source voltage.
3. The power dissipated in any individual resistor must be less than the total power delivered from the voltage source.

To summarize, the procedure used in series-parallel analysis is as follows:

1. Combine all obvious series strings and all obvious parallel banks into their equivalent resistances; redraw the simplified circuit.
2. Add together the series equivalent resistances and again redraw the circuit.
3. Using the product-over-sum or the reciprocal formula, combine the parallel equivalent resistances.
4. By redrawing the circuit as many times as necessary and repeating steps 2 and 3, work your way gradually toward the source until you finally arrive at the value of the total equivalent resistance for the entire circuit.
5. Determine the value of the total source current and then work away from the source in calculating the branch currents and the voltage drops across the individual resistors.
6. Find the powers dissipated in the various resistors and check that the sum of the values obtained equals the total power delivered from the source.

EXAMPLE 4-3 In Fig. 4-3(a), calculate the value of the total equivalent resistance of the entire circuit.

Solution This problem is used to illustrate the difficulties you may encounter when you need to redraw a circuit so that it presents a more conventional appearance and you can isolate the various series and parallel resistor arrangements. This has been done in Fig. 4-3(b), but in order to check its accuracy you should label the corresponding points in the two drawings and then make certain that the same resistors are connected between any two such points.

Starting farthest away from the source, R_6 and R_7 are in series, so that

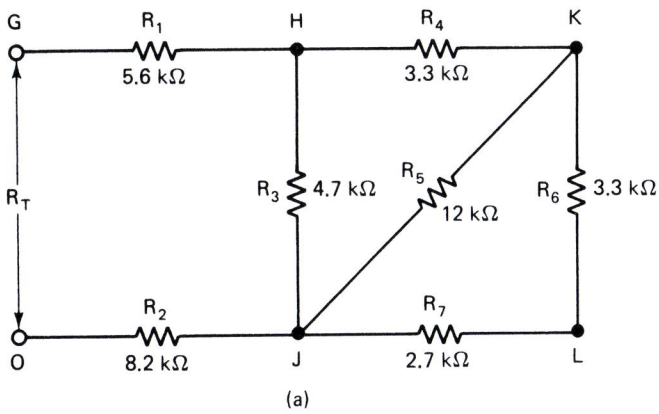

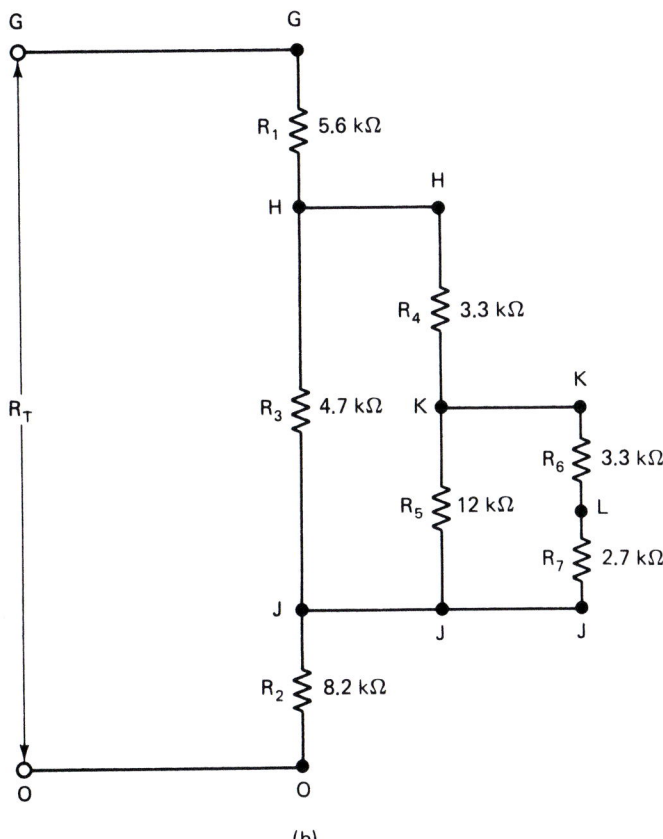

FIGURE 4-3 Circuit for Example 4-3.

their equivalent resistance $R_{67} = 3.3 + 2.7 = 6$ kΩ. R_5 is in parallel with R_{67} and therefore, by using the product-over-sum formula,

$$R_{567} = \frac{12 \times 6}{12 + 6} = 4 \text{ k}\Omega$$

R_4 is in series with R_{567} and consequently, $R_{4567} = 3.3 + 4.0 = 7.3$ kΩ. With R_3 in parallel with R_{4567}, the total resistance of these five resistors is

$$R_{34567} = \frac{7.3 \times 4.7}{7.3 + 4.7} = \frac{7.3 \times 4.7}{12} = 2.86 \text{ k}\Omega$$

Sec. 4-3 Step-by-step Analysis of Complex Series–Parallel Networks

Finally, R_1, R_2, and R_{34567} are in series, so that

$$R_T = 5.6 + 8.2 + 2.86 = 16.7 \text{ k}\Omega \text{ (rounded off)}$$

EXAMPLE 4-4 What arrangement(s) of standard 100-Ω 1-W resistors will produce a total resistance of 100 Ω with a 4-W power rating?

Solution This problem cannot be solved with simple series or parallel arrangements. If four 100-Ω 1-W resistors are connected in series, the power rating is 4 W, but the total resistance is $4 \times 100 = 400 \, \Omega$. Similarly if the four resistors are connected in parallel, the power rating is again 4 W, but the total resistance is only $100/4 = 25 \, \Omega$.

Two possible solutions are shown in Fig. 4-4. In Fig. 4-4(a) the resistance of R_1 and R_2 in parallel is $100/2 = 50 \, \Omega$, and since the two parallel combinations are in series, the total resistance is $2 \times 50 \, \Omega = 100 \, \Omega$; since all four resistors carry the same value of current, the total power rating is $4 \times 1 = 4$ W.

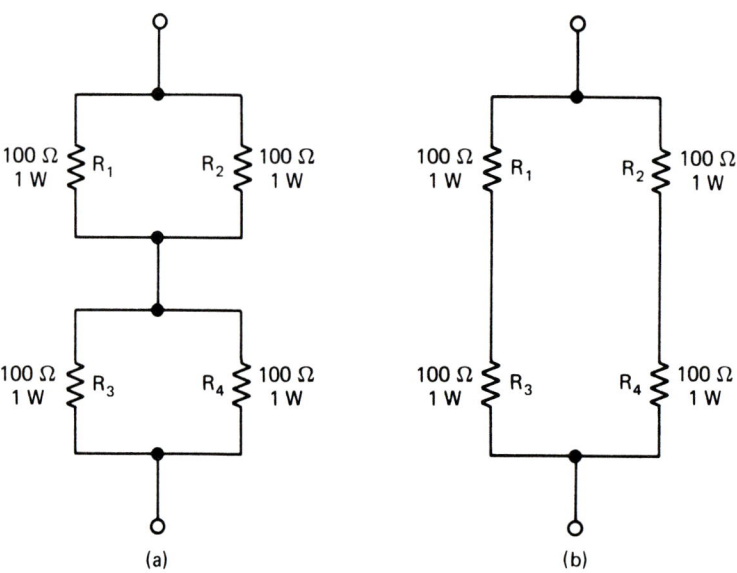

FIGURE 4-4 Circuits for Example 4-4.

Since R_1 and R_3 of Fig. 4-4(b) are a series combination, their equivalent resistance is $2 \times 100 = 200 \, \Omega$. With the R_1, R_3 and R_2, R_4 combinations in parallel, the total resistance is $200/2 = 100 \, \Omega$. Once more the same value of current is flowing through each of the resistors, and therefore the total power rating is again $4 \times 1 = 4$ W.

This example illustrates one use for a series-parallel arrangement of resistors; we can employ standard components to create some required value of resistance with a nonstandard power rating.

4-4

Ground Connections in Series-Parallel Networks

In Section 2-8 we discussed the use of ground as a zero-resistance return path and as a reference level of zero volts. In a series-parallel circuit any one point may be grounded and the voltage at all other points can then be measured relative to ground. As an example we consider the circuit of Fig. 4-5, in which we will ground each of the points in turn and then calculate the values of the potentials at the other points.

Before starting the problem you should check that there is a voltage balance around the various loops.

Loop *PABCDFGN*:

$$V_{AB} + V_{BC} + V_{CD} + V_{DG} + V_{FB} = 60 + 10 + 70 + 20 + 30$$
$$= 190 \text{ V} = V_{PN} = E_T$$

Loop *PABCEFGN*:

$$V_{AB} + V_{BC} + V_{CE} + V_{EF} + V_{FG} = 60 + 10 + 50 + 40 + 30$$
$$= 190 \text{ V} = V_{PN} = E_T$$

Loop *CEFDC*:
Path *CDF* has

$$V_{CD} + V_{DF} = 70 + 20 = 90 \text{ V}$$

Path *CEF* has

$$V_{CE} + V_{EF} = 50 + 40 = 90 \text{ V}$$

G grounded:

potential at $G = 0$ V

potential at $F = +30$ V

potential at $D = +(30 + 20) = +50$ V

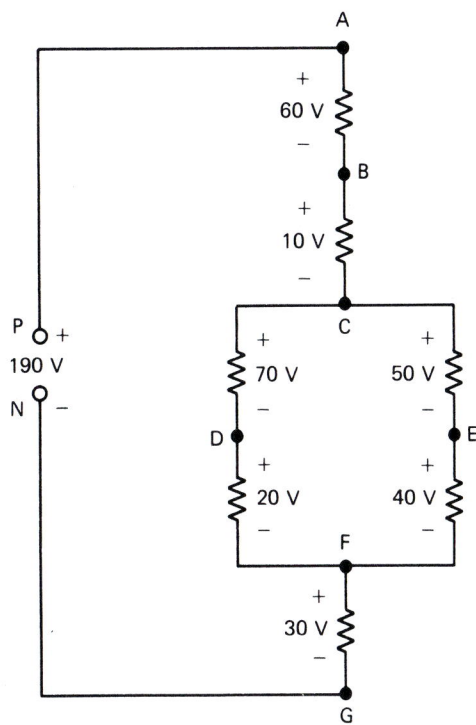

FIGURE 4-5 Measurement of potentials relative to ground.

Sec. 4-4 Ground Connections in Series-Parallel Networks

$$\text{potential at } C = +(30 + 20 + 70) = +120 \text{ V}$$
$$\text{potential at } E = +(30 + 40) = +70 \text{ V}$$
$$\text{potential at } B = +(30 + 20 + 70 + 10) = +130 \text{ V}$$
$$\text{potential at } A = +(30 + 20 + 70 + 10 + 60) = +190 \text{ V}$$

F grounded:

$$\text{potential at } G = -30 \text{ V}$$
$$\text{potential at } F = 0 \text{ V}$$
$$\text{potential at } D = +20 \text{ V}$$
$$\text{potential at } E = +40 \text{ V}$$
$$\text{potential at } C = +(20 + 70) = +(40 + 50) = +90 \text{ V}$$
$$\text{potential at } B = +(20 + 70 + 10) = +(40 + 50 + 10) = +100 \text{ V}$$
$$\text{potential at } A = +(20 + 70 + 10 + 60) = +(40 + 50 + 10 + 60)$$
$$= +160 \text{V}$$

D grounded:

$$\text{potential at } F = -20 \text{ V}$$
$$\text{potential at } G = -(20 + 30) = -50 \text{ V}$$
$$\text{potential at } C = +70 \text{ V}$$
$$\text{potential at } E = +70 - 50 = +20 \text{ V}$$

[Check:
$$\text{potential at } F = +20 - 40 = -20 \text{ V}]$$
$$\text{potential at } B = +(70 + 10) = +80 \text{ V}$$
$$\text{potential at } A = +(80 + 60) = +140 \text{ V}$$

E grounded:

$$\text{potential at } F = -40 \text{ V}$$
$$\text{potential at } G = -(40 + 30) = -70 \text{ V}$$
$$\text{potential at } C = +50 \text{ V}$$
$$\text{potential at } D = +50 - 70 = -20 \text{ V}$$

[Check:
$$\text{potential at } F = -20 - 20 = -40 \text{ V}]$$
$$\text{potential at } B = +(50 + 10) = +60 \text{ V}$$
$$\text{potential at } A = +(50 + 10 + 60) = +120 \text{ V}$$

C grounded:

$$\text{potential at } D = -70 \text{ V}$$
$$\text{potential at } E = -50 \text{ V}$$

potential at $F = -(70 + 20) = -(50 + 40) = -90$ V

potential at $G = -(70 + 20 + 30) = -(50 + 40 + 30) = -120$ V

potential at $B = +10$ V

potential at $A = +(10 + 60) = +70$ V

B grounded:

potential at $C = -10$ V

potential at $D = -(10 + 70) = -80$ V

potential at $E = -(10 + 50) = -60$ V

potential at $F = -(10 + 70 + 20) = -(10 + 50 + 40) = -100$ V

potential at $G = -(10 + 70 + 20 + 30)$

$\qquad = -(10 + 50 + 40 + 30) = -130$ V

potential at $A = +60$ V

A grounded:

potential at $B = -60$ V

potential at $C = -(60 + 10) = -70$ V

potential at $D = -(60 + 10 + 70) = -140$ V

potential at $E = -(60 + 10 + 50) = -120$ V

potential at $F = -(60 + 10 + 70 + 20)$

$\qquad = -(60 + 10 + 50 + 40) = -160$ V

potential at $G = -(160 + 30) = -190$ V

Notice that whichever point is grounded, the voltage difference between D and E is 20 V and that point E is positive with respect to point D.

To summarize, the potential difference between any two points remains the same and is independent of the position of the ground.

4-5 Wheatstone Bridge Circuit

A practical application of a series-parallel network is the Wheatstone bridge circuit, which is used to obtain an accurate measurement of an unknown resistance R_X [Fig. 4-6(a)]. The bridge is made up of four resistor arms R_1, R_2, R_3, R_X, and points P, Q are joined by a center link or "bridge" which contains a sensitive current-indicating device, G. The more conventional drawing of a bridge circuit is shown in Fig. 4-6(b), in which the resistors R_1, R_2, R_3, and R_X form the four arms of the bridge. There are many varieties of the bridge circuit, but they may all be identified by the presence of the four arms and the center link.

The whole of the source voltage, E, must be dropped across R_1R_2 and also R_3R_X. Then, provided that the values of the resistors R_1, R_2, R_3, and R_X satisfy a particular condition, there will be no voltage difference between points P and Q and the current meter will read zero; the bridge is then said to be

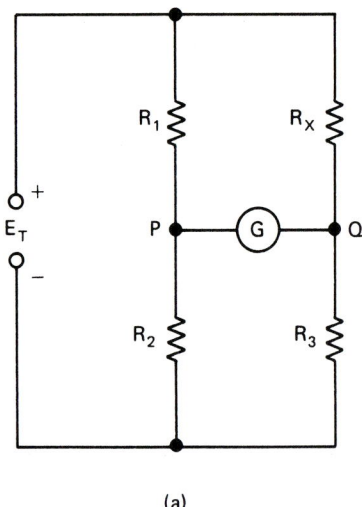
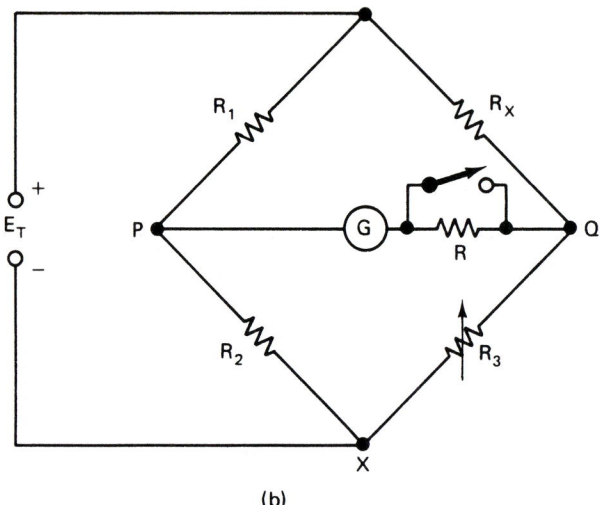

(a) (b)

FIGURE 4-6 Wheatstone bridge.

balanced. If the required balance condition is not fulfilled, G will indicate a current flow, which may be in either direction.

There will be no voltage difference between points P and Q if the ratios $R_1:R_2$ and $R_X:R_3$ are equal. Then

$$\frac{R_X}{R_3} = \frac{R_1}{R_2}, \quad R_X R_2 = R_1 R_3, \quad R_X = R_3 \times \frac{R_1}{R_2} \qquad (4\text{-}5\text{-}1)$$

Notice that in the balanced state, the products ($R_X R_2$ and $R_1 R_3$) of the opposing resistors are equal. This same balance condition can be obtained with a different approach. Ground point X and then apply the voltage division rule [Eq. (2-6-10)]. The potentials V_P and V_Q at points P and Q are

$$V_P = E \times \frac{R_2}{R_1 + R_2} \quad \text{and} \quad V_Q = E \times \frac{R_3}{R_3 + R_X}$$

If the bridge is balanced, $V_P = V_Q$. Then

$$E \times \frac{R_2}{R_1 + R_2} = E \times \frac{R_3}{R_3 + R_X}$$

$$R_2 R_3 + R_2 R_X = R_1 R_3 + R_2 R_3$$

$$R_X = R_3 \times \frac{R_1}{R_2}$$

In measuring the value of R_X, R_3 is an accurately calibrated variable resistor which is used to balance the bridge for a zero reading in the current meter, G. In practice, a protection resistor, R, is commonly included in series with G; a rough balance is then obtained with this resistor present. The resistor is subsequently shorted out and the bridge is accurately balanced.

The high-quality fixed resistors R_1 and R_2 have the possible values of 1 Ω, 10 Ω, 100 Ω, or 1000 Ω. The ratio $R_1:R_2$ can then be set for 1000, 100, 10, 1, 0.1, 0.01, or 0.001. The measured value of R_X may therefore range from 1/1000 of the lowest value of R_3 to 1000 times the highest value of R_3.

To summarize, the Wheatstone bridge circuit consists of four resistor arms and a center link which contains a sensitive current meter. When the bridge is balanced, the current meter reads zero, and it is then possible to obtain an accurate measurement of an unknown resistance.

EXAMPLE 4-5 In Fig. 4-6(b), R_1 is 1 Ω and $R_2 = 100$ Ω. The bridge is then balanced by adjusting R_3 to be 7.4 Ω. What is the value of the unknown resistor, R_X?

Solution When the bridge is balanced,

$$R_X = R_3 \times \frac{R_1}{R_2} \qquad (4\text{-}5\text{-}1)$$

$$= 7.4 \times \frac{1}{100} = 0.074 \text{ Ω}$$

EXAMPLE 4-6 Figure 4-7(a) represents an unbalanced bridge. Calculate the value of the current that flows in the bridge $B'B$.

Solution To find the equivalent resistance the circuit is redrawn in Fig. 4-7(b). Then

$$R_T = \frac{1 \text{ kΩ}}{2} + \frac{6.8 \text{ kΩ} \times 10 \text{ kΩ}}{6.8 \text{ kΩ} + 10 \text{ kΩ}}$$

$$= 500 \text{ Ω} + 4048 \text{ Ω} = 4548 \text{ Ω}$$

$$\text{total current, } I_T = \frac{220 \text{ V}}{4548 \text{ Ω}} = 48.4 \text{ mA}$$

The total current will split equally between the two 1-kΩ resistors and therefore the current flowing from A to B' is 48.4/2 = 24.2 mA. The voltage drop from

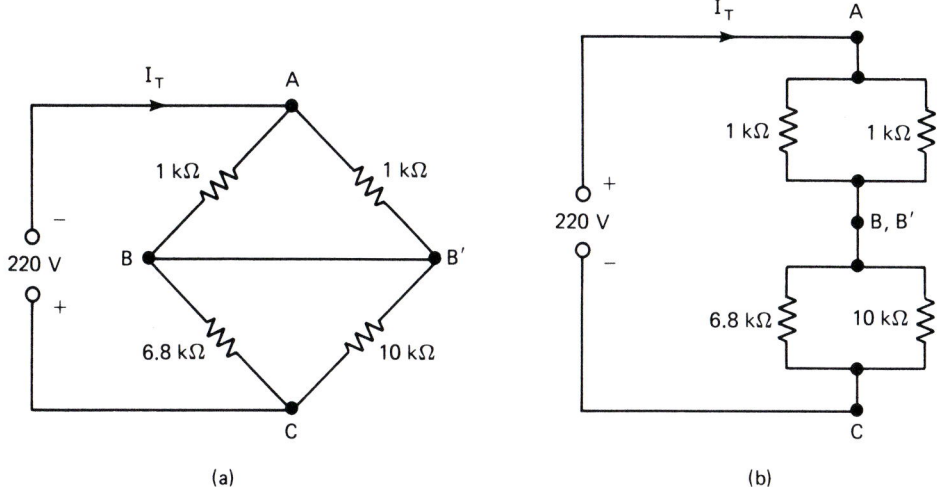

FIGURE 4-7 Unbalanced bridge.

Sec. 4-5 Wheatstone Bridge Circuit

A to B' is 24.2 mA $\times$ 1 kΩ = 24.2 V and the voltage across the 10-kΩ resistor is 220 − 24.2 = 195.8 V. The current flowing from B' to C is 195.8 V/10 kΩ = 19.58 mA, so that, using Kirchhoff's current law for point B', the current in the bridge (center arm) is 24.2−19.58 = 4.6 mA (rounded off). Repeating the current law for point B, the current through the 6.8-kΩ resistor is 24.2 + 4.6 = 28.8 mA (check: the current through the 6.8-kΩ resistor is also 195.8 V/ 6.8 kΩ = 28.8 mA). At point C the sum of the currents through the 6.8-kΩ and 10-kΩ resistors is 19.58 + 28.82 = 48.4 mA, which is the total current, I_T.

Note: This is a good place to mention that not all resistor networks can be analyzed in terms of series-parallel circuits. For example, if in the unbalanced bridge circuit of Fig. 4-7(a) a resistor is included between B and B', this resistor will bear no simple series or parallel relationship to the other four resistors. Such a bridge circuit may only be analyzed using the more sophisticated techniques of Chapter 6.

4-6 Loaded Voltage Divider Circuit

In Section 2-7 we discussed the unloaded voltage divider circuit, which was a single series arrangement, capable of producing a number of different voltages from a single voltage source. However, as soon as loads are connected across the various voltages, the circuit becomes an application of a series-parallel network, as shown in Fig. 4-8. In this loaded voltage divider circuit the three load voltages are V_{L1}, V_{L2}, and V_{L3}, which are associated with their load currents I_{L1}, I_{L2}, I_{L3}, respectively. The bleeder current, I_B, is generally about 10% of the total load current, I_{LT}, which is the sum of I_{L1}, I_{L2}, and I_{L3}. Then

$$I_{LT} = I_{L1} + I_{L2} + I_{L3} \tag{4-6-1}$$

The load voltage, V_{L3}, must equal the voltage drop across R_3 due to the bleeder current, I_B. Therefore,

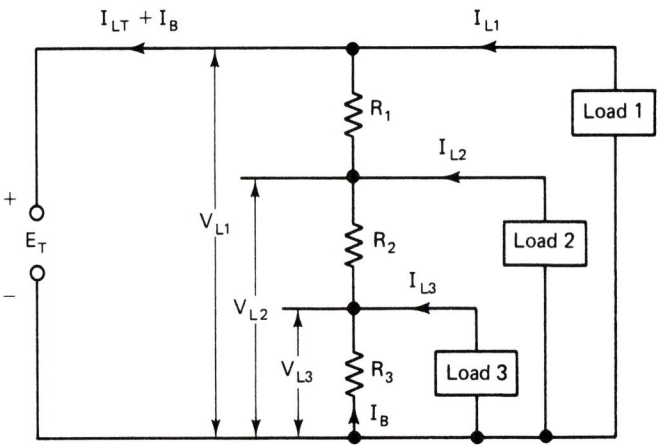

FIGURE 4-8 Loaded voltage divider circuit.

$$V_{L3} = I_B R_3 \text{ and } R_3 = \frac{V_{L3}}{I_B} \tag{4-6-2}$$

The sum of the currents I_{L3} and I_B flows through R_2. This resistor's voltage drop when added to V_{L3} must equal V_{L2}. In equation form,

$$V_{L2} = V_{L3} + R_2(I_{L3} + I_B) \text{ and } R_2 = \frac{V_{L2} - V_{L3}}{I_{L3} + I_B} \tag{4-6-3}$$

In a similar way the sum of the currents of I_{L2}, I_{L3}, and I_B flows through R_1; this resistor's voltage drop when added to V_{L2} must equal V_{L1}, which is the source voltage, E_T. Then

$$E_T = V_{L1} = V_{L2} + R_1(I_{L2} + I_{L3} + I_B) \tag{4-6-4}$$

and

$$R_1 = \frac{V_{L1} - V_{L2}}{I_{L2} + I_{L3} + I_B} \tag{4-6-5}$$

These equations enable the values of the divider resistors R_1, R_2, and R_3 to be calculated, provided that the load voltages with their currents are known and a certain percentage for the bleeder current is assumed. Note that since I_{L1} is drawn directly from the source voltage, its value does not appear in the equations. However, if a dropping resistor, R_D, is included (Fig. 4-9),

$$E_T = V_{L1} + I_T R_D \text{ and } R_D = \frac{E_T - V_{L1}}{I_T} \tag{4-6-6}$$

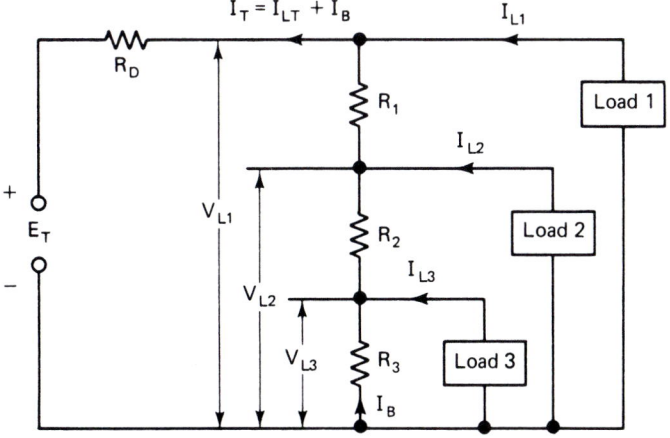

FIGURE 4-9 Loaded voltage divider circuit with dropping resistor.

EXAMPLE 4-7 In Fig. 4-9, $E_T = 50$ V, $V_{L1} = 40$ V, $V_{L2} = 24$ V, $V_{L3} = 9$ V, $I_{L1} = 150$ mA, $I_{L2} = 75$ mA, and $I_{L3} = 50$ mA. If the bleeder current is 10% of the total load current, calculate the values required for R_1, R_2, R_3, and R_D.

Solution

$$\text{total load current, } I_{LT} = I_{L1} + I_{L2} + I_{L3} \tag{4-6-1}$$

$$= 150 + 75 + 50 = 275 \text{ mA}$$

$$\text{bleeder current, } I_B = \frac{10}{100} \times 275 = 27.5 \text{ mA}$$

The voltage divider resistors are

$$R_3 = \frac{V_{L3}}{I_B} \tag{4-6-2}$$

$$= \frac{9 \text{ V}}{27.5 \text{ mA}} = 327 \text{ }\Omega$$

$$R_2 = \frac{V_{L2} - V_{L3}}{I_{L3} + I_B} \tag{4-6-3}$$

$$= \frac{24 \text{ V} - 9 \text{ V}}{50 \text{ mA} + 27.5 \text{ mA}} = 194 \text{ }\Omega$$

$$R_1 = \frac{V_{L1} - V_{L2}}{I_{L2} + I_{L3} + I_B} \tag{4-6-5}$$

$$= \frac{40 \text{ V} - 24 \text{ V}}{75 \text{ mA} + 50 \text{ mA} + 27.5 \text{ mA}} = 105 \text{ }\Omega$$

The power dissipated in these resistors is

$$P_{R1} = 16 \text{ V} \times 152.5 \text{ mA} = 2440 \text{ mW}$$

$$P_{R2} = 15 \text{ V} \times 77.5 \text{ mA} = 1662.5 \text{ mW}$$

$$P_{R3} = 9 \text{ V} \times 27.5 \text{ mA} = 247.5 \text{ mW}$$

Suitable resistors would be R_1, 100 Ω and 5 W; R_2, 200 Ω and 2 W; R_3, 330 Ω and $\frac{1}{2}$ W.

$$\text{dropping resistor, } R_D = \frac{E_T - V_{L1}}{I_T} \tag{4-6-6}$$

$$= \frac{50 \text{ V} - 40 \text{ V}}{275 \text{ mA} + 27.5 \text{ mA}}$$

$$= \frac{10 \text{ V}}{302.5 \text{ mA}} = 33.06 \text{ }\Omega$$

$$\text{its power dissipation, } P_{R_D} = 10 \text{ V} \times 302.5 \text{ mA}$$

$$= 3025 \text{ mW}$$

A suitable resistor would be 33 Ω and 5 W. In voltage divider networks composition resistors are preferred to those of the wirewound variety, which are more liable to open-circuit.

4-7

Effect of Open and Short Circuits in Series–Parallel Networks

In Sections 2-11, 3-9, and 3-10 we discussed fully the effect of open and short circuits on series and parallel networks of resistors. The same principles can be applied to series-parallel circuits, so that if an open circuit occurs, all resistors or combinations of resistors that are directly in series with the open circuit will carry no current. There will be no voltage drop across any of these resistors and they will therefore dissipate no power. As an example, look at Fig. 4-10; when an open circuit occurs between the points X and Y, there will be no current flow through resistors R_4, R_5, R_6, and R_7 and the voltage between X and Y will equal the IR drop across A and B. The total resistance presented to the voltage source will increase, so that the source current and the total power drawn from the source are less. This will cause lower voltage drops across R_1 and R_8, and therefore the voltage across the parallel combination of R_2 and R_3 will be greater. These results are illustrated in Example 4-8.

After reconnecting X and Y so that the open circuit is eliminated, consider the effect of a short circuit between A and B. Since there is zero voltage drop across a short circuit, no current will flow through any of the resistors R_2, R_3, R_4, R_5, R_6, and R_7. The circuit will then be reduced to a simple series arrangement consisting of R_1, the short between A, B, and R_8. The total resistance presented to the source is less and therefore the source current and the total power are greater (see Example 4-9).

To summarize, the effects of open and short circuits on series–parallel networks may be found by investigating which resistors carry no current. The remaining resistors will then make up the circuit, which is effectively across the voltage source.

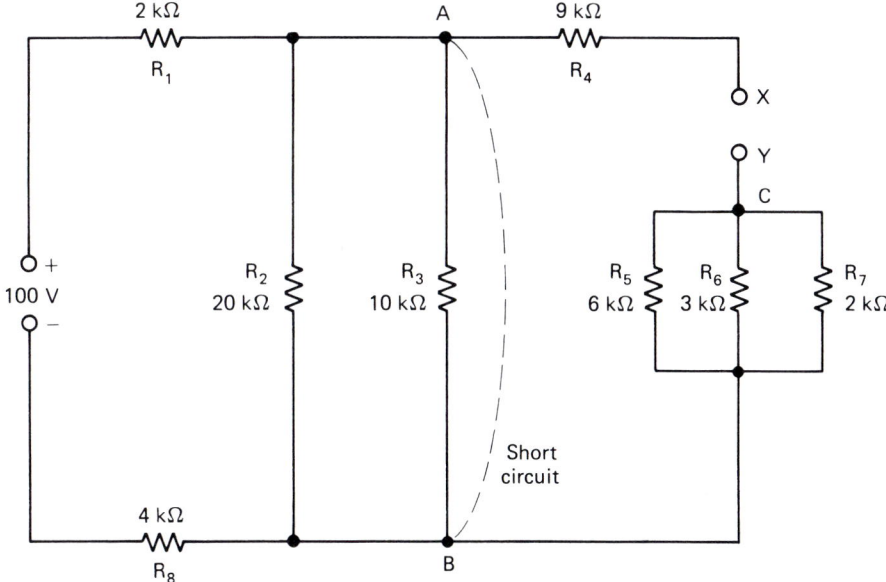

FIGURE 4-10 Effect of open and short circuits in a series–parallel network. Circuit for Examples 4-8 and 4-9.

EXAMPLE 4-8 In Fig. 4-10 calculate the values of R_T, I_T, P_T, V_{R1}, V_{R8}, and V_{AB} (assume that X and Y are connected and that there is no short circuit between A and B). If an open circuit occurs between X and Y, recalculate the values of R_T, I_T, P_T, V_{R1}, V_{R8}, and V_{AB}.

Solution The equivalent resistance, R_{567}, of R_5, R_6, and R_7 is given by

$$R_{567} = \frac{1}{\frac{1}{6} + \frac{1}{3} + \frac{1}{2}} = \frac{1}{1} = 1 \text{ k}\Omega$$

Since R_4 is in series with R_{567}, the total resistance of R_{4567} is $9 + 1 = 10$ kΩ. R_3 is in parallel with R_{4567}, so that R_{34567} equals 10 k$\Omega/2 = 5$ kΩ. Using the product-over-sum formula, the resistance

$$R_{234567} = \frac{20 \times 5}{20 + 5} = \frac{100}{25} = 4 \text{ k}\Omega$$

With R_1 and R_8 in series with R_{234567}, the total equivalent resistance, R_T, is $2 + 4 + 4 = 10$ kΩ.

$$\text{source current, } I_T = \frac{E_T}{R_T} = \frac{100 \text{ V}}{10 \text{ k}\Omega} = 10 \text{ mA}$$

$$\text{total power, } P_T = E_T \times I_T = 100 \text{ V} \times 10 \text{ mA}$$
$$= 1000 \text{ mW} = 1 \text{ W}$$

$$\text{voltage drop, } V_{R1} = I_T \times R_1 = 10 \text{ mA} \times 2 \text{ k}\Omega$$
$$= 20 \text{ V}$$

$$V_{R8} = I_T \times R_8 = 10 \text{ mA} \times 4 \text{ k}\Omega$$
$$= 40 \text{ V}$$

$$V_{AB} = E_T - V_{R1} - V_{R8} = 100 - 40 - 20 = 40 \text{ V}$$

Open Circuit between X and Y
There is now infinite resistance in the branch ACB, so that only R_2 and R_3 are left in parallel. Then

$$R_{23} = \frac{20 \times 10}{20 + 10} = 6.67 \text{ k}\Omega$$

and

$$R_T = 2 + 6.67 + 4 = 12.67 \text{ k}\Omega$$

$$\text{new source current, } I_T = \frac{100 \text{ V}}{12.67 \text{ k}\Omega} = 7.89 \text{ mA}$$

$$\text{total power, } P_T = 100 \text{ V} \times 7.89 \text{ mA} = 789 \text{ mW}$$

$$\text{voltage drop, } V_{R1} = 7.89 \text{ mA} \times 2 \text{ k}\Omega = 15.79 \text{ V}$$

$$V_{R8} = 7.89 \text{ mA} \times 4 \text{ k}\Omega = 31.57 \text{ V}$$

$$V_{AB} = 100 - 15.79 - 31.57 = 52.64 \text{ V}$$

EXAMPLE 4-9 Assume in Fig. 4-10 that X and Y are connected so that there is no open circuit. However, a short circuit occurs across A and B. Calculate the new values of R_T, I_T, P_T, V_{R1}, and V_{R8}.

Solution Since there is zero resistance between A and B, the total equivalent resistance, R_T, consists of R_1 and R_8 in series. Therefore,

$$R_T = R_1 + R_8 = 2 + 4 = 6 \text{ k}\Omega$$

new source current, $I_T = \dfrac{100 \text{ V}}{6 \text{ k}\Omega} = 16.67 \text{ mA}$

total power, $P_T = 100 \times 16.67 \text{ mA} = 1667 \text{ mW}$
$$= 1.667 \text{W}$$

voltage drop, $V_{R1} = 16.67 \text{ mA} \times 2 \text{ k}\Omega = 33.33 \text{ V}$

$V_{R8} = 16.67 \text{ mA} \times 4 \text{ k}\Omega = 66.67 \text{ V}$

4-8
Troubleshooting a Series–Parallel Circuit

We learned in Section 4-7 that if an open occurs in a series–parallel circuit, the total resistance presented to the source is greater and the source current is therefore less. However, the converse is not as simple; if the source current decreases, this might be due to one or more open circuits. However, some short circuits might also be present. The reduction in the source current is merely telling us that the effects of the opens are greater than those of the shorts. Similarly, if the source current increases, you can only deduce that the shorts in the circuit have more effect than the opens.

When troubleshooting more complex networks, you are normally fortunate enough to have the help of a schematic diagram which shows you the anticipated values of resistance and dc voltage at various points in the circuit; all these values are measured with respect to ground. Figure 4-11 illustrates such a schematic. The resistance values shown are those expected when switch S is open. For example, at point A the resistance to ground consists of R_1 in series with $R_4 \| (R_2 + R_3)$ and you should check that this value is, in fact, 15.7

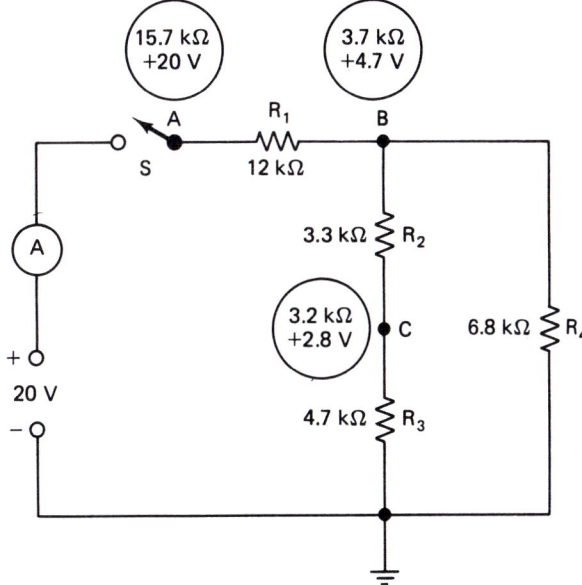

FIGURE 4-11 Schematic diagram of a series–parallel circuit.

kΩ. At point B the resistance to ground is found from $R_4 \| (R_2 + R_3)$, which is calculated to be 3.7 kΩ. Notice that resistor R_1 does not appear in this calculation because of the open circuit represented by switch S. Finally, at point C the resistance to ground is not the obvious 4.7 kΩ because of the parallel path provided by R_2 and R_4; the calculation therefore involves $R_3 \| (R_2 + R_4)$, whose value is equal to 3.2 kΩ.

Let us assume that resistor R_3 has open-circuited. The reading of the milliammeter A then falls, which alerts us to the presence of a fault. Switch S is opened and the black (common) probe of the ohmmeter is connected permanently to the ground. When the movable red probe touches point A, the ohmmeter reading is 18.8 kΩ, which shows the anticipated increase in the total resistance and is the result of connecting R_1 and R_4 in series; we may therefore deduce that either R_2 or R_3 has open-circuited. If the red probe is now shifted to point C, the ohmmeter reading is 3.3 + 6.8 = 10.1 kΩ, not the expected value of 3.2 kΩ. We now know that resistor R_3 has open-circuited and we have located the fault.

With switch S closed, we turn our attention to the dc voltages. At point A the voltage is +20 V with respect to ground, while at point B, the voltage should be +20 V × 3.7 kΩ/15.7 kΩ = +4.7 V (using the voltage division rule). Finally, at point C, the expected voltage is +4.7 V × 4.7 kΩ/(3.3 kΩ + 4.7 kΩ) = +2.8 V.

If the reading of milliammeter A now increases, we know that there is at least one short in the circuit. The black probe of a voltmeter on the 0 to 50 V range is permanently connected to ground while the red probe is used to touch point A; the reading is the expected value of +20 V. However, when the red probe is shifted to point B, the reading is only +3.8 V rather than the expected value of +4.7 V (the voltmeter range has now been changed to 0 to 10 V). A short cannot exist across R_4 since this would cause the reading of the voltage at point B to be zero. Consequently, either R_2 or R_3 must have shorted. If we now shift the red probe to the point C, the voltmeter reading remains at +3.8 V, and therefore the short must exist across resistor R_3.

You might well say that you could have located the short across R_2 by using resistance readings and therefore there seems to be no difference between the resistance and voltage methods. However, in a practical circuit there would probably be devices such as diodes and transistors which operate only when power is applied to the circuit. Consequently, the voltage values cannot be deduced from the resistance readings. The accepted practice for troubleshooting is as follows:

1. Remove power from the circuit and use an ohmmeter to locate the positions of opens and shorts.
2. Apply power to the circuit and use a voltmeter to discover whether the diodes, transistors, and so on, are operating correctly.

The following examples of troubleshooting series–parallel circuits involve the deduction of possible faults from the voltage and resistance readings. However, the main guides to success are your familiarity with a particular circuit and your use of its schematic diagram.

EXAMPLE 4-10 Figure 4-12 illustrates the schematic diagram of a series–parallel circuit. Check that the expected resistance and voltage readings are correct as shown. If the voltmeter's reading is +18 V, what fault, if any, exists in the circuit?

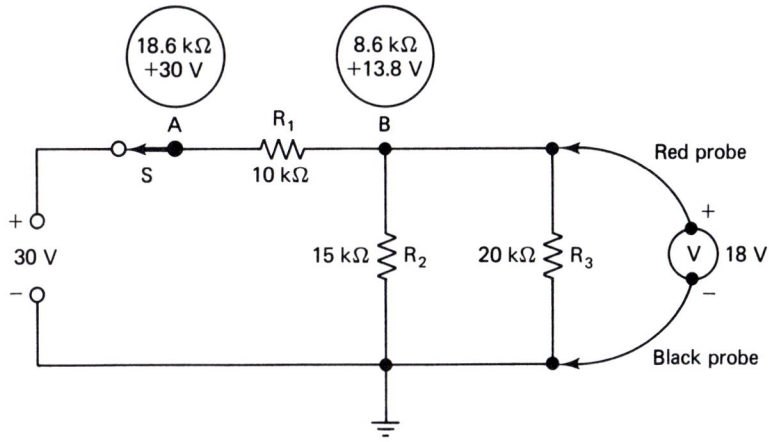

FIGURE 4-12 Schematic diagram for Example 4-10.

Solution The voltmeter reading exceeds the expected value so that it is possible that either R_2 or R_3 has burned out. Assuming that R_2 is open, the voltage across R_3 is 30 V × 20 kΩ/(10 kΩ + 20 kΩ) = 20 V. This is not the same as the voltmeter's reading, and therefore R_3 is the open-circuited resistor.

Check
If R_3 is open, the voltmeter reading is + 30 V × 15 kΩ/(10 kΩ + 15 kΩ) = + 18 V.

EXAMPLE 4-11 The schematic diagram of a series–parallel circuit is shown in Fig. 4-13. Check that the expected resistance and voltage readings are correct as shown. If the voltmeter reading at point B is +20 V, what fault, if any, exists in the circuit?

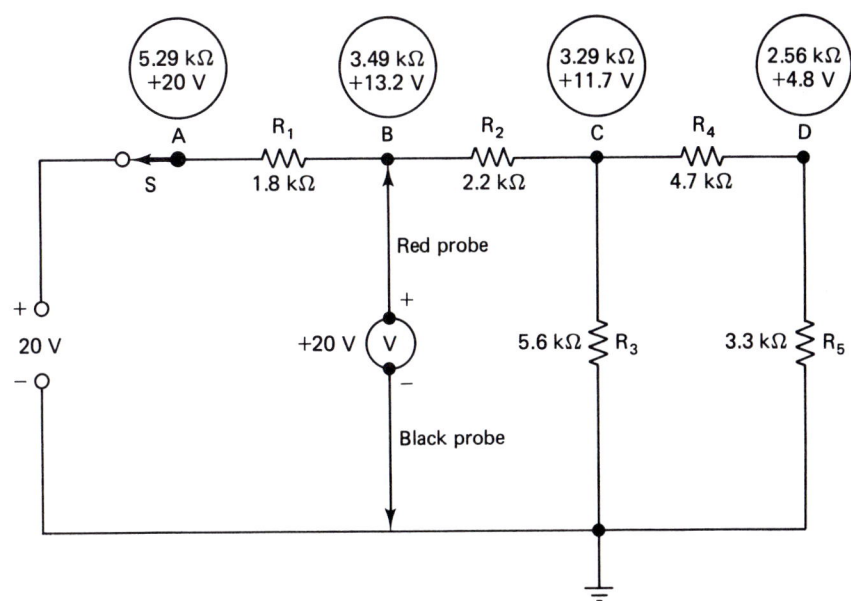

FIGURE 4-13 Schematic diagram for Example 4-11.

Sec. 4-8 Troubleshooting a Series–Parallel Circuit

181

Solution Since the voltmeter reading at point B is the same as the source voltage, there are two possibilities:

1. A short has occurred across the resistor R_1 so that there is no voltage drop between points A and B.
2. Resistor R_2 has open-circuited so that no current is flowing through resistor R_1 across which there will be zero voltage drop.

To check which possibility is correct, shift the red probe to point C. The voltmeter reading drops to zero so that the resistor R_2 is open-circuited (if there were a short across R_1, the voltmeter reading at the point C would merely have been less than $+20$ V).

EXAMPLE 4-12 Figure 4-14 shows the schematic diagram of a series–parallel circuit. Check that the expected resistance and voltage readings are correct as shown. If the voltmeter readings at points B and E are $+16.4$ V and $+10$ V respectively, what faults exist in the circuit?

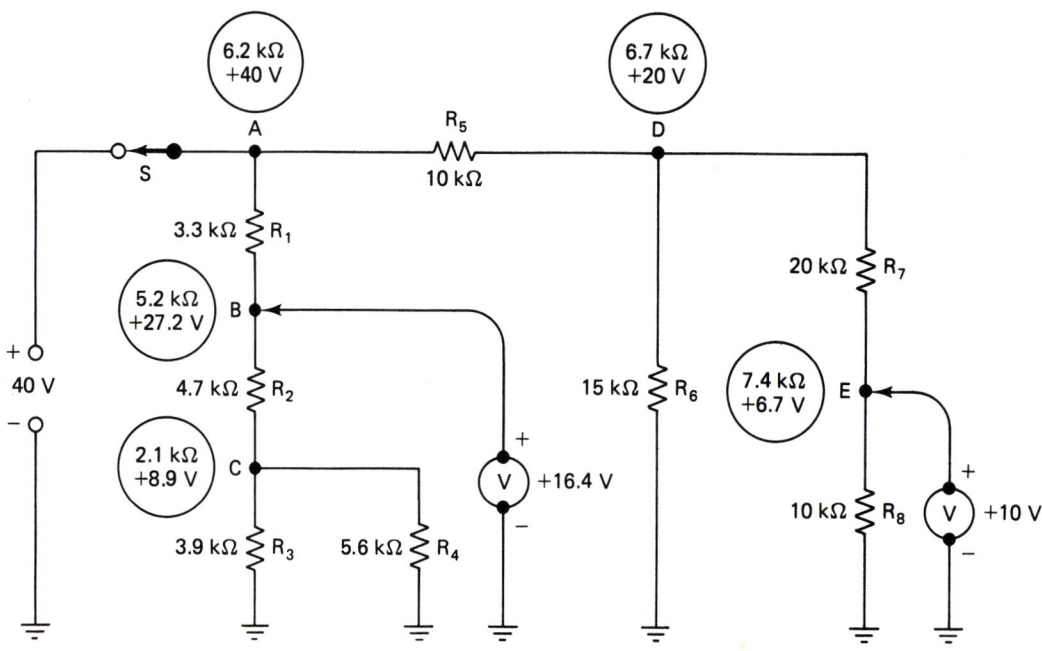

FIGURE 4-14 Schematic diagram for Example 4-12.

Solution Since the voltage at point B has fallen from its expected value, the resistance from B to ground has dropped. There are two possibilities:

1. A short circuit has developed across resistor R_2.
2. A short circuit has developed across the parallel combination of resistors R_3 and R_4.

Move the red probe from point B to point C; the reading remains at $+16.4$V and the short circuit exists across resistor R_2.

Check
The equivalent resistance of $R_3 \| R_4$ is $3.9 \times 5.6/(3.9 + 5.6) = 2.3$ kΩ. With

a short circuit across R_2, the common voltage at points B and C is $+40 \text{ V} \times 2.3 \text{ k}\Omega/(2.3 \text{ k}\Omega + 3.3 \text{ k}\Omega) = +16.4 \text{ V}$.

At point E, the voltage has risen and therefore there are two possibilities:

1. An open circuit in resistor R_6
2. A short circuit across resistor R_5

Shift the red probe from point E to point D. The reading at D is less than $+40$ V and therefore there is no short circuit across R_5. It follows that resistor R_6 must have open-circuited.

Check
If R_6 has open-circuited, the voltage at point E is $+40 \text{ V} \times 10 \text{ k}\Omega/(10 \text{ k}\Omega + 20 \text{ k}\Omega + 10 \text{ k}\Omega) = +10 \text{ V}$.

To summarize, we have used these examples to establish a logical procedure for troubleshooting:

1. Take a voltmeter or ohmmeter reading and then study the schematic diagram to determine the various possibilities that may be causing the fault.
2. Using the schematic diagram, take further readings to eliminate all the possibilities except one. Find the reason for the fault you have discovered and carry out the necessary repair.
3. Continue to take additional readings and check for other faults with the aid of the schematic diagram.

By contrast, do *not* take a single reading and then wade into a lot of mathematical analysis to deduce the location of the fault. This is not troubleshooting in the true sense, and in any case, this method will not work if multiple faults are involved. However, you can, if you wish, use a mathematical check to test the conclusion(s) that you have arrived at through your troubleshooting procedures.

Chapter Summary

4.1. Two resistors in series and a third resistor in parallel [Fig. 4-1(b)]:

$$\text{total resistance, } R_T = \frac{(R_1 + R_2) \times R_3}{R_1 + R_2 + R_3}$$

$$\text{branch current, } I_1 = \frac{E_T}{R_3}$$

$$\text{branch current, } I_2 = \frac{E_T}{R_1 + R_2}$$

$$\text{source current, } I_T = I_1 + I_2$$

$$\text{voltage drop, } V_1 = E_T \times \frac{R_1}{R_1 + R_2}$$

$$\text{voltage drop, } V_2 = E_T \times \frac{R_2}{R_1 + R_2}$$

$$\text{total power, } P_T = P_1 + P_2 + P_3.$$

4.2. Two resistors in parallel and a third resistor in series [Fig. 4-1(d)]:

$$\text{total resistance, } R_T = \frac{R_1R_2 + R_2R_3 + R_3R_1}{R_1 + R_2}$$

$$\text{source current, } I_T = I_1 + I_2$$

$$\text{branch current, } I_1 = I_T \times \frac{R_2}{R_1 + R_2}$$

$$\text{branch current, } I_2 = I_T \times \frac{R_1}{R_1 + R_2}$$

$$\text{voltage drop, } V_1 = I_T R_3$$

$$\text{voltage drop, } V_2 = I_1 R_1 = I_2 R_2$$

$$\text{source voltage, } E_T = V_1 + V_2$$

$$\text{total power, } P_T = P_1 + P_2 + P_3$$

4.3. Analysis of complex series-parallel networks:
Step 1. Find the total equivalent resistance of the entire circuit as presented to the source. This is achieved by a number of simplified circuit drawings in which series and parallel equivalent resistances are alternately combined.
Step 2. Calculate the values of the source current, the individual voltage drops, and the branch currents.
Step 3. Calculate the individual powers dissipated in the resistors and the total power drawn from the voltage source.

4.4. Wheatstone bridge circuit (Fig. 4-6):

$$\text{balanced condition: } R_X = R_3 \times \frac{R_1}{R_2}$$

4.5. Loaded voltage divider circuit (Fig. 4-8):

$$\text{total load current, } I_{LT} = I_{L1} + I_{L2} + I_{L3}$$

$$\text{divider resistor, } R_1 = \frac{V_{L1} - V_{L2}}{I_{L2} + I_{L3} + I_B}$$

$$\text{divider resistor, } R_2 = \frac{V_{L2} - V_{L3}}{I_{L3} + I_B}$$

$$\text{divider resistor, } R_3 = \frac{V_{L3}}{I_B}$$

Self-Review

True–False Questions

1. **T. F.** In calculating the total equivalent resistance of a complex series–parallel circuit, the analysis is carried out in a series of steps by starting at the source.
2. **T. F.** In the analysis of a complex series–parallel circuit, the next step after finding R_T is to calculate the source current, I_T.

3. T. F. The source current of a series–parallel circuit is greater than any one of the branch currents.
4. T. F. In a series–parallel circuit the total equivalent resistance is always greater than the value of any individual resistor.
5. T. F. The total power dissipated by a series–parallel circuit is equal to the sum of the powers dissipated in the individual resistors.
6. T. F. Two equal-value resistors are connected in parallel and this parallel combination is then connected in series with a third resistor of the same value. The total equivalent resistance is 1.5 × the value of each resistor.
7. T. F. Two equal-value resistors are connected in a series combination which is then connected in parallel with a third resistor of the same value. The total equivalent resistance is $\frac{2}{3}$ × the value of one resistor.
8. T. F. The Wheatstone bridge circuit is balanced by adjusting the calibrated resistance for a maximum reading in the current meter.
9. T. F. If an open circuit occurs in a series–parallel network, the total equivalent resistance presented to the source will increase.
10. T. F. If a short circuit occurs in a series–parallel network, the source current will decrease.
11. T. F. In any series–parallel arrangement of resistors the highest current always flows in the lowest-value resistor.
12. T. F. Any complex arrangement of resistors can always be broken down into some combination of series–parallel circuits.
13. T. F. If one load is removed from a loaded voltage divider circuit that includes a series dropping resistor, all the output voltages increase.
14. T. F. The total equivalent resistance of a series–parallel circuit is always less than the lowest-value resistor connected in parallel.
15. T. F. It is possible to obtain output load voltages of +30 V and −20 V from a single 40-V supply.
16. T. F. In a loaded voltage divider circuit the bleeder current must be less than 1% of the total current drawn from the supply.
17. T. F. In Fig. 4-6(b), $R_1 = 1$ kΩ, $R_2 = 10$ kΩ, $R_3 = 5.3$ kΩ, $R_X = 53$ kΩ. The bridge is balanced.
18. T. F. If an open circuit occurs in any series–parallel circuit, the measured voltage across the open circuit is always equal to the source voltage.
19. T. F. In a loaded voltage divider circuit it is never possible to obtain two positive potentials and two negative potentials (all of different values) from a single dc source.
20. T. F. A balanced Wheatstone bridge can be regarded as a series–parallel arrangement of resistors.

Multiple-Choice Questions

21. In Fig. 4-15, the total equivalent resistance presented to the source is
 (a) 60 Ω (b) 130 Ω (c) 13.3 Ω (d) 20 Ω (e) 50 Ω

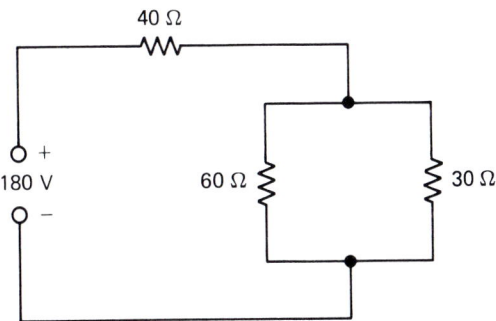

FIGURE 4-15 Circuit for Questions 21 through 25.

22. In Question 21, the voltage drop across the 30-Ω resistor is
    (a) 45 V   (b) 90 V   (c) 30 V   (d) 60 V   (e) 15 V
23. In Question 21, the current flowing through the 60-Ω resistor is
    (a) 3 A   (b) 2 A   (c) 0.5 A   (d) 1 A   (e) 4 A
24. In Question 21, the power dissipated in the 40-Ω resistor is
    (a) 120 W   (b) 540 W   (c) 360 W   (d) 180 W   (e) 90 W
25. In Fig. 4-15, the 30-Ω resistor open-circuits. The total power delivered from the source becomes
    (a) zero   (b) 324 W   (c) 129.6 W   (d) 194.4 W   (e) 820 W
26. In Fig. 4-16, the total source current is
    (a) 3 A   (b) 6.5 A   (c) 4.5 A   (d) 2.5 A   (e) 1.5 A

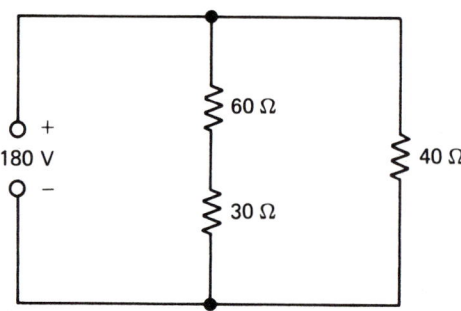

FIGURE 4-16 Circuit for Questions 26 through 30.

27. In Question 26, the voltage drop across the 60-Ω resistor is
    (a) 120 V   (b) 180 V   (c) 60 V   (d) 140 V   (e) 90 V
28. In Question 26, the current flowing through the 30-Ω resistor is
    (a) 6 A   (b) 3 A   (c) 4.5 A   (d) 2 A   (e) 1 A
29. In Question 26, the power dissipated in the 40-Ω resistor is
    (a) 405 W   (b) 810 W   (c) 360 W   (d) 180 W   (e) 1170 W
30. In Fig. 4-16, a short circuit develops across the 30-Ω resistor. The new total equivalent resistance is
    (a) 24 Ω   (b) 100 Ω   (c) 30 Ω   (d) 15 Ω
    (e) 17 Ω (approximately)
31. In Fig. 4-6(a), $R_1 = 180$ Ω, $R_2 = 390$ Ω, $R_3 = 270$ Ω, $R_x = 330$ Ω, $E_T = 15$ V, and meter G is replaced by a short circuit. The total equivalent resistance presented to the source voltage is
    (a) 276 Ω   (b) 1170 Ω   (c) 292 Ω   (d) 67 Ω   (e) 287 Ω
32. In Question 31, the total current drawn from the 15-V source is
    (a) 52.3 mA   (b) 224 mA   (c) 51.4 mA   (d) 12.8 mA   (e) 54.3 mA
33. In Question 31, the current flowing through the 330-Ω resistor is
    (a) 35.1 mA   (b) 18.6 mA   (c) 19.2 mA   (d) 32.8 mA   (e) 18.5 mA
34. In Question 31, the voltage drop across the 180-resistor is
    (a) 807 V   (b) 6.3 V   (c) 3.5 V   (d) 6.1 V   (e) 7.2 V
35. In Question 31, the current flowing through the 270-Ω resistor is
    (a) 19.2 mA   (b) 32.1 mA   (c) 22.2 mA   (d) 35.1 mA   (e) 24.8 mA
36. In Question 31, the voltage drop across the 390-Ω resistor is
    (a) 6.3 V   (b) 8.7 V   (c) 8.9 V   (d) 11.5 V   (e) 7.8 V
37. In Question 31, the direction of the electron flow is
    (a) from P to Q (b) from Q to P
    (c) either from P to Q or from Q to P, according to which of the source terminals is grounded
    (d) the same as the direction of the conventional flow in a bridge circuit
    (e) There is zero current in the short circuit between P and Q.
38. In Question 31, the current flowing in the short circuit between P and Q is
    (a) zero   (b) 54.3 mA   (c) 19.2 mA   (d) 32.1 mA   (e) 12.9 mA
39. In Question 31, the total power dissipated in the circuit is
    (a) 785 mW   (b) 3.36 W   (c) 771 mW   (d) 132 mW   (e) 815 mW
40. In Question 31, the value of R_x is changed so that the bridge is balanced. The new value of R_x is
    (a) 125 Ω   (b) 260 Ω   (c) 585 Ω   (d) 495 Ω   (e) 152 Ω

Practice Problems

41. In Fig. 4-17, what are the values of the current drawn from the 80-V source and the power dissipated in the 3.3-kΩ resistor?

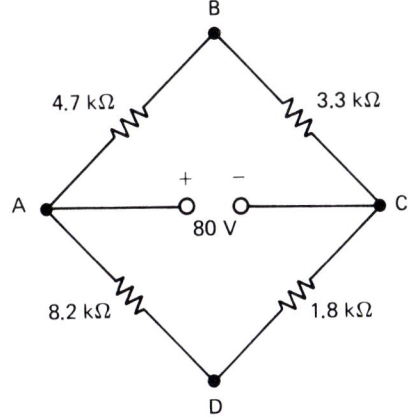

FIGURE 4-17 Circuit for Problems 41 and 42.

42. In Fig. 4-17, points B and D are joined by a connecting wire. What is the new value of the total equivalent resistance presented to the 80-V source?
43. Two loads, of 25 Ω and 4 Ω, are connected in parallel and each requires 100 V for its correct operation. What value of series dropping resistor must be used so that the loads may be correctly operated from a 130-V source? What is the power dissipated in the dropping resistor?
44. What series–parallel arrangement of four 100-kΩ resistors will produce a total equivalent resistance of 75 kΩ? If the circuit is connected across a 300-V source, what is the power dissipated in each of the resistors?
45. In Fig. 4-18, what are the values of the voltage across R_4 and the power dissipated in R_4?
46. In Fig. 4-18, the resistor R_2 open-circuits. What is the new value of the source current?

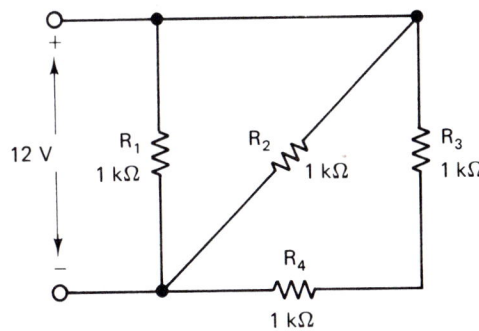

FIGURE 4-18 Circuit for Problems 45, 46, and 47.

47. In Fig. 4-18, a short circuit develops across R_3. What is the new value of the total equivalent resistance?
48. In Fig. 4-6(b), $R_1 = 100$ Ω and $R_2 = 1$ Ω. For the bridge to be balanced so that the reading of G is zero, the value of R_3 is adjusted to be 9.23 Ω. What is the value of the unknown resistance, R_X?
49. In the circuit of Fig. 4-19, what is the value of the potential at point C? If an additional 10-kΩ resistor is connected between points D and G, what is the new value of the potential at C?

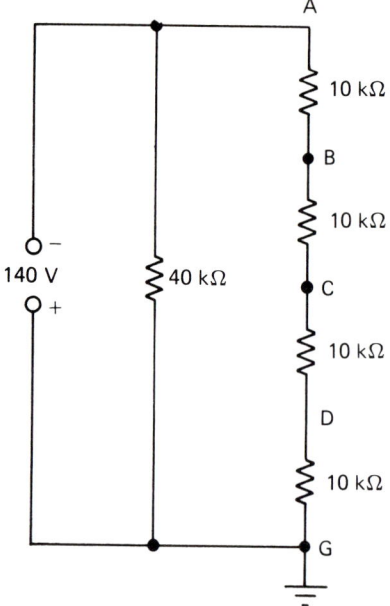

FIGURE 4-19 Circuit for Problem 49.

50. In Fig. 4-8, $V_{L2} = 300$ V, $V_{L3} = 150$ V, $I_{L3} = 40$ mA, and $I_{L2} = 60$ mA. If I_B is 15 mA, what is the value of R_2?

51. A loaded voltage divider circuit has a series dropping resistor of 1.8 kΩ with a supply voltage of 350 V. The load requires 250 V at 40 mA. What is the required value for the bleeder resistor?

52. Two 100-Ω 1-W resistors are connected in parallel. This combination is joined in series with another parallel combination consisting of two 200-Ω 1-W resistors. Assuming that none of the resistors' power ratings is exceeded, what is the maximum power capability of this series–parallel arrangement?

53. In Fig. 4-6(a), $E_T = 12$ V, $R_1 = 1.5$ kΩ, $R_2 = 2.7$ kΩ, $R_3 = 4.7$ kΩ, $R_X = 2.2$ kΩ. What is the current reading of meter G? Ignore any resistance associated with the meter.

54. In Problem 53, resistor R_X open-circuits. What is the new value of the current flowing through meter G?

55. In Problem 53, a short circuit develops across resistor R_2. What is the new value of the current flowing through meter G?

56. A circuit consists of two parallel resistors, 330 Ω and 470 Ω, respectively, connected in series with a 180-Ω resistor. If the current through the 180-Ω resistor is 45 mA, calculate the amount of the total power dissipated in the circuit.

57. A series combination of two resistors, whose values are 220 Ω and 680 Ω, is connected in parallel with an 820-Ω resistor across a dc source. If the current through the 220-Ω resistor is 32 mA, calculate the amount of the total power dissipated in the circuit.

58. Three resistors, whose values are 180 Ω, 270 Ω, and 330 Ω, are connected in parallel; two additional resistors, whose values are 120 Ω and 150 Ω, are also connected in parallel. The two parallel combinations are then joined in series across a dc source. If the current through the 120-Ω resistor is 37 mA, what is the value of the current through the 270-Ω resistor?

59. In Fig. 4-7(a), point B is used as an intermediate ground. What is the value of the potential at point C?

60. In Fig. 4-3(a), a short circuit occurs across resistor R_5. What is the new value of the total equivalent resistance?

Advanced Problems

61. In Fig. 4-20, what is the total equivalent resistance existing between (a) points B and D and (b) points C and D?

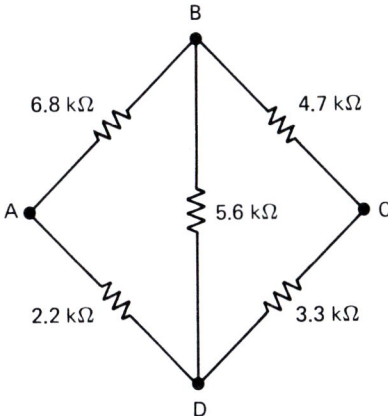

FIGURE 4-20 Circuit for Problems 61 and 62.

62. In Fig. 4-20, the 3.3-kΩ resistor open-circuits. What is the new total equivalent resistance between points A and B?

63. In Fig. 4-21, calculate the voltage drop and the power dissipated for each of the resistors. What is the value of the source voltage, E_T?

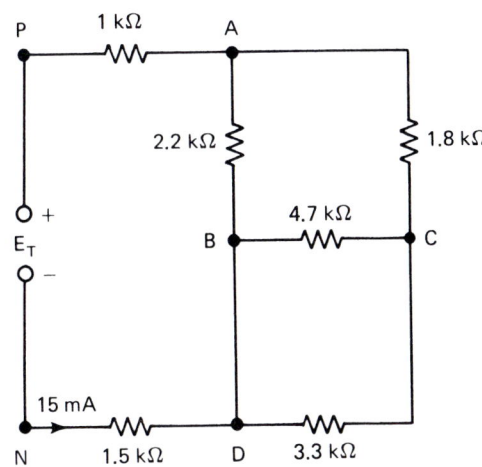

FIGURE 4-21 Circuit for Problems 63, 64, and 65.

64. In Fig. 4-21, a short circuit develops across the 4.7-kΩ resistor. Assuming that the source voltage remains unchanged, recalculate the voltage drop and the power dissipated for each of the resistors. What are the new values of the source current and the total power delivered from the source?

65. In Problem 63, point C is grounded. What is the potential at each of the other points in the circuit?

66. In Fig. 4-9, $R_1 = 470\ \Omega$, $R_2 = 270\ \Omega$, $R_3 = 150\ \Omega$, $I_{L1} = 170$ mA, $I_{L2} = 160$ mA, $I_{L3} = 120$ mA, and $I_B = 50$ mA. Calculate the values of V_{L1}, V_{L2}, and V_{L3}. If $E_T = 250$ V, what is the value of R_D and its power dissipation?

67. In Fig. 4-22, what is the value of the current flowing through the resistor R_5? If the resistor R_1 open-circuits, what is the value of the source current?

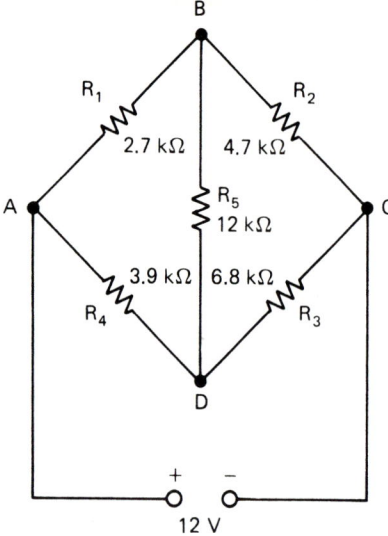

FIGURE 4-22 Circuit for Problem 67.

68. In Fig. 4-23, the power dissipated in the 3.9-kΩ resistor is 2 W. Calculate the value of the source voltage, E_T.

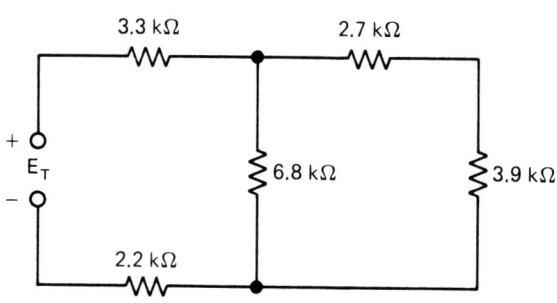

FIGURE 4-23 Circuit for Problem 68.

69. In Fig. 4-24, the IR drop across the 2.2-kΩ resistor is 5 V. Calculate the source current, I_T.

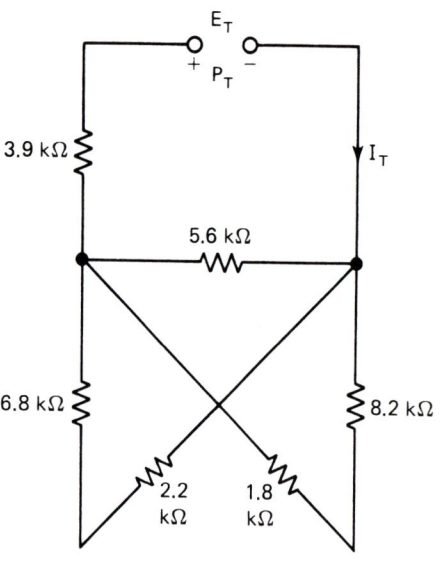

FIGURE 4-24 Circuit for Problems 69 and 70.

70. In Problem 69, what is the value of the total power delivered from the source?
71. In the circuit of Fig. 4-25, determine the appropriate resistor values for R_1, R_2, R_3, and R_4.

FIGURE 4-25 Circuit for Problems 71 and 72.

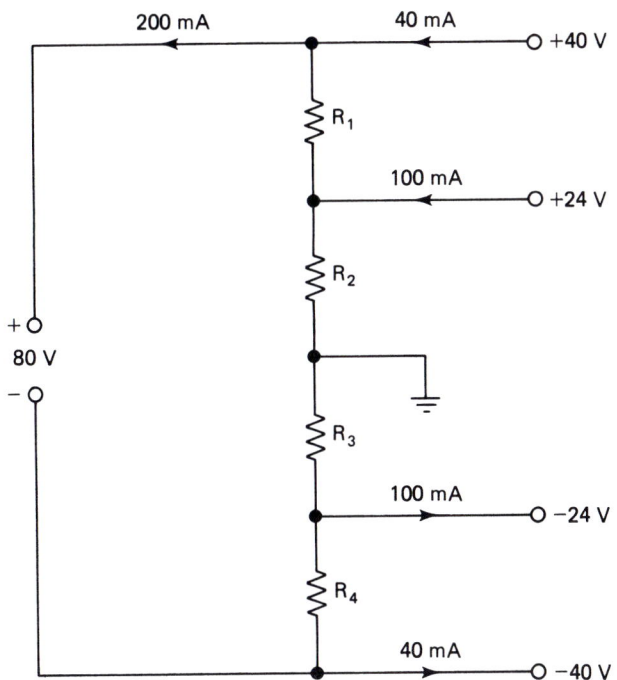

72. In Problem 71, how could the circuit be modified to operate from a 100-V supply?
73. In Fig. 4-26, calculate the values for resistors R_1, R_2, and R_3.

FIGURE 4-26 Circuit for Problems 73 and 74.

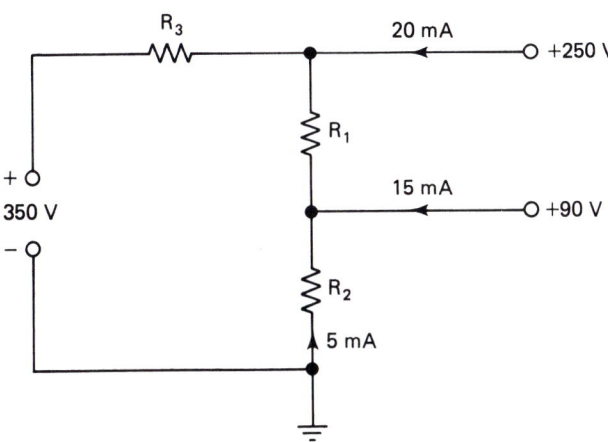

74. In Problem 73, recalculate the values for R_1, R_2, and R_3 if the total supply current is 50 mA. (*Note*: The bleeder current is no longer 5 mA.)
75. In Fig. 4-27, what is the amount of the voltage that exists across the open switch S? When the switch is closed, what is the amount of the current that flows through the switch?

FIGURE 4-27 Circuit for Problem 75.

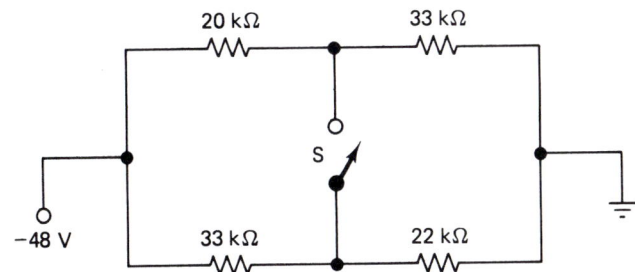

76. In Fig. 4-28, what is the total equivalent resistance between terminals *W* and *X* when (a) terminals *Y* and *Z* are open-circuited as shown, (b) a 100-Ω load is connected between terminals *Y* and *Z*, and (c) terminals *Y* and *Z* are short-circuited?

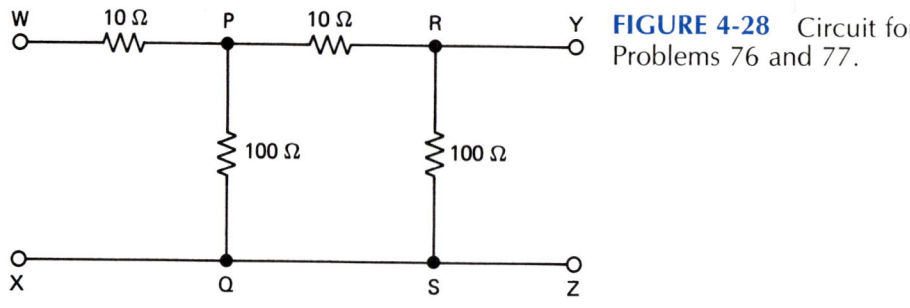

FIGURE 4-28 Circuit for Problems 76 and 77.

77. In Problem 76, a 100-Ω load is connected between terminals *Y* and *Z*, while a 100-V dc source is applied between terminals *W* and *X*. What is the value of the voltage (a) between points *P* and *Q*, (b) between points *R* and *S*, and (c) across the 100-Ω load?

78. Two loads are joined in parallel and then connected to a 120-V dc source by a line that possesses appreciable resistance. The current flowing through one load is 8 A and the resistance of the other load is 10 Ω. If the total voltage drop in the line equals 10% of the source voltage, calculate the values of (a) the power dissipated in the line, (b) the total power delivered to the two loads, and (c) the total power supplied by the source.

79. In Fig. 4-29, calculate the value of *R* if *I* = 240 mA.

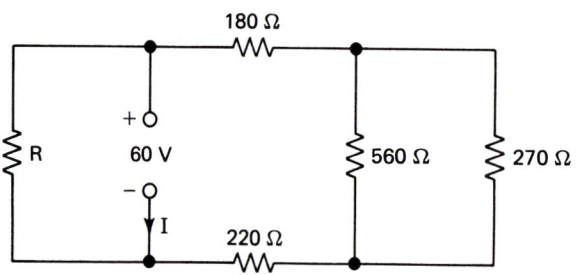

FIGURE 4-29 Circuit for Problems 79 and 80.

80. In Problem 79, a short circuit develops across the 270-Ω resistor. Calculate the new value of *R* if the value of *I* is unchanged.

5

Voltage and Current Sources

In the study of actual electrical sources, you will learn:

1. how primary and secondary cells create their output dc voltages.
2. why all electrical sources possess a certain internal resistance.
3. how an electrical source may be represented by a constant voltage model.
4. how to develop maximum power in the load by the process of matching.
5. that a higher voltage can be obtained by connecting a number of practical sources in series.
6. how an electrical source may be represented by a constant current model.
7. how to reduce the equivalent internal resistance by connecting a number of practical sources in parallel.
8. how to obtain a higher voltage and also reduce the equivalent internal resistance by connecting a number of practical sources in a series–parallel arrangement.

5-0 Introduction

In Sections 2-10 and 3-8 we learned about the properties of cells in series and parallel. However, the cells we discussed were, in a certain sense, idealized since it was assumed that the source voltage remained constant and independent of the value of the load current drawn from the source. This is not true in practice, as we well know from our experience with a car battery. When you have your battery tested at a garage, the voltage is first measured before you start up your car. The car is then started, which is equivalent to placing a load on the battery. Provided that the battery is "good," the measured voltage will fall slightly, but with a "bad" battery, the voltage will fall to a low level and the car will not start.

Most of this chapter is devoted to explaining why it is normal for a source voltage to fall as its load current is increased. The explanation, of course, lies in the voltage source itself, so we will start off by describing in detail the construction of two sources. These are the primary cell (examples of which are the D cells for flashlights) and the lead-acid car battery, which is called a secondary cell. Afterward, we develop two equivalent electrical circuits or models which will simulate the behavior of the practical electrical sources. The models are then loaded so that the equations relating the load's voltage, current, and power can be found. Finally, we examine the results of connecting practical cells in series and parallel.

5-1 Dry Primary Cell

All cells are capable of transforming chemical energy into electrical energy. They basically consist of two electrodes, which are the conductors enabling the electron flow to leave and return to the cell. Between the electrodes is a solution called the *electrolyte*, which acts with the electrodes to produce the electrical energy.

In a primary cell the chemical action eats away one of the electrodes, generally the negative, so that eventually either the electrode must be replaced or the cell must be thrown away. A primary cell of the type used in a flashlight can normally not be recharged, although recently some types have been developed to the point where recharging is possible. We are all familiar with the cordless hand tools whose cells may be recharged with the aid of the charging unit, which is plugged into the household supply.

Figure 5-1 shows the appearance and the construction of a general-purpose primary dry cell. The cylindrical zinc container is the negative electrode, while the positive electrode is the central carbon rod. The electrolyte is in a form of a paste, which is generally a mixture of ammonium chloride (sal ammoniac),

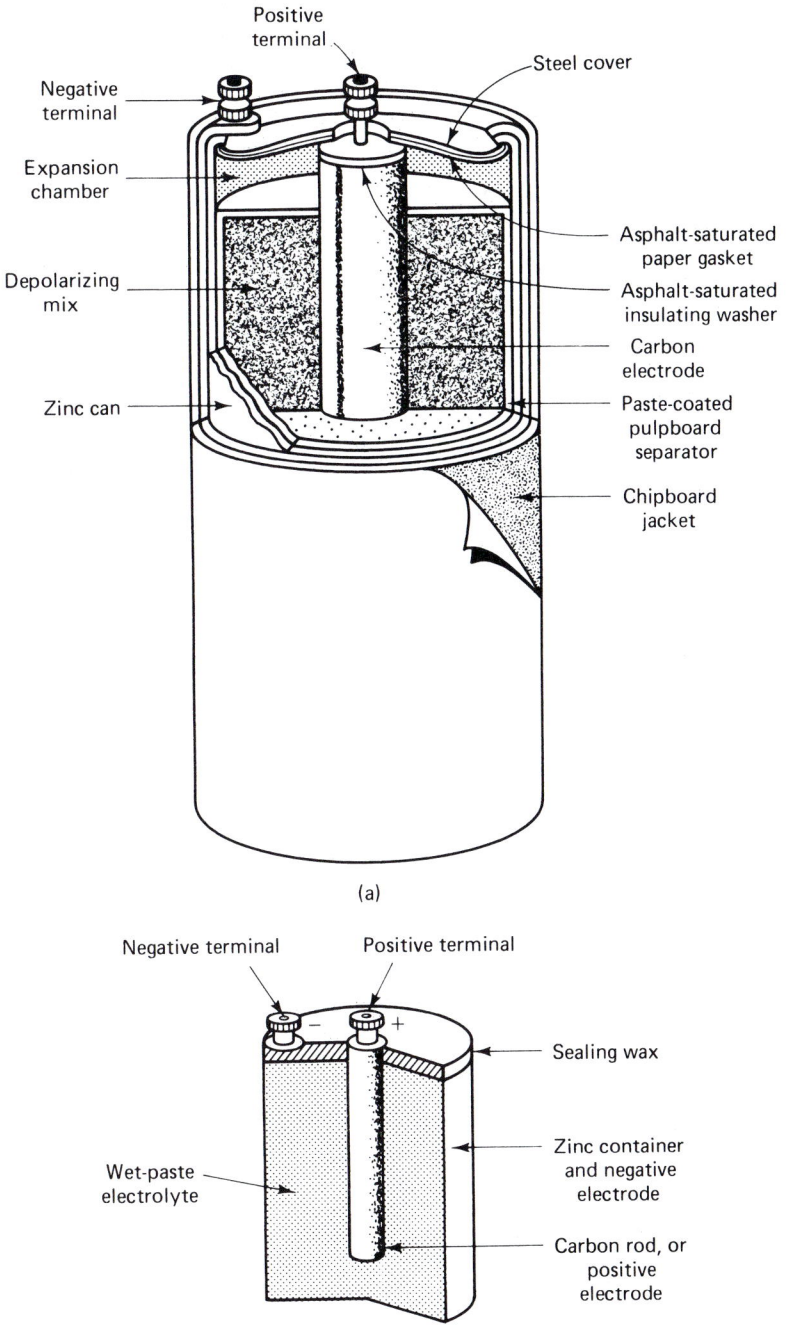

FIGURE 5-1 Primary zinc-carbon dry cell.

powdered coke, ground carbon, graphite, manganese dioxide, zinc chloride, and water.

The pulpboard separator is used to isolate the electrolyte from direct contact with the zinc container; at the same time this type of blotting paper allows the electrolyte to filter slowly through to the container. When the electrolyte is packed in between the carbon rod and the blotting paper, a small expansion space is left at the top of the cell. The terminal posts are then connected to the electrodes and the cell is sealed with asphalt-saturated cardboard to prevent air from entering the cell and drying out the electrolyte.

Sec. 5-1 Dry Primary Cell

Finally, since the zinc container is one of the electrodes, it must be covered with an insulating jacket of chipboard.

The chemical action of the cell involves the zinc and carbon electrodes in conjunction with the sal ammoniac and the water in the electrolyte; the resulting voltage produced between the electrodes is about 1.5 V. As the current is drawn from the cell, the zinc electrode is oxidized and is slowly dissolved in the electrolyte. However, the same chemical action which provides the cell's voltage also produces hydrogen bubbles, which form around the positive electrode until the entire surface may be surrounded; this phenomenon is called *polarization*. It has the result of decreasing the voltage output between the terminals, so we must eliminate the formation of the hydrogen bubbles. In the primary dry cell this is done by introducing a depolarizing agent such as manganese dioxide, which supplies enough free oxygen to combine with all the hydrogen bubbles and create water. Zinc chloride is also added to prevent local action when the cell is idle.

The current that a primary cell may deliver to a load depends on the resistance of the entire circuit, including that of the cell itself. This *internal resistance* of the primary cell depends on the size of the electrodes, their separation, the resistance of the electrolyte, and the degree of any polarization that is present. The larger the electrodes and the less their separation, the lower is the internal resistance and the greater is the current that may be supplied to a load. It follows that the greater the physical size of a primary cell, the less is its internal resistance.

To summarize this section, we have learned that the dry primary cell has a terminal voltage of approximately 1.5 V and an internal resistance which depends on its construction. In addition, most primary cells cannot be recharged and must eventually be discarded.

5-2 Secondary Cell

Secondary cells use the same chemical principles as primary cells, but whereas one of the electrodes in a primary cell is gradually consumed, the substances in a secondary cell are merely converted into another form. The process is reversible, so that a discharged secondary cell may be recharged to its original state by using another source to force a current through the cell in the direction opposite to that during the discharge.

The most common example of the secondary cell is the lead-acid battery, which acts as an electrochemical device for storing energy until it is released as electrical energy. Such a battery contains a number of secondary cells, each producing approximately 2 V. As an example, the 12-V car battery would consist of six cells in series. The construction of this battery is shown in Fig. 5-2. Each of the cell elements has positive and negative lead plates which are insulated from each other by suitable separators and submerged in an electrolyte of a sulfuric acid solution [Fig. 5-2(a)].

The lead electrodes are normally plates which are formed by applying lead oxide pastes to grids made from a lead-antimony alloy [Fig. 5-2(b)]. Each grid is designed to give the plates sufficient mechanical strength, keep the lead oxide in place, and allow conductivity for the electrical current produced by the chemical action.

During manufacture the lead oxide paste is applied to the grids and allowed to dry. An electrochemical forming process then converts the paste of all positive electrodes into lead peroxide and the paste of all negative electrodes into pure sponge lead; both the lead peroxide and the sponge lead offer little

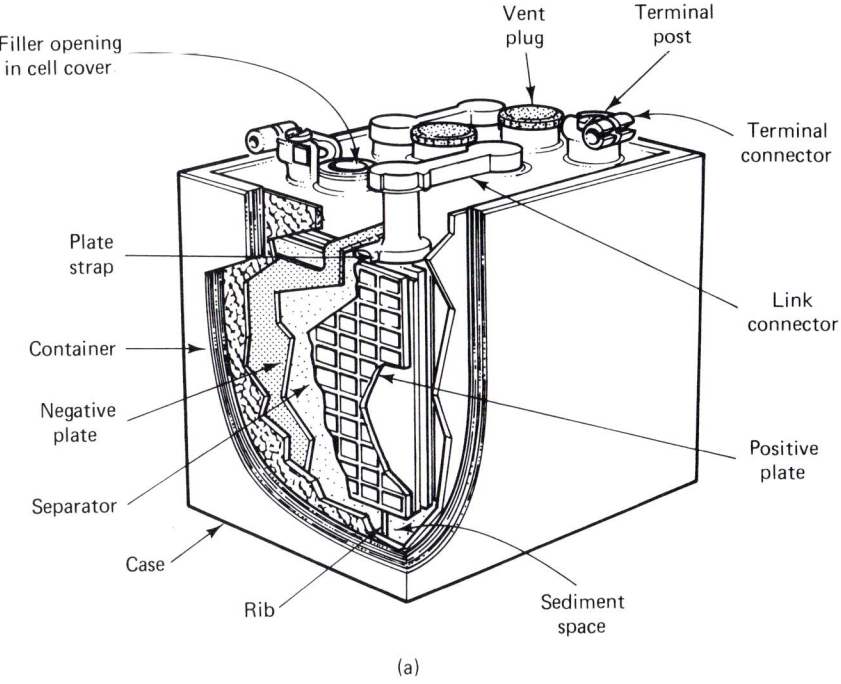

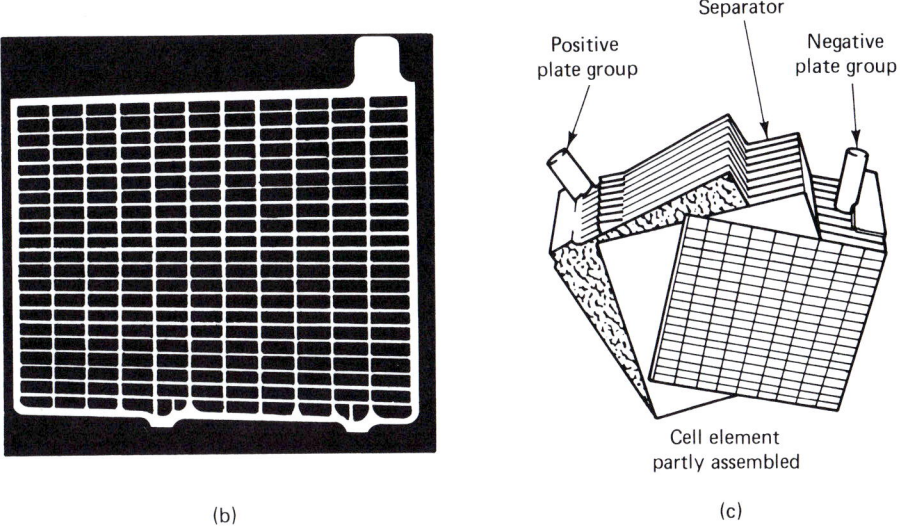

FIGURE 5-2 Secondary lead-acid battery.

resistance. The plates are then built into positive and negative groups, as shown in Fig. 5-2(c). The negative group has one more plate than the positive group, so that when the two sides of the positive plates are acted upon chemically, their expansion and contraction are equal on both sides to prevent buckling. The cell elements are then formed by assembling the positive and negative groups with their separators. These separators are grooved vertically on one side but are smooth on the other side. The grooved side is placed next to the positive plate and allows free circulation of the acid electrolyte around the lead peroxide.

The negative terminal of each cell element is positioned close to the positive terminal of the next cell. The links then join the cell elements in series by providing both an electrical and a physical connection. These link

Sec. 5-2 Secondary Cell

connectors must be sufficiently large so that they do not overheat when the battery is supplying its full-load current. Finally, the output terminals of the battery are distinguished from each other by their markings of "+" and "−."

The covers for the cells and the battery container are generally made from a bituminous composition which is resistant to acid and mechanical shock. Located in the cells' cover are the filler openings, which contain the vent plugs. These plugs prevent leakage or loss of the electrolyte while allowing the escape of gases that form inside the cells.

In its fully charged state [Fig. 5-3(a)] the active elements of the battery are the positive electrodes of lead peroxide, the negative electrodes of pure sponge lead, and the electrolyte, which is roughly composed of one part of concentrated sulfuric acid to four parts of water. The resultant solution has a 1.260 specific gravity, which compares the weight of a given volume of the solution to the same volume of water; in other words, the solution is 1.260 times heavier than water. The value of the specific gravity is measured with a hydrometer and can therefore be adjusted to the required level.

Initially, the pores of both electrodes are filled with the electrolyte. However, as the battery discharges into a load, the acid in contact with the plates is separated from the electrolyte; the results are chemical actions in which both the lead peroxide and the sponge lead are partly converted into lead sulfate. As the discharge continues, more lead sulfate forms on the plates and more acid is taken from the electrolyte [Fig. 5-3(b)]. Consequently, the electrolyte's specific gravity will gradually decrease so that a measurement of the specific gravity can be used as a guide in determining the battery's state of discharge. Ultimately, a point is reached when so much of the positive and negative active material has been converted into lead sulfate that the battery can no longer produce sufficient current and must be recharged from an external source [Fig. 5-3(c)]; the electrolyte's specific gravity is then about 1.110.

During the recharging period [Fig. 5-3(d)] the battery is properly connected, as regards polarity, to a dc source whose voltage is slightly higher than that of the battery. The result is to drive a current through the battery in the opposite direction to the previous discharge current. The acid held in the lead sulfate is then driven back in the electrolyte, so that the specific gravity is raised. At the same time the lead sulfate is converted into lead peroxide for the positive electrode and into sponge lead for the negative electrode. Toward the end of the recharge, hydrogen gas is liberated at the negative plate and oxygen at the positive plate. These gases that escape through the vent plugs are the result of the charging current ionizing the water in the electrolyte when there only remains a small amount of lead sulfate to be converted.

The addition of more sulfuric acid to a discharged battery merely increases the specific gravity of the electrolyte and does not recharge the cell; only a charging current can convert the lead sulfate back into the lead peroxide and the sponge lead.

The capacity of a lead-acid battery is rated in ampere-hours (Ah), which is a measure of the total charge stored (1 Ah = 3600 C). This capacity is primarily determined by the battery's physical size and is equal to the product of the discharge current in amperes and the time in hours during which the battery is supplying the current. Many batteries are rated on a 20-h discharge time, which means that the fully charged battery will be completely discharged after the 20-h period. Therefore, if a battery can deliver a load current of 25 A continuously for 20 h, its capacity will be 25 × 20 = 500 Ah. In equation form:

$$\text{capacity in ampere-hours} = I \times t \quad (5\text{-}2\text{-}1)$$

where I = discharge current (A)
       t = discharge time (h)

The most important feature of the lead-acid battery is its extremely low internal resistance, which is typically less than 0.01 Ω. Such a battery is therefore capable of delivering load currents in excess of 100 A. This low resistance

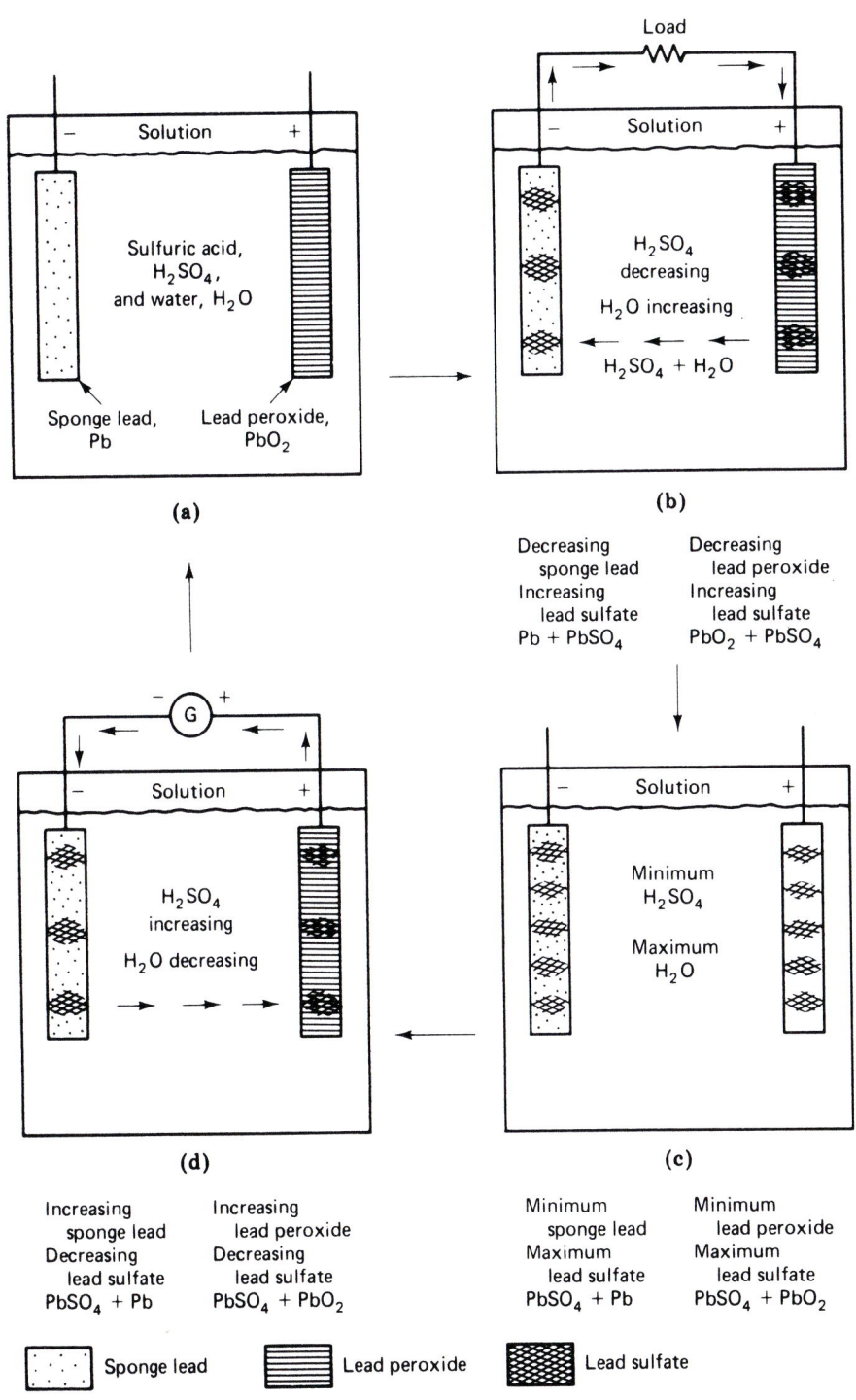

FIGURE 5-3 Charging and discharging processes in the lead acid cell: (a) charged; (b) discharging; (c) discharged; (d) recharging.

Sec. 5-2 Secondary Cell

is due to the large area of the electrodes, their separation, and the high conductivity of the electrolyte. As the battery is discharged, its internal resistance increases so that it is no longer capable of delivering a high current; the internal resistance also increases as a battery ages until ultimately the battery is of no use.

To summarize, a lead-acid cell element is capable of being recharged and produces an EMF of aproximately 2 V. Its capacity is measured in ampere-hours and its internal resistance is extremely low.

EXAMPLE 5-1 A lead-acid battery is rated on a 20-h discharge time. If its capacity is 300 Ah, what is the value of the continuous discharge current?

Solution Using Eq. (5-2-1), we obtain

$$\text{discharge current} = \frac{300 \text{ Ah}}{20 \text{ h}}$$

$$= 15 \text{ A}$$

5-3

Constant Voltage Source

We have learned in Sections 5-1 and 5-2 that both primary and secondary cells contain a certain internal resistance. In general, all electrical sources possess internal resistance and as a result, the voltage existing between the output terminals of an electrical source will fall as the load current is increased. We now need to have some sort of simple model to explain this effect. One such model involves the assumption that the source contains a constant voltage EMF, E, which is in series with the internal resistance, R_i (Fig. 5-4). The constant voltage EMF is then regarded as independent of the load current, although the internal resistance may change for a variety of reasons, such as the loading of the source and its age.

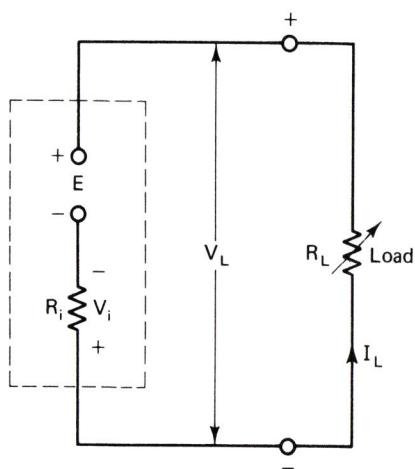

FIGURE 5-4 Constant voltage source.

Let us see how the model explains what happens with an electrical source as a varying load is applied. If the load resistance, R_L, is decreased, the load current, I_L, will rise; this will cause an increase in the voltage drop, V_i, across the internal resistance, R_i, and therefore the terminal voltage, which is the same as the load voltage, V_L, will drop. To summarize, the higher the load current, the lower the terminal voltage, and this is the normal behavior of an electrical source.

Since Fig. 5-4 is actually a series circuit, the load current, I_L, may be found either by dividing V_L by R_L, or dividing V_i by R_i, or dividing E by the total resistance of $R_i + R_L$. In equation form:

$$\text{load current, } I_L = \frac{V_L}{R_L} = \frac{V_i}{R_i} = \frac{E}{R_i + R_L} \qquad (5\text{-}3\text{-}1)$$

$$\text{load voltage, } V_L = I_L R_L \qquad (5\text{-}3\text{-}2)$$

The load voltage may also be found by subtracting the internal voltage drop from the current voltage EMF. Therefore,

$$\text{load voltage, } V_L = E - I_L R_i \qquad (5\text{-}3\text{-}3)$$

$$= E - \frac{E R_i}{R_i + R_L}$$

$$= E \times \frac{R_L}{R_i + R_L} \qquad (5\text{-}3\text{-}4)$$

Equation (5-3-4) is an example of the voltage division rule, in which the source voltage, E, is divided between the load resistance, R_L, and the internal resistance, R_i. To obtain a high voltage across the load will require that the load resistance is large compared with the internal resistance; typically, R_L must be at least 5 to 10 times the value of R_i.

Having derived these equations, we must now have some means of finding the model values of E and R_i for a particular electrical source. The value of E is best measured by disconnecting the load and replacing it with a voltmeter between the terminals. Assuming that the voltmeter draws virtually no current from the source, the voltage drop across the internal resistance is negligible. Consequently, the voltage measured between the terminals will be equal to the constant voltage EMF, E. Since the resistance of the voltmeter is so high, it is practically equivalent to an open circuit and, we can refer to the value of E as the *open-circuit terminal voltage*.

If an ammeter of negligible resistance is connected between the terminals and replaces the load, the value of the effective load resistance is virtually zero and only the source's internal resistance will limit the current. Since the terminals have practically been short circuited, a very large current may flow; its value will be E/R_i, which is normally referred to as the *short-circuit terminal current*. The internal resistance may therefore be calculated from

$$\text{internal resistance, } R_i = \frac{\text{open-circuit terminal voltage, } E}{\text{short-circuit terminal current, } E/R_i} \qquad (5\text{-}3\text{-}5)$$

It may not be practical to measure the short-circuit current, since such a large current may possibly damage the electrical source. A better method of finding the internal resistance is to vary the value of the load resistance, R_L, until the terminal voltage, V_L, is equal to $E/2$; V_i will also be $E/2$ and the measured value of R_L will then equal the internal resistance, R_i.

To illustrate what happens when the load resistance, and therefore the load current, is varied, let us consider an electrical source whose open-circuit terminal voltage is 24 V and whose short-circuit terminal current is 100 A; the source's internal resistance is then 24 V/100 A = 0.24 Ω. With the aid of Eq. (5-3-3), the values of the load voltage, V_L, are calculated for a number of load currents, I_L; the results are shown in Table 5-1. The values of the load voltage and the load current in Table 5-1 have been plotted on the graph illustrated in Fig. 5-5. Because we have assumed that the internal resistance is constant, the shape of the graph is a straight line; we can then say that there is a linear relationship between the load voltage and the load current.

TABLE 5-1 Values of Load Voltage for Various Load Currents

I_L (A)	$V_i = I_L \times R_i$ (V)	$V_L = E - V_i$ (V)	$R_L = V_L/I_L$ (Ω)
0	0	24	Infinite ohms, open circuit
10	2.4	21.6	2.16
20	4.8	19.2	0.96
30	7.2	16.8	0.56
40	9.6	14.4	0.36
50	12	12	0.24
60	14.4	9.6	0.16
70	16.8	7.2	0.103
80	19.2	4.8	0.06
90	21.6	2.4	0.027
100	24	0	Zero ohms, short circuit

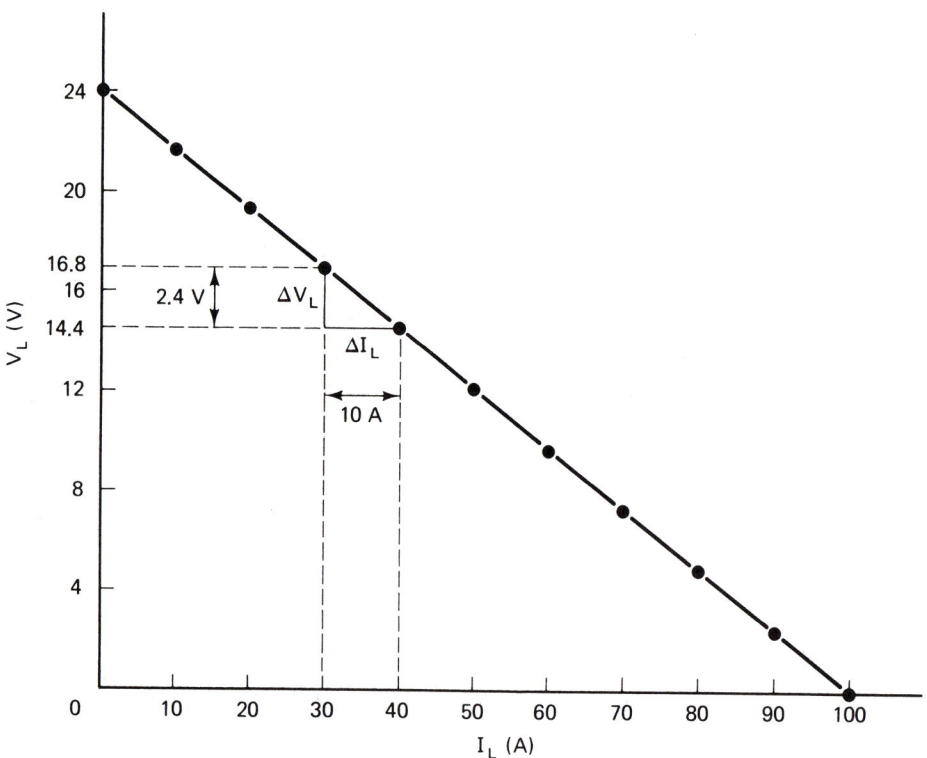

FIGURE 5-5 Graph of load voltage/load current from values shown in Table 5-1.

As shown in Fig. 5-5, the value of the internal resistance may be found from the slope of the line. If the load current increases by a small amount, ΔI_L ("Δ" is a mathematical symbol which means "a small change in"), the rise in the voltage drop across the internal resistance, R_L, is $\Delta I_L \times R_i$, and the load voltage will *fall* by a corresponding amount ΔV_L. Therefore,

$$\Delta V_L = R_i \times \Delta I_L \text{ and } R_i = \frac{\Delta V_L}{\Delta I_L} \qquad (5\text{-}3\text{-}6)$$

From Table 5-1, the change in load current from 30 A to 40 A corresponds to a change in load voltage from 16.8 V to 14.4 V. Then $\Delta I_L = 40 - 30 = 10$ A, $\Delta V_L = 16.8 - 14.4 = 2.4$ V, and $R_i = \Delta V_L/\Delta I_L = 24 \text{ V}/10 \text{ A} = 0.24 \, \Omega$, which is the same as the internal resistance value previously calculated.

Since the load voltage falls as the load current increases, the small changes in V_L and I_L are in opposite directions, with the result that the slope of the V_L/I_L graph is negative. If a graphical relationship between any two quantities looks like "/," the slope is positive, but if the appearance is "\," the slope is negative.

With many electrical sources their internal resistance is not constant but depends on the amount of the load current drawn from the source. Then the load voltage-load current graph will be some form of curve and not a straight line. This curve shows the degree of voltage regulation which the source possesses. The term "voltage regulation" measures the extent to which V_L changes as I_L is varied. Ideal regulation means that the load voltage remains constant and independent of any changes in the load current. For a practical source the degree of voltage regulation is expressed as a percentage and is calculated from

$$\text{percentage voltage regulation} = \frac{V_{NL} - V_{FL}}{V_{FL}} \times 100\% \qquad (5\text{-}3\text{-}7)$$

where V_{NL} = terminal voltage under no-load or minimum-load conditions
V_{FL} = terminal voltage under full-load conditions when the load resistance is R_{FL}

For an ideal source $V_{FL} = E$ and the percentage regulation is zero. The regulation "curve" is then a horizontal line, which shows that the load voltage is constant and independent of the load current.

Under no-load conditions, $V_{NL} = E$, while under full-load conditions, $V_{FL} = E \times R_{FL}/(R_i + R_{FL})$. Then

$$\begin{aligned}
\text{percentage regulation} &= \frac{V_{NL} - V_{FL}}{V_{FL}} \times 100 \\
&= \frac{E - \dfrac{ER_{FL}}{R_i + R_{FL}}}{ER_{FL}/(R_i + R_{FL})} \times 100 \qquad (5\text{-}3\text{-}8) \\
&= \frac{R_i}{R_{FL}} \times 100\%
\end{aligned}$$

In Table 5-1, let us assume that the full-load current is 20 A, which corresponds to a load resistance of 0.96 Ω and a load voltage of 19.2 V. Since the no-load voltage is 24 V, the percentage regulation is

$$\frac{V_{NL} - V_{FL}}{V_{FL}} \times 100 = \frac{24 - 19.2}{19.2} \times 100 = \frac{4.8}{19.2} \times 100 = 25\%$$

Alternatively, the percentage regulation is

$$\frac{R_i}{R_{FL}} \times 100 = \frac{0.24}{0.96} \times 100 = 25\%$$

To summarize, the presence of internal resistance causes a fall in a source's terminal voltage as the load current is increased. The behavior is explained by assuming that the source contains a constant voltage EMF in series with an internal resistance. The condition for a high load voltage then requires that the load resistance is large compared with the internal resistance. The variation of the load voltage with changes in the load current is illustrated by a regulation curve and is measured in terms of the percentage regulation.

EXAMPLE 5-2 The open-circuit terminal voltage and the short-circuit current of an electrical source are 36 V and 50 A. If the full-load resistance is 6 Ω, calculate the full-load voltage and the percentage regulation.

Solution

$$\text{internal resistance, } R_i = \frac{E}{E/R_i} \quad (5\text{-}3\text{-}5)$$

$$= \frac{36 \text{ V}}{50 \text{ A}} = 0.72 \text{ }\Omega$$

Using Eq. (5-3-4), we obtain

$$\text{full-load voltage, } V_{FL} = \frac{36 \text{ V} \times 6 \text{ }\Omega}{6 \text{ }\Omega + 0.72 \text{ }\Omega} = 32.14 \text{ V}$$

From Eq. (5-3-7),

$$\text{percentage regulation} = \frac{36 - 32.14}{32.14} \times 100 = 12\%$$

Check

$$\text{percentage regulation} = \frac{R_i}{R_{FL}} \quad (5\text{-}3\text{-}8)$$

$$= \frac{0.72}{6} \times 100 = 12\%$$

5-4 Maximum Power Transfer and Circuit Efficiency

There are many occasions in communications and electronics where it is necessary to use the condition for maximum power to be developed in the load. One example is the delivery of radio-frequency power from a transmitter (the source) to an antenna (the load); another is the transfer of audiofrequency power from the final stage of a receiver to the loudspeaker load.

In Table 5-1 the load voltage under open-circuit conditions is 24 V, but the corresponding load current is zero. Since the load power is equal to the product of I_L and V_L, its open-circuit value is zero. The load power is also zero for short-circuit conditions, since although I_L is 100 A, V_L is zero. Between these two extreme cases the load power achieves a maximum value that must correspond to a particular load resistance. Using Eq. (5-3-1),

$$\text{load power, } P_L = I_L^2 R_L = \frac{E^2 R_L}{(R_i + R_L)^2} \qquad (5\text{-}4\text{-}1)$$

Referring back to Fig. 5-4, we want to find the condition for P_L to reach its maximum level as we vary the value of R_L. This is normally done by using a calculus solution, but it is possible to obtain the necessary result with the aid of algebra alone. When P_L is at its peak, its reciprocal $1/P_L$ must have its minimum value. Then

$$\frac{1}{P_L} = \frac{(R_i + R_L)^2}{E^2 R_L} = \frac{1}{E^2} \times \left[\frac{(R_L - R_i)^2}{R_L} + \frac{4R_i R_L}{R_L} \right]$$
$$= \frac{(R_L - R_i)^2}{E^2 R_L} + \frac{4R_i}{E^2} \qquad (5\text{-}4\text{-}2)$$

Since the square of a number (whether positive or negative) is always positive, the minimum value of $(R_L - R_i)^2$ must be zero. This will happen when

$$R_L = R_i \qquad (5\text{-}4\text{-}3)$$

This shows that in order to obtain the maximum load power, we must adjust the value of the load resistance until it is equal to the internal resistance of the source. This process is called *matching*, so that when we make R_L equal to R_i, we say that we have matched the load resistance to the internal resistance. As an example, an antena load is matched to the final stage of its transmitter; the radio-frequency power output of that stage is then transferred to the load of the antenna.

When matching is achieved, the maximum load power, $P_{L\max}$, is given by Eq. (5-4-2).

$$\frac{1}{P_{L\max}} = \frac{(R_L - R_i)^2}{E^2 R_L} + \frac{4R_i}{E^2} = 0 + \frac{4R_i}{E^2}$$

when $R_i = R_L$. Then

$$P_{L\max} = \frac{E^2}{4R_i} = \frac{E^2}{4R_L} \qquad (5\text{-}4\text{-}4)$$

and the corresponding load current is

$$I_L = \frac{E}{2R_i} = \frac{E}{2R_L} \qquad (5\text{-}4\text{-}5)$$

This means that to produce maximum power in the load, we can adjust the value of the load resistance until the load current is equal to $E/2R_i$. Apart from maximum power transfer to the load, we must also consider the circuit efficiency; this is the percentage ratio of the load power, P_L, to the total power, P_T, drawn from the constant voltage EMF. Since

$$P_L = I_L^2 R_L, \quad P_T = I_L^2(R_i + R_L)$$

then

$$\begin{aligned}
\text{circuit efficiency} &= \frac{P_L}{P_T} \times 100\% \\
&= \frac{I_L^2 R_L}{I_L^2(R_i + R_L)} \times 100\% \qquad (5\text{-}4\text{-}6) \\
&= \frac{R_L}{R_i + R_L} \times 100\%
\end{aligned}$$

Since from Eq. (5-3-4), $V_L/E = R_L/(R_i + R_L)$, the condition for a large load voltage must be the same as for a high circuit efficiency. However, under matched conditions, $R_L = R_i$ and the powers in the load and the internal resistance are equal. The circuit efficiency for maximum power transfer is only 50%, and therefore in practice it may be necessary to compromise between the load power and the circuit efficiency.

The conditions we have obtained for a large voltage, maximum load power, and a high circuit efficiency can best be illustrated by means of an example. In the circuit of Fig. 5-6 the model of an electrical source has a constant voltage EMF equal to 48 V and its short-circuit terminal current is 6 A; the internal resistance is therefore 48 V/6 A = 8 Ω. The load resistance, R_L is now varied from infinite ohms (open circuit) to zero ohms (short circuit), and for particular load resistances, the corresponding values of I_L, V_L, P_L, and percentage efficiency are calculated as given in Table 5-2.

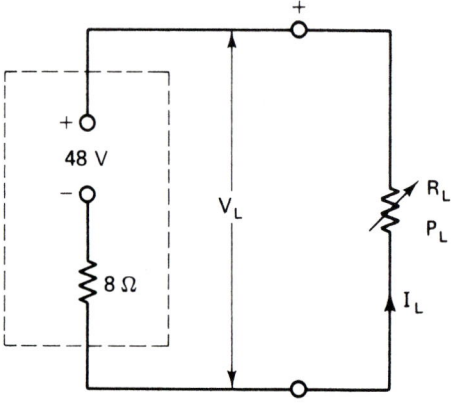

FIGURE 5-6 Circuit related to the values shown in Table 5-2.

TABLE 5-2 Variation of R_L From Infinite to Zero ohms

R_L (Ω)	$I_L = E/(R_i + R_L)$ (A)	$V_L = I_L \times R_L$ (V)	$P_L = I_L \times V_L$ (W)	Efficiency $= P_L/P_T \times 100\%$ $= V_L/E \times 100\%$
Open circuit, infinite (∞) ohms	0	48	0	100
88	0.5	44	22	91.67
40	1	40	40	83.33
24	1.5	36	54	75
16	2	32	64	66.67
8	3	24	72	50
4	4	16	64	33.33
2	4.8	9.6	46.08	20
Short circuit, zero ohms	6	0	0	0

Figure 5-7 shows the graphs of V_L, P_L, I_L, and the circuit efficiency plotted against R_L. When $R_L = 8\,\Omega$ and is therefore matched to the internal resistance, the load power reaches its peak of 72 W, but the load voltage is only 24 V and the circuit efficiency is down to 50%. By contrast, in Table 5-2 when R_L is equal to 88 Ω, which is more than 10 times the internal resistance, the load voltage is 44 V and the circuit efficiency is 91.67%; however, the load power is only 22 W. It is therefore impossible to achieve a high load voltage and a maximum power output simultaneously; as a result, there is the need for the compromise between the load power and the circuit efficiency.

The results of Table 5-2 are also illustrated in Fig. 5-8, which shows the relationships among V_L, P_L, circuit efficiency, and I_L. This is a superior presentation since the load voltage and the percentage efficiency graphs are straight lines and the P_L curve is a symmetrical parabola.

We have learned in this section about one of the most important concepts in the field of electronics, the principle of matching, in which the load resistance is made equal to the internal resistance of the source. The purpose of matching is to achieve maximum power transfer to the load, although it may be necessary to compromise between the load power and the circuit efficiency.

EXAMPLE 5-3 In Fig. 5-4 $E = 12$ V, $R_i = 1.5\,\Omega$, and $R_L = 6\,\Omega$. What is the value of the additional resistance which, when added in parallel with R_L, will result in the maximum power being transferred to the new total load? Calculate the values of the load power before and after the introduction of the additional resistance.

Solution For maximum power transfer, the total load must be matched to the 1.5-Ω internal resistance [Eq. (5-4-3)]. Substituting in Eq. (3-3-4), we obtain

$$\text{value of the additional resistance} = \frac{1.5 \times 6.0}{6.0 - 1.5} = 2\,\Omega$$

$$\text{original load power, } P_L = \frac{E^2 R_L}{(R_i + R_L)^2} \qquad (5\text{-}4\text{-}1)$$

$$= \frac{(12\text{ V})^2 \times 6\,\Omega}{(1.5\,\Omega + 6\,\Omega)^2}$$

$$= 15.36\text{ W}$$

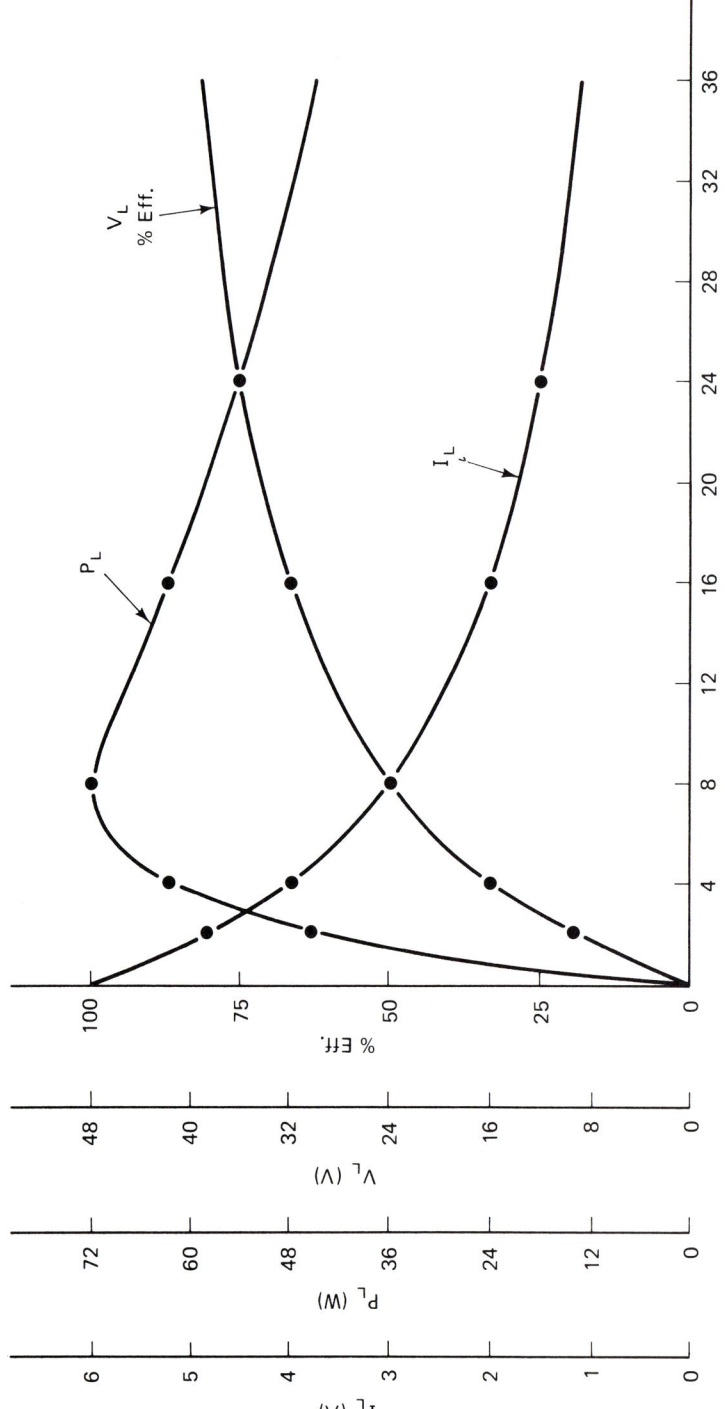

FIGURE 5-7 Graphs of load voltage, load current, load power, and percentage efficiency versus load resistance.

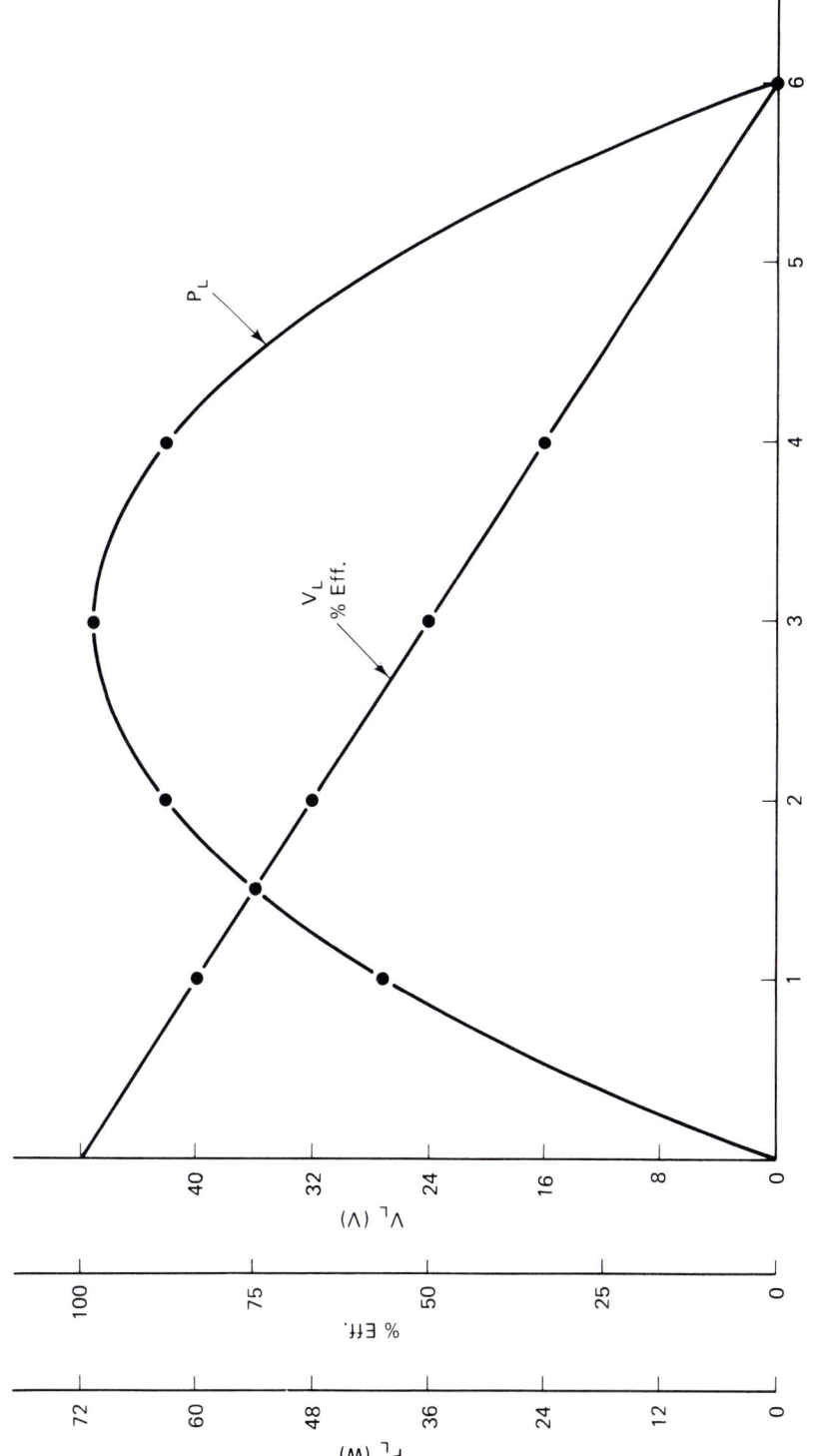

FIGURE 5-8 Relationships among I_L and V_L, P_L, and percentage efficiency.

$$\text{final (maximum) load power } P_{L_{max}} = \frac{E^2}{4R_i} \tag{5-4-4}$$

$$= \frac{(12 \text{ V})^2}{4 \times 1.5 \text{ }\Omega}$$

$$= 24 \text{ W}$$

EXAMPLE 5-4 In Fig. 5-9, the electrical source has a constant voltage EMF of 20 V and an internal resistance of 2.5 Ω. The source is connected to a load, R_L, through an additional series resistance of 1.5 Ω. What is the value of R_L that will permit maximum power transfer to the load? Calculate the maximum load power involved. What would be the new values of load power if the chosen value of R_L were (a) halved and (b) doubled?

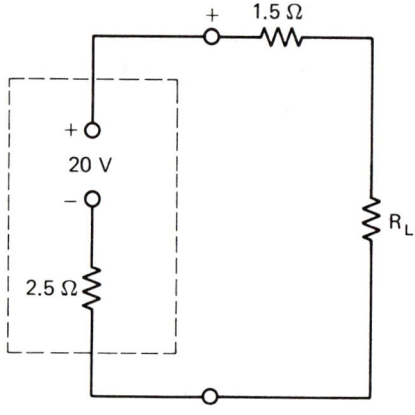

FIGURE 5-9 Circuit for Example 5-4.

Solution For the maximum power transfer to the load, the value of R_L must be matched to the total of all the resistances that are not associated with the load. Therefore, the required value for R_L is 2.5 + 1.5 = 4 Ω. Then

$$\text{maximum load power} = \frac{E^2}{4R_L} = \frac{(20 \text{ V})^2}{4 \times 4 \text{ }\Omega} = 25 \text{ W}$$

(a) When R_L is doubled to 8 Ω, we obtain [Eq. (5-4-1)]

$$\text{load power} = \frac{(20 \text{ V})^2 \times 8 \text{ }\Omega}{(2.5 \text{ }\Omega + 1.5 \text{ }\Omega + 8 \text{ }\Omega)^2} = 22.2 \text{ W}$$

(b) When R_L is halved to 2 Ω, we get

$$\text{load power} = \frac{(20 \text{ V})^2 \times 2 \text{ }\Omega}{(2.5 \text{ }\Omega + 1.5 \text{ }\Omega + 2 \text{ }\Omega)^2} = 22.2 \text{ W}$$

It is worth noting that doubling and halving the matched value of R_L always produces equal load powers.

5-5 Practical Sources in Series

When the cells of Section 2-10 were connected in series-aiding, the total EMF available was the sum of the individual EMFs. However, these cells were idealized since their internal resistances were ignored. When practical dc sources are joined in series and are replaced by their constant voltage models (Fig. 5-10), the total no-load terminal voltage will still be the sum of the individual EMFs. This increase in the voltage available is the usual reason for joining sources in series. However, the internal resistances are also in series and must therefore be added to calculate the total equivalent internal resistance. Consequently, the greater the number of voltage sources in series, the higher is their total internal resistance and the worse is their voltage regulation; this is the real disadvantage of connecting sources in series. Referring to Fig. 5-10,

$$\text{total no-load terminal voltage, } E_T = E_1 + E_2 + E_3 + \ldots + E_N \quad (5\text{-}5\text{-}1)$$

and

$$\text{total internal resistance, } R_{iT} = R_{i1} + R_{i2} + R_{i3} + \ldots + R_{iN} \quad (5\text{-}5\text{-}2)$$

If all the sources are identical and each has a constant voltage EMF, E, with an internal resistance, R_i,

$$E_T = NE \text{ and } R_{iT} = NR_i \quad (5\text{-}5\text{-}3)$$

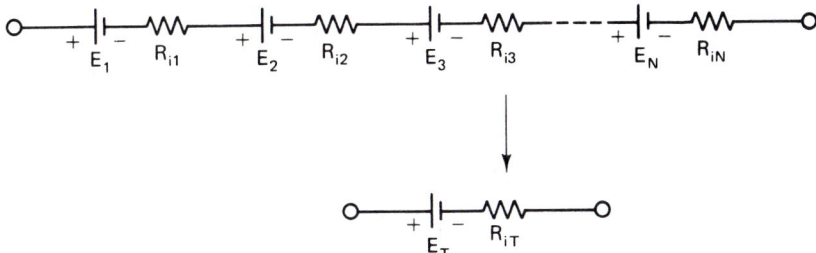

FIGURE 5-10 Cells in series-aiding.

EXAMPLE 5-5 Let Fig. 5-10 represent 10 voltage sources in series, each with an EMF of 2.4 V and an internal resistance of 0.2 Ω. What load resistance should be connected across the sources for maximum power transfer to the load? Calculate the value of the maximum power.

Solution Using Eq. (5-5-3), we have

$$\text{total EMF, } E_T = 10 \times 2.4 \text{ V} = 24 \text{ V}$$

$$\text{total internal resistance, } R_{iT} = 10 \times 0.2 = 2 \text{ Ω}$$

For maximum power transfer to the load,

$$\text{load resistance, } R_L = R_{iT} = 2 \, \Omega$$

$$\text{maximum load power} = \frac{E_T^2}{4R_i} \qquad (5\text{-}4\text{-}4)$$

$$= \frac{(24 \text{ V})^2}{4 \times 2 \, \Omega} = 72 \text{ W}$$

5-6 Constant Current Source

We know from observation that as more load current is drawn from a practical electrical source, the terminal voltage across the load falls. The effect has already been explained in Section 5-4 by assuming that the source contains a *constant voltage* generator *in series* with an internal resistance. But is this the only possible model that can explain the fall in the terminal voltage? An alternative is to consider that the source contains a *constant current* generator which is capable of generating a variable EMF; the value of the constant current is the short-circuit terminal current already discussed in Section 5-3. Across this current generator is a parallel internal resistance which has the same value as the series internal resistance in the constant voltage model.

The comparison between the constant voltage and the constant current generators can best be illustrated by an example in which a practical electrical source has a measured open-circuit voltage of 50 V and a short-circuit current of 10 A; then the internal resistance is 50 V/10 A = 5 Ω. The models for the constant current and constant voltage generators are shown in Fig. 5-11. The arrow convention in the current generator symbol indicates the same electron flow direction as produced by the constant voltage generator.

Constant Voltage Generator

Constant voltage EMF, E = open-circuit terminal voltage

Series internal resistance,

$$R_i = \frac{\text{open-circuit voltage, } E}{\text{short-circuit current, } I}$$

Constant Current Generator

Constant current, I = short-circuit terminal current

Parallel internal resistance,

$$R_i = \frac{\text{open-circuit voltage, } E}{\text{short-circuit current, } I}$$

We shall now connect a load of 20 Ω across each of the generator circuits. With the constant voltage model the load voltage is

$$V_L = 50 \text{ V} \times \frac{20 \, \Omega}{20 \, \Omega + 5 \, \Omega} = 40 \text{ V} \qquad \text{(voltage division rule)}$$

and the load current is 40 V/20 Ω = 2 A. For the constant-current model the load current is

$$I_L = 10 \text{ A} \times \frac{5 \, \Omega}{20 \, \Omega + 5 \, \Omega} = 2 \text{ A} \qquad \text{(current division rule)}$$

and the load voltage, $V_L = 2 \text{ A} \times 20 \, \Omega = 40$ V. By this example we have shown that as far as the external load is concerned, the concepts of constant voltage and constant current generators are equally valid. However, you should

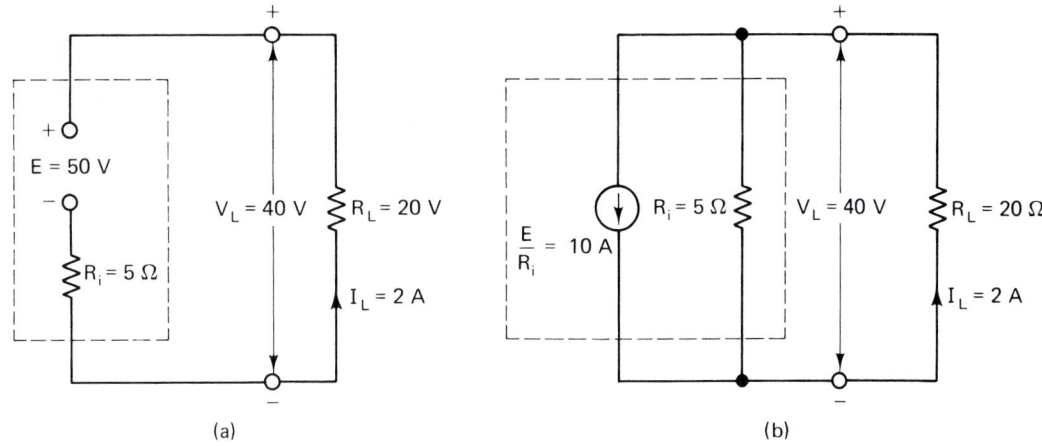

FIGURE 5-11 Equivalent constant voltage and constant current generators.

realize that the generators themselves are not exact equivalents since under open-circuit load conditions, there is power dissipated in the internal resistance of the current generator but not in the internal resistance of the voltage generator.

The question arises: "Why do we need the constant current generator concept when we already have the constant voltage generator?" As shown in Section 5-5, different sources in series can readily be analyzed with the aid of the constant voltage model. However, if these sources are connected in parallel, they cannot be combined in any easy way if equivalent constant voltage generators are used. In Section 5-7 this difficulty will be solved by the introduction of constant current generators which will allow a simple analysis of parallel sources.

We have learned in this section that a practical source may be regarded as equivalent to either a constant voltage or a constant current generator. There are simple relationships between the generators so that you can easily transpose between these two concepts. The constant voltage generator is mainly used for series sources, while the constant current generator is more applicable to sources in parallel.

5-7 Practical Sources in Parallel

In Section 3-8 we combined idealized parallel cells with zero internal resistance and discovered that the total voltage available was equal to the EMF of only one cell. However, these cells in parallel had the advantage of sharing the total load current.

If practical sources are connected in parallel, their equivalent circuit with constant voltage generators is as shown in Fig. 5-12(a). As you will realize in Chapter 6, this circuit is difficult to analyze and will require the use of Kirchhoff's laws. However, if each parallel source is replaced by its equivalent constant current generator [Fig. 5-12(b)], the individual currents may simply be added to produce the total generator current. The equivalent internal resistance can then be calculated by applying the reciprocal formula to the individual parallel internal resistances. When we have obtained the final equivalent current generator, we can transform it, if necessary, into the constant voltage model.

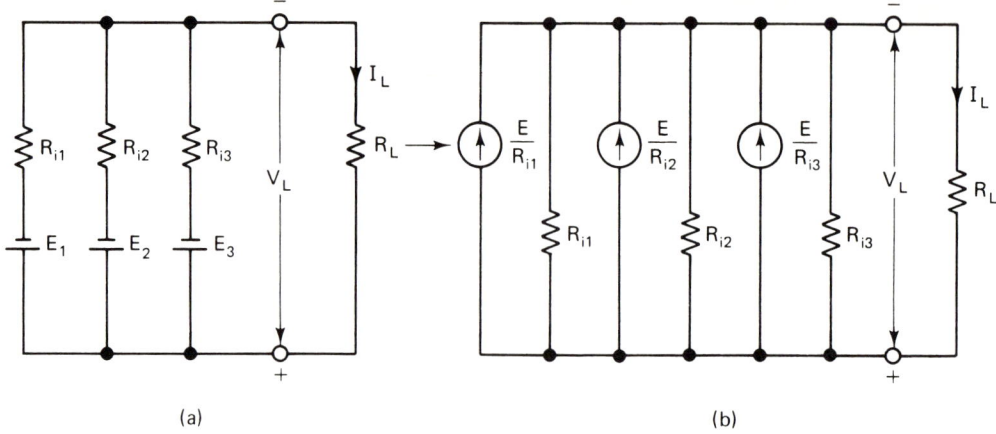

FIGURE 5-12 Constant voltage and constant current equivalent circuits for cells in parallel.

If N practical sources are connected in parallel with all terminals of the same polarity joined together, then

$$I_T = I_1 + I_2 + I_3 + \ldots + I_N \qquad (5\text{-}7\text{-}1)$$

and

$$\frac{1}{R_{iT}} = \frac{1}{R_{i1}} + \frac{1}{R_{i2}} + \frac{1}{R_{i3}} + \ldots + \frac{1}{R_{iN}} \qquad (5\text{-}7\text{-}2)$$

For the total equivalent constant voltage generator,

$$E_T = I_T \times R_{iT} \qquad (5\text{-}7\text{-}3)$$

If the N parallel sources are identical,

$$I_T = NI \qquad (5\text{-}7\text{-}4)$$

$$R_{iT} = \frac{R_i}{N} \qquad (5\text{-}7\text{-}5)$$

and

$$E_T = I_T \times R_{iT} = NI \times \frac{R_i}{N} = I \times R_i \qquad (5\text{-}7\text{-}6)$$

I and R_i are the equivalent constant current and internal resistance of each individual source. Since $E_T = I \times R_i$, the total equivalent voltage is equal to the constant voltage of one source.

The advantage of connecting sources in parallel is not to increase the output voltage but to decrease the equivalent internal resistance. This will produce an improvement in the degree of voltage regulation which is the result of the total load current being shared between the sources.

The advantages of the series and parallel connections can be combined by the use of series-parallel arrangements. If N sources are connected in series and each has a constant voltage EMF, E, with an internal resistance R_i, the total voltage available is NE with an effective internal resistance, of NR_i. However, if the same series arrangement is repeated with M parallel banks (Fig. 5-13), the total voltage is still NE but the internal resistance has been reduced to $(N/M)R_i$. The disadvantage of the series—parallel arrangement is obvious since a total of $N \times M$ cells is required and the complete unit is large, expensive, and cumbersome.

The main purpose of this section has been to show that the best method of combining sources in parallel is to replace the sources by their equivalent constant current generators. The total current of the final generator is then the sum of the generator's separate currents and the total parallel internal resistance is calculated by applying the reciprocal formula to the individual internal resistances.

EXAMPLE 5-6 In Fig. 5-12(a), $E_1 = 15$ V, $E_2 = 20$ V, $E_3 = 16$ V and $R_{i1} = 0.3$ Ω, $R_{i2} = 0.5$ Ω, $R_{i3} = 0.8$ Ω. If the load resistance is 1.8 Ω, what is the value of the load current, I_L?

Solution The values of the constant current generators are

$$I_1 = \frac{15 \text{ V}}{0.3 \text{ Ω}} = 50 \text{ A}$$

$$I_2 = \frac{20 \text{ V}}{0.5 \text{ Ω}} = 40 \text{ A}$$

$$I_3 = \frac{16 \text{ V}}{0.8 \text{ Ω}} = 20 \text{ A}$$

total current for the equivalent generator, $I_T = I_1 + I_2 + I_3$ \hfill (5-7-1)

$$= 50 \text{ A} + 40 \text{ A} + 20 \text{ A}$$

$$= 110 \text{ A}$$

Substitution into Eq. (5-7-2) gives the total equivalent internal resistance:

$$\frac{1}{R_{iT}} = \frac{1}{0.3} + \frac{1}{0.5} + \frac{1}{0.8} = 3.333 + 2.0 + 1.25 = 6.583$$

$$R_{iT} = \frac{1}{6.583} = 0.152 \text{ Ω}$$

By the current division rule,

$$I_L = 110 \text{ A} \times \frac{0.152 \text{ Ω}}{1.8 \text{ Ω}} = 9.3 \text{ A}$$

EXAMPLE 5-7 Ten sources are connected in parallel and each source has a constant voltage EMF of 2.0 V with an internal resistance of 0.15 Ω. What value of load resistance must be connected across the parallel combination for maximum power transfer? Calculate the maximum load power obtained.

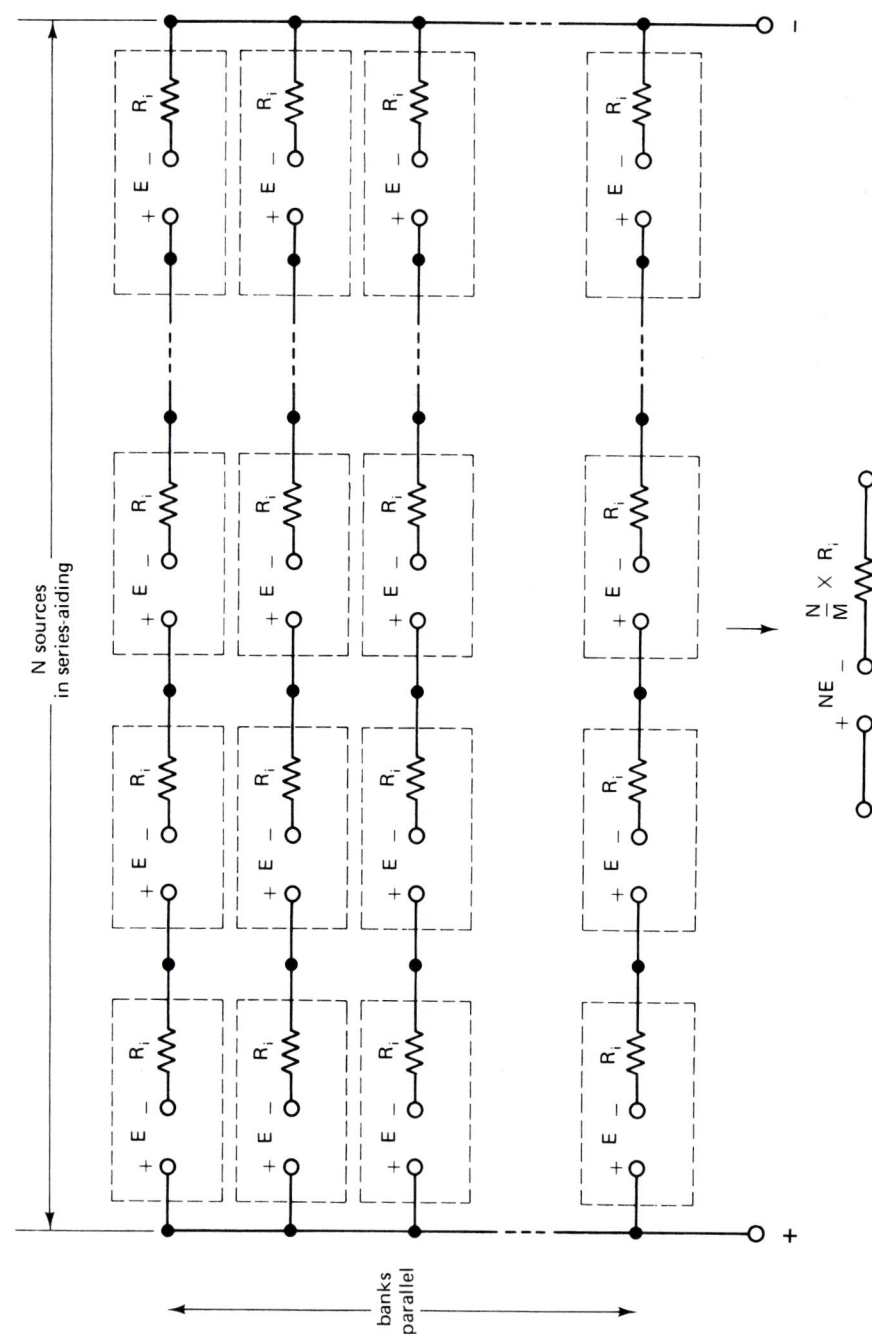

FIGURE 5-13 Series–parallel arrangement of electrical sources.

Solution From Eq. (5-7-6),

$$\text{total EMF, } E_T = \text{voltage of one source} = 2.0 \text{ V}$$

$$\text{total internal resistance, } R_{iT} = \frac{R_i}{N} \quad (5\text{-}7\text{-}5)$$

$$= \frac{0.15 \text{ }\Omega}{10} = 0.015 \text{ }\Omega$$

Using Eq. (5-4-3), we obtain

$$\text{value of load resistance for maximum power transfer} = R_{iT} = 0.015 \text{ }\Omega$$

$$\text{maximum load power} = \frac{E^2}{4R_{iT}} \quad (5\text{-}4\text{-}4)$$

$$= \frac{(2.0 \text{ V})^2}{4 \times 0.015 \text{ }\Omega} = 66.7 \text{ W}$$

EXAMPLE 5-8 Ten cells, each with an EMF of 1.4 V and an internal resistance of 0.08 Ω, are connected in series-aiding. Six such banks are then joined in parallel so that the total number of individual cells is $6 \times 10 = 60$. For such an arrangement calculate the total terminal voltage available and the value of the total internal resistance. Find the maximum load power when the load resistance is matched to the total internal resistance.

Solution

$$\text{total EMF, } E_T = 10 \times 1.4 \text{ V} = 14 \text{ V}$$

$$\text{total internal resistance, } R_{iT} = \frac{10 \times 0.08 \text{ }\Omega}{6} = 0.133 \text{ }\Omega$$

From Eq. (5-4-4),

$$\text{maximum load power} = \frac{(14 \text{ V})^2}{4 \times 0.133 \text{ }\Omega} = 367.5 \text{ W}$$

Chapter Summary

5.1. The dry zinc-carbon primary cell has a terminal voltage of approximately 1.5 V and an appreciable internal resistance. It normally cannot be recharged so that it must eventually be discarded.

5.2. The lead-acid secondary cell element can be recharged and has a terminal voltage of approximately 2 V with a very low internal resistance. Its capacity is measured in ampere-hours (Ah), which is equal to $I \times t$, where I is the discharge current in amperes and t is the discharge time in hours.

5.3. Constant voltage source (Fig. 5-4):

$$\text{constant voltage EMF, } E = \text{open-circuit terminal voltage}$$

$$\text{series internal resistance, } R_i = \frac{\text{open-circuit terminal voltage, } E}{\text{short-circuit terminal current, } E/R_i}$$

$$\text{load current, } I_L = \frac{V_L}{R_L} = \frac{V_i}{R_i} = \frac{E}{R_i + R_L}$$

$$\text{load voltage, } V_L = I_L R_L = E - I_L R_i = \frac{E \times R_L}{R_i + R_L}$$

$$\text{voltage regulation percentage} = \frac{V_{NL} - V_{FL}}{V_{FL}} \times 100\%$$

$$\text{load power, } P_L = I_L V_L = I_L^2 R_L = \frac{V_L^2}{R_L} = \frac{E^2 R_L}{(R_i + R_L)^2}$$

For maximum power transfer to the load, $R_L = R_i$ (matching).

$$\text{maximum load power, } P_{L\max} = \frac{E^2}{4R_i} = \frac{E^2}{4R_L}$$

$$\text{efficiency percentage} = \frac{P_L}{P_T} \times 100\%$$

5.4. Practical sources in series (Fig. 5-10):

$$\text{total no-load terminal voltage, } E_T = E_1 + E_2 + E_3 + \cdots + E_N$$

$$\text{total internal resistance, } R_{iT} = R_{i1} + R_{i2} + R_{i3} + \cdots + R_{iN}$$

If the series sources are identical,

$$E_T = NE \quad \text{and} \quad R_{iT} = NR_i$$

5.5. The constant current source (Fig. 5-11):

$$\text{constant current, } I = \text{short-circuit terminal current, } E/R_i$$

$$\text{parallel internal resistance, } R_i = \frac{\text{open-circuit terminal voltage, } E}{\text{short-circuit terminal current, } E/R_i}$$

$$\text{load current, } I_L = I \times \frac{R_i}{R_i + R_L}$$

5.6. Practical sources in parallel (Fig. 5-12):

$$\text{total generator current, } I_T = I_1 + I_2 + I_3 + \cdots + I_N$$

$$\text{total internal resistance, } R_{iT} = \frac{1}{\frac{1}{R_{i1}} + \frac{1}{R_{i2}} + \frac{1}{R_{i3}} + \cdots + \frac{1}{R_{iN}}}$$

If the parallel sources are identical,

$$I_T = NI$$

$$R_{iT} = \frac{R_i}{N}$$

$$E_T = E \quad \text{(open-circuit voltage of one source)}$$

Self-Review

True–False Questions

1. T. F. Unlike secondary cells, primary cells are always capable of being charged.
2. T. F. The capacity of a lead-acid battery is measured in ampere-hours.
3. T. F. If the load curent drawn from an electrical source is increased, the terminal voltage falls.
4. T. F. A high load voltage requires that the value of the internal resistance is 5 to 10 times the value of the load resistance.
5. T. F. For maximum power transfer to the load, the value of the load resistance must be matched to the internal resistance of the source.
6. T. F. Under the matched condition the circuit efficiency is 100%.
7. T. F. An ideal voltage source has a percentage regulation of 100%.
8. T. F. The internal resistance of a constant current generator is placed in series with the generator.
9. T. F. The disadvantage of connecting electrical sources in series is the increase in the effective internal resistance, so that the percentage regulation increases.
10. T. F. The advantage of connecting electrical sources in parallel is the increase in the available terminal voltage.
11. T. F. The capacity of a cell is measured in amperes per hour (Ah).
12. T. F. When a load is connected, the terminal voltage of a dc source is reduced to one-half of its open-circuit value. The resistance of the load is then equal to the internal resistance of the source.
13. T. F. When a number of cells are connected in series-aiding, their total Ah capacity is equal to the product of one cell's capacity and the number of cells.
14. T. F. A voltage source that has 0% voltage regulation has an effective internal resistance of zero ohms.
15. T. F. The purpose of connecting cells in parallel is to create a power supply with a greater current capacity and a lower internal resistance.
16. T. F. The graph of load power versus load current is symmetrical about the point that represents the maximum power transfer condition.
17. T. F. When the circuit efficiency is 100%, the power developed in the load is always at its maximum value.
18. T. F. The value of the load resistance is equal to the result of dividing the no-load voltage by the load current.
19. T. F. When a load is connected, the terminal voltage is one-third of the open-circuit voltage. The load resistance is then one-half of the internal resistance.
20. T. F. The internal resistance of a constant voltage source has the same value as the internal resistance of the equivalent constant current source.

21. In Fig. 5-14, the voltage of the load resistance is five times the internal resistance. The load voltage is
    (a) 24 V   (b) 20 V   (c) 30 V   (d) 32 V   (e) 28 V

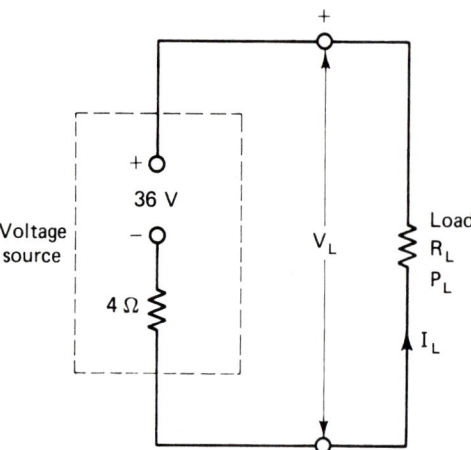

FIGURE 5-14 Circuit for Questions 21 through 27.

22. In Question 21, the value of the load power is
 (a) 45 W   (b) 9 W   (c) 54 W   (d) 36 W   (e) 24 W
23. In Fig. 5-14, the no-load voltage is equal to the open-circuit terminal voltage. When the load resistance is five times the internal resistance, the percentage regulation is
 (a) 5%   (b) 20%   (c) 10%   (d) 15%   (e) 30%
24. In Question 21, the circuit efficiency is
 (a) 90%   (b) 95%   (c) 75%   (d) 70%   (e) 83.3%
25. In Fig. 5-14, the load resistance is changed to match the internal resistance. The maximum load power is
 (a) 81 W   (b) 90 W   (c) 144 W   (d) 72 W   (e) 162 W
26. The constant current generator equivalent of Fig. 5-14 has a current value of
 (a) 32 A   (b) 9 A   (c) 4.5 A   (d) 18 A   (e) 36 A
27. The constant current generator of Question 26 is connected to a load resistance of 8 Ω. The load current is
 (a) 0.75 A   (b) 6 A   (c) 3 A   (d) 1.25 A   (e) 1.5 A
28. Twelve 2-V cells, each with an internal resistance of 0.15 Ω, are connected in series-aiding across a 5.4-Ω load. The load voltage is
 (a) 16 V   (b) 24 V   (c) 18 V   (d) 12 V   (e) 1.5 V
29. Twelve 2-V cells, each with an internal resistance of 0.15 Ω, are connected in parallel across a 0.0875-Ω load. The load current is
 (a) 20 A   (b) 240 A   (c) 1.06 A   (d) 12.7 A   (e) 8.42 A
30. One hundred and twenty 3-V cells, each with an internal resistance of 0.6 Ω, are connected in 12 parallel banks with every bank containing 10 cells in series-aiding. If the series–parallel arrangement is connected across a 1-Ω load, the load power is
 (a) 9 W   (b) 900 W   (c) 600 W   (d) 400 W   (e) 300 W
31. In Fig. 5-12(a), $E_1 = 12$ V, $E_2 = 10$ V, $E_3 = 15$ V, and $R_{i1} = 3$ Ω, $R_{i2} = 2$ Ω, $R_{i3} = 5$ Ω. The constant current generator, which is equivalent to this arrangement of cells, has a current value of
 (a) 12 A   (b) 11 A   (c) 13 A   (d) 14 A   (e) 15 A
32. In Question 31, the value of the total equivalent internal resistance is
 (a) 0.968 Ω   (b) 1.43 Ω   (c) 10 Ω   (d) 1.2 Ω   (e) 1.07 Ω
33. In Question 31, the open-circuit voltage across the output terminals is
 (a) 9.83 V   (b) 15.6 V   (c) 12.3 V   (d) 15.7 V   (e) 11.6 V
34. In Question 31, what is the value of R_L that will allow maximum power transfer to the load?
 (a) 10 Ω   (b) 1.43 Ω   (c) 1.07 Ω   (d) 0.968 Ω   (e) 1.2 Ω
35. In Question 34, the value of the maximum power dissipated in the load is
 (a) 966 W   (b) 227 W   (c) 106 W   (d) 35 W   (e) 205 W
36. In Question 31, a load of 1.8 Ω is connected across the output terminals. The value of the load voltage is
 (a) 8.32 V   (b) 7.54 V   (c) 8.42 V   (d) 7.83 V   (e) 8.13 V

37. In Question 36, the value of the load power is
    (a) 38.5 W   (b) 31.6 W   (c) 39.4 W   (d) 34.1 W   (e) 36.7 W
38. In Question 36, the value of the circuit efficiency is
    (a) 75%   (b) 68%   (c) 65%   (d) 73%   (e) 71%
39. In Question 31, a load of 0.5 Ω is connected across the output terminals. The value of the load power is
    (a) 31.2 W   (b) 60.4 W   (c) 91.7 W   (d) 37.6 W   (e) 42.3 W
40. In Question 39, the value of the circuit efficiency is
    (a) 66%   (b) 34%   (c) 43%   (d) 57%   (e) 27%

Practice Problems

41. The discharge rate for a lead-acid cell is 6 A over a period of 8 h. Calculate in coulombs the amount of charge lost by the battery.
42. In Fig. 5-15, what is the value of R_L that will allow maximum power transfer to the load? Calculate the maximum load power.

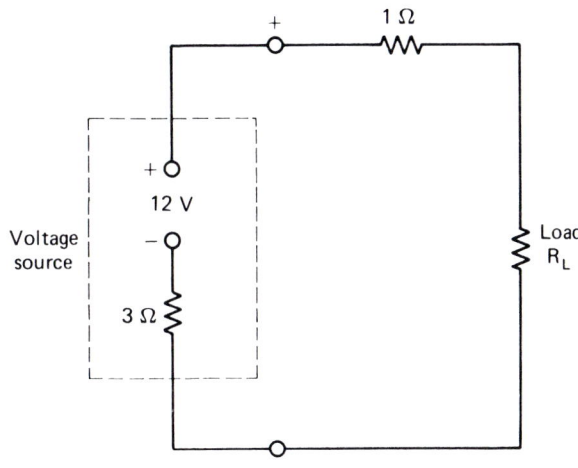

FIGURE 5-15 Circuit for Problem 42.

43. A source has an open-circuit voltage of 24 V. When a load current of 4 A is drawn from the source, the terminal voltage falls to 22 V. Calculate the values of the source's internal resistance, the load resistance, and the short-circuit terminal current. What new value of load resistance will reduce the load voltage to half the value of the open-circuit voltage?
44. For the constant voltage source of Fig. 5-16, derive the equivalent constant current generator and draw its circuit.

FIGURE 5-16 Circuit for Problem 44.

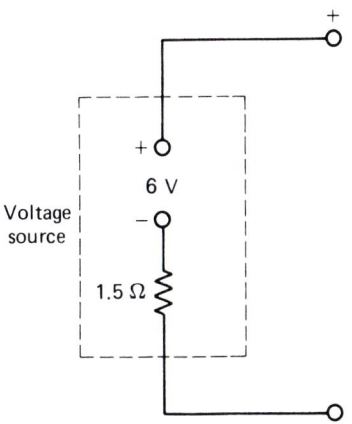

45. For the constant current generator of Fig. 5-17, derive the equivalent constant voltage source and draw its circuit. What is the polarity of the point X?

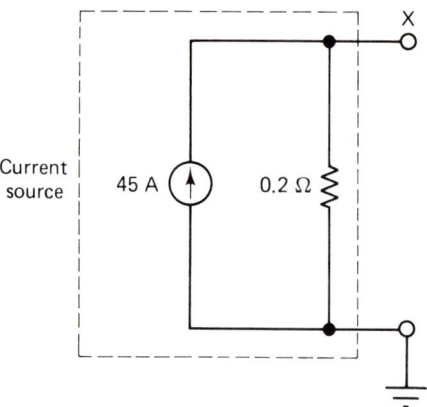

FIGURE 5-17 Circuit for Problem 45.

46. A source's terminal voltage drops from its 80 V no-load value down to 60 V, when it is connected to its full-load resistance of 0.15 Ω. Calculate the values of the (a) load current, (b) source's internal resistance, (c) percentage regulation, and (d) circuit efficiency.

47. Six 1.5-V cells, each with an internal resistance of 0.8 Ω, are connected in series-aiding. What value of a load resistance will allow maximum power transfer to the load? Calculate the value of the maximum load power.

48. Six 1.5-V cells, each with an internal resistance of 0.8 Ω, are connected in parallel. What value of a load resistance will allow maximum power transfer to the load? Calculate the value of the maximum load power.

49. A source has an open-circuit terminal voltage of 24 V and an internal resistance of 4 Ω. Prepare a table of values for I_L, V_L, P_L, and circuit efficiency, with R_L equal in turn to zero, 2, 4, 8, 20, and infinite ohms.

50. Figure 5-18 represents a series–parallel arrangement of electrical sources. What

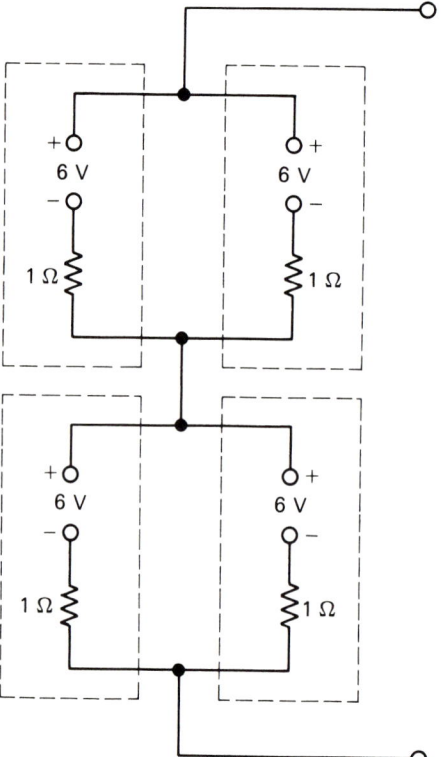

FIGURE 5-18 Circuit for Problem 50.

are the values of the open-circuit terminal voltage and the total internal resistance of the arrangement?

51. A dc source has an open-circuit voltage of 24 V and an internal resistance of 2 Ω. If a current of 1.5 A is drawn from the source, calculate the values of the terminal voltage and the power dissipated in the load.

52. A dc source is used to charge a 12-V cell with an internal resistance of 0.1 Ω. If the charging current is 30 A, what is the required value for the terminal voltage of the source?

53. Three 1.5-V cells are connected in parallel. This combination is joined in series with four 2-V cells that are connected in series-aiding. If all the cells have an internal resistance of 1.5 Ω, derive the equivalent constant current generator for the total arrangement of the cells.

54. A power supply has an open-circuit voltage of 24 V and a percentage regulation of 15%. What is the value of the full-load voltage?

55. One battery has an EMF of 12.8 V with an internal resistance of 0.1 Ω, and a second battery has an EMF of 13.5 V with an internal resistance of 0.08 Ω. If the two batteries are connected in parallel, what are the values of the total equivalent internal resistance of the combination and its terminal voltage on no-load conditions?

56. A dc source has a full-load voltage of 35 V and a percentage regulation of 12%. What is the value of the no-load voltage under open-circuit conditions?

57. One constant current source has a short-circuit current of 4.5 A and an internal resistance of 1.2 Ω. A second constant current source has a short-circuit current of 3.7 A and an internal resistance of 0.8 Ω. If the two sources are correctly paralleled, what are the values of the equivalent open-circuit voltage of the combination and its total equivalent associated resistance?

58. A dc source has an open-circuit voltage of 1.45 V and an internal resistance of 0.5 Ω. When the source is delivering a current of 0.5 A to a resistive load, what is the value of the source's output power?

59. A dc source whose open-circuit voltage is 120 V delivers 50 W to a resistive load when the current is 0.5 A. What is the percentage regulation of the source?

60. A dc source whose open-circuit voltage is 250 V has an internal resistance of 50 Ω. If the circuit efficiency is 65%, what is the required value for the load resistance?

Advanced Problems

61. In Fig. 5-19, what value of R will allow maximum power transfer to the load? Calculate the maximum total load power.

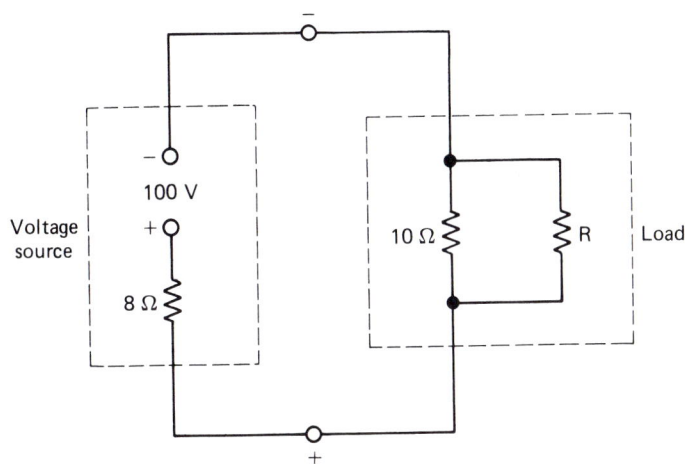

FIGURE 5-19 Circuit for Problem 61.

62. In Fig. 5-20, derive the equivalent constant voltage generator for the three sources in parallel. If R_L is 7 Ω, find the value of the load voltage, V_L.

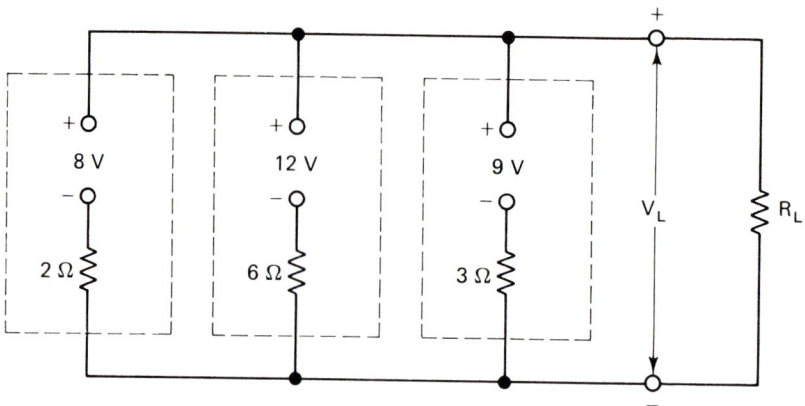

FIGURE 5-20 Circuit for Problem 62.

63. In Fig. 5-21, what is the value and the polarity of the potential at the point X?

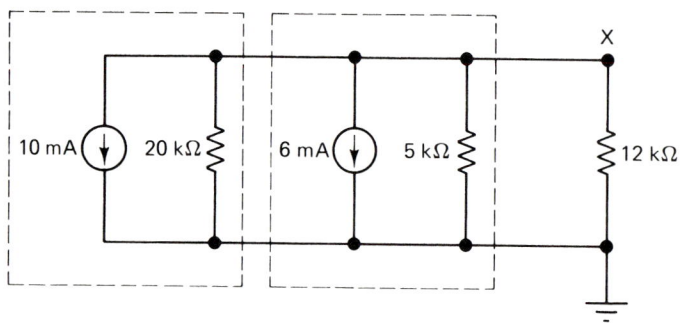

FIGURE 5-21 Circuit for Problem 63.

64. In Fig. 5-22, derive the equivalent constant current generator that can replace the series connected sources between points X and Y.

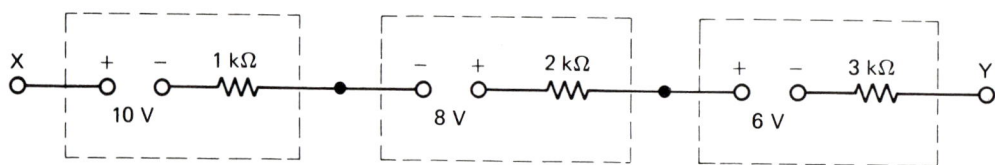

FIGURE 5-22 Circuit for Problem 64.

65. In Fig. 5-23, what is the open-circuit (no-load) voltage between points X and Y? If a full-load resistance of 10 kΩ is now connected between X and Y, what is the new voltage between these two points? Calculate the percentage regulation for the terminal voltage between X and Y.

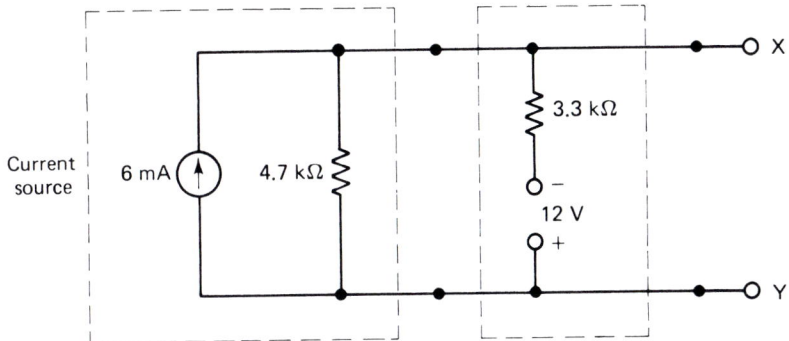

FIGURE 5-23 Circuit for Problem 65.

66. A source has an open-circuit terminal voltage of 45 V, which drops to 30 V when a 2-Ω load is connected. What is the maximum possible value of load power?
67. In Fig. 5-24, derive the equivalent constant current generator between points X and Y. If a 5-kΩ load is connected across X and Y, find the load current.

FIGURE 5-24 Circuit for Problem 67.

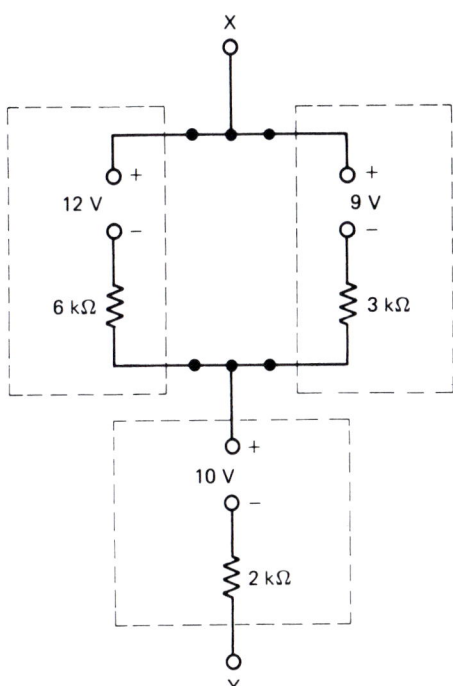

68. Two loads rated at 12 V, 60 W and 12 V, 15 W are properly connected in parallel across a voltage source whose internal resistance is 0.2 Ω. What is the constant voltage EMF of the source?
69. In Fig. 5-25, derive the equivalent constant voltage generator between points X and Y. If a 4-kΩ load is connected across X and Y, calculate the load power and the circuit efficiency.

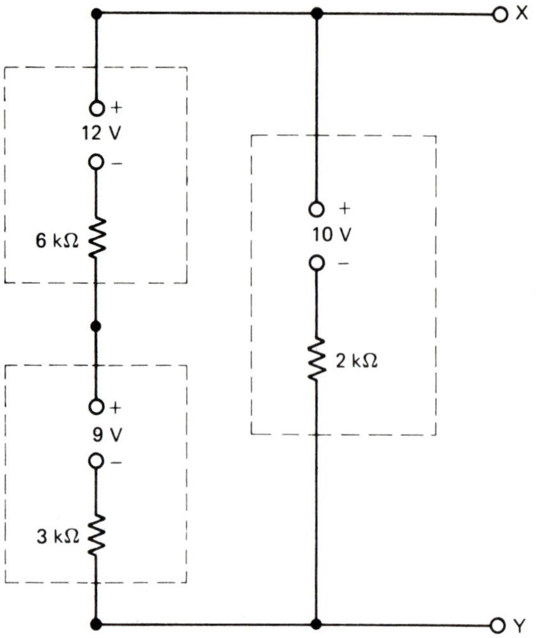

FIGURE 5-25 Circuit for Problem 69.

70. You have 24 cells available, each with a constant voltage EMF of 2 V and an internal resistance of 1 Ω. What arrangement of these cells will produce an open-circuit terminal voltage of 36 V, and what is the effective internal resistance of this arrangement?

71. A dc source has an open-circuit voltage of 30 V and an internal resistance of 5 Ω. What are the two values of load resistance that will provide only 50% of the maximum possible power? In each case, what is the value of the circuit efficiency?

72. A wattmeter is connected between a dc source and a variable resistive load. When the load resistance is adjusted to a value of 8 Ω, the wattmeter indicates a maximum reading of 48 W. What are the values of the source's open-circuit voltage and its percentage regulation?

73. Six cells, each having an open-circuit voltage of 2 V and an internal resistance of 2 Ω, are connected in two groups of three in series-aiding with the two groups joined in parallel. The arrangement is then connected to an external resistance of 30 Ω. What is the value of the current delivered by each cell?

74. A 10-Ω resistive load is connected across the terminals of a battery that consists of four cells in series. Each cell has an open-circuit voltage of 1.5 V and an internal resistance of 1.5 Ω. Calculate the energy in joules absorbed by the resistive load if the circuit is in operation for a period of 2 min.

75. You have a number of identical cells, each with an open-circuit voltage of 3 V and an internal resistance of 1.6 Ω. How many of these cells are required to produce an arrangement whose open-circuit voltage is 27 V and whose internal resistance is 1.8 Ω? If a load is connected across the arrangement for maximum power transfer, what is the value of the current delivered by each cell?

76. A 220-Ω resistor dissipates 10 W when connected to a dc source. When a 330-Ω resistor is connected in series with the 220-Ω resistor, the new dissipation of the 220-Ω resistor falls to 3 W. Calculate the open-circuit voltage of the source.

77. A dc source has an open-circuit voltage of 28 V and an internal resistance of 3 Ω. The resistive load is connected at the end of the line, each wire of which has a resistance of 1 Ω. What is the value of the maximum power that can be developed in the load?

78. A dc source has a terminal voltage of 80 V when a 6-Ω load is connected to its terminals. If the load is changed to 4 Ω, the terminal voltage falls to 7.5 V. Determine the values of the source's internal resistance and its open-circuit voltage.

79. The EHV power supply for the second anode circuit of a television picture tube has an open-circuit voltage of 18 kV and an effective internal resistance of 2.5 MΩ.

What is the second anode voltage when the tube current is 650 μA? If the current is increased by 60%, what is the new value of the second anode voltage? What is the value of the power supply's short-circuit current?

80. In Fig. 5-26, determine the value of R that will allow maximum power transfer to the load. What is the value of the maximum power developed in the load?

FIGURE 5-26 Circuit for Problem 80.

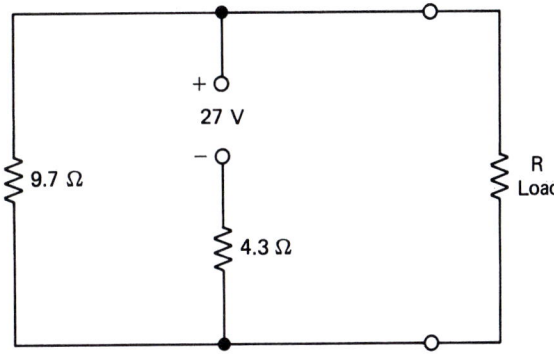

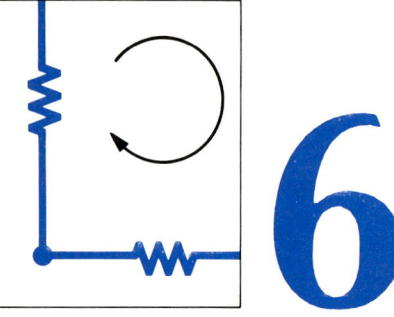

6

Kirchhoff's Laws and the Network Theorems

By studying these aids to circuit analysis, you will learn:

1. how to employ Kirchhoff's laws to analyze complex networks and circuits with more than one voltage source.
2. to apply mesh analysis to solve complex networks as well as circuits with more than one voltage source.
3. to employ nodal analysis in the solution of circuits that contain constant current sources.
4. Millman's theorem and its use in combining practical sources in parallel.
5. the superposition theorem and its application to the solution of circuits with more than one electrical source.
6. how Thévenin's theorem can reduce a complex network to a single constant voltage source.
7. how Norton's theorem can reduce a complex network to a single constant current source.
8. how delta→wye and wye→delta transformations can sometimes be used to simplify complex circuits.

6-0 Introduction

In previous chapters we have solved resistor circuits by applying Ohm's and Kirchhoff's laws. Kirchhoff's voltage law (KVL) was stated as the requirement for a voltage balance around any closed circuit; as we see in this chapter, such a closed circuit may alternatively be called a loop or a mesh. Kirchhoff's current law (KCL) referred to the current balance that must exist at every junction point. Part of this chapter is devoted to a more formal statement of these laws and their application to the analysis of electrical networks containing both resistors and sources. In its simplest form a network consists of a single loop, while more complex networks contain a number of loops which are interdependent.

The voltage across and the current through any resistor in a network may be calculated by the application of Ohm's and Kirchhoff's laws. However, with a complex network the procedure can be cumbersome, because you may have to solve several algebraic simultaneous equations. To simplify the analysis, you can use one or more of a number of network theorems. The main purpose of this chapter is to introduce these theorems, discuss their implication, and apply them to a variety of circuits.

Some of these theorems may be applied to all forms of resistances, whereas others are confined to so-called "linear resistances." A linear resistance is one that obeys Ohm's law, so that the voltage across the resistance is directly proportional to the current through the resistance. Composition, film and wirewound resistors fall into this category, although transistors and other semiconductor devices do not. For example, if you double the voltage across a transistor, its current will increase but will not double. What we are saying is that for a linear resistance its value is constant and independent of its voltage drop. This is not true of a semiconductor device, whose resistance, when measured by an ohmmeter, varies with the resistance range chosen; this is because the various resistance ranges apply different dc voltages to the device.

In this chapter the various methods of circuit analysis and the network theorems are used only with dc, although they are equally applicable to ac. However, with general ac networks it will be necessary to use a type of mathematics known as complex algebra, which is described in Chapter 16.

6-1 Kirchhoff's Voltage and Current Laws

Formally stated, *Kirchhoff's voltage law* says:

> The algebraic sum of the constant voltage EMFs and the voltage drops around any closed electrical loop is always zero.

As soon as you see the words "algebraic sum," you know that positive and negative signs are involved. We must therefore adopt a convention so that each voltage in a loop may be termed either positive or negative. The one normally chosen is to move clockwise around the loop starting at any point and regard a voltage as positive if the negative polarity of that voltage is first encountered. Similarly, if you first meet the positive polarity, the voltage is negative. Let us apply this convention to the circuit of Fig. 6-1. This problem cannot be solved by Ohm's law alone since the sources share a common loop and both are responsible for producing the current through the 10-kΩ resistor. There are, in fact, three interdependent loops and therefore three algebraic equations involving KVL. To derive these equations the following procedure is used:

STEP 1   Identify all points in the circuit that are electrically different and label them with letters of the alphabet. In Fig. 6-1 there are four such points, but other points which are electrically the same have been given the same letter. Then the three loops are $ACDA$ (containing the 12-V source and the 2.4-kΩ and 10-kΩ resistors), $ABCA$ (10-V source, 1.8-kΩ and 10-kΩ resistors), and $ABCDA$ (10-V and 12-V sources, 1.8-kΩ and 2.4-kΩ resistors).

STEP 2   Specify the currents flowing in the various branches of the circuit and insert the polarities of the voltage drops across the resistances. In many cases the direction of the electron flow is obvious, as, for example, the directions of I_1 and I_2 in Fig. 6-1. However, sometimes the direction is doubtful; such is the case with I_3, since the 10-V and 12-V sources tend to drive currents in opposite directions through the 10-kΩ resistor. In fact, this situation does not present any real difficulty since if the wrong direction is chosen, the analysis will reveal that the sign of the current is negative. The direction of I_3 has therefore been arbitrarily chosen to flow from point C to point A.

STEP 3   Select one of the three loops, for example $ACDA$, and a starting point, such as A. Then moving clockwise round the loop, the voltages V_3 and V_1 are negative while the EMF E_1 is positive. Therefore, the first KVL equation is:

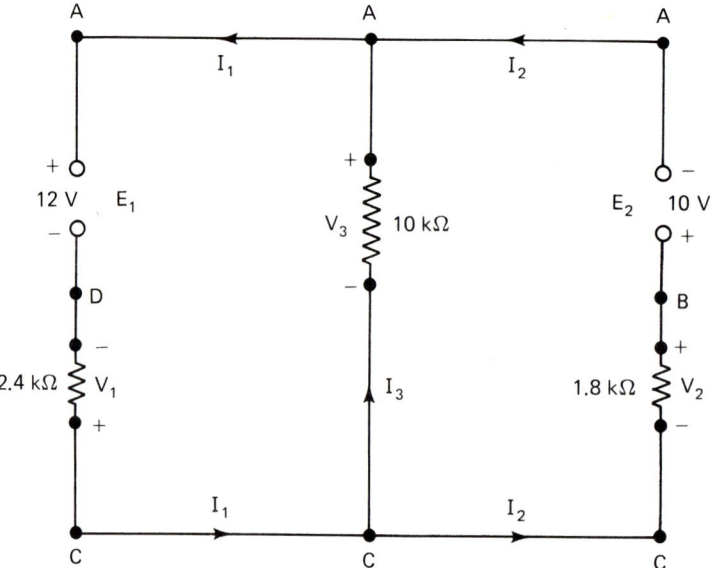

FIGURE 6-1   Circuit illustrating Kirchhoff's laws.

Loop $ACDA$ (starting at A):

$$-V_3 - V_1 + E_1 = 0$$
$$-I_3 \times 10 \text{ k}\Omega - I_1 \times 2.4 \text{ k}\Omega + 12 \text{ V} = 0 \quad \quad (6\text{-}1\text{-}1)$$
$$10I_3 + 2.4I_1 = 12$$

It is assumed that the currents I_1, I_2, and I_3 are all measured in milliamperes. The second KVL equation is:
Loop $ABCA$ (starting at A):

$$+E_2 - V_2 + V_3 = 0$$
$$+10 \text{ V} - I_2 \times 1.8 \text{ k}\Omega + I_3 \times 10 \text{ k}\Omega = 0 \quad \quad (6\text{-}1\text{-}2)$$
$$-10I_3 + 1.8I_2 = 10$$

The third KVL equation is:
Loop $ABCDA$ (starting at A):

$$+E_2 - V_2 - V_1 + E_1 = 0$$
$$+10 \text{ V} - I_2 \times 1.8 \text{ k}\Omega - I_1 \times 2.4 \text{ k}\Omega + 12 \text{ V} = 0 \quad \quad (6\text{-}1\text{-}3)$$
$$2.4I_1 + 1.8I_2 = 22$$

Note that the third equation does not provide extra information since Eq. (6-1-3) can be obtained by merely adding together Eqs. (6-1-1) and (6-1-2). In fact, any one of the three KVL equations may be derived from the other two.

Since there are really only two equations with three unknown currents, it will be necessary to have another equation before the problem can be solved. This is provided by *Kirchhoff's current law*, which, formally stated, is:

> The algebraic sum of the currents existing at any junction point, is zero.

The convention is to assume that those electron flow currents entering the junction point are positive and those leaving are negative. At the junction point, A, I_2 and I_3 are positive and I_1 is negative. Therefore,

$$+I_2 + I_3 - I_1 = 0 \quad \quad (6\text{-}1\text{-}4)$$
$$I_1 = I_2 + I_3$$

Substituting for I_1 in Eq. (6-1-1), we obtain

$$10I_3 + 2.4(I_2 + I_3) = 12$$

or

$$2.4I_2 + 12.4I_3 = 12 \quad \quad (6\text{-}1\text{-}5)$$

Repeating Eq. (6-1-2) yields

$$1.8I_2 - 10I_3 = 10$$

Multiplying Eq. (6-1-5) by 1.8 and Eq. (6-1-2) by 2.4 gives

$$4.32I_2 + 22.32I_3 = 21.6 \qquad (6\text{-}1\text{-}6)$$
$$4.32I_2 - 24.0I_3 = 24.0 \qquad (6\text{-}1\text{-}7)$$

Subtracting Eq. (6-1-7) from Eq. (6-1-6), we get

$$22.32I_3 - (-24.0I_3) = 21.6 - 24.0$$
$$46.32I_3 = -2.4$$
$$I_3 = \frac{-2.4}{46.32} = -0.0518 \text{ mA}$$

The negative sign indicates that the actual direction of the electron flow is from A to C and not from C to A. Substituting the value of I_3 in Eq. (6-1-6) yields

$$4.32I_2 + 22.32 \times (-0.0518) = 21.6$$
$$I_2 = 5.2677 \text{ mA}$$

From Eq. (6-1-4),

$$I_1 = I_2 + I_3$$
$$= 5.2677 + (-0.0518)$$
$$= 5.2159 \text{ mA}$$

Referring back to Fig. 6-1, we have

$$V_1 = 5.2159 \text{ mA} \times 2.4 \text{ k}\Omega = 12.518 \text{ V}$$
$$V_2 = 5.2677 \text{ mA} \times 1.8 \text{ k}\Omega = 9.482 \text{ V}$$
$$V_3 = -0.0518 \text{ mA} \times 10 \text{ k}\Omega = -0.518 \text{ V}$$

The negative sign shows that the polarity of V_3 is actually the reverse of that shown in Fig. 6-1.
Check back with the KVL equations:
Loop *ACDA*:

$$-V_3 - V_1 + E_1 = -(-0.518) - 12.518 + 12 = 0$$

Loop *ABCA*:

$$+E_2 - V_2 + V_3 = 10 - 9.482 + (-0.518) = 0$$

Loop *ABCDA*:

$$+E_2 - V_2 - V_1 + E_1 = +10 - 9.482 - 12.518 + 12 = 0$$

The loop equations therefore check out and the solution is correct. The powers dissipated in the three resistors can be calculated by using the normal power equations.

To summarize, KVL states that the algebraic sum of the constant voltage EMFs and the voltage drops around a closed electrical loop is zero; KCL states that the algebraic sum of the currents at a junction point is zero. To analyze a circuit using these laws requires identifying the individual loops, specifying the various currents flowing in the circuit and then writing down the KVL and KCL equations. Their solutions should be checked with the original loop equations. Finally, you should note that KVL applies only to loops containing constant voltage sources. If a circuit contains a constant current generator, it must be converted into its constant voltage generator equivalent before the Kirchhoff analysis can take place.

EXAMPLE 6-1 In Fig. 6-2, calculate the value of I_L.

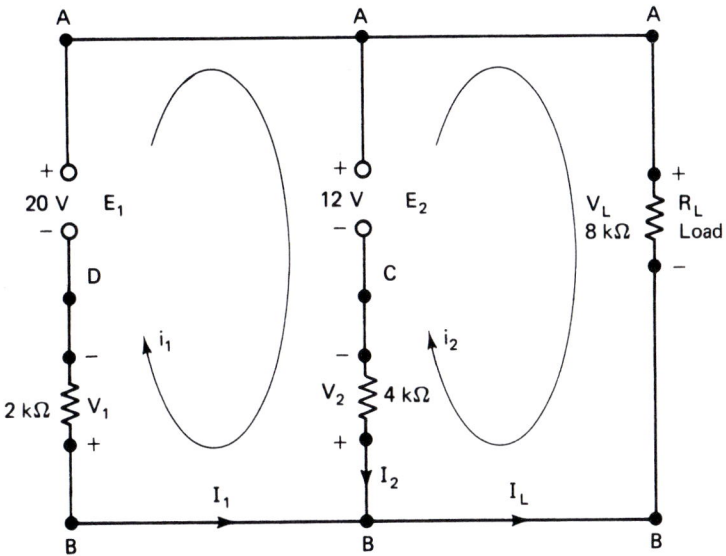

FIGURE 6-2 Circuit for Example 6-1 (also used to compare solutions by Kirchhoff's laws and mesh analysis).

Solution In this solution the arrows labeled i_1 and i_2 are ignored; they will be used in Section 6-2. With the currents measured in milliamperes and as specified, the KVL equations are:

Loop $ABCA$ (starting at A):

$$-V_L + (-V_2) + E_2 = 0$$

$$-I_L \times 8 \text{ k}\Omega + (-I_2 \times 4 \text{ k}\Omega) + 12 \text{ V} = 0$$

$$4I_2 + 8I_L = 12$$

$$2I_2 + 4I_L = 6$$

Loop $ABDA$ (starting at A):

$$-V_L + (-V_1) + E_1 = 0$$

$$-I_L \times 8 \text{ k}\Omega + (-I_1 \times 2 \text{ k}\Omega) + 20 \text{ V} = 0$$

$$2I_1 + 8I_L = 20$$

The KCL equation for the junction point, B, is

$$I_1 + (+I_2) + (-I_L) = 0$$

or

$$I_1 = I_L - I_2$$

Substituting for I_1 in the KVL equation for the loop $ABDA$, we obtain

$$2(I_L - I_2) + 8I_L = 20$$
$$-2I_2 + 10I_L = 20$$

Adding the last equation to the KVL equation for the loop $ABCA$ will eliminate I_2. Then

$$4I_L + 10I_L = 6 + 20 = 26$$

$$I_L = \frac{26}{14} = +1.86 \text{ mA}$$

Substituting back into the original equations leads to $I_1 = 2.57$ mA and $I_2 = -0.714$ mA. This indicates that the actual electron flow is from B to A and not from A to B; the 20-V source is in fact charging the 12-V source.

6-2

Mesh Analysis

Mesh is another term used to describe a closed voltage loop. Using Kirchhoff's laws, a loop can contain a number of different currents, which have to be specified when carrying out step 2 of the procedure. In mesh analysis each of the mesh currents is considered to flow around the entire loop, although an individual resistor may carry one or more mesh currents. One convention is to insert clockwise mesh currents (electron flow) in each of the loops and then write down the KVL equations. This can best be illustrated by the circuit of Fig. 6-3. Notice that the mesh currents, i_1 and i_2, are flowing in opposite

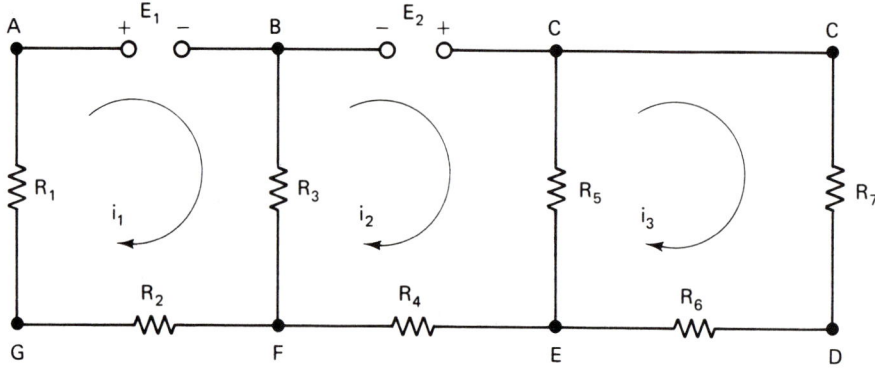

FIGURE 6-3 Circuit illustrating mesh analysis.

directions through the resistor, R_3. In a similar way, resistor R_5 carries the difference of currents i_2 and i_3. Therefore, only the loop *FBCEF* is concerned with all three mesh currents. Writing down the KVL equations:

Mesh *GABFG* (starting at *G*):

$$i_1 R_1 - E_1 + i_1 R_3 + i_1 R_2 - i_2 R_3 = 0$$
$$i_1(R_1 + R_2 + R_3) - i_2 R_3 = E_1 \qquad (6\text{-}2\text{-}1)$$

Mesh *FBCEF* (starting at *F*):

$$i_2 R_3 - i_1 R_3 + E_2 + i_2 R_5 + i_2 R_4 - i_3 R_5 = 0$$
$$-i_1 R_3 + i_2(R_3 + R_4 + R_5) - i_3 R_5 = -E_2 \qquad (6\text{-}2\text{-}2)$$

Mesh *ECDE* (starting at *E*):

$$i_3 R_5 - i_2 R_5 + i_3 R_7 + i_3 R_6 = 0$$
$$-i_2 R_5 + i_3(R_5 + R_6 + R_7) = 0 \qquad (6\text{-}2\text{-}3)$$

The advantages of mesh analysis compared with the method of branch currents are now apparent. The KVL equations can be directly written down by inspection, if the following rules are observed:

1. Add together all the resistances in the particular loop and multiply the result by the mesh current of that loop. This refers to the terms $i_1(R_1 + R_2 + R_3)$, $i_2(R_3 + R_4 + R_5)$, and $i_3(R_5 + R_6 + R_7)$.
2. When a resistor in the loop carries a current related to another mesh, that *IR* drop should be given a negative sign. Such is the case for the terms $-i_2 R_3$, $-i_1 R_3$, $-i_3 R_5$, and $-i_2 R_5$.
3. As far as the signs of the voltage sources are concerned, the normal KVL convention is used.

One other important point: there are three mesh equations with three unknown currents and there is no need to apply KCL to any junction point.

For comparison, let us use mesh analysis to solve for i_2 in the circuit of Fig. 6-2. Notice that i_2 will have the same value as I_L but is flowing in the opposite direction.

Mesh *ACBDA*:

$$(2 \text{ k}\Omega + 4 \text{ k}\Omega)i_1 - 4 \text{ k}\Omega \times i_2 + 20 \text{ V} - 12 \text{ V} = 0$$
$$6i_1 - 4i_2 = -8$$
$$3i_1 - 2i_2 = -4$$

Mesh *ABCA*:

$$-4 \text{ k}\Omega \times i_1 + (4 \text{ k}\Omega + 8 \text{ k}\Omega)i_2 + 12 \text{ V} = 0$$
$$-4i_1 + 12i_2 = -12$$
$$-2i_1 + 6i_2 = -6$$

Sec. 6-2 Mesh Analysis

Multiplying the mesh *ACBDA* equation by 2 and the mesh *ABCA* equation by 3 yields

$$6i_1 - 4i_2 = -8$$
$$-6i_1 + 18i_2 = -18$$

Adding these last two equations together, we get

$$14i_2 = -26$$
$$i_2 = \frac{-26}{14} = -1.86 \text{ mA}$$

The negative sign indicates that the direction of the i_2 flow is in the opposite direction to that shown in Fig. 6-2. This compares with the value of $I_L = +1.86$ mA that we obtained by using Kirchhoff's laws.

We have learned that mesh current analysis is superior to the method of branch currents. Mesh currents can be arbitrarily inserted in a clockwise direction and the only equations required can then be written down directly by carrying out the following steps:

1. Insert all mesh currents in a clockwise direction.
2. In each mesh add together all resistances through which that mesh current flows and determine the associated positive voltage drop. When a resistor in the loop carries a current belonging to an adjacent mesh, that voltage drop is negative.
3. Determine the sign of each voltage source according to its polarity.
4. Write down the KVL equation for each mesh.
5. Solve the simultaneous equations and use the results to check that the voltages balance in each mesh.

Note that the mesh analysis is applicable only to circuits containing voltage sources. If a current generator is present, you must convert it to its equivalent voltage source (Section 5-6).

EXAMPLE 6-2   In Fig. 6-4, determine the value of i_2.

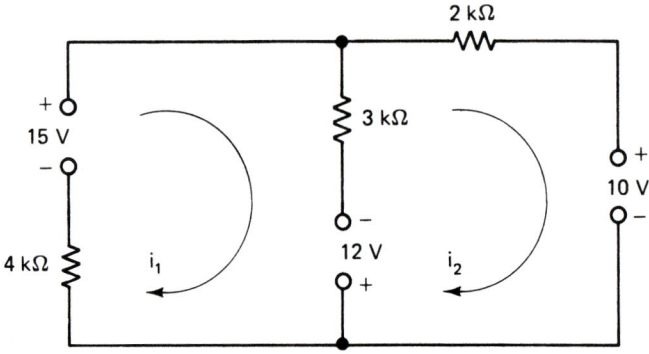

FIGURE 6-4   Circuit for Example 6-2.

Solution The two mesh equations are:

$$i_1(4\ \text{k}\Omega + 3\ \text{k}\Omega) - i_2 \times 3\ \text{k}\Omega + 15\ \text{V} + 12\ \text{V} = 0$$

$$-i_1 \times 3\ \text{k}\Omega + i_2(3\ \text{k}\Omega + 2\ \text{k}\Omega) - 12\ \text{V} - 10\ \text{V} = 0$$

These equations reduce to

$$7i_1 - 3i_2 = -27$$

$$-3i_1 + 5i_2 = 22$$

Multiplying the first mesh equation by 3 and the second mesh equation by 7, we get

$$21i_1 - 9i_2 = -81$$

$$-21i_1 + 35i_2 = 154$$

Adding these last two equations together yields

$$26i_2 = 73$$

$$i_2 = 2.81\ \text{mA}$$

6-3 Nodal Analysis

A *node* is another term for a junction point at which two or more electrical currents exist. This method of analysis requires that the circuit contain only current sources, so that we can then use the KCL to obtain the equation for each node. For some circuits nodal analysis will have an advantage over mesh analysis, since fewer nodal equations will be involved in the solution. For comparison purposes we will apply nodal analysis to the circuit of Fig. 6-2, which has already been solved by Kirchhoff's laws and mesh analysis. As a first step we must convert the constant voltage sources into their equivalent constant current generators with $I_1 = 20\ \text{V}/2\ \text{k}\Omega = 10\ \text{mA}$ and $I_2 = 12\ \text{V}/4\ \text{k}\Omega = 3\ \text{mA}$. The circuit has then been redrawn as in Fig. 6-5, with a ground included as a convenient reference node. The purpose of the analysis will be to find the value of V_L and, as a convention, we shall assume that this voltage at the node, A, is negative with respect to ground; the true polarity will be shown by the sign of the answer for V_L. This convention means that electron currents

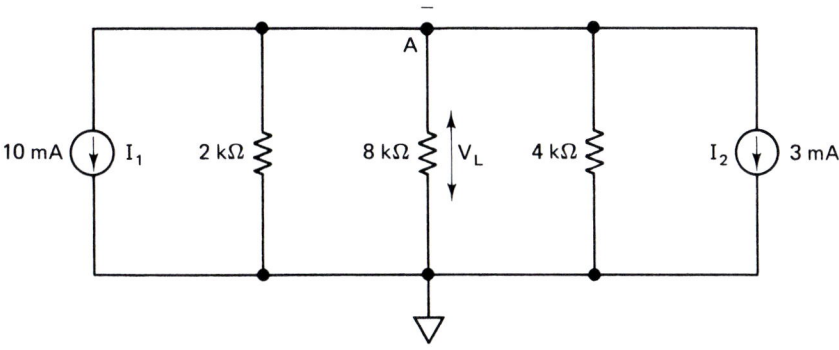

FIGURE 6-5 Circuit illustrating nodal analysis.

must flow from the node through the resistors to ground. Current generators that force currents into the node are then positive, while those driving currents out of the node are negative. The related KCL equation is:

> algebraic sum of generator currents entering the principal node
>
> $\qquad$ = total of currents leaving the node through resistors
>
> (6-3-1)

Since the 10-mA and the 3-mA generators are driving currents out of node A, both these generators will be regarded as negative. The currents leaving the node through the resistors are always considered to be positive and their three values are $V_L/2$ kΩ, $V_L/8$ kΩ, and $V_L/4$ kΩ, all of which are measured in milliamperes. Therefore the equation for the principal node, A, is

$$(-10 \text{ mA}) + (-3 \text{ mA}) = \frac{V_L}{2 \text{ k}\Omega} + \frac{V_L}{8 \text{ k}\Omega} + \frac{V_L}{4 \text{ k}\Omega} = \frac{7V_L}{8 \text{ k}\Omega}$$

$$V_L = -\frac{13 \times 8}{7} = -14.86 \text{ V}$$

Since the sign of the answer for V_L is negative, the voltage at point A is actually positive with respect to ground. The load current, I_L, is equal to 14.86 V/8 kΩ = 1.86 mA and is an electron flow from the ground to point A. This is the same result that we obtained using Kirchhoff's laws and mesh analysis. In this case nodal analysis required only one equation, whereas the other methods needed at least two simultaneous equations. However, this advantage of nodal analysis is not true for all circuits, so you must study each problem carefully and then decide which method of analysis to use.

To summarize, the steps to use in nodal analysis are:

1. If necessary, convert all voltage sources into current sources and then redraw the circuit.
2. Label all the nodes and select the most convenient reference node.
3. For every node, determine the sign of each current generator and then write down the KCL nodal equation.
4. Solving the nodal equations will reveal the node voltages so that the various currents can be calculated.

EXAMPLE 6-3 In the circuit of Fig. 6-6(a), calculate the value of V_N.

Solution Convert the voltage sources into current sources and then redraw as in Fig. 6-6(b). Our principal node is at point N, while the reference node is at ground.

The signs of the 6-mA and 2-mA current generators are positive, while that of the 3-mA generator is negative. Then by using Eq. (6-5-1), we obtain

$$(+6 \text{ mA}) + (+2 \text{ mA}) + (-3 \text{ mA}) = \frac{V_N}{3 \text{ k}\Omega} + \frac{V_N}{9 \text{ k}\Omega} + \frac{V_N}{6 \text{ k}\Omega}$$

$$5 = V_N \times \frac{6 + 2 + 3}{18} = \frac{11 V_N}{18}$$

$$V_N = \frac{5 \times 18}{11} = +8.18 \text{ V}$$

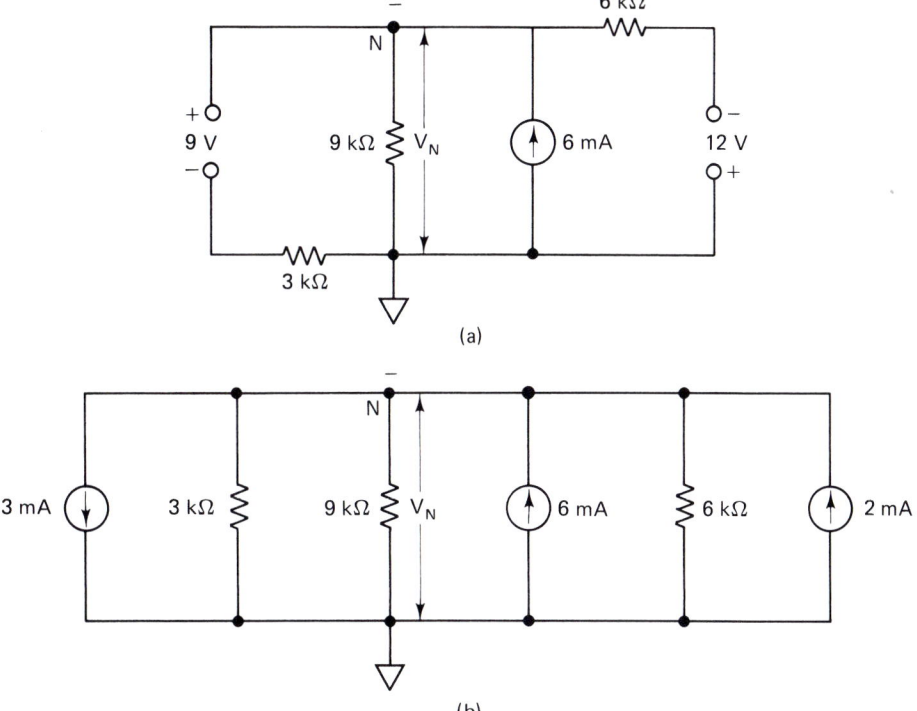

FIGURE 6-6 Circuit for Example 6-3.

Since the value of V_N is positive, the voltage at point N is 8.18 V negative with respect to ground.

6-4
Millman's Theorem

All the circuits that we have so far analyzed could most easily be solved by the use of *Millman's theorem*, a formal statement of which is:

> Any number of constant current sources which are *directly* connected in parallel can be converted into a single current source whose total generator current is the algebraic sum of the individual source currents and whose total internal resistance is the result of combining the individual source resistances in parallel.

Although the statement refers only to current sources, it also can be applied to voltage sources, which must initially be converted into their constant current equivalents. The theorem is also applicable to a mixture of parallel voltage and current sources with the final equivalent generator shown either as a constant current or as a constant voltage source.

As an example, let us solve the circuit of Fig. 6-2 by using Millman's theorem. The first step will be to convert the voltage sources into their equivalent current sources; the circuit is then redrawn as in Fig. 6-5. The algebraic

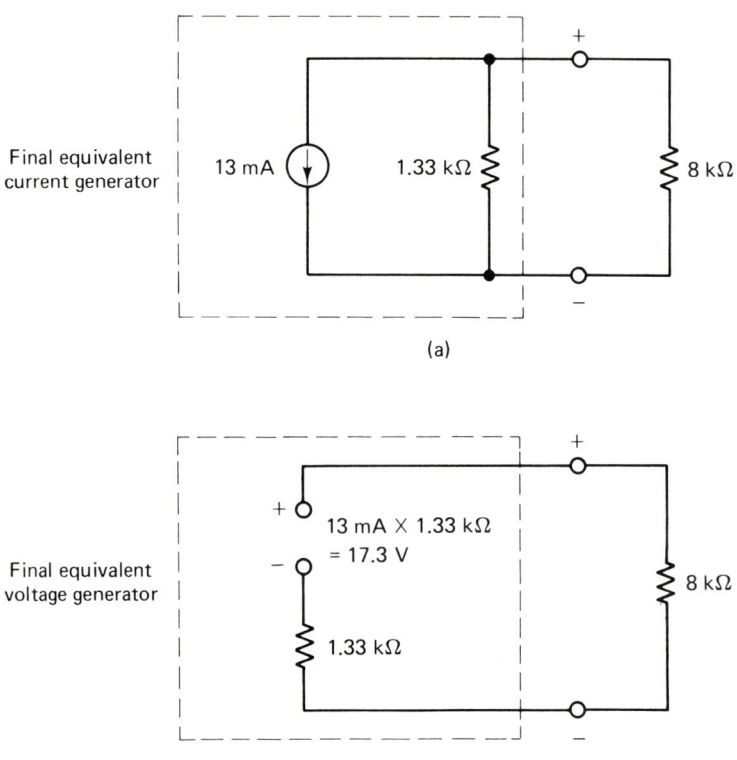

FIGURE 6-7 Illustration of Millman's theorem.

sum of the individual source currents is 10 + 3 = 13 mA and the associated total internal resistance is

$$\frac{2 \times 4}{2 + 4} = 1.33 \text{ k}\Omega$$

This final current generator is then attached to the 8-kΩ resistor [Fig. 6-7(a)]; its equivalent constant voltage source is shown in Fig. 6-7(b). Using the current division rule,

$$I_L = 13 \text{ mA} \times \frac{1.33 \text{ k}\Omega}{1.33 \text{ k}\Omega + 8 \text{ k}\Omega} = 1.86 \text{ mA}$$

This same result was obtained by using Kirchhoff's laws, and mesh and nodal analysis, but with Millman's theorem no algebraic equations were required.

The following is the summary of the steps you must carry out when using Millman's theorem:

1. Convert all voltage sources into their equivalent constant current generators.
2. Calculate the algebraic sum of the individual source currents.
3. Calculate the total internal resistance by combining the individual source resistances in parallel.
4. If necessary, convert the final constant current generator into its constant voltage equivalent.

EXAMPLE 6-4 In Fig. 6-8(a), convert the parallel combination of sources into its equivalent single constant voltage generator.

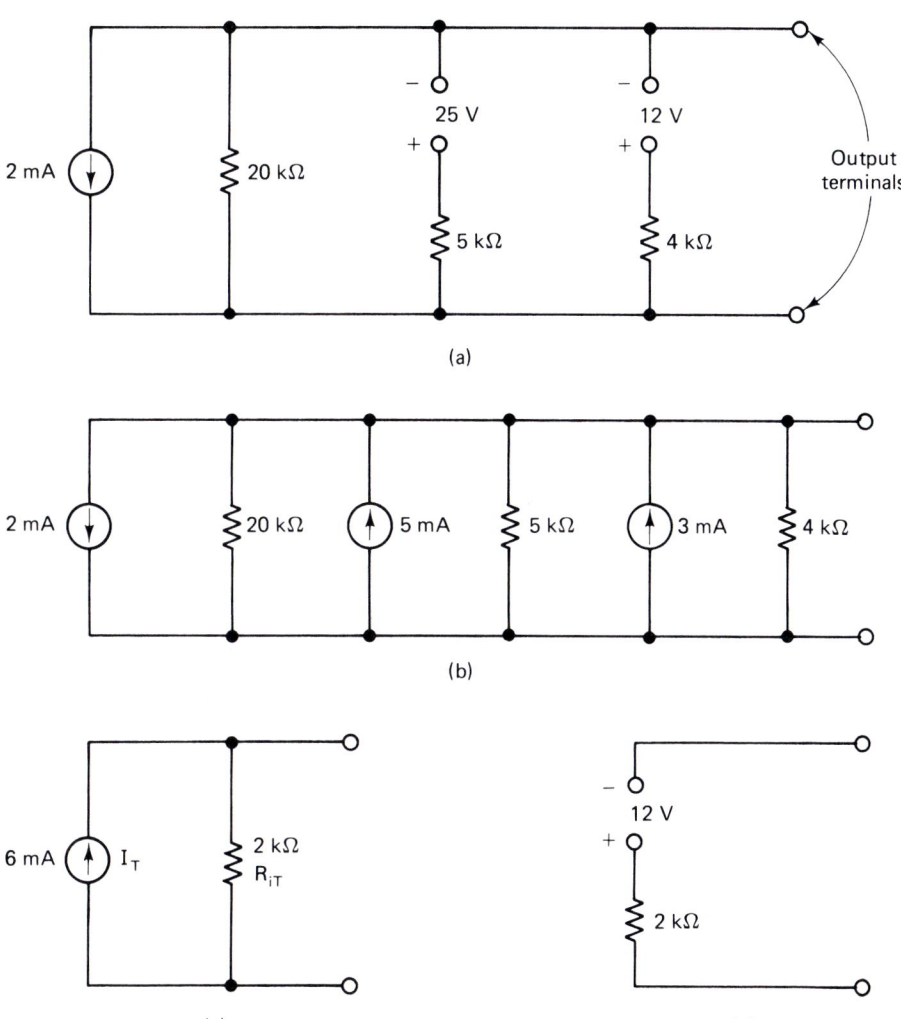

FIGURE 6-8 Circuit for Example 6-4.

Solution Convert the two voltage sources into their equivalent constant current generators and redraw the circuit as in Fig. 6-8(b).

The algebraic sum, I_T, of the individual source currents is

$$I_T = (+5 \text{ mA}) + (+3 \text{ mA}) + (-2 \text{ mA}) = +6 \text{ mA}$$

The total internal resistance, R_{iT}, may be calculated by using the reciprocal formula.

$$\frac{1}{R_{iT}} = \frac{1}{20} + \frac{1}{5} + \frac{1}{4} = \frac{1+4+5}{20} = \frac{10}{20} = \frac{1}{2}$$

Therefore, $R_{iT} = 2$ kΩ and the final constant current and constant voltage generators are shown in Fig. 6-8(c) and (d).

Sec. 6-4 Millman's Theorem

6-5 Superposition Theorem

You may use this theorem to solve a variety of problems involving networks which contain only linear resistances and more than one source. The principle is to consider the currents and voltages produced by each source in turn and then finally combine (superimpose) the results of all sources. The following is a formal statement of the *superposition theorem:*

> In a network of *linear* resistances containing more than one source, the resultant current flowing at any one point is the algebraic sum of the currents which would flow at that point if each source is considered separately while all other sources are replaced by their equivalent internal resistances. This last step is carried out by short-circuiting all sources of constant voltage and open-circuiting all sources of constant current.

Since this theorem allows each source to be considered separately, you can use simple Ohm's law equations to obtain the various currents in the circuit. However, each time that one of the sources is applied to the circuit, a different voltage drop will appear across an individual resistor. The analysis will not work correctly unless the resistances values remain constant and independent of their voltage drops; for this reason all resistors in the network must be linear.

If the circuit contains a large number of sources, the application of the superposition theorem becomes very tedious, since the circuit would have to be solved separately for each source and you would then have to combine all the results. In such cases it would probably be better to use either mesh or nodal analysis.

For comparison purposes we will use the superposition theorem to solve once more the circuit of Fig. 6-2. As the first step we will select the 20-V source and therefore replace the 12-V source by its 4-kΩ internal resistance; the circuit is then as shown in Fig. 6-9(a).

The total resistance presented to the 20-V source is

$$2 + \frac{4 \times 8}{4 + 8} = 4.67 \text{ k}\Omega$$

Then $I'_1 = 20 \text{ V}/4.67 \text{ k}\Omega = 4.29$ mA and, by the current division rule,

$$I'_L = 4.29 \text{ mA} \times \frac{4 \text{ k}\Omega}{4 \text{ k}\Omega + 8 \text{ k}\Omega} = 1.43 \text{ mA}$$

which is the current flowing through the 8-kΩ resistor due to the 20-V source.

Now you turn to the 12-V source and replace the 20-V source by its 2-kΩ internal resistance [Fig. 6-9(b)]. The total resistance offered to the 12-V source is

$$4 + \frac{2 \times 8}{2 + 8} = 5.6 \text{ k}\Omega$$

Therefore, $I''_2 = 12 \text{ V}/5.6 \text{ k}\Omega = 2.14$ mA and

$$I''_L = 2.14 \text{ mA} \times \frac{2 \text{ k}\Omega}{2 \text{ k}\Omega + 8 \text{ k}\Omega} = 0.43 \text{ mA}$$

FIGURE 6-9 Circuit illustrating the superposition theorem.

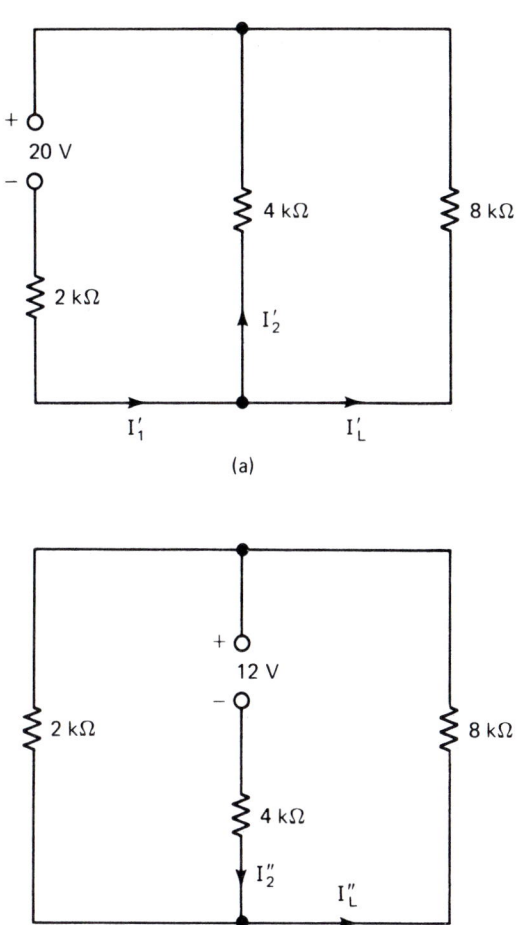

which is the current flowing through the 8-kΩ resistor due to the 12-V source.

Superimposing these results, $I_L = I'_L + I''_L = 1.43 + 0.43 = 1.86$ mA, which is the same value obtained by the other methods.

The following is a summary of the steps you should take when using the superposition theorem:

1. Pick any one source and replace all the other sources by their internal resistances. Remember to short out every constant voltage source and to open circuit every constant current source.
2. Using Ohm's law equations, calculate the values and the directions of all the current components that flow in the circuit as the result of the single source you have selected.
3. Repeat the procedure for each source in turn until all current components have been calculated.
4. For each individual branch, take the algebraic sum of the current components and obtain the actual branch current.

As we mentioned in Section 2-10, the superposition theorem may be applied to the currents and voltages in a network, since these are linearly related by the equation $V = IR$. However, superposition does not apply to the powers dissipated in the resistors. This is because the power-current ($P = I^2R$) and the power-voltage ($P = V^2/R$) relationships are not linear, but involve the squares of the current and the voltage.

Sec. 6-5 Superposition Theorem

EXAMPLE 6-5 Use the superposition theorem to determine the current through the 4-kΩ resistor in Fig. 6-10(a).

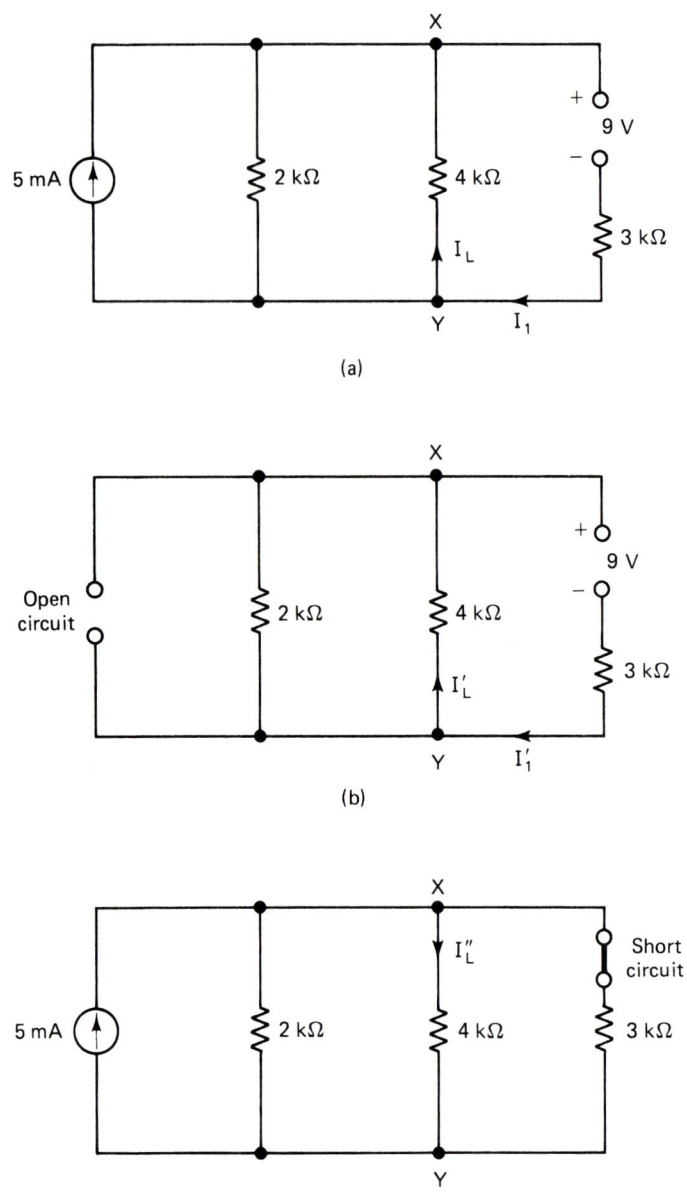

FIGURE 6-10 Circuit for Example 6-5.

Solution Select the 9-V source and open circuit the 5-mA current generator; the circuit is then redrawn as in Fig. 6-10(b). The total resistance presented to the 9-V source is

$$3 + \frac{2 \times 4}{2 + 4} = 4.33 \text{ k}\Omega$$

Then $I_1' = 9\text{ V}/4.33\text{ k}\Omega = 2.08\text{ mA}$ and, by the current division rule,

$$I_L' = 2.08\text{ mA} \times \frac{2\text{ k}\Omega}{2\text{ k}\Omega + 4\text{ k}\Omega} = 0.69\text{ mA}$$

with the electron flow in the direction from point Y to point X.

The next step is to short out the 9-V source so that the circuit now appears as in Fig. 6-10(c). The total resistance of 2 kΩ, 4 kΩ, and 3 kΩ in parallel is

$$\frac{1}{\frac{1}{2} + \frac{1}{4} + \frac{1}{3}} = \frac{12}{13}\text{ k}\Omega$$

and therefore, by the current division rule,

$$I_L'' = 5\text{ mA} \times \frac{\frac{12}{13}\text{ k}\Omega}{4\text{ k}\Omega} = 1.15\text{ mA}$$

with the electron flow in the direction from point X to point Y.

The actual current, I_L, through the 4-kΩ resistor is the algebraic sum of I_L' and I_L'', so that

$$I_L = (+1.15) + (-0.69) = +0.46\text{ mA}$$

Checking this result by Millman's theorem, the 9-V source is converted into a 9 V/3 kΩ = 3 mA current generator. The algebraic sum of the source currents is $(+5\text{ mA}) + (-3\text{ mA}) = +2\text{ mA}$. Then using the current division rule,

$$I_L = 2\text{ mA} \times \frac{\frac{12}{13}\text{ k}\Omega}{4\text{ k}\Omega} = 0.46\text{ mA}$$

6-6 Thévenin's Theorem

This theorem is your most important aid in dealing with complex networks that contain one or more sources. It will enable you to concentrate your analysis on a particular part of a circuit. This part is regarded as being connected between two terminals and is referred to as the *load*. Irrespective of the number of sources and resistors involved, the remainder of the circuit may then be represented by a simple generator with a constant voltage source, E_{TH}, associated with a series internal resistance, R_{TH}. The purpose of Thévenin's theorem is to enable you to calculate E_{TH} and R_{TH} in terms of the values contained in the original circuit. The *Thévenin theorem*, stated formally is:

> The current in a load connected between two output terminals, X and Y, of a complex network of resistors and electrical sources, is the same as if that load were connected across a simple *constant voltage* generator whose EMF, E_{TH}, is the open-circuit voltage measured between X and Y and whose *series* internal resistance, R_{TH}, is the resistance of the network looking back into terminals X and Y with all sources replaced by resistances equal to their internal resistances.

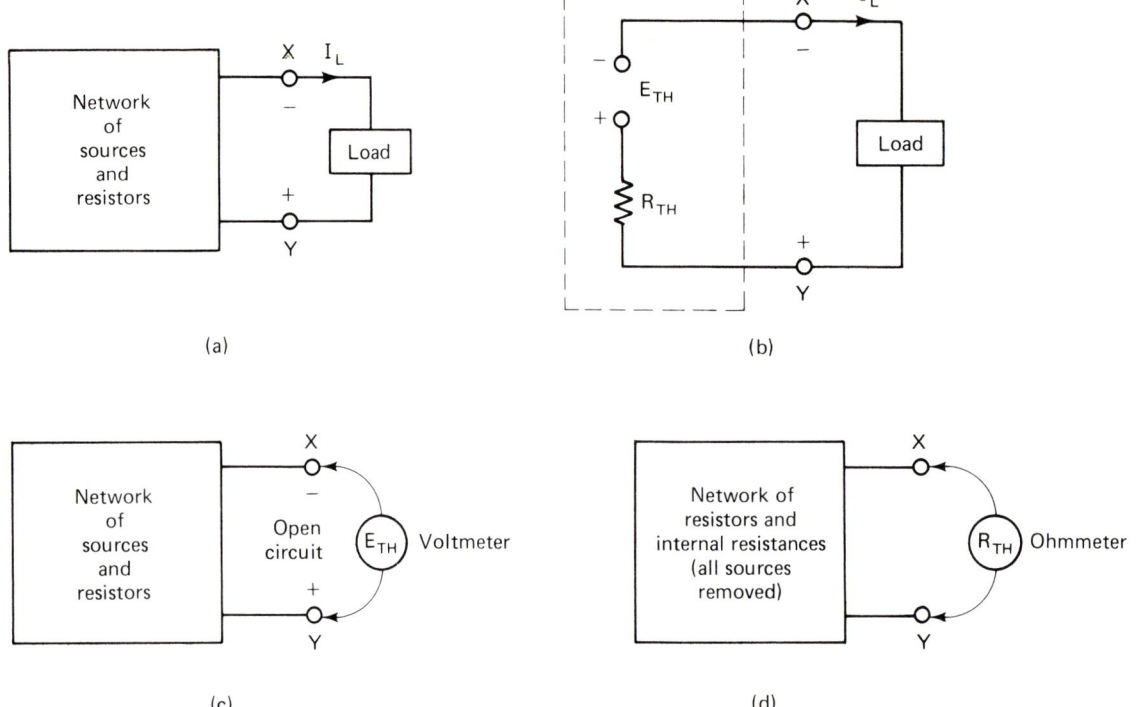

FIGURE 6-11 Illustration of Thévenin's theorem: (a) original circuit; (b) Thévenin equivalent generator; (c) determination of E_{TH}; (d) determination of R_{TH}.

This last step would be achieved by short-circuiting the sources of constant voltage and open-circuiting the sources of constant current.

Figure 6-11 is used to illustrate Thévenin's theorem. First, notice that the polarity of E_{TH} will drive a current through the load in the same direction as in the original circuit. To discover the value of E_{TH}, the load must first be removed and you must then calculate the open-circuit voltage between terminals X and Y. In other words, you must find out what a voltmeter would read when connected between X and Y with the load absent [Fig. 6-11(c)]; this reading is the value of E_{TH}.

To find R_{TH}, the load is again absent and all sources are removed by replacing them with their internal resistances. You eliminate a constant voltage source by placing a *short* circuit across the source while a constant current generator is completely removed to produce an *open* circuit. The value of R_{TH} can then be measured by an ohmmeter connected betweeen X and Y [Fig. 6-11(d)]. The process of obtaining E_{TH} and R_{TH} and then replacing the network of sources and resistors by the single voltage generator is called *Thévenizing* the circuit.

To illustrate Thévenin's theorem we first solve a problem by using Ohm's law equations and then obtain the same answer by Thévenizing the circuit. In Fig. 6-12(a) we are asked to find the voltage drop across the 5-kΩ resistor. Since the 4-kΩ and 5-kΩ resistors are in series and form a branch that is in parallel with the 6-kΩ resistor, the total resistance of the circuit is given by

$$R_T = 3 + \frac{6 \times (4 + 5)}{(4 + 5) + 6} = 3 + \frac{54}{15} = 6.6 \text{ k}\Omega$$

Then

$$I_T = \frac{36 \text{ V}}{6.6 \text{ k}\Omega} = 5.45 \text{ mA}$$

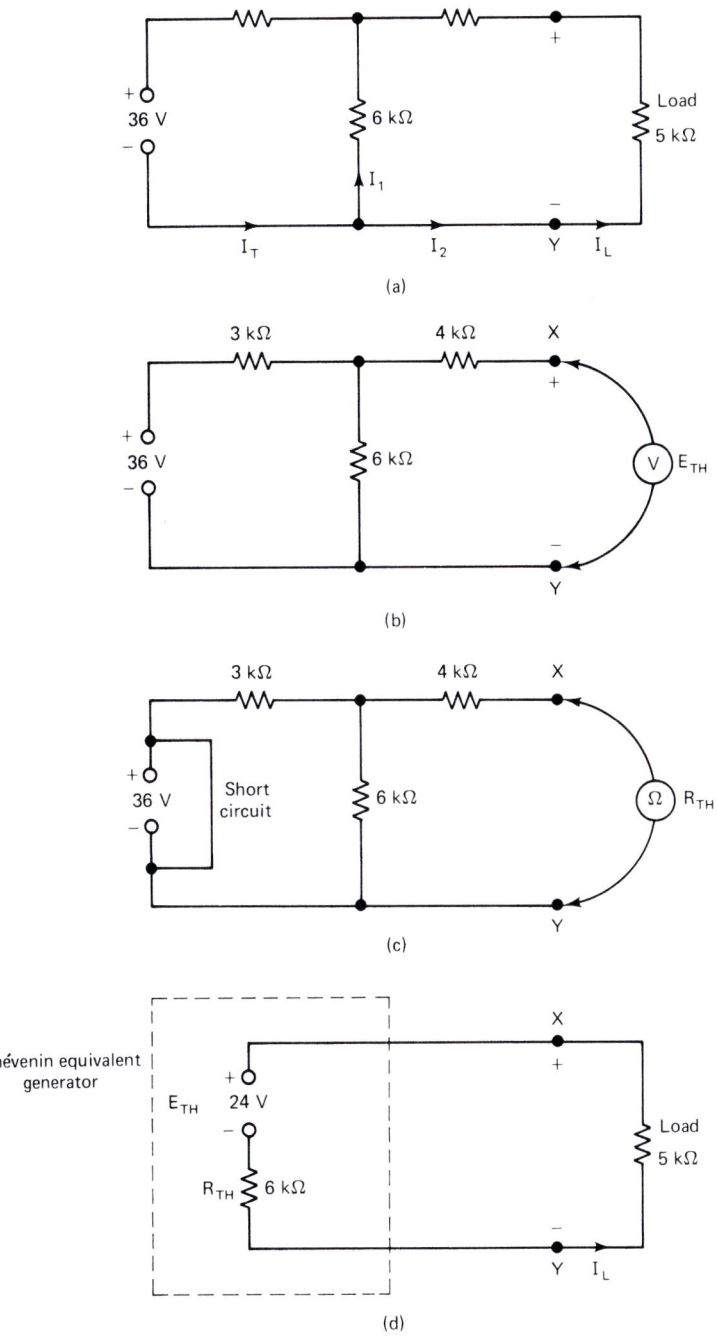

FIGURE 6-12 Circuit to illustrate Thévenin's theorem.

By the current division rule, the load current is

$$I_L = I_2 = 5.45 \text{ mA} \times \frac{6 \text{ k}\Omega}{4 \text{ k}\Omega + 5 \text{ k}\Omega + 6 \text{ k}\Omega}$$

$$= 2.18 \text{ mA}$$

Therefore, the voltage drop across the 5-kΩ resistor is 2.18 mA × 5 kΩ = 10.9 V.

To Thévenize the circuit we will regard the 5-kΩ resistor as the load. The first step is therefore to remove the load and find the Thévenin voltage,

Sec. 6-6 Thévenin's Theorem

E_{TH}, between terminals X and Y [Fig. 6-12(b)]. Since we are measuring the open-circuit voltage between X and Y, there is no current flow through the 4-kΩ resistor and therefore E_{TH} is equal to the voltage drop across the 6-kΩ resistor. By the voltage division rule,

$$E_{TH} = 36 \text{ V} \times \frac{6 \text{ k}\Omega}{6 \text{ k}\Omega + 3 \text{ k}\Omega} = 24 \text{ V}$$

To calculate the value of the Thévenin resistance, R_{TH}, the second step is to remove the load again and then place a short circuit across the 36-V source [Fig. 6-12(c)]. With respect to the position of the ohmmeter connected between X and Y, the 3-kΩ and 6-kΩ resistors are in parallel, so that

$$R_T = 4 + \frac{3 \times 6}{3 + 6} = 6 \text{ k}\Omega$$

The network to the left of X and Y may therefore be replaced by a single voltage generator whose constant EMF is 24 V and whose internal resistance is 6 kΩ. Figure 6-12(d) shows the Thévenin equivalent generator with the load reconnected between terminals X and Y. Then by the voltage division rule, the voltage drop across the 5-kΩ resistor is

$$24 \text{ V} \times \frac{5 \text{ k}\Omega}{5 \text{ k}\Omega + 6 \text{ k}\Omega} = 10.9 \text{ V}$$

Your first reaction is probably one of disappointment! Thévenizing the circuit seems at least as complicated as solving the problem by simple Ohm's law principles. However, with more complex networks, Thévenin's theorem can be a big winner over other methods of analysis. A good example is the unbalanced bridge circuit of Fig. 6-13(a). Here the problem is to find the current through the 2-Ω resistor, which we shall regard as the load. If we attempted to find the answer by mesh or nodal analysis, it would involve three algebraic simultaneous equations, so that the solution would be, to say the least, complicated. Solving the problem by Thévenin's theorem is much simpler and does not require the use of algebra.

To Thévenize the bridge circuit the first step is to remove the 2-Ω load and redraw the circuit as in Fig. 6-13(b). By the voltage division rule, the potential at Y is

$$10 \text{ V} \times \frac{3 \text{ }\Omega}{3 \text{ }\Omega + 5 \text{ }\Omega} = +3.75 \text{ V}$$

while the potential at X is

$$10 \text{ V} \times \frac{7 \text{ }\Omega}{7 \text{ }\Omega + 4 \text{ }\Omega} = +6.36 \text{ V}$$

E_{TH} is therefore $6.36 - 3.75 = 2.61$ V, with the terminal X positive with respect to terminal Y.

Finding R_{TH} is the second step, which is done by removing the load, shorting the 10-V source, and then calculating the value of the resistance between X and Y. With a short across the 10-V source, the junction point of the 4-Ω and 5-Ω resistances is directly connected to ground and may therefore also be labeled as G [Fig. 6-13(c)]. Since we need to know the resistance between X and Y, the circuit has been redrawn in the more conventional manner of Fig. 6-13(d). In redrawing such a circuit you must carefully check that each

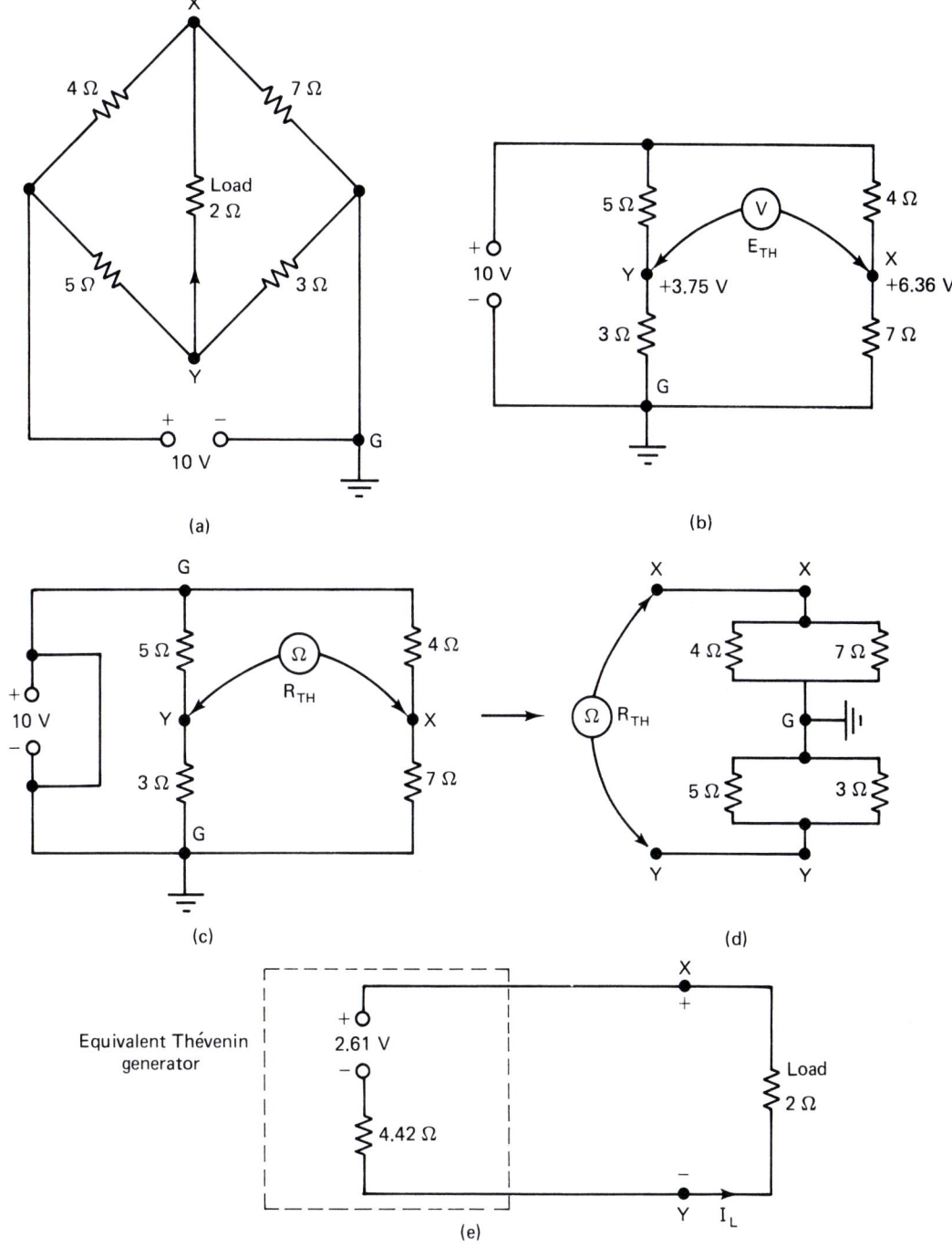

FIGURE 6-13 Analysis of bridge circuit by Thévenin's theorem.

resistor is connected between the same two points in both diagrams. For example, in Fig. 6-13(c) and (d) the 4-Ω and 7-Ω resistors are both connected between points G and X, while the 5-Ω and 3-Ω resistors are joined to G and Y. Using the product-over-sum formula,

$$R_{\text{TH}} = \frac{5 \times 3}{5 + 3} + \frac{4 \times 7}{4 + 7} = 1.875 + 2.545 = 4.42 \, \Omega$$

The Thévenin generator circuit [Fig. 6-13(e)] can now be drawn and the load

Sec. 6-6 Thévenin's Theorem

replaced. Then the load current through the 2-Ω resistor is

$$I_L = \frac{2.61 \text{ V}}{4.42 \text{ Ω} + 2 \text{ Ω}} = \frac{2.61 \text{ V}}{6.42 \text{ Ω}} = 0.41 \text{ A}$$

This section has taught us that in order to Thévenize a circuit for a particular load, you must carry out the following steps:

1. Remove the load and determine the open-circuit voltage, E_{TH}, between the network's output terminals X and Y.
2. With the load still removed, place a *short* circuit across the constant EMF of each *voltage* source, so that only the source's series internal resistance remains. For each *current* generator replace the current source with an *open* circuit, leaving only the parallel internal resistance. Then determine the resistance, R_{TH}, as measured between terminals X and Y.
3. Draw the Thévenin generator consisting of the constant voltage EMF, E_{TH}, in series with the internal resistance, R_{TH}. Replace the load and analyze the circuit to obtain the desired results.

EXAMPLE 6-6 In Fig. 6-14(a) find the value of R_L that will allow maximum power to be developed in the load and calculate the amount of the maximum load power.

Solution

Step 1
Remove the load and calculate the open-circuit voltage, E_{TH}, between the terminals X and Y. To determine E_{TH} we will use the superposition theorem.

Short circuit the 12-V source. By the current division rule the current through the 4.7-kΩ resistor due to the 8-mA generator is

$$8 \text{ mA} \times \frac{3.3 \text{ kΩ} + 2.2 \text{ kΩ}}{(2.2 \text{ kΩ} + 3.3 \text{ kΩ}) + 4.7 \text{ kΩ}} = 8 \times \frac{5.5}{10.2} = 4.31 \text{ mA}$$

with the electron flow in the direction from Y to X.

Remove the 8-mA current source. The direction of the electron flow due to the 12-V source is also from Y to X and the current is

$$\frac{12 \text{ V}}{3.3 \text{ kΩ} + 2.2 \text{ kΩ} + 4.7 \text{ kΩ}} = \frac{12}{10.2} = 1.18 \text{ mA}$$

The total current through the 4.7-kΩ resistor is $4.31 + 1.18 = 5.49$ mA and the open-circuit voltage between X and Y is 4.7 kΩ $\times 5.49$ mA $= 25.8$ V.

Step 2
To find R_{TH} you must place a short circuit across the 12-V source and also remove the 8-mA source to create an open circuit; this step is shown in Fig. 6-14(b). R_{TH} is then the value of ohms which would be measured by an ohmmeter connected between terminals X and Y. With respect to these terminals the 2.2-kΩ and 3.3-kΩ resistors are in series and the 4.7-kΩ resistor is in parallel with this series combination. Using the product-over-sum formula,

$$R_{TH} = \frac{4.7 \times (2.2 + 3.3)}{4.7 + (2.2 + 3.3)} = \frac{4.7 \times 5.5}{10.2} = 2.53 \text{ kΩ}$$

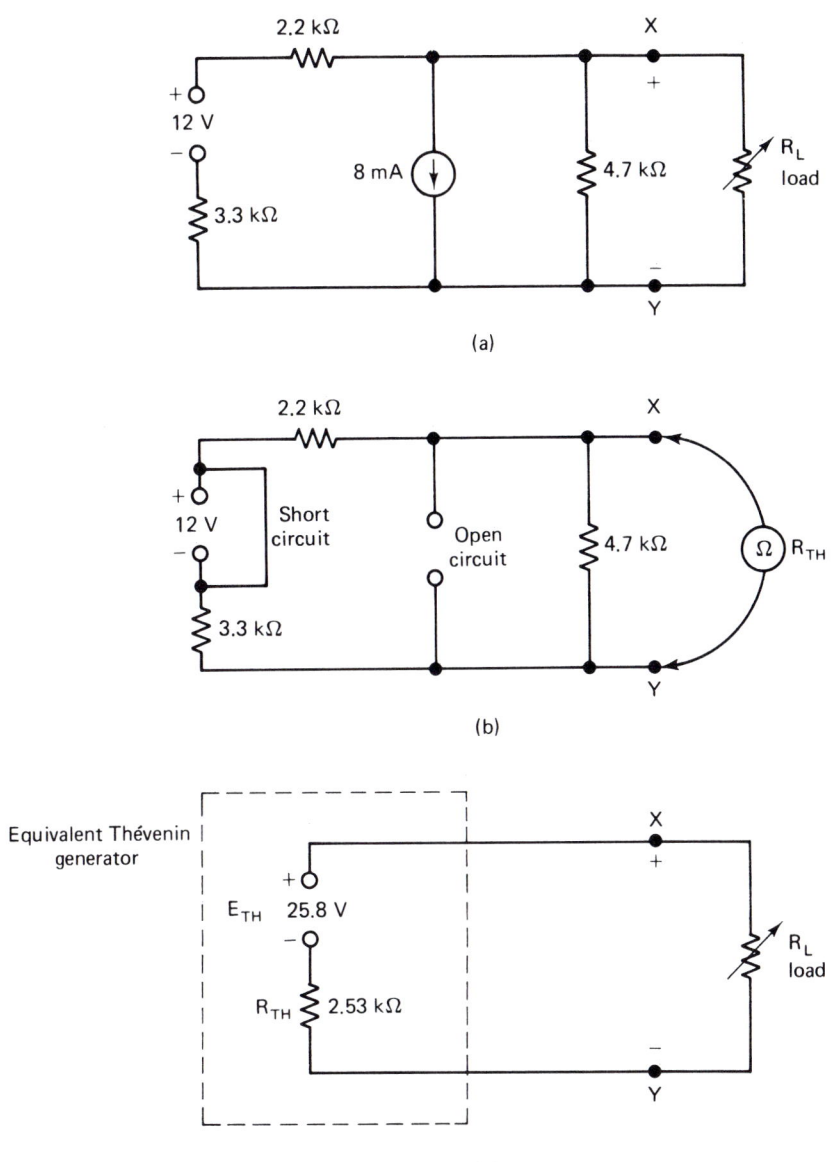

FIGURE 6-14 Circuit for Example 6-6.

Step 3
Draw the Thévenin equivalent generator and replace the load [Fig. 6-14(c)]. For maximum power transfer to the load, the load resistance R_L must be matched to the internal resistance of the Thévenin generator. Therefore, from Eq. (5-4-3),

$$R_L = R_{TH} = 2.53 \text{ k}\Omega$$

Equation (5-4-4) gives the maximum power developed in the load:

$$P_{L_{max}} = \frac{(E_{TH})^2}{4 R_L} = \frac{(25.8 \text{ V})^2}{4 \times 2.53 \text{ k}\Omega} = 65.7 \text{ mW}$$

Sec. 6-6 Thévenin's Theorem

6-7 Norton's Theorem

Norton's theorem is similar to Thévenin's since it enables a complex network of sources and resistors to be replaced by a single generator and a resistance. However, the equivalent Norton generator is of the constant current type with its resistance in parallel, while the equivalent Thévenin generator is of the constant voltage type with its resistance in series (Fig. 6-15). Since the two generators are equivalent to each other as far as the load is concerned, they may be transposed by using the results of Section 5-6. Therefore,

$$\text{Norton current, } I_N = \frac{E_{TH}}{R_{TH}} \qquad (6\text{-}7\text{-}1)$$

$$\text{Thévenin voltage, } E_{TH} = I_N \times R_N \qquad (6\text{-}7\text{-}2)$$

$$\text{Norton resistance, } R_N = \text{Thévenin resistance, } R_{TH} \qquad (6\text{-}7\text{-}3)$$

Norton's theorem, stated formally, is:

> The current in a load connected between two output terminals, X and Y, of a complex network containing electrical sources and resistors is the same as if that load were connected to a *constant current* source whose generator current, I_N, is equal to the *short*-circuit current measured between X and Y. This constant current generator is placed in *parallel* with a resistance, R_N, which is equal to the resistance of the network looking back into terminals X and Y with all sources replaced by resistances equal to their internal resistances.

This last step would involve short-circuiting all sources of constant voltage and open-circuiting all sources of constant current.

Norton's theorem is illustrated in Fig. 6-16. The first thing to notice is that the direction of I_N, as indicated by the arrow in the current generator, is such as to drive an electron flow through the load in the same direction as in the original circuit. The value of I_N is determined by removing the load and

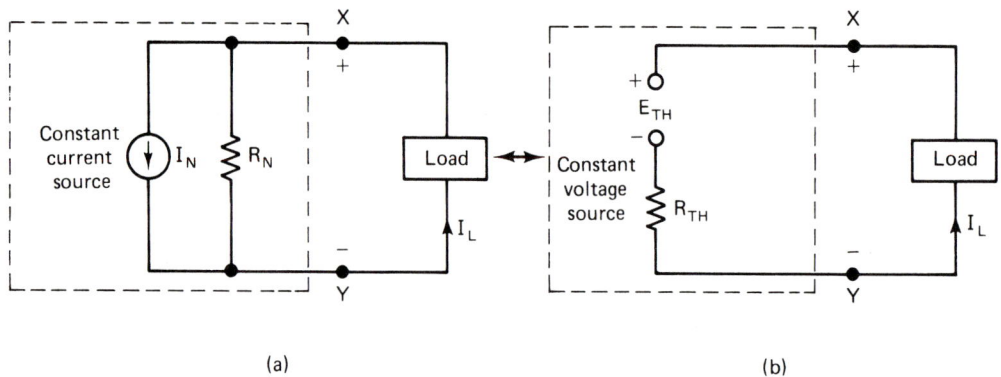

FIGURE 6-15 Relationship between (a) Norton and (b) Thévenin equivalent generators.

FIGURE 6-16 Illustration of Norton's theorem: (a) original circuit; (b) Norton equivalent generator; (c) determination of I_N; (d) determination of R_N.

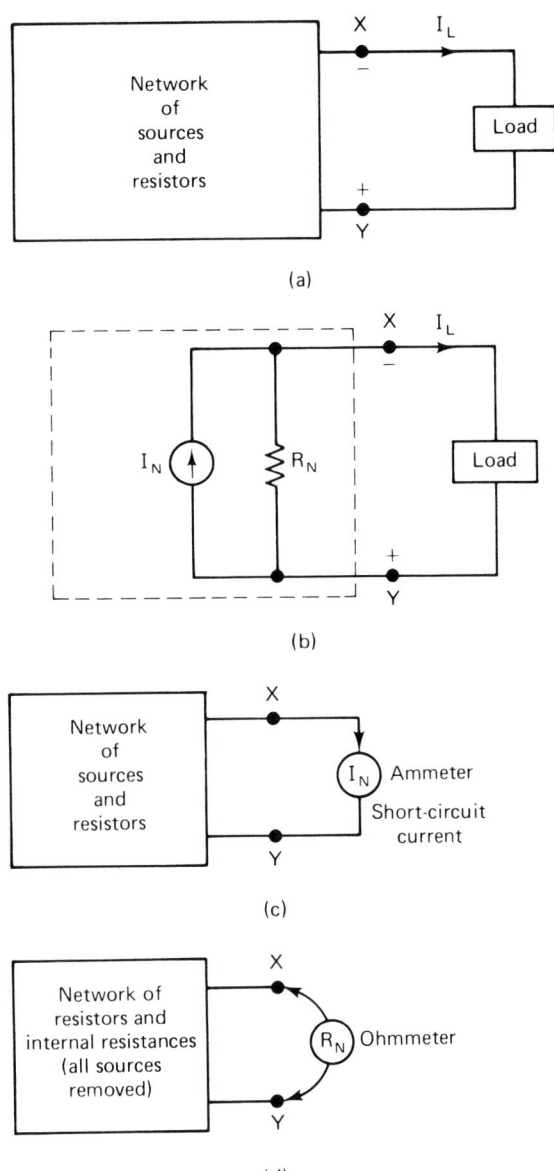

replacing it by a short circuit; the current that flows in this short circuit is then equal to I_N. Therefore, if you use an ammeter to establish the short circuit, the reading of this meter would be the value of I_N [Fig. 6-16(c)].

The procedure for obtaining R_N is exactly the same as the one we used for finding R_{TH} in Thévenin's theorem [Fig. 6-16(d)]. Therefore, as we have already said in Eq. (6-7-3), $R_N = R_{TH}$. The process of obtaining I_N and R_N and then replacing the network of sources and resistors by the single current generator is called *Nortonizing* the circuit.

You may have got the impression that the Thévenin and Norton theorems are so closely related that it does not matter which one you use in analyzing a particular circuit. This is not so. In some circuits it may be far easier to calculate I_N rather than E_{TH}, whereas in other circuits the reverse may be the case. You must therefore study every circuit carefully before deciding on a particular method of analysis.

To compare the use of the Thévenin and Norton theorems, we will use the problem illustrated in Fig. 6-13. The first step will be to remove the 2-Ω

Sec. 6-7 Norton's Theorem

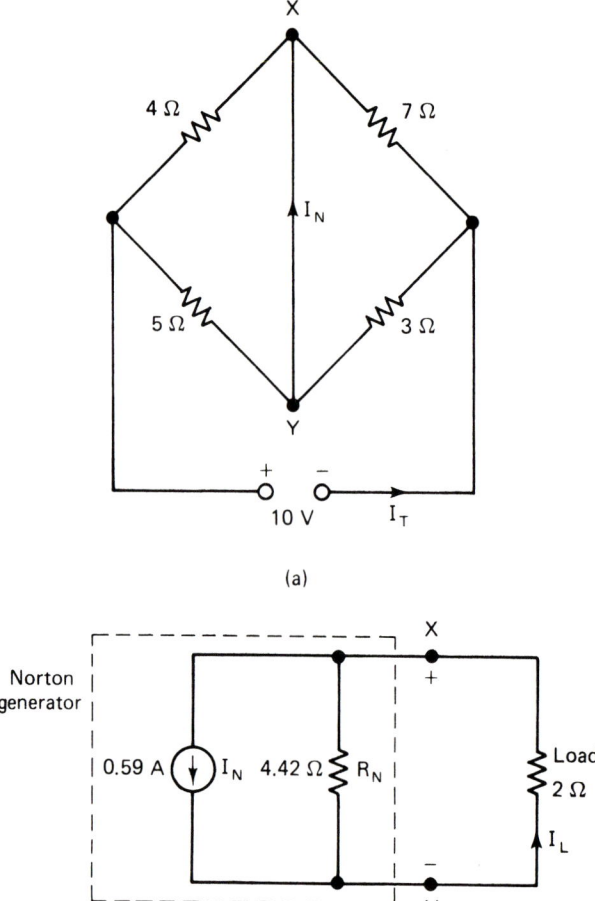

FIGURE 6-17 Analysis of bridge circuit by Norton's theorem.

load and replace it with a short circuit [Fig. 6-17(a)]. The 4-Ω and 5-Ω resistors are in parallel, so that their equivalent resistance is

$$\frac{4 \times 5}{4 + 5} = 2.22 \ \Omega$$

The 7-Ω and 3-Ω resistors are also in parallel, with an equivalent resistance of

$$\frac{7 \times 3}{7 + 3} = 2.10 \ \Omega$$

Since these parallel combinations are in series, the total resistance presented to the 10-V source is 2.22 + 2.10 = 4.32 Ω. Then the total current I_T = 10 V/4.32 Ω = 2.31 A.

Since I_T divides between the 3-Ω and 7-Ω resistors, the current by the CDR rule through the 3-Ω resistor is

$$2.31 \ A \times \frac{7 \ \Omega}{7 \ \Omega + 3 \ \Omega} = 1.62 \ A$$

I_T also divides between the 4-Ω and 5-Ω resistors, so that the current through the 5-Ω resistor is

$$2.31 \text{ A} \times \frac{4 \text{ Ω}}{4 \text{ Ω} + 5 \text{ Ω}} = 1.03 \text{ A}$$

This means that the electron flow entering point Y through the 3-Ω resistor is 1.62 A, while the current leaving point Y through the 5-Ω resistor is 1.02 A. Therefore, there is an electron flow of 1.62 A − 1.03 A = 0.59 A from point Y to point X, and this will be the short-circuit Norton current, I_N. It is obvious that obtaining I_N using Norton's theorem involved more steps than calculating E_{TH} with Thévenin's theorem.

Remembering that $R_N = R_{TH} = 4.42$ Ω, we can draw the equivalent Norton generator and reconnect the 2-Ω load [Fig. 6-17(b)]. By the current division rule, the current through the 2-Ω load is

$$0.59 \text{ A} \times \frac{4.42 \text{ Ω}}{4.42 \text{ Ω} + 2 \text{ Ω}} = 0.41 \text{ A}$$

This is exactly the same result as we obtained using Thévenin's theorem.

As a summary, Nortonizing a circuit for a particular load involves the following steps:

1. Remove the load and determine the short-circuit Norton current, I_N, between the network's output terminals, X and Y.

2. With the load still removed, place a *short* circuit across the constant EMF of each *voltage* source so that only the source's series internal resistance remains. For each *current* generator, replace the current source with an *open* circuit, leaving only the parallel internal resistance. Then determine the Norton resistance, R_N, as measured between terminals X and Y.

3. Draw the Norton generator consisting of the constant current source, I_N, in parallel with the resistance, R_N. Replace the load and analyze the circuit to obtain the desired results.

EXAMPLE 6-7 In Fig. 6-14(a), obtain the equivalent Norton generator between terminals X and Y.

Solution

Step 1
Remove the load and place a short circuit between terminals X and Y [Fig. 6-18(a)]. Using the superposition theorem, short-circuit the 12-V source [Fig. 6-18(b)]; then the electron flow from Y to X due to the current generator is the complete 8 mA, since both the 4.7-kΩ resistor and the series combination of the 2.2-kΩ and 3.3-kΩ resistors have been shorted out. When the 8-mA current generator is removed, the current through the short circuit due to the 12-V source is

$$\frac{12 \text{ V}}{2.2 \text{ kΩ} + 3.3 \text{ kΩ}} = 2.18 \text{ mA}$$

Note that the 4.7-kΩ resistor is still shorted out. Then $I_N = 8 + 2.18 = 10.18$ mA.

Sec. 6-7 Norton's Theorem

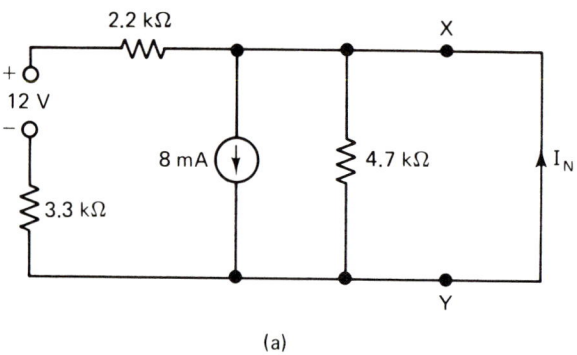

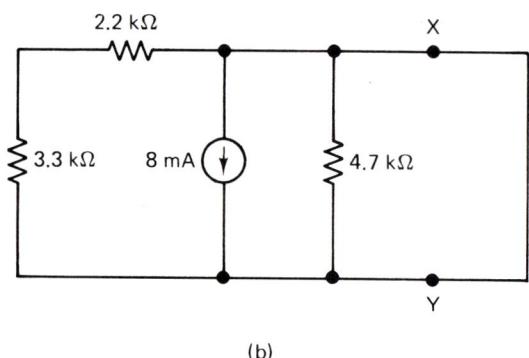

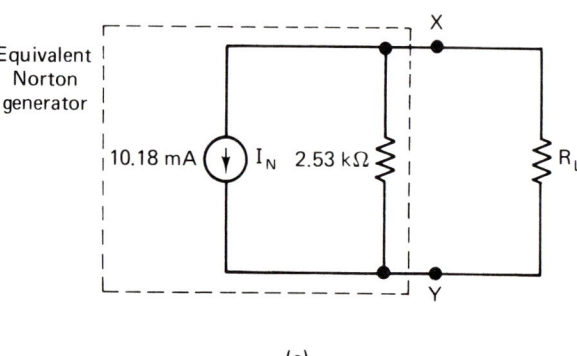

FIGURE 6-18 Circuit for Example 6-7.

(a)

(b)

(c)

Step 2
The procedure for obtaining the Norton resistor, R_N, will be the same as that used for calculating R_{TH} in Example 6-6. Therefore, $R_N = 2.53$ kΩ.

Step 3
Draw the Norton equivalent generator [Fig. 6-18(c)]. Check your results by deriving $E_{TH} = I_N \times R_N = 10.18$ mA $\times$ 2.53 kΩ = 25.8 V. This is the same value for E_{TH} as we obtained in Example 6-6.

The main reason for including this example was to show that in this particular case, it was easier to obtain I_N than E_{TH}.

6-8

Delta-Wye (Π-T) Transformations

The triangular arrangement of resistors shown in Fig. 6-19(a) is commonly known as a *delta connection* (Δ is the Greek capital letter delta). The same

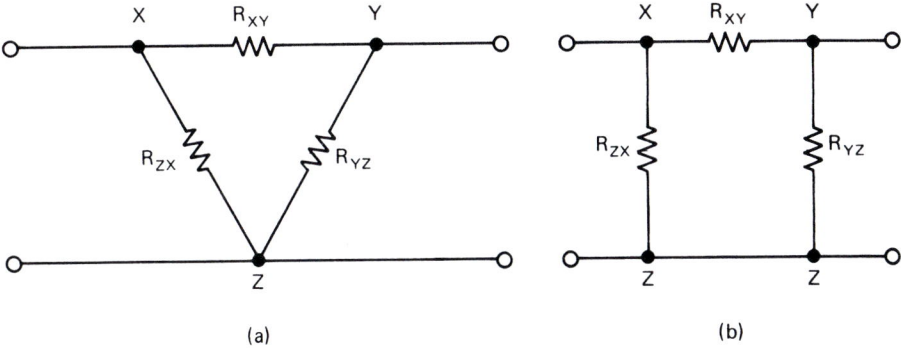

FIGURE 6-19 Delta (Δ) and pi (Π) arrangements.

resistors may also be arranged as in Fig. 6-19(b), so that the configuration now looks like the Greek capital letter *pi*, Π. Pi and delta are therefore alternative names for the same arrangement of resistors.

Figure 6-20(a) shows another resistor configuration, which is referred to as the *wye* (letter Y) connection; the same arrangement has been modified in Fig. 6-20(b) to look like the letter *tee* (T). It follows that either tee or wye may be used to describe these identical resistor networks.

We have already seen that in many instances it is impossible to analyze a circuit directly in terms of series, parallel, or series–parallel connections; a good example of such a network is the unbalanced bridge circuit, where the center arm has no simple series or parallel relationship with the other four resistors. However, in some of these cases the circuit can be changed into a type of series–parallel arrangement if a particular delta formation of resistors is replaced by an equivalent wye connection (or vice versa). Such a change would be called a *delta–wye* (or *wye–delta*) transformation.

For the delta and wye connections to be equivalent, the resistances between the points X, Y, and Z in both arrangements must be the same. Using the product-over-sum formula in Fig. 6-19(a),

total resistance between the points X and Y

$$= \frac{R_{XY}(R_{YZ} + R_{ZX})}{R_{XY} + R_{YZ} + R_{ZX}} \quad (6\text{-}8\text{-}1)$$

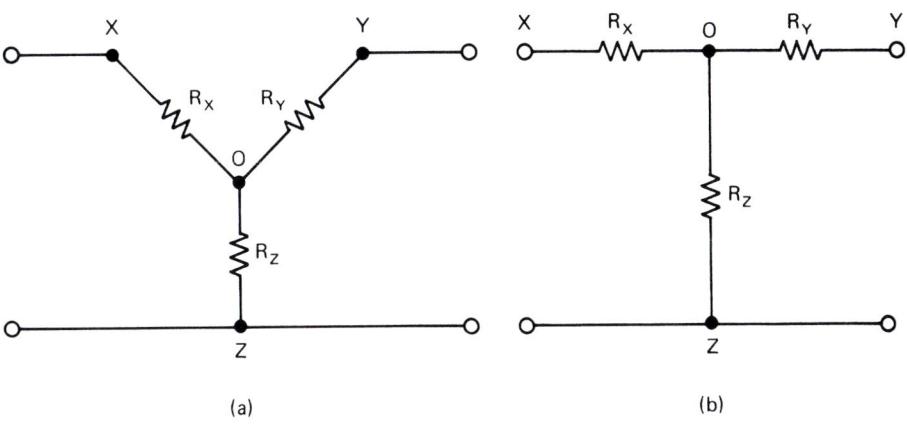

FIGURE 6-20 Wye (Y) and tee (T) arrangements.

Sec. 6-8 Delta-Wye (Π-T) Transformations

In Fig. 6-20(a), R_X and R_Y are directly in series, so that

$$\text{total resistance between } X \text{ and } Y = R_X + R_Y \qquad (6\text{-}8\text{-}2)$$

Therefore,

$$R_X + R_Y = \frac{R_{XY}(R_{YZ} + R_{ZX})}{R_{XY} + R_{YZ} + R_{ZX}} \qquad (6\text{-}8\text{-}3)$$

The symmetry of this relationship will allow us to write down two further equations:

$$R_Y + R_Z = \frac{R_{YZ}(R_{ZX} + R_{XY})}{R_{XY} + R_{YZ} + R_{ZX}} \qquad (6\text{-}8\text{-}4)$$

and

$$R_Z + R_X = \frac{R_{ZX}(R_{XY} + R_{YZ})}{R_{XY} + R_{YZ} + R_{ZX}} \qquad (6\text{-}8\text{-}5)$$

Add Eqs. (6-8-3) and (6-8-5); from this sum subtract Eq. (6-8-4) and then divide by 2. This yields

$$R_X = \frac{R_{XY} R_{ZX}}{R_{XY} + R_{YZ} + R_{ZX}} \qquad (6\text{-}8\text{-}6)$$

Similarly

$$R_Y = \frac{R_{YZ} R_{XY}}{R_{XY} + R_{YZ} + R_{ZX}} \qquad (6\text{-}8\text{-}7)$$

and

$$R_Z = \frac{R_{ZX} R_{YZ}}{R_{XY} + R_{YZ} + R_{ZX}} \qquad (6\text{-}8\text{-}8)$$

You would use these last three equations in a delta-to-wye transformation. For a wye-to-delta transformation you would also need three equations:

$$R_{XY} = \frac{R_X R_Y + R_Y R_Z + R_Z R_X}{R_Z} \qquad (6\text{-}8\text{-}9)$$

$$R_{YZ} = \frac{R_X R_Y + R_Y R_Z + R_Z R_X}{R_X} \qquad (6\text{-}8\text{-}10)$$

$$R_{ZX} = \frac{R_X R_Y + R_Y R_Z + R_Z R_X}{R_Y} \qquad (6\text{-}8\text{-}11)$$

The analysis of a bridge circuit provides a good example for using a delta-to-wye transformation. Once again for comparison purposes we will use the circuit of Fig. 6-13(a), which has been redrawn in Fig. 6-21(a). It is proposed to replace the delta formation of resistors between points X, Y, and Z by its wye equivalent. Then $R_{XY} = 2\ \Omega$, $R_{YZ} = 3\ \Omega$, and $R_{ZX} = 7\ \Omega$, so that by using Eqs. (6-8-6), (6-8-7), and (6-8-8), we obtain

$$R_X = \frac{2 \times 7}{2 + 3 + 7} = \frac{14}{12} = 1.17\ \Omega$$

$$R_Y = \frac{3 \times 2}{2 + 3 + 7} = \frac{6}{12} = 0.50\ \Omega$$

$$R_Z = \frac{7 \times 3}{2 + 3 + 7} = \frac{21}{12} = 1.75\ \Omega$$

To use your calculator most effectively, you add the resistance values, obtain the reciprocal, and then depress the Memory-in ($x \rightarrow M$) key. This transfers the quantity

$$\frac{1}{2 + 3 + 7}$$

to the memory. Subsequent use of the Recall Memory (RM) key enables you to calculate the values of R_X, R_Y, and R_Z. The full procedure is:

Operation	Display	
2 $+$	2	
3 $+$	5	
7 $=$	12	
$1/x$ $x \rightarrow M$ $\times$	0.0833333	
2 $\times$	0.166666	
7 $=$	1.16666	(R_X)
RM $\times$	0.0833333	
3 $\times$	0.25	
2 $=$	0.5	(R_Y)
RM $\times$	0.0833333	
7 $\times$	0.583333	
3 $=$	1.75	(R_Z)

The values of R_X, R_Y, and R_Z have been inserted in Fig. 6-21(b), which is now a relatively simple series–parallel circuit. Combining the series connection of the 4-Ω and the 1.17-Ω resistances, their equivalent resistance is 4 + 1.17 = 5.17 Ω. In the same way, the series arrangement of the 5-Ω and the

Sec. 6-8 Delta-Wye (Π-T) Transformations

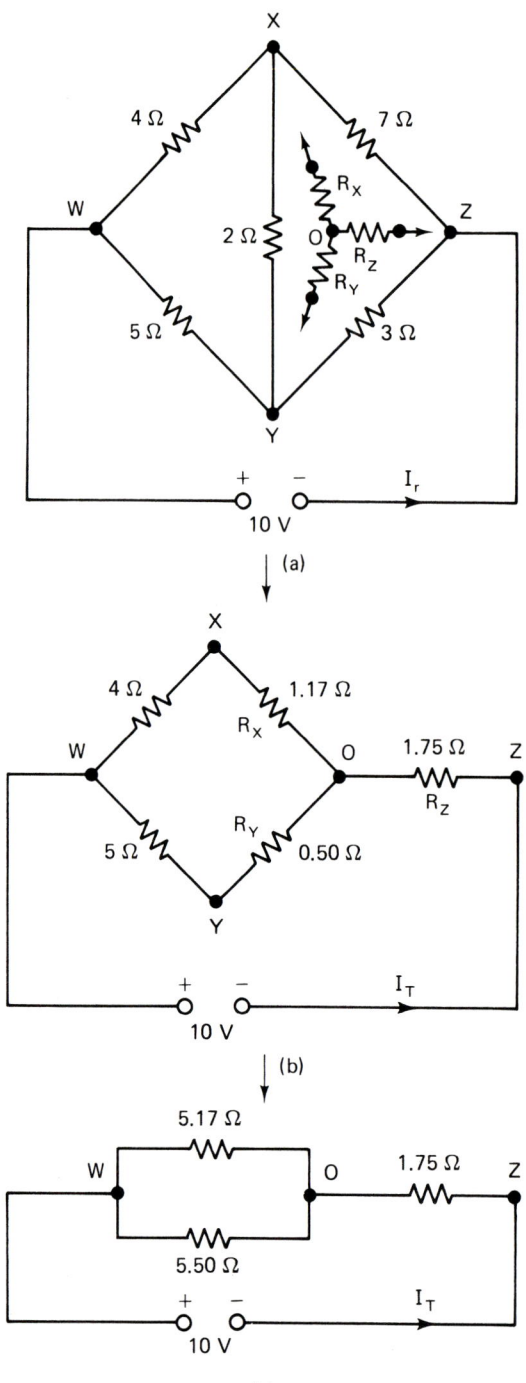

FIGURE 6-21 Delta–wye transformation.

0.5-Ω resistances is equivalent to 5 + 0.5 = 5.5 Ω and the circuit reduces to Fig. 6-21(c). Then

$$\text{resistance of } 5.17\,\Omega\|5.5\,\Omega = \frac{5.17 \times 5.5}{5.17 + 5.5}$$

$$= 2.66\,\Omega$$

Total resistance presented to the 10-V source is 2.66 + 1.75 = 4.41 Ω, so that I_T = 10 V/4.41 Ω = 2.27 A; these results are the same as for the original circuit. Using the current division rule, the current flowing through the 4-Ω resistor of Fig. 6-21(b) is

$$2.27 \text{ A} \times \frac{5 \text{ }\Omega + 0.5 \text{ }\Omega}{(4 \text{ }\Omega + 1.17 \text{ }\Omega) + (5 \text{ }\Omega + 0.5 \text{ }\Omega)} = 1.17 \text{ A}$$

By KCL the current in the 5-Ω resistor is $2.27 - 1.17 = 1.10$ A. Therefore, point X is $4 \text{ }\Omega \times 1.17 \text{ A} = 4.68$ V negative with respect to the positive terminal of the 10-V source, while point Y is $5 \text{ }\Omega \times 1.10 \text{ A} = 5.50$ V negative relative to the same terminal. This means that point Y is negative with respect to point X by $5.50 - 4.68 = 0.82$ V. The current through the 2-Ω resistor is therefore $0.82 \text{ V}/2\Omega = 0.41$ A. This is the same result as we obtained by using both Thévenin's and Norton's theorems.

To make the best use of the delta–wye and wye–delta transformations, you must decide which transformation will help you to convert the circuit you are analyzing into a simple series–parallel arrangement. Carry out the transformation by using Eqs. (6-8-6) through (6-8-11) and then solve the series–parallel circuit to obtain the total source current and the branch currents. Substitute these results into the original circuit but remember that the currents in the delta formation are *not* the same as those in the equivalent wye connection, and vice versa.

EXAMPLE 6-8 In Fig. 6-21(a) the 4-Ω, 7-Ω, and 2-Ω resistors form a T junction. Convert this arrangement into a delta configuration and calculate the total equivalent resistance presented to the 10-V source.

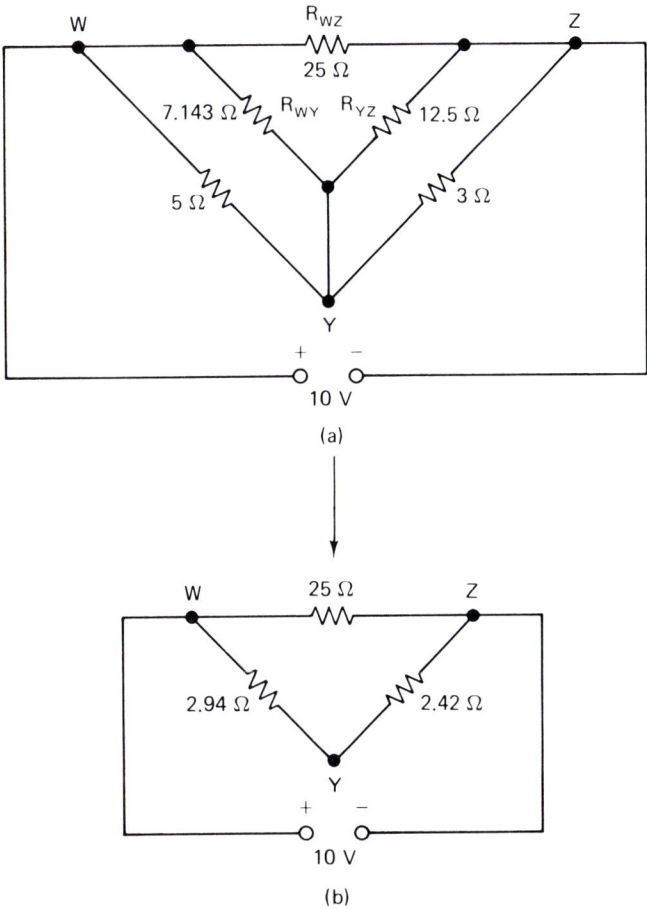

FIGURE 6-22 Wye–delta transformation.

Solution If the 4-Ω, 7-Ω, and 2-Ω resistors are replaced by an equivalent delta formation, the circuit is as shown in Fig. 6-22(a). Using Eqs. (6-8-9), (6-8-10), and (6-8-11), we obtain

$$R_{WZ} = \frac{4 \times 7 + 7 \times 2 + 2 \times 4}{2} = 25 \text{ Ω}$$

$$R_{WY} = \frac{4 \times 7 + 7 \times 2 + 2 \times 4}{7} = 7.143 \text{ Ω}$$

$$R_{YZ} = \frac{4 \times 7 + 7 \times 2 + 2 \times 4}{4} = 12.5 \text{ Ω}$$

The most effective way of finding the equivalent delta resistances is to use a calculator with a Memory Plus, (M+) key. This enables you to add quantities to the memory, which is used to store the value of the quantity "$4 \times 7 + 7 \times 2 + 2 \times 4$." This quantity will then be recalled to be divided in turn by R_Y, R_Z, and R_W. In our example the full procedure is:

Operation	Display	
4 ×	4	
7 x→M	28	
7 ×	7	
2 = M+	14	
2 ×	2	
4 = M+	8	
RM ÷	50	
2 =	25	(R_{WZ})
RM ÷	50	
7 =	7.14286	(R_{WY})
RM ÷	50	
4 =	12.5	(R_{YZ})

The equivalent resistance of R_{WY} in parallel with the 5-Ω resistor is

$$\frac{7.143 \times 5}{7.143 + 5} = 2.94 \text{ Ω}$$

The equivalent resistance of R_{YZ} in parallel with the 3-Ω resistor is

$$\frac{12.5 \times 3}{12.5 + 3} = 2.42 \text{ Ω}$$

[Fig. 6-22(b)]. Since these two parallel combinations are in series, the total resistance in parallel with R_{WZ} is 2.94 + 2.42 = 5.36 Ω. Therefore, the total equivalent resistance presented to the 10-V source is

$$\frac{5.36 \times 25}{5.36 + 25} = 4.41 \text{ Ω}$$

This is the same result as we obtained using the delta–wye transformation.

Chapter Summary

6.1. Kirchhoff's voltage law (KVL):

The algebraic sum of the constant voltage EMFs and the voltage drops around any closed electrical loop (mesh) is always zero.

6.2. Kirchhoff's current law (KCL):

The algebraic sum of the currents existing at any junction point (node) is always zero.

6.3. Nodal analysis:

At a node the algebraic sum of the source currents entering = the sum of the currents leaving through the resistors.

6.4. Millman's theorem:

This theorem may be applied to a number of current and/or voltage sources which are directly connected in parallel. The statement of the theorem is:

Any number of constant current sources that are directly connected in parallel can be converted into a single current source whose total generator current is the algebraic sum of the individual source currents and whose total internal resistance is the result of combining the individual source resistances in parallel.

6.5. Superposition theorem:

In a network of *linear* resistances containing more than one source, the resultant current flowing at any point is the algebraic sum of the currents which would flow at that point if each source is considered separately with all other sources replaced by their equivalent internal resistances.

6.6. Thévenin's theorem:

The current in a load connected between two output terminals X and Y of a complex network of resistors and electrical sources, is the same as if that load were connected across a simple constant voltage generator whose EMF, E_{TH}, is the open-circuit voltage measured between X and Y and whose *series* internal resistance, R_{TH}, is the resistance of the network looking back into terminals X and Y with all sources replaced by resistances equal to their internal resistances.

6.7. Norton's theorem:

The current in a load connected between two output terminals X and Y of a complex network containing electrical sources and resistors, is the same as if that load were connected to a constant current source whose generator current, I_N, is equal to the *short-circuit* current measured between X and Y. This constant current generator is placed in *parallel* with a resistance, R_N, which is equal to the resistance of the network looking back into terminals X and Y with all sources replaced by resistances equal to their internal resistances.

6.8. Relationships between Thévenin's and Norton's generators:

$$E_{TH} = I_N \times R_N \qquad I_N = \frac{E_{TH}}{R_{TH}}$$

$$R_{TH} = R_N$$

6.9. Delta–wye transformation (Figs. 6-19 and 6-20):

$$R_X = \frac{R_{XY}R_{ZX}}{R_{XY} + R_{YZ} + R_{ZX}}, \text{ etc.}$$

6.10. Wye–delta transformation (Figs. 6-19 and 6-20):

$$R_{XY} = \frac{R_X R_Y + R_Y R_Z + R_Z R_X}{R_Z}, \text{ etc.}$$

Self-Review

True–False Questions

1. T. F. Kirchhoff's voltage law states that the algebraic sum of the voltages at a junction point is zero.
2. T. F. Kirchhoff's current law states that the algebraic sum of the currents at a junction point is zero.
3. T. F. In mesh analysis the directions of the mesh currents flowing through a resistor that is common to two meshes, are always in the same direction.
4. T. F. The algebraic sum of the source currents entering a node equals the sum of the currents leaving through the resistors.
5. T. F. Millman's theorem is used to reduce to a single generator a number of voltages and/or current sources which are directly connected in parallel.
6. T. F. Using Thévenin's theorem, E_{TH} is found by removing the load and measuring the open-circuit voltage between the network's output terminals.
7. T. F. The superposition theorem applies to both linear and nonlinear resistances.
8. T. F. Using Norton's theorem, R_N is found by removing the load and then shorting all constant current sources while open-circuiting all constant voltage sources.
9. T. F. Delta and tee are different names for the same formation of resistors.
10. T. F. When comparing the equivalent Norton and Thévenin generators for the same network of sources and resistors, $E_{TH} = I_N R_{TH}$.
11. T. F. Thévenin's theorem can be applied only to a network that cannot be regarded as some form of series–parallel arrangement.
12. T. F. When using mesh analysis, there are always as many independent equations as there are unknown currents.
13. T. F. The superposition theorem is applied only when analyzing circuits with more than one source.
14. T. F. One purpose of a delta-to-wye or a wye-to-delta transformation is to convert a complex network into some relatively simple form of series–parallel circuit.
15. T. F. Two constant current sources have different short-circuit values and different internal resistances. Such sources must never be connected in series.
16. T. F. When using the method of nodal analysis, the circuit must contain only voltage sources.
17. T. F. When using Kirchhoff's laws to analyze a circuit, a negative current solution indicates that the actual direction of the current is opposite to the assumed direction.
18. T. F. Millman's theorem may be adapted to combine series-connected voltage and current sources.
19. T. F. In circuit analysis involving Kirchhoff's laws, a number of different branch currents may flow through one particular resistor.
20. T. F. The Thévenin's equivalent generator for any network of sources and resistors does not change if a different resistor is selected as a load.

Multiple-Choice Questions

21. In Fig. 6-23, the value of the current through the 4-kΩ resistor is found by Kirchhoff's laws to be
(a) 3 mA (b) 2 mA (c) 1 mA (d) 1.5 mA (e) 2.5 mA

FIGURE 6-23 Circuit for Question 21.

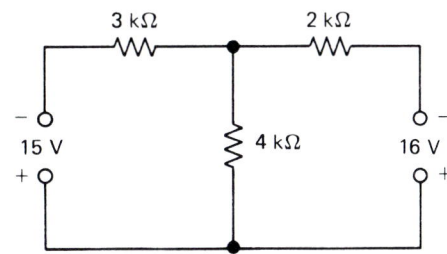

22. In Fig. 6-24, the value of the clockwise mesh current through the 7-kΩ resistor is
(a) 1 mA (b) 2 mA (c) 3 mA (d) 2.5 mA (e) 4 mA

FIGURE 6-24 Circuit for Question 22.

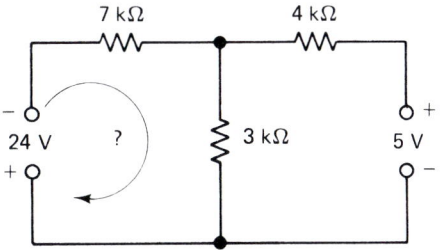

23. In Fig. 6-25, use nodal analysis to determine the voltage at the point N. The value of this voltage is
(a) +24 V (b) −24 V (c) +36 V (d) +18 V (e) +48 V

FIGURE 6-25 Circuit for Question 23.

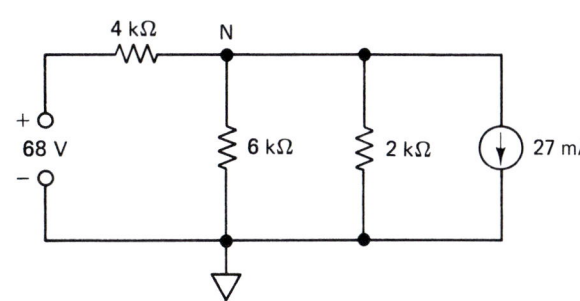

24. Use Millman's theorem to reduce the circuit of Fig. 6-26 to a simple constant current generator. The value of the voltage between the points X and Y is
(a) 12 V (b) 6 V (c) 10 V (d) 11 V (e) 21 V

FIGURE 6-26 Circuit for Question 24.

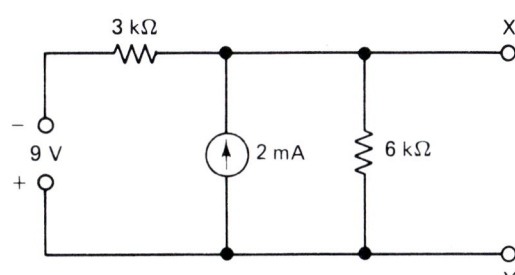

25. In Fig. 6-27, obtain the equivalent Thévenin generator between terminals X and Y. If a 2-Ω load is connected between the terminals, the load voltage is
(a) 4 V (b) 5 V (c) 3 V (d) 6 V (e) 2.5 V

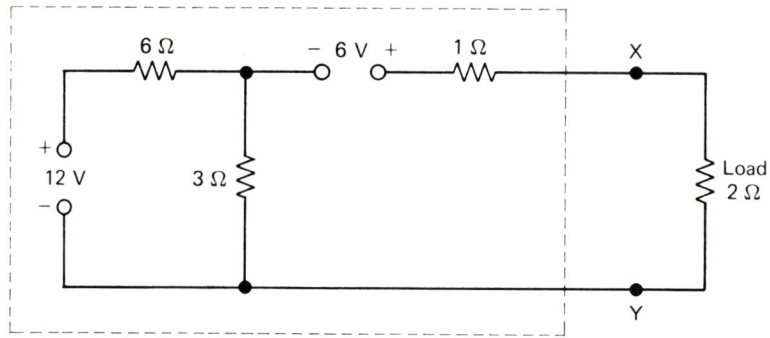

FIGURE 6-27 Circuit for Question 25.

26. Using the superposition theorem in Fig. 6-28, the current flowing through the 5-kΩ resistor is
    (a) 9.9 mA   (b) 6.4 mA   (c) 4.2 mA   (d) 7.2 mA   (e) 3.6 mA
27. In Fig. 6-28, the 5-kΩ resistor is regarded as the load. The value of the Norton current, I_N, between terminals X and Y is
    (a) 10.67 mA   (b) 4 mA   (c) 3.56 mA   (d) 7.53 mA   (e) 5.33 mA
28. In Question 27, the value of the Norton resistance, R_N, between terminals X and Y is
    (a) 32 kΩ   (b) 8 kΩ   (c) 7.5 kΩ   (d) 10.5 kΩ   (e) 12 kΩ

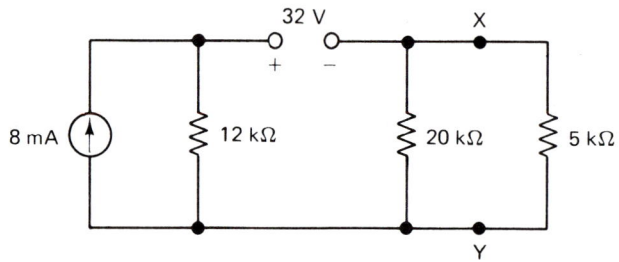

FIGURE 6-28 Circuit for Questions 26, 27, and 28.

29. Figure 6-29 represents equivalent Π and T networks. The value of R_X is
    (a) 1 Ω   (b) 1.25 Ω   (c) 0.8 Ω   (d) 1.67 Ω   (e) 0.6 Ω

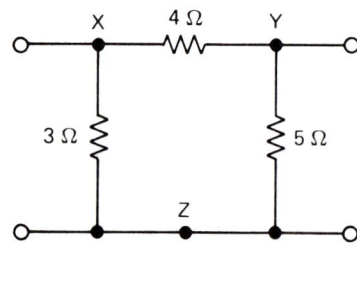

FIGURE 6-29 Circuit for Question 29.

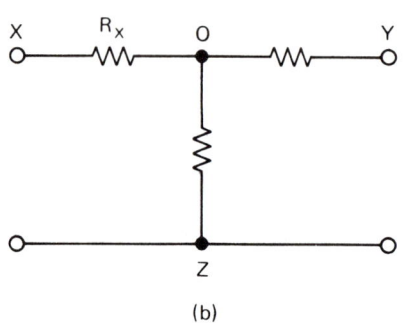

30. Figure 6-30 represents equivalent wye and delta networks. The value of R_{YZ} is
    (a) 11.75 Ω   (b) 9.4 Ω   (c) 10.5 Ω   (d) 6.5 Ω   (e) 15.67 Ω
31. In Fig. 6-23, the value of the 15-V source is changed to 11 V and its polarity is reversed. Using Kirchhoff's laws, the new value of the current through the 2-kΩ resistor is
    (a) 6 mA   (b) 5 mA   (c) 1 mA   (d) 7 mA   (e) 4 mA
32. In Fig. 6-24, the value of the 5-V source is changed to 23 V and its polarity is reversed. If the output of the 24 V source is increased to 36 V, the value of the clockwise mesh current flowing through the 4-kΩ resistor is
    (a) 2 mA   (b) 3 mA   (c) −3 mA   (d) −2 mA   (e) 5 mA
33. In Fig. 6-25, the value of the current generator is changed to 28 mA and its direction is reversed. Using nodal analysis, the value of the voltage at point N is
    (a) −24 V   (b) +24 V   (c) −12 V   (d) +12 V   (e) +16 V
34. In Fig. 6-26, the direction of the current generator is reversed and point Y is grounded. Using Millman's theorem, the value of the voltage at point X is
    (a) +10 V   (b) −10 V   (c) +2 V   (d) +11 V   (e) −2 V
35. In Fig. 6-27, the polarity of the 12-V source is reversed. The new equivalent Thévenin voltage between terminals X and Y is
    (a) 10 V   (b) 6 V   (c) 4 V   (d) 3 V   (e) 2 V
36. In Fig. 6-28, the direction of the current generator is reversed. The new (electron flow) current through the 5-kΩ resistor from point X to point Y is
    (a) −6.4 mA   (b) +6.4 mA   (c) −4.8 mA   (d) +4.8 mA   (e) −3.2 mA
37. In Question 36, the 5-kΩ resistor is regarded as the load. The value of the Norton current between terminals X and Y is
    (a) 10.67 mA   (b) 5.33 mA   (c) 4.8 mA   (d) 3.2 mA   (e) 8.33 mA
38. In Question 36, the value of the Thévenin voltage between terminals X and Y is
    (a) 40 V   (b) 37 V   (c) 42 V   (d) 45 V   (e) 48 V
39. In Fig. 6-29(a), an additional 6-Ω resistor is connected between points X and Z while another additional 20-Ω resistor is connected between points Y and Z. The new equivalent value of R_x in Fig. 6-29(b) is
    (a) 1 Ω   (b) 1.2 Ω   (c) 1.4 Ω   (d) 0.9 Ω   (e) 0.8 Ω
40. In Fig. 6-30(a), an additional 6-Ω resistor is connected between points X and O while a second additional 20-Ω resistor is connected between points O and Z. The new value of the equivalent resistance, R_{YZ}, in Fig. 6-30(b) is
    (a) 15.67 Ω   (b) 16 Ω   (c) 14.67 Ω   (d) 24 Ω   (e) 8 Ω

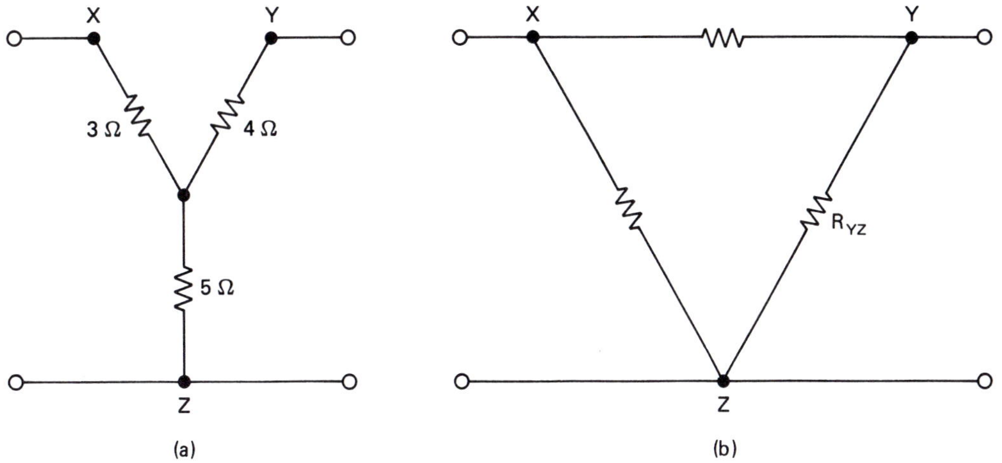

FIGURE 6-30 Circuit for Question 30.

Practice Problems

41. In Fig. 6-31, use Kirchhoff's laws to determine the value of the voltage at point N.

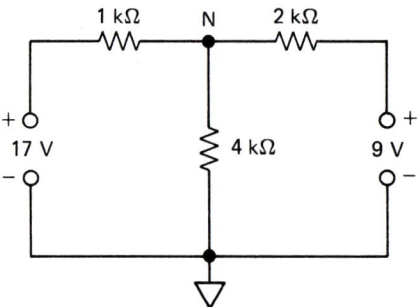

FIGURE 6-31 Circuit for Problem 41.

42. In Fig. 6-32, use mesh analysis to determine the value of the current flowing through the 5-kΩ resistor.

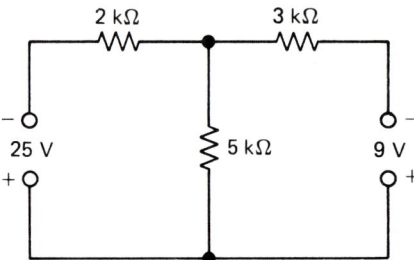

FIGURE 6-32 Circuit for Problem 42.

43. In Fig. 6-33, use nodal analysis to calculate the value and the polarity of the voltage at point N.

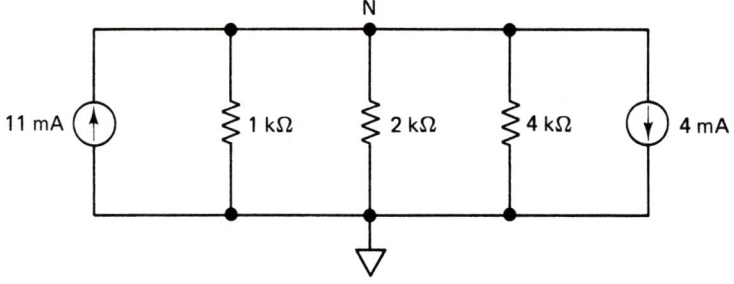

FIGURE 6-33 Circuit for Problem 43.

44. In Fig. 6-34, use Millman's theorem to reduce the circuit to a single constant current source that is connected between the points X and Y.

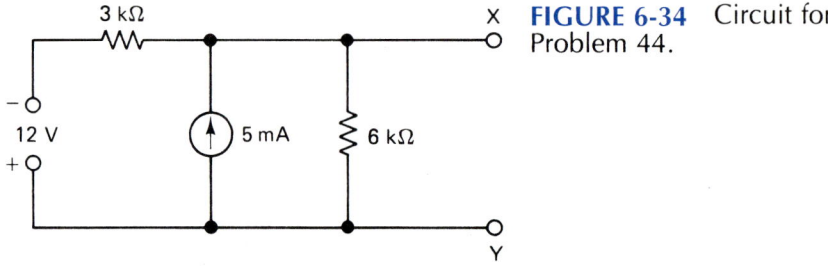

FIGURE 6-34 Circuit for Problem 44.

45. In Fig. 6-35, use Thévenin's theorem to find the current flowing through the 2-kΩ load.

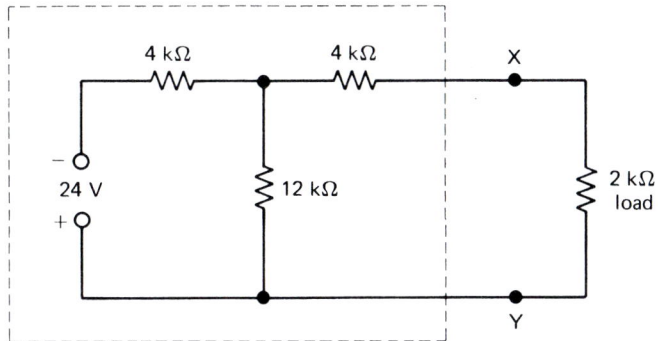

FIGURE 6-35 Circuit for Problem 45.

46. In Fig. 6-36, use Norton's theorem to find I_N and R_N for the equivalent current source that is connected between points X and Y.

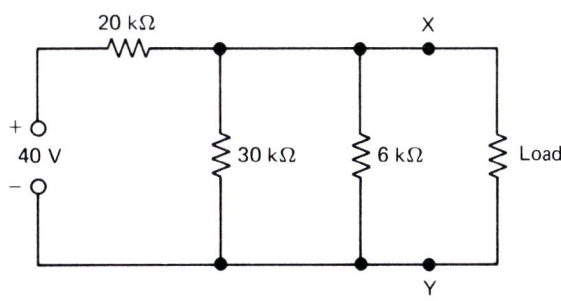

FIGURE 6-36 Circuit for Problem 46.

47. In Fig. 6-37, use the superposition theorem to find the voltage at point N.

FIGURE 6-37 Circuit for Problem 47.

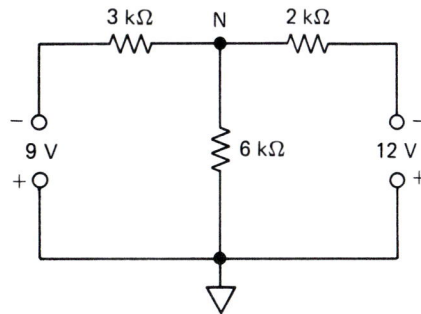

48. In Fig. 6-38, convert the constant current sources into their equivalent constant voltage generators and then calculate the current through the 12-kΩ resistor.
49. The values of the three resistors in a delta formation are 4 Ω, 6 Ω, and 9 Ω. Calculate the values of the three resistances in the equivalent wye arrangement. Draw the two circuits and identify the corresponding points.
50. The values of the three resistors in a T section are 5 Ω, 7 Ω, and 9 Ω. Calculate the values of the three resistors in the equivalent Π section. Draw the two circuits and identify the corresponding points.

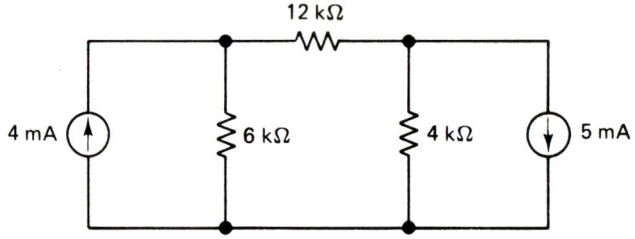

FIGURE 6-38 Circuit for Problem 48.

51. In Fig. 6-39, calculate the values of currents I_1 and I_2.

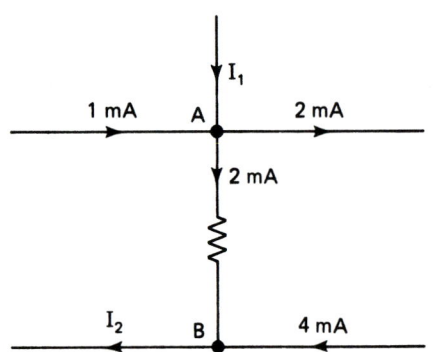

FIGURE 6-39 Circuit for Problem 51.

52. In Fig. 6-40, determine the value of the potential at point X.

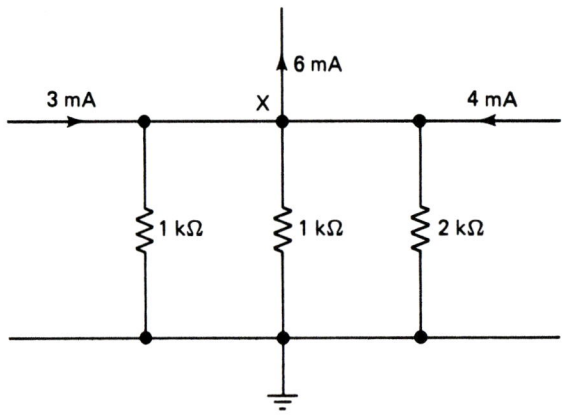

FIGURE 6-40 Circuit for Problem 52.

53. In Fig. 6-41, use mesh analysis to determine the value of the voltage between points X and Y.

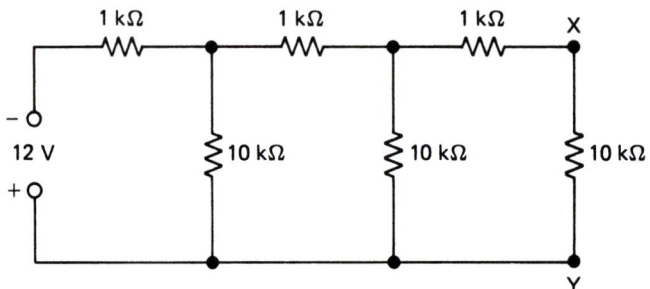

FIGURE 6-41 Circuit for Problem 53.

54. In Fig. 6-42, use nodal analysis to find the value of the node voltage at point N.

FIGURE 6-42 Circuit for Problem 54.

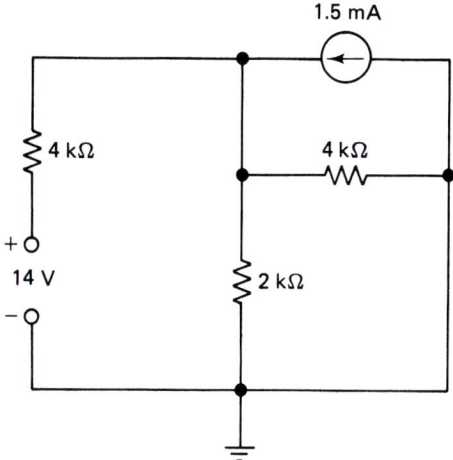

55. In Fig. 6-43, use the superposition theorem to determine the value of the current, I.

FIGURE 6-43 Circuit for Problem 55.

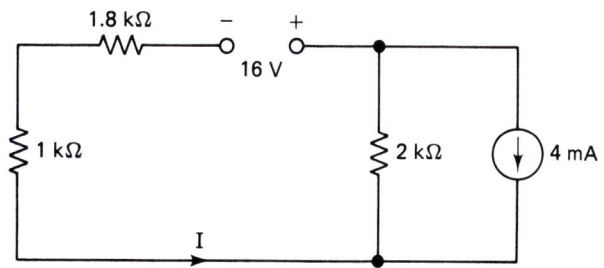

56. Fig. 6-44 shows the circuit of a common-emitter amplifier. Obtain the equivalent Thévenin generator for the bias circuit to the *left* of points X and Y.

FIGURE 6-44 Circuit for Problem 56.

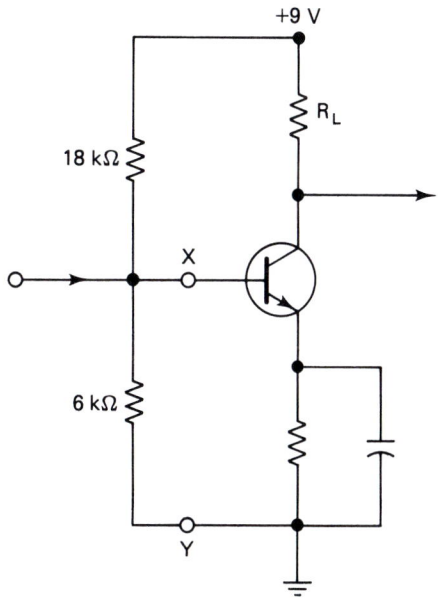

57. In Fig. 6-45, determine the Norton equivalent generator for the circuit to the right of points X and Y.

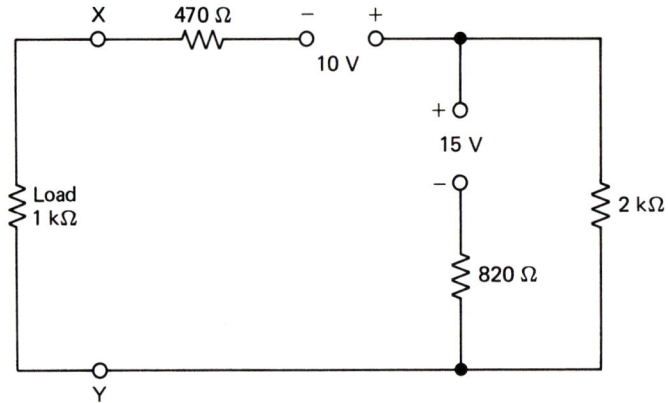

FIGURE 6-45 Circuit for Problem 57.

58. Between two terminals of a resistive network the following results were obtained:

terminal voltage	12 V	0 V
terminal current	0 A	8 A

Obtain the Thévenin and Norton equivalent generators for the resistive network.

59. In Fig. 6-46, convert the circuit between points X, Y, and Z into its wye equivalent.

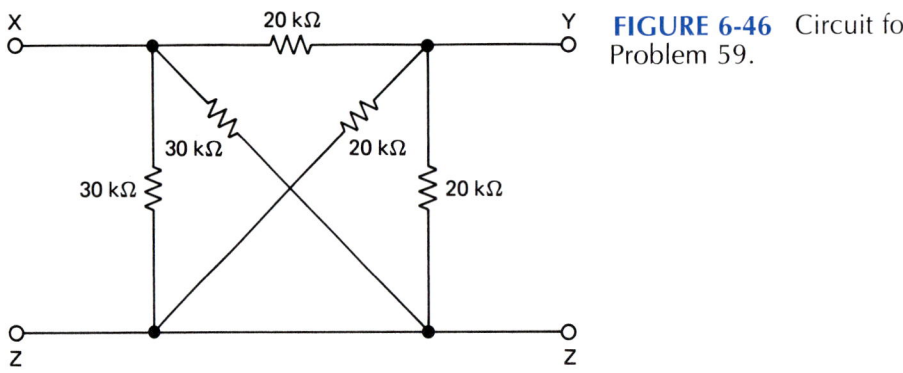

FIGURE 6-46 Circuit for Problem 59.

60. In Fig. 6-47, convert the circuit between points X, Y, and Z into its delta equivalent.

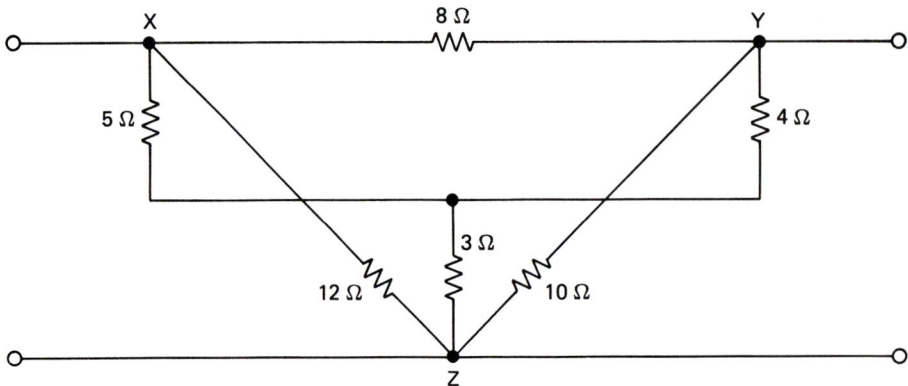

FIGURE 6-47 Circuit for Problem 60.

Advanced Problems

61. Figure 6-48 represents a cube whose sides contain 100-Ω resistors. Use Kirchhoff's laws to find the total equivalent resistance between points X and Y.

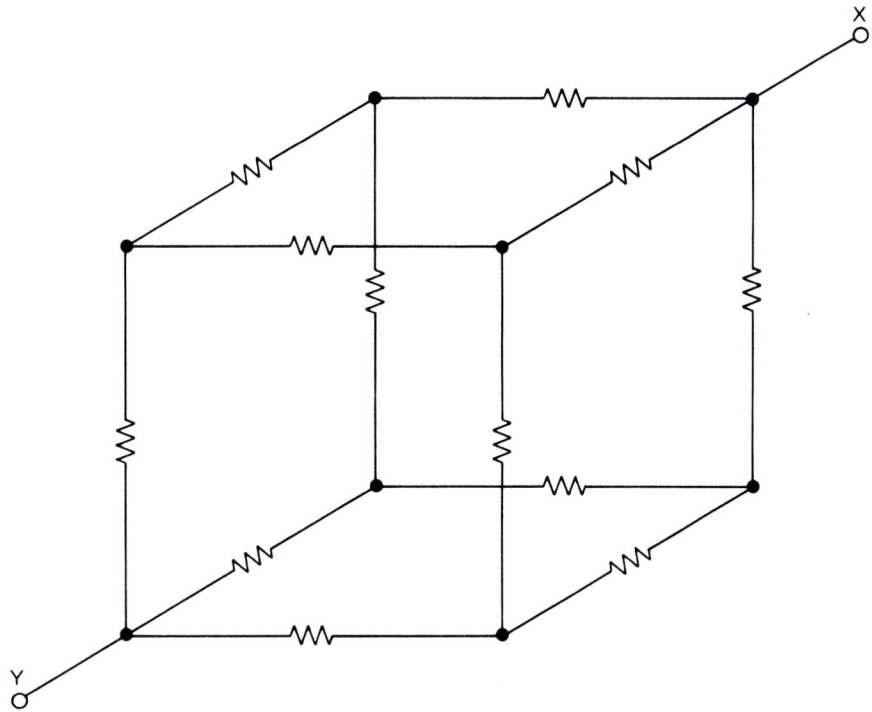

FIGURE 6-48 Circuit for Problem 61.

62. In Fig. 6-49, use mesh analysis to find the value of the current flowing through the 3.3-kΩ resistor.

FIGURE 6-49 Circuit for Problems 62 and 63.

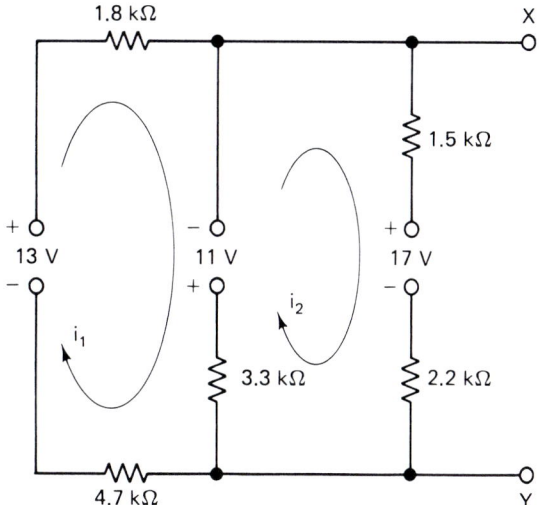

63. In Fig. 6-49, use Millman's theorem to reduce the circuit to a single constant voltage generator that is connected between points X and Y.
64. In Fig. 6-50, use nodal analysis to calculate the value and the polarity of the voltage at point X.

65. In Fig. 6-50, regard the 4.7-kΩ resistor as the load and obtain the values of E_{TH} and R_{TH} for the equivalent Thévenin generator that is connected between points X and Y. Draw the circuit of this generator and identify the X, Y points.

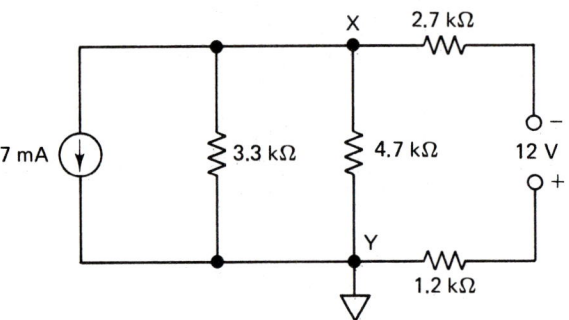

FIGURE 6-50 Circuit for Problems 64 and 65.

66. In Fig. 6-51, use the superposition theorem to find the value and the direction of the current through the 4.7-kΩ resistor.

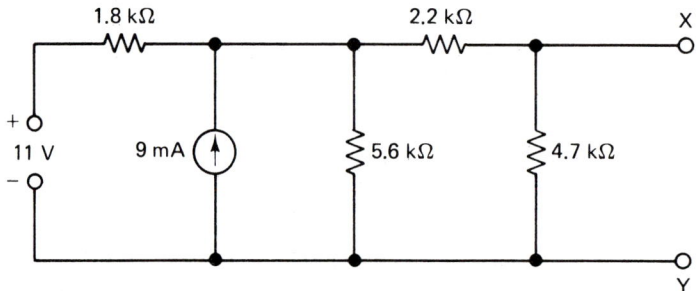

FIGURE 6-51 Circuit for Problems 66 and 67.

67. In Fig. 6-51, calculate the values of I_N and R_N for the equivalent Norton generator between points X and Y. Draw the circuit of this generator and identify the X, Y points.

68. Reduce the circuit of Fig. 6-52 to a single delta formation of three resistances. Draw the arrangement of these resistances and identify points X, Y, and Z.

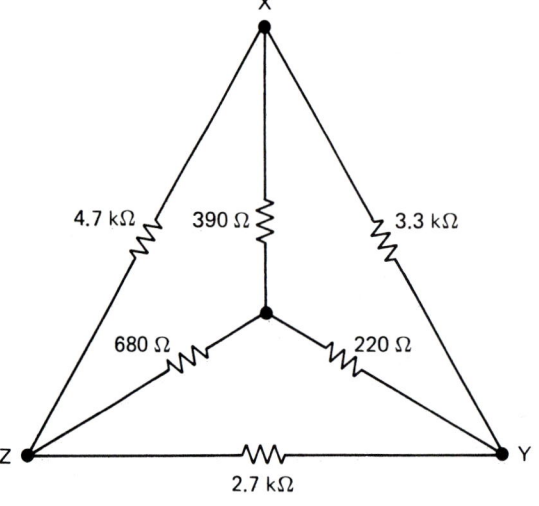

FIGURE 6-52 Circuit for Problem 68.

Kirchhoff's Laws and the Network Theorems Ch. 6

69. In Fig. 6-53, carry out a delta-to-wye transformation for the three resistors connected between points X, Y, and Z. Redraw the circuit after the transformation and calculate the value of the total equivalent resistance between points W and Y.

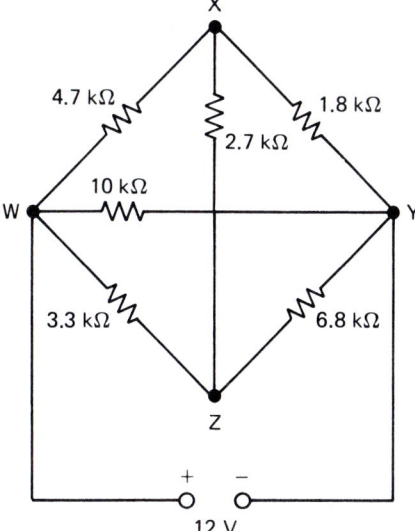

FIGURE 6-53 Circuit for Problems 69 and 70.

70. In Fig. 6-53, Nortonize the circuit between points X and Z and determine the value and the direction of the current flowing through the 2.7-kΩ resistor.

71. In Fig. 6-54, determine the value of the current, I.

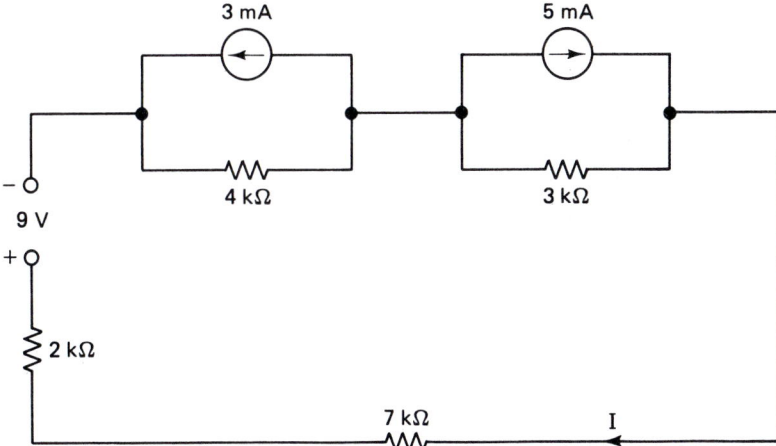

FIGURE 6-54 Circuit for Problem 71.

72. In Fig. 6-55, carry out a delta-to-wye transformation for the three resistors connected between points X, Y, and Z and then determine the value of the load current, I_L.

73. In Fig. 6-21(a), regard the 7-Ω resistor as the load. With this load removed, determine the equivalent Thévenin generator for the remainder of the circuit.

74. In Fig. 6-21(a), regard the 5-Ω resistor as the load. With this load removed, determine the equivalent Norton generator for the remainder of the circuit.

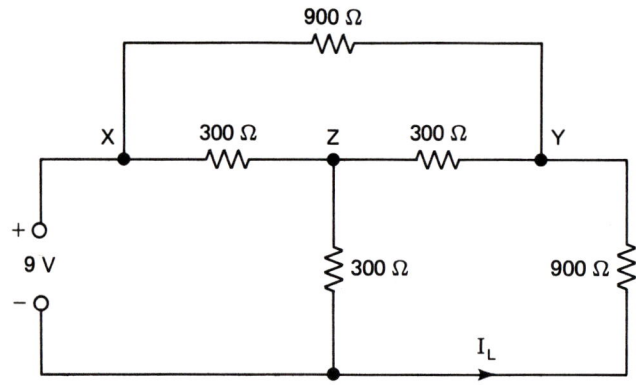

FIGURE 6-55 Circuit for Problem 72.

75. In Fig. 6-56, determine the value of the mesh current, I_1.

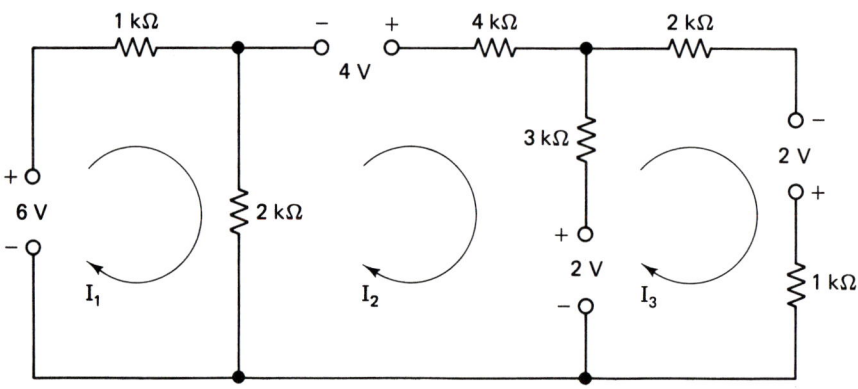

FIGURE 6-56 Circuit for Problem 75.

76. In Fig. 6-57, determine the node voltages at points N_1 and N_2.

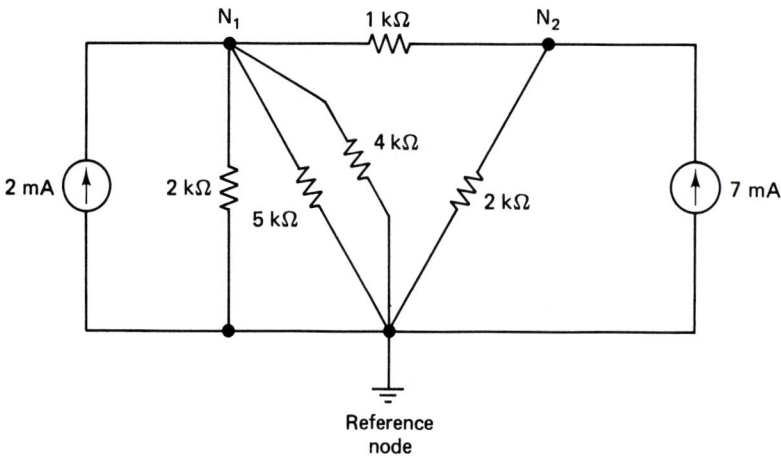

FIGURE 6-57 Circuit for Problem 76.

77. In Fig. 6-58, find the value of *I* by using the superposition theorem.

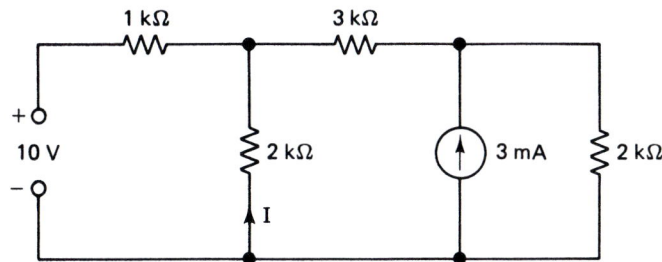

FIGURE 6-58 Circuit for Problem 77.

78. In Fig. 6-59, Thévenize the circuit between terminals *X* and *Y*. What are the values of E_{TH} and R_{TH}?

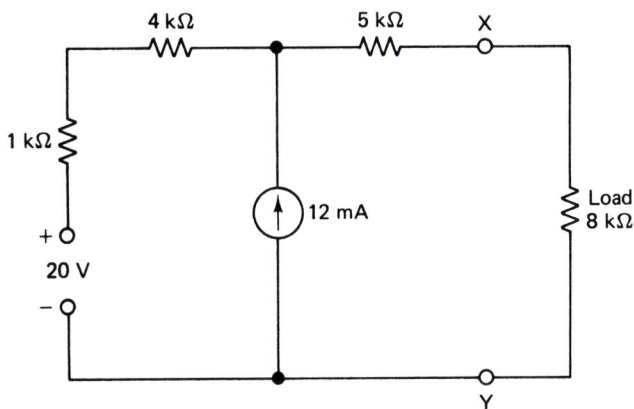

FIGURE 6-59 Circuit for Problem 78.

79. In Fig. 6-60, Nortonize the circuit between the terminals *X* and *Y*. What are the values of I_N and R_N?
80. In Fig. 6-60, use conversions between voltage and current sources to determine the value of the load current, I_L.

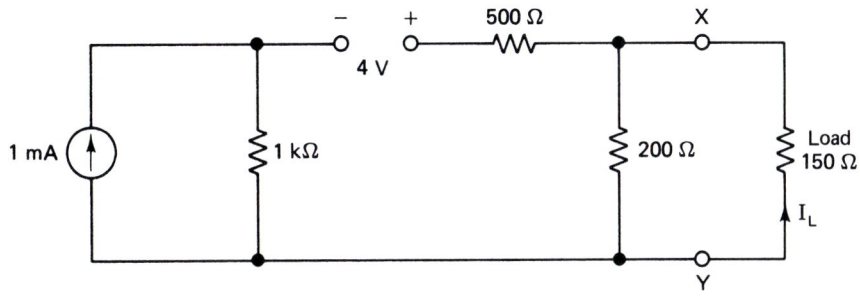

FIGURE 6-60 Circuit for Problems 79 and 80.

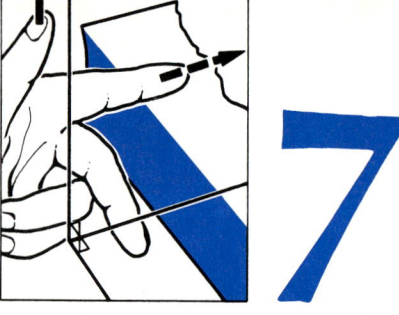

7

Magnetism, Electromagnetism, and Electromagnetic Induction

In our discussion of magnetism, electromagnetism, and electromagnetic induction, you will learn:

1. how a permanent magnet is surrounded by a field pattern of flux lines.
2. how the properties of "hard" iron differ from those of "soft" iron.
3. how the earth behaves as an enormous magnet.
4. about the relationship between magnetic flux and magnetic flux density.
5. about the patterns of the magnetic fields surrounding current-carrying conductors.
6. how a motor converts electrical energy into mechanical energy.
7. about the relationships between magnetomotive force, magnetic field intensity, and flux density when magnetizing a specimen of iron.
8. to use Rowland's law, in which the reluctance of an iron specimen is equal to the ratio of the magnetomotive force to the flux.
9. how the reluctance is related to the permeability of the iron.
10. about the effect of hysteresis in the magnetizing and demagnetizing of an iron specimen.
11. about Faraday's and Lenz's laws, which relate to the production of electrical energy from mechanical energy.
12. how a dc generator produces electrical energy from mechanical energy.

7-0 Introduction

In any study of electricity and electronics, it is vital to have some knowledge of magnetism. Why? As we mentioned in Section 1-5, one of the three electrical properties is inductance, which is just as important as resistance and is strongly associated with magnetism. In fact, the contents of this chapter lead automatically to the discussion of inductance in Chapter 8. Second, at this time the only practical way of generating large quantities of electrical power is through the use of magnetism. Third, many measuring instruments use magnetic principles in their operation; this is fully discussed in Chapter 10.

You are no doubt familiar with a magnet's ability to attract and pick up objects such as pins. This so-called permanent magnet would probably be some form of iron, but there are other materials, such as nickel and cobalt, which have similar magnetic properties. This group is known as the ferromagnetic materials, that is, those materials that have magnetic properties comparable to those of iron.

When a current flows through a conductor, a magnetic field surrounds the conductor. This effect, called electromagnetism, is the production of magnetism through the use of electricity; this principle is used in many measuring instruments and in the electric motor.

If a conductor is moved within or through a magnetic field, a voltage is generated between the ends of the conductor. This phenomenon, called electromagnetic induction, is the principle behind almost all dc generators.

7-1 Bar Magnet. Magnetic Poles. Magnetic Field Patterns. Magnetic Induction

The word "magnet" is derived from magnetite, which is a type of mineral rock to be found in the region of Magnesia, a part of Asia Minor. A lump of this rock is a *natural magnet*, which is capable of picking up steel pins. However, you would find that the pins tended to be concentrated around two regions of the rock; these regions are called the *poles* of the magnet (Fig. 7-1). Early navigators observed that when rocks of this type were conveniently shaped and freely suspended, they would take up an approximate north-south position. Since the rock could therefore indicate a known direction, it was called a *lodestone*, which means "leading stone."

The natural magnets have long since been replaced by more powerful artificial magnets which are manufactured commercially from certain alloys; one of the more common is Alnico, which is made principally from aluminum, nickel, and cobalt. These *artificial* magnets come in a variety of shapes, but at this time we are going to consider the rectangular or bar magnet of Fig. 7-2.

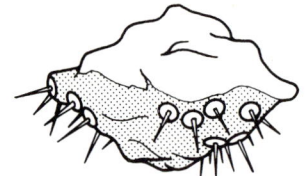

FIGURE 7-1 Natural rock magnet.

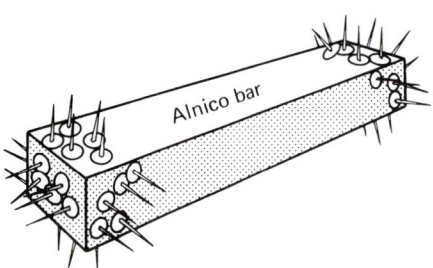

FIGURE 7-2 Artificial permanent magnet.

During its manufacture the Alnico bar was strongly magnetized and it subsequently retained a large amount of its magnetic strength; such a bar is referred to as a "hard-iron" *permanent magnet*. By contrast, there are some alloys, such as silicon steel, which are easily magnetized but afterward keep only a small part of their magnetism; these alloys are used in the manufacture of "soft-iron" *temporary magnets*.

The earth itself behaves as a huge magnet (Fig. 7-3) with its own poles; these are situated close to the geographic north and south poles, which are the ends of the axis on which the earth spins. If a permanent bar magnet is freely suspended, it will take up a position with one of its poles pointing toward the earth's magnetic north pole. This pole of the magnet is called its *north pole*, so that "north" is used in the sense "north seeking." Of course, at the other end of the magnet will be its *south pole*, which is pointing toward the earth's south pole. It is common practice to color-code bar magnets so that the north pole is shown as red and the south pole as blue.

To find the direction of north at any place on the earth's surface you would use a magnetized compass needle, which is light and may be either suspended or pivoted so as to move freely. If such a needle is brought near a bar magnet (Fig. 7-4), it will be found that there is a force of repulsion between two north poles or between two south poles; in other words, like magnetic poles repel each other [Fig. 7-4(a) and (b)]. By contrast, Fig. 7-4(c) illustrates the force of attraction between a north pole and a south pole, so that unlike magnetic poles attract each other. The force of the attraction or repulsion between the poles is proportional to the strength of the poles but is inversely proportional to the square of the distance between them. Consequently, if you double the distance between two poles, the force will be reduced to one-fourth of its original value.

In Fig. 7-2 you see that the pins are concentrated at the poles where the attractive force is greatest; by contrast there is little magnetic effect toward the center of the magnet. If the magnet is now cut into two halves, it will be found that each half behaves as a separate magnet. We can go on subdividing until we get down to the molecules themselves; these are also magnetized with their own poles. As we shall see in Section 7-3, a moving charge creates a magnetic field and consequently an atom can exhibit magnetism due to its orbiting electrons. In ferromagnetic materials atoms whose magnetic effects combine are arranged in groups or domains. With an unmagnetized specimen the magnetic domains are randomly disposed, so there is virtually no overall magnetic effect

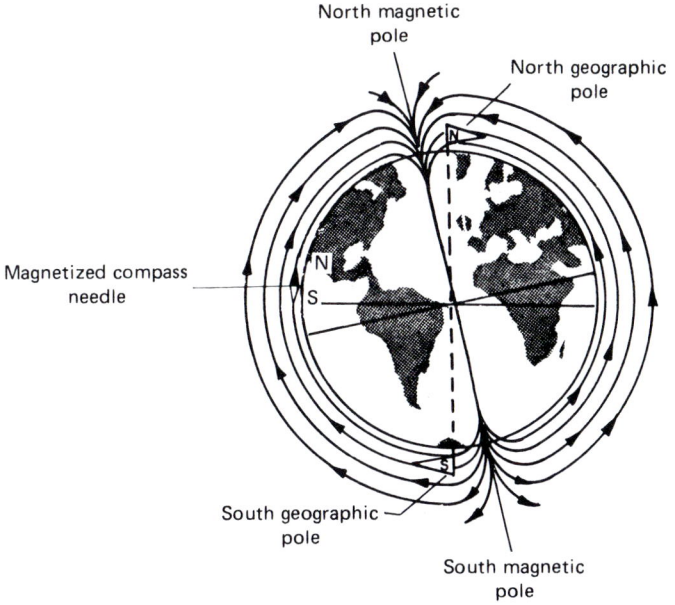

FIGURE 7-3 Earth's magnetic field.

[Fig. 7-5(a)]. When the specimen is magnetized, however, the domains are aligned so that north and south poles appear at the ends [Fig. 7-5(b)]. Subsequent division of the magnet will create a number of smaller magnets, each with its own poles [Fig. 7-5(c)].

Support for this theory comes from the fact that if a permanent magnet is screened from the earth's magnetism and then heated or repeatedly struck with a hammer, it will gradually become demagnetized; this is due to the domain alignment being disarranged. In fact, if a bar magnet is left to itself, it will slowly lose its magnetism; this effect can be considerably reduced by stacking two magnets [Fig. 7-6(a)] so that one magnet provides an iron path for the other. In the case of a horseshoe magnet, its field may be confined by placing a soft iron "keeper" bar between the magnet's poles [Fig. 7-6(b)].

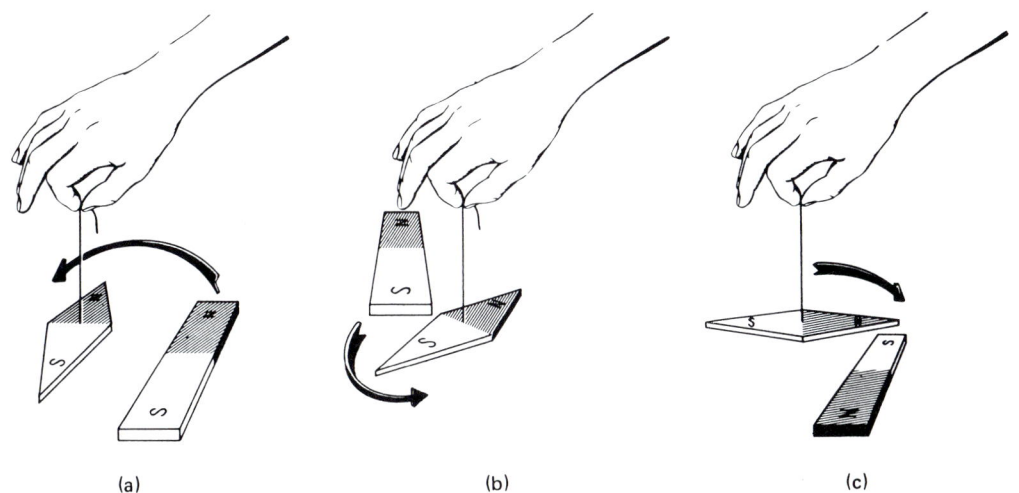

FIGURE 7-4 Attraction and repulsion between magnetic poles: (a), (b) repulsion; (c) attraction.

Sec. 7-1 Bar Magnet. Magnetic Poles. Magnetic Field Patterns. Magnetic Induction

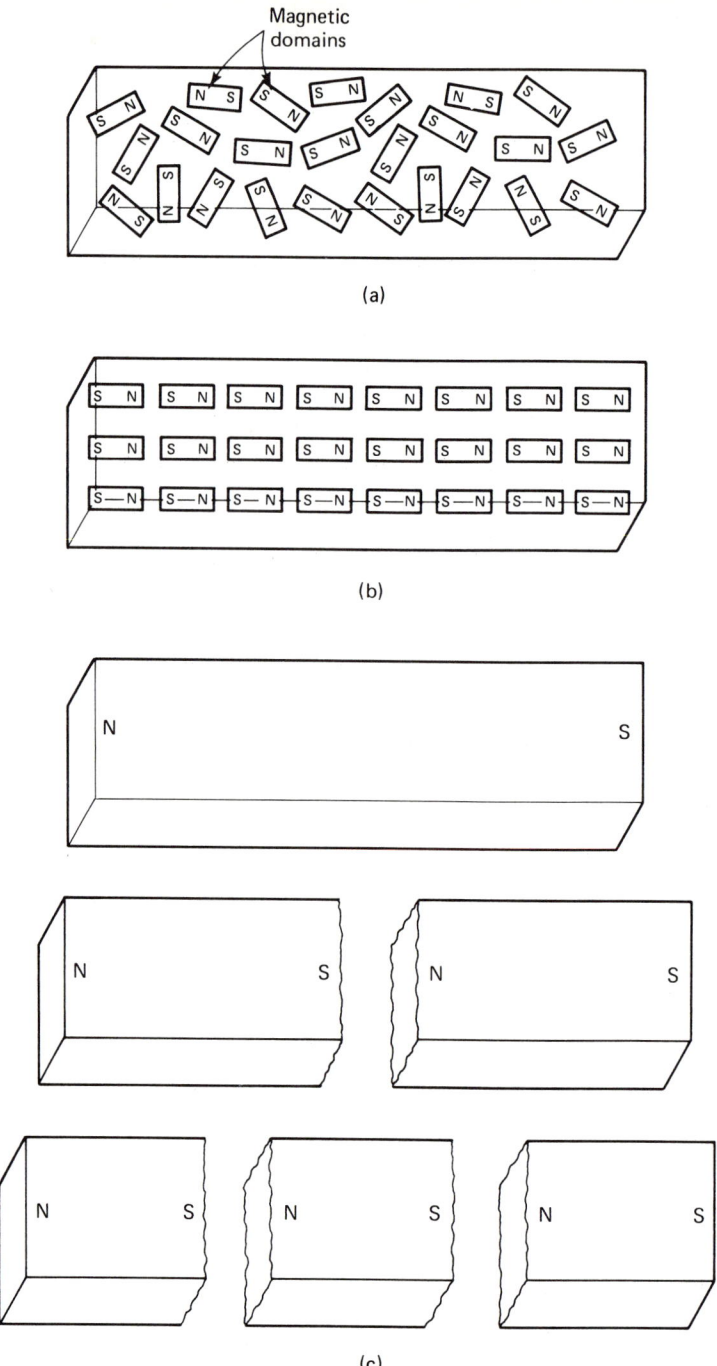

FIGURE 7-5 Domain theory of magnetism.

Magnetic Field Patterns

You are probably familiar with an experiment in which a bar magnet is placed underneath a sheet of paper or glass on top of which iron filings are sprinkled. The paper or glass is then lightly tapped and the iron filings move to arrange themselves in definite paths which form the pattern shown in Fig. 7-7. The iron filings moved as the result of a magnetic force exerted by the magnet. Surrounding the magnet is the space which is the region of influence in which the magnetic force acts; this region of influence is under stress and is called

FIGURE 7-6 Methods of preventing demagnetization: (a) stacked magnets; (b) keeper bar.

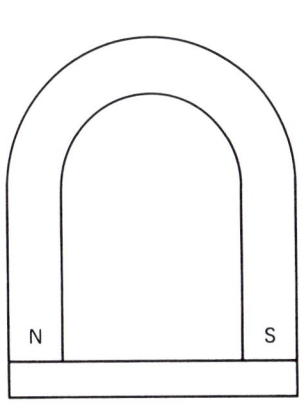

FIGURE 7-7 Magnetic field pattern surrounding a bar magnet.

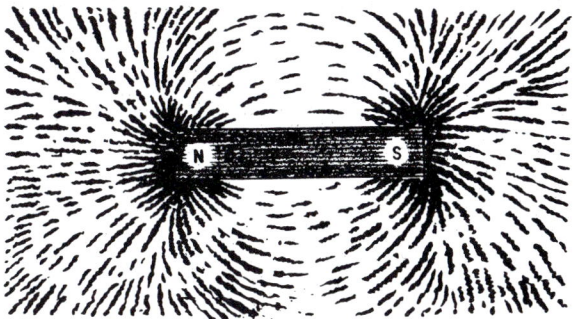

the *magnetic field*. Looking at Fig. 7-7, it appears that the pattern of the magnetic field is made up of a number of lines along which the iron filings lie; these invisible lines are the magnetic lines of the force.

The individual lines of force surrounding a bar magnet may be plotted by placing a pivoted compass needle at several positions in the vicinity of the magnet. At each position the needle will align itself with a magnetic line of force whose direction is assumed to be indicated by the needle's north pole. We can therefore say that a line of force is the path along which a free or isolated north pole would travel in the magnetic field; this is an imaginary concept since in practice a north pole must always be accompanied by a corresponding south pole.

By joining together the directions of various positions, we obtain the pattern of the magnetic lines of force as shown in Fig. 7-8. These lines emanate from the north pole of the bar magnet, pass through the surrounding space, and enter the south pole. We further assume that the lines of force pass from the south pole to the north pole inside the magnet so that each line forms a closed loop. Notice that the lines are concentrated near the poles where the field is strongest, but that as we move away from the poles, the lines spread out or diverge. We can therefore say that the magnetic lines of force repel each other and cannot intersect.

If two bar magnets are placed close to one another, the field patterns surrounding each magnet must be modified since the two sets of lines cannot

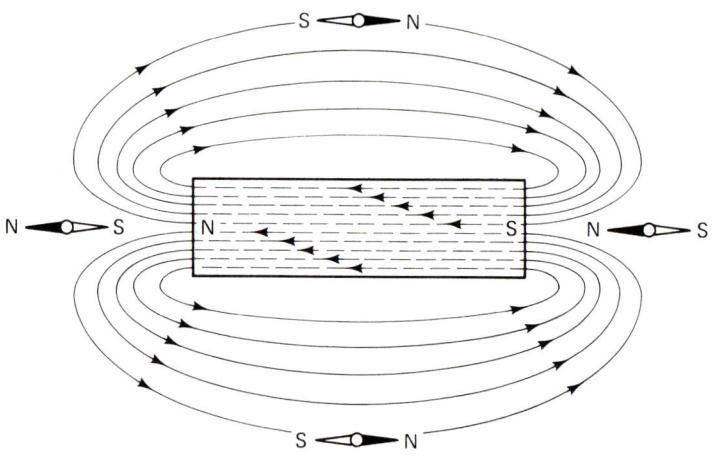

FIGURE 7-8 Magnetic lines of force.

cross. In Fig. 7-9(a) the lines are compressed in the space between the two north poles so that there is a force of repulsion between the magnets. By contrast, Fig. 7-9(b) shows the resultant field pattern which is associated with the attraction between north and south poles.

What we have not explained is the ability of a bar magnet to pick up an unmagnetized object such as a steel pin. The answer lies in magnetic induction. Because the pin lies in the field of the bar magnet, the field pattern is modified

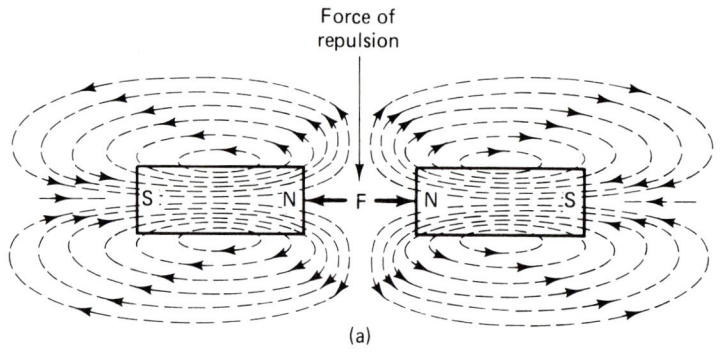

(a)

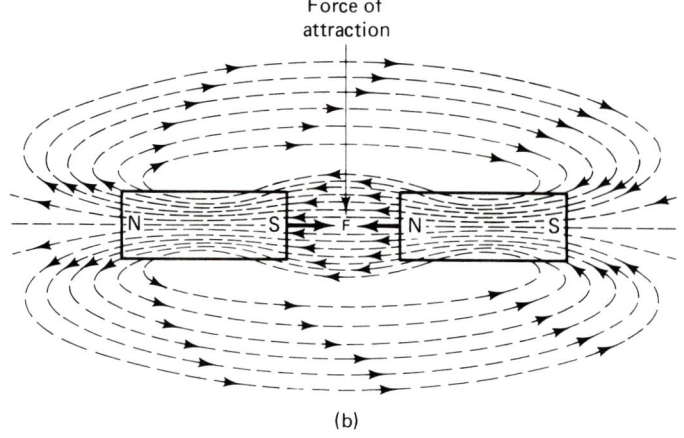

(b)

FIGURE 7-9 Resultant field patterns between neighboring magnets.

284   Magnetism, Electromagnetism, and Electromagnetic Induction   Ch. 7

and there is some alignment of the pin's molecular magnets, so that poles appear at the ends of the pin. The pin's south pole is nearest to the bar magnet's north pole with the result that there is a force of attraction [Fig. 7-10(a)]. The concept of magnetic induction can also be used to magnetize an iron object by stroking it with a bar magnet [Fig. 7-10(b)].

Referring back to Fig. 7-3, the lines of the earth's magnetic field enter the south end of a compass needle and exit from its north end. As shown, the lines must therefore emanate from the earth's south magnetic pole, travel through space, and then enter the north pole. This means that the pole near geographic *north* has the same magnetic properties as the *south* pole of a magnet; in other words, on the principle that unlike poles attract, the earth's north magnetic pole should be color-coded blue.

It is worth mentioning that the angular separation in degrees between the directions of magnetic north and geographic north is called the *variation*; this is not a constant but depends on one's position on the earth's surface. At a particular position the value of the variation changes from year to year at a known rate.

In summation, we have learned in this section about the basic rules of magnetism. Unlike poles attract, like poles repel, and the force of attraction

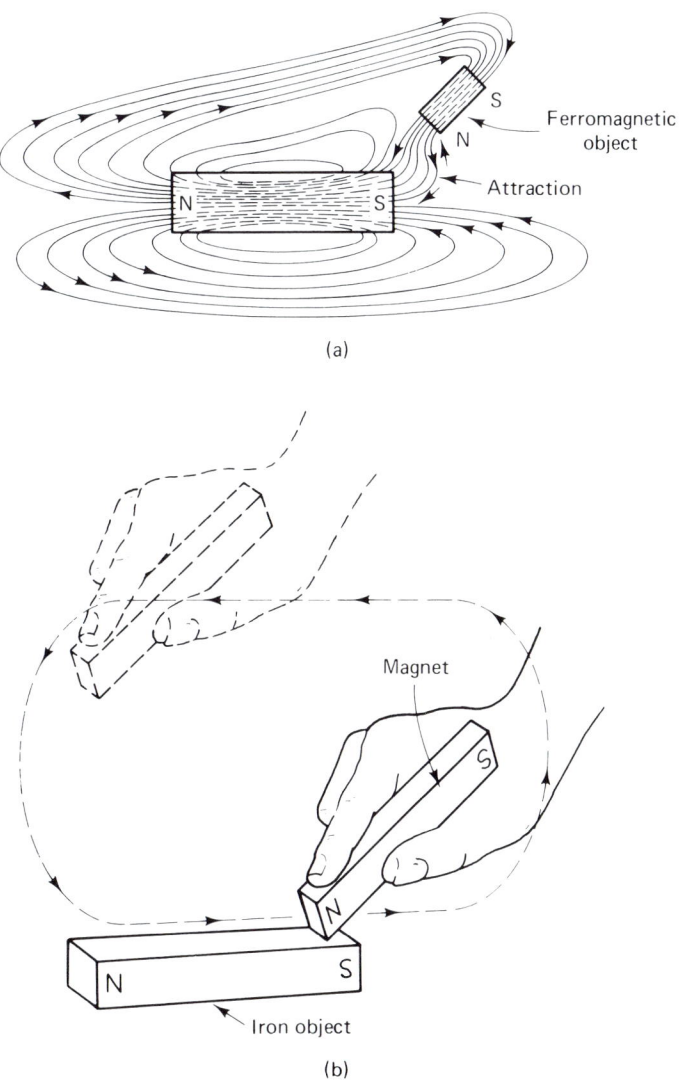

FIGURE 7-10 (a) Magnetic induction; (b) magnetizing by stroking.

Sec. 7-1 Bar Magnet. Magnetic Poles. Magnetic Field Patterns. Magnetic Induction

or repulsion is inversely proportional to the square of the distance between the poles. We regard the field surrounding a magnet to be composed of magnetic lines of force. These lines form closed loops, cannot intersect, and mutually repel each other. The degree of concentration of the lines of force is an indication of the strength of the magnetic field.

EXAMPLE 7-1 Two north poles are situated 2 cm apart and their force of mutual repulsion is 20 N. If their separation is increased to 5 cm, what is the new force of repulsion?

Solution The separation between the poles has been increased by a factor of 5 cm/2 cm = 2.5. Therefore, the force of repulsion must be reduced by a factor of $(2.5)^2 = 6.25$. The new force of repulsion is $20/6.25 = 3.2$ N.

7-2

Magnetic Flux. Magnetic Flux Density

The total number of lines in a magnetic field is referred to as the magnetic flux, whose letter symbol is the Greek lowercase letter ϕ (phi). In the SI system the unit used to measure the magnetic flux is the weber, which is named after Wilhelm Eduard Weber (1804–1890); the unit symbol for the weber is Wb. The phenomenon of electromagnetic induction is used in defining the weber, and this is discussed in Section 7-7. In practice, the weber is a large unit for measuring a typical field and therefore we commonly use microwebers (μWb) and milliwebers (mWb).

We have already said that the concentration of the lines of force is an indication of the strength of a magnetic field. But how is this concentration to be measured? What we are really thinking of is a distribution of the flux, ϕ, over an area, A, which we consider to be at right angles (90°) to the direction of the lines; we can then refer to the flux density, whose letter symbol is B. In the SI system the flux, ϕ, is measured in webers and the area, A, will be in square meters. The flux density will be defined by

$$\text{flux density, } B = \frac{\phi}{A} \quad \text{webers per square meter} \tag{7-2-1}$$

so that

$$\text{flux, } \phi = BA \quad \text{webers} \tag{7-2-2}$$

The value of B measures the amount of flux that is spread over an area of 1 square meter. The SI unit for flux density is the tesla (Nikola Tesla, 1857-1943), whose letter symbol is T. One tesla is equivalent to 1 weber per square meter (Wb/m^2).

Looking again at the magnetic field surrounding a bar magnet (Fig. 7-8), we see that the lines are not spread out uniformly but are instead concentrated near the poles and then diverge as we move away from the poles. This means that the flux density is not constant in the field but varies from one position to another. To get a better idea of flux density, let us take a permanent horseshoe magnet where the ends have been carefully shaped to provide a uniform field which is confined to the air gap between the poles [Fig. 7-11(a)]. In this field the flux lines are evenly spread over the effective area of the pole faces [Fig. 7-11(b)]. The symbol "⊗" is used to indicate that the direction of the lines is into the plane of the paper; consequently, we are assuming that the north pole is placed above the paper while the south pole is beneath. The symbol "⊙" is used to indicate a line that is emerging from the paper toward you.

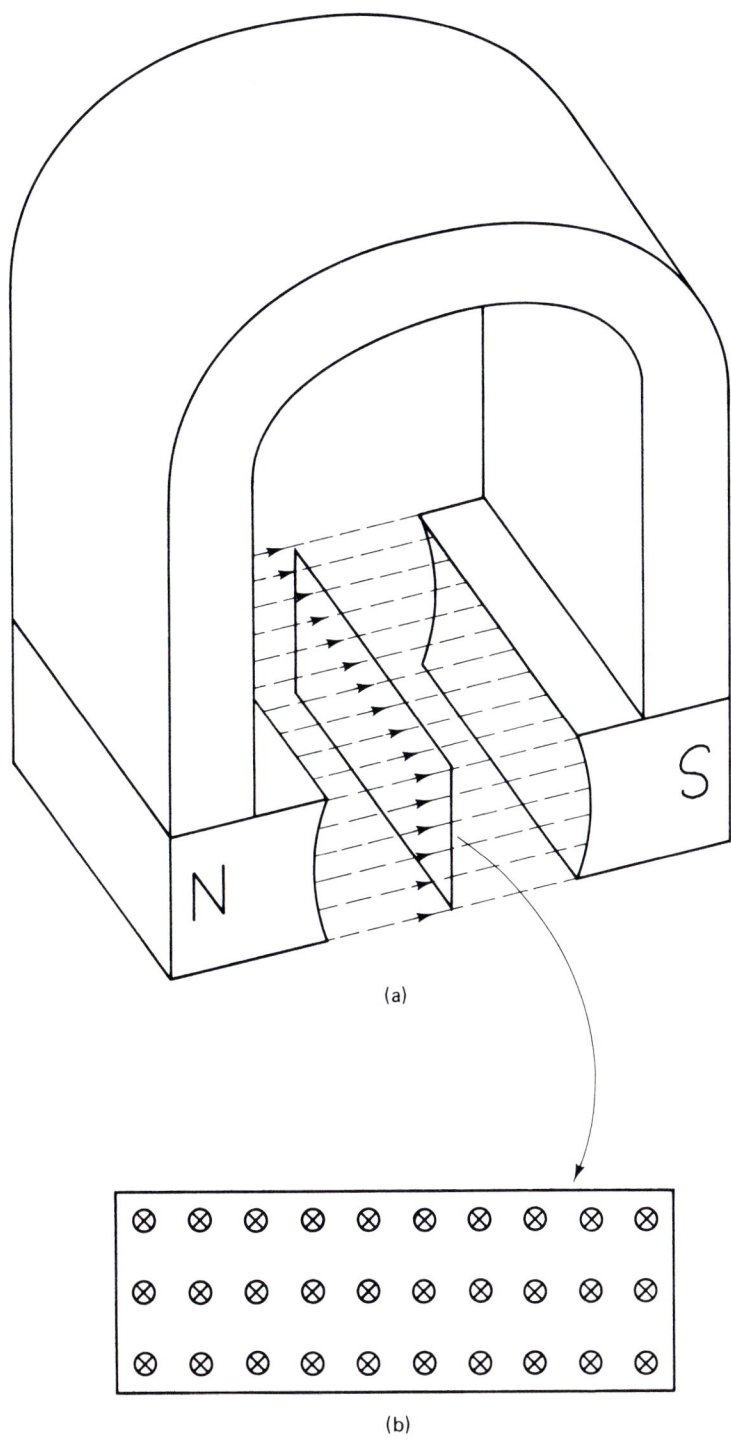

FIGURE 7-11 Flux density in a uniform field.

Sec. 7-2 Magnetic Flux. Magnetic Flux Density

Let us assume that the total flux, ϕ, between the poles is 500 μWb and the effective area, A, of the air gap is 5 cm^2 = 5 × 10^{-4} m^2. Then the flux density, $B = \phi/A$ = 500 μWb/(5 × 10^{-4} m^2) = 500 × 10^{-6} Wb/(5 × 10^{-4} m^2) = 1 Wb/m^2 = 1 T. The flux density of 1 T is the same at all positions in the air gap.

We have covered the important topics of the total magnetic flux, ϕ, which is measured in webers, and the flux density, B, whose unit is the tesla. On the whole we need to pay more attention to the flux density than to the total flux.

EXAMPLE 7-2 A flux density of 0.8 T is constant over a total area of 3.5 cm^2. Calculate the value of the total flux.

Solution

$$A = 3.5 \text{ cm}^2 = 3.5 \times 10^{-4} \text{ m}^2, \quad B = 0.8 \text{ T}$$

$$\text{total flux, } \phi = B \times A \qquad (7\text{-}2\text{-}2)$$
$$= 0.8 \text{ T} \times 3.5 \times 10^{-4} \text{ m}^2$$
$$= 2.8 \times 10^{-4} \text{ Wb} = 280 \text{ } \mu\text{Wb}$$

7-3

Electromagnetism. Motor Effect. Magnetomotive Force

Electromagnetism means the development of magnetism through the use of electricity. If a current flows through a straight conductor, a magnetic field is established around the conductor (which does not have to be manufactured from a ferromagnetic material). In other words, if iron and copper wires, with identical dimensions carry equal values of current, their surrounding magnetic fields will be the same.

The presence of a magnetic field surrounding a current-carrying conductor was demonstrated experimentally by Hans Christian Oersted in 1879. If a magnetic needle is placed near a wire that carries no current [Fig. 7-12(a)], the needle will be aligned with the earth's magnetic field. However if the switch, S, is then closed, current flows through the wire and the needle is deflected. If the field surrounding the conductor is strong compared with the earth's field, the direction of the needle will be at right angles (90°) to the direction of the wire. This means that the magnetic lines of force will be a series of circles whose common center is the conductor; this could be confirmed by an iron-filing experiment. These circles would be close together in the vicinity of the wire but would spread out as the distance from the wire is increased.

The direction of the magnetic lines of force is the direction in which the needle's north pole is pointed. If you think of grasping the conductor with your *left* hand with the thumb used to indicate the direction of the electron flow, the fingers would show the direction of the magnetic lines of force. With the electron flow upward as in Fig. 7-12(b), the direction of the lines of force is clockwise as viewed from above. However, if the polarity of the source voltage, E, is reversed, the lines will be counterclockwise around the conductor. This is further illustrated in Fig. 7-12(c), where the symbols $\otimes$ and $\odot$ are used to indicate the direction of the electron flow.

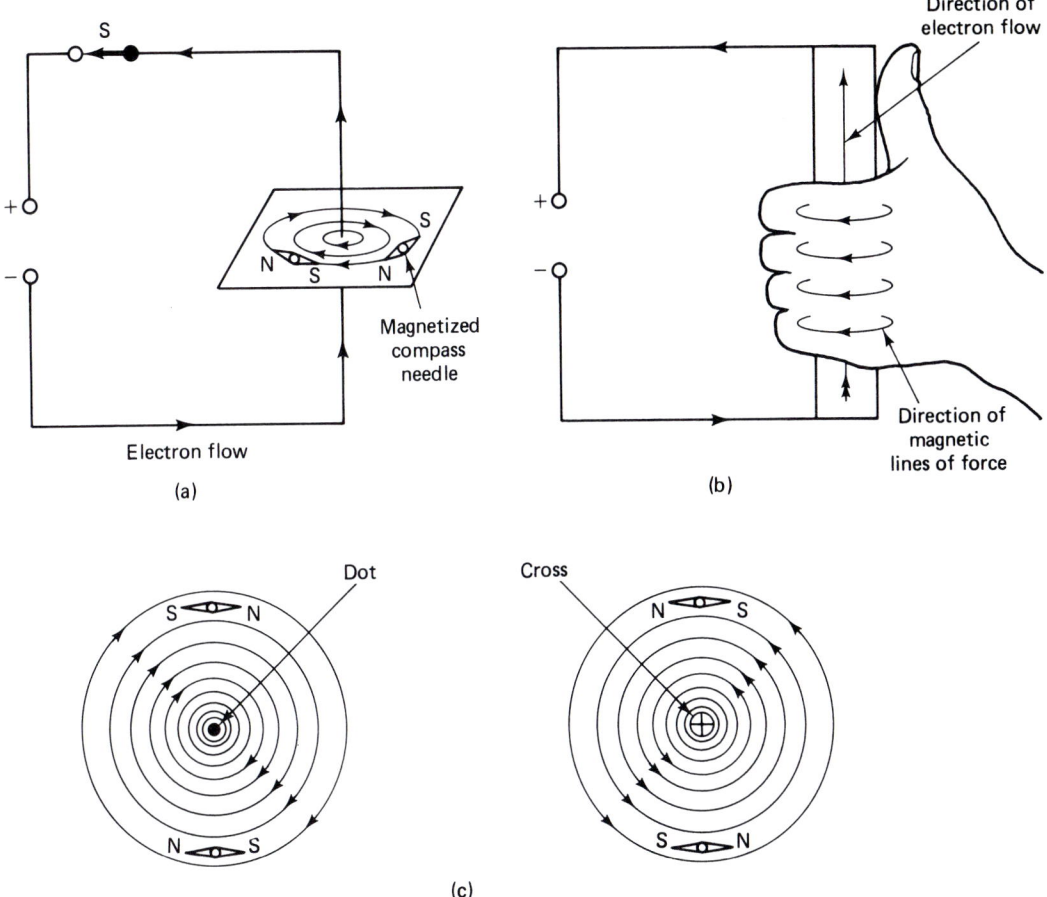

FIGURE 7-12 Magnetic field surrounding a current-carrying conductor.

If two parallel conductors carrying equal currents in the same direction are placed close to one another, the field pattern must be modified, since the two sets of lines cannot cross. The resultant magnetic field is shown in Fig. 7-13(a); the lines of force embrace both conductors, which are therefore drawn together by a force of attraction. If the two adjacent conductors have their currents flowing in opposite directions [Fig. 7-13(b)], the lines of force are crowded together in the space between the conductors, and since these lines mutually repel each other, the tendency will be to push the conductors apart; consequently, there is a force of repulsion between the conductors. Both these results may be demonstrated experimentally; if the conductors are made to be flexible and are carrying currents in the same direction, they will bend inward and toward one another under the force of the attraction. However, if the currents are in the opposite direction, the conductors are repelled from each other and will bend outward.

As mentioned in Section 1-3, the existence of the force between two parallel current-carrying conductors is used to define the ampere, which is the fourth fundamental unit (meter, kilogram, second, ampere) in the SI system. Let two straight parallel conductors of infinite length and negligible cross-sectional area be separated by a distance of 1 m in a vacuum. Then the international ampere is that current which when maintained in each of these two conductors will create a force between the conductors of 2×10^{-7} N per meter length.

The force of attraction or repulsion is the result of placing one current-carrying conductor in the magnetic field surrounding the other conductor. Let

Sec. 7-3 Electromagnetism. Motor Effect. Magnetomotive Force

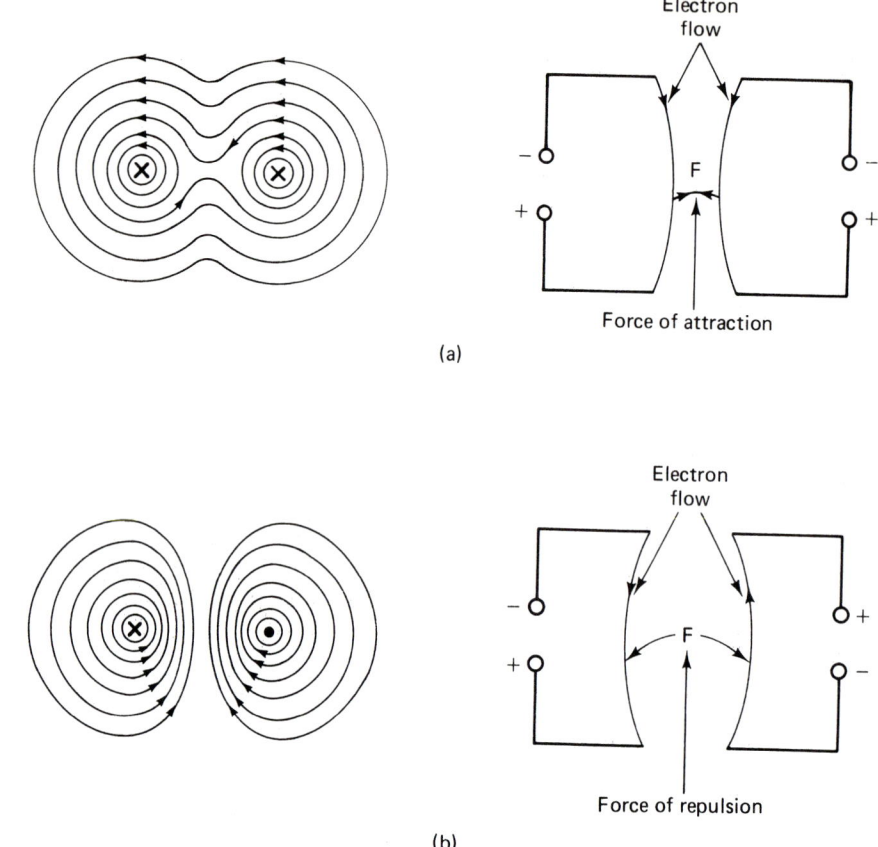

FIGURE 7-13 Forces between adjacent current-carrying conductors.

us see what happens when a conductor with an electron flow of I amperes exiting out of the paper is situated in a uniform magnetic field with a flux density of B teslas [Fig. 7-14(a)]. Below the conductor the two magnetic fields have the same direction and will therefore combine to produce a concentration of the flux lines. However, the fields above the conductor have opposite directions and will tend to cancel, so that the flux lines are spread far apart. The resultant field pattern is shown in Fig. 7-14(b); since the flux lines mutually repel one another, there will be an upward force exerted on the conductor. The direction of this force may be found by means of the *right-hand rule*, which is illustrated in Fig. 7-14(c). The hand is used to show three directions, each of which is at right angles to the other two; the *f*irst finger indicates the direction of the magnetic *f*ield or *f*lux lines, the *se*cond finger, the direction of the electron flow or *c*urrent, and the thu*m*b, the direction of the force or *m*otion if the conductor is allowed to move (corresponding letters have been italicized to help you remember the rule).

The size of the total force, F, exerted on the conductor will clearly depend on the length of the conductor, l meters, the flux density, B teslas, and the current, I amperes. The equation is

$$\text{force, } F = B \times I \times l \quad \text{newtons} \tag{7-3-1}$$

The so-called *motor effect* illustrates how we can convert electrical energy into mechanical energy. We introduce electrical energy in the form of the current

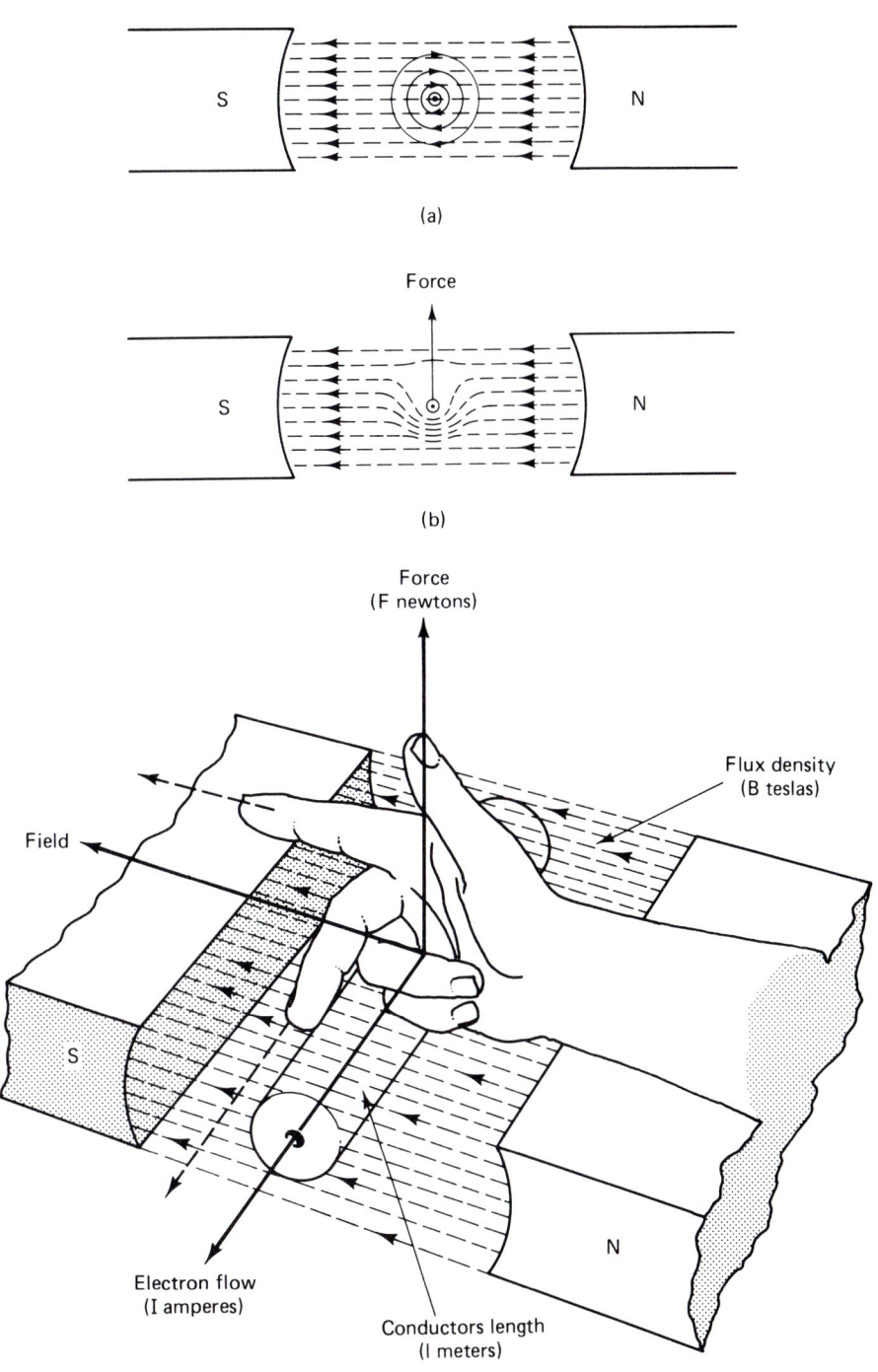

FIGURE 7-14 (a) Force exerted on a current-carrying conductor situated in a magnetic field; (b) resultant field pattern; (c) right-hand motor rule.

flowing through the conductor; the result is the force which is related to mechanical energy. This principle is used in electrical motors, some types of measuring instruments, and in one method of deflecting the electron beam in a cathode ray tube.

When a straight current-carrying conductor is wound on a core into the shape of a coil (Fig. 7-15), its magnetic field pattern is changed. In the partial

Sec. 7-3 Electromagnetism. Motor Effect. Magnetomotive Force

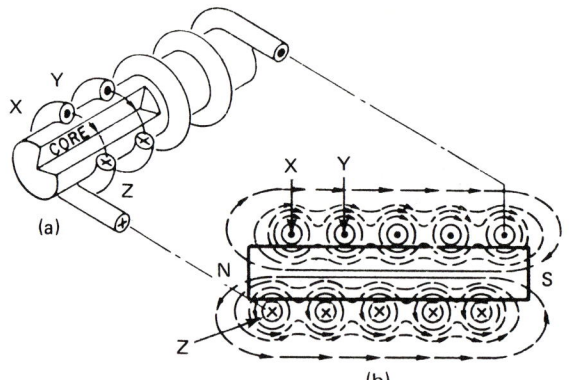

FIGURE 7-15 Magnetic field surrounding a current-carrying coil.

cutaway view of Fig. 7-15(a) we can consider conductors X and Y to be carrying currents in the same direction so that their field pattern is similar to that shown in Fig. 7-13(a). By contrast, conductors X and Z are carrying currents in the opposite directions, so that their field pattern corresponds to Fig. 7-13(b). By combining these two patterns we produce the resultant field, which is illustrated in Fig. 7-15(b). You will immediately notice that the field that is external to the coil is very similar to the field that surrounds a bar magnet. We can regard the coil as possessing an equivalent north pole at one end while at the other end there will be a south pole. If the direction of the electron flow through a coil is known, we can find the location of the north pole by using a *left-hand rule*. Think of grasping the coil in your left hand (Fig. 7-16) with the fingers

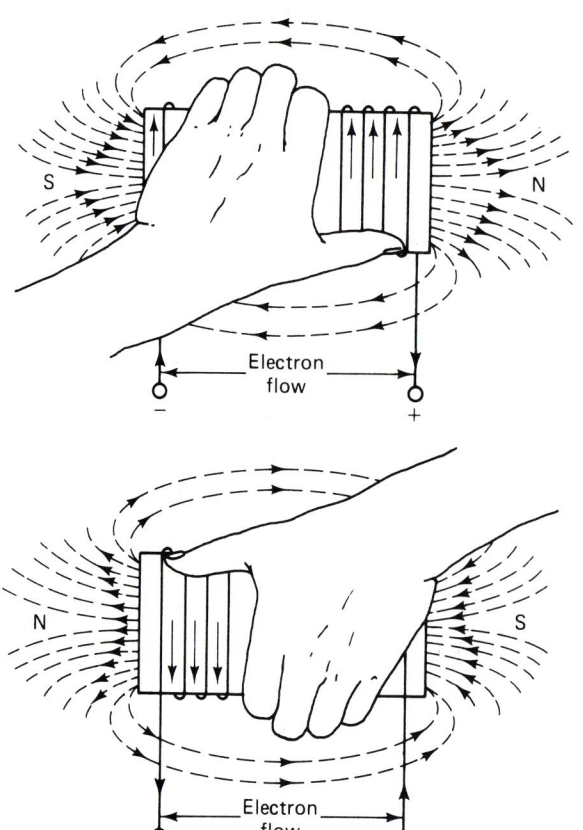

FIGURE 7-16 Left-hand rule to locate the poles of an electro-magnet.

292   Magnetism, Electromagnetism, and Electromagnetic Induction   Ch. 7

wrapped around in the direction of the electron flow; the thumb will then point toward the end where the equivalent north pole exists.

The strength of the magnetic field surrounding a coil will depend on a number of factors:

1. The coil's number of turns
2. The size of the current that is flowing through the coil
3. The ratio of the coil's width to its length
4. The nature of the core

This last factor is extremely important. If the core is made of a ferromagnetic alloy which is easily magnetized, the flow of current through the coil will develop a very strong magnet. However, when the current is removed, most of the magnetism disappears. We refer to such arrangements as *electromagnets*, which are used in electrical generators and motors. Sometimes the coil with its soft-iron core is called a *solenoid*, an example of which is found in relays.

Essentially, the soft-iron core is being magnetized by the flow of current through the coil. The strength of the magnetic field developed will depend on the size of the current and the coil's number of turns. The product of the current, I, in amperes and the number of turns, N, is called the *magnetomotive force* (MMF) whose letter symbol is $\mathscr{F}$. The magnetomotive force is the magnetic equivalent of the electromotive force, in the sense that, as the EMF E, creates a current, I, the MMF, $\mathscr{F}$, establishes a flux, ϕ. We can therefore say that a flux in magnetism is equivalent to a current in electricity but with one important difference; a current flows in an electrical circuit but a flux is merely established in a magnetic circuit and does not flow. The equation for the MMF is

$$\mathscr{F} = IN \quad \text{ampere-turns} \quad (7\text{-}3\text{-}2)$$

Since the number of turns has no units, the basic unit for the MMF is the ampere, while, of course, the basic unit for the EMF is the volt.

In this section we covered the magnetic field patterns surrounding straight conductors and coils and introduced the concept of the MMF, which is responsible for establishing the magnetic flux.

EXAMPLE 7-3 A conductor of length 50 cm is placed at right angles to the direction of a uniform field whose flux density is 0.8 T. If the conductor is carrying a current of 3 A, calculate the force exerted on the conductor.

Solution

$$B = 0.8 \text{ T}, \quad I = 3 \text{ A}, \quad l = 50 \text{ cm} = 50 \times 10^{-2} \text{ m}$$

force exerted on the conductor, $F = B \times I \times l \quad (7\text{-}3\text{-}1)$

$$= 0.8 \text{ T} \times 3 \text{ A} \times 50 \times 10^{-2} \text{ m}$$

$$= 1.2 \text{ N}$$

7-4 Magnetic Field Intensity. Permeability of Free Space

Figure 7-17 shows a toroidal soft-iron ring (shaped like a doughnut) on which is wound a coil of N turns. A current of I amperes is flowing through the coil so that the MMF is IN ampere-turns. A magnetic flux of ϕ webers will be established in the iron ring, and since the cross-sectional area of the ring is A square meters, the flux density, B, is ϕ/A teslas.

We are assuming that the flux is entirely confined to the iron; however, in practice some flux will exist in the surrounding air and this is referred to as the *leakage flux*. Notice that the flux lines are continuous around the ring and therefore no poles exist. However, if a radial cut is made in the ring to cause an air gap, a north pole exists on one side of the gap while a south pole appears on the other side.

Suppose that we keep the same values of I, N, and A but we make the ring larger by increasing its mean circumference, which is the average of the inner and outer circumferences. The turns will be spread farther apart and the flux density inside the iron ring will be reduced. The value of the flux density therefore depends not only on the current, I, and the number of turns, N, but also on the length, l, of the magnetic path over which the flux is established. These three factors are contained in the *magnetic field intensity*, H (sometimes called the *magnetizing force*), which can be regarded as equivalent to the MMF per unit length and is therefore measured in ampere-turns per meter (AT/m) or, more basically, in amperes per meter (A/m). The equation is:

$$\text{magnetic field intensity, } H = \frac{IN}{l} \quad \text{ampere-turns per meter} \quad (7\text{-}4\text{-}1)$$

and therefore,

$$\text{magnetomotive force, } \mathscr{F} = IN = H \times l \quad \text{amperes} \quad (7\text{-}4\text{-}2)$$

What we are saying is that the magnetic field intensity, H, is directly responsible for producing the magnetic flux density, B; consequently, there must be some relationship between B and H so that if we know the value of H, we can find the value of B.

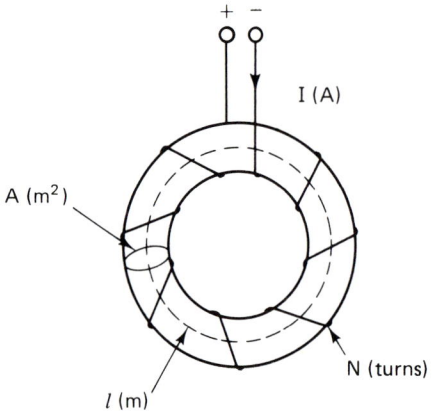

FIGURE 7-17 Toroidal ring magnetic circuit.

294   Magnetism, Electromagnetism, and Electromagnetic Induction   Ch. 7

Permeability of Free Space (VACUUM)

Figure 7-18 shows the cross section of a long straight conductor, A, which is situated in a vacuum and carries an electron current of 1 A emerging from the paper. One of the circular lines of force (C) has a radius of 1 m, so that the length of its magnetic path is $2\pi \times 1$ m $= 2\pi$ meters (length of a circle's circumference $= 2\pi \times$ radius, where $\pi \approx 3.142$). The MMF acting on this path is 1 A, so that the corresponding magnetic field intensity, H, is $1/2\pi$ ampere per meter.

A' is a second conductor which is parallel to and 1 m away from the first conductor. If the second conductor also carries a current of 1 A, we can use Eq. (7-3-1) to find the force exerted on 1 m length of the second conductor. This force is

$$F \text{ (newtons)} = B \times I \times l = B \times 1 \text{ A} \times 1 \text{ m}$$

where B is the flux density in teslas at the position of the second conductor. But from the definition of the ampere, this force must also be 2×10^{-7} N. Consequently,

$$F = B \times 1 \text{ A} \times 1 \text{ m} = 2 \times 10^{-7} \text{ N}$$

or

$$B = 2 \times 10^{-7} \text{ T}$$

We now know that, at the same position, a field intensity, H, of $1/2\pi$ ampere per meter develops a flux density of 2×10^{-7} T. Therefore, in a vacuum, which is sometimes referred to as *free space*, the ratio of B to H is

$$\frac{B}{H} = \frac{2 \times 10^{-7} \text{ T}}{1/2\pi \text{ A/m}} = 4\pi \times 10^{-7} \qquad (7\text{-}4\text{-}3)$$
$$= 12.57 \times 10^{-7} \text{ SI units}$$

The ratio of B to H in a vacuum is called the *permeability* of free space, whose letter symbol is μ_0. As we shall see in Chapter 8, μ_0 is measured in henrys per meter. The word "permeability" is a measure of the ability of a medium (for example, vacuum, air, iron) to allow the establishment of a flux density by applying a magnetic field intensity. For a vacuum only,

$$\mu_0 = \frac{B}{H}, \qquad B = \mu_0 H, \qquad H = \frac{B}{\mu_0} \qquad (7\text{-}4\text{-}4)$$

FIGURE 7-18 Two parallel conductors in a vacuum.

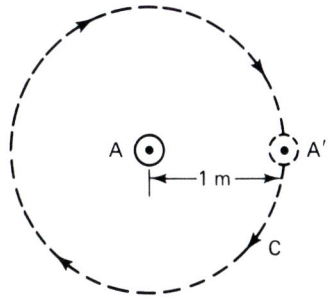

Sec. 7-4 Magnetic Field Intensity. Permeability of Free Space

Most materials are nonmagnetic and have virtually the same permeability as free space.

The magnetic field intensity, H, is responsible for producing the magnetic flux density, B. In this section we have discovered that the relationship between B and H is the permeability of the medium in which the magnetic flux is established.

EXAMPLE 7-4 Calculate the MMF required to establish a magnetic flux of 600 μWb in an air gap whose length is 0.5 mm and whose cross-sectional area is 4 cm².

Solution Air is nonmagnetic and therefore its permeability is the same as that of free space.

$$A = 4 \times 10^{-4} \text{ m}^2, \quad l = 0.5 \times 10^{-3} \text{ m}, \quad \phi = 600 \times 10^{-6} \text{ Wb}$$

$$\text{flux density, } B = \frac{\phi}{A} \tag{7-2-1}$$

$$= \frac{600 \times 10^{-6} \text{ Wb}}{4 \times 10^{-4} \text{ m}^2}$$

$$= 1.5 \text{ T}$$

$$\text{field intensity, } H = \frac{B}{\mu_0} \tag{7-4-4}$$

$$= \frac{1.5}{12.57 \times 10^{-7}}$$

$$= 1.19 \times 10^6 \text{ AT/m}$$

$$\text{magnetomotive force, } \mathscr{F} = H \times l \tag{7-4-2}$$

$$= 1.19 \times 10^6 \text{ AT/m} \times 0.5 \times 10^{-3} \text{ m}$$

$$= 596 \text{ AT}$$

7-5

Relative Permeability. Reluctance. Rowland's Law

From the principle of the electromagnet we already know that the flux density inside a coil is intensified when a soft-iron core is inserted. If we keep the magnetic field intensity the same, the ratio of the flux density established in the material of the core to the flux density produced in a vacuum is called the *relative permeability*, μ_r, which has no units and is just a number. For air and any other nonmagnetic core, $\mu_r = 1$. However, there are some so-called *paramagnetic materials* (for example, aluminum and platinum) whose permeability is slightly better than that of free space. These substances have a small degree of magnetism with a relative permeability which is marginally greater than 1. Yet other elements, such as silver and copper, are diamagnetic and have a relative permeability of just less than 1; diamagnetic materials therefore offer slightly more opposition than does free space to a magnetic flux. However, both effects are so small that in practice we can regard both paramagnetic and diagmagnetic materials as possessing a permeability equal to that of free space.

A complete contrast is provided by ferromagnetic materials, such as certain nickel-iron alloys which have extremely high relative permeabilities ranging up to 100,000. However, as we shall see in Section 7-6, μ_r is not a constant for a particular alloy since the relative permeability varies with the value of the magnetic field intensity. From the definition of μ_r,

$$\text{relative permeability, } \mu_r = \frac{\text{flux density with a magnetic core}}{\text{flux density with a vacuum core}} \quad (7\text{-}5\text{-}1)$$

From Eq. (7-4-3) the flux density for a vacuum core is $\mu_0 H$. Therefore,

$$\text{flux density with a magnetic core, } B = \mu_0 \mu_r H \quad (7\text{-}5\text{-}2)$$

The product $\mu_0 \mu_r$ is called the *absolute permeability*, μ, of the magnetic material. As we shall see in Chapter 8, the absolute permeability is measured in henrys per meter. Therefore,

$$\mu = \mu_0 \mu_r \quad \text{henrys per meter} \quad (7\text{-}5\text{-}3)$$

and

$$B = \mu H, \quad \mu = \frac{B}{H} \quad (7\text{-}5\text{-}4)$$

The fact that the absolute permeability is directly related to the relative permeability means that the value of μ is also not constant but depends on the magnetic field intensity.

Referring back to the iron ring of Fig. 7-17, we have already determined that

$$\text{total flux, } \phi = B \times A \quad \text{webers}$$

and

$$\text{magnetomotive force, } \mathcal{F} = H \times l \quad \text{ampere-turns}$$

Therefore, when we divide the MMF (the magnetic equivalent of the EMF) by the flux (the magnetic equivalent of the current), the result is

$$\frac{\text{MMF}, \mathcal{F}}{\text{flux}, \phi} = \frac{H \times l}{B \times A}$$

But $B = \mu_0 \mu_r H = \mu H$. Therefore,

$$\frac{\text{MMF}, \mathcal{F}}{\text{flux}, \phi} = \frac{H \times l}{\mu_0 \mu_r H \times A} = \frac{l}{\mu_0 \mu_r A} = \frac{l}{\mu A} \quad (7\text{-}5\text{-}5)$$

Sec. 7-5 Relative Permeability. Reluctance, Rowland's Law

This relationship, sometimes called *Rowland's law*, is the magnetic equivalent of the Ohm's law equation

$$\frac{\text{EMF}, E}{\text{current}, I} = \text{resistance}, R$$

$$= \frac{\text{length}, l}{\text{conductivity}, \sigma \times \text{cross-sectional area}, A}$$

The quantity $l/\mu A$ is therefore comparable with the resistance and is called the reluctance, $\mathcal{R}$, of the magnetic circuit. Then

$$\mathcal{R} = \frac{l}{\mu A} = \frac{\mathcal{F}}{\phi} \tag{7-5-6}$$

The reluctance is a measure of the magnetic circuit's opposition to the establishment of the flux. There are no special SI units for reluctance, but it can be referred to in terms of ampere-turns per weber or as the reciprocal of the henry unit, which is defined in Chapter 8. By comparing $R = l/\sigma A$ and $\mathcal{R} = l/\mu A$, you can see that the electrical conductivity, σ, is equivalent to the magnetic permeability, μ. Since the reluctance of a ferromagnetic specimen is inversely proportional to the absolute permeability, the value of the reluctance will not be a constant but will depend on the magnetic field intensity. The principle of magnetic shielding relies on the fact that the reluctance of soft iron is very low compared with that of air. We can protect an object such as a meter movement from stray magnetic fields by enclosing the object in a soft-iron case, which has a high value of relative permeability. The case then offers a low reluctance for the magnetic flux so that only a few lines can penetrate through to the shielded space (Fig. 7-19).

Just as the conductance in siemens is the reciprocal of the resistance in ohms, the reciprocal of the reluctance is the permeance, $\mathcal{P} = \mu A/l$. The SI unit for permeance is the henry or the weber per ampere-turn.

Many of the principles we have used in the analysis of series and parallel

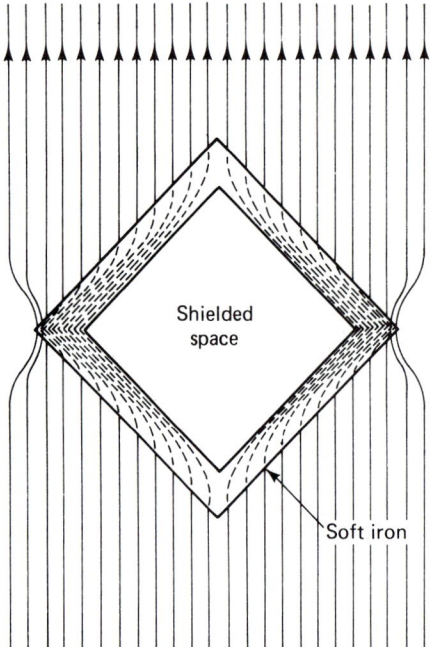

FIGURE 7-19 Magnetic shield.

resistor circuits can be applied to magnetic circuits. For example, we can make a cut in the iron ring of Fig. 7-17 and cause a narrow air gap. We then assume that the same flux is established in both the air and the iron, which is electrically equivalent to saying that the current is the same throughout a series circuit. A certain number of ampere-turns would be necessary to establish the flux in the air gap and a different number of ampere-turns would develop the same flux in the iron. By summing the two numbers of ampere-turns we would obtain the total required MMF; this would be equivalent to finding the source voltage by adding together the individual voltage drops across two resistors in series.

One important difference between magnetic and electrical circuits lies in the concept of power. Whereas energy must be continuously supplied to an electrical circuit that contains resistors, a magnetic flux, once it is established, does not require any further energy from the source. In the case of the magnetized iron ring, the only energy supplied after the magnetic flux is set up will be dissipated by the coil's resistance in the form of heat.

We discussed in this section the magnetic properties of various materials in terms of their relative permeabilities. In addition, we examined the close relationship between the properties of electrical and magnetic circuits; a detailed comparison is shown in the chapter summary.

EXAMPLE 7-5 A certain magnetic circuit has an air gap which has a length of 1.5 mm and a cross-sectional area of 8 cm². Calculate the gap's reluctance and the MMF required to establish a flux of 800 µWb across the gap.

Solution

$$\mu_r = 1, \quad A = 8 \times 10^{-4} \text{ m}^2, \quad l = 1.5 \times 10^{-3} \text{ m}$$

reluctance of the gap, $\mathcal{R} = \dfrac{l}{\mu A}$ (7-5-6)

$$= \dfrac{l}{\mu_0 \mu_r A}$$

$$= \dfrac{1.5 \times 10^{-3}}{12.57 \times 10^{-7} \times 8 \times 10^{-4}}$$

$$= 1.49 \times 10^6 \text{ SI units}$$

required MMF, $\mathcal{F} = \mathcal{R} \times \phi$

$$= 1.49 \times 10^6 \times 800 \times 10^{-6}$$

$$= 1192 \text{ AT}$$

EXAMPLE 7-6 An iron ring has a uniform cross-sectional area of 6 cm² and a mean circumference of 50 cm. The flux density established in the iron is 1.5 T and the corresponding relative permeability is 2500. Calculate the reluctance of the ring and the MMF required to establish the flux.

Solution

$$l = 50 \times 10^{-2} \text{ m}, \quad A = 6 \times 10^{-4} \text{ m}^2$$

$$\mu = \mu_0 \mu_r = 12.57 \times 10^{-7} \times 2500 \text{ SI units}$$

reluctance of the ring, $\mathcal{R} = \dfrac{l}{\mu A}$ (7-5-6)

$$= \dfrac{50 \times 10^{-2}}{12.57 \times 10^{-7} \times 2500 \times 6 \times 10^{-4}}$$

$$= 2.65 \times 10^5 \text{ SI units}$$

magnetic flux, $\phi = B \times A$ (7-2-2)

$$= 1.5 \text{ T} \times 6 \times 10^{-4} \text{ m}^2 = 9 \times 10^{-4} \text{ Wb}$$

magnetomotive force, $\mathcal{F} = \phi \times \mathcal{R}$

$$= 9 \times 10^{-4} \times 2.65 \times 10^5$$

$$= 240 \text{ AT (rounded off)}$$

7-6 Magnetization Curve. Hysteresis

We have already pointed out that the value of the relative permeability, μ_r, for a ferromagnetic material is not a constant but depends on the levels of the magnetic field intensity, H, and the corresponding magnetic flux density, B; consequently, there is no simple relationship between these two variables. It is therefore common practice to use a magnetization curve which enables us to obtain directly the related values of B and H. This B,H curve is obtained experimentally by using the circuit of Fig. 7-20. If the slider of the potentiometer is set to point O, there will be zero current flow through the coil which is wound on the ferromagnetic specimen; as a result, the magnetic field intensity will be zero. If the slider is gradually shifted toward point X, the current, I, and the field intensity, H, will increase in one direction and magnetize the ring; a flux density meter will then record the corresponding values of B. By moving the slider back from X to O, I and H will gradually be reduced to zero. If we then move the slider from O to Y, both I and H will again be increased but their directions will be reversed. Finally, by returning the slider from Y to O, I and H will together decrease to zero.

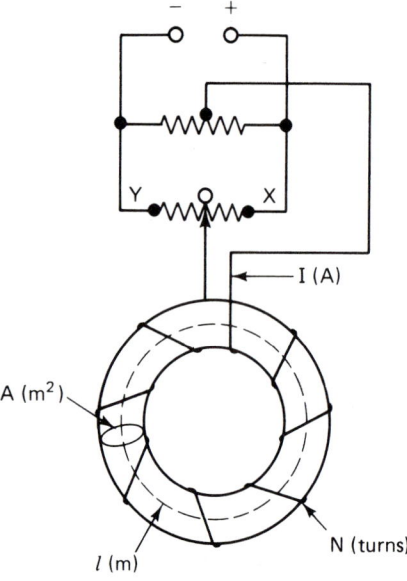

FIGURE 7-20 Circuit for obtaining a magnetization curve.

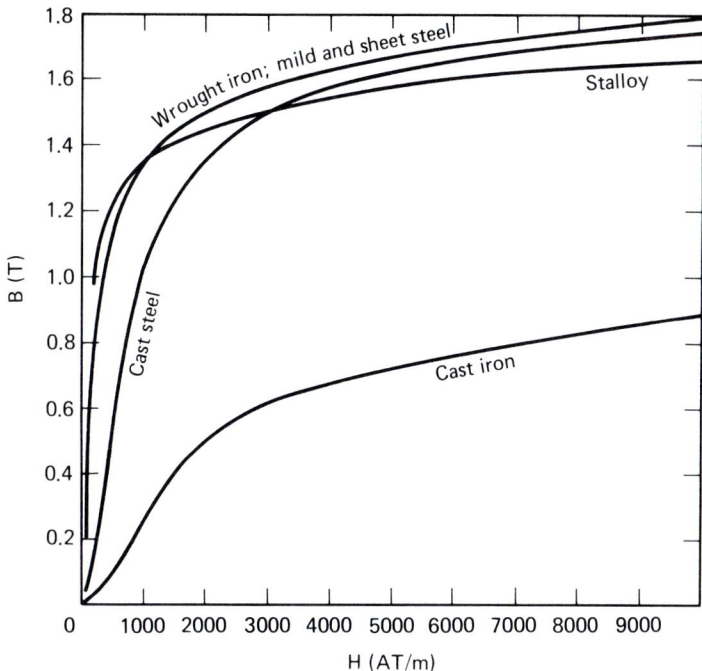

FIGURE 7-21 Magnetization curves.

Examples of magnetization curves, obtained by moving the slider from O to X, are shown in Fig. 7-21. The curves differ considerably, but in each case the ferromagnetic material is initially assumed to be completely demagnetized so that its magnetic domains are oriented in a random manner. At the start of each experiment there is a short curved region in which the magnetic domains are starting to align in the required direction. This process is subsequently speeded up, which corresponds to the steep portions of the B,H curves. Then the curves start to bend over; this indicates saturation in which virtually all of the magnetic domains have been aligned with the magnetizing field produced by the coil. Finally, the B,H curves become straight lines with only gradual slopes; the flux density is then increasing with the magnetic field intensity in the same way as it would for any nonmagnetic material.

Since $\mu = B/H$, we can divide the flux density by the corresponding value of the magnetic intensity and obtain the absolute permeability for any point on the curves. This will enable us to calculate the relative permeability by using the equation $\mu_r = \mu/\mu_0$. The results are displayed in the μ_r/H curves of Fig. 7-22. As you would expect, the relative permeability in every case rises rapidly prior to saturation and then falls gradually away.

In Fig. 7-23, let us assume that we have already moved the slider from O to X and have therefore reached the corresponding point, X', on the B,H curve with the flux density represented by the line AX'. If we now return the slider from X to O, we do not follow the same curve back to the origin. This is because most ferromagnetic materials have the property of retentivity, so that when the magnetic field intensity is completely removed, we have returned along the curve $X'B$ and there is still some flux density remaining as measured by the line OB. This value of B is called the *residual* or *remanent magnetism*. To remove this flux density we must move the slider some distance from O toward Y so that I and H are reversed. The amount of reverse field intensity required to eliminate the residual magnetism is called the *coercive force*, which is shown by the line OC.

By further shifting the slider all the way to Y, we move from C to Y' on the curve and the flux density increases to the level represented by $Y'D$. Re-

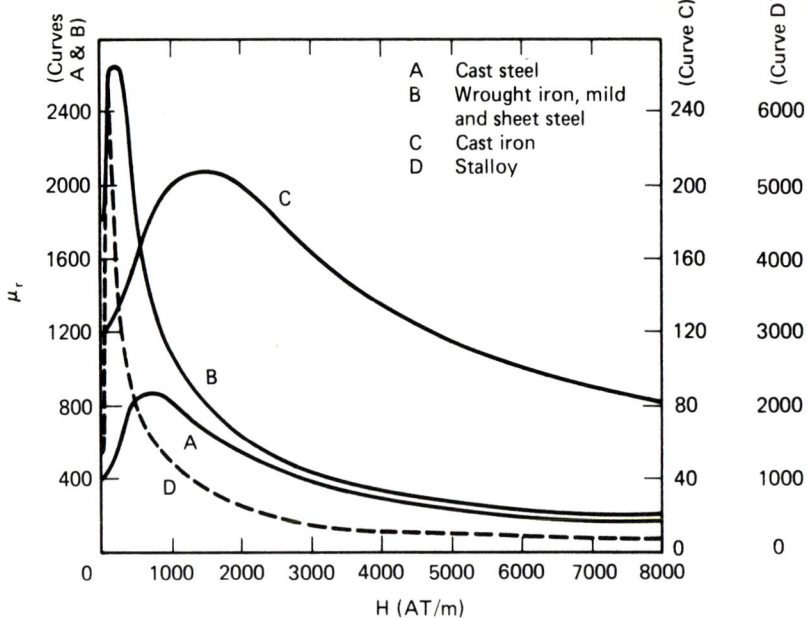

FIGURE 7-22 Relationship between relative permeability and magnetic field intensity.

versing the slider back to O will remove the field intensity, so that once again there is only the residual magnetism with its flux density measured by OE. This is eliminated by the coercive force, OF, which is developed by moving the slider some distance from O back toward X. Finally, we can return the slider to its starting point, X, so that we will again have arrived at the point X' on the B,H curve.

We have seen that when H is removed, the flux density, B, which was developed by the flux intensity, does not fall to zero. This means that the flux density, which is the "effect," is lagging behind the field intensity, the "cause"; this phenomenon of lagging is called *hysteresis*. If we gradually and contin-

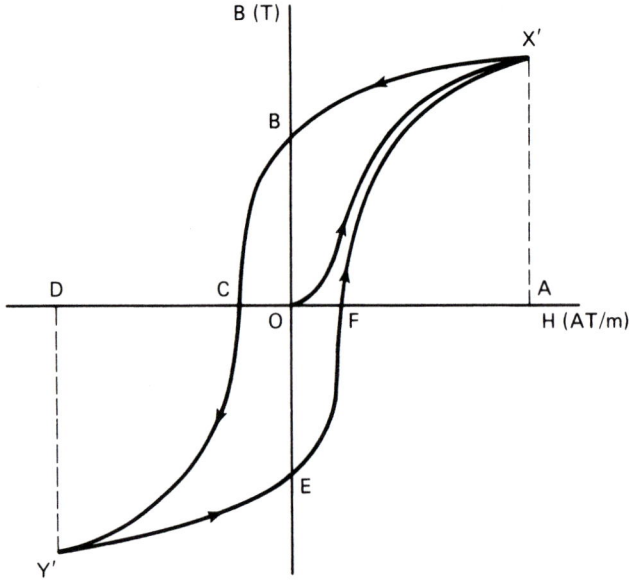

FIGURE 7-23 Hysteresis loop.

uously shift the slider from X to O to Y to O to X, we will correspondingly move from X' to B to C to Y' to E to X' on the closed B,H graph, which is called the *hysteresis loop*. As we shall see in Section 15-1, the effect of hysteresis is to introduce an energy loss in the iron core of an alternating current transformer.

You may well ask: "Are we permanently condemned to move around on the hysteresis loop so that we can never demagnetize the ferromagnetic specimen?" The answer is "no"—by reversing the magnetic intensity back and forth while gradually decreasing the current to zero, we can ultimately return to the origin of the magnetization curve.

In this section we discussed the magnetization curve, which directly relates the flux density to the field intensity; at any point on the curve we derive the value of the relative permeability. The phenomenon of hysteresis is the result of ferromagnetic materials retaining some of their magnetism when the field intensity is removed.

EXAMPLE 7-7 A cast-iron ring has a cross-sectional area of 4 cm² and a mean circumference of 60 cm. What is the current required in a magnetizing coil of 400 turns to establish a total flux of 160 μWb?

Solution

$$\phi = 160 \times 10^{-6} \text{ Wb}, \quad A = 4 \times 10^{-4} \text{ m}^2, \quad l = 60 \times 10^{-2} \text{ m}$$

$$\text{flux density, } B = \frac{\phi}{A} \quad (7\text{-}2\text{-}1)$$

$$= \frac{160 \times 10^{-6} \text{ Wb}}{4 \times 10^{-4} \text{ m}^2} = 0.4 \text{ T}$$

In Fig. 7-21 a flux density of 0.4 T in cast iron corresponds to a field intensity of 1500 AT per meter.

$$\text{magnetomotive force, } \mathscr{F} = H \times l \quad (7\text{-}4\text{-}2)$$

$$= 1500 \text{ AT/m} \times 60 \times 10^{-2} \text{ m}$$

$$= 900 \text{ AT}$$

From Eq. (7-3-2),

$$\text{magnetizing current, } I = \frac{\mathscr{F}}{N} = \frac{900}{400} = 2.25 \text{ A}$$

EXAMPLE 7-8 A cast steel toroidal ring has a cross-sectional area of 5 cm² and a mean circumference of 50 cm. A radial cut has been made in the ring to create an air gap of thickness 1 mm. Assuming no flux leakage, what is the total MMF required to establish a flux density of 1.5 T in the air gap?

Solution This problem involves a magnetic circuit in which we regard the steel and the air gap as being in series. We will therefore carry out separate calculations to find the MMF required to establish the common flux in the air gap and the steel. The two MMFs are then added to give the total MMF.

From Fig. 7-21, a flux density of 1.5 T in cast steel requires a field intensity of $H_1 = 3000$ AT/m.

$$H_1 = 3000 \text{ AT/m}, \quad l_1 = 50 \times 10^{-2} \text{ m}$$

The MMF required to establish the flux in the steel ring is

$$\mathscr{F}_1 = H_1 \times l_1 \qquad (7\text{-}4\text{-}2)$$
$$= 3000 \text{ AT/m} \times 50 \times 10^{-2} \text{ m} = 1500 \text{ AT}$$

For the air gap the magnetic field intensity is [Eq. (7-4-3)]

$$H_2 = \frac{B}{\mu_0} = \frac{1.5}{12.57 \times 10^{-7}} = 1.19 \times 10^6 \text{ AT/m}$$

Then

$$H_2 = 1.19 \times 10^6 \text{ AT/m}, \quad l_2 = 1 \times 10^{-3} \text{ m}$$

The MMF required to establish the flux in the air gap is

$$\mathscr{F}_2 = H_2 \times l_2 = 1.19 \times 10^6 \text{ AT/m} \times 1 \times 10^{-3} \text{ m} \qquad (7\text{-}4\text{-}2)$$
$$= 1190 \text{ AT}$$

$$\text{total MMF}, \mathscr{F}_T = \mathscr{F}_1 + \mathscr{F}_2 = 1500 + 1190 = 2690 \text{ AT}$$

7-7

Electromagnetic Induction. Faraday's Law. Lenz's Law

The phenomenon of electromagnetic induction is the link between magnetism and the generation of electricity. Faraday discovered in 1831 that when a conductor cuts or is cut by a magnetic field, an EMF is produced or, as we say, *induced* in the conductor. Notice that either the conductor or the field may be moving; it is necessary only that there be relative motion between the conductor and the magnetic flux.

Figure 7-24 shows a moving conductor which is cutting the flux lines at right angles (90°). The ends of the conductor are connected to a sensitive voltmeter. The movement of the conductor in the magnetic field requires mechanical energy, while the induced EMF and the subsequent current flow through the instrument represent electrical energy. We are therefore converting mechanical energy into electrical energy, which is the principle behind the electrical generator. Most power stations provide excellent examples of converting from one form of energy into another. Coal, oil, nuclear fission, and, in the future, nuclear fusion all initially produce heat; this is used to raise the temperature of water to produce steam, which then drives a turbine. The mechanical energy at the end of the turbine shaft is used to produce some form of motion between conductors and a magnetic flux; the voltages are then induced in the conductors so that electrical energy is available.

Now why does this effect occur? If the conductor is stationary, its free electrons are moving in random directions. However, if we cause the conductor

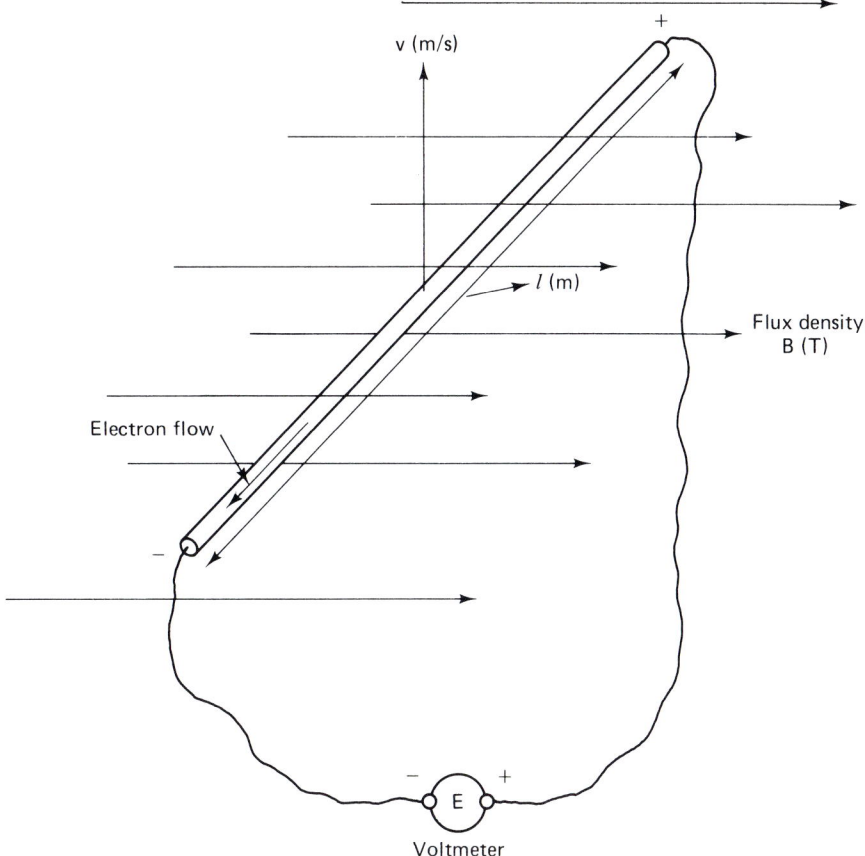

FIGURE 7-24 Faraday's law. Generator effect.

to cut the flux, we give the electrons a movement in a particular direction. A force will then be exerted on the moving charge and the free electrons will be driven along the conductor. Consequently, one end of the conductor will be negatively charged while the other end is positively charged, so that a voltage exists between the ends of the conductor.

The next question is: "What factors determine the size of the induced voltage?" Faraday's law states that the magnitude of the EMF depends on the rate of the cutting of the flux, ϕ, which means the number of lines of magnetic force cut by the conductor in 1 second. As an equation,

$$E = -\frac{\Delta\phi}{\Delta T} \qquad (7\text{-}7\text{-}1)$$

where E is the induced EMF in volts and $\Delta\phi/\Delta T$ represents the rate of the cutting of the flux; the meaning of the minus sign will be explained in our discussion of Lenz's law. Equation (7-7-1) is used in defining the weber, which is the SI unit of magnetic flux. The induced EMF is 1 V if a conductor cuts a flux of 1 Wb in a time of 1 s.

If there are a total of N conductors, each of which is cutting the same flux,

$$E = -N\frac{\Delta\phi}{\Delta T} \qquad (7\text{-}7\text{-}2)$$

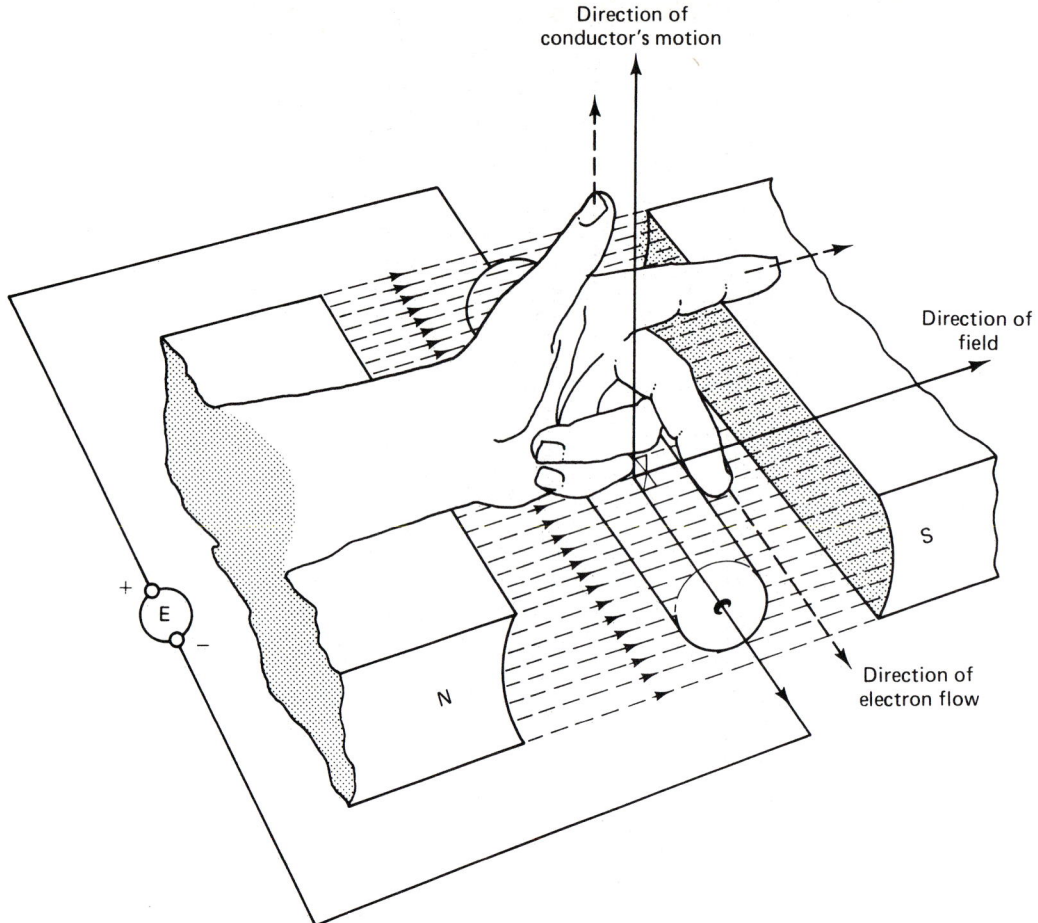

FIGURE 7-25   Left-hand generator rule.

The direction of the induced EMF and the subsequent electron flow through the instrument is found by the left-hand rule; remember that it is the right-hand rule for the "motor effect," but the left-hand rule for the "generator effect." However, for both rules the assigned directions are the same. The *f*irst finger indicates the direction of the *f*lux, the se*c*ond finger is the direction of the induced EMF or electron *c*urrent flow, and the thu*m*b shows the direction of the conductor's *m*otion; the italicized letters are to help you remember the rule, which is illustrated in Fig. 7-25. Notice that if you reverse the direction

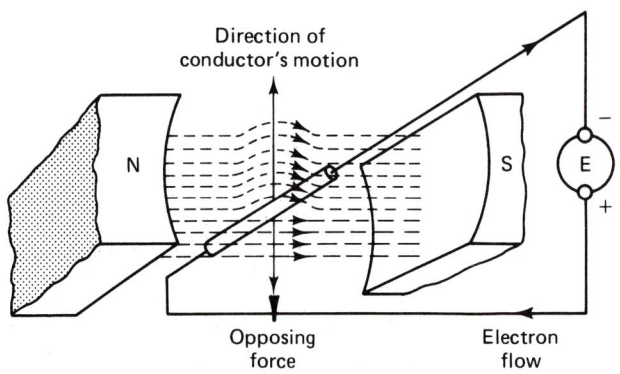

FIGURE 7-26   Generator effect and Lenz's law.

in which you move the conductor, the direction of the electron flow will also be reversed.

We already know that the magnitude of the induced EMF is equal to the rate at which the conductor cuts (or is cut by) the flux. Let us assume that, in Fig. 7-24, the uniform flux density is B teslas, the conductor's length is l meters, and the conductor is moving downward at right angles (90°) to the field with a velocity of v meters per second. In 1 s (ΔT) the conductor will have covered a vertical distance of v meters and will have swept out an area, $\Delta A = v \times l$ square meters. Then using Eq. (7-7-1), we obtain

flux cut by conductor in 1 s = $B \times \Delta A = Blv$ webers

$$\text{value of the induced EMF, } E = \frac{\Delta \phi}{\Delta T}$$
$$= \frac{Blv \text{ (Wb)}}{1 \text{ s}} \quad (7\text{-}7\text{-}3)$$
$$= Blv \quad \text{volts}$$

The value of the induced EMF is therefore directly proportional to the flux density, the conductor's length, and its speed. This means that the same induced EMF can be generated either by moving a conductor slowly in a strong magnetic field or by moving the same conductor rapidly in a weaker field.

Now there's a good rule in life: "You don't get something for nothing." So what price are we paying to produce this electrical energy? As soon as we move the conductor and cut the magnetic flux, the EMF is induced and there is a current flow through the conductor. We therefore have a current-carrying conductor situated in a magnetic field which reacts with the field surrounding the conductor. As a result, the conductor will experience a force whose direction is opposite to the motion of the conductor (the direction of this force may be found by the right-hand "motor rule"). We must therefore provide mechanical energy in order to move the conductor against the opposing force (Fig. 7-26). All this is summed up in Lenz's law (H. Lenz, 1804–1865), which states: "The direction of the induced EMF is such as to oppose the change which originally developed the EMF." Let us examine this statement in detail. The "change" is the conductor's motion, which develops the EMF by cutting the flux. The EMF produces a current flow in a particular direction so that a magnetic field appears around the conductor. The reaction between the two fields causes a force to be exerted on the conductor and the direction of this force is such as to oppose the original motion of the conductor (Fig. 7-26). Because of this opposition, E is given a negative sign in the equation

$$E = -\frac{\Delta \phi}{\Delta T}$$

A good example of Lenz's law is illustrated in Fig. 7-27. A coil is connected to a sensitive current meter which has a scale with a center zero so that the meter can indicate a current flow in either direction. When you move the bar magnet downward toward the coil, the field emanating from the north pole will cut the turns of the coil. By Faraday's law, an EMF is induced into the coil so that a current will flow through the meter and deflect the needle in one direction away from the zero position. Due to the flow of the current through the coil, a north pole will appear at the end that is closest to the north pole of the magnet. As a result, there will be a force of repulsion between the two poles and you must therefore do work against this force of repulsion in order

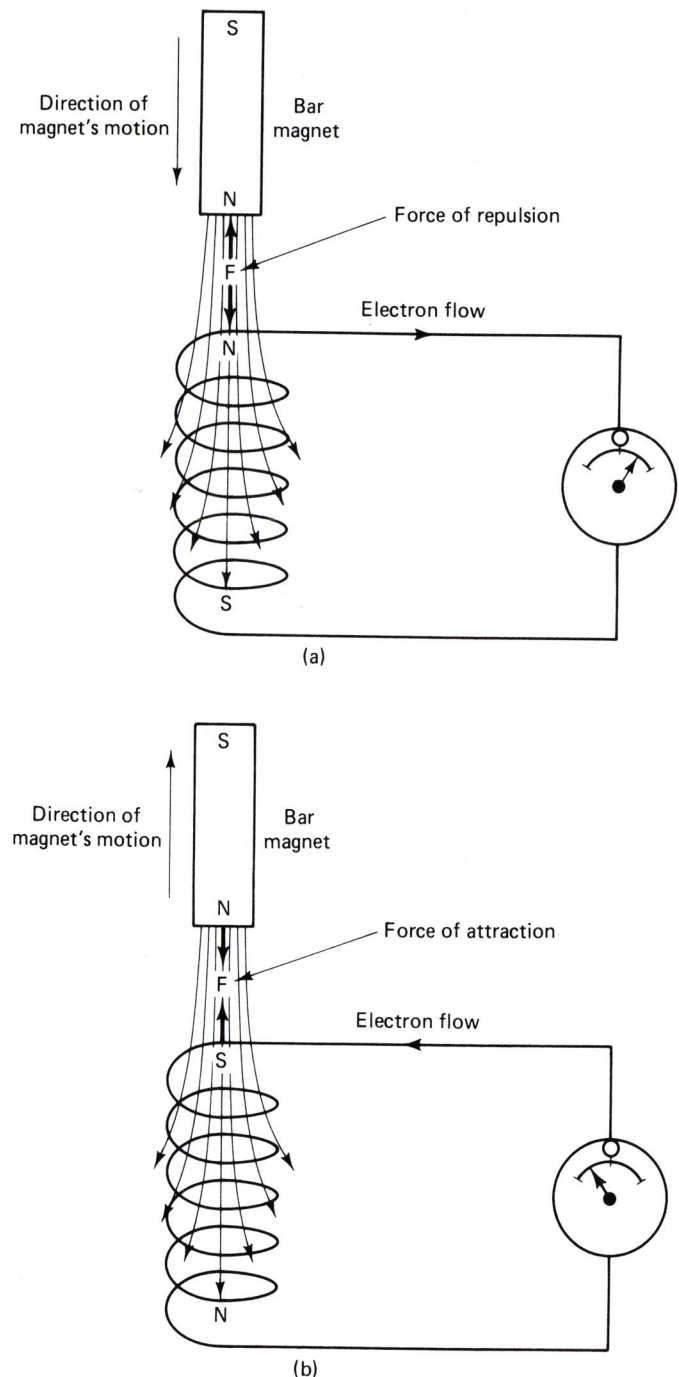

FIGURE 7-27 Illustration of Lenz's law.

to generate the electricity in the coil. In terms of Lenz's law, the "change" is the motion of the bar magnet toward the coil. The "direction of the induced EMF" and its associated current flow creates poles at the ends of the coil, so that there is a force of repulsion between the coil and the magnet. This force "opposes" the motion of the magnet.

If you move the magnet upward, the direction of the EMF and the current are reversed, so that the needle is deflected to the other side of the zero position. The south pole of the coil will then be nearest to the north pole of the magnet.

There will be a force of attraction that will oppose the motion of the magnet away from the coil.

As we have described the experiment, the conductor (coil) is stationary while the flux (magnet) is moving; the results would be similar if we kept the magnet in a fixed position and moved the coil instead.

In this section we saw how a combination of magnetism and mechanical energy can be used to produce electrical energy. Faraday's law revealed the factors that determine the size of the induced EMF. Lenz's law connected the directions of the induced voltage and its associated current with the opposition to the change that originally developed the EMF.

EXAMPLE 7-9 A conductor cuts a flux of 500 μWb in a time of 0.25 ms. What is the value of the induced EMF?

Solution

$$\text{flux change, } \Delta\phi = 500 \times 10^{-6} \text{ Wb}$$
$$\text{time change, } \Delta T = 0.25 \times 10^{-3} \text{ s} \quad (7\text{-}7\text{-}1)$$
$$\text{value of induced EMF, } E = (-)\frac{\Delta\phi}{\Delta T}$$
$$= \frac{500 \times 10^{-6} \text{ Wb}}{0.25 \times 10^{-3} \text{ s}} = 2\text{V}$$

EXAMPLE 7-10 A conductor of length 50 cm is moving at right angles to the direction of a uniform magnetic field whose flux density is 0.8 T. If the conductor's velocity is 120 m/s, what is the value of the induced EMF?

Solution

$$B = 0.8 \text{ T}, \quad l = 50 \times 10^{-2} \text{ m}, \quad v = 120 \text{ m/s}$$
$$\text{induced EMF, } E = Blv \quad (7\text{-}7\text{-}3)$$
$$= 0.8 \text{ T} \times 50 \times 10^{-2} \text{ m} \times 120 \text{ m/s} = 48 \text{ V}$$

7-8
Principle of DC Electrical Generator

The dc generator or dynamo is a practical application of electromagnetic induction. In its simplest form the generator consists of a single turn or loop which is mounted on a shaft; this shaft is then turned by some external form of mechanical energy. When the loop is rotated at constant speed in a uniform magnetic field, the flux is cut by the loop's conductors T and T' (Fig. 7-28); by electromagnetic induction a voltage will then appear at the ends of the loop. In position B one conductor T is moving vertically upward while the other conductor T' is moving downward. Using the left-hand rule for the generator effect, one end of the loop (point X) is negative with respect to the other end (point Y). However, after one-half revolution of the loop (position D), the

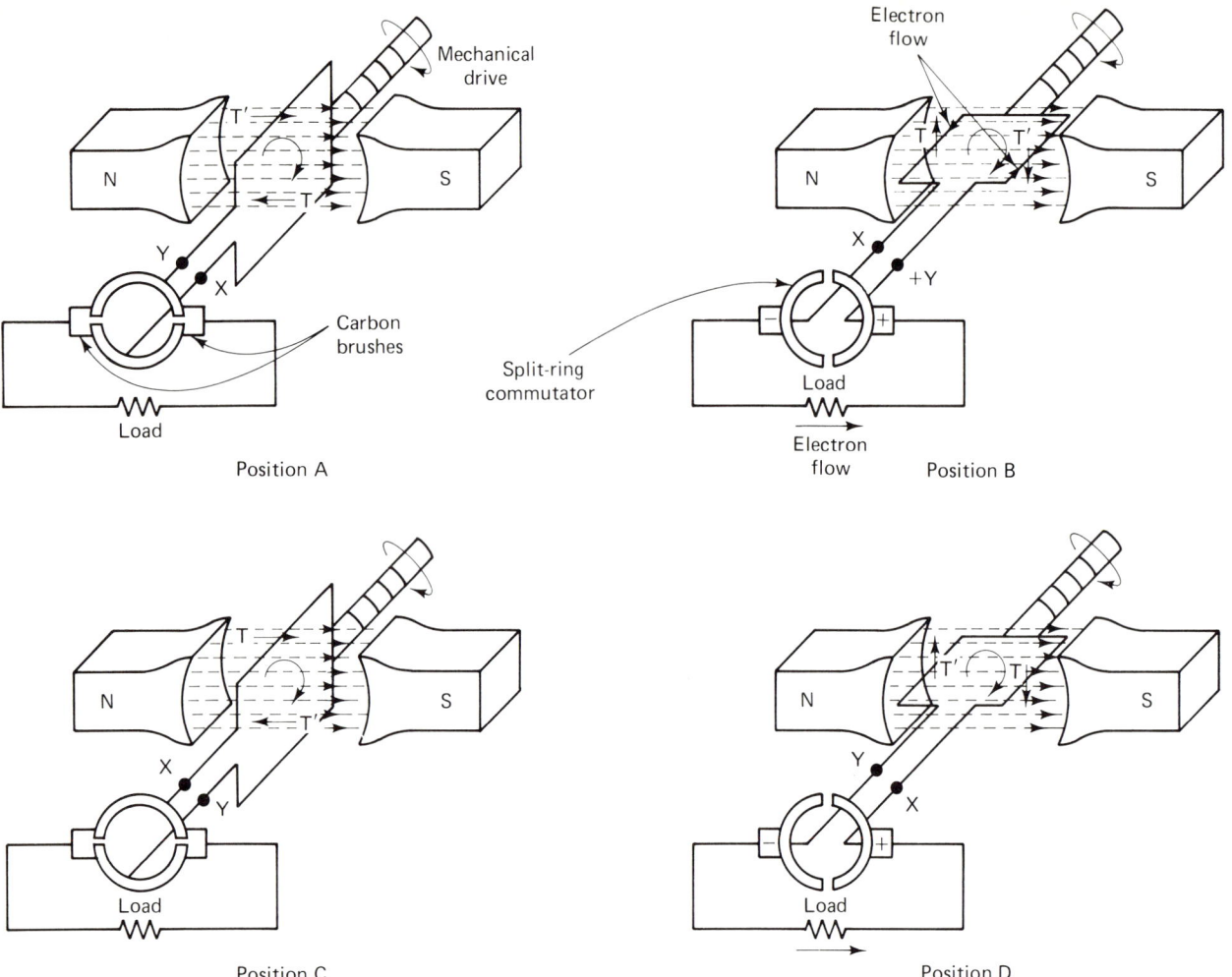

FIGURE 7-28 Principle of the dc generator.

movements of the two conductors are reversed, so that the polarities at X and Y are changed. This means that a loop which rotates in a magnetic field naturally generates an alternating voltage.

As a means of delivering this ac voltage to a load, the ends of the loop are taken to two brass slip rings, which rotate with the shaft and make contact with two stationary carbon brushes. These brushes are then joined to the generator's external terminals across which the load is connected.

In order to produce a dc output voltage, the two separate slip rings are replaced by a single ring, split into two segments which are insulated from each other. This split ring is called a *commutator*, which mechanically reverses the coil connections to the load at the same instant that the polarity of the generated voltage reverses in the loop. By this means the alternating voltage which is naturally generated by the loop is converted into a form of dc voltage to be applied to the load. A device that converts ac to dc is called a *rectifier*, so that the commutator is behaving as a type of mechanical rectifier.

As the loop is rotated clockwise from position A to position B, the loop is cutting the flux at an increasing angle, so that the EMF is rising and driving an electron flow from left to right through the load. When we reach position B the conductors of the loop are instantaneously cutting the flux at right angles, so that the generated EMF reaches its maximum value. As the rotation is continued from position B to position C, the angle at which the loop's con-

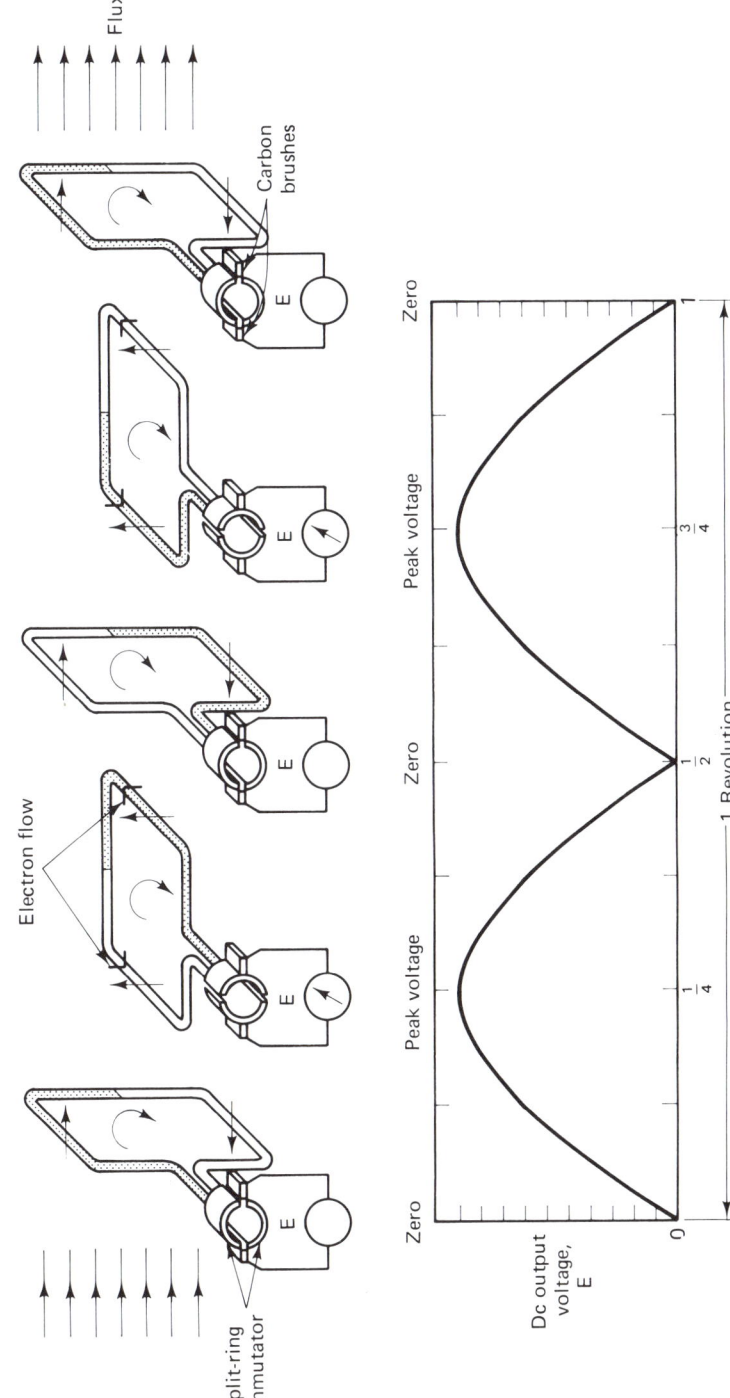

FIGURE 7-29 Variation of dc output voltage during one revolution of the loop.

ductors are cutting the flux is decreasing, so that the output voltage across the load falls. In position C the conductors are momentarily moving parallel with the flux, so that no lines are cut and the generated EMF is zero. At this same instant each of the carbon brushes makes contact with both segments of the split ring, so that the loop is temporarily short-circuited. As we move from position C to position D the output voltage again rises from zero to a maximum, but the polarities of the voltages induced in the loop's conductors are reversed. However, since the segments of the commutator have rotated with the loop and are now connected to opposite carbon brushes, the direction of the electron flow through the load remains unchanged. We describe the output voltage across the load as fluctuating but unidirectional since its polarity never reverses.

Figure 7-29 shows the graph of the fluctuating dc output voltage for one complete rotation of the loop. This type of dc voltage could be used for charging a battery such as a lead-acid secondary cell, but it is unsuitable for most applications. Practical dc generators create a steady dc voltage which is accompanied by a small fluctuation or ripple. The ripple represents the variation in the output voltage as the conductors are rotated; it is of course undesirable and must be reduced to an acceptable level.

To increase the output voltage the single loop can be replaced by a coil consisting of a number of turns. This coil is mounted on a slotted cylindrical armature which is made of sheet steel laminations. To reduce the large ripple generated by a single coil, two separate coils can be connected to a commutator which is split into four segments (Fig. 7-30). When the voltage generated by one coil falls to the level represented by E_1, the brushes break contact with the pair of opposite segments connected to this coil. At the same instant the brushes make contact with the other pair of segments, which are connected to the coil whose generated EMF is rising toward the maximum level, E_2. The

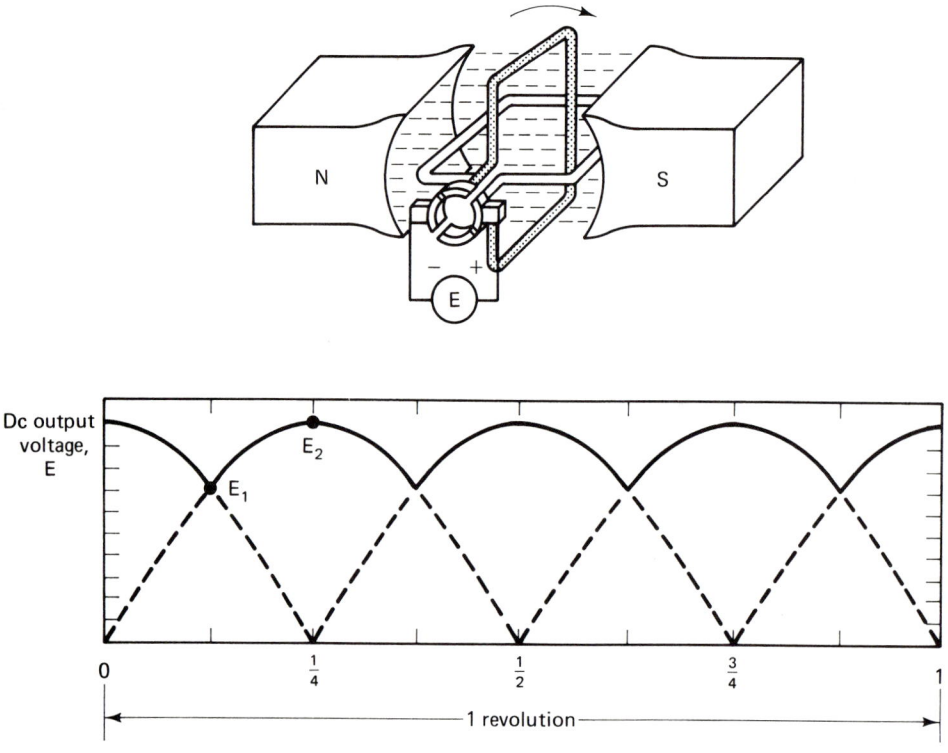

FIGURE 7-30 Dc output voltage generated by two coils.

variation in the output voltage is now restricted to the difference between the E_1 and E_2 levels. The ripple may obviously be further reduced by distributing more coils around the solid armature and splitting the commutator into a greater number of segments.

We must emphasize that we have only touched on the principle of the dc generator. For example, there has been no discussion on the ways in which the conductors may be wound on the armature and of the losses in the armature itself. Such a detailed discussion is best left to a text on electricity rather than one on electronics.

Chapter Summary

7.1. Like poles repel, unlike poles attract. The force between two poles is directly proportional to the product of the pole strengths but is inversely proportional to the square of their distance apart. The north pole of a magnet means the "north-seeking" pole.

7.2. A magnetic field is made up of lines of flux whose direction is from a north pole of a magnet toward the south pole. These lines indicate the strength of the field, cannot intersect, and mutually repel each other.

7.3. Magnetic flux, ϕ, is measured in webers (Wb).

$$\text{flux density, } B = \frac{\phi}{A} \quad \text{teslas (T)}$$

7.4. Magnetomotive force (MMF), $\mathscr{F} = IN$ ampere-turns (AT)

Magnetic field intensity or magnetizing force, H

$$= \frac{\text{MMF}}{\text{length of magnetic path}} = \frac{IN}{l} \quad \text{ampere-turns per meter (AT/m)}$$

or amperes per meter (A/m)

For free space (vacuum),

$$\frac{\text{flux density, } B \text{ (T)}}{\text{magnetic field intensity, } H \text{ (AT/m)}} = \mu_0$$

absolute permeability of free space, $\mu_0 = 4\pi \times 10^{-7} = 12.57 \times 10^{-7}$ SI units

7.5. Relative permeability, μ_r

$$= \frac{\text{flux density with material as the core}}{\text{flux density with a vacuum core}}$$

Flux density in the core material, $B = \mu_0 \mu_r H$ teslas

Absolute permeability of the core material, $\mu = \frac{B}{H} = \mu_0 \mu_r$ SI units

7.6. Rowland's law:

$$\frac{\text{magnetomotive force } (\mathscr{F})}{\text{flux}(\phi)} = \text{reluctance } (\mathscr{R})$$

$$= \frac{l}{\mu A} \quad \text{SI units}$$

7.7. Faraday's law (left-hand rule):
Single conductor:

$$\text{induced EMF, } E = -\frac{\text{change of flux, } \Delta\phi \text{ (Wb)}}{\text{change of time, } \Delta T \text{ (s)}}$$

$$= -\frac{\Delta\phi}{\Delta T} \quad \text{volts}$$

Coil of N turns:

$$\text{induced EMF, } E = -N\frac{\Delta\phi}{\Delta T} \quad \text{volts}$$

7.8. Motor effect (right-hand rule):

Force exerted on a current-carrying conductor, $F = BIl$ newtons

7.9. Comparison between electrical and magnetic units:

Electric Circuit: Ohm's law, resistance = EMF/current		Magnetic Circuit: Rowland's law, reluctance = MMF/flux	
Quantity	SI Unit	Quantity	SI Unit
EMF, E	Volt, V	MMF, $\mathscr{F}$	Ampere-turn (AT), or ampere (A)
Current, I	Ampere, A	Flux, ϕ	Weber, Wb
Current density	Ampere per square meter, A/m²	Flux density, B	Weber per square meter, Wb/m², or tesla, T
Resistance, R	Ohm, Ω	Reluctance, $\mathscr{R}$	Ampere-turn per weber, AT/Wb
Conductance, G	Siemens, S	Permeance, $\mathscr{P}$	Henry, H, or weber per ampere-turn, Wb/AT
Conductivity, σ	Siemens per meter, S/m	Permeability, μ	Henry per meter, H/m
Electric field intensity, $\mathscr{E}$	Volt per meter, V/m	Magnetic field intensity, H	Ampere-turn per meter, AT/m, or ampere per meter, A/m

Self-Review

True–False Questions

1. **T. F.** Magnetic lines of force mutually attract and tend to concentrate.
2. **T. F.** Due to the force of attraction, the north pole of a suspended bar magnet points toward the earth's south pole.
3. **T. F.** The magnetomotive force is measured in amperes.
4. **T. F.** For a particular specimen of soft iron, the ratio of the flux density to the magnetic field intensity is a constant.
5. **T. F.** Rowland's law states that MMF/flux density = reluctance.
6. **T. F.** The right-hand rule is used for the motor effect provided that the direction of the current is the direction of the electron flow.

7. **T. F.** The permeability of free space has a value of 2×10^{-7} SI unit.
8. **T. F.** If a flux of 50 mWb is cut by a conductor in 2 ms, the induced EMF is 25 V.
9. **T. F.** There is a force of attraction between two parallel conductors which are carrying currents in the same direction.
10. **T. F.** The hysteresis loop shows the relationship between the flux density and the magnetic field intensity for a ferromagnetic specimen.
11. **T. F.** Copper and silver are examples of ferromagnetic materials.
12. **T. F.** A magnetic flux line is the path that is followed by an electron when traveling from a north pole to a south pole.
13. **T. F.** Like magnetic poles repel each other, while unlike poles attract.
14. **T. F.** The right-hand rule is used to locate the poles of an electromagnet.
15. **T. F.** When the current is removed from an electromagnet, most of the magnetism disappears.
16. **T. F.** The voltage induced in a conductor is greatest when the motion of the conductor is parallel to the direction of the flux lines.
17. **T. F.** The SI unit of the flux density is the tesla per square meter.
18. **T. F.** When a current-carrying conductor is situated in a magnetic field, the force exerted on the conductor is proportional to the product of the current and the flux density.
19. **T. F.** The ratio of the magnetic flux to the magnetic field intensity is equal to the permeability.
20. **T. F.** The left-hand rule is used to find the direction of the electron flow in the generator effect.

Multiple-Choice Questions

21. The distance between a north pole and a south pole is halved. The force of the attraction between the poles is
    (a) halved   (b) doubled   (c) divided by 4
    (d) multiplied by 4   (e) multiplied by 8
22. A flux of 400 μWb is uniformly distributed over an area of 5 cm². The flux density is
    (a) 0.8 T   (b) 2 T   (c) 8 T   0.2 T   (e) 0.08 T
23. A conductor of length 40 cm is carrying a current of 3 A and is lying at right angles to the direction of a magnetic field whose flux density is 1.2 T. The force in newtons exerted on the conductor is
    (a) 9   (b) 1.44   (c) 3.6   (d) 90   (e) 2.88
24. A MMF of 500 AT establishes a flux of 200 μWb in a magnetic circuit. The reluctance of the circuit in SI units is
    (a) 1×10^5 (b) 3×10^6 (c) 4×10^{-7} (d) 2.5×10^6 (e) 1×10^6
25. A conductor of length 50 cm is moving with a velocity of 20 m/s at right angles to a magnetic field with a flux density of 1.6 T. The voltage induced in the conductor is
    (a) 16 V   (b) 8 V   (c) 4 V   (d) 80 V   (e) 20 V
26. A magnetic field intensity of 800 AT/m establishes a flux density of 2 T in a specimen of soft iron. The iron's absolute permeability in SI units is
    (a) 400   (b) 1600   (c) 0.0025   (d) 3200   (e) 2000
27. In Question 26, the iron's relative permeability is approximately
    (a) 0.0025   (b) 400   (c) 2000   (d) 1600   (e) 3200
28. A coil of 200 turns cuts a flux of 400 μWb in a time of 2 ms. The total EMF induced in the coil is
    (a) 80 V   (b) 20 V   (c) 40 V   (d) 8 V   (e) 200 V
29. An iron ring has a cross-sectional area of 2 cm² and a mean circumference of 40 cm. If the iron has an absolute permeability of 0.002, the reluctance of the ring in SI units is
    (a) 1×10^6 (b) 0.5×10^6 (c) 2.0×10^6 (d) 12.57×10^6
    (e) 8.0×10^5

30. A coil of 200 turns is wound on an iron ring with a mean circumference of 20 cm. If the current in the coil is 2 A, the magnetic field intensity in AT/m is
    (a) 1000   (b) 500   (c) 4000   (d) 2500   (e) 2000
31. A coil of 500 turns is wrapped on a magnetic core whose length is 10 cm. The current flowing through the coil is 0.4 A. The magnetomotive force is
    (a) 200 AT   (b) 0.4 A   (c) 2000 AT/m   (d) 20 AT   (e) 50 AT
32. If the flux density in the core of Question 31 is 0.2 T, the permeability in SI units of the core is
    (a) 2.5×10^{-3} (b) 1×10^{-4} (c) 5×10^{-5} (d) 2×10^{-4} (e) 1×10^{-3}
33. A magnetomotive force of 500 AT is applied to a magnetic circuit with a reluctance of 2×10^6 SI units. The flux established in the circuit is
    (a) 250 µWb   (b) 1 mWb   (c) 0.4 mWb   (d) 400 µWb   (e) 4 mWb
34. The flux density in a mild steel ring is 1.6 T. Referring to Fig. 7-21, the required magnetic field intensity is
    (a) 3500 AT/m   (b) 4750 AT/m   (c) 5500 AT/m   (d) 1600 AT/m
    (e) 2500 AT/m
35. In Question 34, refer to Fig. 7-22. The value of the ring's relative permeability is
    (a) 380   (b) 165   (c) 300   (d) 275   (e) 400
36. A long straight conductor is situated in air and is carrying a current of 200 A. If the return conductor is far away, the value of the magnetizing force at a radius of 10 cm is
    (a) 159 AT/m   (b) 628 AT/m   (c) 314 AT/m   (d) 318 AT/m   (e) 400 AT/m
37. In Question 36, the flux density at a radius of 10 cm from the conductor is
    (a) 4×10^{-3} T   (b) 2×10^{-4} T   (c) 3.18×10^{-4} T   (d) 4×10^{-4} T
    (e) 2×10^{-5} T
38. Two straight parallel conductors are situated in air with a spacing of 10 cm between their centers. If these conductors carry currents of 1000 A each in opposite directions, the value of the magnetic field intensity at a position midway on a straight line joining and at right angles to the two conductors is
    (a) 3180 A/m   (b) 1590 A/m   (c) 6370 A/m   (d) 3140 A/m   (e) 2000 A/m
39. In Question 38, the value of the flux density at a position 3 cm from one conductor on a straight line joining and at right angles to the two conductors is
    (a) 0.008 T   (b) 0.00951 T   (c) 0.00381 T   (d) 0.00318 T   (e) 0.00985 T
40. In Question 38, the direction of the current in one of the conductors is reversed. The value of the magnetic field intensity at a position 3 cm from one conductor on a straight line joining and at right angles to the two conductors is
    (a) 3030 A/m   (b) zero   (c) 7580 A/m   (d) 6370 A/m   (e) 1590 A/m

Practice Problems

41. An air gap has a length of 1 mm and a cross-sectional area of 5 cm². Calculate the reluctance of the air gap.
42. In Problem 41, what is the value of the MMF required to establish a flux of 600 µWb in the air gap?
43. A soft-iron ring has a relative permeability of 1200, a cross-sectional area of 2 cm², and a mean circumference of 15 cm. Calculate the reluctance of the ring.
44. In Problem 43, the ring is wound with 400 turns to establish a flux of 1 mWb in the iron. What is the value of the direct current that is flowing through the coil?
45. A magnetic field intensity of 1600 AT/m produces a flux density of 1.2 T in a specimen of iron. Under these conditions, what is the iron's absolute permeability?
46. A conductor cuts a magnetic flux of 800 µWb in a time of 0.02 ms. What is the value of the EMF induced in the conductor?
47. A conductor of length 40 cm is moving at 100 m/s through a magnetic field whose flux density is 0.9 T. If the conductor is cutting the flux lines at 90°, what is the value of the EMF induced in the conductor?
48. In Problem 47, the conductor is connected to a 12-Ω load. Calculate the value of the force exerted on the conductor.
49. Two radial cuts are made in an iron ring and the gaps are filled with nonmagnetic

material. Each gap requires an MMF of 2500 AT to establish the required flux while an MMF of 800 AT is necessary in order to develop the same flux in the iron. If the ring is wound with a coil of 500 turns, what is the current flowing in the coil (ignore any flux leakage)?

50. An iron ring has a mean circumference of 30 cm and is wound with a coil of 500 turns. When a current of 4 A is flowing through the coil, calculate the value of the magnetic field intensity. If the iron's relative permeability is 1200, what is the value of the flux density in the ring?

51. An air-cored inductor has a length of 0.18 m and a diameter of 6 cm. If this coil has 650 turns, what value of current is required to provide a magnetic field intensity of 4500 AT/m?

52. A coil with an air core consists of 1200 turns and is 35 cm long. If the flux density in the air core is 0.18 Wb/m^2, what is the value of the current flowing through the coil?

53. When a magnetomotive force of 550 AT is applied to a magnetic circuit, the resultant flux is 720 mWb. Calculate the value of the circuit's reluctance.

54. A cylindrical magnetic core has a diameter of 2.3 cm. If the flux established in the core is 2×10^{-4} Wb, determine the value of the flux density in teslas.

55. A magnetic circuit has a reluctance of 3400 SI units. If a magnetomotive force of 820 AT is applied to the circuit, calculate the value of the flux.

56. The dimensions of an air gap with a rectangular cross section are 2.4 cm long, 5.3 cm high, and 3.3 cm wide. If the flux across the air gap is 0.17 Wb, calculate the required value of the magnetic field intensity.

57. A metal toroidal ring with a square cross-sectional area has an internal radius of 8 cm and an external radius of 10 cm. A coil of 400 turns is wrapped around the core and carries a current of 3 A. If the flux established in the core is 670 μWb, what are the values of the magnetic field intensity and the core's relative permeability?

58. In Problem 57, a 1.5-mm air gap is cut into the toroid. Determine the increased value of the current required to maintain the original flux in the core.

59. A metal toroidal ring with a circular cross-sectional area has an internal diameter of 10 cm and an external diameter of 14 cm. A coil of 250 turns is wrapped around the ring and carries a current of 800 mA. If the metal's permeability is 7.0×10^{-4} SI units, determine the values of the ring's reluctance and the flux established in the ring.

60. A coil of wire consists of 10 rectangular turns whose dimensions are 25 cm long and 20 cm wide. The long dimension is at right angles to the direction of a magnetic field whose flux density is 0.6 T. If the current through the coil is 7 A, calculate the value in N·m of the torque that can be exerted on the coil about its axis of rotation.

Advanced Problems

61. Figure 7-31 shows a cast steel ring with a cross-sectional area of 5 cm^2 and a mean circumference of 45 cm. There are two radial cuts at diametrically opposite points and these cuts are filled with nonmagnetic material to a thickness of 1 mm. If

FIGURE 7-31 Circuit for Problem 61.

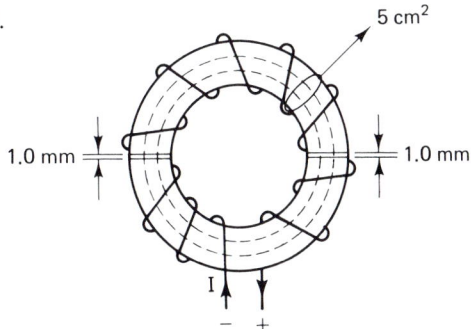

there is no magnetic leakage, calculate the total MMF required to establish a flux density of 1.5 T. (*Hint:* Use the information on cast steel in Fig. 7-21.)

62. Figure 7-32 shows a cast iron magnetic core whose center limb's cross-sectional area is 12 cm². If the cross-sectional area of each outside limb is 10 cm², calculate the total MMF required to establish a flux of 0.2 T in the air gap. Neglect any flux leakage. (*Hints:* This is a series–parallel magnetic circuit. Use the information on cast iron in Fig. 7-21.)

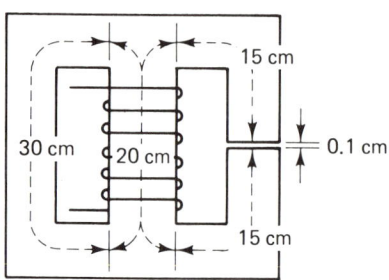

FIGURE 7-32 Circuit for Problem 62.

63. A cast iron ring has a mean circumference of 40 cm and a cross-sectional area of 5 cm². What is the current required in a coil of 250 turns to establish a total flux of 200 μWb? (*Hint:* Use the information on cast iron in Fig. 7-21.)

64. An iron ring has a mean circumference of 50 cm and a uniform cross-sectional area of 4 cm². When the flux density in the iron is 1.3 T, its relative permeability is 1100. For this particular value of flux density, calculate the reluctance of the ring and the required MMF.

65. A coil of 400 turns carries a current of 5 A and is uniformly wound over a non-magnetic toroidal ring with a mean circumference of 30 cm and a cross-sectional area of 2.5 cm². Calculate the values of the magnetic field intensity, the flux density, and the total flux.

66. The length of a loop is 40 cm and its width is 20 cm. If the loop is rotated at 2000 rpm in a uniform magnetic field whose flux density is 1.2 T, calculate the maximum value of the loop's output voltage.

67. A copper loop has a length of 20 cm and a width of 10 cm. The loop carries a current of 4 A and is situated in a uniform magnetic field whose flux density is 1.5 T. Calculate the maximum torque on the loop in newton-meters.

68. During an experiment to obtain the magnetization curve of a ferromagnetic specimen it is found that a field intensity of 2000 AT/m produces a flux density of 2.5 T. What is the specimen's relative permeability at this point on the curve?

69. In Fig. 7-33 the soft-iron magnetic circuit has a cross-sectional area of 5 cm² and the mean length of the magnetic path is 40 cm. If the relative permeability of the soft iron is 900, calculate the current, I, required to establish a flux of 800 μWb.

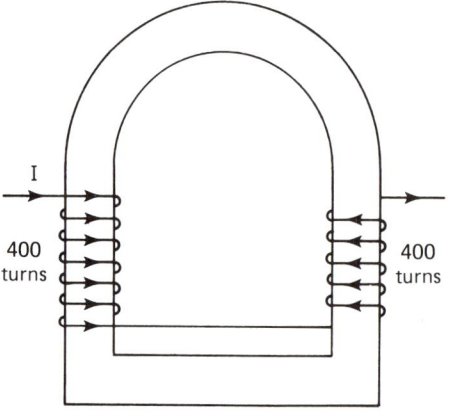

FIGURE 7-33 Circuit for Problem 69.

70. Figure 7-34 shows a cast steel ring that is made in two sections. It is required to establish a flux density of 1.0 T in the section whose mean magnetic path is 20 cm long. Calculate the total MMF required to establish the same flux in both sections. (*Hints:* Ignore any flux leakage. Use the information on cast steel in Fig. 7-21.)

FIGURE 7-34 Circuit for Problem 70.

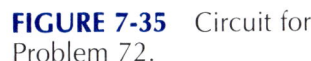

71. The cylindrical core of a coil has a flux of 74 mWb and a flux density of 1.5 Wb/m². What is the diameter of the core?

72. Figure 7-35 illustrates the dimensions of a nonmagnetic toroidal core which is completely wound with a coil of thin wire. If a current of 80 mA flows through the coil and establishes a flux density of 0.01 T, calculate the required number of turns.

FIGURE 7-35 Circuit for Problem 72.

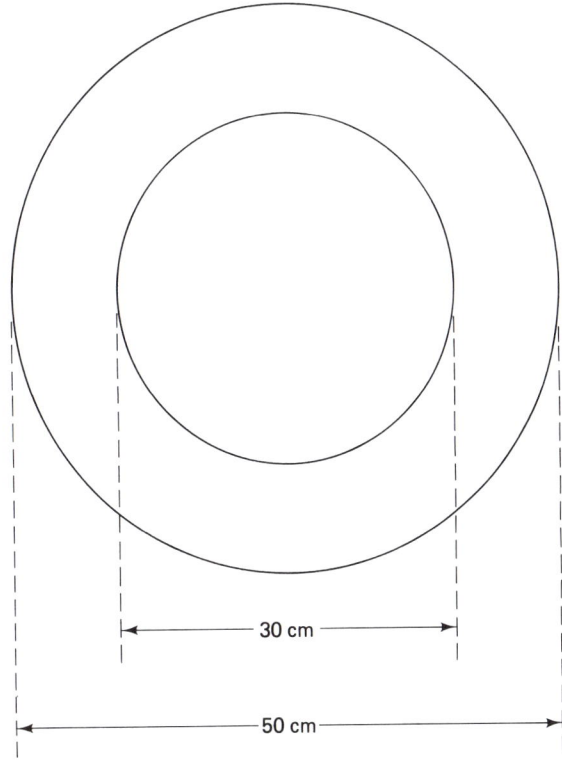

73. In Fig. 7-36, a current of 340 mA flows through the coil. What is the value in teslas of the flux density established in the nonmagnetic cylindrical core?

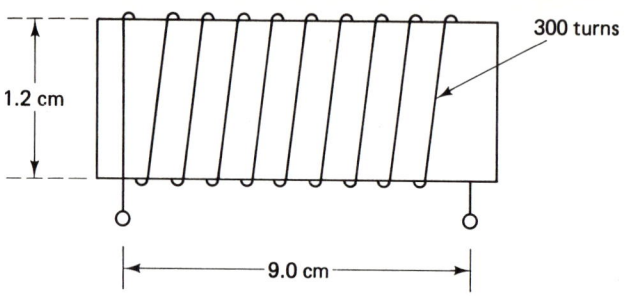

FIGURE 7-36 Circuit for Problem 73.

74. In the magnetic circuit of Fig. 7-37, the flux density in the X section is 0.6 T. What is the value of the flux density in the Y section?

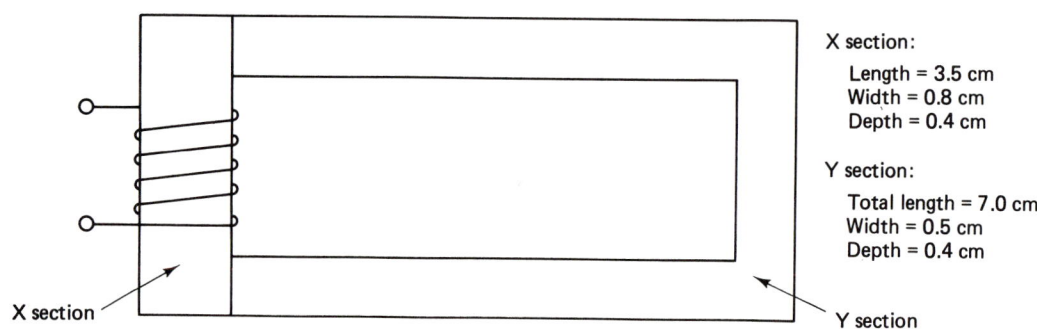

FIGURE 7-37 Circuit for Problem 74.

75. Figure 7-38 shows a cast steel toroidal ring in which the flux density is 0.95 T. If the current flowing through the coil is 800 mA, what is the number of turns wound on the ring?

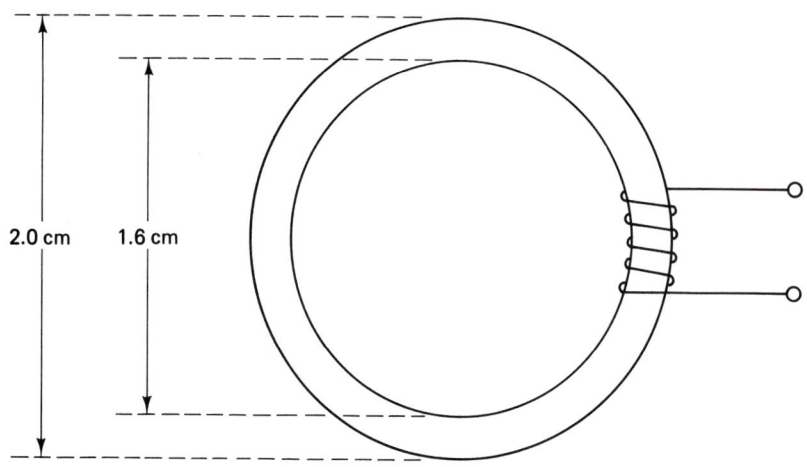

FIGURE 7-38 Circuit for Problem 75.

76. In Fig. 7-39, there are 900 turns, which are wound on the center limb of the cast iron specimen. If the current is 570 mA, what is the flux density in the center limb?

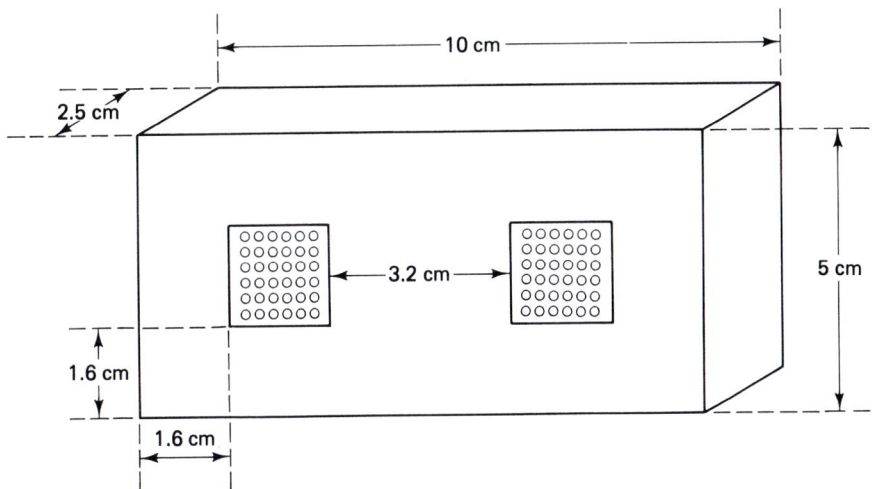

FIGURE 7-39 Circuit for Problem 76.

77. In the series magnetic circuit of Fig. 7-40, the flux across the air gap is 58 μWb. Calculate the value of the magnetic field intensity that is required to establish this value of flux.

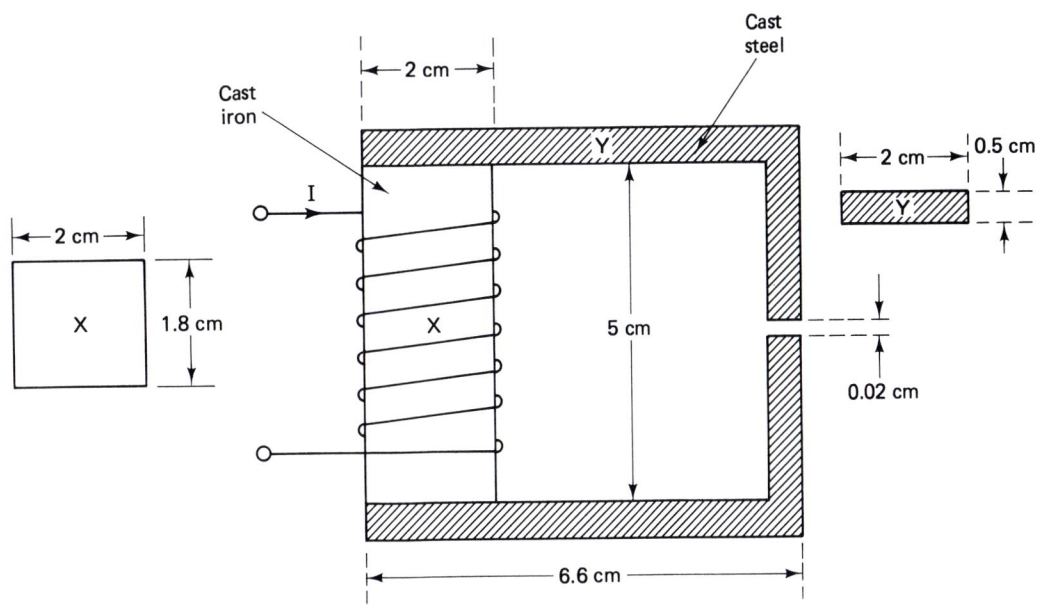

FIGURE 7-40 Circuit for Problem 77.

78. In the circuit of Fig. 7-41, calculate the value of the magnetic flux.

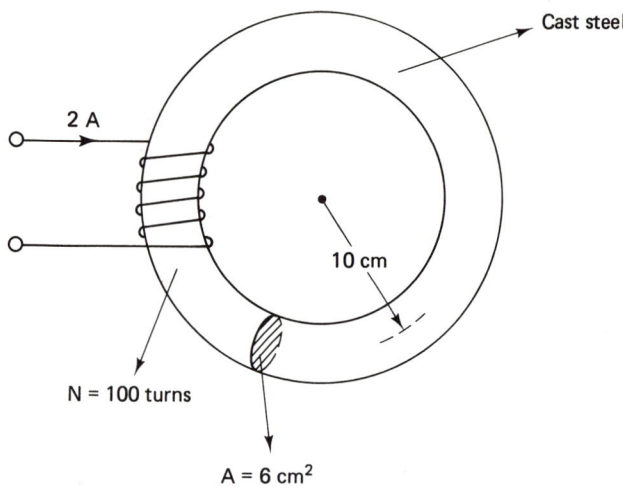

FIGURE 7-41 Circuit for Problem 78.

79. The flux in the air gap of Fig. 7-42 is 100 μWb. Determine the value of the current, I_2.

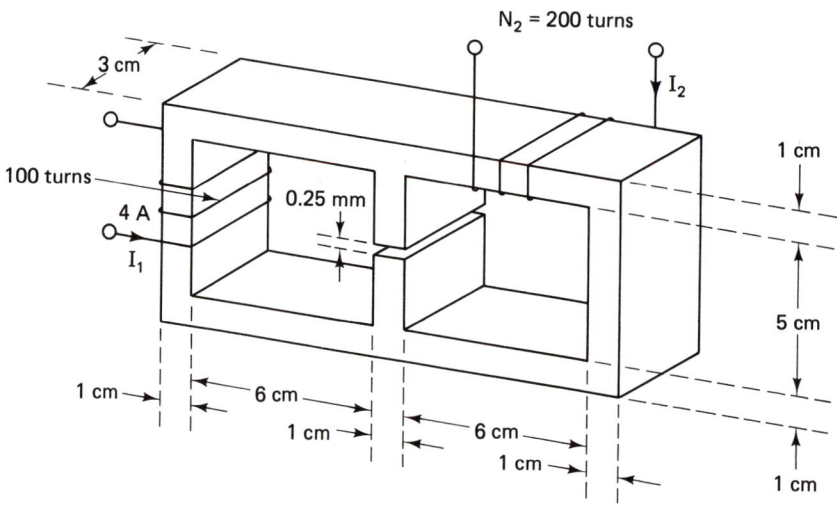

FIGURE 7-42 Circuit for Problem 79.

80. In Fig. 7-43, the flux in the wrought iron is 700 μWb. How many turns of wire are wound on the cast steel section? Assume that there is no flux leakage at the transition between the wrought iron and the cast steel.

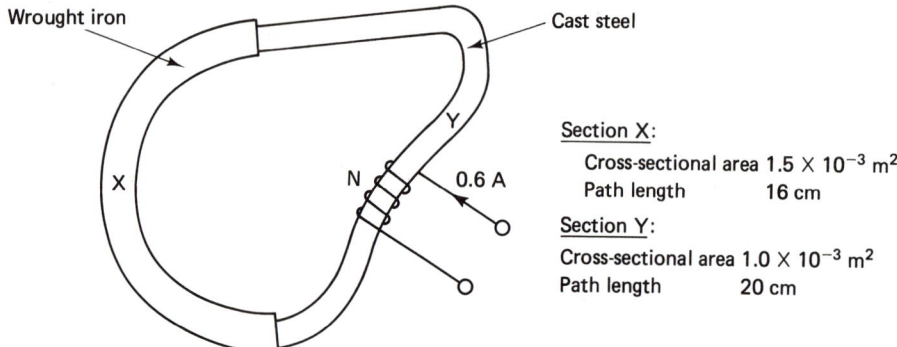

Section X:
 Cross-sectional area 1.5×10^{-3} m^2
 Path length   16 cm

Section Y:
 Cross-sectional area 1.0×10^{-3} m^2
 Path length   20 cm

FIGURE 7-43 Circuit for Problem 80.

Inductance in DC Circuits

In studying the effects of self-inductance, you will learn:

1. how to compare the electrical properties of resistance and inductance.
2. the definition of the unit of inductance.
3. which physical factors determine a coil's self-inductance.
4. about the reasons for using the various types of inductors.
5. how to calculate the amount of energy stored in the magnetic field surrounding an inductor.
6. what occurs when a number of inductors are connected in series.
7. what occurs when a number of inductors are connected in parallel and series–parallel.
8. about the concept of the time constant, L/R seconds, and its relation to the duration of the transient state.
9. how to troubleshoot for defects in inductors.

8-0

Introduction

In previous chapters we have examined the property of resistance, which opposes and therefore limits the flow of current. However, as well as resistance, a circuit in general may possess two other properties: inductance and capacitance (Chapter 9). This chapter is devoted to a study of self-inductance and its effect in a dc circuit. Inductance is defined as that electrical property which prevents any sudden or abrupt *change* in current and limits the rate of change of current. In other words, if an electrical circuit contains inductance, the current in that circuit cannot rise or fall instantaneously.

It is worth mentioning that the mechanical equivalent of inductance is mass. When a force is applied to a mass, the mass accelerates from rest, so that its velocity cannot change instantaneously. In the same way a moving mass cannot stop instantaneously but will decelerate toward rest.

A straight piece of wire possesses some inductance, but the property is more marked in a coil, which is called an inductor. As we saw in Chapter 7, a current flowing through a coil creates a magnetic field around the coil. Consequently, inductance may also be regarded as that property which draws energy from a source and *stores* it in the form of a magnetic field; by contrast, resistance takes energy from a source and *dissipates* it in the form of heat.

If the current increases or decreases in a coil, the magnetic field will expand or collapse and will cut the turns of the coil itself. On the principle of Faraday's law involving electromagnetic induction, a voltage will be induced into the coil and this voltage will depend on the *change* of the current. Therefore, due to the current flow, the coil creates its own flux, which cuts its own turns and induces the voltage across the coil—this is the reason for referring to this property of the coil as its *self*-inductance, with L as the letter symbol.

8-1

Comparison Between Electrical Properties of Resistance and Inductance

It is extremely important to understand the differences between the properties of resistance and inductance. In Fig. 8-1 we have a resistor and an inductor (coil), connected in parallel across a dc source. The circuit symbol for the inductor is ⎯⎯⎮⎮⎮⎯⎯. We are assuming that the material from which the coil is made has negligible resistance, so that the coil only has the property of inductance. By contrast, a practical coil would possess both inductance and resistance.

As soon as the switch, S, is closed, the current in the resistor immediately rises from zero to a constant value of $I_R = E/R$ amperes, so that there has been an instantaneous change in the current; with the values shown, $I_R = 15$

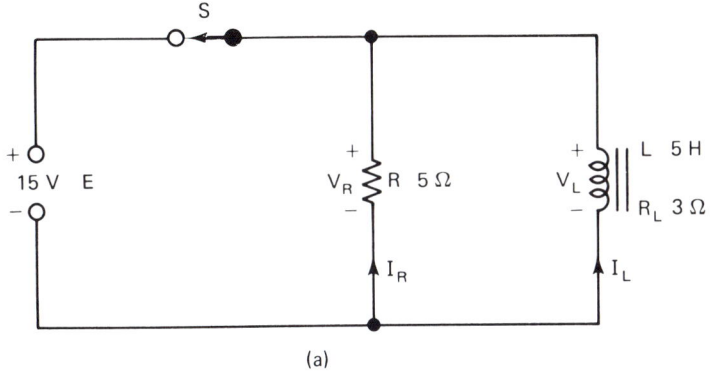

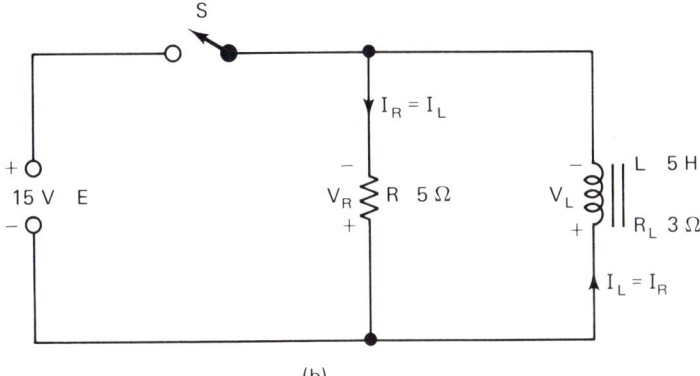

FIGURE 8-1 Comparison between the properties of resistance and inductance.

V/5 Ω = 3 A [Fig. 8-2(a)]. The voltage drop across R, V_R (15 V) must exactly balance (oppose) the source voltage, E (15 V). We can therefore write that

$$V_R = -I_R \times R \qquad (8\text{-}1\text{-}1)$$

The minus sign is used to indicate that the polarity of V_R is opposing the polarity of E when we consider the complete loop. By Kirchhoff's voltage law (KVL),

$$E + V_R = 0 \qquad (8\text{-}1\text{-}2)$$

Combining Eqs. (8-1-1) and (8-1-2) leads to

$$E = I_R \times R \qquad (8\text{-}1\text{-}3)$$

which is the normal Ohm's law equation.

With the inductor the situation is totally different. As soon as S is closed, the current, I_L, starts to grow in the coil. This creates a magnetic flux which, as it expands outward, cuts the turns of the coil itself. By Faraday's law, an EMF, V_L, will be self-induced into the inductor and this voltage, like V_R, must exactly balance (oppose) the source voltage, E. Since the polarity of V_L opposes the polarity of E around the loop, it is common to refer to V_L as a *counter EMF*.

If the value of I_L were constant, the magnetic flux would be stationary and would not cut the turns of the coil; consequently, the magnitude of V_L

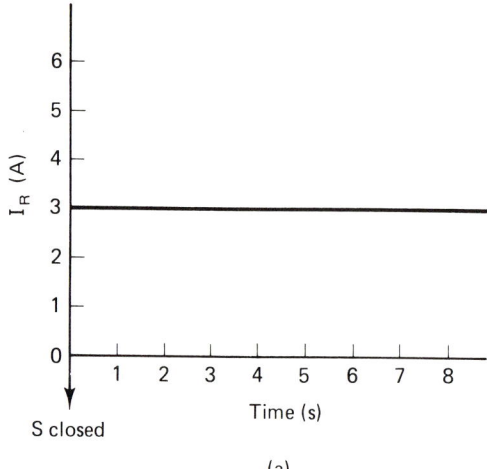

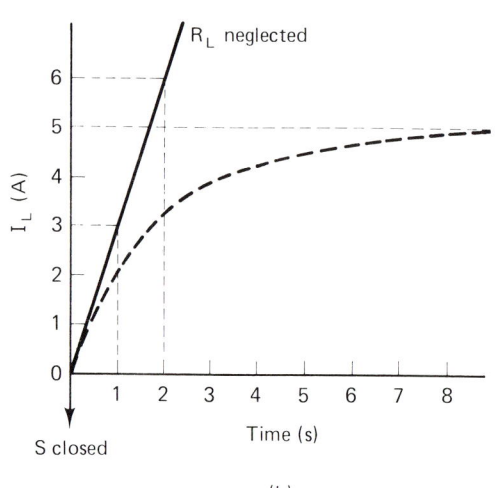

FIGURE 8-2 Variation of I_R (a) and I_L (b) with time in the circuit of Fig. 8-1.

would be zero. The value of V_L that is necessary to balance the source voltage can only be achieved by a varying flux that will require a changing current. This is an illustration of Lenz's law (Section 7-7) since the polarity of the induced EMF is such as to oppose the change producing the EMF, namely the change in the current. Since the values of V_L and E are fixed, the current must start at zero and then grow at a constant rate. This rate of current growth would be measured in amperes per second, so that if, for example, the rate of current growth was 3 A/s, the current after 1 s would be 3 A, after 2 s 6 A, after 3 s 9 A, and so on. Remember that the inductor's current, I_L, is zero before the switch S is closed and is still zero just after S is closed; this means that unlike the resistor current, I_R, the inductor current, I_L, does not change instantaneously.

Although the counter EMF must be proportional to the rate of change of the current, V_L must also depend on some factor associated with the coil. This factor is called the *inductance*, L, which is determined by the coil's number of turns, nature of the core, cross-sectional area, and length (Section 8-2). In equation form,

$$V_L = -L \times \frac{\Delta I_L}{\Delta T} \tag{8-1-4}$$

where V_L = counter EMF, with the negative sign indicating that the polarity of V_L opposes the polarity of E

L = coil's inductance, which is measured in henrys (H); this unit is named after Joseph Henry (1797–1878)

$\dfrac{\Delta I_L}{\Delta T}$ = rate of change of the current (A/s)

Remember that the symbol "Δ" means "a change in," so that ΔI_L is the change in the inductor current and ΔT is the corresponding change in time. For example, if, from the time S is closed, the values of I_L after 3 s and 5 s are, respectively, 9 A and 15 A, then $\Delta I_L = 15 - 9 = 6$ A and $\Delta T = 5 - 3 = 2$ s. The rate of change of the current, $\Delta I/\Delta T = 6$ A/2 s $= 3$ A/s. Equations (8-1-1) and (8-1-4) should be carefully compared.

$$V_R = -R \times I_R, \qquad V_L = -L \times \frac{\Delta I_L}{\Delta T}$$

The voltage across a resistor is directly proportional to the current flowing through the resistor, but the voltage across an inductor is directly proportional to its rate of change of current.

Using KVL in the loop containing the inductor and the source voltage yields

$$E + V_L = 0 \qquad (8\text{-}1\text{-}5)$$

Combining Eqs. (8-1-4) and (8-1-5), we get

$$E = L \times \frac{\Delta I_L}{\Delta T} \qquad (8\text{-}1\text{-}6)$$

We can use this equation to define the henry unit. The inductance is 1 H if when the current is changing at the rate of 1 A/s, the induced or counter EMF is 1 V.

Rearranging Eq. (8-1-6), we obtain

$$\frac{\Delta I_L}{\Delta T} = \frac{E}{L \text{ (HENRYS)}} \approx \text{A/s} \qquad (8\text{-}1\text{-}7)$$

and

$$L = \frac{E}{\Delta I_L / \Delta T} \quad \text{henrys} \qquad (8\text{-}1\text{-}8)$$

Referring to the values shown in Fig. 8-1, $\Delta I_L/\Delta T = 15$ V/5 H $= 3$ A/s. The current starts at zero and then grows at the rate of 3 A/s. The graph of the inductor's current variation with time is shown in Fig. 8-2(b). It appears that the current is increasing toward infinity, and that obviously is impossible. However, you must remember that we have neglected the coil's resistance, R_L. In practice, the inductor current will take an appreciable time to reach a final constant value of E/R_L amperes (this so-called "transient state" is discussed more fully in Section 8-5); with the values shown, the inductor current eventually levels off at 15 V/3 Ω = 5 A [Fig. 8-2(b)], and the final total current

$I_T = 5 + 3 = 8$ A. When the steady-state conditions have been reached, the magnetic flux is stationary and the counter EMF is zero; the source voltage is then opposed by the voltage drop across the coil's resistance.

When the inductor current has reached its final steady value, let us see what happens when S is opened (Fig. 8-1(b)). Initially, there is no change in the inductor's current, but as I_L starts to decay, the magnetic flux begins to collapse and will induce a counter EMF whose polarity will oppose the decrease in I_L. This is another illustration of Lenz's law (Section 7-7), which states that the polarity of the induced EMF is such as to oppose the change producing the EMF. The return path for the inductor current is now through the resistor, so that the current is the same through both components and will again take an appreciable time to decay to zero.

If a dc source is suddenly switched across an inductor and the current reaches its final steady value, what happens if the switch is opened and there is no resistor through which the inductor's current can flow? In this case there will be a rapid collapse of the magnetic flux, which can induce a counter EMF many times greater than the original dc source voltage. The result is an inductive flash (arc) across the contacts of the switch.

To summarize, the inductance is defined as that circuit property which opposes any abrupt change in the current, whether it be growth or decay, and limits the rate of change of current. If a dc voltage is suddenly switched across an inductor with negligible resistance, the current will start at zero and then grow at a constant rate which is measured in amperes per second. The value of the inductance, L, is measured in henrys and will determine the rate at which the current changes.

EXAMPLE 8-1 A 10-V dc source is suddenly switched across a 2-H inductor whose resistance is negligible. What are the value of the initial current, the initial rate of growth of current, the counter EMF, and the current after 0.5 s?

Since the inductor opposes any sudden change of current, the initial current is zero.

The initial growth is $E/L = 10$ V/2 H $= 5$ A/s [Eq. (8-1-7)].

If the resistance is negligible, the counter EMF will exactly balance the source voltage and must therefore equal 10 V, [Eq. (8-1-5)].

The inductor current after 0.5 s is 5 A/s $\times$ 0.5 s $= 2.5$ A.

EXAMPLE 8-2 In a time of 50 ms, the current flowing through a coil changes from 3 A to 5 A. If the coil's self-inductance is 2.5 H, what is the value of the induced EMF?

Solution

$$\text{change in the current, } \Delta I = 5\,\text{A} - 3\,\text{A} = 2\,\text{A} \qquad (8\text{-}1\text{-}4)$$

corresponding change in the time, $\Delta T = 50 \times 10^{-3}$ s

$$\text{value of the induced EMF} = L \times \frac{\Delta I}{\Delta T}$$

$$= 2.5\,\text{H} \times \frac{2\,\text{A}}{50 \times 10^{-3}\,\text{s}} = 100\,\text{V}$$

EXAMPLE 8-3 The counter EMF induced in a coil is 50 V when the current changes by 20 mA in a time of 4 μs. What is the value of the coil's self-inductance?

Solution

$$\text{self-inductance, } L = \frac{E}{\Delta I_L/\Delta T} \qquad (8\text{-}1\text{-}8)$$

$$= \frac{50 \text{ V}}{20 \times 10^{-3} \text{ A}/(4 \times 10^{-6} \text{ s})}$$

$$= \frac{50 \times 4 \times 10^{-6}}{20 \times 10^{-3}} = 0.01 \text{ H} = 10 \text{ mH}$$

8-2 Physical Factors That Determine a Coil's Self-Inductance. Types of Inductors

Consider a long coil of N turns with a cross-sectional area of A square meters and a length of l meters (Fig. 8-3). We assume that the current in this coil changes by ΔI amperes in a corresponding time change of ΔT seconds. As shown, the coil is uniformly wound on a magnetic core with a relative permeability, μ_r (Section 7-5). Regarding each turn as a conductor and using Eq. (7-7-2) for Faraday's law in conjunction with Eq. (8-1-6), the source voltage is given by

$$E = N\frac{\Delta \phi}{\Delta T} = L\frac{\Delta I}{\Delta T} \qquad (8\text{-}2\text{-}1)$$

where $\Delta \phi$ is measured in webers and is the change of flux produced by the

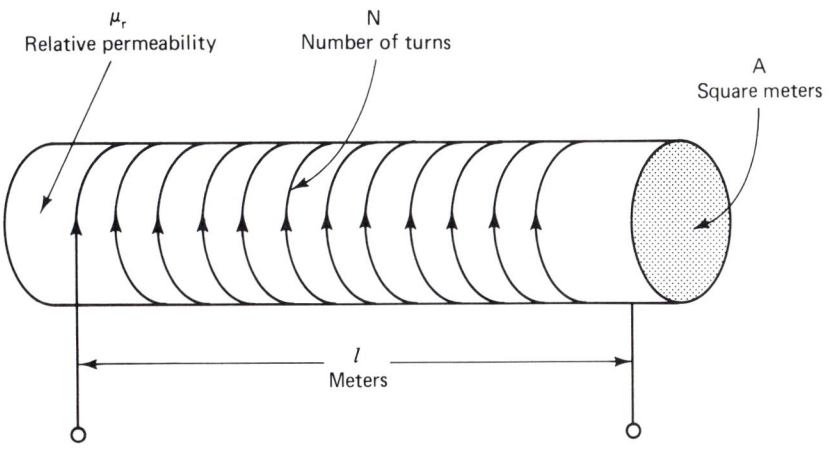

FIGURE 8-3 Factors determining the self-inductance of a coil.

current change, ΔI. Rearranging Eq. (8-2-1), yields

$$L = N\frac{\Delta \phi}{\Delta I} = \frac{\text{change in flux linkages}}{\text{change in current}} \quad (8\text{-}2\text{-}2)$$

The term "flux linkages" means the product of the flux and the number of turns through which the flux passes or with which the flux is linked. We can then define the henry unit in another way; a coil's inductance is 1 H if a current change of 1 A creates a change in flux linkages of 1 Wb-turn.

Using Eq. (7-2-2), $\phi = BA$, where B is the flux density in the core and is measured in teslas. Then $\Delta\phi = A \times \Delta B$, where ΔB is the change in the flux density due to the change in the flux, $\Delta\phi$. Substituting for $\Delta\phi$ in Eq. (8-2-2), we get

$$\begin{aligned} L &= N\frac{\Delta\phi}{\Delta I} \\ &= NA \times \frac{\Delta B}{\Delta I} \end{aligned} \quad (8\text{-}2\text{-}3)$$

From Eq. (7-5-2), $B = \mu_0\mu_r H$; therefore, $\Delta B = \mu_0\mu_r\Delta H$, where ΔH, in ampere-turns per meter, is the change in the magnetic field intensity associated with the change, ΔB, in the flux density. As discussed in Section 7-4, μ_0 is the absolute permeability of free space and has a value of $4\pi \times 10^{-7} = 12.57 \times 10^{-7}$ SI units. Substituting for ΔB in Eq. (8-2-3), we get

$$L = NA \times \frac{\Delta B}{\Delta I} = NA \times \mu_0\mu_r\frac{\Delta H}{\Delta I} \quad (8\text{-}2\text{-}4)$$

Equation (7-4-1) states that $H = NI/l$, so that $\Delta H = N \times \Delta I/l$, where ΔH is the change in the magnetic field intensity produced by the change in the current, ΔI. Substituting for ΔH in Eq. (8-2-4), we obtain

$$\begin{aligned} L &= NA \times \mu_0\mu_r\frac{\Delta H}{\Delta I} \\ &= NA \times \mu_0\mu_r \times \frac{N}{l} \times \frac{\Delta I}{\Delta I} \\ &= \frac{\mu_0\mu_r N^2 A}{l} \\ &= 12.57 \times 10^{-7} \times \frac{\mu_r N^2 A}{l} \quad \text{henrys} \end{aligned} \quad (8\text{-}2\text{-}5)$$

The self-inductance is therefore directly proportional to the square of the number of turns, the cross-sectional area, and the core's relative permeability, but is inversely proportional to the coil's length.

For a coil with a nonmagnetic core such as air or plastic, the value of μ_r is virtually 1, so that the equation for the self-inductance is

$$L = 12.57 \times 10^{-7} \times \frac{N^2 A}{l} \quad \text{henrys} \quad (8\text{-}2\text{-}6)$$

The values of the inductors used in electronics range from henrys to microhenrys. Those with a value of several henrys are referred to as *chokes* [Fig. 8-4(a)]; they are manufactured with a large number of copper turns which are wound on an iron core; the two lines to the side of the coil symbol indicate the core's presence. The physical size of a choke is determined primarily by its current rating since the current develops the magnetic field, which represents the energy drawn from the source. By contrast, you will recall that the physical size of a resistor is governed mainly by its power rating.

Although a choke's inductance cannot normally be varied by external means, the value of L does depend on the relative permeability μ_r, which varies with the amount of direct current flowing through the coil (Section 7-5). The fixed value of L is therefore normally quoted for a particular dc current level.

At the other end of the scale, a coil with a value of a few microhenrys is made up of a number of turns which are wound on a nonmagnetic coil [Fig. 8-4(b)]. The inductance may then be varied by means of an adjustable tap which is in contact with the turns of the coil. As you will learn in Chapter 14, one of the principal uses of an inductor occurs in the selection of a radio station or a TV channel; for example, each of channels 2 through 13 requires a separate inductor whose value is of the order of microhenrys. The fixed values of some small inductors may be indicated by a color code which is described in Appendix B.

In the AM broadcast band the inductor's value is a few hundred microhenrys and is therefore of the order of millihenrys [Fig. 8-4(c)]. Such an inductor uses an iron-dust core in which granules are insulated from each other and then compressed to form a solid *slug*. The value of the inductance may be increased by inserting the slug deeper into the coil. In the circuit symbol the dashed lines represent the powdered iron core; if the slug is movable in order to vary the inductance, an arrow is shown at the end of the dashed lines.

Another type of core, which has the same symbol as powdered iron, is a ceramic material known as ferrite. The magnetic properties of ferrite are similar to those of iron, but unlike iron, ferrite is an insulator and this helps to reduce one type of energy loss associated with the core (Section 15-1).

We have learned that the value of a coil's inductance depends on the square of the number of turns, the cross-sectional area, the length, and the nature of the core; by controlling these factors a range of inductance values may be manufactured. The use of a powdered-iron or ferrite slug allows the inductance to be varied.

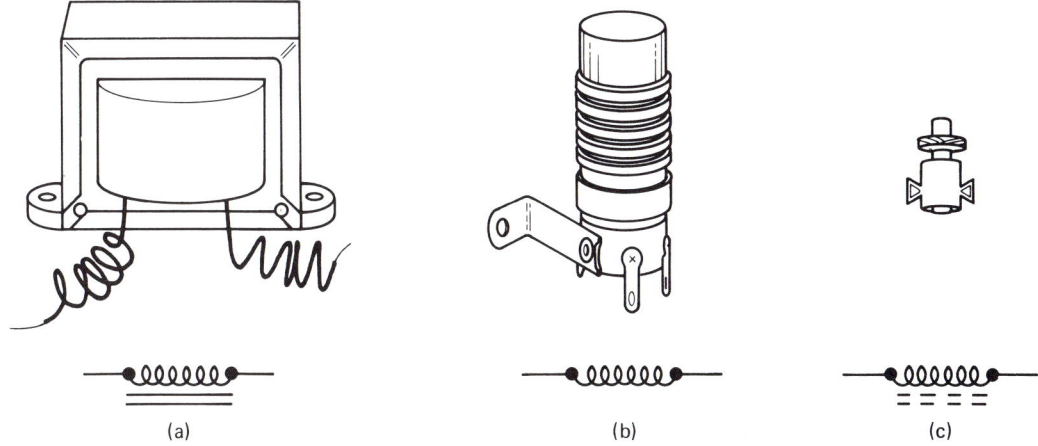

FIGURE 8-4 Types of inductor with their schematic symbols: (a) iron core inductor or choke; (b) air core inductor; (c) powdered iron core inductor. (Photographs from Abraham Marcus and Charles Thomson, *Electricity for Technicians*, Third Edition, Prentice Hall, Englewood Cliffs, N.J., 1982, p. 172.)

EXAMPLE 8-4 A coil of 400 turns is wound on a nonmagnetic core. If a current of 5 A flowing through the coil establishes a flux of 250 μWb, what is the inductance of the coil and the average value of the induced EMF if the current is reversed in 20 ms?

Solution

$$\text{inductance, } L = N \times \frac{\Delta\phi}{\Delta I} \tag{8-2-2}$$

$$= \frac{400 \times 250 \times 10^{-6} \text{ Wb}}{5 \text{ A}}$$

$$= 20 \times 10^{-3} \text{ H} = 20 \text{ mH}$$

If the flux is reversed, the *change* in the flux $= 2 \times 250 \times 10^{-6}$ Wb.

$$\text{average induced EMF, } E = N \times \frac{\Delta\phi}{\Delta T} \tag{8-2-1}$$

$$= \frac{400 \times 2 \times 250 \times 10^{-6} \text{ Wb}}{20 \times 10^{-3} \text{ s}} = 10 \text{ V}$$

Alternatively,

$$E = L \times \frac{\Delta I}{\Delta T}$$

$$= 20 \times 10^{-3} \text{ H} \times \frac{2 \times 5 \text{ A}}{20 \times 10^{-3} \text{ s}} = 10 \text{ V}$$

EXAMPLE 8-5 A coil is wound with 600 turns on an air core. If the length of the coil is 5 cm and the cross-sectional area is 4 cm², calculate the value of the coil's inductance. If an iron core is inserted with a relative permeability of 900, what is the new inductance value?

Solution

$$\text{length of coil, } l = 5 \text{ cm} = 5 \times 10^{-2} \text{ m}$$

$$\text{cross-sectional area of coil, } A = 4 \text{ cm}^2 = 4 \times 10^{-4} \text{ m}^2$$

$$\text{inductance, } L = 12.57 \times 10^{-7} \times \frac{N^2 A}{l} \tag{8-2-6}$$

$$= 12.57 \times 10^{-7} \times \frac{600^2 \times 4 \times 10^{-4}}{5 \times 10^{-2}} \text{ H}$$

$$= 3.62 \text{ mH}$$

When the iron core is inserted, the new inductance is 900×3.62 mH $= 3.26$ H [by Eq. (8-2-5)].

8-3 Energy Stored in Magnetic Field of Inductor

The magnetic field surrounding an inductor represents a form of energy that is developed by the current flowing through the coil. This energy has been delivered by the source from which current is drawn. In Fig. 8-5(a) we assume that the coil has negligible resistance but has a constant inductance of L henrys. After the switch, S, is closed, the current grows at a constant rate from zero to I amperes in a time of t seconds [Fig. 8-5(b)]. The rate of growth of current is I/t amperes per second, so that the source voltage, E, is $L \times I/t$ volts. The average current is $I/2$, so that:

$$\begin{aligned}
\text{average power delivered from the source to the magnetic field} \\
= E \times \text{average current} \\
= L \times \frac{I}{t} \times \frac{I}{2} \\
= \frac{1}{2} \frac{LI^2}{t} \quad \text{watts}
\end{aligned} \quad (8\text{-}3\text{-}1)$$

The total energy stored in the magnetic field must be equal to the product of the average power and the time interval. Therefore,

$$\begin{aligned}
\text{total energy stored} &= \frac{1}{2} \frac{LI^2}{t} \times t \\
&= \frac{1}{2} LI^2 \quad \text{joules}
\end{aligned} \quad (8\text{-}3\text{-}2)$$

When the switch, S, is opened, the magnetic field collapses rapidly and a large counter EMF is induced in the coil. The energy that had been previously stored in the magnetic field is now dissipated in the arc which occurs between the contacts of the switch.

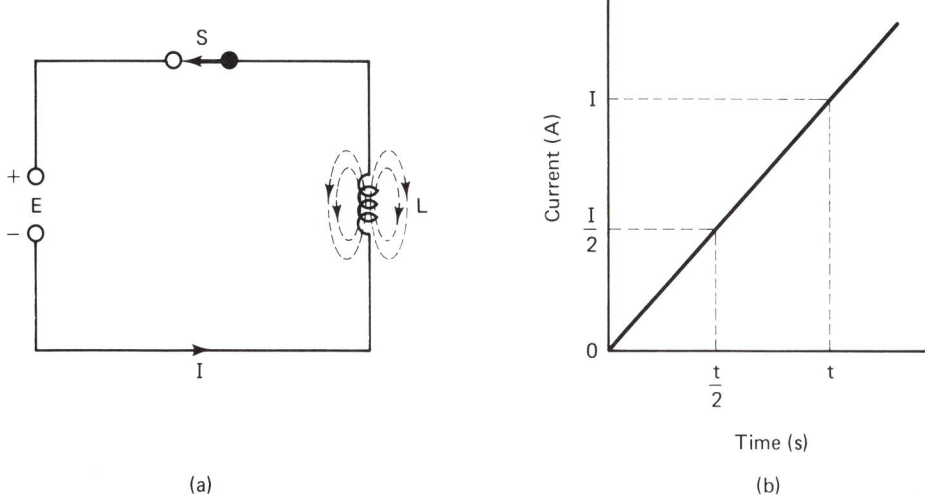

FIGURE 8-5 Energy stored in a magnetic field.

EXAMPLE 8-6 A coil with an inductance of 5 H and a resistance of 30 Ω is suddenly switched across a 90-V source. Find (a) the initial rate of growth of the current, (b) the value of the final steady current, and (c) the energy stored in the magnetic field when the final current is flowing through the coil.

Solution (a) From Eq. (8-1-7), the initial rate of growth of the current is

$$\frac{E}{L} = \frac{90 \text{ V}}{5 \text{H}} = 18 \text{ A/s}$$

(b) The final steady current is

$$\frac{E}{R_L} = \frac{90 \text{ V}}{30 \text{ Ω}} = 3 \text{ A}$$

In looking at the answers to (a) and (b) you must realize that the current reaches its final value in less than 1 s and that the rate of growth of the current is initially at its maximum value but subsequently decreases.

(c) The final energy stored in the magnetic field [Eq. (8-3-2)] is

$$\frac{1}{2} LI^2 = \frac{1}{2} \times 5 \text{ H} \times (3 \text{ A})^2 = 22.5 \text{ J}$$

8-4 Inductors in Series and Parallel

Series Arrangement

Figure 8-6 represents N inductors which are joined end to end and are therefore in series. The same rate of growth of the current, $\Delta I/\Delta T$, is associated with each inductor, so that

$$V_1 = -L_1 \frac{\Delta I}{\Delta T}, \quad V_2 = -L_2 \frac{\Delta I}{\Delta T}, \quad V_3 = -L_3 \frac{\Delta I}{\Delta T}, \quad \ldots, \quad V_N = -L_N \frac{\Delta I}{\Delta T}$$

From Kirchhoff's voltage law,

$$E + V_1 + V_2 + V_3 + \cdots + V_N = 0$$

These equations lead to

$$E = \frac{\Delta I}{\Delta T}(L_1 + L_2 + L_3 + \cdots + L_N) \qquad (8\text{-}4\text{-}1)$$

If L_T is the total equivalent self-inductance,

$$E = \frac{\Delta I}{\Delta T} \times L_T \qquad (8\text{-}4\text{-}2)$$

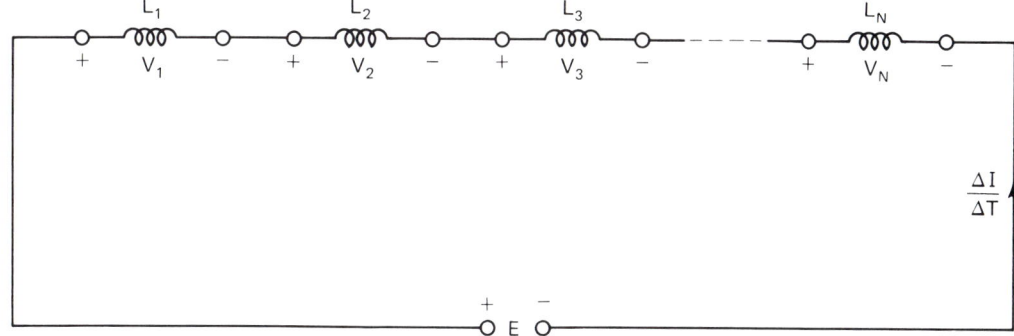

FIGURE 8-6 Inductors in series.

Therefore,

$$L_T = L_1 + L_2 + L_3 + \cdots + L_N \qquad (8\text{-}4\text{-}3)$$

If the N inductors all have the same inductance, L,

$$L_T = NL \qquad (8\text{-}4\text{-}4)$$

The formulas for series inductors and for series resistors are therefore identical.

Parallel Arrangement

Figure 8-7 shows N inductors connected between two common lines so that the inductors are in parallel. The total rate of growth of the current drawn from

FIGURE 8-7 Inductors in parallel.

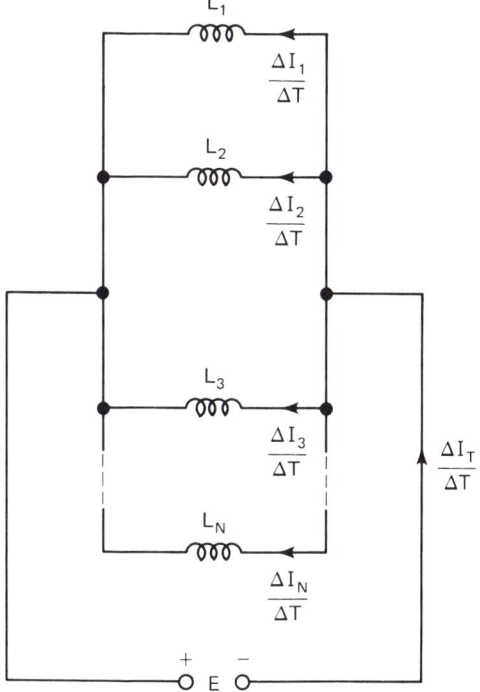

Sec. 8-4 Inductors in Series and Parallel

the source is equal to the sum of the rates of growth in the individual inductors. Then from Eq. (8-1-7),

$$\frac{\Delta I_1}{\Delta T} = \frac{E}{L_1}, \quad \frac{\Delta I_2}{\Delta T} = \frac{E}{L_2}, \quad \frac{\Delta I_3}{\Delta T} = \frac{E}{L_3}, \quad \ldots, \quad \frac{\Delta I_N}{\Delta T} = \frac{E}{L_N}$$

and

$$\frac{\Delta I_T}{\Delta T} = \frac{\Delta I_1}{\Delta T} + \frac{\Delta I_2}{\Delta T} + \frac{\Delta I_3}{\Delta T} + \cdots + \frac{\Delta I_N}{\Delta I_T} \qquad (8\text{-}4\text{-}5)$$

These equations lead to

$$\frac{\Delta I_T}{\Delta T} = E \times \left(\frac{1}{L_1} + \frac{1}{L_2} + \frac{1}{L_3} + \cdots + \frac{1}{L_N} \right) \qquad (8\text{-}4\text{-}6)$$

If L_T is the total equivalent self-inductance,

$$\frac{\Delta I_T}{\Delta T} = \frac{E}{L_T} = E \times \frac{1}{L_T} \qquad (8\text{-}4\text{-}7)$$

Comparing Eqs. (8-4-6) and (8-4-7), we get

$$\frac{1}{L_T} = \frac{1}{L_1} + \frac{1}{L_2} + \frac{1}{L_3} + \cdots + \frac{1}{L_N}$$

or

$$L_T = \frac{1}{\frac{1}{L_1} + \frac{1}{L_2} + \frac{1}{L_3} + \cdots + \frac{1}{L_N}} \qquad (8\text{-}4\text{-}8)$$

This is the familiar reciprocal formula, which is identical to the formula for resistances in parallel. It follows that, for two parallel inductors,

$$L_T = \frac{L_1 \times L_2}{L_1 + L_2} \quad \text{(product-over-sum formula)} \qquad (8\text{-}4\text{-}9)$$

If the N inductors in parallel are identical and each has a self-inductance of L henrys,

$$L_T = \frac{L}{N} \qquad (8\text{-}4\text{-}10)$$

To summarize, the total equivalent inductance of inductors in series, parallel, and series-parallel may be found by using the same formulas and principles as we have used for resistors in Chapters 2, 3, and 4.

EXAMPLE 8-7 In Fig. 8-8, $L_1 = 20$ H, $L_2 = 5$ H, $L_3 = 4$ H, and $E = 58$ V. Find the values of the voltages V_1, V_2, and V_3 and the rate of the growth of the current. After 3 s from the time the switch, S, is closed, calculate the amount of energy stored in each inductor and the total energy drawn from the source (neglect the resistances of the inductors).

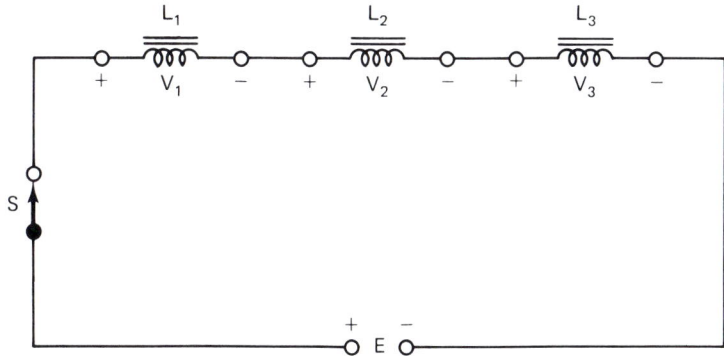

FIGURE 8-8 Circuit for Example 8-7.

Solution

$$\text{total self-inductance, } L_T = L_1 + L_2 + L_3 \qquad (8\text{-}4\text{-}3)$$
$$= 20 + 5 + 4 = 29 \text{ H}$$

Substitution into Eq. (8-4-2) yields

$$\text{rate of growth of current} = \frac{E}{L_T} = \frac{58 \text{ V}}{29 \text{ H}} = 2 \text{ A/s}$$

From Eq. (8-1-4),

$$\text{counter EMF, } V_1 = 2 \text{ A/s} \times 20 \text{ H} = 40 \text{ V}$$
$$\text{counter EMF, } V_2 = 2 \text{ A/s} \times 5 \text{ H} = 10 \text{ V}$$
$$\text{counter EMF, } V_3 = 2 \text{ A/s} \times 4 \text{ H} = 8 \text{ V}$$

Voltage Check

$$E = V_1 + V_2 + V_3 = 40 + 10 + 8 = 58 \text{ V}$$

The current, I, after 3 s is 2 A/s × 3 s = 6 A.
From Eq. (8-3-2),

$$\text{energy stored in the 20-H inductor, } W_1 = \tfrac{1}{2} L_1 I^2$$
$$= \tfrac{1}{2} \times 20 \text{ H} \times (6 \text{ A})^2$$
$$= 360 \text{ J}$$

$$\text{energy stored in the 5-H inductor, } W_2 = \tfrac{1}{2}L_2I^2$$
$$= \tfrac{1}{2} \times 5\text{ H} \times (6\text{ A})^2$$
$$= 90\text{ J}$$
$$\text{energy stored in the 4-H inductor, } W_2 = \tfrac{1}{2}L_3I^2$$
$$= \tfrac{1}{2} \times 4\text{ H} \times (6\text{ A})^2$$
$$= 72\text{ J}$$
$$\text{total energy stored, } W_T = W_1 + W_2 + W_3$$
$$= 360 + 90 + 72$$
$$= 522\text{ J}$$

Energy Check

$$\text{total energy drawn from the source} = \tfrac{1}{2} \times L_T \times I^2$$
$$= \tfrac{1}{2} \times 29\text{ H} \times (6\text{ A})^2$$
$$= 522\text{ J}$$

The total energy may be checked in another way. The average current during the 3-s interval is 6 A/2 = 3 A. Therefore, the total energy drawn from the source is 58 V × 3 A × 3 s = 522 J.

EXAMPLE 8-8 In Fig. 8-9, $L_1 = 20$ H, $L_2 = 5$ H, $L_3 = 4$ H, and $E = 58$ V. Find the rates of growth of the currents in the individual branches and the rate of growth of the total current drawn from the source. After 3 s from the time the switch is closed, calculate the amount of energy stored in each inductor and the total energy drawn from the source (neglect the resistances of the inductors).

Note: For comparison purposes, the values in this example are the same as those chosen in Example 8-7.

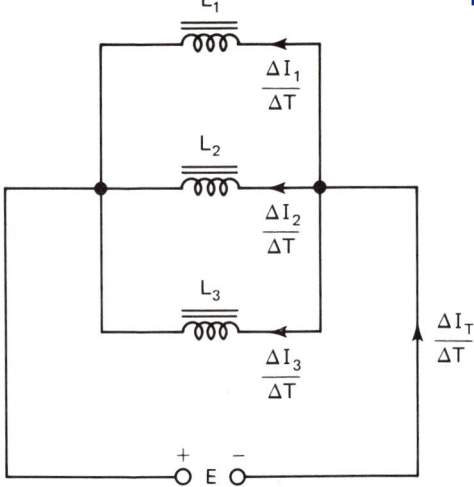

FIGURE 8-9 Circuit for Example 8-8.

Solution The total self-inductance, L_T, is given by [Eq. (8-4-8)]

$$\frac{1}{L_T} = \frac{1}{20} + \frac{1}{5} + \frac{1}{4} = \frac{1 + 4 + 5}{20} = \frac{10}{20} = \frac{1}{2}$$

Therefore, $L_T = 2$ H.

Use of Eq. (8-1-7) yields

$$\text{rate of growth of the current in the 20-H inductor} = \frac{\Delta I_1}{\Delta T}$$

$$= \frac{58 \text{ V}}{20 \text{ H}}$$

$$= 2.9 \text{ A/s}$$

$$\text{rate of growth of the current in the 5-H inductor} = \frac{\Delta I_2}{\Delta T}$$

$$= \frac{58 \text{ V}}{5 \text{ H}}$$

$$= 11.6 \text{ A/s}$$

$$\text{rate of growth of the current in the 4-H inductor} = \frac{\Delta I_3}{\Delta T}$$

$$= \frac{58 \text{ V}}{4 \text{ H}}$$

$$= 14.5 \text{ A/s}$$

rate of growth of the total current drawn from the source,

$$\frac{\Delta I_T}{\Delta T} = \frac{\Delta I_1}{\Delta T} + \frac{\Delta I_2}{\Delta T} + \frac{\Delta I_3}{\Delta T} \tag{8-4-5}$$

$$= 2.9 + 11.6 + 14.5 = 29.0 \text{ A/s}$$

Check on the Rate of Growth of the Total Current

Rate of growth of the total current drawn from the source is also $E/L_T = 58$ V/ 2 H = 29.0 A/s. After 3 s the branch currents I_1, I_2, and I_3 are, respectively, $3 \times 2.9 = 8.7$ A, $3 \times 11.6 = 34.8$ A, and $3 \times 14.5 = 43.5$ A. Substitution into Eq. (8-3-2) gives

$$\text{energy stored in the 20-H inductor, } W_1 = \tfrac{1}{2} \times L_1 \times I_1^2$$

$$= \tfrac{1}{2} \times 20 \text{ H} \times (8.7 \text{ A})^2$$

$$= 756.9 \text{ J}$$

$$\text{energy stored in the 5-H inductor, } W_2 = \tfrac{1}{2} \times L_2 \times I_2^2$$

$$= \tfrac{1}{2} \times 5 \text{ H} \times (34.8 \text{ A})^2$$

$$= 3027.6 \text{ J}$$

$$\text{energy stored in the 4-H inductor, } W_3 = \tfrac{1}{2} \times L_3 \times I_3^2$$

$$= \tfrac{1}{2} \times 4 \text{ H} \times (43.5 \text{ A})^2$$

$$= 3784.5 \text{ J}$$

The total energy stored in the three inductors is

$$W_T = W_1 + W_2 + W_3 = 756.9 + 3027.6 + 3784.5 = 7569 \text{ J}$$

Energy Check
The total energy drawn from the source,

$$W_T = \tfrac{1}{2} L_T I_T^2 = \tfrac{1}{2} \times 2 \text{ H} \times (3 \times 29 \text{ A})^2 = 7569 \text{ J}$$

Alternatively, the average current during the 3 s is $(29 \times 3)/2 = 43.5$ A. Therefore, the total energy drawn from the source is

$$W_T = 58 \text{ V} \times 43.5 \text{ A} \times 3 \text{ s} = 7569 \text{ J}$$

8-5

Time Constant of *LR* Circuit

We already know that the property of inductance prevents any sudden change of current and also limits the rate of change of the current. In other words, if a dc voltage is suddenly switched across the inductive circuit of Fig. 8-10(a), the current must take a certain time before reaching a final steady value, which is limited by the circuit's resistance. *R* may either represent the resistance of the coil or may be a combination of an actual series resistor and the coil's resistance. The purpose of this section is to discuss the factors that control the duration of the so-called transient or changing state; this is the interval between the time the switch, *S*, is closed in position 1 and the time at which the final or steady-state conditions are reached.

The behavior of the circuit immediately after *S* is closed in position 1 is called the *transient response*. Since the current I_1 is zero immediately before *S* is closed, it must initially remain at zero due to the inductance, which will prevent any sudden change of the current. Therefore, the voltage drop across *R* is zero and the source voltage must be exactly balanced by the counter EMF induced in *L*. This counter EMF must be created by the initial rate of the current growth, which, by Eq. (8-1-6), must equal E/L amperes per second.

To summarize, the conditions at the beginning of the transient state are:

$$I_1 = 0, \quad V_R = 0, \quad V_L = E$$

$$\text{initial rate of growth, } \frac{\Delta I_1}{\Delta T} = \frac{E}{L} \text{ A/s} \qquad (8\text{-}5\text{-}1)$$

A question frequently asked by students is: "How can there be a rate of current growth when the current iself is zero?" As an analogy, think of the beginning of the Indianapolis 500; as a result of the lap preceding the race, a car may cross the starting line at 100 mi/h (rate of change of distance with time), but the amount of the race distance covered at that time is zero. In the same way a current may have an initial rate of growth equal to 2 A/s, but until time has elapsed, the current is zero.

As the current increases and causes the magnetic field to expand, V_R rises and V_L falls, since at all times the sum of the voltage drop across *R* and the counter EMF induced in *L* must exactly balance the source voltage, *E*. When the counter EMF decreases, the rate of current growth is correspondingly

FIGURE 8-10 Time constant of an *LR* circuit.

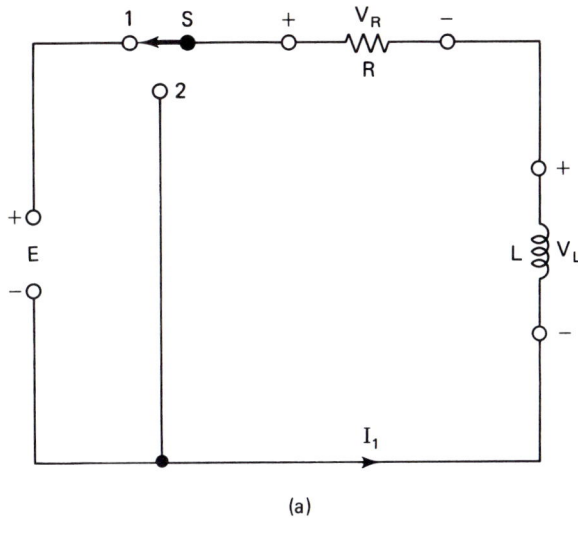

(a)

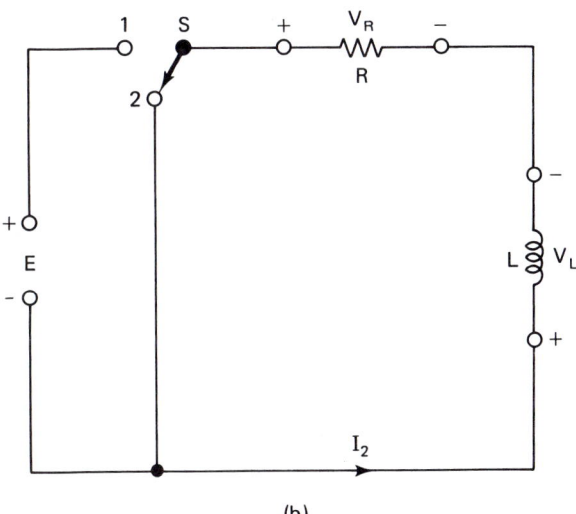

(b)

reduced, so we have a situation in which the greater the value of the current, the less is its rate of growth; for an everyday analogy, the taller a tree grows, the less is the rate of its height increase. Theoretically, it would take infinite time for the current to reach the final steady value, E/R, which is limited by the resistance. However, as we shall see, the current reaches its final value to within 1% after a time interval that is determined by the values of the inductance and the resistance.

The initial rate of current growth is E/L amperes per second and the final current value is E/R amperes. If the inductance is increased but the resistance and the source voltage are unchanged, the initial rate of the current growth will be reduced while the final steady current will be the same. The transient state will therefore last longer, so that its duration depends directly on the inductance; in other words, the greater the inductance, the longer it will take the current to reach its final value to within 1%. By contrast, let us keep the inductance and the source voltage the same but raise the value of the resistance. The initial rate of the current growth will be the same, but the value of the final steady current will be reduced. The transient state will not last as long and therefore its duration is inversely dependent on the value of the resistance.

When these results are combined, we are led to the conclusion that the interval of the transient state is determined by the $L{:}R$ ratio. The quantity

TABLE 8-1 *L/R* Time Constant

Inductance, L	Resistance, R	Time Constant, L/R
H	Ω	seconds
H	kΩ	milliseconds
H	MΩ	microseconds
mH	Ω	milliseconds
mH	kΩ	microseconds
mH	MΩ	nanoseconds
μH	Ω	microseconds
μH	kΩ	nanoseconds
μH	MΩ	picoseconds

L/R is called the *time constant*, which is measured in seconds if L is in henrys and R is in ohms. For example, a 20-H inductor whose resistance is 100 Ω, will have a time constant of 20 H/100 Ω = 0.2 s. For other values of L and R, you may estimate the order of the time constant by using Table 8-1.

How is the value of the time constant related to the transient response of the *LR* circuit? Mathematically, it may be shown that after one time constant the current has reached 63.2% or 0.632 of its final value, E/R. Consequently, after L/R seconds or one time constant from the instant S is closed in position 1, the conditions in the circuit are:

$$I_1 = 0.632 \frac{E}{R}, \qquad V_R = 0.632E,$$

$$V_L = E - V_R = E - 0.632E = 0.368E \qquad (8\text{-}5\text{-}2)$$

$$\text{rate of current growth} = \frac{V_L}{L} = \frac{0.368E}{L} \text{ A/s}$$

The factor 0.632 is extremely important. After a further interval of one time constant or a total of $2L/R$ seconds from the instant S is closed, the current completes 63.2% of its *remaining* climb toward its final steady value. Since after one time constant the remaining percentage climb is 100 − 63.2 = 36.8%, the current increase during the second time constant is 0.632 × 36.8 = 23.26%, so that the total percentage climb after two time constants is 63.2 + 23.26 = 86.5% (rounded off). During the third time constant, the remaining percentage climb is 100 − 86.5 = 13.5%, the increase is 13.5 × 0.632 = 8.532%, and the total percentage climb is 86.5 + 8.532 = 95.0% (rounded off). Repeating the procedure for the fourth and fifth time constants, you can verify that the total percentage climbs are respectively 98.2% and 99.3%. This means that after five time constants or $5L/R$ seconds, the current has reached a level that is within 1% of its final steady value. We then assume that the circuit's transient state has been completed and that the final or steady-state conditions have been reached. These conditions are:

$$I_1 = \frac{E}{R}, \qquad V_R = E, \qquad V_L = 0$$

$$\text{rate of current growth} = 0 \qquad (8\text{-}5\text{-}3)$$

We should mention that there is another interpretation of the time constant. Since the final current is E/R amperes and the *initial* rate of the current growth is E/L amperes per second, the current would have reached its final value in

L/R seconds or *one* time constant *if* the initial rate of the current growth had been maintained. Of course, the rate of the current growth does, in fact, fall off, so that *five* time constants are required for the current to reach its final level.

Notice that the initial voltage and current conditions (immediately after the switch, S, is closed in position 1) may be predicted by regarding the inductor as an *open* circuit; the final steady-state conditions can be obtained by considering the inductor to be a *short* circuit.

The way in which the current has increased in this circuit is referred to as *exponential growth*. This is graphically illustrated in Fig. 8-11(a); notice that the vertical axis is marked in percentage and the horizontal time axis is measured with intervals of L/R seconds. The graph is then a universal curve which can be used to solve exponential growth problems by converting all current or voltage values to percentages of their final levels and expressing all intervals in terms of the time constant. The variations of I_1, V_R, and V_L with time are shown in Fig. 8-12(a); you can clearly see that I_1 and V_R follow exponential growth curves but V_L falls away with time and represents an exponential decay, which we discuss next in more detail.

Assuming that the steady-state conditions have been reached, we will now switch S to position 2 [Fig. 8-10(b)]. Remembering that the current cannot change instantaneously, the direction and value of I_2 must initially be the same as the final direction and value of I_1, namely E/R amperes. V_R will still equal E but in order for the voltages to balance around the loop, V_L must also be equal to E. However, the polarity of V_L is reversed compared with the polarity during the growth of the current. This is because the current is starting to decay, so that the magnetic field is collaping rather than expanding. Summarizing, the circuit conditions immediately after S is closed in position 2, are:

$$I_2 = \frac{E}{R}, \quad V_R = E, \quad V_L = -E$$

initial rate of change of current, $\quad \frac{\Delta I_2}{\Delta T} = -\frac{E}{L} \quad$ A/s

(8-5-4)

The minus signs represent the reversal of V_L's polarity and the fact that the current change is a decay and not a growth.

As the current decays, V_R and therefore V_L must fall together, so that the voltage balance around the loop is maintained. The decrease in V_L means that the rate of change in the current is lower, so we have a situation in which the smaller the current, the less is its rate of decay. Theoretically, the current would take infinite time to decay completely; however, the fall is exponential and the current will have dropped to below 1% of its original value after five time constants. Once again the 0.632 factor is all-important. After one time constant or a time interval of L/R seconds from the instant that S is switched to position 2, the current, I_2, has *lost* 63.2% and has therefore fallen to 36.8% of its original value.

Then the circuit conditions are:

$$I_2 = 0.368 \frac{E}{R}, \quad V_R = 0.368E, \quad V_L = -0.368E,$$

$$\frac{\Delta I_2}{\Delta T} = -\frac{0.368E}{L} \quad \text{A/s}$$

(8-5-5)

Using the results we obtained from the growth curve, the current level after two, three, four, and five time constants will be, respectively, $100 - 86.5 =$

Sec. 8-5 Time Constant of *LR* Circuit

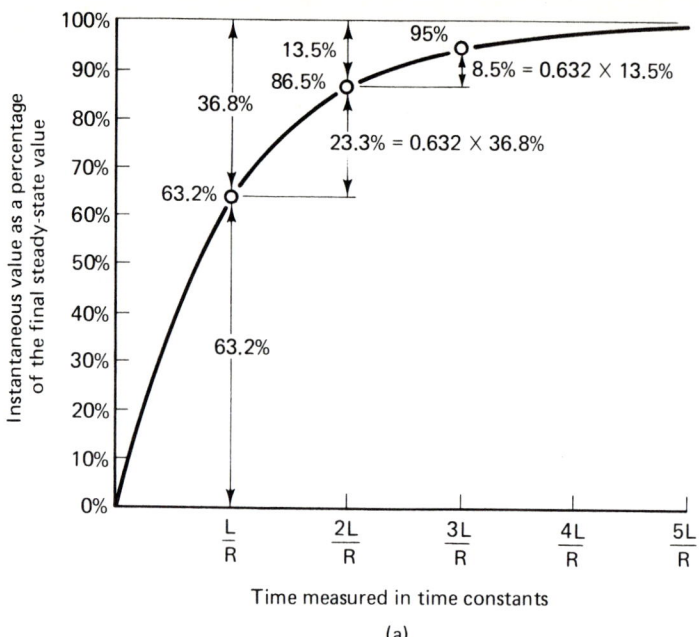

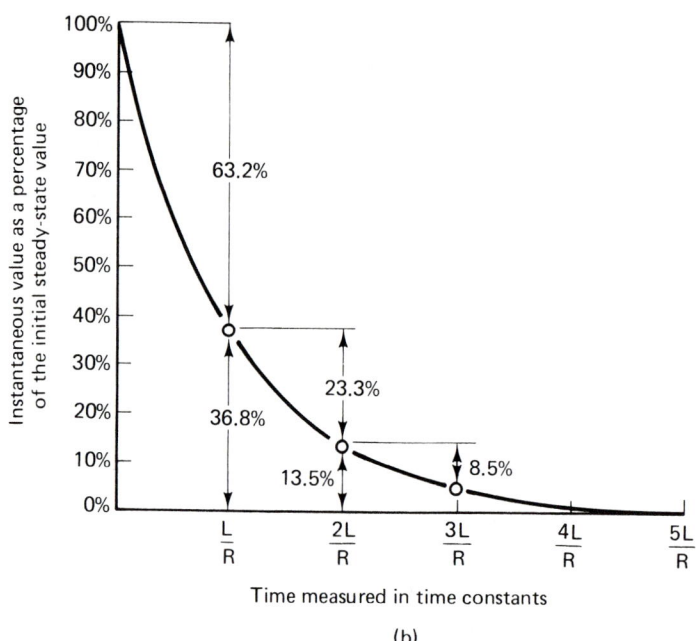

FIGURE 8-11 (a) Universal exponential growth curve; (b) universal exponential decay curve.

13.5%, $100 - 95.0 = 5.0\%$, $100 - 98.2 = 1.8\%$, and $100 - 99.3 = 0.7\%$ of its original value. After five time constants we assume that the transient decay state has been concluded, with the values of I_2, V_R, and V_L all essentially equal to zero. The universal decay curve is illustrated in Fig. 8-11(b) and the graphs for I_2, V_R, and V_L are shown in Fig. 8-12(b). Notice that I_2 would have fallen to zero in *one* time constant if the initial rate of the current decay had been maintained.

The importance of the time constant lies in its control of the transient response for an *LR* circuit during both the growth and the decay of the current.

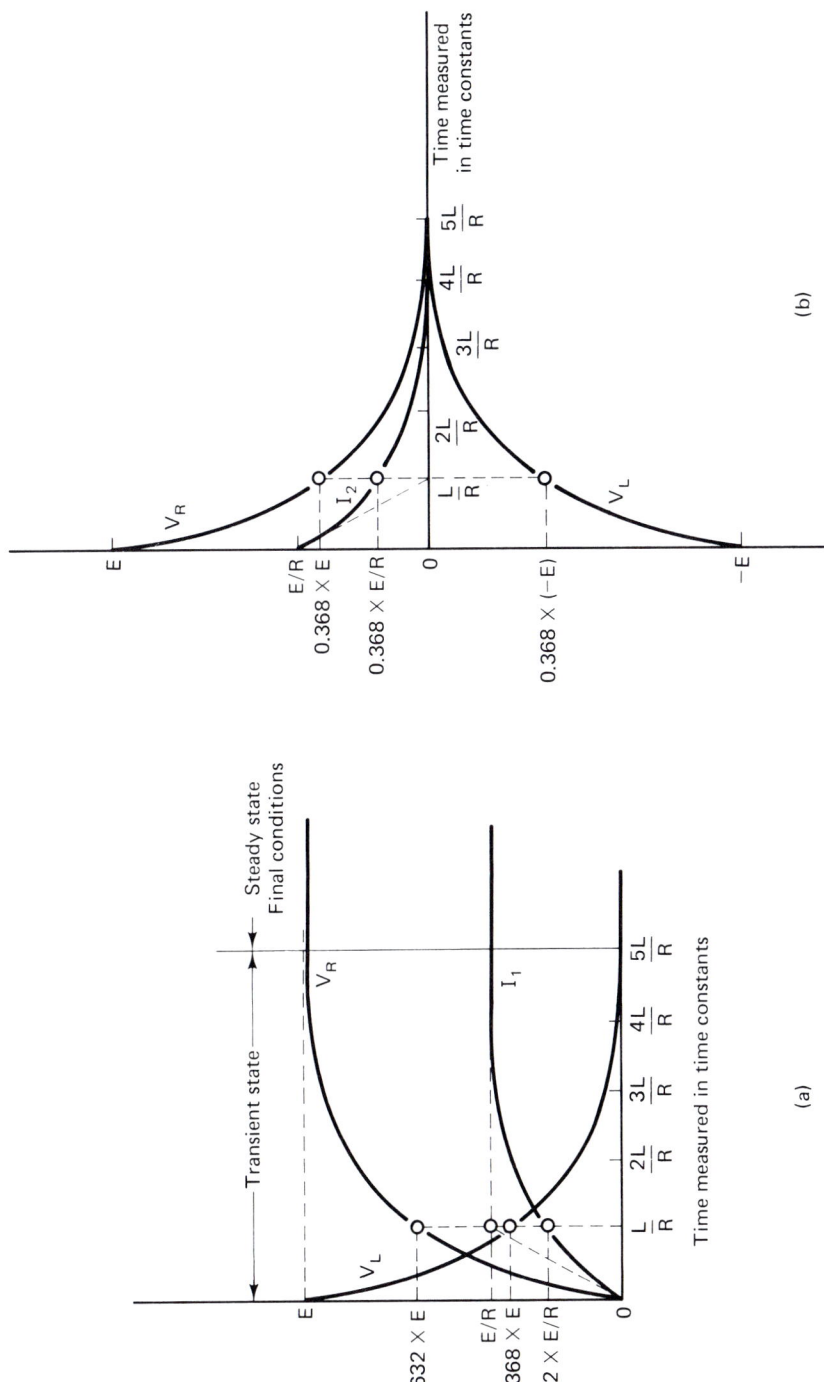

FIGURE 8-12 (a) Growth of current in the *LR* circuit; (b) decay of current in the *LR* circuit.

TABLE 8-2 Percentage and *L/R* Time Constant Values

Time Constant, *L/R* (seconds)	Percentage Growth	Percentage Decay
0.2	18.1	81.9
0.5	39.3	60.7
0.7	50.3	49.7
1.0	63.2	36.8
2.0	86.5	13.5
3.0	95.0	4.98
4.0	98.2	1.83
5.0	99.3	0.67

Using the universal curves, corresponding values of percentage and time constant are shown in Table 8-2.

EXAMPLE 8-9 In Fig. 8-10(a), $R = 3$ kΩ, $L = 30$ mH, and $E = 90$ V. Find the value of the circuit's time constant. With S closed in position 1, calculate the initial values of I_1, V_R, and V_L and their values after time intervals of $t = 7$, 10, 23, and 60 μs.

Solution From Table 8-1,

$$\text{time constant}, \frac{L}{R} = \frac{30 \text{ mH}}{3 \text{ k}\Omega} = 10 \ \mu\text{s}$$

Use of Eqs. (8-5-1) yields:
Initial values for t = 0:

$$I_1 = 0 \text{ A}, \quad V_R = 0 \text{ V}, \quad V_L = 90 \text{ V}$$

and

$$\text{rate of current growth}, \frac{E}{L} = \frac{90 \text{ V}}{30 \text{ mH}} = \frac{90 \text{ V}}{30 \times 10^{-3} \text{ H}} = 3000 \text{ A/s}$$

t = 7 μs
Since the time constant is 10 μs, 7 μs = 7/10 × time constant = 0.7 time contant. On the universal growth curve of Fig. 8-11(a), 0.7 time constant corresponds approximately to 50%.

$$\text{final current}, \frac{E}{R} = \frac{90 \text{ V}}{3 \text{ k}\Omega} = 30 \text{ mA}$$

Therefore,

$$\text{current after 7 } \mu\text{s} = \frac{50}{100} \times 30 = 15 \text{ mA}$$

$$V_R = 15 \text{ mA} \times 3 \text{ k}\Omega = 45 \text{ V} \quad \text{and} \quad V_L = 90 - 45 = 45 \text{ V}$$

t = 10 μs
The interval of 10 μs is equal to the time constant of the circuit.

Equations (8-5-2) yield

$$\text{current after } 10 \ \mu s = 0.632 \times 30 = 18.96 \text{ mA}$$

$$V_R = 18.96 \text{ mA} \times 3 \text{ k}\Omega = 56.9 \text{ V} \quad \text{and} \quad V_L = 90 - 56.9 = 33.1 \text{ V}$$

t = 23 μs
The interval of 23 μs is equivalent to 23/10 = 2.3 time constants. On the universal growth curve, 2.3 time constants corresponds approximately to 90%.

$$\text{current after 2.3 time constants} = \frac{90}{100} \times 30 = 27 \text{ mA}$$

$$V_R = 27 \text{ mA} \times 3 \text{ k}\Omega = 81 \text{ V} \quad \text{and} \quad V_L = 90 - 81 = 9 \text{ V}$$

t = 60 μs
This interval is more than five time constants, so that the final steady state has been reached.
Equations (8-5-3) yield

$$\text{current after } 60 \ \mu s = \text{final value of 30 mA}$$

$$V_R = 90 \text{ V} \quad \text{and} \quad V_L = 0 \text{ V}$$

EXAMPLE 8-10 In Fig. 8-10(b), $R = 3$ kΩ, $L = 30$ mH, and $E = 90$ V. A time interval of 60 μs has elapsed from the closing of the switch, S, in position 1. S is then switched to position 2. What are the initial values of I_2, V_R, V_L, and their subsequent values after time intervals of $t = 7$, 10, 23, and 60 μs?

Solution Equations (8-5-4) yield:

Initial values for t = 0:

$$I_2 = 30 \text{ mA}, \quad V_R = 90 \text{ V}, \quad V_L = -90 \text{ V}$$

t = 10 μs
From Eqs. (8-5-5),

$$I_2 = 0.368 \times 30 = 11.04 \text{ mA}, \quad V_R = 11.04 \text{ mA} \times 3 \text{ k}\Omega = 33.12 \text{ V},$$

$$V_L = -33.12 \text{ V}$$

t = 23 μs
From the universal decay curve of Fig. 8-11(b), the percentage corresponding to 23 μs or 2.3 time constants is approximately 10%. Therefore,

$$I_2 = \frac{10}{100} \times 30 \text{ mA} = \text{mA}, \quad V_R = 3 \text{ mA} \times 3 \text{ k}\Omega = 9 \text{ V},$$

$$V_L = -9 \text{ V}$$

t = 60 μs
The interval of 60 μs exceeds five time constants, so that the current, I_2, is virtually zero. Therefore, $V_R = V_L = 0$ V.

8-6
Trouble Shooting for Faults in Inductors

We have already seen that a practical inductor possesses the property of resistance, R, as well as that of inductance, L. This resistance is normally represented as being in series with the inductance, so that the inductor's equivalent electrical circuit is as shown in Fig. 8-13. In Chapter 9 we shall see that the practical inductor also possesses the property of distributed capacitance, C, which is in parallel with the inductance-resistance combination.

Inductors suffer from two problems: open circuits and short circuits in the coil winding. Let us first discuss the open circuit. If an inductor is open, it will possess zero self-inductance, since no magnetic field can be established to induce a counter EMF. At the same time an ohmmeter will record infinite resistance when connected between the ends of the coil [Fig. 8-14(a)]; this

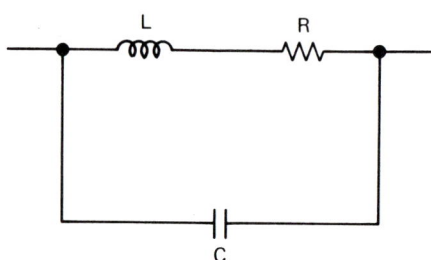

FIGURE 8-13 Equivalent electrical circuit of practical inductor.

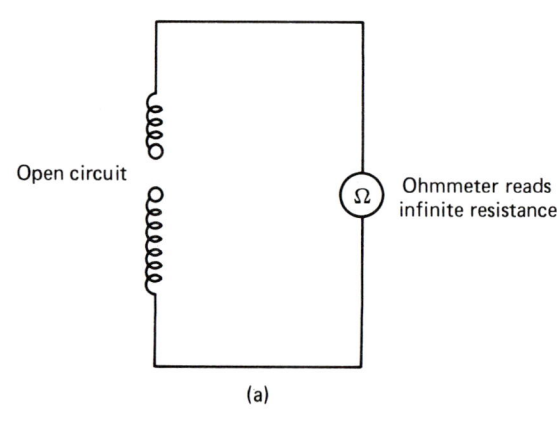

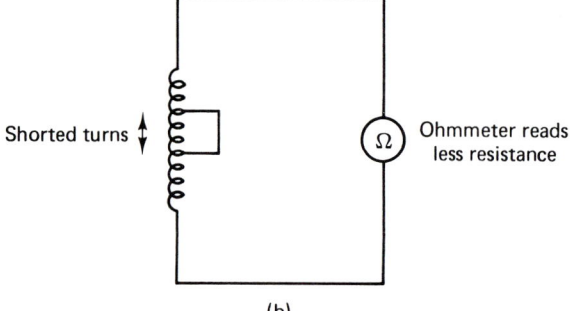

FIGURE 8-14 Ohmmeter showing (a) open circuit and (b) short circuit in the coil winding.

Inductance in DC Circuits Ch. 8

means that you can use the same continuity tests and trouble-shooting techniques with inductors as we used for resistors. Again we must be careful to look out for any resistance paths that are in parallel with the coil, since these will affect the reading of the ohmmeter.

As far as short circuits are concerned, the turns of an inductor must naturally be insulated from each other. This is achieved by covering the copper wire with fiber or plastic; alternatively, the turns may be coated with an insulating varnish. However, for a variety of reasons such insulation may break down, so that some of the inductor's turns are shorted out [Fig. 8-14(b)]. As a result, the coil has a lower self-inductance and a lower resistance. Consequently, if the reading of an ohmmeter connected across the ends of the coil is substantially less than the rated resistance value, you can suspect that there is a partial short across some of the turns. In the extreme case there can be a short across the entire coil so that the ohmmeter reading is zero. Typically, the rated resistance values for chokes with self-inductances of several henrys range from tens to hundreds of ohms. For radio broadcast or TV channel coils with self-inductances from a few microhenrys or less up to several millihenrys, the dc resistances lie between 1 Ω (or less) and tens of ohms.

Chapter Summary

8.1. Counter EMF, $V_L = -L \dfrac{\Delta I}{\Delta T} = -N \dfrac{\Delta \phi}{\Delta T}$ volts

Inductance, $L = N \dfrac{\Delta \phi}{\Delta I}$ henrys

Rate of change of current, $\dfrac{\Delta I}{\Delta T} = \dfrac{E}{L}$ amperes per second

8.2. Inductance of coil, $L = \dfrac{\mu_0 \mu_r N^2 A}{l}$ henrys

8.3. Energy stored in a magnetic field $= \tfrac{1}{2} L I^2$ joules

8.4. Inductors in series:

$$L_T = L_1 + L_2 + L_3 + \cdots + L_N \quad \text{henrys}$$

8.5. Inductors in parallel:

$$L_T = \dfrac{1}{\dfrac{1}{L_1} + \dfrac{1}{L_2} + \dfrac{1}{L_3} + \cdots + \dfrac{1}{L_N}} \quad \text{henrys}$$

8.6. Time constant of LR circuit $= \dfrac{L}{R}$ seconds

Duration of transient state (time required to reach steady-state conditions)
$= \dfrac{5L}{R}$ seconds

Final steady-state current $= \dfrac{E}{R}$ amperes

Self-Review

True–False Questions

1. T. F. Inductance is that property of an electrical circuit which opposes the flow of current.
2. T. F. The energy drawn from an electrical source by an inductor is stored in the form of a magnetic field.
3. T. F. The value of self-inductance is directly proportional to the inductor's length.
4. T. F. The value of self-inductance is directly proportional to the square root of the number of turns.
5. T. F. If an air core of an inductor is replaced by a core of soft iron, the value of the self-inductance is greatly increased.
6. T. F. The energy stored in the magnetic field of an inductor is equal to LI^2 watts.
7. T. F. Three chokes with self-inductances of 6 H, 3 H, and 2 H are connected in series. Their total equivalent self-inductance is 11 H.
8. T. F. Three chokes with self-inductances of 20 H, 5 H, and 4 H are connected in parallel. Their total equivalent self-inductance is 2 H.
9. T. F. In an RL circuit, the time constant is the time for the current to reach 100% of its final steady-state value.
10. T. F. A 1-H inductor has a resistance of 20 Ω. The time constant of this combination is 0.05 s.
11. T. F. A conductor must be in the form of a coil before it can possess the property of self-inductance.
12. T. F. The inductance of a coil with a magnetic core is always a constant.
13. T. F. The polarity of a coil's self-induced EMF is always such that it opposes any change in the current flowing through the coil.
14. T. F. Certain variable types of inductor use an adjustable core to alter the effective permeability.
15. T. F. The equations for determining the total equivalent self-inductance of a series–parallel arrangement of inductors are the same as those for an equivalent arrangement of resistors.
16. T. F. On the universal exponential decay curve a time interval of $3L/2R$ seconds corresponds to a percentage of approximately 68%.
17. T. F. A practical inductor possesses the properties of inductance, resistance, and distributed impedance.
18. T. F. If the diameter of an inductor's cross-sectional area is doubled without changing any of the other physical properties, the value of the self-inductance is also doubled.
19. T. F. The property of inductance does not allow any instantaneous change in the current.
20. T. F. When a dc voltage is suddenly switched across a series combination of an inductor and a resistor, the inductor initially behaves as an open circuit.

Multiple-Choice Questions

21. A 20-V dc source is suddenly switched across a 5-H choke. The initial rate of current growth is
    (a) 100 A/s (b) 25 A/s (c) 15 A/s (d) 5 A/s (e) 4 A/s
22. The current flowing through a 3-H choke is decaying at the rate of 6 A/s. The value of the counter EMF is
    (a) 2 V (b) 9 V (c) 3 V (d) 18 V (e) 54 V

23. A coil has a self-inductance of 50 mH. If the number of turns is doubled without altering either the inductor's length or its cross-sectional area, the value of the new self-inductance will be
    (a) 100 mH   (b) 200 mH   (c) 25 mH   (d) 12.5 mH   (e) 50 mH

24. The air core of a 150-μH inductor is replaced by an iron core with a relative permeability of 800. The new value of the self-inductance is
    (a) 0.12 H   (b) 12 mH   (c) 1.2 H   (d) 0.012 H
    (e) both (b) and (d)

25. An 8-H inductor with a resistance of 100 Ω is carrying a current of 2 A. The amount of energy stored in the magnetic field surrounding the inductor is
    (a) 16 J   (b) 32 J   (c) 200 J   (d) 400 J   (e) 64 J

26. Three coils with self-inductances of 500 μH, 300 μH, and 200 μH are switched in series across a 50-V dc source. The initial counter EMF induced in the 300-μH inductor is
    (a) 10 V   (b) zero   (c) 15 V   (d) 25 V   (e) 30 V

27. Three chokes with self-inductances of 10 H, 5 H, and 2 H are connected in parallel. If this combination is switched across a 30-V dc source, the initial total rate of growth of the current drawn from the source is
    (a) 1.76 A/s   (b) 3 A/s   (c) 24 A/s   (d) 15 A/s   (e) 37.5 A/s

28. A 2-mH inductor with negligible resistance is connected in series with a 1-kΩ resistor. If this combination is suddenly switched across a 20-V dc source, the initial rate of growth of the current is
    (a) 10 mA/s   (b) 20 mA/s   (c) 10 A/s   (d) 10000 A/s   (e) 100 A/s

29. In the circuit of Question 28, the value of the time constant is
    (a) 2 s   (b) 0.5 s   (c) 2 ms   (d) 1 ms   (e) 2 μs

30. In the circuit of Question 28, the value of the steady current flowing at the end of the transient state is
    (a) 20 mA   (b) 10 mA   (c) 10 A   (d) 20 μA   (e) 1000 A

31. The current in an 8-H inductor increases from 1.6 A to 2.4 A in a period of 200 ms. The rate of growth of current is
    (a) 4 A/s   (b) 8 A/s   (c) 1.2 A/s   (d) 12 A/s   (e) 0.8 A/s

32. The current in a 150-μH coil decreases from 200 mA to 120 mA in a period of 1.6 μs. The value of the induced EMF is
    (a) 75 mV   (b) 750 mV   (c) 7.5 mV   (d) 7.5 V   (e) 30 V

33. A 20-H inductor is connected in parallel with a 5-H inductor. This parallel combination is then joined in series with a 4-H inductor. If the complete circuit is switched across a 24-V dc source, the initial EMF induced in the 5-H inductor is
    (a) 8 V   (b) 12 V   (c) 6 V   (d) 4 V   (e) 16 V

34. In Question 33, assume that the resistances of the coils are negligible. If 2 s has elapsed from the time the circuit was switched across the dc source, the energy stored in the magnetic field surrounding the 4-H inductor is
    (a) 72 J   (b) 36 J   (c) 144 J   (d) 12 J   (e) 54 J

35. A 400-μH inductor with negligible resistance is connected in series with a 20-Ω resistor. If this combination is suddenly switched across a 30-V dc source, the initial voltage across the inductor is
    (a) 30 V   (b) 20 V   (c) 15 V   (d) 10 V   (e) zero

36. In Question 35, a period of 120 μs has elapsed since the circuit was suddenly switched across the dc source. The voltage across the resistor is then
    (a) 30 V   (b) 20 V   (c) 15 V   (d) 10 V   (e) zero

37. In Question 36, the current flowing through the inductor is
    (a) zero   (b) 0.5 A   (c) 0.75 A   (d) 1 A   (e) 1.5 A

38. A 30-mH inductor is connected in parallel with a 60-mH inductor. If the rate of change of current through the 30-mH inductor is 5 A/s, the total rate of change of current for both inductors is
    (a) 2.5 A/s   (b) 10 A/s   (c) 15 A/s   (d) 7.5 A/s   (e) 12.5 A/s

39. On the universal exponential growth curve, 80% corresponds to an interval that is equal to the time constant multiplied by
    (a) 0.22   (b) 1.6   (c) 1.4   (d) 1.8   (e) 1.3

40. A coil of 500 turns is wound on a nonmagnetic core. If a current of 6 A flowing through the coil establishes a flux of 300 μWb, the inductance of the coil is
    (a) 20 mH   (b) 25 mH   (c) 35 mH   (d) 15 mH   (e) 40 mH

Practice Problems

41. The current flowing through a 200-mH inductor decays from 8 A to 5 A in 12 ms. What is the average value of the induced counter EMF?
42. A 6-V source is suddenly switched across a 3-mH inductor. If the inductor's resistance is neglected, what is the value of the inductor's current after 5 ms?
43. A current of 8 A establishes a flux of 120 mWb in an iron-cored inductor that is wound with 1000 turns. What is the value of the self-inductance?
44. A coil is wound with 600 turns on an air core and has a self-inductance of 2 mH. If the number of turns is increased to 900 without changing either the coil's cross-sectional area or its length, what is the new value of the self-inductance?
45. A 3-H inductor is suddenly switched across a 12-V dc source. If the inductor's resistance is neglected, what is the energy stored in the inductor after a time interval of 0.5 s?
46. Two chokes with self-inductances of 3 H and 5 H are joined in series and are then suddenly switched across a 12-V dc source. What are the initial values of the rate of the current growth and the counter EMF induced in the 5-H inductor?
47. Two chokes with self-inductances of 20 H and 5 H are connected in parallel and then suddenly switched across a dc source. If the total rate of the current growth drawn from the source is 2 A/s, what is the value of the source voltage?
48. In Fig. 8-15, calculate the value of the time constant and the initial rate of current growth after the switch, S, is closed.

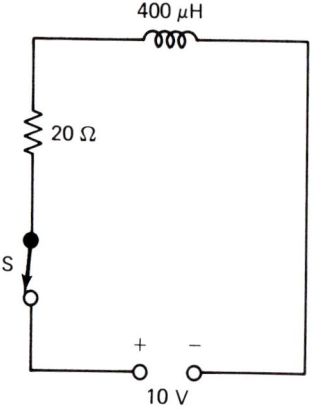

FIGURE 8-15 Circuit for Problems 48, 49, and 50.

49. In Problem 48, what is the duration of the transient state that occurs after S is closed? Calculate the value of the final steady-state current that flows through the inductor.
50. In Problem 48, what are the values of the circuit current after intervals of (a) 20 µs and (b) 60 µs, from the time the switch is closed?
51. An average voltage of 15 V is induced in a coil of 120 turns as a result of a change in the magnetic flux. If this event occurs in a total time of 180 ms, what is the amount of the total flux change?
52. An inductor has a length of 7.5 cm and a diameter of 1 cm. How many turns are required to produce a self-inductance of 135 µH?
53. A coil has a self-inductance value of 120 mH. What value of resistance must be placed in series with the coil to produce a time constant of 30 µs?
54. A 50-mH coil whose resistance is negligible has a terminal voltage of 40 V. Determine the rate of change of current in the coil.
55. An 8-H inductor is manufactured by wrapping a number of turns of copper wire around an iron ring with a circular cross-sectional area. The inner and outer diameters of the ring are 8 cm and 10 cm, and the relative permeability of the iron is 320. Calculate the number of turns required.

56. The current in a 10-mH coil increases uniformly from 1.0 mA to 1.8 mA in a period of 12 μs. Subsequently, the current remains at the 1.8 mA level for 24 μs and then decays uniformly to zero in 3 μs. Determine the waveform of the voltage induced in the coil and indicate the voltage values involved.

57. Three coils have self-inductances of 10 mH, 20 mH, and 30 mH. How may these coils be connected to provide a total self-inductance of 22 mH?

58. Two coils have a total self-inductance of 25 H when connected in series but only 4 H when connected in parallel. What are the values of the individual self-inductances?

59. When a 12-V battery is switched across a coil, the current increases from zero to a final level of 240 mA in 160 ms. Calculate the values of the coil's resistance, its self-inductance, and the time constant.

60. A coil of 1500 turns surrounds a magnetic circuit with a reluctance of 6×10^6 SI units. What is the value of the self-inductance?

Advanced Problems

61. In Fig. 8-16, what is the value of the total equivalent self-inductance between points P and Q?

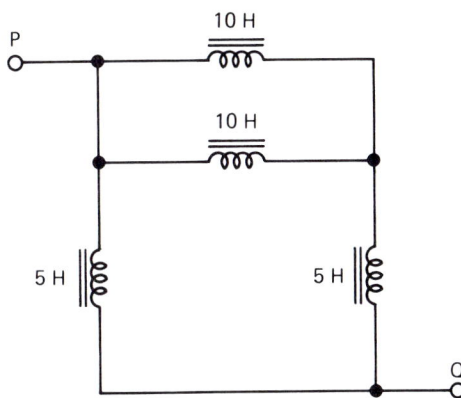

FIGURE 8-16 Circuit for Problem 61.

62. The average counter EMF induced in a 500-μH coil is 10 V when the current changes by 40 mA. Over what interval of time does the current change?

63. A coil of 400 turns is wound on a nonmagnetic core. A flux of 250 μWb is established by a current of 5 A flowing through the coil. What is the value of the coil's inductance? What is the average value of the induced EMF if the current is reversed in 20 ms?

64. A cast iron ring has a mean circumference of 40 cm and a cross-sectional area of 5 cm². The ring is uniformly wound with a coil of 400 turns. Calculate the inductance if currents of (a) 2.5 A and (b) 10 A are reversed in the coil. (*Hint:* Find the magnetic field intensity and then use Fig. 7-21 to obtain the corresponding flux density for cast iron. Notice how the value of the inductance changes with the value of the current.)

65. A coil of 700 turns is wound on a magnetic core whose relative permeability is 1100. If the coil's length is 5.5 cm and its cross-sectional area is 2.5 cm², what is the value of the coil's self-inductance? If the core is replaced by one made from nonmagnetic material, what is the new value of the self-inductance?

66. A current of 3.5 A flows through a coil of 450 turns and establishes a flux of 650 μWb. What is the value of the self-inductance, and what is the amount of energy stored in the magnetic field surrounding the inductor?

67. In Fig. 8-17, the switch, S, is closed in position 1. Find the value of the time constant and the values of the inductor's current after time intervals of 50, 80, 120, 200, and 300 ms. (*Hint:* Use the exponential growth curve.)

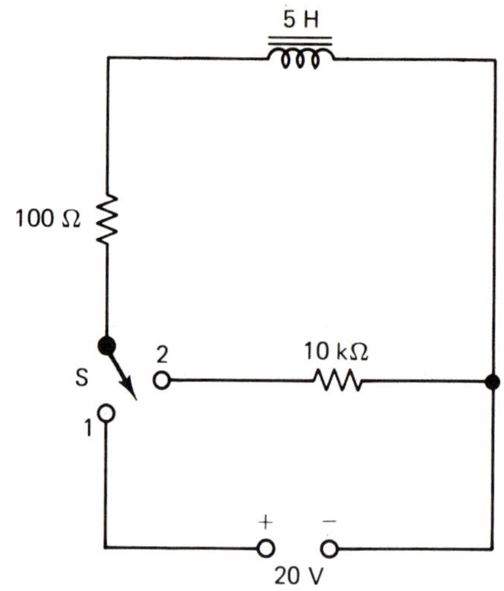

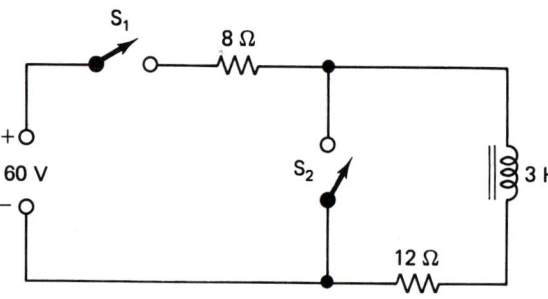

FIGURE 8-17 Circuit for Problems 67 through 70.

68. In Problem 67, what are the values of the inductor's current and its counter EMF at the end of the transient state? What is the amount of the energy then stored in the magnetic field surrounding the inductor?

69. In Problem 67, the circuit has reached its final steady-state condition with the switch in position 1. S is now switched to position 2. Calculate the initial value of the induced counter EMF and the rate of the current's decay. (*Hint:* Remember that the current cannot change its value instantaneously. Find the voltage drop across the total resistance.)

70. In Problem 69, what is the value of the circuit's time constant with the switch, S, in position 2? How long will it take for the current to decay to zero after S is switched to position 2?

71. In Fig. 8-18, switch S_1 is closed while switch S_2 is left open. How long does it take for the inductor current to reach 1 A?

FIGURE 8-18 Circuit for Problems 71 and 72.

72. In Problem 71, assume that switch S_1 has been closed for so long that the current through the inductor is limited only by the circuit's resistance. If S_2 is now closed, how long does it take for the inductor current to fall to 1 A?

73. A mild steel ring is wound with 200 turns. When the current is reduced from 7 A to 5 A, the flux decreases from 800 μWb to 760 μWb. What is the value of the coil's effective self-inductance over this range of current variation?

74. In Fig. 8-19, what are the initial values of I_1, I_2, and I_T *immediately* after switch S is closed?

FIGURE 8-19 Circuit for Problems 74 and 75.

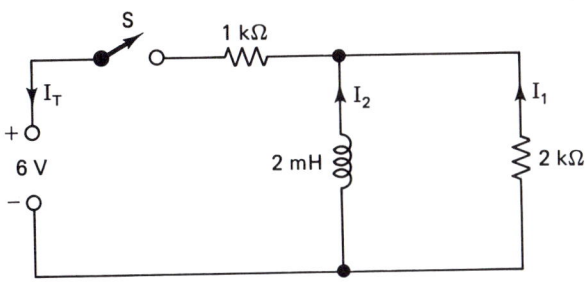

75. In Problem 74, assume that the transient state has been concluded. What are the steady-state values of I_1, I_2, and I_T?
76. In Fig. 8-20, what are the initial values of I_1, I_2, I_T, V_1, V_2, and V_3 *immediately* after switch S is closed?

FIGURE 8-20 Circuit for Problems 76 and 77.

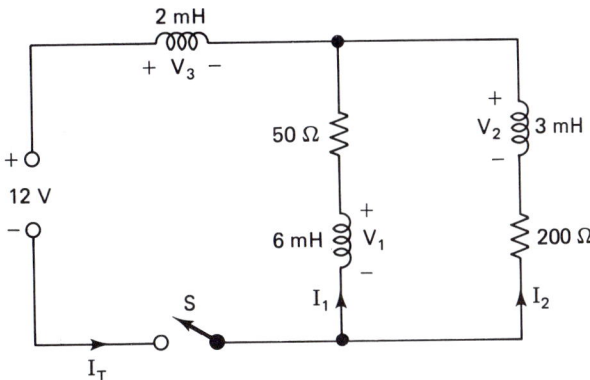

77. In Problem 76, assume that the transient state has been concluded. What are the steady-state values of I_1, I_2, I_T, V_1, V_2, and V_3?
78. A coil with 500 turns wrapped around a magnetic core has a diameter of 2 cm and a length of 12 cm. If the relative permeability of the core is 450 and the current through the coil is 6.5 A, what is the amount of energy stored in the magnetic field surrounding the coil?
79. In Problem 78, the current is decreased from 6.5 A to 5.0 A. How much energy is returned to the source?
80. A relay coil with a resistance of 500 Ω and an inductance of 7.5 H, operates when its current level reaches 40 mA. The relay is suddenly switched across a 25-V dc source. How soon will the relay operate after the switch is closed? [*Hint:* Refer to the exponential growth curve of Fig. 8-11(a).]

9

Capacitance and Electrostatics

When studying the subjects of electrostatics and capacitance, you will learn:

1. how objects are charged and are surrounded by electric field patterns that consist of flux lines.
2. how the charges and their separation feature in Coulomb's law.
3. about the relationship between the charge density, the electric field intensity, and the permittivity.
4. about the electrical properties of capacitance and the definition of its unit.
5. which physical factors determine the capacitance of a capacitor.
6. about the reasons for using the various types of capacitor.
7. how to calculate the amount of energy stored in the electric field associated with a capacitor.
8. what occurs when a number of capacitors are connected in series.
9. what occurs when a number of capacitors are connected in parallel and in series–parallel.
10. about the concept of the time constant and how it relates to the charging and discharging of a capacitor through a resistor.
11. how to troubleshoot for faults in capacitors.

9-0
Introduction

Now we come to the third and last property that an electrical circuit can possess—capacitance. Remembering that resistance opposes and limits the flow of current while inductance prevents any sudden change of *current*, what does capacitance do? Capacitance (previously called "capacity") is that property of an electrical circuit which prevents any sudden change of *voltage* and limits the rate of change of voltage; in other words, the voltage across a capacitance, whose letter symbol is C, cannot change instantaneously. As we shall see, the property of capacitance is associated with stationary charges, which is the subject of electrostatics, briefly discussed in Section 1-2. We start this chapter by studying electrostatics and its SI units in greater depth.

The device that possesses the property of capacitance is the capacitor (previously called a "condenser"), which consists basically of two conducting surfaces (for example, aluminum, copper, tin foil), separated by an insulator, also referred to as the dielectric (for example, air, mica, ceramic). Between these two surfaces is established an electric field or flux, so that we can alternatively think of capacitance as that property which draws energy from a source and *stores* it in the form of an electric field. Compare this with inductance, which stores energy in a magnetic field, and resistance, which dissipates energy as heat. Mechanically inductance is equivalent to mass, and resistance is equivalent to friction. Although there is a mechanical equivalent to capacitance, it is too complex to help in our understanding of the property.

9-1
Electric Flux. Coulomb's Law

As we learned in Section 1-2, an object or body that loses some electrons from its atoms is positively charged. Similarly, if the body gains electrons, it will become negatively charged. When a negatively charged body comes into contact with a positively charged body, electrons will leave the negative body and enter the positive body. This electron flow will continue until the charges on the two bodies have equalized.

If two charged bodies are not brought into contact but are only positioned near to one another, there will be no electron flow but an electric force will exist between the bodies. This phenomenon is called *static electricity* or *electrostatics*, since the charges cannot move.

The simplest way of producing a static charge is by the friction or triboelectric method. If two bodies are rubbed together, electrons will literally be "wiped off" from one body on to the other. This will not be successful if the bodies are good conductors, since equalizing currents will flow easily and prevent any appreciable charge from accumulating on the bodies. Good results are obtained by using two insulators, one of which is a hard rod and the other

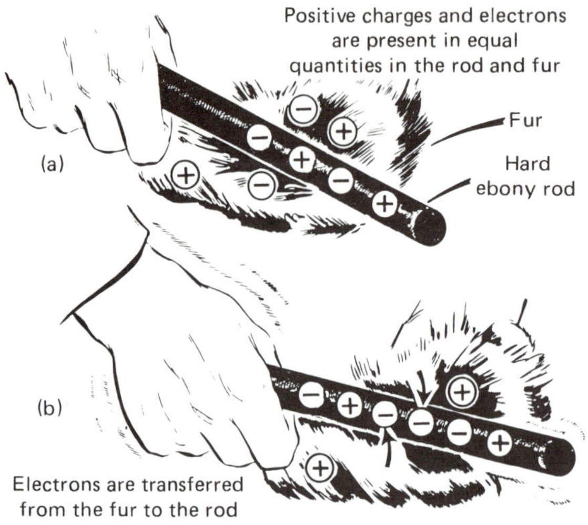

FIGURE 9-1 Use of friction to generate static electricity.

some form of fluffy fur. This is illustrated in Fig. 9-1, where an ebony rod is rubbed against a piece of cat's fur. Electrons are transferred from the fur to the rod so that the fur becomes positively charged and the rod negatively charged. If sufficient charge is eventually accumulated on both bodies, equalizing currents will flow in the form of sparks, which produce a crackling sound.

A fundamental law of electrostatics is that charges with the same polarity (*like* charges) *repel* each other, whereas charges with the opposite polarity (*unlike* charges) *attract* each other. This may be demonstrated by an experiment in which two light paper pulp (pith) balls are suspended close to one another but are initially separated [Fig. 9-2(a)]. An ebony rod is negatively charged and then brought into contact with the left-hand ball, which will therefore acquire a negative charge. The right-hand ball will be positive with respect to the left-hand ball, so that the two balls will be attracted toward one another and will be brought into contact [Fig. 9-2(b)]. This is the result of electrostatic induction, which can be compared with magnetic induction (Section 7-1). The negative charge will then be shared between the balls [Fig. 9-2(c)], so that they will then be repelled from each other and will swing apart [Fig. 9-2(d)].

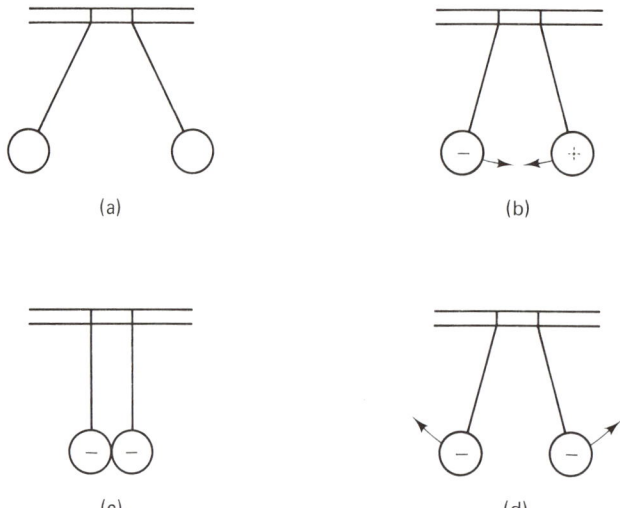

FIGURE 9-2 Forces of attraction and repulsion between charged bodies.

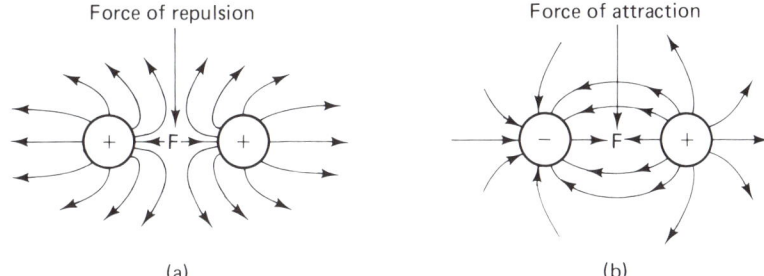

FIGURE 9-3 Electric fields between charged bodies: in (a), F is the force of repulsion; in (b), F is the force of attraction.

The influence that a charged body exerts on its surroundings is called its *electric* or *electrostatic field*. Such a field always terminates on a charged body and extends between positive and negative charges. It can exist in a vacuum or in such materials as air, glass, and waxed paper.

For convenience, an electric field is represented by lines of force which indicate the direction and the strength of the field, so that a strong field would be shown as a large number of lines close together. The total number of lines is referred to as the *electric flux*, whose letter symbol is the Greek lowercase letter psi, ψ. In the SI system a single line is assumed to emanate from a positive charge of 1 C and to terminate on an equal negative charge. The number of lines is therefore equal to the number of coulombs and we do not have to invent a new unit in order to measure the electric flux. By contrast, we did invent a new SI unit, the weber, with which to measure the magnetic flux, ϕ.

The electric flux patterns for the force of repulsion between like charges and the force of attraction between unlike charges are shown in Fig. 9-3. In 1785, Coulomb discovered through experiments that the magnitude of the force between two charged bodies was directly proportional to the product of the charges but inversely proportional to the square of their separation. If the charges are placed in a vacuum, the full equation for Coulomb's law is

$$\text{force, } F = \frac{Q_1 Q_2}{4\pi\epsilon_0 d^2} = \frac{kQ_1 Q_2}{d^2} \quad \text{newtons} \tag{9-1-1}$$

where Q_1, Q_3 = charges, measured in coulombs
d = distance in meters between the charges

and

$$k = \frac{1}{4\pi\epsilon_0} = \frac{1}{4 \times 3.1416 \times 8.85 \times 10^{-12}}$$
$$= 8.99 \times 10^9 \text{ SI units}$$

The meaning of ϵ_0, the permittivity of free space, will be explained in Section 9-2.

To summarize, the electric fields that surround charged bodies are illustrated by lines of force which show the direction and strength of the field. The total number of lines is the electric flux measured in coulombs. Unlike charges attract, like charges repel, and the force of the attraction or repulsion is directly proportional to the product of the charges but inversely proportional to the square of their separation.

Sec. 9-1 Electric Flux. Coulomb's Law

EXAMPLE 9-1 In a vacuum a positive charge of 5 μC is positioned 10 cm away from a negative charge of 5 μC. What are the values of the total electric flux and the force exerted on the charges? State whether the force is one of attraction or repulsion.

Solution The total electric flux, ψ, is 5 μC. Substitution into Eq. (9-1-1) yields

$$Q_1 = Q_2 = 5 \times 10^{-6} \text{ C}, \quad d = 10 \times 10^{-2} \text{ m}$$

$$\text{force, } F = \frac{8.99 \times 10^9 \times (5 \times 10^{-6})^2}{(10 \times 10^{-2})^2}$$

$$= 8.99 \times 25 \times 10^{-1} = 22.5 \text{ N}$$

Since the charges are unlike, the force will be one of attraction.

9-2 Charge Density. Electric Field Intensity. Permittivity of Free Space

Figure 9-4 represents two rectangular conducting plates and for each plate the area of one side is A square meters. These parallel plates are d meters apart and are positioned in a vacuum. We have charged the plates to a potential difference of V_C volts, with the result that one plate carries a positive charge of Q coulombs and the other plate has an equal negative charge. The total electrical flux, ψ, therefore contains Q lines of force and, consequently,

$$\psi = Q \quad \text{coulombs} \tag{9-2-1}$$

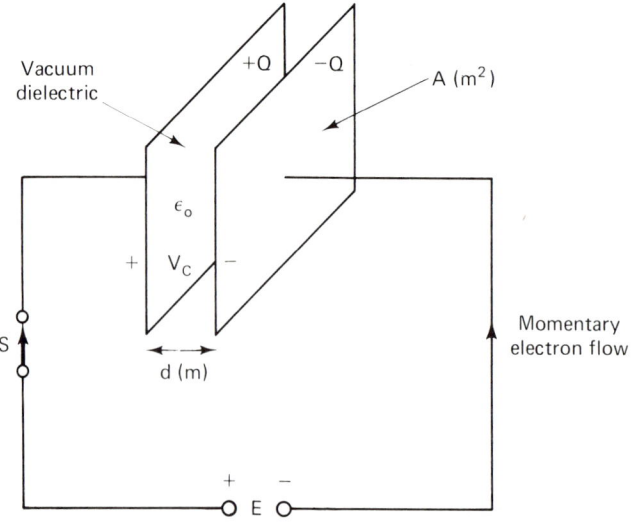

FIGURE 9-4 Basic capacitor.

The lines of force are spread over an area of A square meters. The electric flux density, whose letter symbol is D, will be given by

$$D = \frac{\psi}{A} = \frac{Q}{A} \quad \text{coulombs per square meter} \quad (9\text{-}2\text{-}2)$$

You should compare this with the magnetic flux density,

$$B = \frac{\phi}{A} \quad \text{webers per square meter}$$

The total potential difference of V_C volts is uniformly distributed along the lines of force which extend over the distance of d meters. Increasing the value of V_C and reducing the distance, d, will produce a stronger electric field. We can therefore refer to the electric field intensity, or electric field strength, whose letter symbol is $\mathscr{E}$. This will be expressed in terms of the voltage gradient, V_C/d, which is measured in volts per meter. In equation form:

$$\text{electric field intensity, } \mathscr{E} = \frac{V_C}{d} \quad \text{volts per meter} \quad (9\text{-}2\text{-}3)$$

This is comparable with the magnetic field intensity,

$$H = \frac{NI}{l} \quad \text{amperes per meter}$$

If a negative charge in coulombs is positioned in the vacuum, it will be repelled from the negative plate and attracted toward the positive plate. The force exerted in newtons will depend on the amount of the charge and the value of the electric field intensity. Therefore,

force in newtons = (charge in coulombs)

× (electric field intensity in volts per meter)

or

$$\text{electric field intensity (V/m)} = \frac{\text{force (N)}}{\text{charge (C)}} \quad (9\text{-}2\text{-}4)$$

The electric field intensity is 1 V/m if when a charge of 1 C is placed in the field, the force exerted on the charge is 1 N. Consequently, electric field intensity may be measured in either volts per meter or newtons per coulomb.

In electromagnetism the magnetic field intensity, H, acting in a vacuum, resulted in the creation of a flux density B. Their ratio, B/H, was equal to the permeability of free space, μ_0, whose value was $4\pi \times 10^{-7}$ H/m in terms of SI units. Similarly, in electrostatics an electric field intensity, $\mathscr{E}$, acting in a vacuum, will create an electric flux density, D. Their ratio of $D/\mathscr{E}$ is called the permittivity of free space, whose letter symbol is ϵ_0. As an equation,

$$\frac{D \text{ (C/m}^2\text{)}}{\mathscr{E} \text{ (V/m)}} = \epsilon_0 \quad (9\text{-}2\text{-}5)$$

The value of ϵ_0 is 8.85×10^{-12} SI unit. These units may either be expressed in coulombs per volt-meter or, as we shall see later, in farads per meter. It is worth mentioning that since $\mu_0 = 4\pi \times 10^{-7}$ SI unit and $\epsilon_0 = 8.85 \times 10^{-12}$ SI unit,

$$\frac{1}{\sqrt{\mu_0 \epsilon_0}} = 3 \times 10^8 \text{ m/s} \qquad (9\text{-}2\text{-}6)$$

This is the velocity in free space of all electromagnetic waves which consist of electric and magnetic fields; examples are light and radio waves. A famous Scottish mathematician, Clerk Maxwell, discovered this and other relationships in 1865, with the result that he was able to predict the existence of radio waves some 20 years before they were successfully demonstrated through the experiments of Heinrich Hertz.

This section has covered the concept of the electric field intensity, $\mathscr{E}$, which is measured in terms of the potential gradient (volts per meter). In a vacuum the electric field intensity, $\mathscr{E}$, is related to the charge density, D (coulombs per square meter) by the permittivity of free space, ϵ_0.

EXAMPLE 9-2 In Fig. 9-5 the area of one side of each plate is 5 cm² and their separation in a vacuum is 2 cm. Calculate the values of the electric field intensity, the electric flux density, and the charge on each plate.

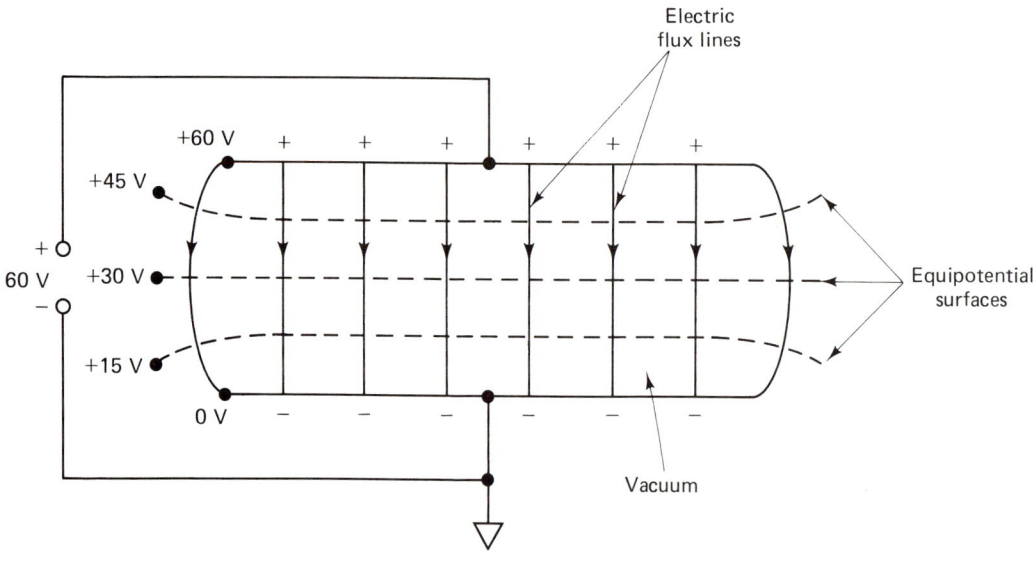

FIGURE 9-5 Circuit for Example 9-2.

Solution

$$A = 5 \times 10^{-4} \text{ m}^2, \quad d = 2 \times 10^{-2} \text{ m}, \quad V_C = 60 \text{ V}$$

electric field intensity, $\mathscr{E} = \dfrac{V_C}{d}$ \hfill (9-2-3)

$$= \frac{60 \text{ V}}{2 \times 10^{-2} \text{ m}} = 3000 \text{ V/m}$$

$$\text{flux density}, D = \mathcal{E} \times \epsilon_0$$
$$= 3000 \times 8.85 \times 10^{-12} \qquad (9\text{-}2\text{-}5)$$
$$= 2.655 \times 10^{-8}\, \text{C/m}^2$$
$$\text{charge on each plate}, Q = \text{flux}, \psi = D \times A \qquad (9\text{-}2\text{-}2)$$
$$= 2.655 \times 10^{-8} \times 5 \times 10^{-4}\,\text{C} = 13.3\,\text{pC}$$

9-3 Factors that Determine the Capacitance. Relative Permittivity. Dielectric Strength. Types of Capacitors

The property of capacitance is possessed by a capacitor, which is a device for storing charge and consists, as we have already said, of two conducting surfaces which are separated by an insulator, also called the *dielectric*. Figure 9-4 therefore represents a capacitor with a vacuum as the dielectric. When a dc voltage, E, is switched across the capacitor, there is a momentary current flow during which electrons are drawn off the left-hand plate, flow through the voltage source, and are then deposited on the right hand plate. This electron flow quickly ceases when the voltage, V_C, between the plates exactly balances the source voltage, E. The charge then stored by the capacitor depends on the applied voltage and also on some constant of the capacitor called the *capacitance*. The letter symbol for capacitance is C and its unit is the farad, F (named after Michael Faraday). In equation form:

$$V_C(\text{volts}) = -\frac{Q(\text{coulombs})}{C(\text{farads})} \qquad (9\text{-}3\text{-}1)$$

The negative sign indicates that the voltage, V_C, across the capacitor exactly balances the source voltage, E. Using KVL,

$$E + V_C = 0$$

Therefore,

$$E = \frac{Q}{C} \qquad (9\text{-}3\text{-}2)$$

where E = source voltage (V)
Q = charge stored (C)
C = capacitance (F)

It follows that the capacitance is 1 F if when the voltage applied between the capacitor plates is 1 V, the charge stored by the capacitor is 1 C. Unfortunately, the farad is far too large a unit for practical purposes, so that capacitances are normally measured in either microfarads (1 μF = 1 × 10^{-6} F) or picofarads (1 pF = 1 $\mu\mu$F = 1 × 10^{-6} μF = 1 × 10^{-12} F); a picofarad may alternatively be referred to as a micromicrofarad ($\mu\mu$F). However when substituting a value of capacitance into an equation, that value must always be in farads. From Eqs. (9-3-1), (9-2-5), (9-2-2), and (9-2-3),

$$\text{permittivity of free space, } \epsilon_0 = \frac{D}{\mathscr{E}} = \frac{Q/A}{E/d} = \frac{Q}{E} \times \frac{d}{A} = \frac{C \times d}{A} \quad (9\text{-}3\text{-}3)$$

With C in farads, d in meters, and A in square meters, the SI units of ϵ_0 are farads per meter. Transposing Eq. (9-3-3) yields

$$C = \epsilon_0 \times \frac{A}{d} \quad \text{farads} \quad (9\text{-}3\text{-}4)$$

Therefore, the capacitance is directly proportional to the area of the plates but is inversely proportional to their distance apart. This distance will be equal to the thickness of the dielectric.

Since $V_C = -Q/C$, it is impossible to alter the voltage across the capacitor's plates without changing the value in coulombs of the stored charge. From Eq. (1-3-1), $Q = I \times T$, and therefore Q, the charge in coulombs, cannot be changed unless a current, I (in amperes or coulombs per second) shall flow and either charge or discharge the capacitor over a period of time, T seconds. This means that a capacitor can neither charge nor discharge instantaneously, and therefore capacitance is that electrical property which prevents any sudden change in the voltage and limits the voltage's rate of change.

Now let us look at it another way. Since charge, Q = capacitance, C × volts, E, it follows that

$$\text{rate of change of charge with time, } \frac{\Delta Q}{\Delta T}$$
$$= \text{capacitance, } C, \times \text{rate of change of voltage, } \frac{\Delta E}{\Delta T} \quad (9\text{-}3\text{-}5)$$

But $\Delta Q/\Delta T$ is measured in coulombs per second and is therefore the current, I. As a result,

$$I = C \times \frac{\Delta E}{\Delta T} \quad (9\text{-}3\text{-}6)$$

or

$$\text{rate of change of voltage, } \frac{\Delta E}{\Delta T} = \frac{I}{C} \quad \text{volts per second} \quad (9\text{-}3\text{-}7)$$

Since $\Delta E/\Delta T$ is inversely proportional to C, the capacitance will limit the rate of change of the voltage. This should be compared with Eq. (8-1-7), $\Delta I/\Delta T = E/L$ amperes per second, which showed that the inductance limited the rate of change of the current.

But what does all this mean as far as capacitance is concerned? Let us keep the figures simple by assuming that a 1-F capacitor, initially uncharged, is connected to a constant current generator whose output is 1 A (1 C/s). After 1 s the capacitor will have acquired 1 C and the potential difference between its plates will be 1 C/F = 1 V. When 2 s have passed, the potential difference will be 2 C/F = 2 V, so that the potential difference is increasing at the rate of 1 V/s. Now let us start again, but this time we will increase the capacitance to 2 F. After 1 s the potential difference is only 1 C/2 F = 0.5 V, after 2 s the potential difference is 2 C/2 F = 1 V, and therefore the rate of change of the voltage has fallen to 0.5 V/s. Consequently, the capacitance limits the rate of change of the voltage since doubling the value of C has halved the corresponding value of $\Delta E/\Delta T$.

If the vacuum between the capacitor's plates is replaced by some type of insulating material, the capacitance is increased. The ratio of a capacitor's capacitance with an insulating material as the dielectric, to its capacitance with a vacuum dielectric, is equal to the material's *relative permittivity* (sometimes referred to as the *dielectric constant*). Since the ratio of two capacitances is involved, ϵ_r is just a number and has no units. Therefore,

$$\text{relative permittivity of an insulating material, } \epsilon_r = \frac{\text{capacitance with the insulating material as the dielectric}}{\text{capacitance with a vacuum dielectric}} \quad (9\text{-}3\text{-}8)$$

The values of the relative permittivity for some commonly used dielectrics are shown in Table 9-1. When the space between the capacitor plates is filled by a dielectric with a relative permittivity, ϵ_r,

$$\text{capacitance, } C = \frac{\epsilon_0 \epsilon_r A}{d} \quad \text{farads} \quad (9\text{-}3\text{-}9)$$

It follows that

$$\frac{\text{electric flux density, } D}{\text{electric field intensity, } \mathscr{E}} = \epsilon_0 \epsilon_r = \epsilon \quad (9\text{-}3\text{-}10)$$

The quantity ϵ is called the dielectric's *absolute permittivity*, which will be measured in farads per meter. Notice that since ϵ_r for air is 1.0006, the absolute permittivity for air (and other gases) is very near to that of free space (vacuum).

Equation (9-3-9) should be compared with Eqs. (7-5-3) and (7-5-4), which state that

$$\frac{\text{magnetic flux density, } B}{\text{magnetic field intensity, } H} = \mu_0 \mu_r = \mu$$

where μ is the absolute permeability.

TABLE 9-1 Relative Permittivity Values of Some Dielectrics

Dielectric Material	Relative Permittivity, ε_r	Dielectric Strength (kV/mm)
Air	1.0006	3
Mica	3–6	50
Waxed paper	3.5	20
Glass	5–10	100

From the formula

$$C = \frac{\epsilon_0 \epsilon_r A}{d} = \frac{\epsilon A}{d}$$

it would appear that we could indefinitely increase the capacitance (without raising the capacitor's physical size) by reducing the value of d, or, in other words, by bringing the conducting plates closer together. However, we must remember that when we charge a capacitor, there is a potential difference between the capacitor plates, so that if the dielectric is too thin, it can break down and arcing will occur between the plates. The capacitor is then short-circuited, which may cause damage to other parts of the circuit. The arcing is due to electrons being torn out of their orbits within the dielectric and is the result of an excessive electric field intensity. The ability of a particular dielectric to withstand a high electric field intensity is measured by its dielectric strength, which is most conveniently expressed in kilovolts per millimeter. The dielectric strengths of the common insulators used in capacitors are shown in Table 9-1. Because of differences in manufacture, the quoted figures are average values. As an example, the average dielectric strength of mica is 50 kV/mm; this means that in order to avoid breakdown, not more than 50,000 V should be applied across a 1-mm thickness of a mica dielectric.

Because of its insulator's dielectric strength, each capacitor has a voltage rating as recommended by the manufacturer. This is the dc working voltage (WVDC) and is the highest dc voltage that should be applied across the capacitor without risking a breakdown of the dielectric. In this respect we must point out that dc and ac voltages must be regarded differently; as we shall see, an ac voltage is commonly given in terms of its effective value, which is only about 70% of its peak value. Consequently, if an ac voltage with an effective value of 100 V is applied across a capacitor whose rating is 100 WVDC, dielectric breakdown may occur.

The voltage rating is important since it determines a capacitor's size; if the same dielectric is used, a 0.1-μF capacitor with a 1-kV WVDC rating will be larger in appearance than another 0.1-μF capacitor with only a 100-V WVDC rating. In this sense the voltage rating of a capacitor may be compared with the power rating of a resistor and the current rating of an inductor.

So far we know that a capacitor has a value in μF or pF as determined by its physical construction, uses a particular type of dielectric, and has a voltage rating. These factors should be borne in mind as we discuss the various types of practical capacitors which have fixed values of capacitance. The schematic symbol for most such capacitors is ─||─ or ─)|─; the symbol with the curved line is more widely used since it avoids confusion with the symbol for open-circuit relay contacts. The curved line itself usually represents the outer conducting surface of the capacitor and is normally connected to the negative side of an electrical circuit. The common types of fixed-value capacitors are:

PAPER CAPACITOR [Fig. 9-6(a)] The conducting surfaces are flat thin strips of metal foil separated by a paper dielectric. These are rolled together into a tubular form and then impregnated with wax or resin to exclude moisture. The outer covering may be either cardboard, molded hard plastic, or metal. The available values of capacitance commonly range from 100 pF to 4 μF, while the voltage ratings lie between 200 and 600 V.

MICA CAPACITOR [Fig. 9-6(b)] A number of metal plates are sandwiched between mica sheets and the whole assembly is then covered in molded plastic. The values of capacitance typically range from a few picofarads up to 0.05 μF; the larger values are limited by the capacitor's high cost. Mica is an excellent

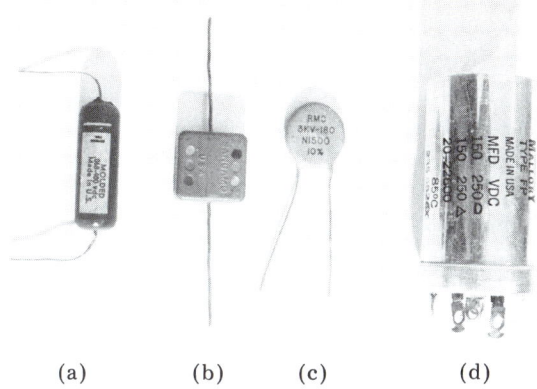

FIGURE 9-6 Types of fixed-value capacitor; (a) paper capacitor; (b) mica capacitor; (c) ceramic capacitor; (d) electrolytic capacitor.

insulator and can stand higher voltages than paper; consequently, the voltage ratings can extend up to several kilovolts.

CERAMIC CAPACITOR One such capacitor uses a hollow ceramic cylinder as the dielectric. Thin metal films are then deposited on the inner and outer surfaces of the cylinder. A second type of ceramic capacitor [Fig. 9-6(c)] is manufactured in the shape of a disk. After leads are attached to the metal surfaces on each side of the capacitor, it is completely covered with an insulating moisture-proof coating. The range of capacitance values extend from a few picofarads to 0.1 μF, with voltage ratings up to several kilovolts. If the value of capacitance is shown on the coating as a whole number, that value is in picofarads; however, values in microfarads will contain a decimal point.

Ceramic capacitors are extremely stable and their change of capacitance with change in temperature is predictable. This temperature coefficient is measured in parts per million per degree Celsius (ppm/°C) and may be either positive or negative. As an example, a 0.01-μF ceramic capacitor may be marked "P 100 (+100 ppm/°C)"; since 0.01 μF = 10,000 pF, the capacitance would increase by (10,000 × 100)/1,000,000 = 1 pF for every 1°C rise in temperature.

The values of paper, mica, and ceramic capacitors may be marked directly on the cover or indicated by a color code, which is described in Appendix C.

ELECTROLYTIC CAPACITOR [Fig. 9-6(d)] This type is capable of producing very high capacitance values but is much smaller in size than the paper, mica, and ceramic capacitors. For the higher voltage ratings which exceed 450 V, the range of capacitances available is typically from 4 to 200 μF, but with the lower voltage ratings, the range is extended to 600 μF. In some computer circuits, you can find capacitance values up to 0.2 F! Generally speaking, electrolytic capacitors are found in circuits (for example, power supplies), where it is not important if the actual value of capacitance differs appreciably from the rated value.

The common type of electrolytic capacitor contains two aluminum surfaces, one of which is usually the external can or housing. The other surface is aluminum foil which is in contact with an electrolyte of ammonium borate. During manufacture a current is passed through the capacitor and this forms a thin aluminum oxide film on the foil surface. This oxide film, which is approximately 0.00001 cm thick, is the capacitor's dielectric. The fact that the dielectric is so thin accounts for the very high capacitance values.

Unlike the other types of capacitor, you must observe polarity when connecting electrolytic capacitors into a circuit. For this reason the terminals of electrolytic capacitors are clearly marked "+" and "−"; internally, the +

terminal is connected to the foil and the − terminal to the can. Externally, the + and − terminals must be connected to the positive and negative sides, respectively, of the circuit's dc voltage source. If this is not done, there will be a flow of current through the capacitor in the opposite direction to the flow that originally formed the dielectric film. As a result, the chemical action will be reversed, the film will be destroyed, and the capacitor is short circuited; in addition, gas is released during the destruction of the film, so that pressure builds up inside the unit.

Figure 9-6(d) shows an electrolytic capacitor which has, for convenience, four common negative terminals connected directly to the can. However, there are two separate positive terminals, which means that there are two electrolytic capacitors housed in the unit. As indicated on the information pasted to the side of the container, these are 150-µF capacitors, each with a 250 WVDC rating. The normal identifying marks for the positive terminals are the half-moon, triangle, and square, which are stamped on the bottom of the container and are related to the information on the side of the unit.

Radio receivers usually contain a *variable capacitor*, which allows you to tune in to the various stations. The most common type is illustrated in Fig. 9-7(a); it consists of two sets of metal plates (aluminum or brass) separated by air as the dielectric. One set of plates is called the *stator*, which is rigidly attached to an insulating frame. The other set is the *rotor*, which is mounted on a shaft with bearings; at the end of the shaft is a knob to control the capacitance. As the rotor is moved, its plates slide in between the stator plates; this changes the effective areas of the opposing sets of plates and therefore varies the capacitance. Because of the limitation of physical size, variable capacitors provide only small values of capacitance. On the low side the range is typically 3 to 30 pF, while on the high side the range can be from 500 to 1000 pF. The symbol for a variable capacitor is ⇥⊬, ⇥⊢, or ⇥⊬ with the curved line generally used to indicate the rotor.

The same basic type of construction is used in the adjustable capacitor, which is found in circuits where, on occasions, it is necessary to reset the value of the capacitance. Here the rotor shaft is shortened and slotted so that a screwdriver can control the value; the rotor's position is then secured by a locking nut that fits over the shaft.

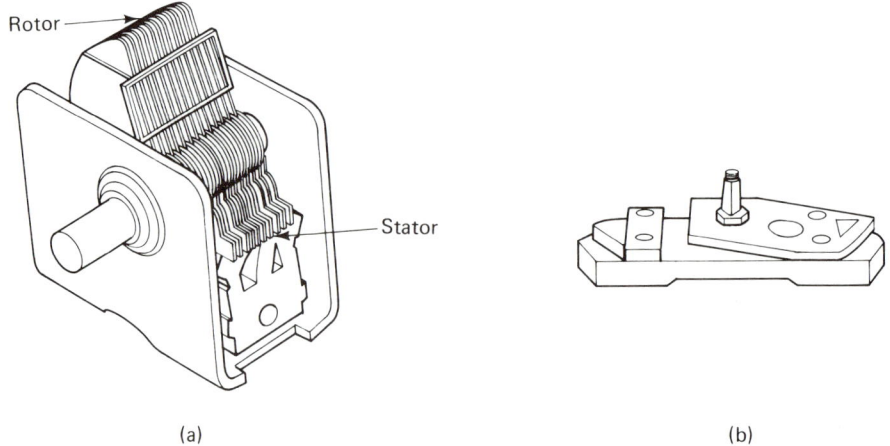

(a)   (b)

FIGURE 9-7 Variable and adjustable capacitors. (From Abraham Marcus and Charles Thomson, *Electricity for Technicians*, Third Edition, Prentice Hall, Englewood Cliffs, N.J., 1982, p. 192.)

Another type of adjustable capacitor [Fig. 9-7(b)] changes the distance between the plates in order to reset the capacitance. This is known as a *compression capacitor* and uses a mica dielectric. A screw then threads through a springy phosphor bronze plate and the insulating frame. The technician uses a screwdriver to vary the pressure on the unit; this will alter the effective distance between the plates and change the capacitance. The values of adjustable capacitors are of the order of a few picofarads.

In this section we covered the following aspects of capacitance:

1. Capacitance prevents any sudden change in voltage and is measured in farads. Also, capacitance is the ratio of the charge stored in the capacitor to the voltage applied between the capacitor's plates ($C = Q/E$).
2. Since $C = \epsilon_0\epsilon_r A/d$, the capacitance is directly proportional to the area of the plates and the relative permittivity but is inversely proportional to the thickness of the dielectric.
3. Because of the insulator's dielectric strength, each capacitor has a voltage rating (WVDC), which primarily determines its physical size.

EXAMPLE 9-3 Two parallel aluminum plates are separated by 1 mm in air. If the area of one side of each plate is 100 cm², calculate the value of the capacitance.

Solution

$$d = 1 \times 10^{-3} \text{ m}, \quad A = 100 \times 10^{-4} \text{ m}^2$$

$$\text{capacitance, } C = \frac{\epsilon_0 \epsilon_r A}{d} \quad\quad (9\text{-}3\text{-}9)$$

$$= \frac{8.85 \times 10^{-12} \times 1.0006 \times 100 \times 10^{-4} \text{ F}}{1 \times 10^{-3}}$$

$$= 89 \text{ pF} \quad \text{(rounded off)}$$

This example illustrates the low value of capacitance provided by an air capacitor even though its physical size is large.

EXAMPLE 9-4 A capacitor consists of two sheets of metal foil, each side of which has an area of 2000 cm². The sheets are separated by a 0.1-mm thickness of paper dielectric whose relative permittivity is 2.5. Calculate the value of the capacitance. If the capacitor is charged to 100 V, what are the values of the (a) charge stored, (b) electric flux density, and (c) electric field intensity?

Solution Substitution into Eq. (9-3-9) gives

$$d = 0.1 \times 10^{-3} \text{ m}, \quad A = 2000 \times 10^{-4} \text{ m}^2, \quad \epsilon_r = 2.5$$

$$\text{capacitance, } C = \frac{8.85 \times 10^{-12} \times 2.5 \times 2000 \times 10^{-4} \text{ F}}{0.1 \times 10^{-3}}$$

$$= 0.044 \text{ }\mu\text{F} \quad \text{(rounded off)}$$

(a) From Eq. (9-3-2),

$$\text{charge stored, } Q = CE = 0.044 \; \mu\text{F} \times 100 \text{ V} = 4.4 \; \mu\text{C}$$

(b) From Eq. (9-2-2),

$$\text{electric flux density, } D = \frac{Q}{A} = \frac{4.4 \; \mu\text{C}}{2000 \times 10^{-4} \text{ m}^2} = 22 \; \mu\text{C/m}^2$$

(c) From Eq. (9-2-3),

$$\text{electric field intensity, } \mathscr{E} = \frac{V_C}{d} = \frac{100 \text{ V}}{0.1 \times 10^{-3} \text{ m}} = 1 \times 10^6 \text{ V/m}$$

You can check these results by using Eqs. (9-2-5) and (9-3-10):

$$\frac{D}{\mathscr{E}} = \frac{22 \; \mu\text{C/m}^2}{1 \times 10^6 \text{ V/m}} = 22 \times 10^{-12} \text{ F/m}$$

and

$$\frac{D}{\mathscr{E}} = \epsilon = \epsilon_0 \epsilon_r = 8.85 \times 10^{-12} \times 2.5 = 22 \times 10^{-12} \text{ F/m}$$

9-4   Energy Stored in an Electric Field

SL.

In Fig. 9-8, a capacitor (initially uncharged) with a capacitance of C farads is being charged by a constant current, I amperes, for a period of t seconds. At the end of this time the final charge stored is Q coulombs, which, from Eq. (1-3-1) is equal to $I \times t$. The final voltage, V_C, between the capacitor plates is Q/C [Eq. (9-3-1)] and the average voltage during the charging period is $V_C/2$. Therefore,

$$\begin{aligned}\text{average power} &= \text{average voltage} \times \text{charging current} \\ &= \frac{V_C}{2} \times I \quad\quad\quad\quad (9\text{-}4\text{-}1) \\ &= \frac{V_C}{2} \times \frac{Q}{t}\end{aligned}$$

and

$$\begin{aligned}\text{energy stored} &= \text{average power} \times \text{time} \\ &= \frac{V_C}{2} \times \frac{Q}{t} \times t \quad\quad\quad (9\text{-}4\text{-}2) \\ &= \frac{1}{2} \times QV_C \quad \text{joules}\end{aligned}$$

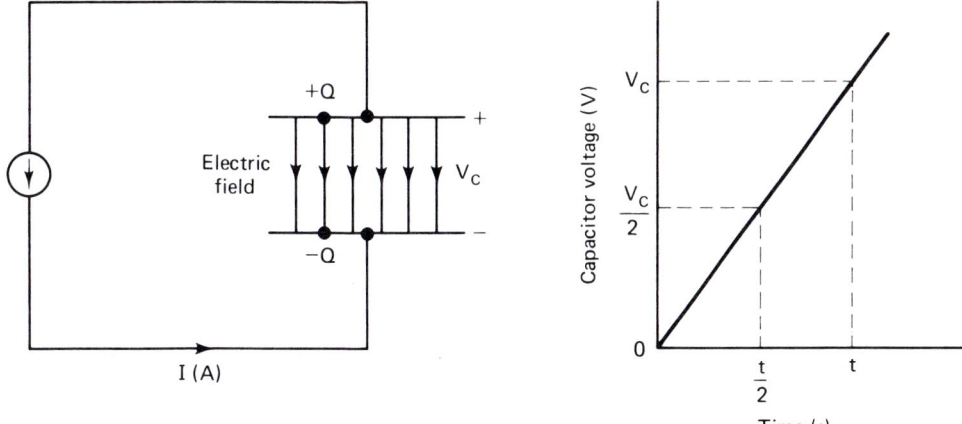

FIGURE 9-8 Energy stored in a capacitor.

Since $Q = CV_C$ [Eq. (9-3-1)],

$$\text{energy stored} = \frac{1}{2} Q \times \left(\frac{Q}{C}\right) = \frac{1}{2} \times \frac{Q^2}{C} \quad \text{joules} \quad (9\text{-}4\text{-}3)$$

$$= \frac{1}{2} \times (CV_C) \times V_C = \frac{1}{2} \times C \times V_C^2 \quad \text{joules} \quad (9\text{-}4\text{-}4)$$

The quantity $\frac{1}{2} CV_C^2$ for the energy stored in the electric field between the capacitor plates should be compared with $\frac{1}{2} LI^2$, which is the expression for the energy stored in the magnetic field surrounding an inductor.

EXAMPLE 9-5 A 4-µF capacitor is charged by a 50-V dc source. Calculate the amount of energy stored in the electric field between the capacitor's plates.

Solution

$$\text{energy stored} = \frac{1}{2} \times C \times V_C^2 = \frac{1}{2} \times 4 \text{ µF} \times (50 \text{ V})^2 = 5000 \text{ µJ}$$

9-5 Capacitors in Series and in Parallel

Capacitors in Series

In Fig. 9-9, the capacitors $C_1, C_2, C_3, \ldots, C_N$ are connected in series. When the switch, S, is closed, these capacitors are rapidly charged by the source voltage, E_T. The momentary charging current must be the same throughout the circuit, and therefore at the end of the charging interval, each capacitor must store the same charge, Q. Since there is no electron flow through the

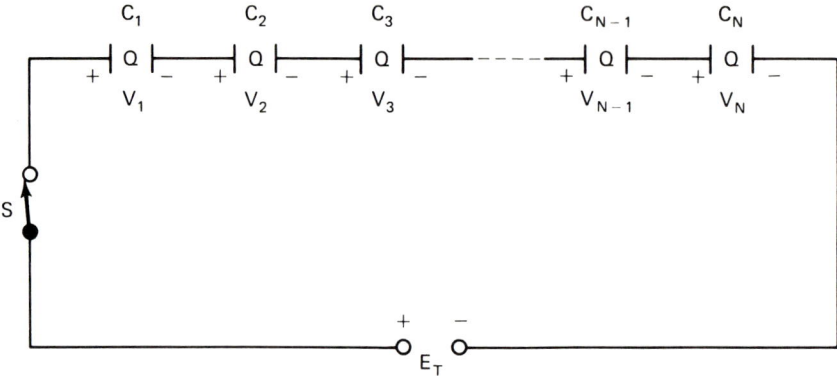

FIGURE 9-9 Capacitors in series.

dielectric, the capacitors $C_2, C_3, \ldots, C_{N-1}$ are charged by electrostatic induction. From Eq. (9-3-1),

$$V_1 = -\frac{Q}{C_1}, \quad V_2 = -\frac{Q}{C_2}, \quad V_3 = -\frac{Q}{C_3}, \ldots, V_N = -\frac{Q}{C_N}$$

By KVL, $E_T + V_1 + V_2 + V_3 + \cdots + V_N = 0$. Therefore,

$$E_T = \frac{Q}{C_1} + \frac{Q}{C_2} + \frac{Q}{C_3} + \cdots + \frac{Q}{C_N}$$
$$= Q \times \left(\frac{1}{C_1} + \frac{1}{C_2} + \frac{1}{C_3} + \cdots + \frac{1}{C_N}\right) \tag{9-5-1}$$

If C_T is the total equivalent capacitance that will store the same charge, Q, with the same applied voltage, E_T,

$$E_T = \frac{Q}{C_T} = Q \times \frac{1}{C_T} \tag{9-5-2}$$

Comparison of Eqs. (9-5-1) and (9-5-2) yields

$$\frac{1}{C_T} = \frac{1}{C_1} + \frac{1}{C_2} + \frac{1}{C_3} + \cdots + \frac{1}{C_N}$$

or

$$C_T = \frac{1}{\frac{1}{C_1} + \frac{1}{C_2} + \frac{1}{C_3} + \cdots + \frac{1}{C_N}} \tag{9-5-3}$$

This is our well-known reciprocal formula, but this time it applies to a series, not a parallel, arrangement. Connecting capacitors in series will reduce, not increase, the capacitance; the total capacitance itself will be less than the value of the smallest series capacitance. Fundamentally, a series arrangement creates an increase in the distance between the end plates which are connected to the source voltage and store the charge, Q. Referring to Section 9-3, this raising of the distance between the effective plates of a capacitor will reduce the capacitance. So why not use a lower-value capacitance in the first place and dispense with the series arrangement? Because a capacitor has a dc working voltage rating, one purpose of the series arrangement is to divide the source voltage between the capacitors so that the voltage across an individual capacitor does not exceed its rating. Since $E_T = Q/C_T$ and $Q = C_1V_1 = C_2V_2 = \cdots = C_NV_N$,

$$\frac{V_1}{V_2} = \frac{C_2}{C_1}, \quad \text{etc.} \tag{9-5-4}$$

and

$$E_T = \frac{Q}{C_T} = V_1 \times \frac{C_1}{C_T} = V_2 \times \frac{C_2}{C_T} = V_3 \times \frac{C_3}{C_T} = \cdots = V_N \times \frac{C_N}{C_T} \tag{9-5-5}$$

Therefore,

$$V_1 = E_T \times \frac{C_T}{C_1}, \quad V_2 = E_T \times \frac{C_T}{C_2},$$
$$V_3 = E_T \times \frac{C_T}{C_3}, \ldots, V_N = E_T \times \frac{C_T}{C_N} \tag{9-5-6}$$

Equations (9-5-4), (9-5-5), and (9-5-6) show how the source voltage is divided between the series capacitors. If the N series capacitors all have the same capacitance, C, then

$$C_T = \frac{C}{N} \tag{9-5-7}$$

For two capacitors in series,

$$C_T = \frac{C_1C_2}{C_1 + C_2} \quad \text{(product-over-sum formula)} \tag{9-5-8}$$

and

$$V_1 = E_T \times \frac{C_2}{C_1 + C_2}, \quad V_2 = E_T \times \frac{C_1}{C_1 + C_2} \tag{9-5-9}$$

Capacitors in Parallel

In Fig. 9-10, the capacitors $C_1, C_2, C_3, \ldots, C_N$ are connected in parallel and each capacitor is therefore charged by the source voltage, E_T. Since the capacitances are different, the charges stored by the individual capacitors are not the same and are given by

$$Q_1 = C_1E_T, \quad Q_2 = C_2E_T,$$
$$Q_3 = C_3E_T, \ldots, Q_N = C_NE_T \tag{9-5-10}$$

Sec. 9-5 Capacitors in Series and in Parallel

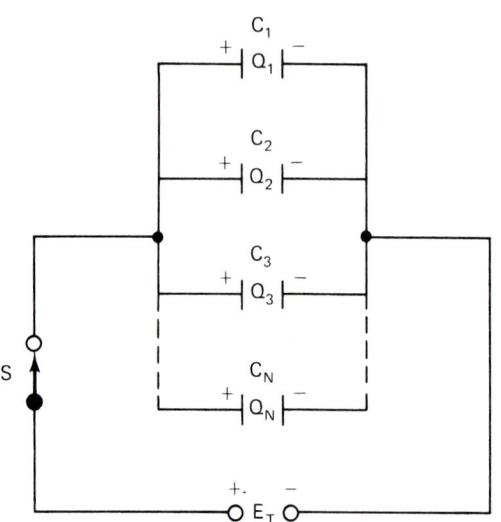

FIGURE 9-10 Capacitors in parallel.

or

$$\frac{Q_1}{Q_2} = \frac{C_1}{C_2}, \text{ etc.} \qquad (9\text{-}5\text{-}11)$$

The total charge, Q_T, is the sum of the individual charges stored in the capacitors. Therefore,

$$\begin{aligned} Q_T &= Q_1 + Q_2 + Q_3 + \cdots + Q_N \\ &= C_1 E_T + C_2 E_T + C_3 E_T + \cdots + C_N E_T \\ &= E_T \times (C_1 + C_2 + C_3 + \cdots + C_N) \end{aligned} \qquad (9\text{-}5\text{-}12)$$

If C_T is the total capacitance,

$$Q_T = E_T \times C_T \qquad (9\text{-}5\text{-}13)$$

Comparing Eqs. (9-5-12) and (9-5-13), we obtain

$$C_T = C_1 + C_2 + C_3 + \cdots + C_N \qquad (9\text{-}5\text{-}14)$$

This shows that the total equivalent capacitance of capacitors in *parallel* is the sum of their individual capacitances (this is the same way in which we combined resistances or inductances in *series*). A parallel arrangement of capacitors increases the effective surface area and therefore raises the capacitance, which is directly proportional to the area.

When the N parallel capacitors all have the same capacitance, C,

$$C_T = NC \qquad (9\text{-}5\text{-}15)$$

If a multiplate capacitor contains a total of N plates with alternate plates connected as in Fig. 9-11, there are $(N - 1)$ parallel capacitors, each with a capacitance of $C = \epsilon_0 \epsilon_r A / d$. The total equivalent capacitance is

$$C_T = (N - 1)C = \frac{\epsilon_0 \epsilon_r (N - 1) A}{d} \quad \text{farads} \qquad (9\text{-}5\text{-}16)$$

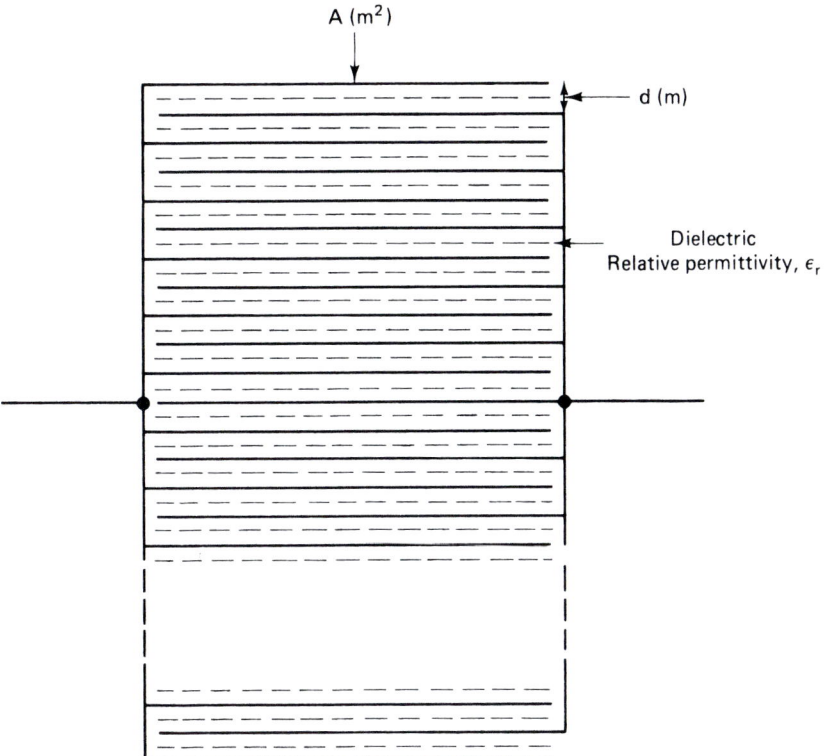

FIGURE 9-11 Multiplate capacitor.

We have learned that capacitors in series carry the same charge and that we must use the reciprocal or product-over-sum formula to calculate the total capacitance. With capacitors in parallel the total charge is equal to the sum of the charges stored in the individual capacitors and the total capacitance is equal to the sum of their individual capacitances.

EXAMPLE 9-6 In Fig. 9-12(a), $C_1 = 20$ μF, $C_2 = 5$ μF, $C_3 = 4$ μF, and $E_T = 58$ V. Calculate the values of C_T, V_1, V_2, and V_3. What are the amounts of charge and energy stored in each capacitor?

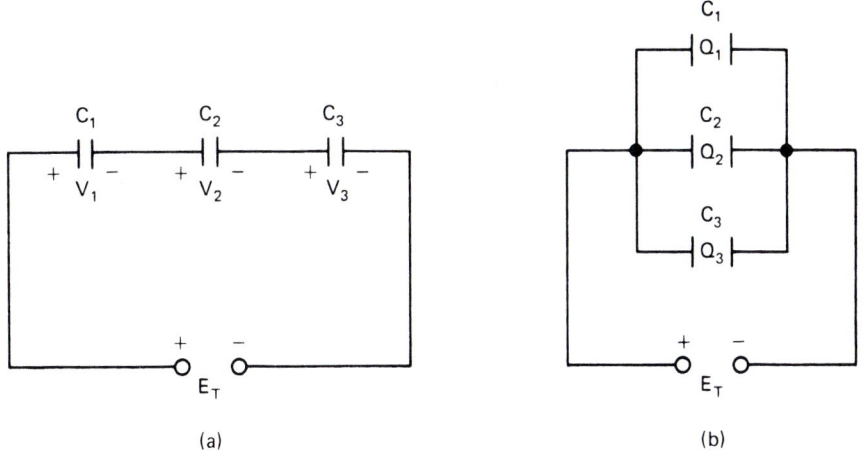

FIGURE 9-12 Circuits for Examples 9-6 and 9-7.

Sec. 9-5 Capacitors in Series and in Parallel

Solution The capacitors are in series and we must therefore use the reciprocal formula to calculate the value of C_T. Substitute in Eq. (9-5-3):

$$\text{total equivalent capacitance, } C_T = \frac{1}{\frac{1}{20} + \frac{1}{5} + \frac{1}{4}}$$

$$= \frac{20}{1 + 4 + 5} = \frac{20}{10} = 2 \text{ }\mu\text{F}$$

From Eq. (9-5-2),

$$\text{charge stored in each capacitor, } Q = C_T E_T = 2 \text{ }\mu\text{F} \times 58 \text{ V} = 116 \text{ }\mu\text{C}$$

and from Eq. (9-3-1),

$$\text{potential difference, } V_1 = \frac{Q}{C_1} = \frac{116 \text{ }\mu\text{C}}{20 \text{ }\mu\text{F}} = 5.8 \text{ V}$$

$$\text{potential difference, } V_2 = \frac{Q}{C_2} = \frac{116 \text{ }\mu\text{C}}{5 \text{ }\mu\text{F}} = 23.2 \text{ V}$$

$$\text{potential difference, } V_3 = \frac{Q}{C_3} = \frac{116 \text{ }\mu\text{C}}{4 \text{ }\mu\text{F}} = 29.0 \text{ V}$$

Notice that the largest value of capacitance has the smallest potential difference and the lowest value of capacitance has the highest potential difference.

Voltage Check

$$E_T = V_1 + V_2 + V_3 = 5.8 + 23.2 + 29.0 = 58.0 \text{ V}$$

From Eq. (9-4-2),

energy stored in the 20-μF capacitor,

$$W_1 = \tfrac{1}{2} \times Q \times V_1 = \tfrac{1}{2} \times 116 \text{ }\mu\text{C} \times 5.8 \text{ V} = 336.4 \text{ }\mu\text{J}$$

energy stored in 5-μF capacitor,

$$W_2 = \tfrac{1}{2} \times Q \times V_2 = \tfrac{1}{2} \times 116 \text{ }\mu\text{C} \times 23.2 \text{ V} = 1345.6 \text{ }\mu\text{J}$$

energy stored in the 4-μF capacitor,

$$W_3 = \tfrac{1}{2} \times Q \times V_3 = \tfrac{1}{2} \times 116 \text{ }\mu\text{C} \times 29.0 \text{ V} = 1682.0 \text{ }\mu\text{J}$$

The highest value of capacitance stores the least energy and the lowest value of capacitance stores the most energy.

$$\text{total energy stored, } W_T = W_1 + W_2 + W_3 = 336.4 + 1345.6 + 1682.0$$

$$= 3364 \text{ }\mu\text{J}$$

Energy Check

$$\text{total energy stored} = \tfrac{1}{2} \times C_T \times E_T^2 \qquad (9\text{-}4\text{-}4)$$

$$= \tfrac{1}{2} \times 2 \text{ }\mu\text{F} \times (58 \text{ V})^2 = 3364 \text{ }\mu\text{J}$$

EXAMPLE 9-7 In Fig. 9-12(b), $C_1 = 20\ \mu F$, $C_2 = 5\ \mu F$, $C_3 = 4\ \mu F$, and $E_T = 58$ V. Calculate the values of C_T, Q_1, Q_2, Q_3, and Q_T. What is the amount of energy stored in each capacitor?

Note: For comparison purposes we have used the same values of capacitance as we did in Example 9-6.

Solution The capacitors are in parallel and we must therefore add up the individual capacitances to obtain the value of C_T.

Substitute first in Eq. (9-5-14):

total capacitance, $C_T = C_1 + C_2 + C_3 = 20 + 5 + 4 = 29\ \mu F$

Then from Eq. (9-5-10),

charge stored in the 20-μF capacitor,

$$Q_1 = C_1 E_T = 20\ \mu F \times 58\ V = 1160\ \mu C$$

charge stored in the 5-μF capacitor,

$$Q_2 = C_2 E_T = 5\ \mu F \times 58\ V = 290\ \mu C$$

charge stored in the 4-μF capacitor,

$$Q_3 = C_3 E_T = 4\ \mu F \times 58\ V = 232\ \mu C$$

Notice that the highest-value capacitor stores the most charge, while the lowest-value capacitor stores the least charge.

$$\text{total charge stored, } Q_T = Q_1 + Q_2 + Q_3 \qquad (9\text{-}5\text{-}12)$$
$$= 1160 + 290 + 232 = 1682\ \mu C$$

Charge Check

$$\text{total charge, } Q_T = E_T \times C_T \qquad (9\text{-}5\text{-}13)$$
$$= 58\ V \times 29\ \mu F = 1682\ \mu C$$

Substituting in Eq. (9-4-2) gives

energy stored in the 20-μF capacitor,

$$W_1 = \tfrac{1}{2} \times Q_1 \times E_T = \tfrac{1}{2} \times 1160\ \mu C \times 58\ V = 33{,}640\ \mu J$$

energy stored in the 5-μF capacitor,

$$W_2 = \tfrac{1}{2} \times Q_2 \times E_T = \tfrac{1}{2} \times 290\ \mu C \times 58\ V = 8410\ \mu J$$

energy stored in the 4-μF capacitor,

$$W_3 = \tfrac{1}{2} \times Q_3 \times E_T = \tfrac{1}{2} \times 232\ \mu C \times 58\ V = 6728\ \mu J$$

total energy stored, $W_T = W_1 + W_2 + W_3$
$$= 33{,}640 + 8410 + 6728 = 48{,}778\ \mu J$$

This time it is the highest-value capacitor that stores the most energy, while the lowest-value capacitor stores the least energy.

Energy Check

$$\text{total energy stored} = \tfrac{1}{2} \times C_T \times E_T^2 \qquad (9\text{-}4\text{-}4)$$
$$= \tfrac{1}{2} \times 29\ \mu F \times (58\ V)^2 = 48{,}778\ \mu J$$

EXAMPLE 9-8 Two 0.1-μF capacitors are connected in parallel and are charged from a 150-V source. The supply is then disconnected and is replaced by a third uncharged 0.1-μF capacitor. What is the voltage across the third capacitor? Calculate the total energy stored before and after the third capacitor is connected.

Solution By Eqs. (9-3-2) and (9-5-12), the total capacitance before the connection of the third capacitor was $2 \times 0.1 = 0.2$ μF, and the total charge stored in this capacitance was $0.2\ \mu F \times 150\ V = 30\ \mu C$.

By Eq. (9-4-4), total energy stored before the connection of the third capacitor, was $\tfrac{1}{2} \times 0.2\ \mu F \times (150\ V)^2 = 2250\ \mu J$.

By Eqs. (9-3-2) and (9-5-12), the total capacitance after the connection of the third capacitor is $3 \times 0.1 = 0.3$ μF, and since the total charge stored is still 30 μC, the voltage across the third capacitor (and also across the other two capacitors) is $30\ \mu C/0.3\ \mu F = 100\ V$.

By Eq. (9-4-4), the total energy stored after the connection of the third capacitor is $\tfrac{1}{2} \times 0.3\ \mu F \times (100\ V)^2 = 1500\ \mu J$.

Notice that an energy amount of $2250 - 1500 = 750\ \mu J$ has been lost. If we neglect the resistance of the connecting wires, the lost energy appeared in the form of the spark that occurred when the third capacitor was connected.

EXAMPLE 9-9 In Fig. 9-11, a capacitor is manufactured with 11 metal plates. The area of one side of each plate is 125 cm² and adjacent plates are separated by sheets of mica which are 0.5 mm thick and have a relative permittivity of 5. What is the value of the total capacitance?

Solution

$$N = 11, \quad A = 125 \times 10^{-4}\ m^2, \quad d = 0.5 \times 10^{-3}\ m, \quad \epsilon_r = 5$$

total capacitance, C_T

$$= \frac{\epsilon_0 \epsilon_r (N-1) A}{d} \qquad (9\text{-}5\text{-}16)$$

$$= \frac{8.85 \times 10^{-12} \times 5 \times (11-1) \times 125 \times 10^{-4}}{0.5 \times 10^{-3}}\ F$$

$$= 0.0111\ \mu F \quad \text{(rounded off)}$$

9-6 Time Constant of *CR* Circuit

We are now going to look at the charging and discharging of a capacitor through a resistor; the capacitor is assumed to be ideal and is therefore without losses.

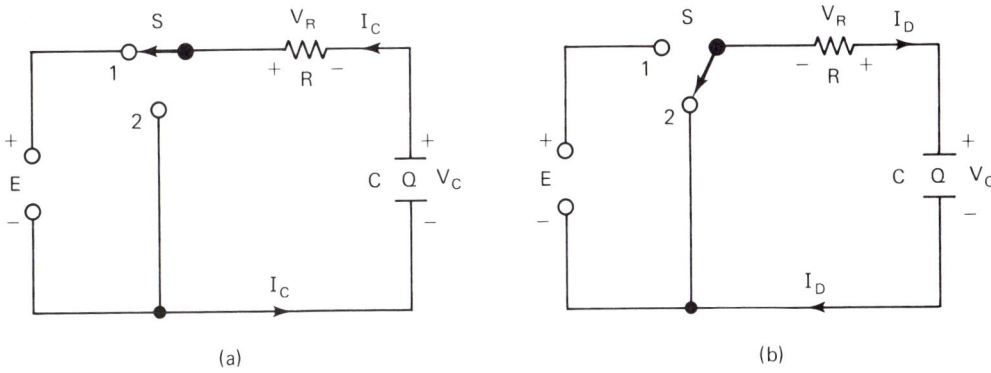

FIGURE 9-13 Charge and discharge of a capacitor through a resistor.

The results we obtain should be carefully compared with those for the LR circuit of Section 8-5.

Let us start by analyzing the conditions at the beginning of the transient state; as already discussed in Section 8-5, the transient state is the interval between the time the switch, S, is closed in position 1 and the time at which the final or steady-state conditions are reached.

When the dc voltage is switched across the CR circuit of Fig. 9-13(a), we must remember that the capacitor cannot charge instantaneously. Therefore, since the capacitor was uncharged just before the switch, S, was closed in position 1, it must still be uncharged immediately after S is closed; consequently, the voltage between the capacitor plates, V_C, must initially be zero. The voltage drop across the resistor, V_R, is at its maximum value, which must exactly balance the source voltage, E. The initial charging current, I_C, is also at its maximum value, which is equal to E/R amperes.

To summarize, the conditions at the beginning of the transient state are

$$Q = 0, \quad V_C = 0, \quad V_R = E, \quad I_C = \frac{E}{R} \quad (9\text{-}6\text{-}1)$$

These same conditions could be obtained by assuming that the capacitor initially behaves as a *short* circuit.

As the current flows and charges the capacitor, V_C rises and therefore V_R falls, since at all times the sum of the voltage drop across the resistor and the potential difference between the capacitor plates must exactly balance the source voltage. As V_R decreases, the charging current, I_C ($= V_R/R$) is less and we therefore have a situation in which the higher the voltage to which the capacitor is charged, the lower is its rate of charging. Theoretically, it would take infinite time for the capacitor to be fully charged; however, the capacitor acquires more than 99% of its final charge after a time interval which is determined by the values of the capacitance and the resistance.

Assuming that the capacitor is fully charged, the potential difference between the capacitor's plates is E volts and the charge stored is $Q = CE$ coulombs. If the capacitance is increased but the resistance and the source voltage are unchanged, the capacitor will have to store more charge, but the initial charging current (E/R amperes) will remain the same. The capacitor will take longer to charge and therefore the duration of the transient state depends directly on the capacitance. Again, let us keep the capacitance and the source voltage the same but increase the value of the resistance. The capacitor will have to store the same final charge, but the initial value of the charging current

Sec. 9-6 **Time Constant of CR Circuit**

will be less. The transient state will last longer and therefore its duration is also directly dependent on the value of the resistor.

By combining these results, we are led to the conclusion that the interval of the charging transient state is determined by the product of C and R. The quantity CR is called the *time constant*, which is measured in seconds if C is in farads and R is in ohms. For example, a 0.5-μF capacitor and a 10-kΩ resistor will have a time constant of 0.5 μF $\times$ 10 kΩ = $0.5 \times 10^{-6} \times 10 \times 10^3$ s = 5000 μs = 5 ms. For other values of C and R you may estimate the order of the time constant by using Table 9-2. The CR time constant should be compared with the L/R time constant for the inductor and the resistor (Section 8-5).

How is the value of the time constant related to the transient response of the CR circuit? Mathematically, it can be shown that after one time constant the capacitor has acquired 63.2% or 0.632 of its final charge, so that V_C has also reached 0.632 of its final steady-state value, E. Consequently, after CR seconds or one time constant from the instant S is closed in position 1, the conditions in the circuit are:

$$Q = 0.632CE, \quad V_C = 0.632E, \quad V_R = E - 0.632E = 0.368E$$

$$\text{and} \quad I_C = \frac{V_R}{R} = \frac{0.368E}{R} \qquad (9\text{-}6\text{-}2)$$

As with the LR circuit, the factor 0.632 is all-important. After a further interval of one time constant or a total of 2 CR seconds from the instant S is closed, V_C completes 63.2% of its *remaining* climb toward its final steady-state value. Since after one time constant the *remaining* percentage climb is $100 - 63.2 = 36.8\%$, the increases in V_C and Q during the second time constant are $0.632 \times 36.8 = 23.26\%$, so that the total percentage climb after two time constants is $63.2 + 23.26 = 86.5\%$ (rounded off). During the third time constant, the remaining percentage climb is $100 - 86.5 = 13.5\%$, the increase is $13.5 \times 0.632 = 8.532\%$, and the total percentage climb is $86.5 + 8.532 = 95.0\%$ (rounded off). Repeating the procedure for the fourth and fifth time constants you can verify that the total percentage climbs are 98.2% and 99.3%, respectively. This means that after five time constants or 5 CR seconds, V_C and Q have reached levels that are within 1% of their final steady-state values. We then assume that the circuit's transient state has been completed and that the final or steady-state conditions have been reached. These conditions are:

$$Q = CE, \quad V_C = E, \quad V_R = 0, \quad I_C = 0 \qquad (9\text{-}6\text{-}3)$$

TABLE 9-2

C	R	Time Constant, CR
F	Ω	seconds
μF	Ω	microseconds
μF	kΩ	milliseconds
μF	MΩ	seconds
nF	Ω	nanoseconds
nF	kΩ	microseconds
nF	MΩ	milliseconds
pF	Ω	picoseconds
pF	kΩ	nanoseconds
pF	MΩ	microseconds

You can predict these final conditions by regarding the capacitor as an *open* circuit.

If an open circuit occurs in a series arrangement of resistors (Section 2-11), the ends between which the open circuit exists will behave as conducting surfaces, while the open circuit itself will act as an air dielectric. The open circuit therefore represents a very small capacitance which will be charged to the value of the source voltage. Moreover, the time constant of the charge path will be short, so that the source voltage appears almost instantaneously across the open circuit.

The variations of Q, V_C, V_R, and I_C with time are shown in Fig. 9-14(a); the way in which V_C and Q rise is referred to as *exponential growth*, while the manner in which V_R and I_C fall is *exponential decay*. For convenience the universal growth and decay curves of Fig. 8-11 have been reproduced in Fig. 9-15. The vertical axis is again marked in percentage, but now the horizontal time axis is measured with intervals of CR seconds. Using these universal curves, corresponding values of percentage (rounded off to three significant figures) and time constant are shown in Table 9-3.

There is another interpretation of the time constant. Since the final charge stored by the capacitor is CE coulombs while the initial charging current is E/R amperes (coulombs per second), the capacitor would have taken a time of

$$\frac{CE \text{ coulombs}}{E/R \text{ coulombs per second}} = CR \text{ seconds} \quad \text{(one time constant)}$$

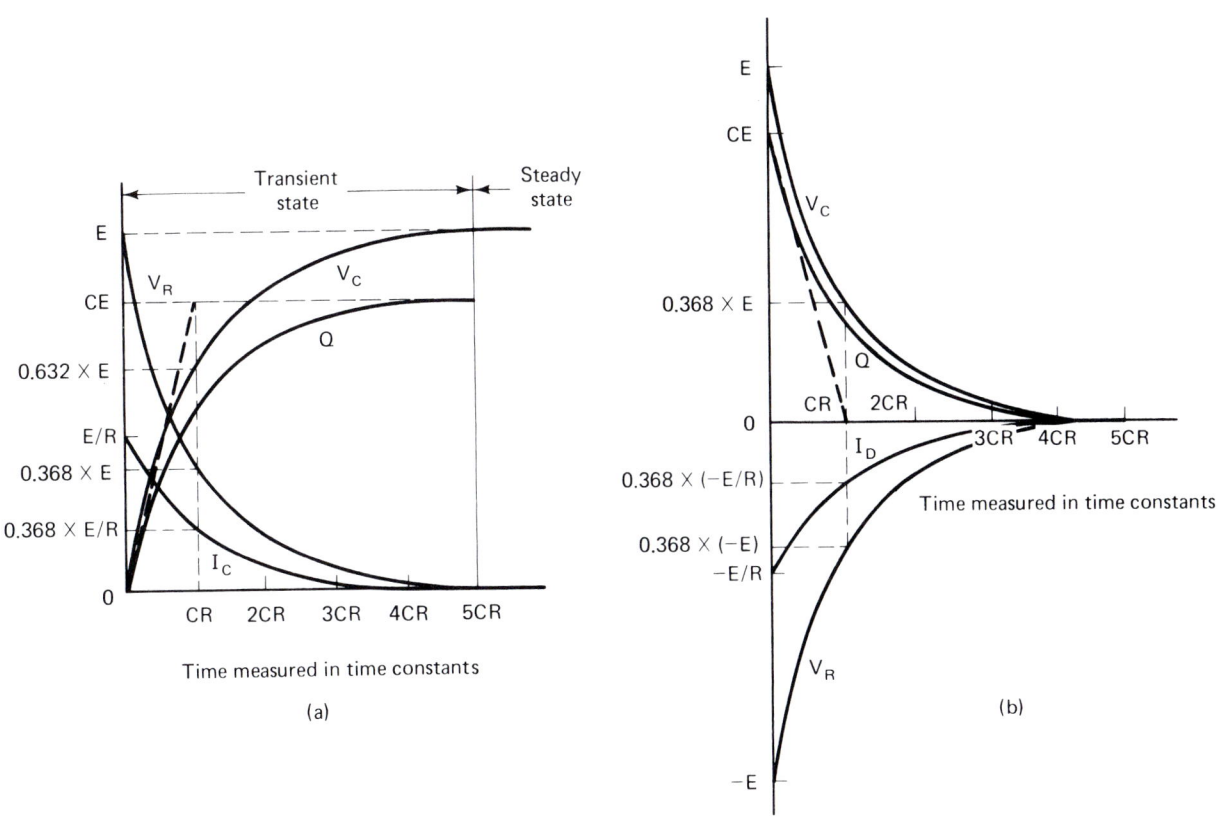

FIGURE 9-14 (a) Charging of a capacitor through a resistor; (b) discharge of a capacitor through a resistor.

Sec. 9-6 Time Constant of *CR* Circuit

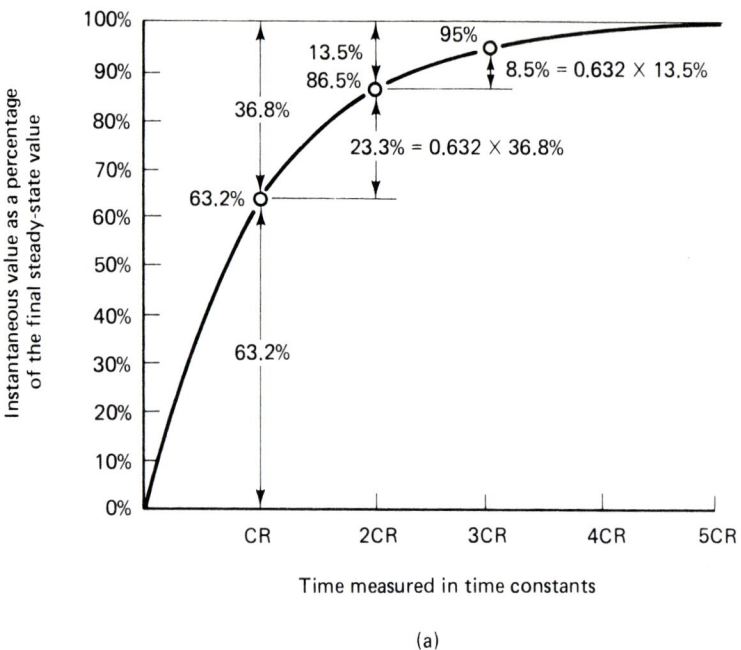

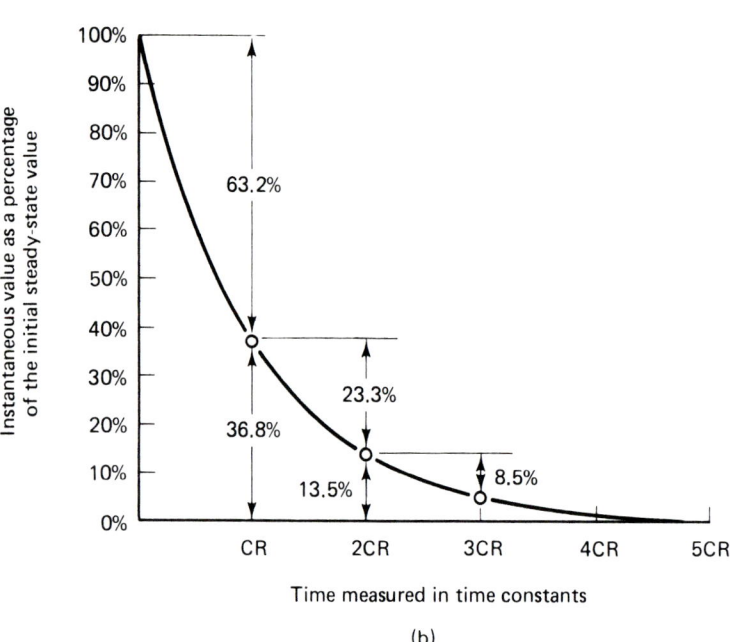

FIGURE 9-15 (a) Universal exponential growth curve; (b) universal exponential decay curve.

to charge fully *if* the *initial* rate of charging had been maintained; this is indicated by the dotted line in Fig. 9-14(a). Of course, this does not happen in practice, since the more the capacitor is charged, the less is the value of I_C and the lower is the rate of charging.

Assuming that the steady-state conditions have been reached in the circuit of Fig. 9-13(a), we will now switch S to position 2 [Fig. 9-13(b)]. Remembering that the capacitor cannot discharge instantaneously, the values of V_C and Q must remain exactly the same. Since V_C is still equal to E, V_R must also equal E in order for the voltages to balance around the loop. However, the polarity

TABLE 9-3 Percentage and CR Time Constant Values

Time Constant (CR seconds)	Percentage Growth	Percentage Decay
0.2	18.1	81.9
0.5	39.3	60.7
0.7	50.3	49.7
1.0	63.2	36.8
2.0	86.5	13.5
3.0	95.0	4.98
4.0	98.2	1.83
5.0	99.3	0.674

of V_R is now reversed when compared with the polarity during the charging of the capacitor. This is because the capacitor is starting to discharge and the discharge current, I_D, is in the opposite direction to the previous charging current, I_C.

To summarize, the circuit conditions immediately after S is closed in position 2 are

$$Q = CE, \quad V_C = E, \quad V_R = -E, \quad I_D = -\frac{E}{R} \quad (9\text{-}6\text{-}4)$$

The minus signs indicate the reversals of V_R's polarity and the direction of the current.

As the capacitor discharges, V_C and therefore V_R fall together, so that the voltage balance around the loop is maintained. The decrease in V_R means that the discharge current is less, so we have a situation in which the more the capacitor is discharged, the less is its rate of discharge. Theoretically, the capacitor would take infinite time to discharge completely but Q, V_C, V_R, and I_D all drop to below 1% of their original values after five time constants. Once again the 0.632 factor is the key to the exponential decay. After one time constant or a time interval of CR seconds from the instant that S is switched to position 2, Q and V_C have lost 63.2% and have therefore fallen to 36.8% of their original values. Then the circuit conditions are:

$$Q = 0.368CE, \quad V_C = 0.368E, \quad V_R = -0.368E, \quad I_D = \frac{-0.368E}{R} \quad (9\text{-}6\text{-}5)$$

Using the results we obtained from the growth curve, the levels of Q, V_C, V_R, and I_D after two, three, four, and five time constants will be $100 - 86.5 = 13.5\%$, $100 - 95.0 = 5.0\%$, $100 - 98.2 = 1.8\%$, and $100 - 99.3 = 0.7\%$, respectively, of their original values; more exact percentages are shown in Table 9-3. After five time constants we assume that the transient decay state has been concluded with the values of Q, V_C, V_R, and I_D all essentially equal to zero. The variations of these quantities with time are shown in Fig. 9-14(b); the dotted line indicates that the capacitor would have fully discharged in *one* time constant if the *initial* rate of discharge, I_D, had been maintained.

We have learned about the importance of the time constant in controlling the transient response of a CR circuit during both the charging and the discharging of a capacitor through a resistor. We should mention that if there is more than one resistor or one capacitor in the CR circuit, you must first calculate the total equivalent resistance and capacitance and use these values to obtain the time constant.

EXAMPLE 9-10 In Fig. 9-13(a), $R = 100$ kΩ, $C = 50$ pF, and $E = 50$ V. Calculate the value of the time constant. From the instant S is closed in position 1, what are the values of Q, V_C, V_R, and I_C after time intervals, $t = 0$, 3.5, 5.0, 15.0, and 30.0 μs? What is the amount of energy stored in the capacitor after 30 μs?

Solution By Table 9-2,

$$\text{time constant} = C \times R = 50 \text{ pF} \times 100 \text{ k}\Omega$$
$$= 5000 \text{ ns} = 5 \text{ }\mu\text{s}$$

Substituting in Eqs. (9-6-1) yields:

$t = 0$ (start of the transient state)

$$Q = 0, \quad V_C = 0, \quad V_R = 50 \text{ V}, \quad I_C = 50 \text{ V}/100 \text{ k}\Omega = 0.5 \text{ mA}$$

Use Table 9-3 to obtain the following growth and decay percentages.

$t = 3.5 \text{ }\mu\text{s} = 3.5/5 \text{ time constant} = 0.7 \text{ time constant}$

$$\text{growth percentage} = 50.3\%, \quad \text{decay percentage} = 49.7\%$$

These results could also have been obtained from the universal growth and decay curves. Therefore,

$$V_R = \frac{49.7}{100} \times 50 \text{ V} = 24.85 \text{ V}$$

$$V_C = \frac{50.3}{100} \times 50 \text{ V} = 25.15 \text{ V}$$

$$Q = 24.85 \text{ V} \times 50 \text{ pF} = 1242.5 \text{ pC}$$

$$I_C = \frac{24.85 \text{ V}}{100 \text{ k}\Omega} = 0.2485 \text{ mA}$$

$t = 5.0 \text{ }\mu\text{s} = 1 \text{ time constant}$

$$\text{growth percentage} = 63.2\%, \quad \text{decay percentage} = 36.8\%$$

$$V_C = \frac{63.2}{100} \times 50 \text{ V} = 31.6 \text{ V}$$

$$V_R = \frac{36.8}{100} \times 50 \text{ V} = 18.4 \text{ V}$$

$$Q = 31.6 \text{ V} \times 50 \text{ pF} = 1580 \text{ pC}$$

$$I_C = \frac{18.4 \text{ V}}{100 \text{ k}\Omega} = 0.184 \text{ mA}$$

$t = 15.0 \ \mu s = 3$ time constants

$$\text{growth percentage} = 95.0\%, \quad \text{decay percentage} = 4.9\%$$

$$V_C = \frac{95.0}{100} \times 50 \ V = 47.5 \ V$$

$$V_R = \frac{4.9}{100} \times 50 \ V = 2.5 \ V$$

$$Q = 47.5 \ V \times 50 \ pF = 2375 \ pC$$

$$I_C = \frac{2.5 \ V}{100 \ k\Omega} = 0.025 \ mA$$

$t = 30 \ \mu s = 6$ time constants
The transient state is now ended and the final conditions [Eqs. (9-6-3)] are

$$V_C = 50 \ V, \quad V_R = 0, \quad Q = 50 \ V \times 50 \ pF = 2500 \ pC, \quad I_C = 0$$

By Eq. (9-4-1)

energy stored in the capacitor after 30 μs
$$= \tfrac{1}{2} \times 2500 \ pC \times 50 \ V = 625{,}000 \ pJ = 0.625 \ \mu J$$

EXAMPLE 9-11 In Example 9-10 the circuit has reached its final steady-state condition. S is now switched to position 2 [Fig. 9-13(b)]. From that instant, what are the values of V_C, V_R, Q, and I_D after time intervals of $t = 0$, 3.5, 5.0, 15.0, and 30.0 μs?

Solution By Eqs. (9-6-4):

$t = 0$ (start of the discharging transient state):

$$V_C = 50 \ V, \quad V_R = -50 \ V, \quad Q = 2500 \ pC, \quad I_D = -\frac{50 \ V}{100 \ k\Omega}$$
$$= -0.5 \ mA$$

Table 9-3 yields the following decay percentages.

$t = 3.5 \ \mu s = 0.7$ time constant

$$\text{decay percentage} = 50.3\%$$

$$V_C = \frac{50.3}{100} \times 50 \ V = 25.15 \ V$$

$$V_R = -25.15 \ V$$

$$Q = 25.15 \ V \times 50 \ pF = 1257.5 \ pC$$

$$I_D = -\frac{25.15 \ V}{100 \ k\Omega} = -0.2515 \ mA$$

$t = 5.0 \ \mu s = 1$ *time constant*

$$\text{decay percentage} = 36.8\%$$

$$V_C = \frac{36.8}{100} \times 50 \text{ V} = 18.4 \text{ V}$$

$$V_R = -18.4 \text{ V}$$

$$Q = 18.4 \text{ V} \times 50 \text{ pF} = 920 \text{ pC}$$

$$I_D = -\frac{18.4 \text{ V}}{100 \text{ k}\Omega} = -0.184 \text{ mA}$$

$t = 15.0 \ \mu s = 3$ *time constants*

$$\text{decay percentage} = 4.9\%$$

$$V_C = \frac{4.9}{100} \times 50 \text{ V} = 2.5 \text{ V}$$

$$V_R = -2.5 \text{ V}$$

$$Q = 2.5 \text{ V} \times 50 \text{ pF} = 125 \text{ pC}$$

$$I_D = -\frac{2.5 \text{ V}}{100 \text{ k}\Omega} = -0.025 \text{ mA}$$

$t = 30.0 \ \mu s = 6$ *time constants*
The decay transient state is now concluded, so that the capacitor is completely discharged.

$$V_C = 0, \quad V_R = 0, \quad Q = 0, \quad I_D = 0$$

EXAMPLE 9-12 In Fig. 9-16, the capacitor is initially uncharged and the switch, S, is then closed. At the start and the finish of the transient state, what are the values of I_T, I_1, I_2, and the voltage at the point P?

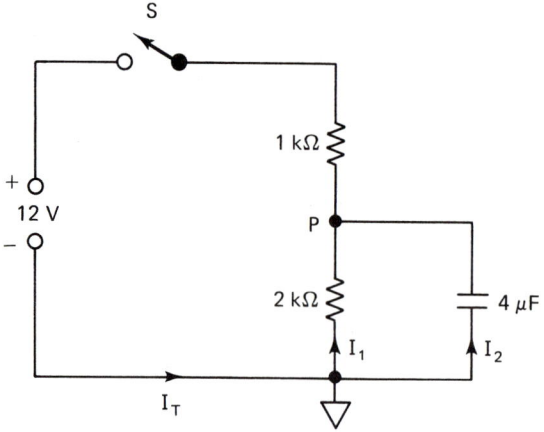

FIGURE 9-16 Circuit for Example 9-12.

Solution The capacitor cannot charge instantaneously and therefore the initial voltage at the point P is zero. Consequently, I_1 is also zero and $I_2 = I_T = 12\text{V}/1 \text{ k}\Omega = 12$ mA. These results could have been obtained by regarding the capacitor as a *short* circuit.

At the end of the transient state the capacitor is fully charged, so that $I_2 = 0$. Then $I_T = I_1 = 12\text{ V}/3\text{ k}\Omega = 4\text{ mA}$, and the voltage at P is $+4\text{ mA} \times 2\text{ k}\Omega = +8\text{ V}$. You could have predicted these values by considering the capacitor to be an *open* circuit.

9-7 Features of a Practical Capacitor

So far in this chapter we have assumed that all the capacitors used were ideal or perfect in the sense that only the property of capacitance was involved and there were no energy losses. For example, we assumed that the dielectric possessed infinite resistance, so that when the capacitor was being charged, there was no electron flow between the conducting surfaces. Furthermore, after such a capacitor was charged, it would retain its charge indefinitely since the dielectric was perfect. This is not true in practice; only some capacitors can retain a high level of charge for an appreciable time after the charging source is removed; such capacitors may prove dangerous unless safety measures are taken.

The purpose of this section is to discuss the various ways in which the practical capacitor differs from the ideal.

Leakage Resistance

A capacitor's dielectric, such as mica, ceramic, or paper has a resistance which is very large but not infinite. The insulator's resistance is called its *leakage value*, so that when a potential difference is established between the capacitor's plates, a small leakage current will flow in the dielectric. As a result, energy will be dissipated as heat in the insulator and this is a power loss within the capacitor. The effect is represented by the high value of leakage resistance, R, which is in parallel with the capacitance [Fig. 9-17(a)].

Electrolytic capacitors have very high leakage currents, which may be up to 1 mA or so; the corresponding leakage resistance is of the order of 1 MΩ. This high leakage current is the main disadvantage of electrolytic capacitors. There is a moderate leakage current in paper capacitors, where the leakage

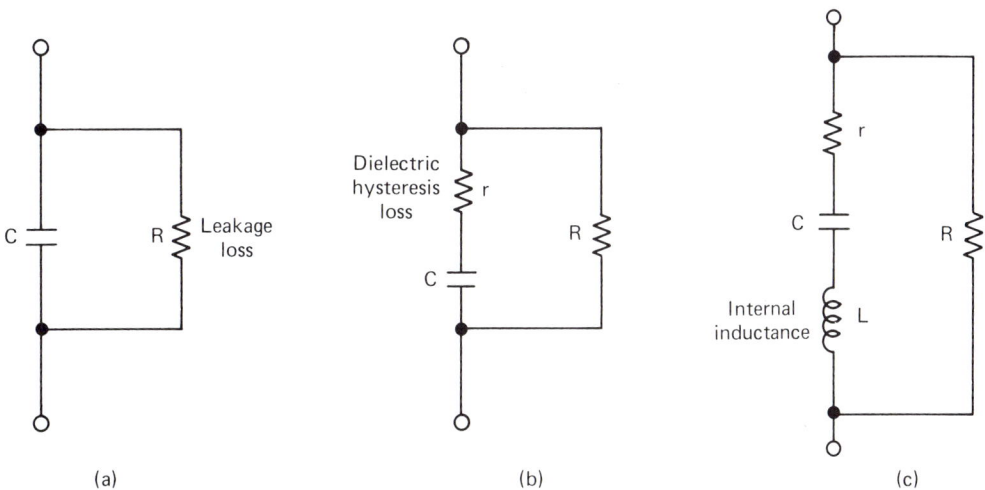

FIGURE 9-17 Practical capacitor.

resistance is about 100 MΩ. However, the leakage current in mica and ceramic capacitors is very low; their leakage resistances are of the order of 1000 MΩ.

The power losses are measured by a capacitor's power factor, which is just a number. An ideal capacitor would have a power factor of zero, so that a high-quality practical capacitor should have a very low power factor. The power factor for a capacitor with an air dielectric is virtually zero, while the approximate power factors for mica and paper capacitors are 0.0005 and 0.01, respectively. The power factor can be measured by an item of test equipment known as a capacitor checker. The checker will also enable you to measure the capacitance and to test the voltage rating.

Dielectric Hysteresis

If we reverse the polarity of the source voltage applied to a capacitor, the electric field intensity, $\mathscr{E}$, will also be reversed and this should cause a reversal of the electric flux density, D. However, if the polarity of the source voltage is reversed, the flux density cannot respond and will lag behind the change in the electric field intensity. This effect, called *dielectric hysteresis*, is the electrostatic equivalent of magnetic hysteresis, which we discussed fully in Section 7-6 and illustrated in Fig. 7-23. Dielectric hysteresis will cause a further energy loss which is represented by a low value of resistance (of the order of 1 Ω) in series with the capacitance [Fig. 9-17(b)]. This type of loss is particularly severe in microwave circuits, where special insulators such as polyethylene, polystyrene, and teflon have to be used.

Internal Inductance

Capacitors with tubular construction naturally use curved conducting surfaces. As a result, such capacitors will possess a certain amount of internal inductance, whose value is determined mainly by the capacitor's physical size. This inductance, L, is regarded as in series with the capacitance [Fig. 9-17(c)].

We should point out that the connecting wires in a circuit possess some self-inductance, which is typically less than 1 μH. In addition, these wires can be positioned close to the metal chassis on which the circuit is built so that capacitance will exist between the wires and the chassis; the value of this capacitance is normally a few picofarads. These properties of connecting wires are known as *stray inductance and capacitance*, which may have to be taken into account in the design of some circuits. Another effect to be considered is the internal capacitances contained in the active devices such as transistors.

We are led to the conclusion that resistors, inductors, and capacitors *each* possess to a certain degree the properties of capacitance, resistance, and inductance. The property that dominates determines the name which we give to a particular component; for example, if the resistance of a component is its dominant property, we will call that component a resistor; and so on.

9-8 Troubleshooting for Faults in Capacitors

Compared with resistors and inductors, capacitors are most likely to suffer from defects. These defects are in the form of open circuits, short circuits, or partial short circuits. In the first two cases the capacitor is entirely useless since it cannot store any charge. With the partial short some charge may be stored, but the circuit containing the capacitor will probably suffer from poor performance.

Although a capacitor can be completely tested only by a capacitor checker, its defects can be found with an ohmmeter, such as a volt-ohmmeter (VOM), which is switched to the R × 1 MΩ range (Section 10-6). When connecting a capacitor across the ohmmeter it is only necessary to observe polarity with the electrolytic type; here the red (plus) probe of the VOM should be attached to the capacitor's + terminal and the black (minus) probe to the − terminal.

If the capacitor to be tested has not been removed from the circuit, the usual care must be taken to avoid the effect of any parallel resistance paths across the capacitor. This can be done by entirely disconnecting one end of the capacitor from the ground side of the circuit. One further point before carrying out any test; you must make certain that the capacitor is completely discharged; you can do this by connecting a conductor, such as a clip lead, across the capacitor's terminals.

A normal capacitor will behave initially as a short circuit, so that immediately after the capacitor is connected, the needle of the VOM will move swiftly across to the right and indicate zero ohms. After that the battery inside the VOM will charge the capacitor and the needle will move quite slowly to the left; this is the normal "capacitor action." In the case of air, mica, ceramic, and paper capacitors, the needle will finally come to rest at infinite ohms (symbol $\infty \, \Omega$), since the leakage resistances of these capacitors should be extremely high. However, with electrolytic capacitors the normal leakage resistance can be less than 1 MΩ.

We will now look at the results of testing a defective capacitor with a VOM used as an ohmmeter.

Open Circuit

With mica, ceramic, and paper capacitors an open connection may occur between a conducting surface and its external lead. In the case of electrolytic capacitors, the resistance of the electrolyte may increase to a very high value due to age and excessive temperatures.

When a capacitor with an open circuit is tested by an ohmmeter, the needle will not initially move to the right but will instead remain steady at infinite ohms. One word of caution—if you are trying to test a low-value capacitor (less than 100 pF), there will only be a slight initial movement of the needle to the right because the charging of the capacitor will have a very short time constant.

Short Circuit

A short between its external leads may, of course, occur with any capacitor. However, paper and electrolytic capacitors tend to suffer most from "shorts." Their dielectrics deteriorate with age as a result of the continuous or repeated application of the potential difference between the conducting surfaces; this effect is made worse by excessive temperatures.

With a short-circuited capacitor, the needle of the ohmmeter will move across to the right and stay in that position. There will be no charging from the ohmmeter's battery and no normal capacitor action.

Partial Short Circuit

The dielectrics of paper electrolytic capacitors may deteriorate slowly with age so that their leakage resistances fall and their leakage currents rise; the result is a partial short circuit.

If a capacitor with a partial short circuit is tested with a VOM, the needle will initially move to the right of the ohm's scale and then move back to the left. However, its final position will indicate a value of leakage resistance which is much lower than normal.

9-9 Differentiator Circuits

These circuits use the concept of time constant to achieve their objectives. The process of differentiation itself is related to the subject of calculus in mathematics. Let us assume that we have two waveforms which we will identify as waveform A and waveform B. Then if waveform B is the result of differentiating waveform A, the instantaneous value of any point in waveform B is equal to the value of the *slope* at the corresponding point in waveform A. In this case the word "corresponding" means "occurring at the same moment in time."

As an example, let us consider a mass that is falling as the result of gravity. In Section 1-1 we learned that the acceleration under the force of the earth's gravitational pull is 9.81 m/s². Consequently, if the mass is dropped from rest with an initial velocity of zero, its velocity after 1 s will be 9.81 m/s, so that its average velocity during this interval is 9.81/2 = 4.905 m/s and the distance through which the mass has fallen is equal to 4.905 m/s × 1 s = 4.905 m.

After 2 s the velocity of the mass is 2 × 9.81 = 19.62 m/s, the average velocity over the 2-s interval is 19.62/2 = 9.81 m/s, and the total distance through which the mass has fallen is 9.81 m/s × 2 s = 19.62 m. Notice that when the time is doubled (for example, from 1 s to 2 s), the velocity is doubled, but the distance fallen has quadrupled (from 4.905 m to 19.62 m). This suggests that the distance fallen is directly proportional to the *square* of the elapsed time.

Table 9-4 shows corresponding values of time, velocity, and distance fallen during the first 4 seconds. The symbol v represents the velocity acquired by the mass at the *end* of each time interval and s indicates the corresponding total distance through which the mass has fallen.

The equations relating these results are

$$\text{acceleration,}\ a = 9.81\ \text{m/s}^2 \qquad (9\text{-}9\text{-}1)$$

$$\text{velocity,}\ v = 9.81 \times t\ \text{m/s} \qquad (9\text{-}9\text{-}2)$$

Equation (9-9-2) states that velocity = acceleration × time [Eq. (1-1-1)].

$$\begin{aligned}\text{total distance fallen,}\ s &= \text{average velocity} \times \text{time} \\ &= \frac{v \times t}{2} \\ &= \frac{9.81 \times t}{2} \times t = 4.905 t^2\ \text{m}\end{aligned} \qquad (9\text{-}9\text{-}3)$$

Equation (9-9-3) indicates (as we suspected) that the distance fallen is directly proportional to the square of the elapsed time.

The three graphs of distance, velocity, and acceleration versus time are illustrated in Fig. 9-18. The distance versus time graph is a parabolic curve which is the result of the "square-law" relationship between the two variables

TABLE 9-4

Time, t (s)	Velocity, v (m/s)	Distance, s (m)	Acceleration, a (m/s²)
0	0	0	9.81
1	9.81	4.905	9.81
2	19.62	19.62	9.81
3	29.43	44.14	9.81
4	39.24	78.48	9.81

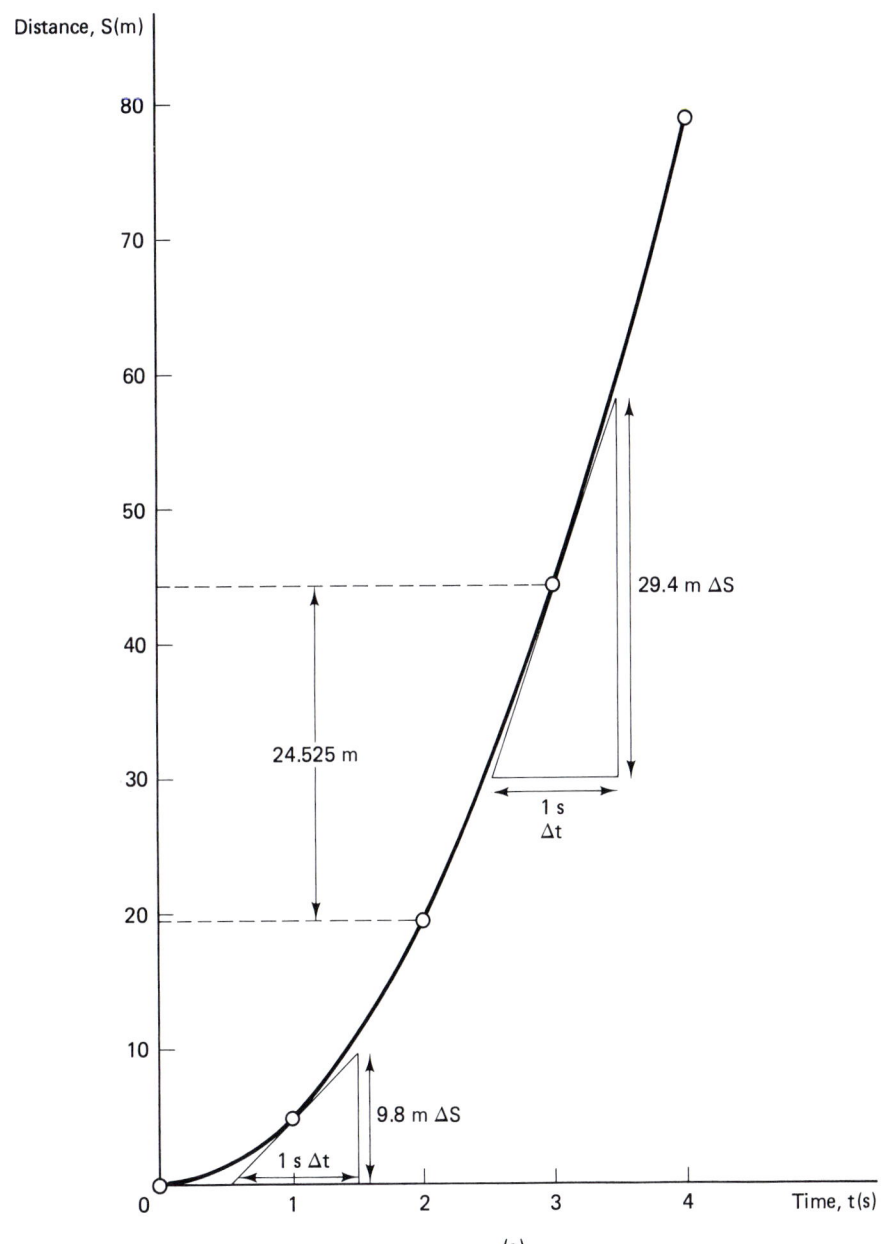

FIGURE 9-18 Graphs of (a) distance, (b) velocity, and (c) acceleration versus time.

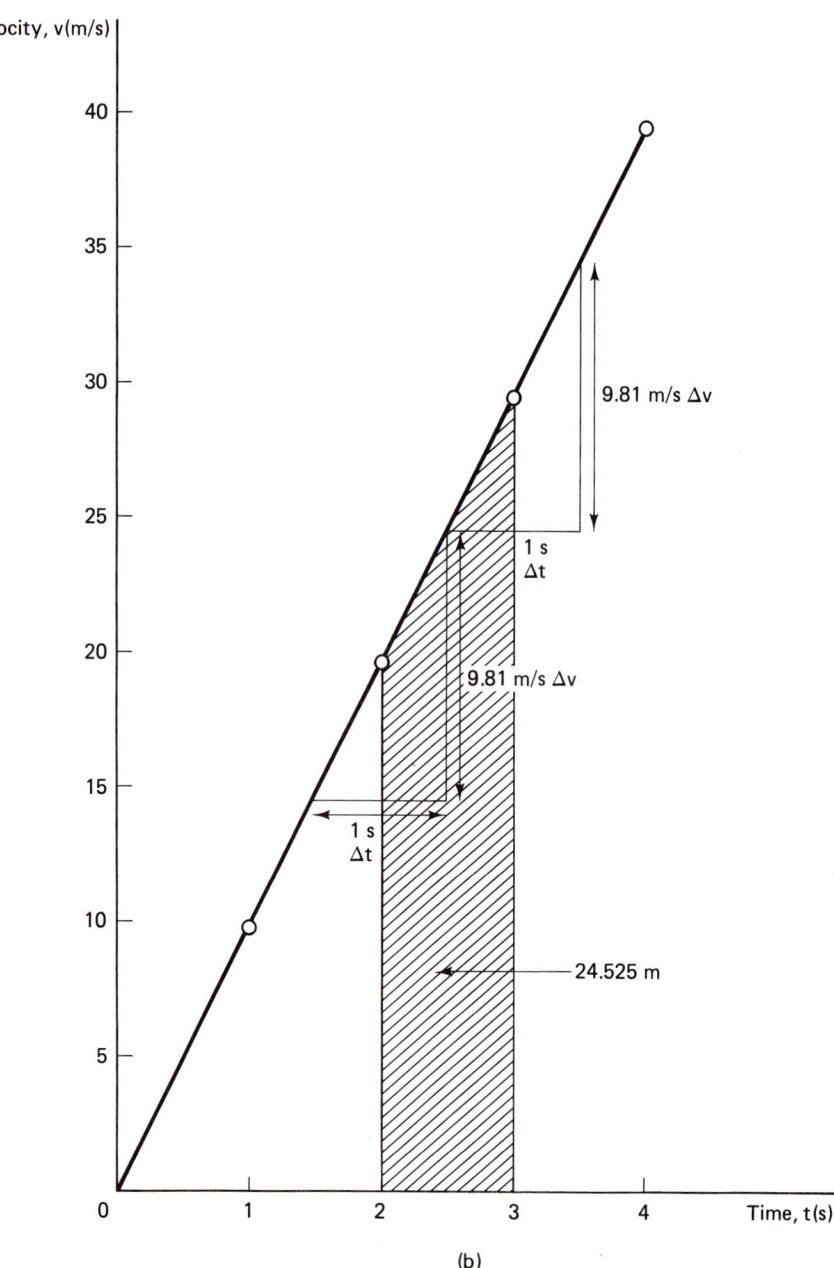

FIGURE 9-18 continued

(distance, $s = 4.905t^2$ m). By contrast, there is a "linear" equation connecting the velocity and the time (velocity, $v = 9.81 \times t$ m/s) so that their graph is a straight line [Fig. 9-18(b)]. The gravitational acceleration remains constant ($a = 9.81$ m/s^2), and therefore the graph of Fig. 9-18(c) is a *horizontal* straight line.

At the point specified by $t = 1$ s, $s = 4.905$ m, we will draw (by estimation) a tangent that touches the parabolic curve as shown in Fig. 9-18(a). A right-angled triangle is then constructed with the tangent line as the hypotenuse. For convenience the horizontal side of the triangle is chosen to represent 1 s and the corresponding vertical side is equivalent to 9.8 m approximately. The slope of the tangent can then be found by dividing the vertical side of 9.8 m by the horizontal side of 1 s. As a result, the slope is equal to 9.8 m/1 s = 9.8 m/s, which is the velocity of the mass corresponding to $t = 1$ s. Repeating

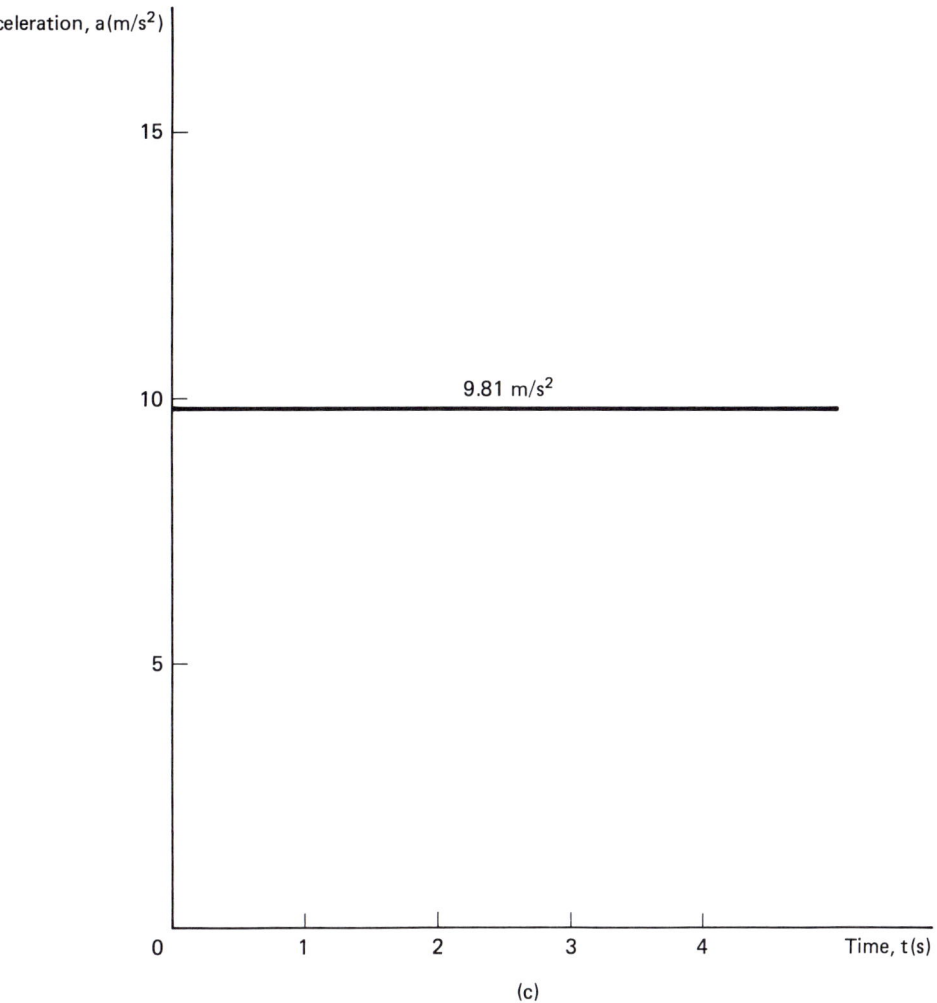

FIGURE 9-18 continued

the process for the point $t = 3$ s, $s = 44.14$ m, the slope of the tangent is approximately 29.4 m/1 s = 29.4 m/s. In other words, the slope of the tangent measures the ratio of the variation in the distance fallen to the variation in the corresponding time interval; the value of this ratio is equal to the *instantaneous* velocity of the mass. In calculus terminology the result of *differentiating* the distance with respect to the time is equal to the velocity, so that the differentiation of the distance versus time graph produces the velocity versus time graph.

We will now turn our attention to the velocity versus time graph. This is an inclined straight line, so that its slope is the same at all points. In Fig. 9-18(b), triangles have been constructed for the two points specified by $t = 2$ s, $v = 19.62$ m/s, and $t = 3$ s, $v = 29.43$ m/s; in each case the slope is (9.8 m/s)/(1 s) = 9.8 m/s², which is the value of the gravitational acceleration. It follows that the result of differentiating the velocity/time graph is the acceleration/time graph.

Parabolic curves are not commonly encountered in electronics, which is more concerned with square waves, sine waves, and "ramp" or "sawtooth" waveforms. For example, what would we expect the resultant waveform to look like if we differentiated a square wave? But to begin with, how would we generate a square wave such as that shown in Fig. 9-19(b)? When we look at the circuit of Fig. 9-19(a), let us assume that switch S is closed in position

Sec. 9-9 Differentiator Circuits

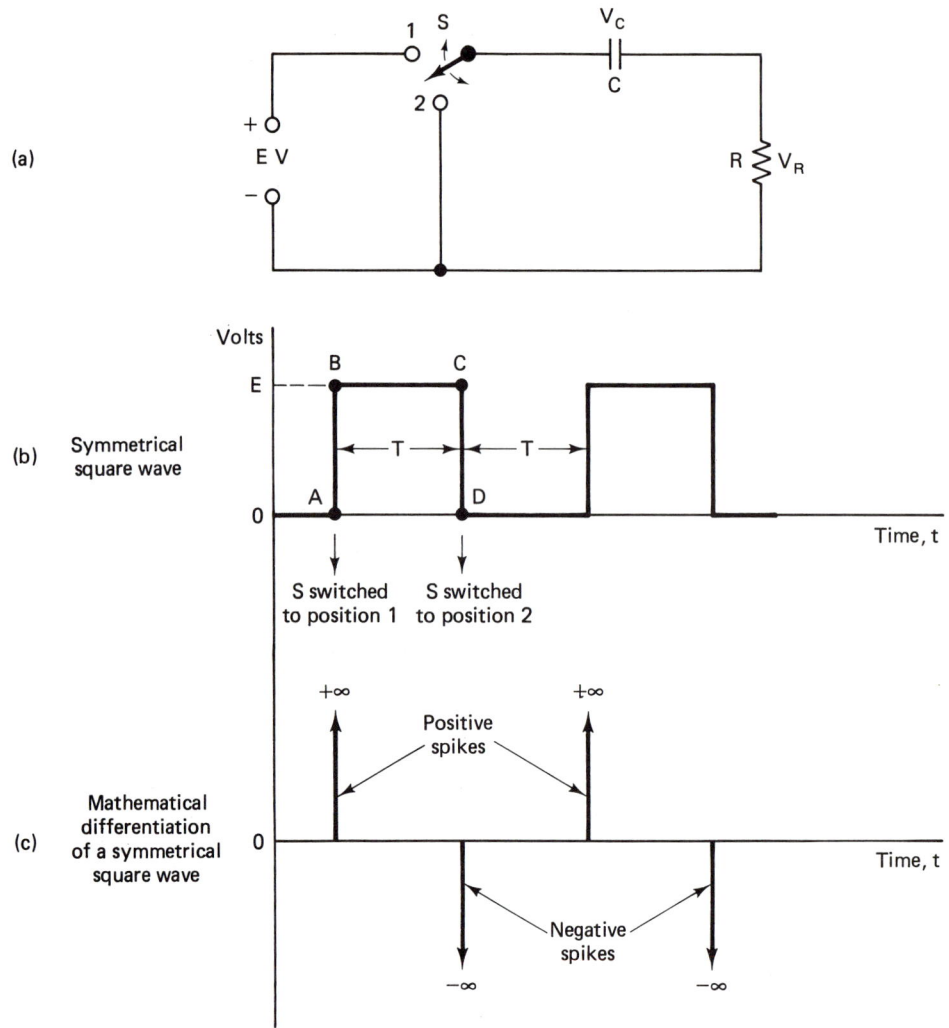

FIGURE 9-19 Mathematical differentiation of a symmetrical square wave.

1; as a result, E volts is applied across the series combination of the capacitor C and the resistor R. After an appreciable time (T), S is switched *instantaneously* to position 2. The source voltage is now replaced by a short circuit that is equivalent to applying zero volts. After the same time interval, T, the switch is instantaneously returned to position 1 and the sequence is repeated. The applied voltage is then represented by the symmetrical square wave of Fig. 9-19(b). Of course, this square wave is idealized since, in practice, the switch cannot move instantaneously from position 1 to position 2, or vice versa. There must be a certain "rise" time in the movement of the voltage from point A to point B and also a "fall" time in its decrease from point C to point D.

To differentiate the square wave, we must consider the slope of the waveform at every point during its entire sequence. When the voltage rises instantaneously from zero volts to E volts, the time involved is ideally zero, so that the slope is theoretically infinite. Moreover, this slope is *positive* since the waveform is increasing from the low level of zero volts to the high level of E volts. Consequently, the mathematical result of differentiating the waveform from point A to point B is an infinite positive "spike," as shown in Fig. 9-19(c).

As we move from point B to point C, the voltage remains unchanged, so that this section of the square wave is a horizontal line whose slope is zero. Between point C and point D, the level of the square wave falls instantaneously from E volts to zero volts, so that the differentiated result is an infinite *negative* spike.

Summarizing, when we differentiate a square wave, we produce alternate positive and negative spikes which are separated from one another by a zero level.

CR Differentiator Circuit

Electronically, we cannot create the precise waveform of Fig. 9-19(c). But how can we obtain a close approximation? Here we must refer back to the curves of Fig. 9-14. The V_R curve of Fig. 9-14(a) looks something like a positive spike, while the V_R curve of Fig. 9-14(b) resembles a negative "spike." It therefore appears that we could apply a symmetrical square wave to a series CR combination and obtain the differentiated output from across the resistor [Fig. 9-20(a)]. But when we join the two V_R curves together [Fig. 9-20(b)], it is clear that the "spikes" are far too broad and must be *shortened* to improve the degree of approximation.

In the waveform of Fig. 9-20(b) the decay time of $5CR$ seconds associated with each V_R curve is equal to the interval of T seconds for the symmetrical square wave. A practical differentiator circuit normally uses a time constant

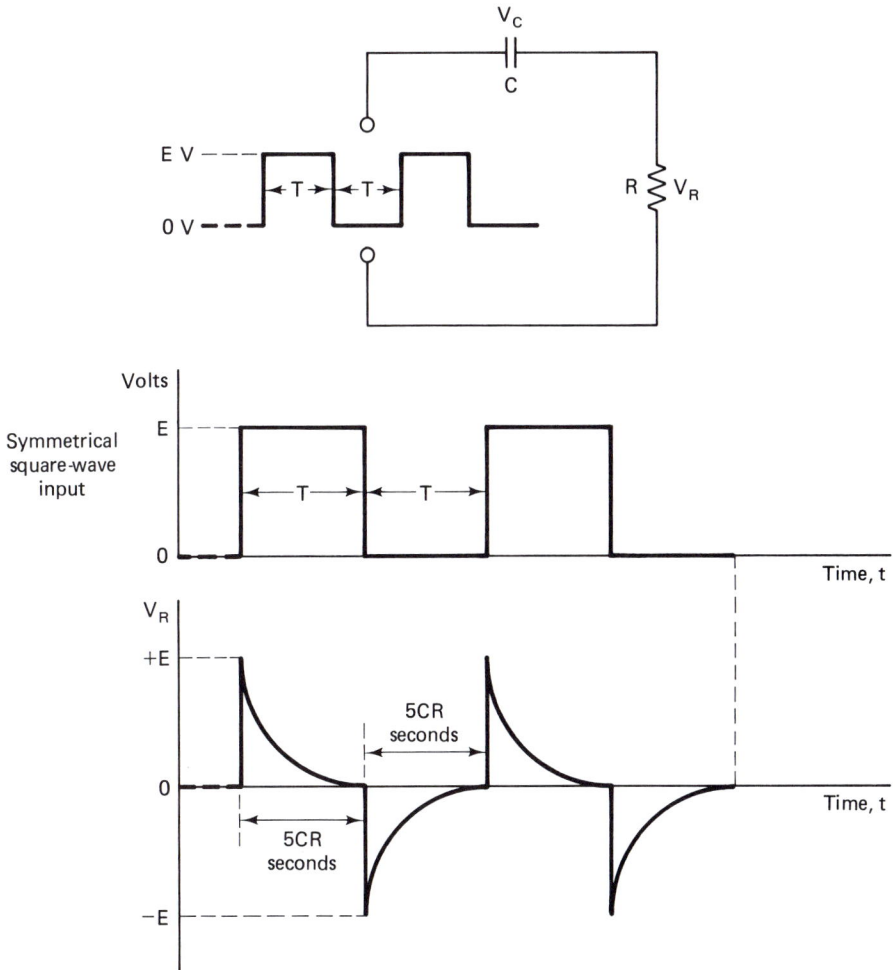

FIGURE 9-20 Application of a symmetrical square wave where time, $T = 5CR$ seconds.

Sec. 9-9 Differentiator Circuits

which is much less than one-tenth of T seconds; this means that the decay time of $5CR$ seconds is less than $T/2$, so that the capacitor has either charged or discharged completely before one-half of each "T" interval has elapsed.

As an example, let us assume that the square wave is repeating at the rate of 500 times per second, so that one complete sequence involves a time of $1/500$ s $= 2000$ μs. The time interval, T, is 1000 μs, and therefore the basic differentiator circuit would require a time constant of less than $1000/10 = 100$ μs. In the circuit of Fig. 9-21(a), $R = 500$ kΩ, $C = 100$ pF, so that the time constant is 100 pF $\times$ 0.5 MΩ $= 50$ μs. We have therefore fulfilled the condition that $CR < T/10$ and the capacitor will either charge or discharge completely after one-fourth of each "T" interval has elapsed. The differentiated output voltage across the resistor then consists of narrow positive and negative "spikes" and is a reasonable approximation to the ideal waveform of Fig. 9-19(c).

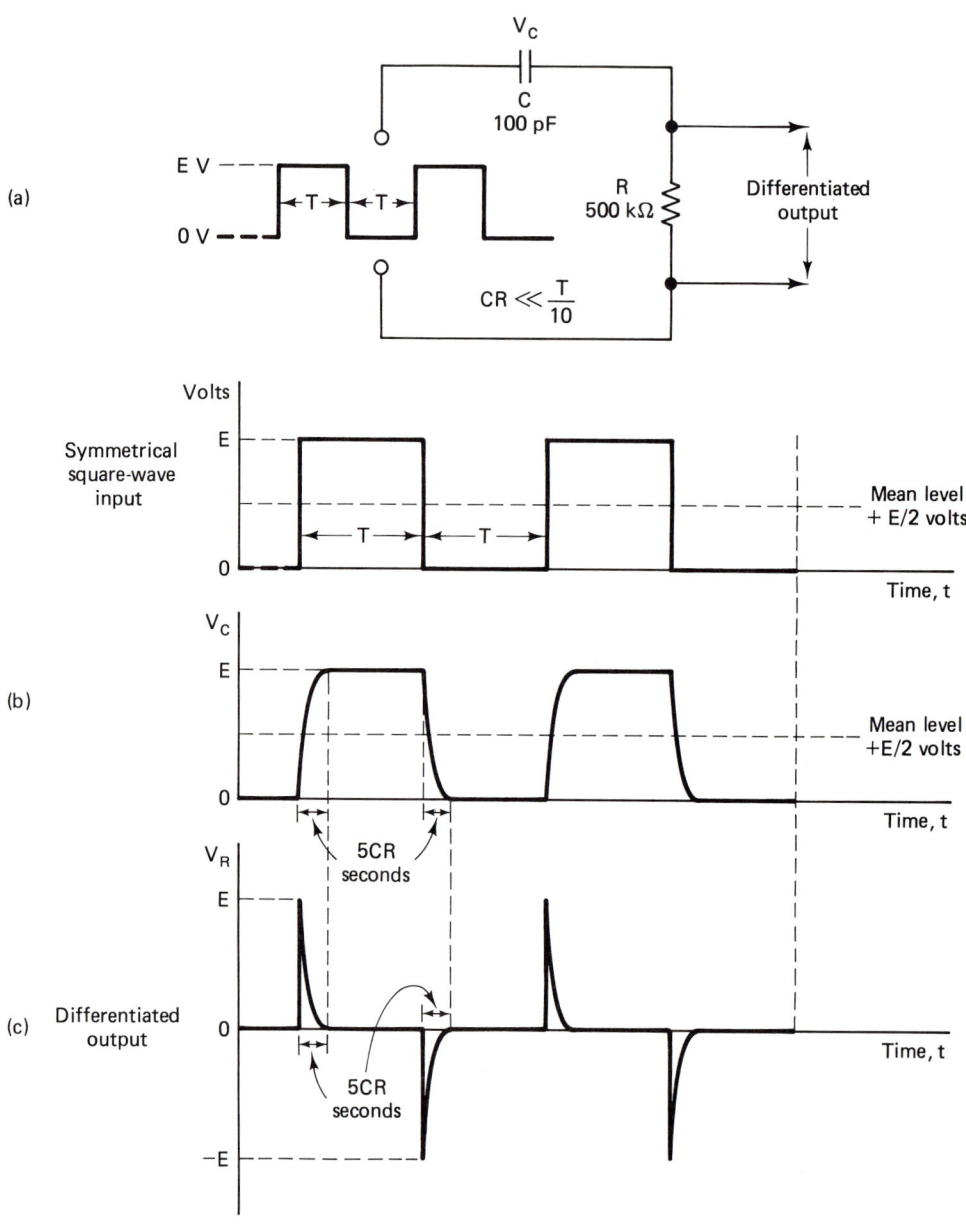

FIGURE 9-21 Practical *RC* differentiator circuit.

For the circuit of Fig. 9-21(a) the waveforms of V_C and V_R are shown in Fig. 9-21(b) and (c). Notice that the mean level of the V_R waveform is zero, while the mean level of the V_C waveform is $E/2$ volts (which is the same as the mean level of the source voltage). To obey Kirchhoff's voltage law at all times, the sum of the instantaneous values of the voltages V_R and V_C must equal the instantaneous value of the source voltage.

You might well ask: "Why don't we reduce the values of C and R still further so that we can obtain an even better approximation to the ideal differentiated waveform?" Let's first consider the effect of reducing the value of the resistance. The source of the square wave must possess a certain internal resistance so that the abrupt rise from zero to E volts is divided between the internal resistance and the differentiating resistor. If the value of R is reduced to the point where it becomes comparable with the internal resistance, the output positive spike will be decreased by the amount dropped across the internal resistance; the differentiated waveform will then be distorted. In a similar way, if the value of C is reduced to the same order as that of the stray capacitance in the circuit, further distortion will occur.

To summarize, one purpose of differentiation is to create short-interval spikes or pulses from a square-wave input of comparatively long duration. This is achieved by using a series CR combination whose time constant is small compared with the duration of the square wave; the differentiated output is taken from across the resistor and then one set of pulses (either the positive group or the negative group) can be used to "trigger" (control) the operation of the following circuitry.

LR Differentiator Circuit

Let us refer back to the graphs displayed in Fig. 8-12. These show the growth and decay of current in the series LR circuit of Fig. 8-10. The two graphs of V_L versus time have the same appearance as the two V_R versus time graphs for the series RC circuit (Fig. 9-14). It therefore seems that differentiation can be provided by either a CR circuit or an LR circuit; in the latter case the time constant L/R seconds would be less than $T/10$ seconds and the differentiated output would be taken from across the inductor (Fig. 9-22).

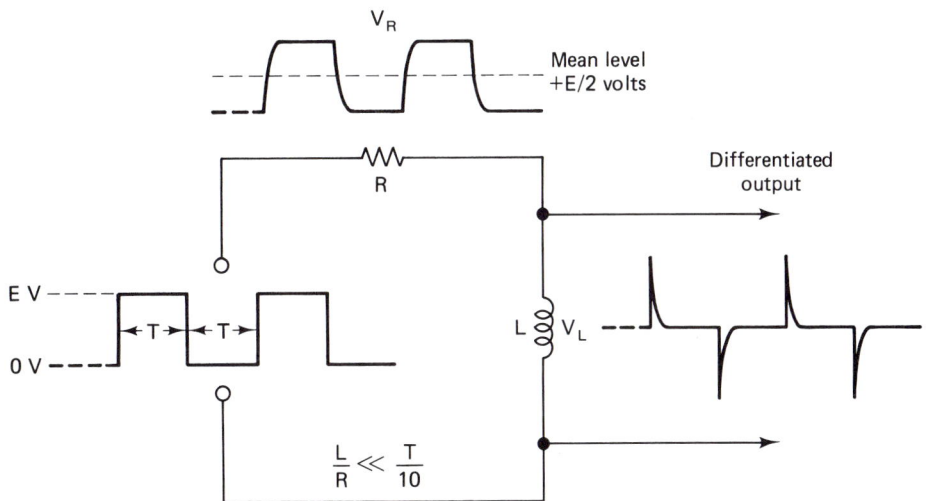

FIGURE 9-22 *LR* differentiator circuit.

The *LR* differentiator circuit has two disadvantages compared with the *CR* arrangement:

1. Inductors normally possess appreciable resistance so that their losses are greater than those of capacitors. As a result, the differentiated output from the *LR* circuit is lower and is more distorted.

2. As discussed in Section 8-6, an inductor must be considered to have a distributed self-capacitance. When the square wave is applied, it is possible for the inductance and the self-capacitance to set up an oscillation which is referred to as "ringing." The effect of ringing is to introduce further distortion into the waveform of the differentiated output voltage.

Because of these disadvantages, the *LR* differentiator circuit is not commonly found in electronic circuits.

Differentiation of a Sawtooth Waveform

The symmetrical sawtooth waveform of Fig. 9-23(a) is frequently encountered in electronics and communications. For example, a sawtooth voltage is applied between the *X* plates of an electrostatic cathode ray tube in order to move the spot of light at uniform speed across the face of the tube (Chapter 20).

Between points *A* and *B* the slope of the sawtooth is fixed at *m* volts per second, so that the mathematically differentiated waveform has a constant value over the interval and will therefore be represented by a horizontal line [Fig. 9-23(b)]. As the sawtooth drops from point *B* to point *C* in theoretically zero time, the slope is infinitely steep and the differentiated result is an instantaneous negative spike. Over the interval between points *C* and *D* the value of the sawtooth is always zero, so that the differentiated waveform is a horizontal line that is also at the zero level.

When the sequence is repeated, the mathematically differentiated waveform [Fig. 9-23(b)] has the appearance of a symmetrical square wave. It follows that the result of differentiating a sawtooth waveform is to create a square wave.

The basic *CR* differentiator circuit is shown in Fig. 9-23(c). Once again we will assume that the repetition rate of the sawtooth waveform is 500 times per second; consequently,

$$T = \frac{1}{2 \times 500} \text{ s} = 1000 \text{ }\mu\text{s}$$

The value of $C \times R$ is 100 pF × 0.5 MΩ = 50 μs and the requirement for a short time constant ($CR < T/10$) is fulfilled.

After a brief transient time, the circuit has reached its steady-state condition, in which the rate at which the voltage across the capacitor is increasing must be the same as the rate at which the sawtooth voltage is rising: namely, *m* volts per second. Since the charge, q = voltage, V_C × capacitance, C, it follows that the capacitor is being charged at the rate of $m \times C$ coulombs per second. Therefore, the charging current is mC amperes and the voltage across the resistor has a constant value of mCR volts. The waveforms of V_C and V_R are illustrated in Fig. 9-23(d).

If we assume that m = 0.1 V/μs, the peak value of the sawtooth is 0.1 V/μs × 1000 μs = 100 V and the steady voltage across the resistor is mCR = 0.1 V/μs × 100 pF × 0.5 MΩ = 5 V.

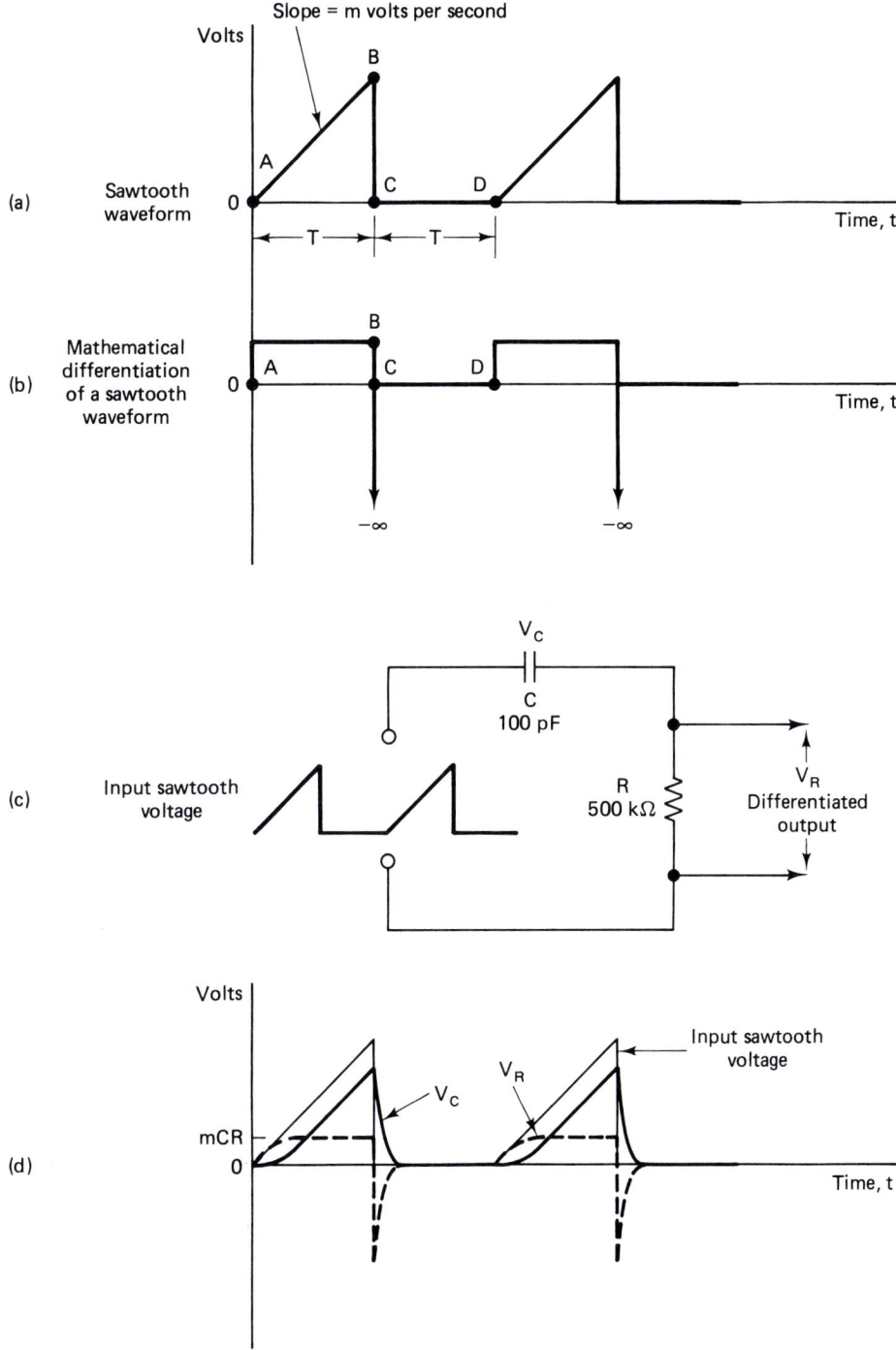

FIGURE 9-23 Differentiation of a sawtooth waveform.

Applications of the Differentiator Circuit

The use of a differentiator circuit is not uncommon in electronics and communications. For example, in the transmitter of a radar system a sine-wave source may be used to create a square wave. Differentiation of the square wave produces positive and negative spikes of short duration. The negative spikes are eliminated, while the positive spikes are used to activate or "trigger" the action of a modulator unit. This stage is responsible for determining the

Sec. 9-9 Differentiator Circuits

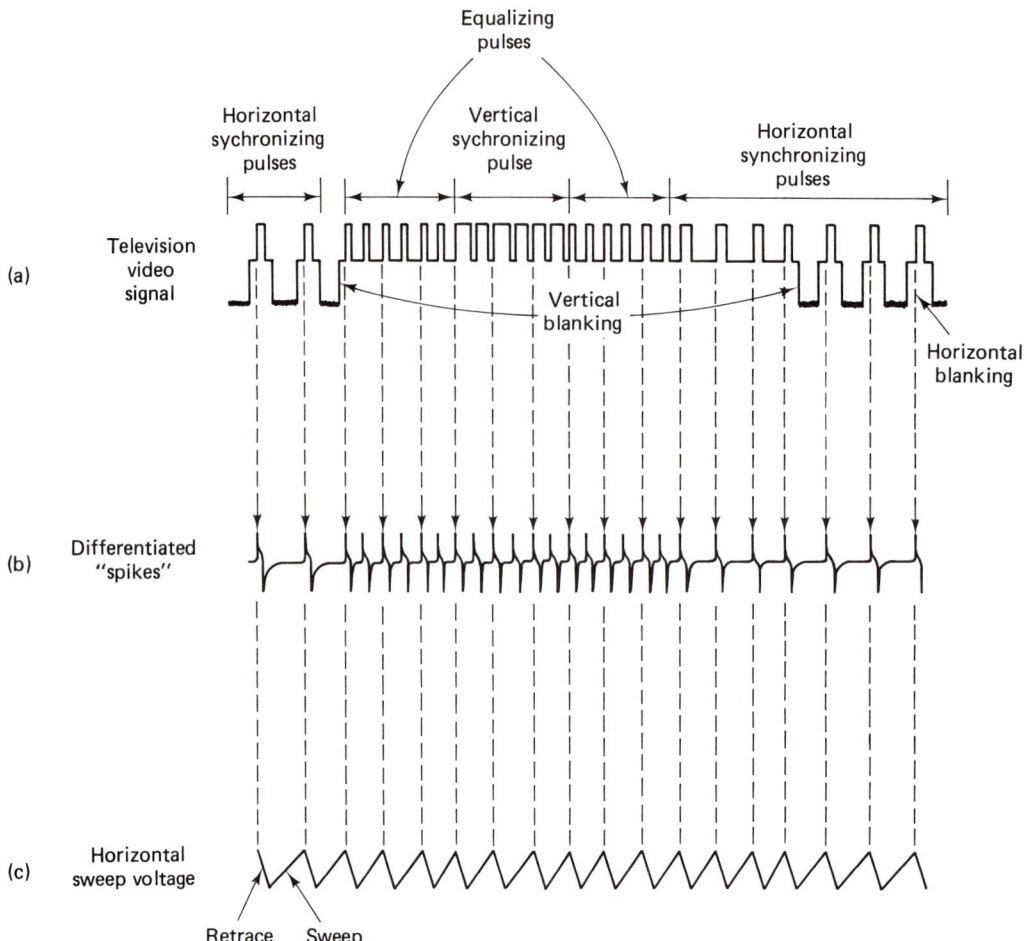

FIGURE 9-24 Differentiation of synchronizing and equalizing pulses in the television receiver.

duration of the radio-frequency (RF) pulse which finally emanates from the antenna system.

In another application a TV transmitter sends out horizontal (H) and vertical (V) synchronizing pulses, so that the sweeping of the receiver's television tube is in step with the action of the studio camera. At the receiver the horizontal synchronizing pulses are differentiated and the resulting positive spikes are used to trigger the horizontal AFC (automatic frequency control) circuit, which then initiates the horizontal retrace and the subsequent sweep [Fig. 9-24(a)]. When the serrated synchronizing pulse arrives to start off the vertical retrace, it is essential that the horizontal scanning circuits continue to be correctly triggered (although the retrace is blacked out). To achieve this result, the serrated pulses are correctly spaced so that when differentiation occurs, *every other* positive spike triggers the horizontal AFC circuit. In Fig. 9-24(b) these active positive spikes are indicated by arrows which are lined up with the start of the retrace, as shown in the horizontal sweep voltage waveform [Fig. 9-24(c)]. The same differentiating action applies to the equalizing pulses that occur before and after the vertical synchronizing pulse.

EXAMPLE 9-13 In Fig. 9-21, assume that the symmetrical square-wave input has a repetition rate of 1000 times per second. If the chosen value for R is

100 kΩ and the time constant is 5% of T seconds, what is the required value of the capacitor for differentiation to occur?

Solution

$$\text{time interval, } T = \frac{1}{2 \times 1000} \text{ s} = 500 \text{ μs}$$

$$\text{time constant} = \frac{5}{100} \times 500 = 25 \text{ μs}$$

$$\text{capacitance, } C = \frac{\text{time constant}}{R}$$

$$= \frac{25 \text{ μs}}{100 \text{ kΩ}}$$

$$= 250 \times 10^{-12} \text{ F}$$

$$= 250 \text{ pF}$$

9-10 Integrator Circuits

CR Integrator Circuit

As far as calculus is concerned, integration is the reverse process to differentiation. In terms of the falling mass, we discovered that the velocity was the result of differentiating the distance with respect to the time. It therefore follows that the distance is the result of integrating the velocity with respect to the time. With regard to waveforms, if waveform B is the result of differentiating waveform A, then waveform A is the result of integrating waveform B. For example, we found out that when we differentiated a sawtooth waveform, we obtained a square wave. Reversing the process means that if we integrate a square wave, we will produce a sawtooth waveform.

Looking back to Fig. 9-19(a), we have already discussed the creation of a square wave by the rapid movement of the switch between positions 1 and 2. But which of the waveforms in Fig. 9-14(a) represents a reasonable approximation to the integrated sawtooth output? Only the graph of V_C versus time shows an increasing voltage, but this graph is a curve of exponential growth, whereas we need a linear ("straight line") rise of constant slope. However, we should observe that the curve shows good linearity between the times of zero and $CR/10$ seconds. Consequently, if we choose values of C and R such that $CR > 10T$, the V_C waveform will be a close approximation to the required integrated output. The basic integrator circuit is shown in Fig. 9-25(a). Once again we shall assume that the input square wave is repeating at the rate of 500 times per second; the time interval, T, is 1000 μs, and therefore the circuit will require a time constant greater than $10 \times 1000 = 10{,}000$ μs. Since $C = 0.02$ μF and $R = 1$ MΩ, the time constant is 0.02 μF $\times 1$ MΩ $= 20{,}000$ μs and the necessary condition is fulfilled. The integrated sawtooth output is taken from across the capacitor, as indicated by the waveforms of Fig. 9-25(b).

Integration in calculus is also a method of determining the area beneath a curve. Referring back to Fig. 9-18(a) and (b), we will integrate the velocity versus time graph between the times $t = 2$ s and $t = 3$ s. The result is the

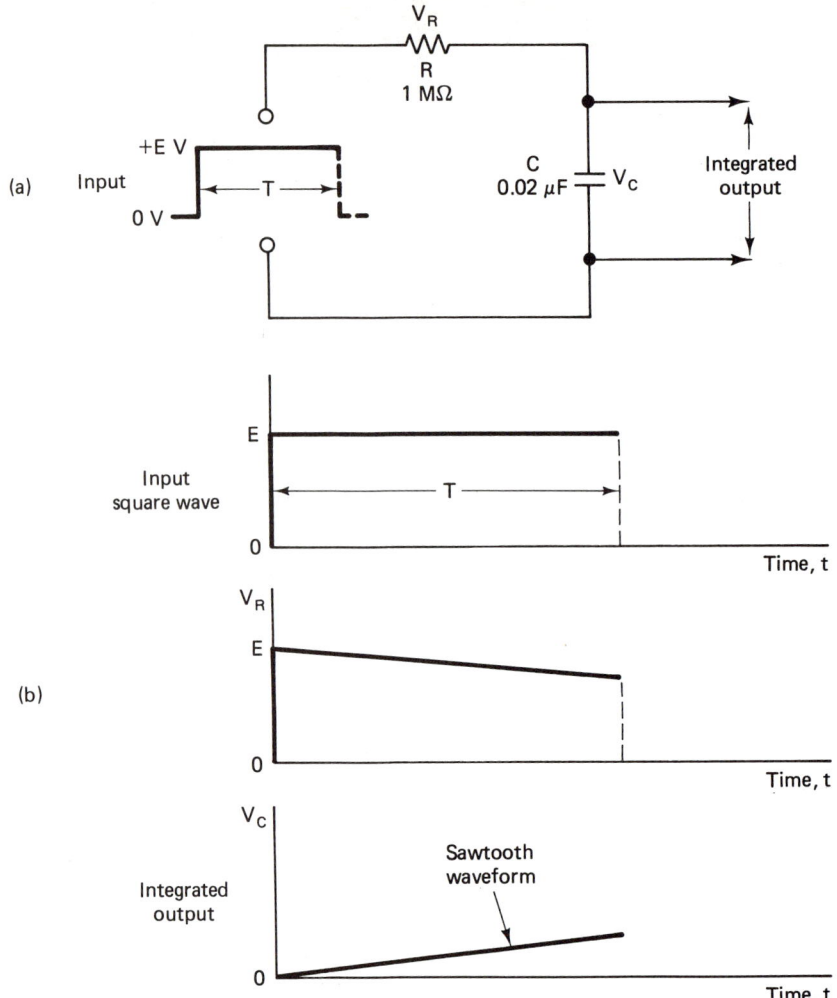

FIGURE 9-25 Basic integrator circuit.

shaded area, which has a value of $\frac{1}{2} \times (19.62 + 29.43)$ m/s $\times (3 - 2)$ s = 24.525 m. Remembering that the distance/time graph is the integral of the velocity/time graph, this area should have the same value as the distance covered between $t = 2$ s and $t = 3$ s; from Table 9-4 this distance is $44.145 - 19.02 = 24.525$ m and therefore the anticipated result has been verified.

Let us assume that a repeating symmetrical square wave is applied to the integrating circuit of Fig. 9-26(a). When steady-state conditions have been reached, the capacitor will charge during the time that the square wave is at the E-volt level. However, there will be only a slow "charging" rate because of the long time constant. In a similar way, the capacitor will only discharge slowly when the square wave is at the zero-volt level. The V_C waveform of Fig. 9-26(b) fluctuates slightly about the $+E/2$-volt level, while the average value of the V_R waveform is zero. At all times the sum of the instantaneous V_R and V_C values must equal the instantaneous value of the input square wave. The total of the shaded areas for the square wave is equal to the shaded area shown in the V_C waveform; the circuit therefore satisfies the area concept of integration.

If we idealize the V_C waveform, it becomes a steady level of $+E/2$ volts; by contrast the corresponding V_R waveform is a square wave with levels between $+E/2$ and $-E/2$ volts. The capacitor has in fact charged to the mean level of the input square wave, while the square wave itself appears across the resistor.

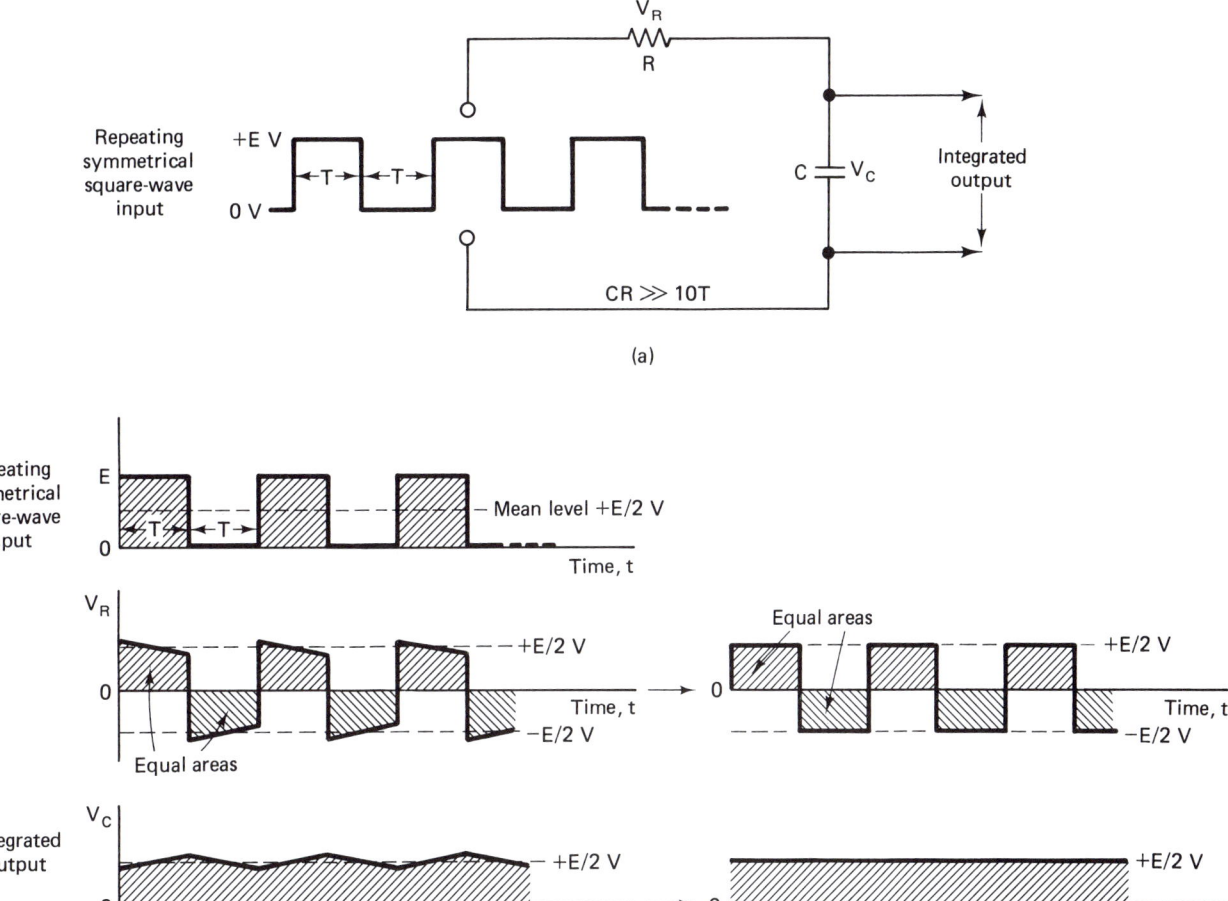

FIGURE 9-26 Application of a symmetrical square wave to an integrator circuit.

These results are used when coupling a signal from one circuit to another (Fig. 9-27). The mean level of the input signal is "blocked" by the capacitor, while the voltage across the resistor has an average value of zero, but the shape of the waveform is virtually undistorted. This will only be true provided that the time constant of the coupling components is long compared with the time taken for one complete sequence of the signal. To summarize, a CR integrator circuit requires a high time constant and the integrated output is developed across the capacitor.

Asymmetrical Square-Wave Input

If the symmetrical square wave of Fig. 9-26(a) is followed by the asymmetrical square wave of Fig. 9-28(b), the mean level is no longer equal to $+E/2$ volts. Assuming that $T_1 = 3T_2$, the new mean level is

$$+ \frac{T_1}{T_1 + T_2} \times E = + \frac{3T_2 \times E}{4T_2} = + \frac{3E}{4} \text{ volts}$$

The integrated output, V_C, will therefore climb from $+E/2$ to $+3E/4$ and the new waveforms are as shown in Fig. 9-28(b). We have deduced that the value

Sec. 9-10 Integrator Circuits

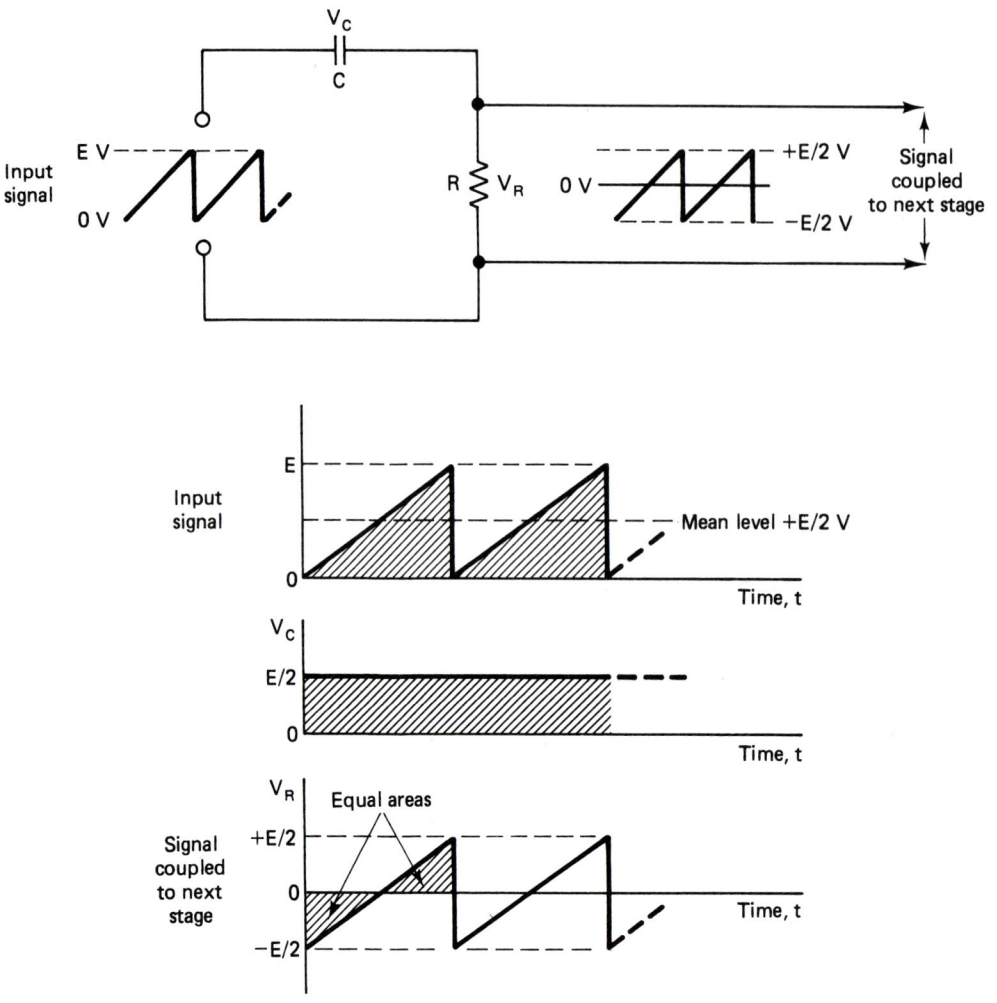

FIGURE 9-27 Principle of the CR coupling circuit.

of the integrated output depends on the degree of asymmetry in the square-wave input. Once again the sum of the areas under the input square wave is equal to the area under the integrated output. The V_R waveform has equal areas above and below the zero line, so that its average value is again zero.

LR Integrator Circuit

This circuit is illustrated in Fig. 9-29. To achieve the integrating action we shall require a long time constant so that $L/R > 10T$. The integrated output is then developed across the resistor, R whose value is limited by the effective load, which is represented by the following circuitry; this load is in parallel with R and consequently the total equivalent resistance can never exceed the value of the load.

Application of the Integrator Circuit

To initiate the vertical retrace in a TV receiver, the vertical synchronizing pulse (Fig. 9-30) is integrated by a CR circuit with a long time constant. Since this vertical pulse has narrow serrations, its integrated value will be higher than that of the preceding six equalizing pulses, which establish a low starting level.

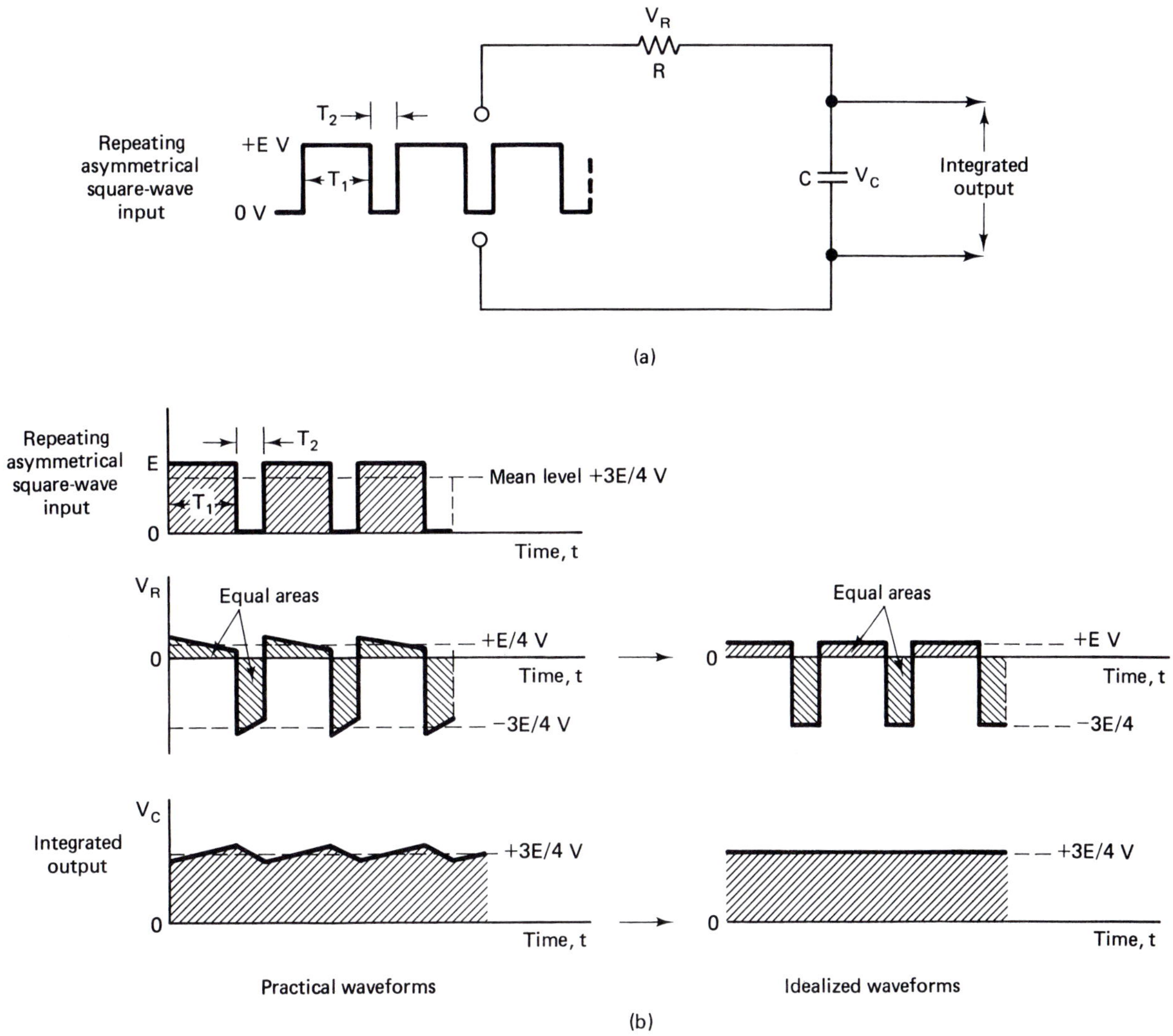

FIGURE 9-28 Application of an asymmetrical square wave to an integrator circuit.

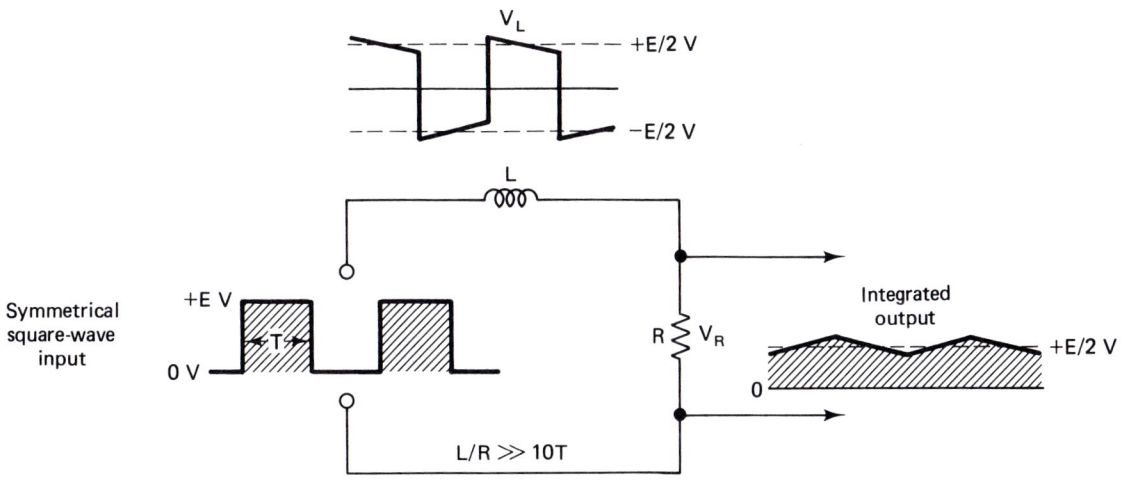

FIGURE 9-29 LR integrator circuit.

Sec. 9-10 Integrator Circuits

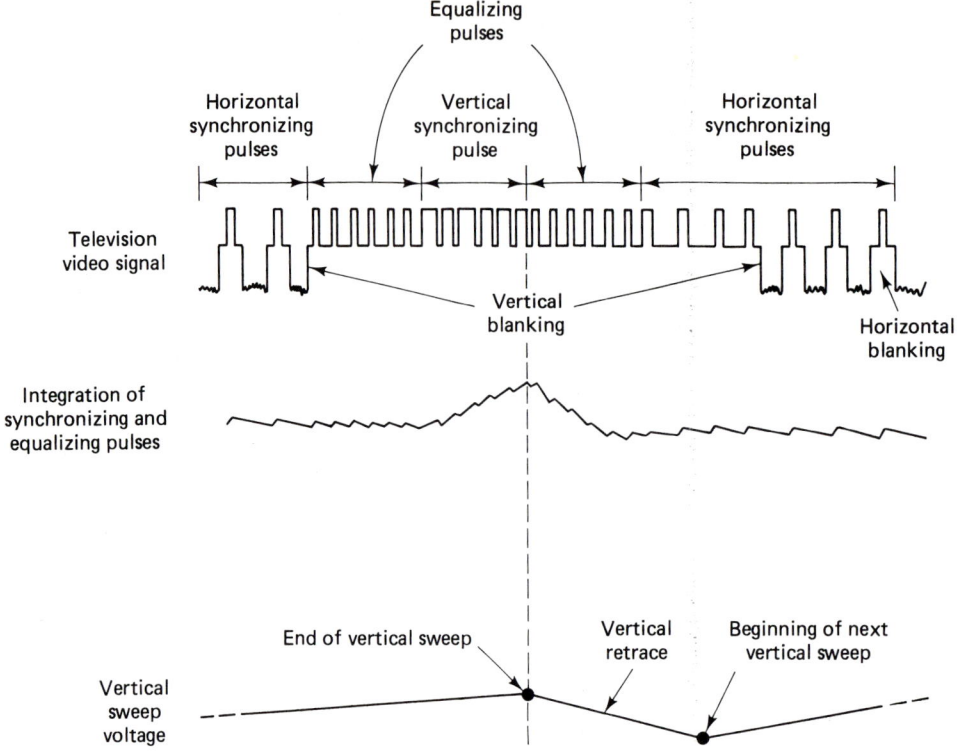

FIGURE 9-30 Integration of synchronizing and equalizing pulses in the television receiver.

You will remember that the serrations are also present, so that the differentiated output can continue to trigger the horizontal retrace.

When the vertical synchronizing pulse is applied to the CR circuit, the integrated output across the capacitor climbs to the point where the retrace is initiated. This causes the blanked-out trace to move from the bottom of the TV screen back to the top. After the vertical synchronizing pulse has ended there are another six equalizing pulses which serve to restore the integrated output to its original level.

EXAMPLE 9-14 In Fig. 9-24(a), assume that the symmetrical square wave input has a repetition rate of 1000 times per second. If the value of R is 1 MΩ and the time constant is $20 \times T$ seconds, what capacitor value is required for integration to occur?

Solution

$$\text{time interval, } T = \frac{1}{2 \times 1000} \text{ s} = 500 \text{ }\mu\text{s}$$

$$\text{time constant} = 20 \times T = 10{,}000 \text{ }\mu\text{s}$$

$$\text{capacitance, } C = \frac{\text{time constant}}{R}$$

$$= \frac{10{,}000 \text{ }\mu\text{s}}{1 \text{ M}\Omega}$$

$$= 0.01 \text{ }\mu\text{F}$$

Chapter Summary

9.1. Electrostatic equations:

$$\text{Coulomb's law: } F = \frac{Q_1 Q_2}{4\pi\epsilon_0 d^2} \quad \text{newtons}$$

$$\text{permittivity of free space, } \epsilon_0 = 8.85 \times 10^{-12} \text{ F/m}$$

$$\text{electric flux, } \psi = Q \quad \text{coulombs}$$

$$\text{electric field intensity, } \mathcal{E} = \frac{E}{d} \quad \text{volts per meter or newtons per coulomb}$$

$$\text{force, } F = Q\mathcal{E} \quad \text{newtons}$$

$$\text{electric flux density, } D = \frac{Q}{A} \quad \text{coulombs per square meter}$$

$$\text{charge, } Q = CE \quad \text{coulombs}$$

$$\text{absolute permittivity, } \epsilon = \frac{D}{\mathcal{E}} = \epsilon_0 \epsilon_r \quad \text{farads per meter}$$

relative permittivity (dielectric constant) of an insulator, ϵ_r

$$= \frac{\text{capacitance with the insulator as the dielectric}}{\text{capacitance with a vacuum dielectric}}$$

$$\text{capacitance, } C, = \frac{\epsilon_0 \epsilon_r A}{d} = \frac{\epsilon A}{d} \quad \text{farads}$$

$$\text{energy stored in an electric field} = \frac{1}{2} CE^2 = \frac{1}{2} QE = \frac{1}{2} \frac{Q^2}{C} \quad \text{joules}$$

9.2. Capacitors in series:

$$C_T = \frac{1}{\frac{1}{C_1} + \frac{1}{C_2} + \frac{1}{C_3} + \cdots + \frac{1}{C_N}} \quad \text{farads}$$

$$Q_1 = Q_2 = Q_3 = \cdots = Q_N = Q_T = C_T E_T \quad \text{coulombs}$$

$$V_1 = E_T \times \frac{C_T}{C_1} \quad \text{volts,} \quad \text{etc.}$$

Equal-value capacitors:

$$C_T = \frac{C}{N} \quad \text{farads}$$

Two capacitors in series:

$$C_T = \frac{C_1 C_2}{C_1 + C_2} \quad \text{farads}$$

$$V_1 = E_T \times \frac{C_2}{C_1 + C_2} \quad \text{volts}$$

$$V_2 = E_T \times \frac{C_1}{C_1 + C_2} \quad \text{volts}$$

9.3. Capacitors in parallel:

$$C_T = C_1 + C_2 + C_3 + \cdots + C_N \quad \text{farads}$$

$$Q_T = Q_1 + Q_2 + Q_3 + \cdots + Q_N = C_T E_T \quad \text{coulombs}$$

$$Q_1 = C_1 E_T \quad \text{coulombs, etc.}$$

Equal-value capacitors:

$$C_T = NC \quad \text{farads}$$

Multiplate capacitor:

$$C = \frac{\epsilon_0 \epsilon_r (N - 1) A}{d} \quad \text{farads}$$

9.4. Charge and discharge of a capacitor through a resistor:

$$\text{time constant} = CR \text{ seconds}$$

$$\text{total time for the transient state} = 5CR \text{ seconds}$$

9.5. Comparisons among resistance, inductance, and capacitance:

	R	**L**	**C**
Property	Opposes the current	Opposes the rate of change of the current	Opposes the rate of change of the voltage
Unit	Ohm, Ω	Henry, H	Farad, F
Voltage–current relationship	$E = R \times I$	$E = L \times \dfrac{\Delta I}{\Delta T}$	$I = C \times \dfrac{\Delta E}{\Delta T}$
Physical factors	$R = \dfrac{\rho l}{A}$	$L = \dfrac{\mu_0 \mu_r N^2 A}{l}$	$C = \dfrac{\epsilon_0 \epsilon_r A}{d}$
Energy (joules)	*Dissipated* as heat: $W = I^2 RT$	*Stored* as a magnetic field: $W = \tfrac{1}{2} LI^2$	*Stored* as an electric field: $W = \tfrac{1}{2} CE^2$
Series arrangement	$R_T = R_1 + R_2 + \cdots + R_N$	$L_T = L_1 + L_2 + \cdots + L_N$	$C_T = \dfrac{1}{\dfrac{1}{C_1} + \dfrac{1}{C_2} + \cdots + \dfrac{1}{C_N}}$
Parallel arrangement	$R_T = \dfrac{1}{\dfrac{1}{R_1} + \dfrac{1}{R_2} + \cdots + \dfrac{1}{R_N}}$	$L_T = \dfrac{1}{\dfrac{1}{L_1} + \dfrac{1}{L_2} + \cdots + \dfrac{1}{L_N}}$	$C_T = C_1 + C_2 + \cdots + C_N$

9.6. A *CR* differentiator circuit has a *short* time constant compared with the duration of the input waveform. The differentiated output is taken from across the resistor.

9.7. A *CR* integrator circuit has a *long* time constant when compared with the duration of the input waveform. The integrated output is taken from across the capacitor.

Self-Review

True–False Questions

1. T. F. Capacitance is that property of an electrical circuit which opposes voltage.
2. T. F. Electric flux is measured in coulombs.

3. T. F. With a mica dielectric the ratio of the electric flux density to the electric field intensity is the permittivity of free space, ϵ_0.
4. T. F. If the thickness of a dielectric is doubled, the capacitance is halved.
5. T. F. If a number of capacitors are connected in series, the same charge is stored in each capacitor.
6. T. F. The force of repulsion between two unlike charges is inversely proportional to the square of their separation.
7. T. F. If capacitances of 1 µF, 0.02 µF, and 5000 pF are connected in parallel, their total capacitance is 1.07 µF.
8. T. F. The time constant of a 500-pF capacitor and a 10-MΩ resistor is 5 ms.
9. T. F. A 0.1 µF capacitor is charged to 50 V. The energy stored in the electric field is 250 µJ.
10. T. F. A 1-µF capacitor charged to 100 V is discharged through a 10-kΩ resistor. The time constant of the discharge is 0.01 s.
11. T. F. For a parallel-plate capacitor the charge existing on one plate is exactly the same as the charge on the other plate.
12. T. F. The permittivity of an insulator is equal to the product of the relative permittivity and the permittivity of free space.
13. T. F. The dielectric strength of air is greater in value than that of waxed paper.
14. T. F. The maximum working voltages of a number of capacitors connected in parallel is determined by the highest individual working voltage.
15. T. F. If the voltage across a capacitor is doubled, the energy stored in the capacitor is also doubled.
16. T. F. Only electrolytic capacitors normally provide capacitance values of several microfarads.
17. T. F. A parallel-plate capacitor with an air dielectric is charged to a certain voltage. If the source is removed and the mica dielectric is inserted between the plates, the electric field intensity and the voltage across the capacitor are both increased.
18. T. F. A *CR* integrator circuit requires a high time constant and the integrated output is taken from across the resistor.
19. T. F. The relative permittivity of an insulator is equal to the ratio of the capacitance value with the insulator as its dielectric to the capacitance value with a free-space dielectric.
20. T. F. A *CR* differentiator circuit requires a short time constant and the differentiated output is taken from across the resistor.

Multiple-Choice Questions

21. A 200-pF capacitor is charged to 40 V. The charge stored is
    (a) 800 pC   (b) 200 pC   (c) 16,000 pC   (d) 0.008 µC   (e) 0.002 µC
22. An air capacitor consists of two parallel plates that are 0.2 mm apart; the area of one side of each plate is 20 cm². If the capacitor contains a charge of 0.004 µC, the flux density in microcoulombs per square meter between the plates is
    (a) 2   (b) 4   (c) 0.2   (d) 20   (e) 40
23. In Question 22, the value of the capacitance is
    (a) 177 pF   (b) 88.5 pF   (c) 17.7 pF   (d) 354 pF   (e) 8.85 pF
24. In Question 22, the voltage gradient in volts per meter between the plates is
    (a) 432,000   (b) 113,000   (c) 226,000   (d) 43,200   (e) 86,400
25. If a charge of 20 µC is placed between the plates of the capacitor of Question 22, the force in newtons exerted on the charge is
    (a) 2.3   (b) 4.5   (c) 9.0   (d) 1.1   (e) 11.0
26. A 30-µF capacitor and a 6-µF capacitor are connected in series across an 84-V source. The voltage across the 6-µF capacitor is
    (a) 84 V   (b) 14 V   (c) 70 V   (d) 81.2 V   (e) 2.8 V

27. A 4-µF capacitor and a 20-µF capacitor are connected in parallel across a 48-V source. The total charge stored is
    (a) 0.5 µC (b) 2 µC (c) 2.5 µC (d) 14.4 µC (e) 1152 µC
28. In Question 27, the total energy stored in the two capacitors is
    (a) 1152 µJ (b) 576 µJ (c) 4608 µJ (d) 23,040 µJ (e) 27,648 µJ
29. The combination of a 220-pF capacitor and a 200-kΩ resistor has a time constant of
    (a) 4.4 ms (b) 11 µs (c) 44 µs (d) 440 µs (e) 4400 ns
30. A mica dielectric has a relative permittivity of 5. Its absolute permittivity in farads per meter is
    (a) 5×10^{-12} (b) 8.85×10^{-12}
    (c) 1.77×10^{-12} (d) 44.3×10^{-12}
    (e) 13.85×10^{-12}
31. The total equivalent capacitance of two capacitors in series is 4 µF. If the value of one capacitor is 20 µF, the value of the other capacitor is
    (a) 16 µF (b) 24 µF (c) 0.2 µF (d) 5 µF (e) 8 µF
32. The energy stored in a 4-µF capacitor is 800 µJ. The voltage across the capacitor is
    (a) 400 V (b) 20 V (c) 800 V (d) 28.28 V (e) 14.14 V
33. The maximum voltage that can be applied across a 0.5-mm thickness of mica is
    (a) 10 kV (b) 25 V (c) 5 kV (d) 25 kV (e) 2500 V
34. A *CR* differentiator circuit requires a time constant of 50 µs. If the value of the resistor is 200 kΩ, the required value for the capacitor is
    (a) 0.01 µF (b) 40 µF (c) 400 pF (d) 0.004 µF (e) 250 pF
35. Two 0.1-µF capacitors are charged in series from a 120-V source. When the charging is completed, the source is carefully removed and the two free leads are connected together. The subsequent voltage across each of the capacitors is
    (a) 60 V (b) 120 V (c) zero (d) 80 V (e) 240 V
36. Three capacitors, whose values are 40 µF 100 WVDC, 20 µF 150 WVDC, and 25 µF 125 WVDC, are connected in series. The working voltage of the combination is
    (a) 150 WVDC (b) 100 WVDC
    (c) 345 WVDC (d) 125 WVDC
    (e) 375 WVDC
37. In Question 36 the three capacitors are reconnected in parallel. The new maximum working voltage is
    (a) 150 WVDC (b) 100 WVDC
    (c) 375 WVDC (d) 125 WVDC
    (e) 40 WVDC
38. A *CR* integrator circuit requires a time constant of 80 µs. If the value of the capacitor is 0.0005 µF, the value of the resistor is
    (a) 160 kΩ (b) 16 kΩ (c) 40 kΩ (d) 400 kΩ (e) 4 MΩ
39. The voltage across a 200-pF capacitor is increasing uniformly at the rate of 20 V/ms. The value of the charging current is
    (a) 1 mA (b) 1 µA (c) µA (d) 4 mA (e) 4 µA
40. In Question 39, assume that a time interval of 4 ms has elapsed. The additional charge stored by the capacitor is
    (a) 0.016 µC (b) 1.6 µC (c) 0.16 µC (d) 0.08 µC (e) 0.8 µC

Practice Problems

41. A parallel-plate capacitor has an air dielectric. If the area of one side of each plate is 0.004 m² and the separation between the plates is 0.0005 m, calculate the value of the capacitance.

42. If the capacitor of Problem 41 is charged to 80 V, what is the amount of the charge stored? Calculate the force of attraction between the plates.

43. In Problem 42, what are the values of the electric field intensity and the electric flux density?

44. A 6-μF capacitor and a 3-μF capacitor are connected in series across a 180-V dc source. What are the values of the charge and the energy stored in each capacitor? What is the value of the voltage across each capacitor?

45. A 6-μF capacitor and a 3-μF capacitor are connected in parallel across a 180-V dc source. What are the values of the charge and the energy stored in each capacitor?

46. If we close the switch, S, in the circuit of Fig. 9-31, what are the initial and final values of the potential at point X? How long will it take for the circuit to reach its steady-state condition?

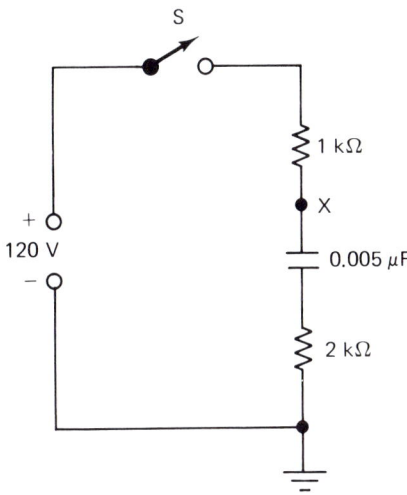

FIGURE 9-31   Circuit for Problem 46.

47. If we close the switch, S, in the circuit of Fig. 9-32, what are the initial and final values of the potentials at points X and Y? What is the value of the circuit's time constant?

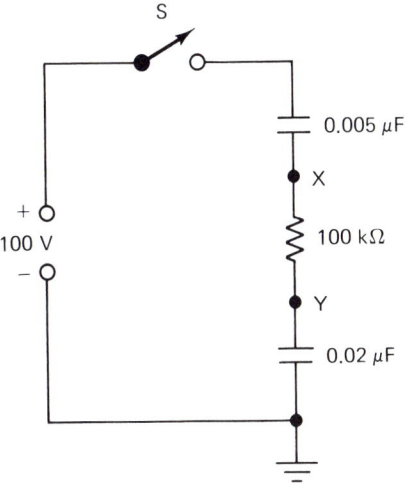

FIGURE 9-32   Circuit for Problems 47 and 48.

48. In Problem 47, the 0.005-μF capacitor short-circuits. What are the new values of the final potentials at points X and Y?

49. In Fig. 9-33, what is the value of the total equivalent capacitance between points P and Q?

Self-Review

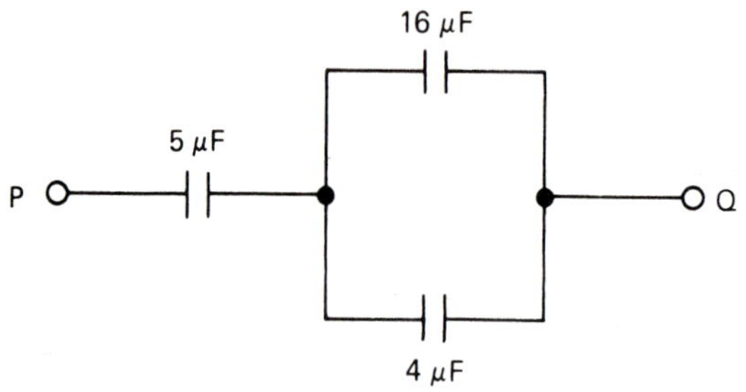

FIGURE 9-33 Circuit for Problem 49.

50. Two 0.1-μF capacitors are charged in series from a 240-V source. After the charging is completed, the source is removed. The capacitors are then separated and carefully paralleled. Finally, a third, uncharged 0.1-μF capacitor is connected across the parallel combination. What would be the voltage across the third capacitor?

51. A 0.05-μF capacitor is connected in series with a 0.1-μF capacitor. A 0.03-μF capacitor is joined in parallel across the series combination. What is the total equivalent capacitance of the complete network?

52. The initial voltage across a 0.03-μF capacitor is 50 V, but after an interval of 5 ms this voltage has risen to 95 V. What is the average value of the charging current?

53. In Problem 52, the capacitor is charged to 95 V but is then uniformly discharged by a current of 1.8 mA. How long will it take for the capacitor to discharge completely?

54. A variable capacitor with an initial capacitance of 800 pF is charged to 100 V. The plates of the capacitor are then separated until the capacitance is reduced to 200 pF. What is the value of the new voltage across the capacitor? Calculate the values of the energy stored before and after the plates are separated. Explain the difference between these energy values.

55. Three capacitors of identical dimensions possess dielectrics whose relative permittivities are 1, 3, and 5, respectively. If these capacitors are connected in series across a 360-V dc source, calculate the value of the voltage across each capacitor.

56. A slice of insulating material with a thickness of 4 mm is inserted between the plates of a parallel-plate capacitor. To restore the value of the capacitance to its original value, it is necessary to increase the spacing between the plates by 2 mm. Calculate the value of the relative permittivity for the insulating material.

57. To what voltage must a 0.01-μF capacitor be charged in order that the energy stored is 50 μJ?

58. Three capacitors whose capacitances and voltage ratings are 40 μF 25 WVDC, 25 μF 40 WVDC, and 50 μF 20 WVDC, respectively, are connected in parallel. What is the maximum charge that can be stored by this combination without exceeding the voltage rating of any of the capacitors?

59. The complete sequence of a symmetrical square wave repeats at a rate of 1000 times every second. This square wave is applied to a CR differentiator circuit in which the value of the resistor is 100 kΩ. What is the capacitor's maximum value that will allow adequate differentiation to occur?

60. The same square wave of Problem 59 is applied to an RC integrating circuit in which the value of the resistor is 1 MΩ. What is the capacitor's minimum value that will allow adequate integration to occur?

Advanced Problems

61. A paper (relative permittivity = 3.5) capacitor has two aluminum foil surfaces which are 0.15 mm apart. The area of one side of each surface is 80 cm². If the capacitor is charged to a potential difference of 120 V, what are the values of the electric flux, the electric field intensity, and the electric flux density?

62. In Problem 61, calculate the values of the capacitance and the charge stored. What is the amount of the energy stored in the electric field? If a charge of 50 µC is placed between the surfaces, find the force that is exerted on this charge.

63. A 16-µF capacitor is charged from a 200-V source, while a 25-µF capacitor is separately charged from a 250-V source. If the charged capacitors are disconnected from their sources and correctly paralleled, what is the final voltage across the combination? Calculate the total energy stored before and after the parallel connection is made.

64. If the switch, S, is closed in Fig. 9-34, what are the initial and final values of the potential at point X? What is the duration of the transient state?

FIGURE 9-34 Circuit for Problem 64.

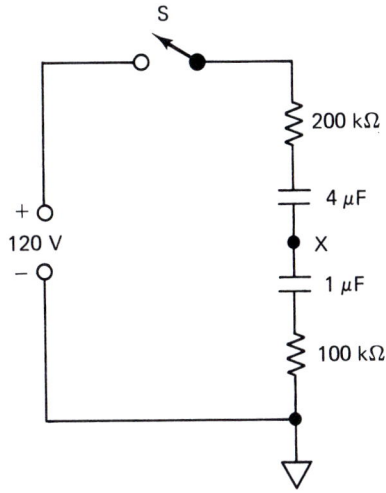

65. In Fig. 9-35, points P and Q are connected to the terminals of a 100-V dc source. What is the amount of the total energy stored?

FIGURE 9-35 Circuit for Problem 65.

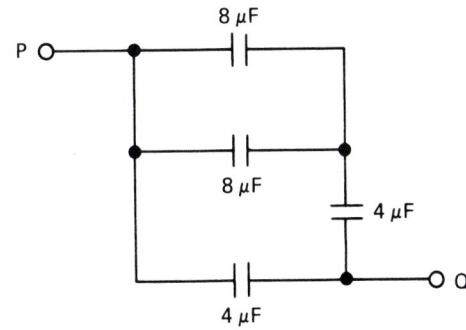

66. In Fig. 9-36, what is the value of the circuit's time constant? (*Hint:* Use Thévenin's theorem.)

FIGURE 9-36 Circuit for Problem 66.

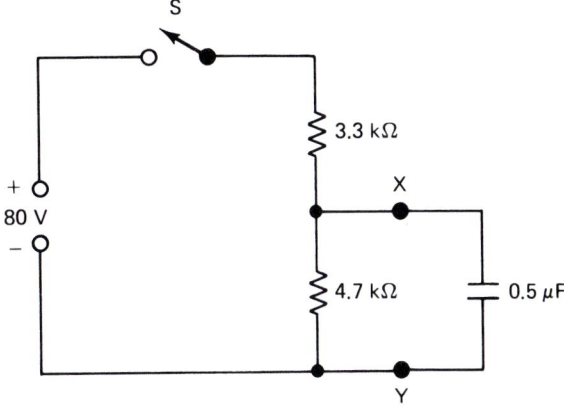

67. If we close the switch, S, in the circuit of Fig. 9-37, what are the values of I_T, I_1, and I_2 at the beginning and the end of the transient state? What is the amount of the final charge stored in the 4-μF capacitor?

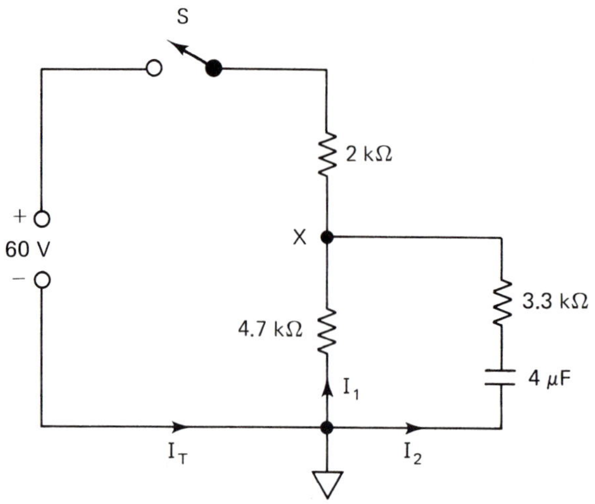

FIGURE 9-37 Circuit for Problem 67.

68. There are number of 0.1-μF capacitors available, each with a rating of 100 WVDC. Using a minimum number of these capacitors, sketch an arrangement which has a total capacitance of 0.1 μF and a rating of 200 WVDC. (*Note:* There is more than one solution.)

69. Two capacitors, whose values are 0.2 μF and 0.05 μF, are connected in series across a 120-V dc source. After the capacitors have been charged, the source is disconnected and replaced by a third, uncharged 0.1-μF capacitor. What is the final voltage across the third capacitor?

70. A variable capacitor with an air dielectric has a rotor and a stator with a total of 23 aluminum plates. The effective area of one side of each plate is 8 cm^2. If the separation between adjacent plates is 0.5 mm, calculate the *maximum* capacitance that the capacitor can provide.

71. In Fig. 9-38, use the universal exponential growth curve to determine the value of V_C when S_1 has been closed for a time interval of 200 ms. From the moment S_1 is closed, what is the time taken for V_{R1} to fall to the level of 50 V?

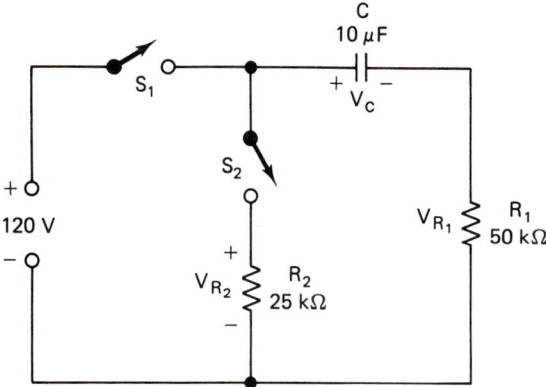

FIGURE 9-38 Circuit for Problems 71 and 72.

72. In Problem 71, S_1 remains closed until the capacitor has charged to 120 V. S_1 is then opened and, simultaneously, S_2 is closed. Using the universal exponential decay curve, determine the value of V_{R2} after a time of 300 ms has elapsed. What is the time required for V_C to fall to a level of 40 V? What is the value of the discharge current when S_2 has been closed for a time of 800 ms?

73. In Fig. 9-39, switch S_1 is closed and then after a time of 2 ms, switch S_2 is closed. Using the universal exponential growth curve, determine the value of V_C after an interval of 1 ms has elapsed from the time the switch S_2 is closed.

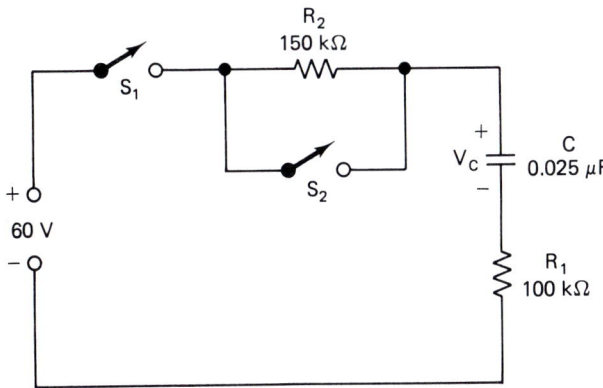

FIGURE 9-39 Circuit for Problems 73 and 74.

74. In Problem 73, switch S_1 is initially closed and when the value of V_c reaches 35 V, switch S_2 is closed. What is the time interval between the closing of switch S_1 and the closing of switch S_2? When the value of V_C reaches 55 V, how much time has elapsed from the closing of switch S_2?

75. The voltage to which a capacitor is charged rises from 30 V to 50 V in a time of 2.5 μs. If the average value of the charging current is 0.8 A, calculate the value of the capacitance.

76. When a capacitor is connected across a 400-V dc source, the energy stored is 0.3 J. Calculate the values of the capacitance and the charge stored by the capacitor.

77. A parallel-plate capacitor has a capacitance of 300 pF. Each rectangular plate has a length of 4 cm and a width of 3 cm, while the separation between adjacent plates is filled by mica with its relative permittivity of 5. If there is a total of 9 plates, determine the thickness of the mica.

78. A 0.002-μF capacitor with a waxed paper dielectric (Table 9-1) has a 600 WVDC rating, but it is necessary to operate with a safety factor of 2. Calculate the required area for each conducting surface on one side.

79. A CR circuit consists of a 250-pF capacitor in series with a 200-kΩ resistor. If a symmetrical square wave is applied to the circuit, what is the square wave's highest repetition rate that will allow an adequate differentiated output to appear across the resistor? What is the lowest repetition rate that will allow an adequate integrated output to appear across the capacitor?

80. In Fig. 9-40, the same input waveform, V_{in}, is applied to each of the CR circuits. For both circuits sketch the V_R and V_C waveforms and include the voltage levels with the time intervals involved.

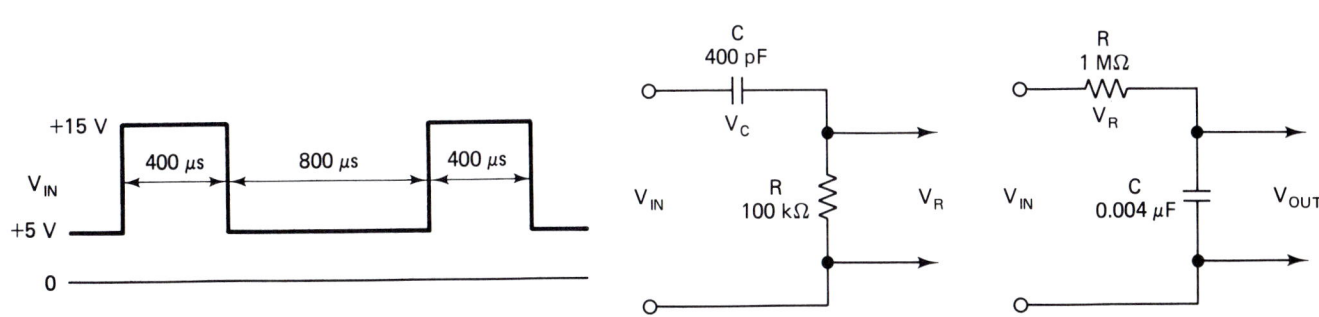

FIGURE 9-40 Circuits for Problem 80.

Self-Review

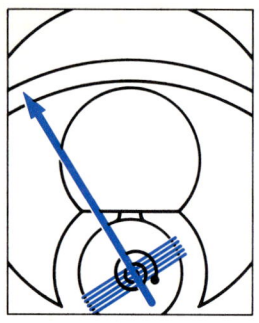

10

DC Electrical Measurements

When studying the different types of meters, you will learn:

1. about the construction of a moving-coil meter movement and how it employs the motor effect to achieve its deflection.
2. how the moving-coil instrument is adapted to behave as an ammeter for measuring the current.
3. how the moving-coil instrument is adapted to behave as a voltmeter to measure potential differences.
4. the effect that the resistance of a voltmeter has on voltage values being measured.
5. how the moving-coil instrument is adapted to behave as an ohmmeter to measure resistance.
6. about the principles of moving-iron ammeters and voltmeters as well as their advantages and disadvantages with respect to moving-coil instruments.
7. the use of the thermocouple instrument to measure a current.
8. how an electrodynamometer meter movement is used to measure power.

10-0

Introduction

In our study of dc we have been concerned mainly with four electrical quantities: voltage (E), current (I), resistance (R), and power (P). This chapter is devoted to those test instruments (voltmeter, ammeter, ohmmeter, and wattmeter) which measure these quantities in terms of their units. At this stage we will describe the use of these meters in recording dc values and only mention the subject of ac measurements, which is covered in Chapter 20 and in the accompanying laboratory manual.

The most common type of indicating instrument is fitted with a needle which shows on a scale the value of the quantity being measured. Such an instrument has a moving system which is carried by a hardened steel spindle with its ends tapered and highly polished. The tapered ends form pivots which rest in hollow-ground sapphire bearings set in steel screws. These indicating instruments all possess three essential parts:

1. A deflecting device, which enables a mechanical force to be exerted by the voltage, current, or power being measured
2. A controlling device, which ensures that the amount of the deflection is dependent on the magnitude of the measured quantity
3. A damping device, which prevents oscillation of the moving system and enables the final position to be reached quickly

The various devices used depend on the particular type of instrument and its purpose.

10-1

Moving-Coil (D'Arsonval) Meter Movement

One of the most common items of test instrument used in electronics is the voltmeter-ohmmeter-milliammeter or VOM (Fig. 10-1), which is a multimeter capable of measuring either voltage or current or resistance; it does not require an external power supply for its operation. The VOM contains a moving-coil or electromechanical meter movement which depends on the motor effect (Section 7-3) to produce the necessary deflection. This type of meter movement is also used in panels, switch gear, and many other applications besides the VOM.

Referring to Fig. 10-2, the deflecting device of the moving-coil meter movement consists of a rectangular coil, which is made of thin insulated copper wire wound on a light aluminum frame. This frame is carried by a spindle which pivots in jeweled bearings. Current is led into and out of the coil by the spiral hairsprings, H, which are the controlling device and therefore provide

FIGURE 10-1 VOM multimeter.

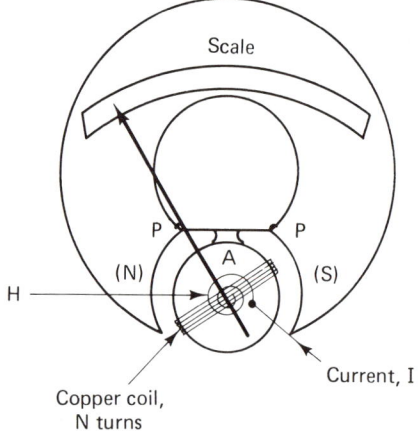

FIGURE 10-2 Moving-coil meter movement.

the restoring torque. The coil is free to move in the gaps between the permanent magnet pole pieces, P, and a soft-iron cylinder, A, which is normally carried by a nonmagnetic bridge attached to P. The purpose of the soft-iron cylinder is to:

1. Concentrate the magnetic flux by reducing the amount of the air gap.
2. Produce a radial magnetic field with a uniform flux density; this is also achieved by special shaping of the pole pieces, P. A radial field is one in which the flux lines are directed toward a central region (Fig. 10-3). As a result, the position of the coil in relation to the flux lines is the same over a wide arc.

Let the coil consist of N turns, each with a length of l meters and a width of d meters. In the radial field the torque, T, exerted on the coil is due to the motor effect. From Eqs. (1-1-5) and (7-3-1),

$$T = BINld = BINA \quad \text{newton-meters} \tag{10-1-1}$$

DC Electrical Measurements   Ch. 10

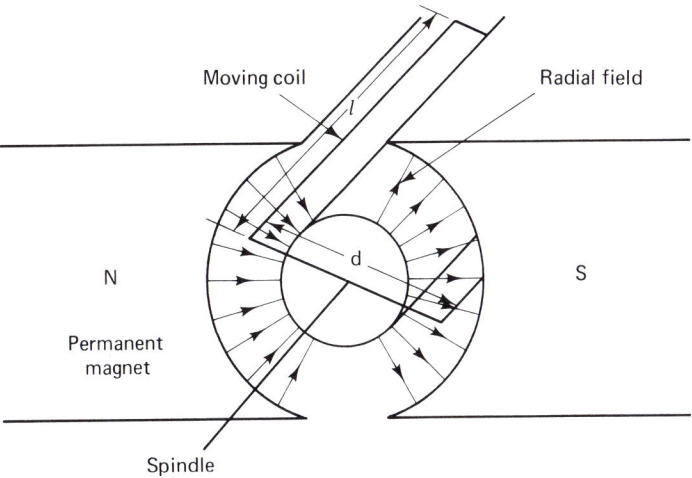

FIGURE 10-3 Position of the moving coil in a radial magnetic field.

where I = current flowing through the coil (A)
B = flux density (T)
$A = ld$ = area of the coil (m²)
N = number of turns in the coil

The restoring torque, T', provided by the hairsprings, is directly proportional to the amount of the deflection, $\theta°$. Therefore, $T' = k\theta°$, where k is a constant for the system of the hairsprings. In the final rest position of the needle, the two torques must exactly balance, so that

$$T = T'$$

Therefore,

$$BINA = k\theta°$$

and

$$I = \frac{k}{BAN}\theta = K\theta° \qquad (10\text{-}1\text{-}2)$$

where K is the constant of proportionality for the entire meter movement. Since the deflection is directly proportional to the current, the scale is linear with the divisions evenly spaced.

As the coil rotates toward the equilibrium position, it could overswing so that the needle would oscillate before finally coming to rest. The required damping of this unwanted oscillation is introduced by winding the coil on the aluminum frame. As the coil rotates, currents are induced into the frame and by Lenz's law the effect of these currents will be such as to oppose the motion of the deflection system. When the coil reaches its equilibrium position, it has no energy to swing farther and since the deflecting and restoring torques are then balanced, there is no tendency for the needle to oscillate.

The sensitivity of the moving-coil meter movement will be determined by the amount of current required to provide full-scale deflection; the less is the value of current, the greater is the movement's sensitivity. High sensitivity will be achieved by using a powerful permanent magnet to produce a large flux density, which in turn will create a strong deflection torque. In addition, a high sensitivity will require many turns of fine copper wire to be wound on a light aluminum frame and then attached to delicate hairsprings. The full-scale deflection current for such a meter movement would be 20 μA or less. By using fewer turns of a larger wire size together with stronger hairsprings, the full-scale deflection current can be increased to a few milliamperes; the limitation on the current then becomes the physical size of the meter movement.

Since the sensitivity and the full-scale deflection current are inversely related, the sensitivity is directly measured in terms of the current's reciprocal. With a current equivalent to volts divided by ohms, the unit of sensitivity will be ohms per volt. For example, if the meter movement's full-scale deflection current is 50 μA, the sensitivity is

$$\frac{1}{50 \ \mu A} = \frac{1}{50 \times 10^{-6} \ A} = 20{,}000 \ \Omega/V$$

Therefore,

$$\text{sensitivity (ohms per volt)} = \frac{1}{\text{full-scale deflection current (amperes)}}$$

and

$$\text{full-scale deflection current (amperes)} = \frac{1}{\text{sensitivity (ohms per volt)}} \quad (10\text{-}1\text{-}3)$$

Since the full-scale deflection currents typically range from 20 μA to 1 mA, the moving-coil meter movement is basically a microammeter or milliammeter. The resistance of the coil itself is of the order of tens of ohms, and therefore the meter movement can only be calibrated to measure up to a few millivolts.

The passage of the current through the coil raises its temperature and alters its resistance. It is normal practice to add a swamping resistor in series with the coil. The swamping resistor has a negative temperature coefficient which counteracts the positive temperature coefficient of the copper coil; it is also used to bring the total resistance (coil resistance plus swamping resistance) to a suitable value, such as 50 Ω. It must be remembered that in order to measure current at a particular point, the circuit must be broken and the meter inserted in the break. The total 50-Ω resistance of the meter movement will be in series with the remainder of the circuit and will therefore affect the reading to a certain degree. The amount of the error will depend on the value of the circuit's total resistance in comparison with the 50-Ω resistance of the meter movement.

The meter accuracy is given as the error percentage with full-scale deflection. The specified accuracy is normally of the order of ±2%, so that for a 0–1 mA meter movement the maximum error is ±(2/100) × 1 = ±0.02 mA.

This same error is maintained throughout the scale, so that the lower the reading the worse is the percentage accuracy. For example, if the reading is 0.1 mA, the accuracy is ±0.02 mA and the percentage error increases to ±(0.02/0.1) × 100 = ±20%.

The main advantages of the moving-coil meter movement are:

1. Uniform scale
2. High sensitivity

3. High degree of shielding from stray magnetic fields
4. High degree of accuracy

The principal disadvantages are:

1. Suitable only for dc measurement
2. Relatively expensive when compared with the moving-iron instrument (Section 10-6)
3. Fragile construction

We have examined the construction of the moving-coil meter movement and have discussed the factors that affect its ohms per volt sensitivity. In addition, we know that the amount of the deflection is directly proportional to the current flowing through the coil, so that the meter's scale markings are evenly spaced.

EXAMPLE 10-1 The sensitivity of a moving-coil meter movement is 1000 Ω/V. What is the value of the full-scale deflection current?

Solution By Eq. (10-1-3),

$$\text{full-scale deflection current} = \frac{1}{1000 \ \Omega/V}$$

$$= \frac{1}{1000} \ A = 1 \ mA$$

EXAMPLE 10-2 A 1.5-V cell with negligible internal resistance is connected across a 3-kΩ resistor. The current is to be measured by a 0.1-mA moving-coil meter movement which has a total resistance of 50 Ω. After the circuit is broken to insert the meter movement, what is the value of the recorded current?

Solution Before the meter movement is inserted, the circuit current is 1.5 V/3 kΩ = 0.5 mA. With the meter movement connected, the total resistance is 3 kΩ + 50 Ω = 3.05 kΩ, and the recorded current is 1.5 V/3.05 kΩ = 0.4918 A. The difference of 0.5 − 0.4918 = 0.0082 mA is due to the loading effect of the meter on the circuit. The worst loading effect will occur with low voltages and low resistances.

10-2

Ammeter

To convert the basic moving-coil meter movement into an instrument capable of measuring hundreds of milliamperes or even amperes, a shunt resistor, R_{sh}, is connected across the series combination of the meter movement and the swamp resistor (Fig. 10-4). This shunt is a low-value precision-type resistor (Section 1-8) which is usually made from constantan wire, with its negligible

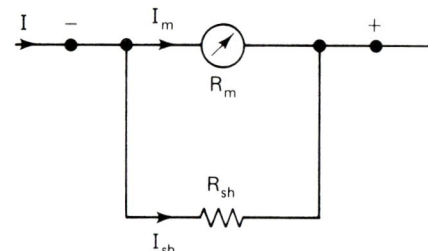

FIGURE 10-4 Ammeter and meter shunt.

temperature coefficient. Because of the shunt's low resistance, most of the current to be measured will be diverted through R_{sh} and only a small fraction will pass the meter movement to provide the required deflection. Using KCL, we have

$$I = I_m + I_{sh} \qquad (10\text{-}2\text{-}1)$$

Since there is the same voltage drop across the shunt resistor and the meter movement,

$$I_{sh} \times R_{sh} = I_m \times R_m \qquad (10\text{-}2\text{-}2)$$

Therefore,

$$R_{sh} = \frac{I_m \times R_m}{I_{sh}} = \frac{I_m \times R_m}{I - I_m} \qquad (10\text{-}2\text{-}3)$$

where I = current to be measured
  I_m = current flowing through the meter movement
  R_{sh} = shunt resistance
  R_m = total meter movement resistance

Figure 10-5 shows an instrument that has a number of shunts for different current ranges. However, there is a danger that when switching from range to range, the meter movement will be placed directly in the circuit without the protection of the shunt; this could cause damage to the meter movement. The

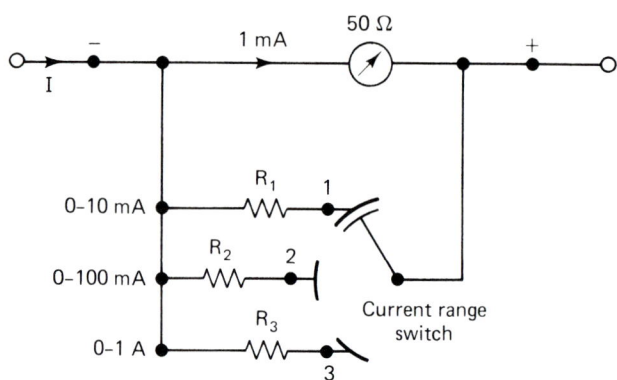

FIGURE 10-5 Multirange ammeter.

danger may be avoided by making the switch of a make-before-break type. Alternatively, the instrument may use an Ayrton or ring shunt (Fig. 10-6), which ensures that the meter cannot be placed in the circuit without the presence of a shunt.

Because of its very low resistance, the ammeter must *never* be placed directly across a voltage source. To avoid the danger of such damage to a VOM, you should always switch to a high voltage range after you have finished taking your current measurements. When measuring an unknown current, the meter should initially be switched to the highest range. If the reading is then too small, the meter can be switched to a lower range; in this way the meter movement is protected from excessive current.

We have learned about the use of the shunt in extending the current range of a moving-coil meter movement. The ammeter must always be placed directly in the path of the current to be measured.

EXAMPLE 10-3 In Fig. 10-5 the full-scale deflection current of the meter movement is 1 mA and its resistance is 50 Ω. Calculate the values of the shunt resistors for the ranges (a) 0–10 mA, (b) 0–100 mA, and (c) 0–1 A.

Solution (a) Use Eqs. (10-2-1) and (10-2-2). The voltage across the meter movement = 1 mA × 50 Ω = 50 mV. For the full scale current of 10 mA, the shunt current is 10 mA − 1 mA = 9 mA. The value of the shunt resistor is

$$\frac{50 \text{ mV}}{9 \text{ mA}} = 5.555 \ldots \Omega$$

The total resistance of the instrument is now

$$\frac{50 \times \frac{50}{9}}{50 + \frac{50}{9}} = 5 \text{ }\Omega$$

When the current to be measured is 5 mA, the voltage across the instrument is 5 mA × 5 Ω = 25 mV, and the current through the meter movement is 25 mV/50 Ω = 0.5 mA. The pointer therefore correctly indicates half of the full-scale deflection.
(b) Substitution in Eq. (10-2-3) yields

$$I = 100 \text{ mA}, \quad I_m = 1 \text{ mA}, \quad R_m = 50 \text{ }\Omega$$

$$R_{sh} = \frac{1 \text{ mA} \times 50 \text{ }\Omega}{100 \text{ mA} - 1 \text{ mA}} = 0.505 \text{ }\Omega$$

(c) From Eq. (10-2-3),

$$I = 1 \text{ A} = 1000 \text{ mA}$$

$$R_{sh} = \frac{1 \text{ mA} \times 50 \text{ }\Omega}{1000 \text{ mA} - 1 \text{ mA}} = 0.05005 \text{ }\Omega$$

These shunts whose values we have calculated could normally be placed inside the ammeter's casing. For currents greater than 20 A, the shunts are too large to be included within the instrument and are therefore connected externally.

EXAMPLE 10-4 In Fig. 10-6, an Ayrton shunt is used for the same meter movement and current ranges as in Example 10-3. Calculate the required values for R_1, R_2, and R_3.

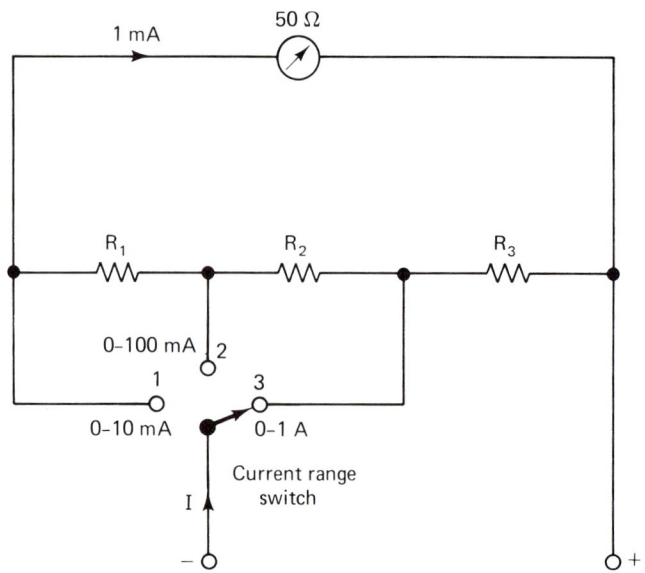

FIGURE 10-6 Multirange ammeter with an Ayrton shunt.

Solution The problem is best solved by using the current division rule.

0–10 mA Range
While 1 mA flows through the 50-Ω meter movement, 9 mA is flowing through the series combination R_1, R_2, R_3. Therefore,

$$\frac{R_1 + R_2 + R_3}{50} = \frac{1 \text{ mA}}{9 \text{ mA}}$$

$$R_1 + R_2 + R_3 = \frac{50}{9}$$

$$R_2 + R_3 = \frac{50}{9} - R_1$$

0–100 mA Range
1 mA is flowing through R_1 and the 50 Ω in series while the current through R_2, R_3 is 99 mA. Then

$$\frac{R_2 + R_3}{R_1 + 50} = \frac{1 \text{ mA}}{99 \text{ mA}}$$

$$R_2 + R_3 = \frac{R_1 + 50}{99}$$

Equating the two expressions for $R_2 + R_3$, we get

$$\frac{50}{9} - R_1 = \frac{R_1 + 50}{99}$$

$$550 - 99 R_1 = R_1 + 50$$

$$R_1 = 5\ \Omega$$

Then

$$R_2 + R_3 = \frac{50}{9} - 5 = \frac{5}{9} = 0.5\dot{5}\ \Omega$$

and

$$R_2 = \frac{5}{9} - R_3$$

The symbol $0.5\dot{5}$ means $0.5555\ldots$, which is normally called "0.5 recurring."

0–1 A Range
999 mA flows through R_3 and 1 mA flows through the series combination of R_1, R_2 and the 50 Ω. It follows that

$$\frac{R_3}{R_1 + R_2 + 50} = \frac{1\ \text{mA}}{999\ \text{mA}}$$

$$999 R_3 = R_2 + 55$$

$$999 R_3 - R_2 = 55$$

Substituting $R_2 = \frac{5}{9} - R_3$ yields

$$1000 R_3 = 55 + \frac{5}{9} = 55.5\dot{5}$$

$$R_3 = 0.05\dot{5}\ \Omega$$

Then

$$R_2 = 0.5\dot{5} - 0.05\dot{5} = 0.5\ \Omega$$

The required values are therefore

$$R_1 = 5\ \Omega,\quad R_2 = 0.5\ \Omega,\quad R_3 = 0.05\dot{5}\ \Omega$$

10-3 Voltmeter

As mentioned in Section 10-1, the moving-coil meter movement is basically a millivoltmeter or microvoltmeter. To adapt the movement for higher voltage ranges it is necessary to connect a series multiplier resistor. Most of the voltage to be measured is then dropped across this high-value series resistor and only

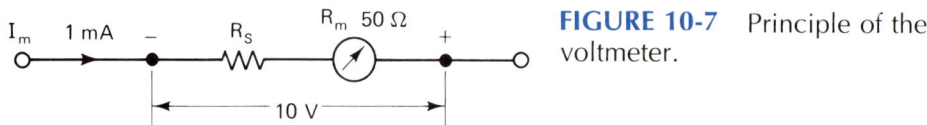

FIGURE 10-7 Principle of the voltmeter.

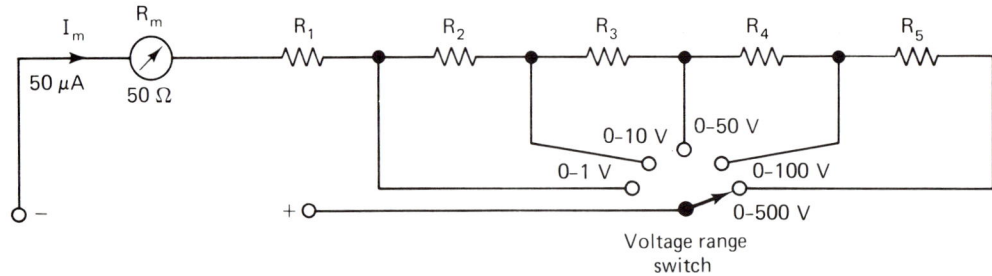

FIGURE 10-8 Multirange voltmeter.

a small part appears across the meter movement to provide the necessary deflection.

In the example of Fig. 10-7, the meter movement once again has a total resistance of 50 Ω and requires a full-scale deflection current of 1 mA. To adapt the instrument for the 0–10 V range, the total resistance of the voltmeter must be 10 V/1 mA = 10 kΩ and the value of the multiplier resistor is 10 kΩ − 50 Ω = 9950 Ω. In equation form,

$$R_s = \frac{V}{I_m} - R_m \qquad (10\text{-}3\text{-}1)$$

or

$$R_s = V \times (\text{sensitivity in ohms per volt}) - R_m \qquad (10\text{-}3\text{-}2)$$

where R_s = value of the series multiplier resistor
V = voltage indicated by the full-scale deflection
I_m = full-scale deflection current
R_m = total resistance of the meter movement

In a multirange voltmeter it would be possible to use a separate multiplier resistor for each range. However, it is simpler to use a series of resistors, as shown in Fig. 10-8.

In contrast with the low resistance of the ammeter, the voltmeter has a high resistance and is placed across (in parallel with) the resistor whose voltage drop is to be measured. The "loading" effect of the voltmeter is discussed in the next section.

EXAMPLE 10-5 In Fig. 10-8 the meter movement has a resistance of 50 Ω and a sensitivity of 20,000 Ω/V. The voltage ranges are 0–1 V, 0–10 V, 0–50 V, 0–100 V, and 0–500 V. Calculate the required values for the resistors R_1, R_2, R_3, R_4, and R_5.

Solution Use Eq. (10-3-2).

0–1 V Range

$$R_1 = (20{,}000 \text{ }\Omega/\text{V} \times 1 \text{ V}) - 50 \text{ }\Omega = 19{,}950 \text{ }\Omega$$

0–10 V Range

$$R_1 + R_2 = (20{,}000 \times 10) - 50 = 199{,}950 \text{ }\Omega$$
$$R_2 = 199{,}950 - 19{,}950 = 180{,}000 \text{ }\Omega$$

0–50 V Range

$$R_1 + R_2 + R_3 = (20{,}000 \times 50) - 50 = 999{,}950 \text{ }\Omega$$
$$R_3 = 999{,}950 - 199{,}950 = 800{,}000 \text{ }\Omega$$

0–100 V Range

$$R_1 + R_2 + R_3 + R_4 = (20{,}000 \times 100) - 50 = 1{,}999{,}950 \text{ }\Omega$$
$$R_4 = 1{,}999{,}950 - 999{,}950 = 1{,}000{,}000 \text{ }\Omega$$

0–500 V Range

$$R_1 + R_2 + R_3 + R_4 + R_5 = (20{,}000 \times 500) - 50 = 9{,}999{,}950 \text{ }\Omega$$
$$R_5 = 9{,}999{,}950 - 1{,}999{,}950 = 8{,}000{,}000 \text{ }\Omega$$

10-4 Loading Effect of a Voltmeter

The voltmeter is placed across (in parallel with) the component whose dc voltage drop is to be measured. The ideal voltmeter should have infinite resistance, so that the instrument does not load the circuit being monitored. However, the resistance of a moving-coil voltmeter is far from infinite and its value changes with the particular voltage range selected. For example, if the meter movement has a sensitivity of 20,000 Ω/V, the instrument's resistance in the 0–10 V range is 20,000 Ω/V $\times$ 10 V = 200 kΩ, while in the 0–100 V range, the resistance increases to 20,000 Ω/V $\times$ 100 V = 2 MΩ. The worst loading effect by such a voltmeter will therefore occur in high-resistance, low-voltage circuits, as shown in Fig. 10-9(a).

The instrument should measure a voltage drop of 6 V between the points X and Y; the voltmeter is therefore switched to the 0–10 V range and connected across the lower 200-kΩ resistor [Fig. 10-9(b)]. However, the voltmeter's resistance is also 200 kΩ, so that the total resistance of the voltmeter and the resistor in parallel is only 200 kΩ/2 = 100 kΩ. Using the voltage division rule, the reading of the voltmeter is

$$12 \text{ V} \times \frac{100 \text{ k}\Omega}{200 \text{ k}\Omega + 100 \text{ k}\Omega} = 4 \text{ V}$$

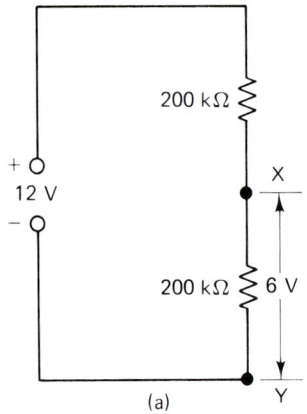
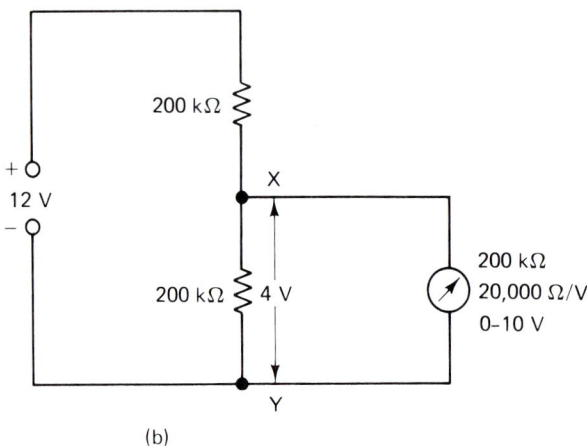

FIGURE 10-9 Loading effect of a voltmeter.

The voltmeter loading effect has therefore reduced the voltage between X and Y from 6 V to 4 V.

The vacuum-tube voltmeter (VTVM) and the field-effect transistor (FET) voltmeter have high resistances (several megohms) which are independent of the voltage range chosen; such voltmeters have a minimum loading effect.

EXAMPLE 10-6 In Fig. 10-9(a), the voltage drop between X and Y is measured by a voltmeter whose sensitivity is 1000 Ω/V. What is the reading of the voltmeter when switched to the 0–10 V range?

Solution

$$\text{resistance of voltmeter} = 1000 \ \Omega/\text{V} \times 10 \ \text{V}$$
$$= 10{,}000 \ \Omega$$
$$= 10 \ \text{k}\Omega$$

Total resistance of the voltmeter and the lower 200-kΩ resistor in parallel is

$$\frac{10 \times 200}{10 + 200} = 9.524 \ \text{k}\Omega$$

Using the voltage division rule, the voltmeter reading is

$$12 \text{ V} \times \frac{9.524 \text{ k}\Omega}{200 \text{ k}\Omega + 9.524 \text{ k}\Omega} = 0.545 \text{ V}$$

This result clearly illustrates the disastrous loading effect of a low-sensitivity voltmeter.

10-5
Ohmmeter

The moving-coil meter movement may be adapted for resistance measurements by the inclusion of additional resistors and one or more primary cells, Since these cells alone must provide the necessary deflection, it is *essential* that resistance measurements are only made with all power removed from the circuit.

A basic ohmmeter could consist of a 1.5-V primary cell, a rheostat, and a 0–1 mA moving-coil meter movement with its 50-Ω total resistance. The resistance, R, to be measured is then connected between the ohmmeter probes [Fig. 10-10(c)].

Initially, the probes are placed apart so that an open circuit (infinite resistance) exists and no current flows [Fig. 10-10(a)]. The needle is therefore on the left-hand side of the scale and indicating zero current but infinite resistance; the symbol for infinity is "∞," which is often marked on the ohms scale. If the needle is not on the zero current mark, you can mechanically adjust the torque provided by the hairsprings.

The next step is to short the probes together (zero resistance) and adjust the rheostat ("zero ohms" control) until a current of 1 mA flows [Fig. 10-10(b)]. If the terminal voltage of the primary cell is actually 1.5 V, the circuit's total resistance is 1.5 V/1 mA = 1500 Ω. The necessary value for the rheostat would then be 1500 − 50 = 1450 Ω. Differences in the cell's voltage can be accommodated by small adjustments of the rheostat. However, if you are unable to bring the current up to 1 mA, it probably means that the cell has too low a terminal voltage and must be replaced.

Since the ohms scale has infinity marked on the left and zero on the right, the value of ohms increases from right to left; this is called a *back-off scale*, since as the resistance increases, the current drops back (backs off) from its full-scale deflection value.

If a 1.5-kΩ resistor is connected between the probes as in Fig. 10-10(c), the circuit current would be 1.5 V/(1450 Ω + 50 Ω + 1500 Ω) = 0.5 mA. The needle would then be in the center of the 1.0 mA scale. Consequently, we have infinite ohms on the left, zero ohms on the right, and 1500 Ω in the center. Since the current is *inversely* proportional to the resistance, the scale is nonlinear, so that it is expanded on the low-resistance side but crowded on the high-resistance end. Such a scale could not be used for accurate measurements of either low resistances (less than 1 Ω) or high resistances (more than 10 kΩ).

Figure 10-11(a) shows the circuit of a practical ohmmeter which uses a meter movement with a sensitivity of 20,000 Ω/V and a resistance of 2000 Ω. The full-scale deflection current will therefore be 50 μA, which must flow through the meter movement when the red and black probes are shorted together so that the measured resistance is zero ohms. Shorting the probes

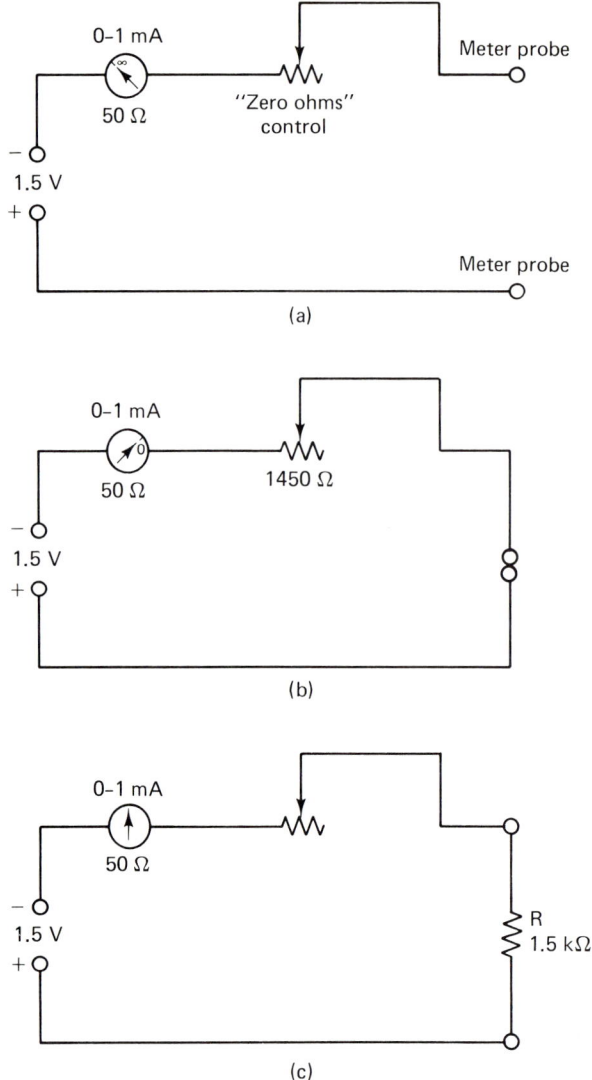

FIGURE 10-10 Basic ohmmeter.

together is the initial step in the use of the ohmmeter; the value of the "zero adjust" resistor, R_6, is set to produce the full-scale deflection current, which will coincide with the reading of zero ohms on the right-hand side of the resistance scale [Fig. 10-11(b)].

With the range switch in the R × 1 position the value of R_6 is 1.5 V/ 50 μA − 1138 Ω − 21,850 Ω − 2000 Ω = 30 kΩ − 24,988 Ω = 5012 Ω. The value of R_1 is significant since 11.5 Ω is the value in the center of the scale, so that when the unknown resistance, R_x, is 11.5 Ω, the current through the meter movement must be 50 μA/2 = 25 μA. On the left side of the scale the deflection is zero and therefore R_x is an open circuit of infinite ohms. This means that the resistance scale is nonlinear, as shown in Fig. 10-11(b). The R × 1 range will be used to measure resistances between zero and about 200 Ω, although it is difficult to measure accurately on the left-hand side of the scale.

When R_x is 11.5 Ω, the cell voltage of 1.5 V will divide equally between R_x and R_1 (ignoring the shunting effect of $R_M + R_2 + R_4 + R_6 = 30$ kΩ across R_1). The voltage across R_1 is therefore 1.5 V/2 = 0.75 V, which will drive the necessary current of 0.75 V/30 kΩ = 25 μA through the meter movement.

In the R × 100 range position, the value of R_6 must be reset; R_3 is then included so that when R_x is 11.5 × 100 = 1150 Ω, the current through the

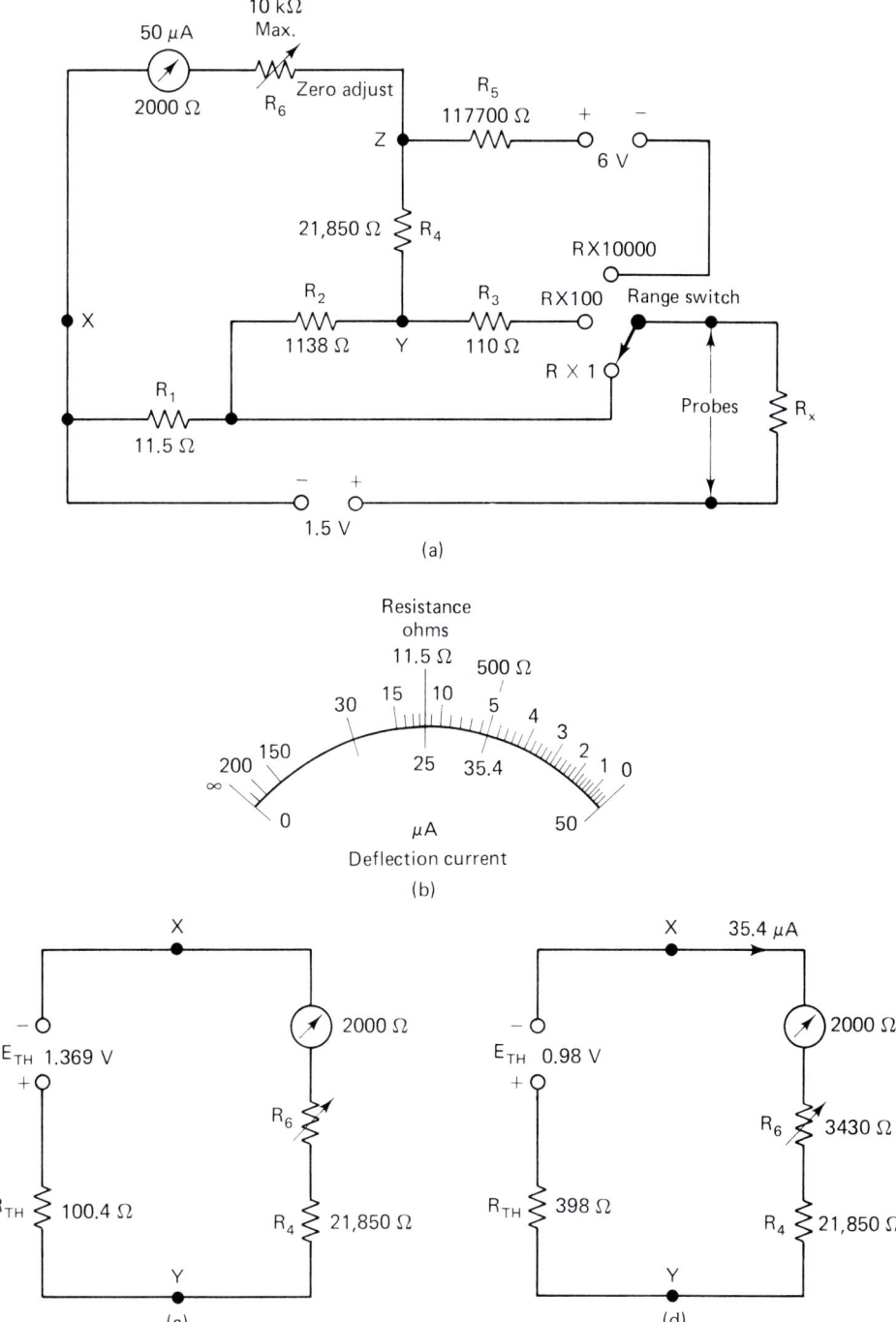

FIGURE 10-11 Practical ohmmeter and its equivalent circuits.

meter movement is 25 μA. This range will be used to measure values of R_x between 200 Ω and 20,000 Ω.

On the R × 10,000 range (R_x between 20,000 Ω and 2 MΩ), the additional 6-V cell has to be added in order to drive the required current through the meter movement. The value of R_5 is such that when R_x is 11.5 × 10,000 = 115,000 Ω, the deflection current is again 25 μA.

When measuring the value of an unknown resistance, start with the R × 1 range and then, if necessary, switch to the range that brings the needle roughly toward the center of the scale.

Sec. 10-5 Ohmmeter

EXAMPLE 10-7 In Fig. 10-11(a), calculate the value for R_6 on the R × 100 range. If R_x is 500 Ω, what is the value of the deflection current through the meter movement?

Solution With the probes shorted, Thévenize the circuit between points X and Y [Fig. 10-11(c)]. Then by the voltage division rule,

$$E_{TH} = 1.5 \text{ V} \times \frac{11.5 \text{ Ω} + 1138 \text{ Ω}}{11.5 \text{ Ω} + 1138 \text{ Ω} + 110 \text{ Ω}}$$

$$= \frac{1.5 \times 1149.5}{1259.5} = 1.369 \text{ V}$$

and

$$R_{TH} = \frac{110 \times 1149.5}{110 + 1149.5} = 100.4 \text{ Ω}$$

Then

$$R_6 = \frac{1.369 \text{ V}}{50 \text{ μA}} - 21{,}850 \text{ Ω} - 2000 \text{ Ω} - 100.4 \text{ Ω}$$

$$= 3430 \text{ Ω}$$

When the resistance of 500 Ω is connected between the probes [Fig. 10-11(d)],

$$E_{TH} = 1.5 \text{ V} \times \frac{11.5 \text{ Ω} + 1138 \text{ Ω}}{11.5 \text{ Ω} + 1138 \text{ Ω} + 110 \text{ Ω} + 500 \text{ Ω}}$$

$$= 1.5 \times \frac{1149.5}{1759.5} = 0.98 \text{ V}$$

and

$$R_{TH} = \frac{610 \times 1149.5}{610 + 1149.5} = 398 \text{ Ω}$$

The deflection current is

$$\frac{0.98 \text{ V}}{398 \text{ Ω} + 3430 \text{ Ω} + 2000 \text{ Ω} + 21{,}850 \text{ Ω}} = 35.4 \text{ μA}$$

This result is illustrated in Fig. 10-11(b).

10-6

Moving-Iron Ammeter and Voltmeter

These instruments can be divided into two types:

1. The attraction type, in which a piece of soft iron is attracted toward a solenoid
2. The repulsion type, in which two parallel soft-iron vanes are magnetized inside a solenoid and therefore repel each other

Attraction Moving-Iron Meter

This instrument is illustrated in Fig. 10-12(a). The current to be measured flows through the coil and sets up a magnetic field. This field magnetizes the soft-iron disk and draws it into the center of the coil. The force acting on the iron depends on the flux density and on the coil's magnetic field intensity, both of which are proportional to the current. The torque and the deflection are therefore not proportional to the current alone (as in the moving-coil meter movement) but to the square of the current. The result is a nonlinear scale (Fig. 10-14) which is cramped at the low end but open at the high end. However, by careful shaping of the iron disk, the scale's linearity can be improved.

The restoring torque of the controlling device is supplied by a spring or sometimes by gravity. The damping device is commonly obtained by the use of an air piston.

Repulsion Moving-Iron Meter

Two iron vanes are situated axially in a short solenoid as shown in Fig. 10-12(b). One vane is fixed while the other is movable and attached to the pivot, which also carries the needle. When the current flows through the coil, the vanes are equally magnetized and repel each other. The repulsion creates a deflection on the scale, which is calibrated for direct reading. The force of repulsion is dependent on the flux density of each iron vane and is directly proportional to the square of the current; therefore, the scale is again nonlinear. The restoring and damping systems are similar to those of the attraction type.

Since the strength of the deflection torque depends on the number of ampere turns for the coil, it is possible to arrange for various ranges by winding different numbers of turns on the coil. Also, by varying the type of wire used, the meter's resistance can be changed so that there is no need for shunt and swamping resistances.

The moving-iron milliammeter may be converted to a voltmeter by adding a suitable noninductive series resistor.

Compared with the moving-coil meters, the moving-iron instruments are relatively cheap and robust; however, they have low sensitivity, are affected by stray magnetic fields, and are liable to hysteresis error. Their accuracy is normally ±5% of the full-scale deflection value. Unlike the moving-coil meters, the moving-iron instruments can be used for direct ac measurements.

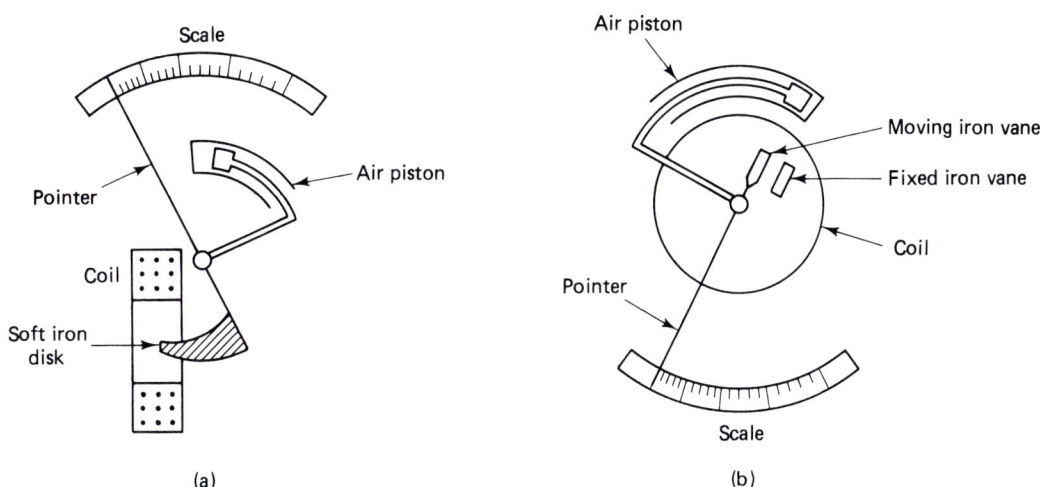

FIGURE 10-12 Moving-iron instruments.

EXAMPLE 10-8 A moving-iron meter needs 450 ampere-turns to provide full-scale deflection. How many turns on the coil will be required for the 0–10 A range? For a voltage scale of 0–300 V with a full-scale deflection current of 15 mA, how many turns are required, and what is the total resistance of the instrument?

Solution

$$\text{number of turns for the 0–10 A range} = \frac{450}{10}$$

$$= 45 \text{ turns}$$

$$\text{number of turns for the 0–300 V range} = \frac{450}{0.015}$$

$$= 30{,}000 \text{ turns}$$

$$\text{total resistance} = \frac{300 \text{ V}}{15 \text{ mA}} = 20 \text{ k}\Omega$$

10-7

Thermocouple Meter

The principle of this type of meter is based on the Seebeck effect, which was originally discovered in 1821. If a circuit consisting of different metals is at the same temperature throughout, there is no resultant EMF. However, if a junction between two dissimilar metals is maintained at a different temperature from the rest of the circuit, a "thermoelectric" EMF is generated and a current can then be driven through a conventional moving-coil meter movement.

The construction of the meter is shown in Fig. 10-13. The current to be measured passes between X and Y, raising the temperature of the heater wire. Attached to the center of this wire is the thermocouple junction, J; bismuth and antimony are commonly used as the dissimilar metals, although many other combinations are possible. As the temperature rises, the thermoelectric EMF will increase and drive a greater current through the meter movement. The amount of deflection on the scale depends on a heating effect which is proportional to the square of the measured current. The meter scale is therefore

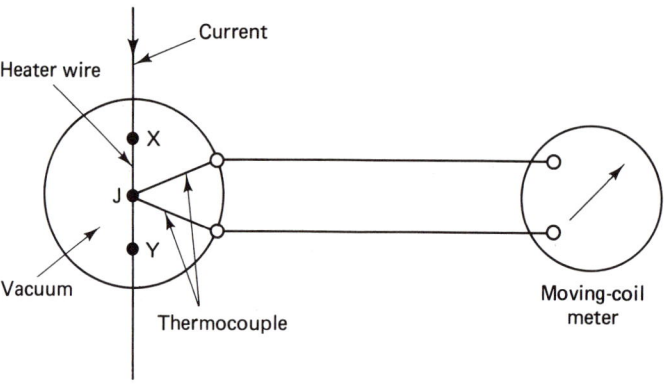

FIGURE 10-13 Thermocouple meter.

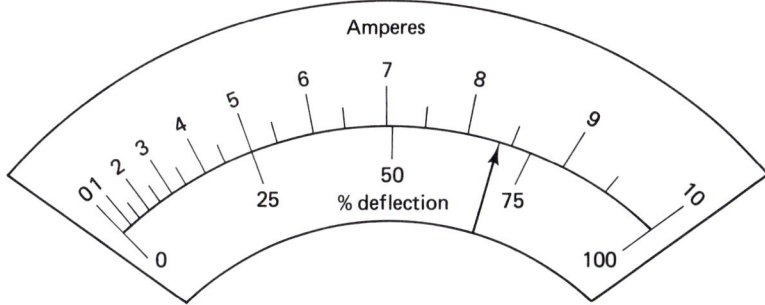

FIGURE 10-14 Current-squared meter.

nonlinear, so that it is cramped at the low end and open at the high end (Fig. 10-14).

This type of "current-squared" meter is suitable for reading both dc and ac. Its particular importance lies in the measurement of radio-frequency currents such as occur in antenna systems of broadcast transmitters. Once this type of meter has been calibrated, the calibration is accurate from dc up to very high frequencies.

EXAMPLE 10-9 A dc current of 10 A provides full-scale deflection of a thermocouple ammeter. What percentage of full-scale deflection corresponds to a current of 4 A?

Solution The amount of deflection is proportional to the square of the current. Therefore, the deflection corresponding to 4 A is

$$100 \times \left(\frac{4 \text{ A}}{10 \text{ A}}\right)^2 = 16\% \text{ of the full-scale deflection}$$

EXAMPLE 10-10 A dc current of 5 A corresponds to the full-scale deflection on a thermocouple ammeter. What value of current will produce 60% of the full-scale deflection?

Solution The current is proportional to the square root of the deflection. Therefore, the current corresponding to 60% deflection is

$$5 \text{ A} \times \sqrt{\frac{60}{100}} = 3.87 \text{ A}$$

10-8
Electrodynamometer Movement. Wattmeter

In the moving-coil meter movement the flux associated with the permanent magnet is fixed in direction; consequently, this type of meter can only be directly used for dc measurements. However, if the permanent magnet is replaced by

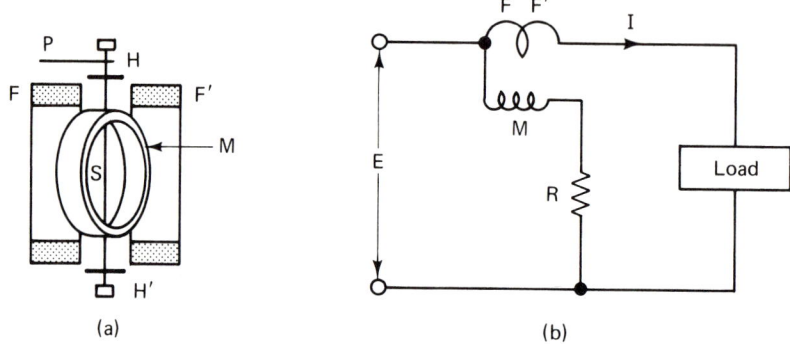

FIGURE 10-15 Electrodynamometer movement and wattmeter.

an electromagnet, the direction of the flux may be reversed. This is shown in the electrodynamometer movement of Fig. 10-15(a), in which two fixed coils, F and F', provide the electromagnet. The moving coil, M, is then carried by a spindle and the controlling torque is exerted by spiral hairsprings H and H', which also serve to lead the current into and out of the moving coil.

For dc measurements the electrodynamometer movement has no advantages over the D'Arsonval type. Compared with ac moving-iron instruments, dynamometer ammeters and voltmeters are less sensitive and more expensive, so that they are rarely used. However, the dynamometer wattmeters are important since they are the most common way of measuring power directly in ac circuits. Figure 10-15(b) shows the way in which the wattmeter is connected into the circuit. The fixed coils are joined in series with the load so that they carry the load current, I. The moving coil is in series with a high-value multiplier resistor, so that the current through the moving coil is proportional to the source voltage, E. The torque on the moving coil is proportional to the product of the current through the fixed (current) coils and the current through the moving (voltage) coil, and is therefore directly proportional to the power taken by the load.

The controlling torque provided by the hairsprings depends directly on the deflection, which is therefore proportional to the power; consequently, electrodynamometer wattmeters have linear scales, with the markings evenly spaced.

When used for ac measurements, the moving coil cannot follow the rapid fluctuations in the instantaneous power, so that it assumes a mean or average position which corresponds to the true power in the circuit. Consequently, the electrodynamometer movement automatically takes into account the ac circuit's power factor (Section 13-1).

Chapter Summary

10.1. Moving-coil meter movement:

$$\text{torque exerted on coil} = BINld = BINA \quad \text{newton-meters}$$

$$\text{current}, I = K \times \theta° \text{ (deflection)}$$

$$\text{sensitivity (ohms per volt)} = \frac{1}{\text{full-scale deflection current (amperes)}}$$

10.2. Ammeter:

$$\text{measured current, } I = I_m + I_{sh}$$

$$\text{voltage drop across shunt resistor and meter movement} = I_{sh} \times R_{sh} = I_m \times R_m$$

$$\text{shunt resistor value, } R_{sh} = \frac{I_m \times R_m}{I_{sh}} = \frac{I_m \times R_m}{I - I_m}$$

10.3. Voltmeter:

$$\text{multiplier resistor value, } R_s = \frac{V}{I_m} - R_m$$

10.4. Current-squared meter:

$$\text{meter deflection is proportional to (current)}^2$$

$$\text{current is proportional to } \sqrt{\text{meter deflection}}$$

10.5. Wattmeter: torque is directly proportional to the product of source voltage and source current.

Self-Review

True–False Questions

1. **T. F.** The sensitivity in ohms per volt is directly proportional to the full-scale deflection current.
2. **T. F.** For a moving-coil meter movement a high-sensitivity requires a powerful electromagnet.
3. **T. F.** To measure the current, the circuit is broken and the low-resistance ammeter is inserted in the path of the current.
4. **T. F.** A voltmeter has a high resistance and is placed across the resistor whose voltage drop is to be measured.
5. **T. F.** A 0–50 V voltmeter has a resistance of 50 kΩ. Its sensitivity is 1000 Ω/V.
6. **T. F.** A moving-iron ammeter has a linear scale with the markings evenly spaced.
7. **T. F.** The scale of a thermocouple meter is crowded at the low-value end and expanded at the high end.
8. **T. F.** Before taking a resistance reading, all power must be removed from the circuit.
9. **T. F.** The electrodynamometer movement used in wattmeters contains a total of four coils.
10. **T. F.** The higher the current range of a multimeter, the higher is the value of the shunt resistance.
11. **T. F.** A moving-coil meter responds only to direct current.
12. **T. F.** The electrodynamometer movement has a nonlinear scale.
13. **T. F.** When you shift to a higher current range on a VOM, the resistance of the instrument is increased.
14. **T. F.** The loading effect of a VOM is reduced when using a higher range. This is true whether the instrument is being used either as a voltmeter or as an ammeter.
15. **T. F.** The use of a moving-coil meter as an ohmmeter requires a zero-ohms control.
16. **T. F.** The thermocouple instrument is used primarily for measuring direct currents with low values.

17. **T. F.** In the ohmmeter circuit of Fig. 10-10, the cell voltage falls due to aging. This causes the resistance reading to be too low.
18. **T. F.** In circuits where the loading of the instrument does not present a problem, a range is chosen so that the reading is close to the full-scale deflection value.
19. **T. F.** The high current ranges of a VOM require low values for the shunt resistors.
20. **T. F.** If it is impossible to zero an ohmmeter on the R × 1 range, its low voltage cell must be replaced.

Multiple-Choice Questions

21. A moving-coil meter movement has a sensitivity of 50,000 Ω/V. Its full-scale deflection current is
    (a) 50 μA (b) 0.2 mA (c) 20 μA (d) 0.5 mA (e) 5 μA
22. A moving-coil meter movement has a full-scale deflection current of 200 μA and a total resistance of 100 Ω. The meter movement's sensitivity is
    (a) 100 Ω/V (b) 200 MΩ/V (c) 20,000 Ω/V (d) 50,000 Ω/V (e) 5000 Ω/V
23. In Question 22, the maximum voltage that the meter movement can measure is
    (a) 2 mV (b) 200 μV (c) 20 mV (d) 200 mV (e) 2 V
24. The meter movement of Question 22 is adapted to read a full-scale current of 1 mA. The resistance value of the required shunt is
    (a) 0.04 Ω (b) 25 Ω (c) 20 Ω (d) 2 Ω (e) 0.05 Ω
25. The meter movement of Question 22 is adapted to read 10 V full-scale. The value of the series multiplier resistor is
    (a) 50 kΩ (b) 19,900 Ω (c) 20,000 Ω (d) 49.9 kΩ (e) 4900 Ω
26. A simple ohmmeter contains a 1.5-V cell and the meter movement of Question 22. The "zero ohms" rheostat has a value of
    (a) 7.5 kΩ (b) 7.4 kΩ (c) 7.6 kΩ (d) 750 Ω (e) 74.9 kΩ
27. The ohmmeter of Question 26 is used to measure a resistance of 22.5 kΩ. The current through the meter movement is
    (a) 25 μA (b) 50 μA (c) 75 μA (d) 100 μA (e) 150 μA
28. A "current-squared" meter has a full-scale (100%) deflection of 10 A. The current corresponding to 50% deflection is
    (a) 7.07 A (b) 5 A (c) 3.535 A (d) 2.5 A (e) 4 A
29. In Question 28, a current of 8 A corresponds to a deflection of
    (a) 80% (b) 40% (c) 88% (d) 28.3% (e) 64%
30. A D'Arsonval meter has a full-scale (100%) deflection of 10 A. The current corresponding to 50% deflection is
    (a) 5 A (b) 7.07 A (c) 3.535 A (d) 2.5 A (e) 4 A
31. On the 0–10 V range a 50,000-Ω/V VOM has a reading of 4 V. The total resistance of the instrument is
    (a) 50 kΩ (b) 200 kΩ (c) 300 kΩ (d) 20 kΩ (e) 2 MΩ
32. A voltmeter contains a 50-μA, 2-kΩ moving-coil meter movement. The lowest possible voltage range of the instrument is
    (a) 0–50 V (b) 0–25 V (c) 0–100 V (d) 0–50 A (e) 0–100 V
33. The accuracy of a VOM on the 0–10 A range is 2% of its full-scale deflection value. If the reading is 4 mA, the maximum percentage error is
    (a) ±0.8% (b) ±2% (c) ±4% (d) ±5% (e) ±10%
34. To double the current range of a 50-μA, 2-kΩ moving-coil meter movement, the required value for the shunt resistor is
    (a) 1 kΩ (b) 2 kΩ (c) 3 kΩ (d) 4 kΩ (e) 5 kΩ
35. The accuracy of a voltmeter on the 0–50 V range is 2% of the full-scale deflection value. The accuracy of a current meter on the 0–10 mA range is 3% of the full-scale deflection value. The reading of the voltmeter when connected across a resistor, is 25 V. When connected in series with the resistor, the reading of the current meter is 5 mA. The maximum percentage error in the calculated value of the resistance is
    (a) 2.5% (b) 5% (c) 7% (d) 8% (e) 10%
36. The deflection of a 100-μA, 2-kΩ moving-coil meter movement is half of its full-scale value. The voltage drop across the instrument is
    (a) 200 V (b) 100 V (c) 100 mV (d) 200 mV (e) 0.02 V

37. In Question 36, a 998-kΩ resistor is connected in series with the moving-coil meter movement. The highest voltage that can be measured is
    (a) 1 V   (b) 10 V   (c) 100 V   (d) 1 kV   (e) 100 mV
38. A wattmeter is connected to a 120-V dc source and is operated on the 0–100 W range. If the source current is 600 mA, the percentage of full-scale deflection is
    (a) 72%   (b) 12%   (c) 60%   (d) 20%   (e) 7.2%
39. The current flowing through a meter with a current-squared scale is doubled. The percentage of the deflection is multiplied by
    (a) 2   (b) 4   (c) 8   (d) $\sqrt{2}$ (e) $2\sqrt{2}$
40. The flux density for a moving-coil meter movement is doubled, the number of the turns is doubled, but the area of each turn remains the same. If the diameter of the wire is halved, the torque exerted on the coil is multiplied by
    (a) 2   (b) 4   (c) 8   (d) 0.25   (e) 0.5

Practice Problems

41. A moving-coil meter movement has a full-scale deflection current of 100 μA and a total resistance of 50 Ω. Calculate the required value of the shunt resistance for the 0–10 mA range. When the shunt is included, what is the total resistance of the instrument?
42. The meter movement of Problem 41 is to be used as a voltmeter with a 0–100 V range. What is the value of the required series multiplier resistor and the total resistance of the voltmeter?
43. Two 1-MΩ resistors are joined in series across a 120-V dc source. If the voltmeter of Problem 42 is connected across one of the resistors, calculate the reading of the voltmeter.
44. A moving-iron ammeter is wound with 40 turns and gives a full-scale deflection with 5 A. How many turns would be required to give a full-scale deflection with 20 A?
45. A coil of a moving-coil instrument is wound with $30\frac{1}{2}$ turns on an aluminum frame which has an effective length of 3 cm and an effective width of 2 cm. The flux density in the gap is 0.075 T. Calculate the torque exerted on the coil when the current flowing through the coil is 20 mA.
46. A basic ohmmeter circuit consists of a 1.5-V primary cell, a rheostat, and a 0–100 μA meter movement whose total resistance is 100 Ω. When the measured resistance is 10 kΩ, what is the current flowing through the meter movement?
47. A 10,000-Ω/V meter movement is adapted for use as a voltmeter with a 0–10 V range. What additional value of resistance must be connected in series when switching to the 0–100 V range?
48. A current-squared meter has a full-scale deflection of 5 A. What percentage of the full-scale deflection corresponds to a current of 2 A? What value of current corresponds to 80% of the full-scale deflection?
49. An electrodynamometer wattmeter has a full-scale deflection of 100 W. What value of power will correspond to 40% of the full-scale deflection?
50. A 0–100 μA meter movement with a total resistance of 100 Ω is adapted as an ammeter with a 0–1 A range. When the measured current is 0.7 A, what is the value of the current flowing through the shunt resistor?
51. An ammeter has a resistance of 0.1 Ω and a voltmeter has a resistance of 15 kΩ. The voltmeter is connected directly across a resistor and the ammeter is joined in series with this combination. If the readings are 60 mA and 150 V, calculate the value of the resistor before and after allowing for the loading effects of the instruments.
52. A basic ohmmeter consists of a 0–100 μA meter movement in series with a 3-V cell and a rheostat. Find the extreme resistances measured by the ohmmeter when the deflection of the needle is (a) one-fourth, (b) one-half, and (c) three-fourths of its full-scale value.
53. An unknown resistor is connected in series with an ammeter whose resistance is 0.2 Ω. A voltmeter, whose resistance is 15 kΩ, is connected across the series

combination. If the readings of the ammeter and the voltmeter are 150 mA and 150 V, respectively, calculate the value of the unknown resistor after allowing for the loading effects of the instruments.

54. In a basic ohmmeter circuit a 1.5-V cell is connected in series with a 22-kΩ precision resistor, a rheostat, and a moving-coil meter movement whose sensitivity is 20,000 Ω/V. What is the rheostat's value required for full-scale deflection to be obtained when the test probes are in contact?

55. In Problem 54, a current of 10 μA is flowing through the meter movement. What is the value of the external resistance being measured?

56. In Problem 54, the voltage of the cell falls by 10% due to aging. Calculate the new value to which the rheostat must be reset.

57. A component with a high value of resistance is connected in *series* with a voltmeter whose resistance is 150 kΩ. When 120 V dc is applied across this series combination, the reading of the voltmeter is 36 V. Calculate the value of the component with the high resistance.

58. A moving-coil meter movement has a resistance of 50 Ω and is used in an ammeter that is switched to the 0.5 A scale. If the internal shunt resistance is 0.5 mΩ, calculate the value of the full-scale deflection current.

59. An ammeter whose resistance is 0.5 Ω is connected in series with a 24-V dc source and a load that normally requires a current of 24 A from the source. What is the value of the current indicated by the ammeter?

60. A dc source with an open-circuit voltage of 12 V has an internal resistance of 100 kΩ. The terminal voltage is checked by a 20,000-Ω/V VOM on its 0–50 V range. What is the reading of the voltmeter?

Advanced Problems

61. A moving-coil meter with a total resistance of 3000 Ω has a scale with a total of 10 divisions. If the meter is connected across a 120-V dc source, the needle is deflected full-scale. When the same instrument is connected in series with an additional resistor across a 150-V dc source, the needle's deflection is eight divisions. What is the value of the additional resistor?

62. A dc voltmeter has a resistance of 28,600 Ω. When connected in series with an external resistance across a 480-V dc source, the voltmeter reads 220 V. What is the value of the external resistance?

63. A 1000-Ω/V meter movement is adapted as a voltmeter with a 0–50 V range. Two 200-kΩ resistors are joined in series across an 80-V dc source. If the voltmeter is connected across one of the resistors, what will be its reading? If a second identical voltmeter is then connected across the other resistor, what will be the reading of each voltmeter?

64. A moving-coil meter has an accuracy of $\pm 2\%$ on the 0–100 mA scale. What is the percentage error when the reading is 25 mA?

65. In Fig. 10-16, calculate the values of the resistors R_1 and R_2.

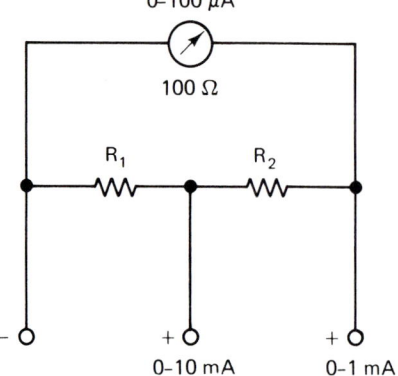

FIGURE 10-16 Circuit for Problem 65.

66. In Fig. 10-17, calculate the values of the resistors R_1 and R_2.

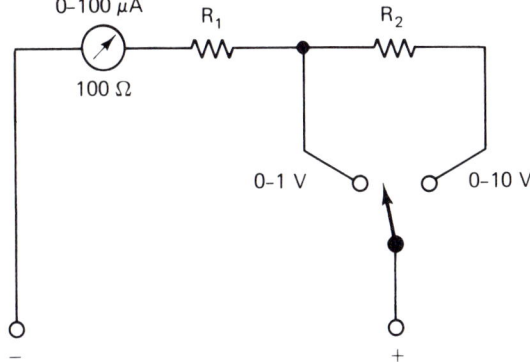

FIGURE 10-17 Circuit for Problem 66.

67. Two resistors whose values are 1 MΩ and 2 MΩ are connected in series across a 12-V dc source. The voltage across the 2-MΩ resistor is measured by a 0–100 μA meter movement which has been adapted for use as a voltmeter. What is the reading of the voltmeter on the (a) 0–10 V scale and (b) 0–100 V scale?

68. In Fig. 10-18, calculate the values of the "zero ohms" control for the R $\times$ 1 and R $\times$ 10 ranges. For each range what values of resistance are measured in the center of the ohms scale?

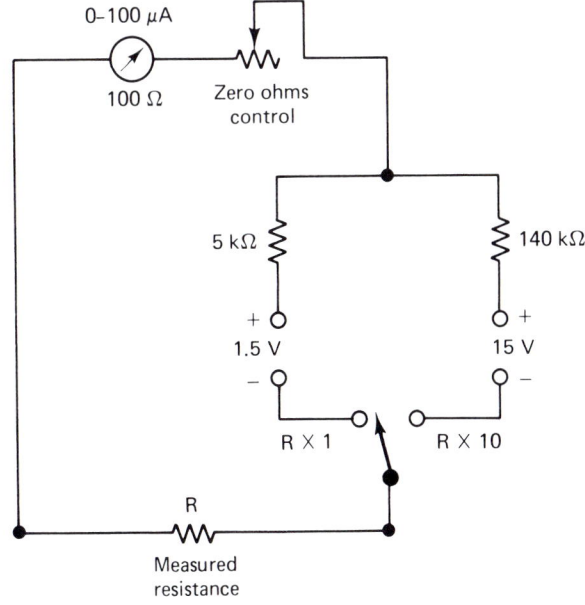

FIGURE 10-18 Circuit for Problem 68.

69. A 1.5-V cell is connected in series with a 20-kΩ resistor and a 0–100 μA meter movement whose total resistance is 100 Ω. Calculate the percentage error in the current reading due to the "loading" effect of the meter.

70. In Fig. 10-11, find the required value of R_6 for the R $\times$ 10,000 range.

71. A D'Arsonval meter movement has a rectangular moving coil that consists of $30\frac{1}{2}$ turns. The effective axial length of the magnetic field is 2 cm and the coil's radius is 0.8 cm. The flux density in the gap is 0.12 T and the controlling torque provided by the hairsprings is 0.5×10^{-6} N·m per degree of deflection. Calculate the amount of current necessary to provide a deflection of 60°.

72. A D'Arsonval meter movement has a resistance of 50 Ω, a full-scale deflection current of 100 μA, and is adapted for use as a voltmeter on the 0–250 V range.

If two resistors with values of 220 kΩ and 330 kΩ are connected in series across a 300-V supply, determine the value of the reading when the voltmeter is placed across each of the resistors in turn.

73. A milliammeter and a voltmeter are used in measuring the value of an unknown resistor. When the voltmeter, whose sensitivity is 10,000 Ω/V, is connected directly across the resistor, the readings of the instruments are 196 µA and 240 V. When the milliammeter, whose resistance is 50 Ω, is joined directly in series with the resistor (the voltmeter is in parallel with this combination), the readings are 100 µA and 240 V. If the voltmeter is on the 0–250 V range, what is the measured resistance for each set of readings, and which of the two measurements is more accurate? Determine the actual value of the resistance after allowing for the loading effects of the instruments.

74. A basic ohmmeter uses a 0–100 µA meter movement whose resistance is 30 Ω. This meter movement is connected in series with a 4-V cell and a rheostat. Across the meter movement is a variable shunt resistor whose value is 30 Ω. To what value is the rheostat set? Find the resistances measured when the deflection of the needle is (a) one-third, (b) one-half, and (c) two-thirds of its full-scale value. If the cell's voltage falls to 3 V, calculate the new resistance value to which the variable shunt must be adjusted.

75. A moving-coil meter movement has a full-scale deflection current of 100 µA and a resistance of 5 Ω. The flux density of the permanent magnet is then increased by 50%, the number of turns is doubled, and the diameter of the wire is halved. If the area of the moving coil is increased by $33\frac{1}{3}\%$, what is the sensitivity (in Ω/V) of the new meter movement?

76. In Problem 75, assume that although the area is increased, the perimeter of each turn remains the same. What is the new resistance of the instrument?

77. A moving-coil meter movement has a resistance of 2 kΩ and a sensitivity of 20,000 Ω/V. What are the required values for an Ayrton shunt that will provide ranges of 0–200 µA, 0–1 mA, and 0–5 mA?

78. A moving-coil meter movement has a resistance of 2 kΩ and a sensitivity of 10,000 Ω/V. This movement is used in a multirange voltmeter whose ranges are 0–1 V, 0–10 V, and 0–50 V. If the multiplier resistors are series connected, what are the required values of these resistors?

79. A current meter on the 0–50 mA range has an accuracy of 4% of its full-scale deflection value. If the readings are in turn 10 mA, 30 mA, and 50 mA, determine in each case (a) the possible error in milliamperes, (b) the possible range of measured values for the true current, and (c) the percentage error in each reading.

80. A 1-MΩ resistor is connected in series with a 470-kΩ resistor across a 12-V dc source whose internal resistance is negligible. An 11-MΩ electronic voltmeter (EVM) on the 0–5 V range is connected across the 470-kΩ resistor; what is the reading of the EVM? A 20,000-Ω/V VOM on the 0–5 V range is then placed in parallel with the EVM. What are the new readings of the instruments? If both instruments are now switched to their 0–50 V ranges, what are the revised readings for the instruments?

Alternating Currents

SEE NOTES

The objectives of this chapter are as follows:

1. To be able to identify periodic waveforms and calculate both the period and the frequency.
2. To recognize the characteristics of sine-wave voltages and currents in resistive circuits.
3. To be able to calculate numerical values of sine and cosine functions with given triangles.
4. To establish the relationship between the rotation of conducting loops and the magnetic field in the simple ac generator.
5. To calculate the instantaneous power level and effective power levels for sine-wave voltages and currents in resistive circuits.
6. To convert measured peak sine-wave levels to effective values.
7. To determine average levels after rectification of the sine wave.
8. To establish current and voltage amplitude and phase relationships for sine waves in inductive circuits.
9. To perform calculations involving voltage, current, and inductive reactance.
10. To calculate inductive reactance with known inductance values and frequency.
11. To represent sine-wave voltages and currents using phasors containing amplitude and phase information.
12. To calculate and apply the capacitive reactance concepts in circuit analysis.
13. To apply phasors in the analysis of ac circuits with capacitance.
14. To represent the sine wave in full mathematical form having amplitude, frequency, phase angle, and time keyed to variable and constant elements within the mathematical expression.

11-0

Introduction

Beginning in this chapter the basic theory of electricity will be extended to include alternating currents, currents that regularly change direction of flow. The most elementary form of alternating-current generator using a switch in what would otherwise be a simple dc circuit is explored first. Frequency and period are defined, after which sine-wave alternating currents and voltages are examined. Consideration of power with ac allows the relationship between peak values and effective values to be stated.

Next, the flow of alternating current in an inductor is analyzed. It will be seen that an opposition to current flow, called inductive reactance, must be accounted for. In a similar way there is capacitive reactance for a capacitor.

11-1

Alternating-Current Fundamentals

In previous chapters we have been concerned with current that flows in one direction only. When a resistor is connected across the terminals of a battery, electrons are forced through the resistor. Electrons come out of the negative battery terminal, pass through the resistor, and reenter the positive terminal of the battery. We know that the amount of current is governed by Ohm's law.

The current flow may be stopped by disconnecting the leads at the battery terminals. If we now reverse the leads and reconnect them, dc current will again flow. However, because of the reversal of the leads, the current will now flow through the resistor in the opposite direction. Electrons will pass through the resistor in the reverse direction from the original direction. We have caused the current to alternate directions.

It is possible to cause the current to alternately flow in one direction, then the reverse direction, and back again by manually disconnecting, reversing, and reconnecting the leads at the battery terminal as often as we choose. It might be better to connect a double-pole, double-throw (DPDT) switch in the circuit, as shown in Fig. 11-1. This makes the current reversal easy. When the swinger arm is thrown to the right, point *A* is connected to the lead attached to the negative battery terminal as shown in Fig. 11-2. This results in the direction of electron current flow indicated by the arrows. The current will flow downward through the resistor from *A* to *B*.

The current in the resistor can be reversed by throwing the swinger arm to the left (see Fig. 11-3). The arrows on the schematic diagram now indicate that current is passing upward through the resistor from *B* to *A*.

Suppose that we now direct the person who operates the switch to move the swinger arm from one side to the other side each 5 s. The sweep second hand of a watch or clock will be used. The switch will be thrown to the right

FIGURE 11-1 Current reversing circuit.

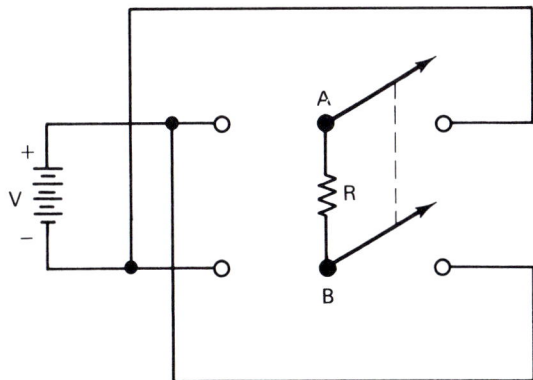

FIGURE 11-2 Electron flow from A to B.

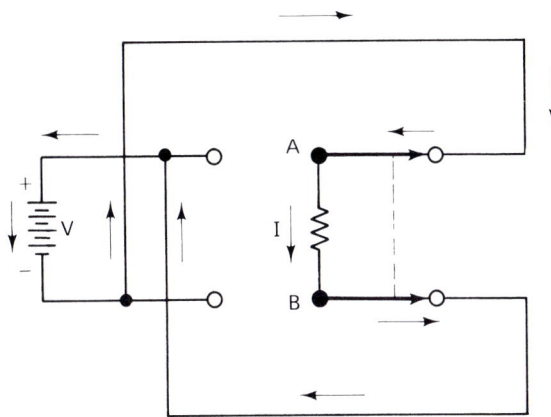

FIGURE 11-3 Electron flow from B to A.

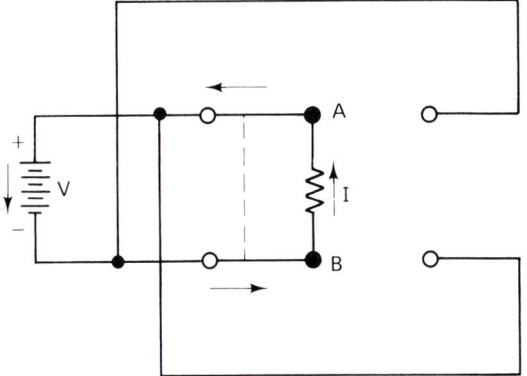

when the second hand is a 0 (straight up). It will then be thrown in the opposite direction when the sweep second hand reaches 5 s. After another 5 s, when the second hand reaches 10 s, the switch will again be thrown. The switch operator has been directed to throw the switch at 0, 5, 10, 15, 20 s, and so on. Each time the switch is thrown, the current through the resistor reverses or alternates. We now have a simple form of alternating current.

EXAMPLE 11-1 Refer to Fig. 11-1. The battery supplies 24 V and $R = 12$ kΩ. Assume the timing conditions of the preceding paragraph. Find the current and indicate the direction of flow at 1 s and 6 s after the switch is thrown to the right (Fig. 11-2).

Sec. 11-1 Alternating-Current Fundamentals

Solution We will start with the clock second hand at 0. This means that the electron current will then flow down from A to B (A-B) for 5 s. At $t = 1$ s,

$$I = \frac{V_B}{R} = \frac{24}{12} = 2 \text{ mA}$$

However, the current will reverse directions when the clock hand reaches 5 s and remain constant in the opposite direction from $t = 5$ s to $t = 10$ s. Therefore, at $t = 6$ s,

$$I = \frac{V_B}{R} = -\frac{24}{12} = -2 \text{ mA}$$

It has been shown that an alternating current can be produced in a resistor using a voltage source and a DPDT switch as connected in the circuit of Fig. 11-1. However, the switch must be thrown over and over again after regular time changes on a timing device such as a clock. During one time interval the current will flow in one direction, which we will call the *positive* direction. During a second time interval the current will be reversed. We customarily refer to the reversed current as being in the *negative* direction. A picture in the form of a graph shows exactly what is happening. One scale in the down-up (vertical) direction shows the current magnitude and direction at any moment. A second scale in the left-right (horizontal) direction shows the passage of time. The alternating current that we produce using our circuit and throwing the switch initially and then after equal intervals of time appears in Fig. 11-4. It will be seen that the circuit values and time intervals of Example 11-1 have been used.

Exactly what is the meaning of this graph? Imagine that a bug starts to crawl to the right from the zero position on the horizontal axis of Fig. 11-4. His position is marked at 5-s intervals. At any time after starting, the bug looks up and can see the graph above. However, if he looks down, he will see nothing. The conclusion: Resistor current is 2 mA from the time the bug starts to crawl until he has traveled for 5 s. Between 5 s and 10 s, if the bug looks up, he will see nothing. If he looks down, he sees the graph. The conclusion: Resistor current is -2 mA between 5 and 10 s.

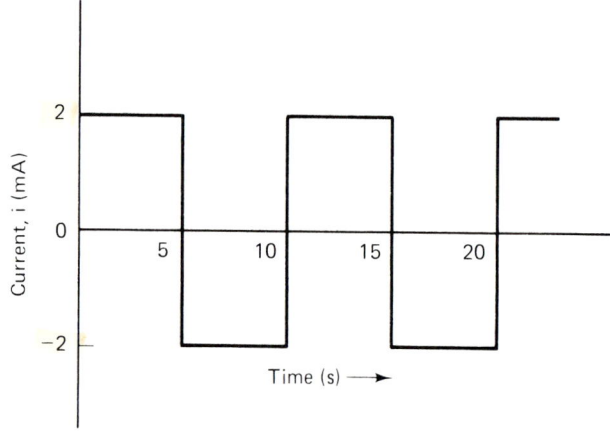

FIGURE 11-4 Graph for Example 11-1.

Looking carefully at Fig. 11-4, we notice that the graph repeats itself after a 10-s time interval. The shape of the graph appearing between the 0 point and the 10-s point is the same as that between 10 s and 20 s. To further illustrate this, notice that the current goes from $+2$ mA to -2 mA at $t = 5$ s and again at $t = 15$ s and still another time at $t = 25$ s. More generally, anything that occurs during the first 10 s will be repeated at any multiple of 10 s later. Multiples of 10 s are 20 s, 30 s, 50 s, 70 s, and 150 s, for example. We know that the current is 2 mA at 3 s. Therefore, the current will be 2 mA at 23 s, 33 s, 53 s, 73 s, and 153 s.

EXAMPLE 11-2 Find the current for the graph of Fig. 11-4 at $t = 7$ s, $t = 17$ s, $t = 27$ s, $t = 87$ s, and $t = 107$ s.

Solution Notice that 17 s, 27 s, 87 s, and 107 s are equal to 7 s added to 10 s or a multiple of 10 s.

$$17 = 7 + 10$$
$$27 = 7 + 20$$
$$87 = 7 + 80$$
$$107 = 7 + 100$$

Therefore, the current will be -2 mA at 7 s, 17 s, 27 s, 87 s, and 107 s.

A wave that repeats after fixed time intervals is called a "periodic" wave. The wave in Fig. 11-4 is a periodic wave. Moreover, the time interval for repeating is named the *period*. The period of Fig. 11-4 is 10 s. We customarily use capital T as the symbol for the period. For Fig. 11-4,

$$T = 10 \text{ s}$$

EXAMPLE 11-3 Find the period, T, for Fig. 11-5.

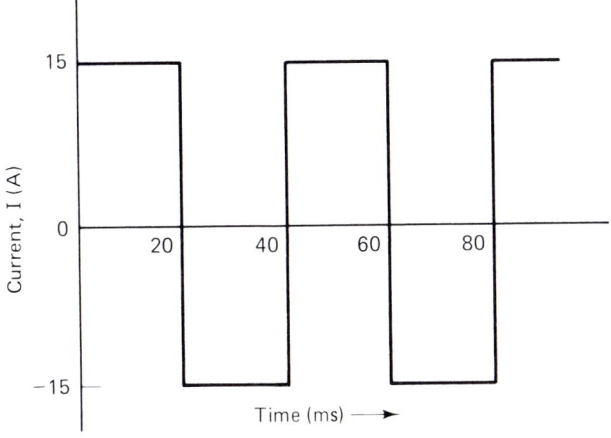

FIGURE 11-5 Graph for Example 11-3.

Sec. 11-1 Alternating-Current Fundamentals

Solution We first compare the graph of Fig. 11-5 with that previously seen in Fig. 11-4. The second differs from the first two ways. First, the current positive value is 15 A and the negative value is −15 A. This means only that the currents result from a new combination of supply voltage and resistance. The different current value has no bearing on the period.

Second, the changes in current direction occur at 20-ms intervals. The resistor current goes positive at 0, negative at 20 ms, and positive again at 40 ms. This means that

$$T = 40 \text{ ms}$$

There is another important factor that is of importance in alternating current circuits. We are interested in how often or how frequently the waveform of the graph is repeated. This factor, called the *frequency*, is closely related to the period. If a nonrepeating part of the graph is called a *cycle*, the frequency is the number of cycles produced in a unit of time. More specifically, we are interested in the number of cycles per second. Since a cycle occupies a time interval equal to a period, we could say that the frequency is the number of periods per second. Let us join these several related ideas by referring to Fig. 11-6. In addition to the time scale in seconds, points along the time scale are indicated as A, B, C, D, E, and F. From the previous discussion we know that the graph or waveform from A to C is one cycle, as is the waveform from C to E. The time interval from A to C, 0.1 s, is one period. Similarly, the time interval from C to E, 0.1 s, is one period.

If we extend the time scale of Fig. 11-6 all the way to 1 s, we will see that a total of 10 cycles and 10 periods will be connected. The frequency is the number of cycles in 1 s. Therefore, the frequency for the waveform of Fig. 11-6 is 10 cycles per second. One cycle per second has been given a special

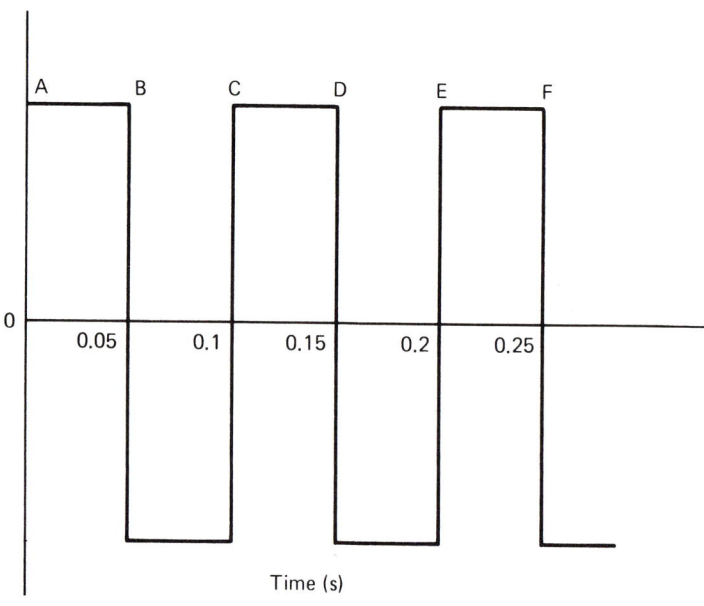

FIGURE 11-6 Graph of 10-Hz waveform.

name, the hertz, abbreviated Hz. Also, the term "frequency" is abbreviated as f. For Fig. 11-6, the frequency is

$$f = 10 \text{ cycles/second}$$
$$= 10 \text{ Hz}$$

The frequency can be calculated more directly from the relationship

$$f = \frac{1}{T} \tag{11-1-1}$$

To illustrate the use of this "reciprocal" equation, let us use the same information and find the frequency. Remember that $T = 0.1$ s. Therefore,

$$f = \frac{1}{T} = \frac{1}{0.1} = 10 \text{ Hz}$$

It is proper to say that f and T are reciprocals of one another. When this is so, another equation is also true.

$$T = \frac{1}{f} \tag{11-1-2}$$

We will make use of Eq. (11-1-2) in finding the period when the frequency is known.

EXAMPLE 11-4 Find the frequency for the waveform of Fig. 11-7.

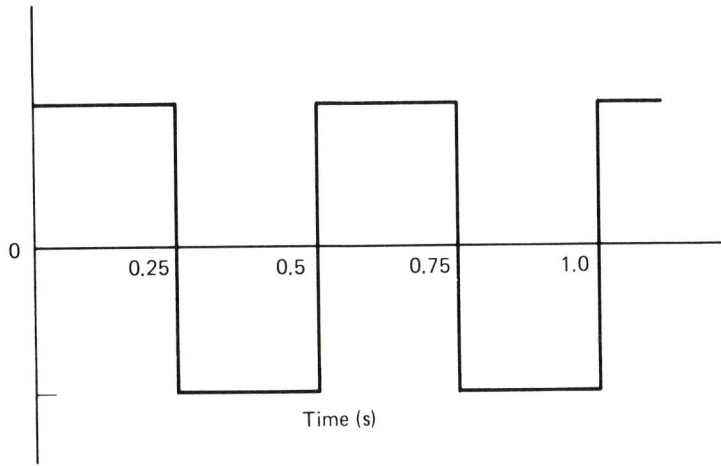

FIGURE 11-7 Graph for Example 11-4.

Solution One cycle occupies the interval between 0 and 0.5 s and a second cycle between 0.5 and 1.0 s. Since there are two cycles in 1.0 s of time, the frequency is given by

$$f = 2 \text{ Hz}$$

Alternatively, using Eq. (11-1-1) and knowing that $T = 0.5$ s, we obtain

$$f = \frac{1}{T} = \frac{1}{0.5} = 2 \text{ Hz}$$

EXAMPLE 11-5 Find the period for an alternating voltage with a frequency of 500 Hz.

Solution

$$T = \frac{1}{f} \tag{11-1-2}$$

$$= \frac{1}{500} = 0.002 \text{ s} = 2 \text{ ms}$$

In all of the preceding examples the square wave has changed from negative to positive at $t = 0$. This has been so because the time origin was selected to make this happen. The position of the time origin does not affect the period or the frequency.

EXAMPLE 11-6 Calculate the period and the frequency for the waveform of Fig. 11-8.

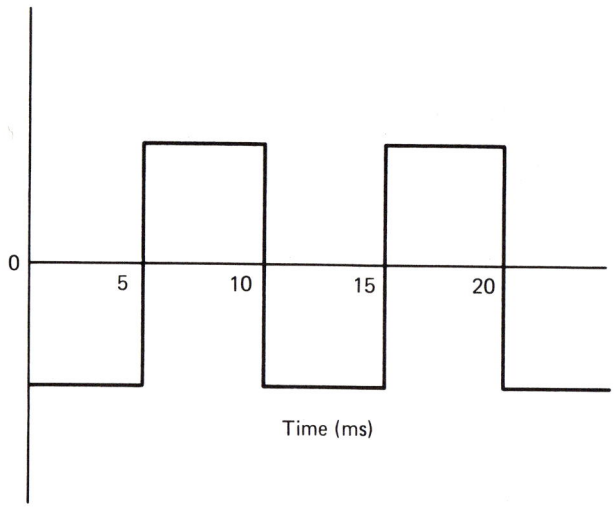

FIGURE 11-8 Graph for Example 11-6.

Solution The waveform changes polarity at 5 ms and again at 10 ms. It is seen that the waveform has the same shape between 10 and 20 ms as it has from 0 to 10 ms. Therefore, it is periodic with a period of 10 ms.

$$f = \frac{1}{T} \tag{11-1-1}$$

$$= \frac{1}{10^{-2}} = 10^2 = 100 \text{ Hz}$$

EXAMPLE 11-7 A waveform has the shape shown in Fig. 11-9. What are the values of the period and the frequency?

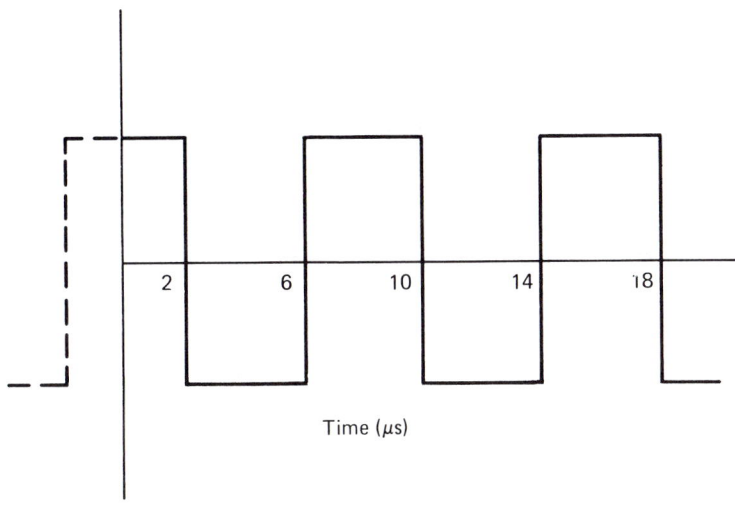

FIGURE 11-9 Graph for Example 11-7.

Solution At first glance the waveform does not appear to be periodic. The spacing from 0 to 2 μs is different than that from 2 μs to 6 μs. Since we do not know what happened before $t = 0$, it is better to start at $t = 2$ μs and then search for the period.

From $t = 2$ μs to $t = 6$ μs, there is a span of 4 μs. Then, from $t = 6$ μs to $t = 10$ μs there is another span of 4 μs. Starting with 10 μs and moving to 18 μs, there is a similar pattern. We see, then, one cycle from 2 μs to 10 μs, a span of 8 μs. A second, identical cycle appears from 10 μs to 18 μs with the same time span. The period is 8 μs.

$$T = 8 \text{ μs} = 8 \times 10^{-6} \text{ s}$$

The frequency is found using Eq. (11-1-1):

$$f = \frac{1}{T}$$

$$= \frac{1}{8 \times 10^{-6}} = 1.25 \times 10^5 \text{ Hz} = 125 \text{ kHz}$$

Sec. 11-1 Alternating-Current Fundamentals

Note: The waveform is periodic only if the waveform is extended to negative time, as indicated by the dashed lines in Fig. 11-9.

In this section we have considered an alternating current of a particular waveform. It is called a *square wave*. There are other alternating current waveforms. Each reaches a maximum positive value and a minimum negative value. And, like the square wave, every alternating current waveform has a period and a frequency which are related as reciprocals.

11-2 Sine-Wave Alternating Current

Most electrical power throughout the world originates as ac (alternating current) in large rotating machines. Because of the nature of these machines, the alternating current that flows has a unique waveform known as a sine wave. Figure 11-10 shows two cycles of a sine-wave current. One cycle occupies the time from 0 to 0.02 s. The second cycle runs from 0.02 s to 0.04 s. The period, which is a constant, is $T = 0.02$ s. Therefore, the frequency is found using Eq. (11-1-1):

$$f = \frac{1}{T}$$

$$= \frac{1}{0.02} = 50 \text{ Hz}$$

Unlike the square wave, whose waveform is flat on top and flat on the bottom, the ac waveform is a gentle curve rising and falling almost continuously at every moment. Only at the peak does it pause. Similarly, at the lowest extreme it hesitates at the minimum (negative peak) values. In Fig. 11-10 the positive peak value is 10 A and the negative peak value is −10 A.

Figure 11-11 shows an ac circuit consisting of an ac voltage generator and a 5-Ω resistor. The current in this circuit is that previously described and graphed in Fig. 11-10. In this figure no DPDT switch is required as was used

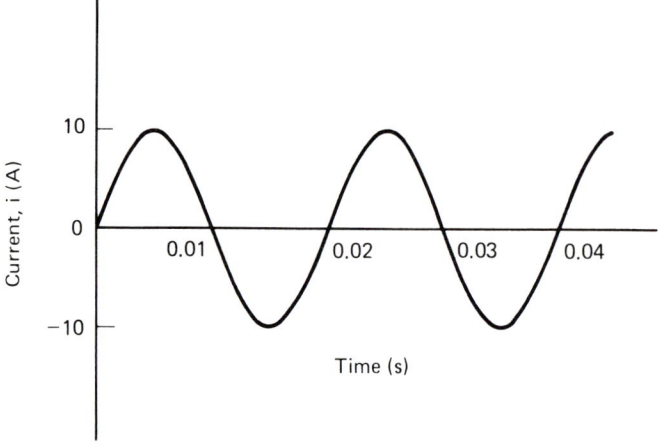

FIGURE 11-10 Sine-wave alternating current.

FIGURE 11-11 Alternating-current circuit.

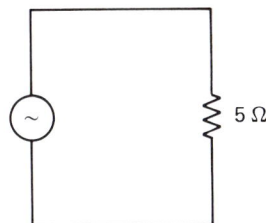

in Section 11-1. The ac voltage generator automatically reverses the voltage polarity to match the current direction of Fig. 11-10.

Moreover, the amount of voltage at every moment is such that Ohm's law governs. For example, if $I_p = 10$ A (the peak value) at a particular time (0.005 s), then the voltage supplied by the generator will also be maximum. By Ohm's law this peak voltage will be

$$V_p = I_p R \qquad (11\text{-}2\text{-}1)$$
$$= 10 \times 5 = 50 \text{ V}$$

Subscripts are used to indicate the peak voltage V_p and the peak current I_p. Both V_p and I_p are constants. We can think of climbing a mountain whose elevation is 5000 ft. On the way up the mountain the climber will be at different elevations (heights above sea level). On the way down the climber will also be at different elevations. But the top or peak is at one constant or fixed elevation, 5000 ft.

At every instant of time Ohm's law applies and the generator voltage can be found. Thus, if $i = 5$ A, then

$$v = iR \qquad (11\text{-}2\text{-}2)$$
$$= 5 \times 5 = 25 \text{ V}$$

Here we use lowercase v and i to indicate a changing or variable value. Similarly, if $i = 0$, then

$$v = iR \qquad (11\text{-}2\text{-}2)$$
$$= 0 \times 5 = 0 \text{ V}$$

If we examine the generator voltage waveform point by point, we must conclude that it must be equal to the iR drop. This is not to say that the upper terminal of the resistor in Fig. 11-11 will always be positive. However, when it is positive, the upper terminal of the generator will also be positive. Similarly, when the upper terminal of the resistor is negative, the upper terminal of the generator will also be negative.

The voltage across and the current through a resistor are said to be *in phase*. In particular, the two reach positive peaks at the same instant, negative peaks at the same instant, and pass through zero at the same instant. Drawn to the same time scale, the voltage and current will have the time relationships of Fig. 11-12.

We stated earlier that the form of the sine wave of current or voltage is determined by the generator. However, the word "sine" is the name of a trigonometric function. Trigonometry deals with triangles. This branch of mathematics is useful in the analysis of ac devices and circuits. In our use of trigonometry, only the right triangle will be used for most purposes. A right triangle has one interior angle of 90° (see Fig. 11-13). Our right triangle in this figure has a hypotenuse whose length is one unit. (We are accustomed

Sec. 11-2 Sine-Wave Alternating Current

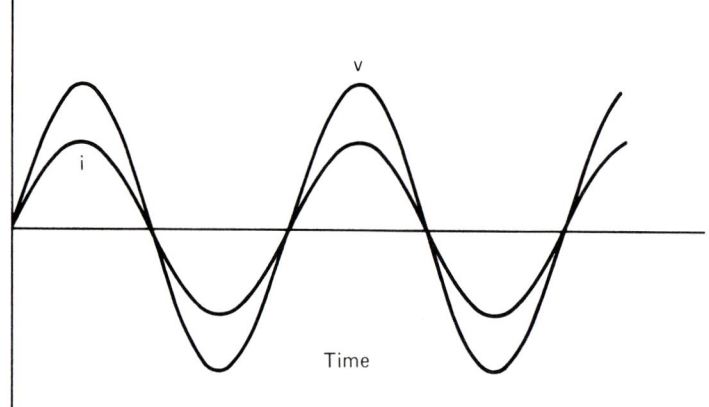

FIGURE 11-12 In-phase alternating voltage and current.

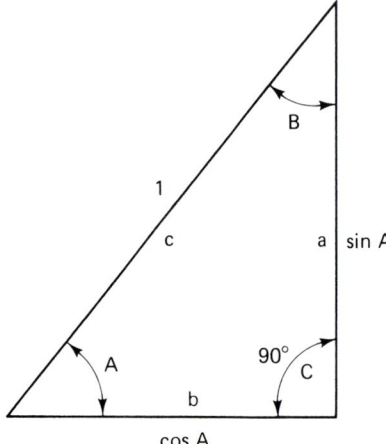

FIGURE 11-13 Right triangle.

to length units of feet, yards, miles, or meters.) Whatever the unit is for the hypotenuse, we should use the same unit for the sides. For the present we will use a, b, c for the triangle side lengths and angles A, B, C with C equal to $90°$, so that a right triangle is shown. When the hypotenuse has a length of 1 unit, then $a = $ sine A, customarily written as $a = \sin A$. This may seem strange. What it means is that a depends on the value of A in a way described as the sine of angle A.

EXAMPLE 11-8   If $A = 0$, what is a?

Solution   When $A = 0$, the hypotenuse and side b are one and the same, and $a = 0$. We might refer to the triangle as being squashed.

EXAMPLE 11-9   If $A = 90°$, find $\sin 90°$.

Solution   We have an odd situation here. When $A = 90°$ the hypotenuse is vertical. This means that $b = 0$ and $c = a$. Since $c = a$, $a = 1$ and $\sin 90° = 1$.

At angles between 0 and 90°, sin A will be between 0 and 1. At angles greater than 90° and less than 180°, the angle appears as shown in Fig. 11-14. Here the angle with positive horizontal axis has the value θ (Greek lowercase theta).

$$\sin \theta = \sin A = a$$

When θ lies between 180° and 270° as seen in Fig. 11-15, the sine is negative.

$$\sin \theta = -\sin A$$

With θ between 270° and 360° (Fig. 11-16),

$$\sin \theta = -\sin A$$

Figure 11-17 illustrates the graph of sin θ from 0 to 360°. By comparison with Fig. 11-12 we conclude that a sine-wave voltage or current when plotted against time has the same shape as the sine function plotted against angle. A complete cycle of sine-wave voltage or current occupies one period of time. The corresponding angle for the trigonometric function of Fig. 11-17 is 360°.

Consider the simplified ac generator of Fig. 11-18. As the loop rotates counterclockwise from the initial position shown, the horizontal side of the loop cuts the flux lines between the magnet poles. This causes an induced voltage in the loop which starts at 0, rises to its maximum value after 90° of rotation, and falls to 0 again after 180° of rotation. As rotation continues, the induced voltage will built to a negative maximum at 270° and to zero again at 360°. The waveform is a sine wave and the time of one rotation is the period.

FIGURE 11-14 Right triangle with θ between 90 and 180°.

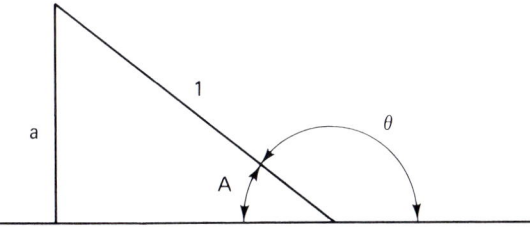

FIGURE 11-15 Right triangle with θ between 180 and 270°.

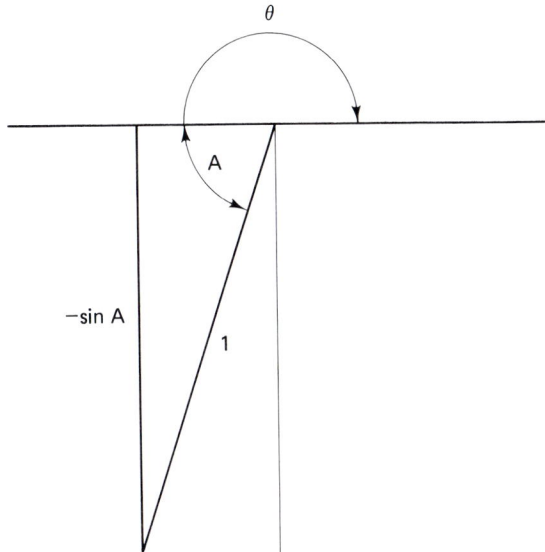

Sec. 11-2 Sine-Wave Alternating Current

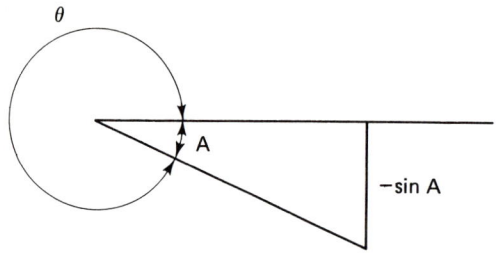

FIGURE 11-16 Right triangle with θ between 270 and 360°.

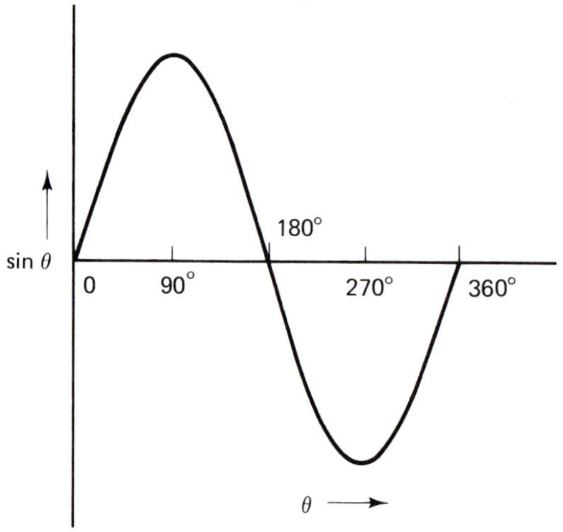

FIGURE 11-17 Graph of sine wave.

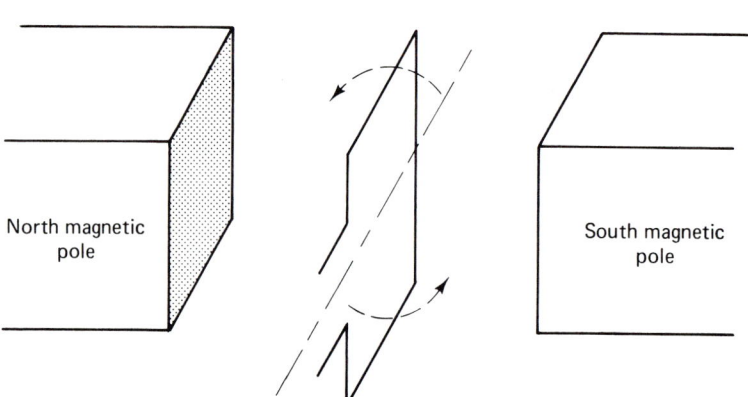

FIGURE 11-18 Simple ac generator.

Let us pause for a little while and reconsider the action of the ac generator in Fig. 11-18. We start with the loop in the position shown in the figure, that is, with the loop contained in a vertical plane. Even though the loop is rotating with a steady (constant) rotation rate, at this particular moment the top and bottom conductors, the sides of the loop, are moving parallel to the magnetic-field flux lines. The top side is moving to the left and the bottom side is moving to the right. Neither side is cutting magnetic flux lines, so zero electromotive force is induced in the sides at this moment.

As the loop rotates beyond the position of Fig. 11-18, the motion of the

456   Alternating Currents   Ch. 11

sides is no longer parallel to the magnetic flux lines. Therefore, an electromotive force (EMF) will be induced in each side of the loop. The polarities of these EMFs will be such as to be additive around the loop, giving twice the voltage of either across the loop terminals. The ends of the loop contribute no EMF since neither cuts magnetic flux lines.

As the coil rotates just past the starting position the induced EMF is initially quite small, but it builds up as the angle of rotation increases. This happens because the motion of the loop sides causes larger number of magnetic flux lines to be cut for each degree of rotation. More magnetic flux lines are cut for each tiny, but equal, interval of time. The induced EMF reaches its peak at the moment that each side is closest to the neighboring magnetic pole. We say that the loop is in a horizontal plane and that it has rotated 90° from the initial position.

If your understanding of the generator action during the first 90° is a bit fuzzy, go back three paragraphs and reread the explanation. Otherwise, let us continue as the loop rotates past the 90° position. As you probably will reason, when the angle continues to increase, the induced EMF decreases. This occurs because the sides move more nearly parallel to the magnetic flux lines as the angle increases from 90° to 180°. At 180° the loop is once again in a vertical plane and the induced EMF is again zero, as it was at the start. However, we must remember that the positions of the sides have been interchanged. The side that was on top in the beginning is now at the bottom.

The rotation of the loop through the first 180° has created one *alternation*, one half-cycle with the same polarity of EMF, for a sinusoidal waveform. As you will surely reason, rotation through the next half-circle, from 180° to 360°, provides us with a second alternation starting with zero volts, reaching a maximum at 270°, and then back to zero volts again at 360°. However, as viewed at the loop terminals, the polarity of this second half-cycle will be the opposite of that for the first half-cycle. A little thought about the direction of movement, up or down, for the respective sides should confirm this conclusion.

11-3

Power and Effective Value

We have learned that an alternating current may have a waveform described as a sine wave. When this current flows through a resistor, power will be dissipated in accordance with the formulas discussed in Section 1-5. To reflect the changing nature of the current and power, we write Eq. (1-5-2) as

$$p = i^2 R \qquad (11\text{-}3\text{-}1)$$

where i is the instantaneous value of current and p is the instantaneous value of power. Of course, when the unit of current is the ampere, the power unit is watts for resistance R measured in ohms when applying Eq. (11-3-1).

Both the sine-wave current and the power appear in Fig. 11-19. As might have been predicted, the power curve is zero at the points where the current is zero. This occurs when i crosses the horizontal time axis when going upward or downward. Also, the power curve peaks at the same point that the current is maximum. It also peaks when the current is greatest in the negative direction. This occurs because the square of a negative quantity is positive. Therefore, the instantaneous power will peak twice during one cycle of the alternating

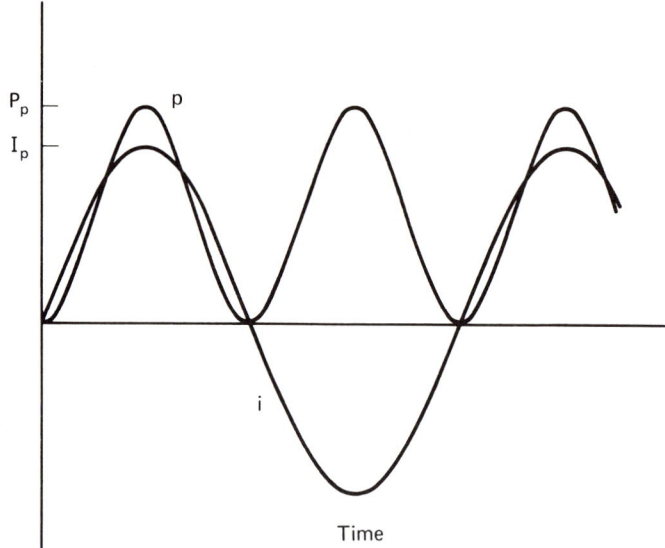

FIGURE 11-19 Sine-wave current and power.

current that produces it. This relationship between peak power P_p and peak current I_p is expressed in Eq. (11-3-2).

$$P_p = I_p^2 R \qquad (11\text{-}3\text{-}2)$$

EXAMPLE 11-10 An alternating current having a peak value of 3 A passes through a 10-Ω resistor. What is the peak power?

Solution

$$P_p = I_p^2 R \qquad (11\text{-}3\text{-}2)$$
$$= 3^2 \times 10 = 90 \text{ W}$$

EXAMPLE 11-11 Find the resistance that has a peak power dissipation of 700 mW for a peak current of 24 mA.

Solution

$$R = \frac{P_p}{I_p^2} \qquad (11\text{-}3\text{-}2)$$
$$= \frac{0.7}{0.024^2} = 1215 \text{ }\Omega$$

Note that the given quantities were converted to basic units before the calculation was made.

EXAMPLE 11-12 Corresponding to a peak power of 2 kW and 150 Ω of resistance, what is the peak current?

Solution Using Eq. (11-3-2), it is possible to solve for I_p^2 and then for I_p by taking the square root.

$$I_p^2 = \frac{P_p}{R} \qquad (11\text{-}3\text{-}2)$$

$$= \frac{2000}{150} = 13.333 \ldots \text{A}^2$$

$$I_p = \sqrt{13.333\ldots} = 3.65 \text{ A}$$

Note that the value of I_p^2 would not be written as is done here. Normally, there would be no purpose in doing so. However, we have written it here as an intermediate result for the student who is following the solution.

EXAMPLE 11-13 At some moment in time the instantaneous current in a 3.3-kΩ resistor is 20 mA. What is the power at the particular instant?

Solution

$$p = i^2 R \qquad (11\text{-}3\text{-}1)$$

$$= 0.02^2 \times 3300 = 1.32 \text{ W}$$

In this problem as stated, there is no way to identify the peak power or the peak current. The information given will not allow the peak values to be found. We can only find the power at the particular moment for which the changing current is known.

Referring back to Fig. 11-19, it is possible to find the *average power*. If a line (dashed line in the figure) is drawn midway between the horizontal time axis (see Fig. 11-20) and the peak power P_p, we have the average power. Because of the shape of the power curve, the area above the dashed line under the curve will just fill in the area under the dashed line above the curve. This is why the dashed line represents the average power.

We now come to the idea of the *effective* value of current. An effective ampere of ac will have the same heating result while flowing through a given resistor as will 1 A of dc. If we can measure, or otherwise find, the effective current, we can apply any of the power equations from Section 1-5. We will not have to learn new equations for calculations involving power.

Referring again to Fig. 11-20, a dc current would have to produce the same power dissipation as the average power level to be equivalent to the effective ac current. In other words, the power must be $I_p^2 R/2$:

$$I^2 R = \frac{I_p^2 R}{2}$$

$$I = \frac{I_p}{\sqrt{2}} = 0.707 I_p \qquad (11\text{-}3\text{-}3)$$

Sec. 11-3 Power and Effective Value

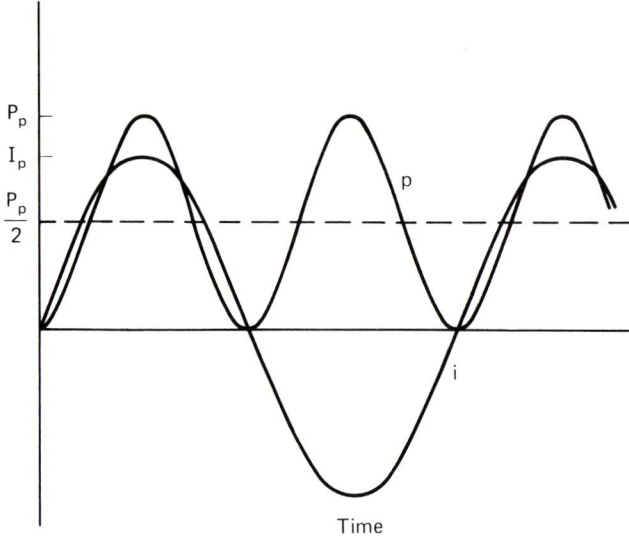

FIGURE 11-20   Peak and average power.

In looking at the meaning of effective ac current, we see that Eq. (11-3-3) also relates the effective ac current to the peak ac current.

EXAMPLE 11-14   A sine-wave current has a peak value of 100 mA. What is the effective current?

Solution

$$I = 0.707 I_p \qquad (11\text{-}3\text{-}3)$$

$$= 0.707 \times 100 = 70.7 \text{ mA}$$

EXAMPLE 11-15   A standard ammeter measures 1.75 A (effective). If the current is a sine wave, what is the peak current?

Solution   Solving for I_p using Eq. (11-3-3), we obtain

$$I_p = \frac{I}{0.707} = \frac{1.75}{0.707} = 2.48 \text{ A}$$

The same relationship also applies for sine-wave voltages; that is,

$$V = 0.707 V_p \qquad (11\text{-}3\text{-}4)$$

where V_p is the peak value and V is the effective value.

There is another way of stating the effective value when referring to ac voltage or current. We often refer to the *root-mean-square* (rms) voltage or current, which are the same as the effective values. Voltmeters and current meters measure effective or rms values unless specifically designed to do otherwise. Look at the panel, meter face, or instruction manual of the instrument if there is any doubt. Some instruments may read peak value or peak-to-peak value. In electronics it is common to mean the effective or rms values unless some other measurement is intended. Therefore, when applying any of the forms that the power equations (11-3-5), (11-3-6), and (11-3-7) can take, it is understood that effective values must be used.

$$P = IV \qquad (11\text{-}3\text{-}5)$$

$$P = I^2R \qquad (11\text{-}3\text{-}6)$$

$$P = \frac{V^2}{R} \qquad (11\text{-}3\text{-}7)$$

EXAMPLE 11-16 The voltage in a resistive circuit is measured to be 120 V and the current 3.5 A. What is the power?

Solution Effective voltage and current are given in the problem statement. Use Eq. (11-3-5).

$$P = IV = 3.5 \times 120 = 420 \text{ W}$$

EXAMPLE 11-17 Three milliamperes (peak) passes through a 4700-Ω resistor. Calculate the power.

Solution Use Eq. (11-3-3) and then Eq. (11-3-6).

$$I = 0.707I_p = 0.707 \times 3 = 2.12 \ldots \text{ mA}$$

$$P = I^2R = 0.00212\ldots^2 \times 4700 = 0.0211 \text{ W}$$

EXAMPLE 11-18 The peak-to-peak voltage across a 560-Ω resistor is measured to be 2.5 V. Find the power.

Solution Remember first to convert the peak-to-peak voltage of 2.5 V to 1.25 V (peak), then use Eq. (11-3-4), and finally apply Eq. (11-3-7).

$$V = 0.707V_p = 0.707 \times 1.25 = 0.883\ldots \text{ V}$$

$$P = \frac{V^2}{R} = \frac{(0.833\ldots)^2}{560} = 0.00139 \text{ W}$$

Before we leave this subject, another type of average needs to be introduced. We should recognize that the average value of a sine wave is zero. The positive loop area is just equal to the negative loop area (see Fig. 11-21). These loops are more accurately referred to as *alternations*. The area above the time axis is the same as the area below for a whole number of cycles.

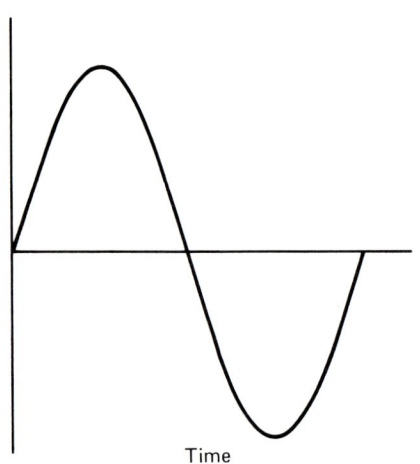

FIGURE 11-21 Sine wave.

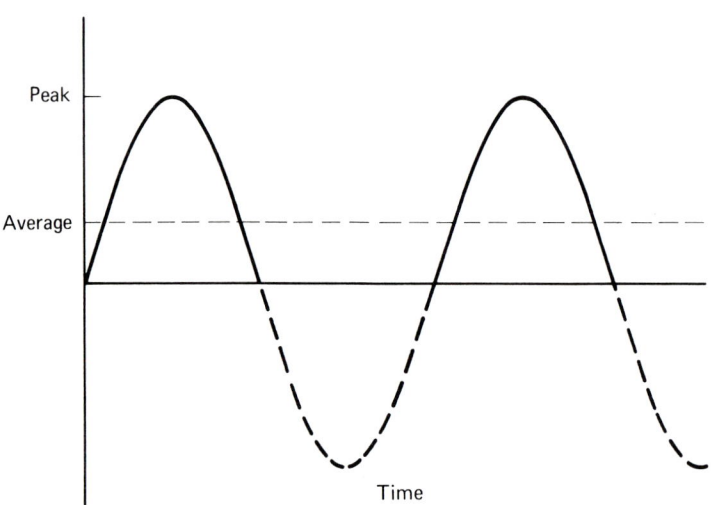

FIGURE 11-22 Output of half-wave rectifier.

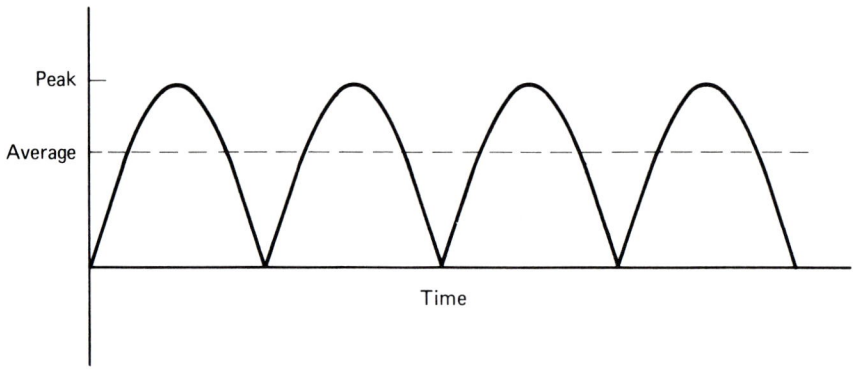

FIGURE 11-23 Output of full-wave rectifier.

However, it is sometimes necessary to consider a single alternation which has an average value given by

$$I_{av} = 0.637 I_p \qquad (11\text{-}3\text{-}8)$$

$$V_{av} = 0.637 V_p \qquad (11\text{-}3\text{-}9)$$

These equations have application in examining rectifiers which are a part of the power supplies used in our laboratory and in almost every item of electronic equipment. As two examples of these waveforms, Fig. 11-22 shows the output of a half-wave rectifier which chops off the negative alternation and the full-wave rectifier of Fig. 11-23, which inverts the negative alternation. It should be recognized that Eq. (11-3-8) or (11-3-9) applies directly to a waveform of Fig. 11-23. However, the results so obtained for Fig. 11-22 must be halved as the number of alternations are halved.

EXAMPLE 11-19 Find the average value of a single alternation of current whose peak is 4.72 mA.

Solution Use Eq. (11-3-8).

$$I_{av} = 0.637 \times I_p = 0.637 \times 4.72 = 3.00 \text{ mA}$$

EXAMPLE 11-20 If the average voltage measured at the output of a full-wave rectifier is 100 V, what is the peak voltage?

Solution Solve for V_p using Eq. (11-3-9).

$$V_p = \frac{V_{av}}{0.637} = \frac{100}{0.637} = 157 \text{ V}$$

11-4 Alternating Current in an Inductor

We previously learned that the property of inductance does not cause opposition to the flow of direct current. We now consider the flow of alternating current in an inductor and will discover that inductance does present opposition to flow.

It will be recalled that inductance is associated with magnetism caused by the flow of electrical current. A single continuous electrical circuit has inductance because a magnetic field is produced. When the conductor in such a circuit is coiled, the magnetic field is intensified and the flux representing the field is increased. Moreover, the flux now passes through a larger number of turns. The greater field and the larger number of turns both increase the inductance. Adding a core of magnetic material, such as iron, increases the inductance even more.

Immediately after an inductor is connected into a complete electrical circuit with a dc source, the flow of current is zero. Eventually, the current will be limited only by the circuit resistance, but not in the beginning. Why does this happen?

The reason is that time is required to transfer energy from the source to the magnetic field of the inductor. Circuit current must build up to its final value. As long as the current is changing in building up, there is a counter electromotive force across the inductor terminals which is opposing the growth of the current. This EMF or voltage is always greater, as the rate of current change is greater. Eventually, the current reaches the final value and the induced EMF falls to zero.

When alternating current passes through the inductor, the voltage is greatest when the rate of current change is the greatest. Stated another way, the rate of current change is largest when the applied voltage is maximum. The rate of current change is zero when the applied voltage is zero.

We can demonstrate the rate of change by looking at the graph of current in Fig. 11-24(a). One complete cycle of ac current is shown. The rate of change of current with respect to time is the slope of the graph. If we think of the sine wave as the elevation of the terrain, the graph represents how high or low a climber will be with respect to a starting point (point A). At point A the climber is midway in elevation between the valley (point D) and the

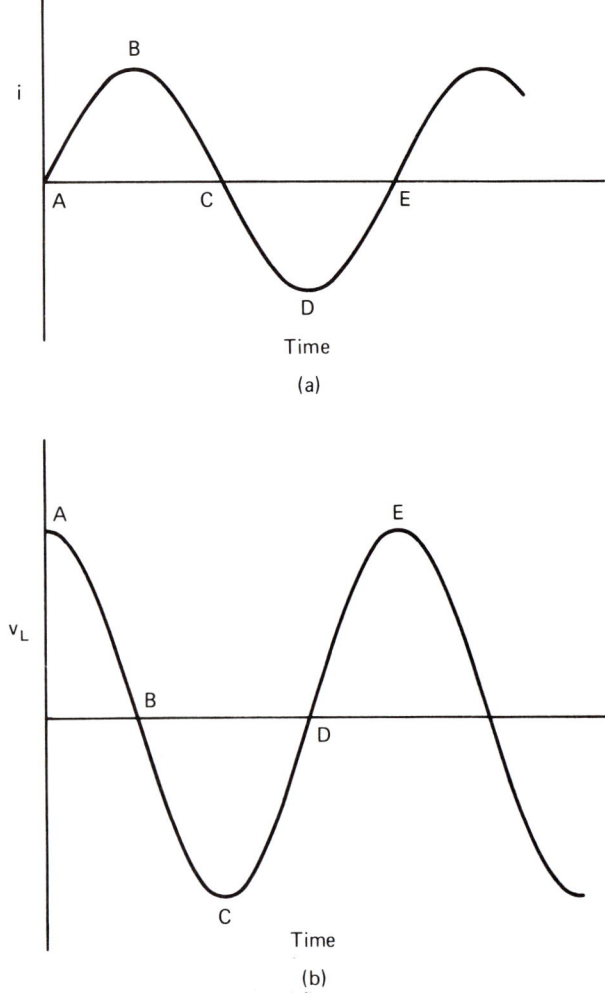

FIGURE 11-24 Current and inductor voltage.

peak (point B). The climbing effort at point A is greatest because the slope is the greatest there. At point B the climber has reached the top and the slope is zero. When the path is downhill, the slope is negative. At point C the slope reaches the greatest negative value. At D the slope is again zero.

We are now prepared to show the time relationship between the current, i, and the voltage, v_L. It is for this reason that Fig. 11-24(b) has been positioned immediately below Fig. 11-24(a). The top sine wave is the current and the bottom sine wave is the voltage. Notice that the voltage is at its positive maximum value when the current "uphill slope" is greatest. This occurs at point A and again at point E. At points B and D, the current slope is zero. These are the points where the voltage crosses the time axis. At point C the current "downhill slope" is the most, and the voltage is the greatest in the negative direction. Since the voltage reaches its maximum value at A before the current reaches its maximum at B, we say that the voltage *leads* the current. Also, it is proper to think that the current *lags* the voltage. Either way is correct. It is easy to see that the difference in lead or lag amounts to one-quarter of a complete cycle. Remember that one cycle is created by a 360° rotation of the two-pole generator described in Section 11-2. For one-quarter of a cycle the rotation will be 90°. Therefore, the voltage leads the current by 90°. Also, the current lags the voltage by 90°.

There is one more fact to be brought out about the behavior of ac current and voltage in a purely inductive circuit. The maximum voltage and the maximum current (peak values of voltage and current) can be related by a simple extension of Ohm's law. Since we are talking about an opposition to ac current flow caused by inductance, resistance is not used. This how it works.

Referring back to the introduction of inductance in Section 8-1, the voltage across the inductance, the EMF is given by the product

$$v_L = L \times \text{slope of current}$$

Recall that v_L is in volts when L is expressed in henrys and slope of current is in amperes per second. We recognize that the slope is really the time rate of current change. This slope has the peak value of current as one factor. It also has the rate of angle change as a second factor. However, the angle measure must be in a unit called the radian, not the degree. A circle consists of 360°, as we know. A circle also consists of 2π radians. The *pi* is a constant expressed as 3.14159 to an accuracy of six digits.

The radian, as an angular measure, requires further thought. Consider the radius of a circle, which may be regarded as one spoke in a wheel. Suppose that we take a second spoke and bend it around the rim of the wheel. The angle, as seen from the axle of the wheel, between the ends of the second spoke, is 1 radian. We refer to the curved item, formerly a spoke, as an *arc*. The arc length being the same as the radius (original spoke length) is the requirement for the angle being equal to 1 radian. Since the distance around the circle is slightly greater than three times the length of one radius, the number of radians in a circle is the same. This common number is represented as the Greek lowercase letter π, whose approximate value was quoted in the preceding paragraph.

Since the angle changes at the rate of 2π radians for a cycle and the number of cycles per second is the frequency, f, the time rate of angle change is $2\pi f$ radians per second.

One rotation is a cycle, or 360° or 2π radians. The rotation frequency, turns per second, or cycles per second, is usually referred to as "frequency." It is also useful to think of angle rates. In place of rotations per second, we might want to express the change as "degrees per second" or, even better for some purposes, "radians per second." This last angle rate of change is $2\pi f$.

Let us see where we stand. Let I_p be the peak or maximum value for the current. Then

$$\text{slope of current} = (I_p)(2\pi f)$$

Finally, when we multiply the slope of current by L to get the peak voltage,

$$V_p = (L)(I_p)(2\pi f)$$
$$= (I_p)(2\pi f L)$$

Parentheses are retained here for emphasis. If we now compare the product of I_p and $2\pi f L$ with Ohm's law, $V = IR$, we see that $2\pi f L$ should have the unit of ohms. It does. We call $2\pi f L$ the *inductive reactance*, which is the official name for inductive opposition to current flow. Instead of R as a symbol we use X, and as a further memory aid, the subscript L.

$$X_L = 2\pi f L \quad (11\text{-}4\text{-}1)$$

This equation will be used in its present form when f and L are known and X_L is to be found. It can be changed to allow L to be calculated when X_L and f are given.

$$L = \frac{X_L}{2\pi f} \quad (11\text{-}4\text{-}2)$$

Still another variation of Eq. (11-4-1) is possible. Suppose that the values of X_L and L are known. It is then possible to obtain the frequency, f.

$$f = \frac{X_L}{2\pi L} \quad (11\text{-}4\text{-}3)$$

Perhaps the most important points to remember about the formulas for inductive reactance are the following. Whenever any two of the three quantities f, L, and X_L are known, the third can be calculated. The fundamental unit for frequency, f, is the hertz, abbreviated as Hz, which is a cycle per second. For inductance, L is the henry (H), and for inductive reactance, X_L is the ohm (Ω).

Inductive reactance has one property in common with resistance: Ohm's law applies. The product of current and inductive reactance is voltage. As we shall see shortly, there are also differences in comparing inductive reactance and resistance. For example, inductive reactance depends on the frequency. Doubling the frequency automatically doubles the number of inductive reactance ohms.

EXAMPLE 11-21 A 0.1-H inductor is connected into a 60-Hz circuit. Calculate the inductive reactance, X_L.

Solution

$$X_L = 2\pi f L \quad (11\text{-}4\text{-}1)$$
$$= 2\pi(60)(0.1)$$
$$= 37.7 \, \Omega$$

EXAMPLE 11-22 Find the inductive reactance of a 10-mH inductor at 50 kHz.

Solution

$$X_L = 2\pi f L \qquad (11\text{-}4\text{-}1)$$
$$= 2\pi(50 \times 10^3)(10 \times 10^{-3})$$
$$= 3140 \ \Omega$$

EXAMPLE 11-23 A winding has 2500 Ω of inductive reactance for a frequency of 10,000 Hz. Find the inductance.

Solution

$$L = \frac{X_L}{2\pi f} \qquad (11\text{-}4\text{-}2)$$
$$= \frac{2500}{2\pi \times 10{,}000}$$
$$= 0.0398 \ H = 39.8 \ mH$$

EXAMPLE 11-24 Determine the frequency for which a 10-mH inductor has 1 MΩ of reactance.

Solution

$$f = \frac{X_L}{2\pi L} \qquad (11\text{-}4\text{-}3)$$
$$= \frac{10^6}{2\pi \times 10^{-2}}$$
$$= 1.59 \times 10^7 \ Hz = 15.9 \ MHz$$

We now intend to look at sine waves in a slightly different way. This will also allow us to see the voltage and current of an inductor in a new and useful form. This approach will be the basis for the tools used in analyzing ac circuits for the remainder of the book.

We first think of a wheel rolling on the ground. As the wheel turns, the axle moves relative to the earth. Another way of doing almost the same thing is shown in Fig. 11-25. We have the same wheel as before. However, the axle is now fixed and the wheel has a rope wrapped around the outside of the rim (circumference). The wheel is turned by pulling the rope to the right. We also see the axes of a graph, shown as dashed lines in Fig. 11-25. The horizontal axis is at the same level as the axle of the wheel at fixed point A. A point P on the rim of the wheel is at the same level. We have carefully placed the end of the rope at the starting line for the horizontal axis.

As the rope is pulled to the right and with the axle, A, fixed in position, the wheel turns. We can see that the rotation of the wheel will cause P to go

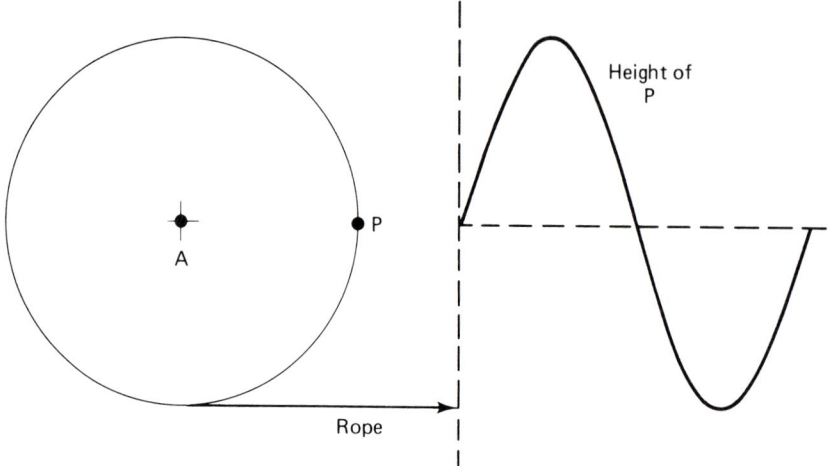

FIGURE 11-25 Generation of sine wave by rotating wheel.

up. The wheel will rotate in the counterclockwise direction. This means that the rotation is in the direction just opposite to that of a clock's hands.

For each position of the rope end, there will be a position of point *P*. The end of the rope moves horizontally to the right. Position *P* changes in both the horizontal and vertical directions. We are interested only in the up or down (vertical) movement. That vertical motion, when combined with the horizontal motion of the rope end, gives us a graph which is a sine wave. When the rope has been pulled to the right enough to turn the wheel exactly one-quarter revolution, point *P* will be at the highest position and directly above *A*. For this condition the rotation has been 90°. A sine-wave peak has been reached.

An addition rotation of 90° brings point *P* to the original level, but on the left side of *A*. Beyond this point during an additional rotation of 180°, point *P* will be below the starting level until the total angular rotation will be 360°. A cycle has been completed.

This may seem to be a very complicated way to generate a sine wave. Be patient. The reason for doing this will be explained in a moment. Before continuing it is best to review what is happening as the rope is pulled to the right. You should understand this before continuing. If the picture in your mind is somewhat fuzzy, it might be well to discuss it with a fellow student or your instructor.

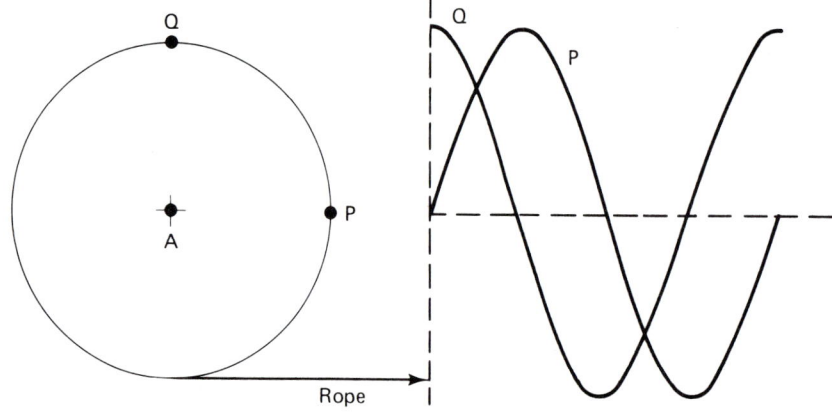

FIGURE 11-26 Generation of two sine waves by rotating wheel.

FIGURE 11-27 Phasor diagram.

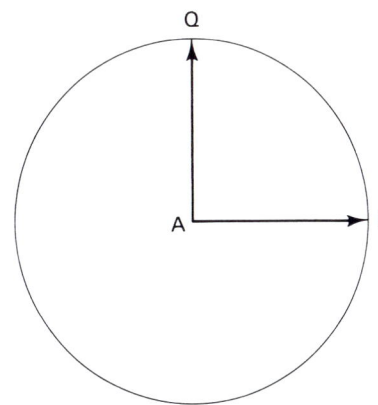

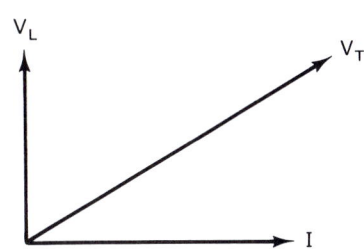

FIGURE 11-28 Typical phasor diagram, including voltage and current.

We now get to the reason for using a wheel and a rope. Suppose that two points are marked on the rim of the wheel (see Fig. 11-26). The second point is labeled Q and Q is directly above A. Now as the wheel turns with the movement of the rope end to the right, the height of P generates a sine wave as before. Q will also generate a sine wave, with a different starting point. The sine wave generated by Q leads the sine wave generated by P by 90°. Stated differently, the sine wave generated by P lags the sine wave generated by Q by 90°. We say that the *phase difference* is 90°.

It is possible to indicate phase difference using marks on a circle as was done in Figs. 11-25 and 11-26. However, we commonly furnish the same information by drawing the spokes of the wheel connecting P to A and Q to A (see Fig. 11-27). Arrowheads are used to make sure which end of the line is connected to the circle. Such directed lines representing electrical quantities are called *phasors*. Two or more phasors may appear on the phasor diagram similar to Fig. 11-27. These may be ac voltages, ac currents or a mixture of ac voltages and ac currents. In Fig. 11-27, the two voltages have the same peak value since the phasor lengths are the same. In general, the phasor lengths are not the same. A typical phasor diagram has been drawn for Fig. 11-28. We see in it a voltage phasor labeled V_L and a current phasor I which lags V_L by 90°. These are exactly the phase relationship for the voltage across an inductor and the current flow in an inductor. Voltage V_T leads the current I by an angle that is less than 90°. In Chapter 12 you will learn that V_T represents the total voltage for a series circuit containing both R and X_L.

11-5

Alternating Current in a Capacitor

A capacitor placed in a dc circuit will become charged. Current will flow in the branch containing the capacitor as long as the capacitor is not fully charged. After the capacitor is fully charged, current will no longer flow in that branch.

In an ac circuit a capacitor will continue to charge and discharge as long as the circuit is activated. Current must flow first in one direction and then in the other, so that the charge makes the capacitor terminal voltage match the source voltage. For a simple series circuit with a capacitor connected across an ac voltage source,

$$v_C = \frac{q}{C} \tag{11-5-1}$$

We are looking at the instantaneous values of voltage and charge where v_C is the capacitor voltage in volts, q is the capacitor charge in coulombs, and C is the capacitance in farads.

With a sine waveform, the voltage v_C reaches a maximum once each cycle. Charge q, which is also a sine wave, must be maximum at the same moment. At that particular moment the current will be zero. The current is also zero when the voltage reaches the negative maximum once each cycle. When the current is zero, the capacitor voltage is not changing, which occurs at the maximum and minimum (negative maximum) points. We must conclude that there is a 90° phase angle between the capacitor voltage and the capacitor current.

Current is the time rate of charge movement. Charge will accumulate on the capacitor plates during the positive-current half-cycle. Charge of the opposite polarity will accumulate during the negative-current half-cycle. Therefore, the capacitor current leads the voltage by 90° (see Fig. 11-29). This is exactly the opposite phase relationship of that of the inductor. Current in an inductor lags the voltage by 90°.

We now reexamine Eq. (11-5-1) and recall that $q = It$, where I is a current and t is the time the current is flowing. This suggests that the peak capacitor voltage V_p is a function of (1) the peak current I_p, (2) the time that charging takes place, which will be some fraction of the period, T, and (3) the value of C. The functional relationship is

$$V_p = \frac{I_p T}{2\pi C} \tag{11-5-2}$$

Here the 2π mysteriously appears because the charging current varies as a sine wave rather than being constant and charging occurs for half a period, $T/2$. Since the period and the frequency are reciprocals, Eq. (11-5-2) can be written in a form that is more commonly used.

$$V_p = \frac{I_p}{2\pi f C} \tag{11-5-3}$$

Rearranging Eq. (11-5-3), a new form of Ohm's law appears.

$$\frac{V_p}{I_p} = \frac{1}{2\pi f C} \tag{11-5-4}$$

We refer to $1/2\pi f C$ as the *capacitive reactance* of the capacitor.

$$X_C = \frac{1}{2\pi f C} \tag{11-5-5}$$

When C is given in farads and f in Hz, the unit for X_C is ohms.

FIGURE 11-29 Current and capacitor voltage.

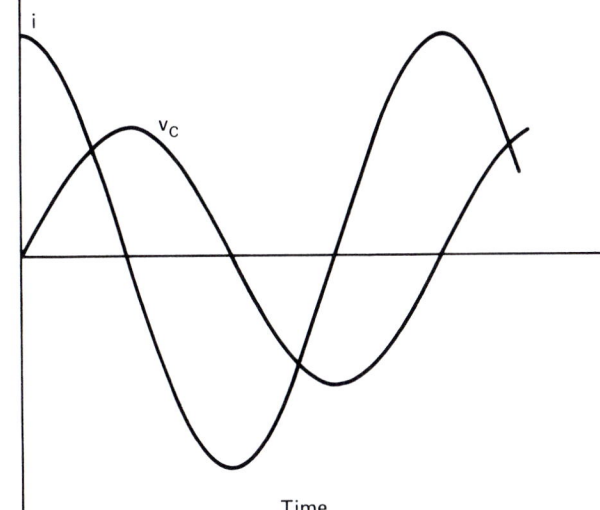

The whole idea of capacitive reactance is a little hard to take. The trouble occurs because we depend on a series of mathematical statements (equations). Since these are piled one upon the other as a child constructs castles from blocks of different shapes, the placement of each level is critical to the stability of the structure. Fortunately, our understanding is not dependent on the understanding of each step in the mathematical development. We learn in the laboratory, too. Moreover, the fundamental facts about capacitive reactance are quickly discovered on the job. Despite these consoling thoughts for the student struggling with the mathematical foundations, it is essential that several points be understood.

1. Capacitive reactance presents opposition to the flow of alternating current. This opposition is measured in ohms.
2. Capacitive reactance is dependent on the frequency. This is the message of Eq. (11-5-5). The greater the frequency, the less the capacitive reactance.

EXAMPLE 11-25 If $f = 100$ Hz and $C = 0.001$ F, find the value of X_C.

Solution

$$X_C = \frac{1}{2\pi f C} \tag{11-5-5}$$

$$= \frac{1}{2\pi \times 100 \times 0.001} = 1.59 \ \Omega$$

EXAMPLE 11-26 Find X_C for $f = 10$ kHz and $C = 0.01$ μf.

Solution

$$X_C = \frac{1}{2\pi f C} \tag{11-5-5}$$

$$= \frac{1}{2\pi \times 10^4 \times 10^{-8}} = 1590 \ \Omega$$

Sec. 11-5 Alternating Current in a Capacitor

EXAMPLE 11-27 When $X_C = 100$ kΩ and $C = 100$ pF, what is the frequency?

Solution

$$X_C = \frac{1}{2\pi f C} \quad (11\text{-}5\text{-}5)$$

$$f = \frac{1}{2\pi X_C C} = \frac{1}{2\pi \times 10^5 \times 10^{-10}} \text{Hz} = 15.9 \text{ kHz}$$

EXAMPLE 11-28 Calculate C for which the capacitive reactance is 1.5 MΩ for a frequency of 250 Hz.

Solution

$$X_C = \frac{1}{2\pi f C} \quad (11\text{-}5\text{-}5)$$

$$C = \frac{1}{2\pi f X_C} = \frac{1}{2\pi \times 2.5 \times 10^2 \times 1.5 \times 10^6}$$

$$= 4.24 \times 10^{-10} \text{ F} = 424 \text{ pF}$$

11-6

Mathematical Representation of Sine Wave

We have seen that a sine wave of voltage or current may be associated with the rotation of a wheel. At any particular moment the exact quantity depends on the angle of rotation, which accounts for the characteristic shape of the graph. Moreover, the size of the wheel, as measured by the radius (length of a spoke), determines the peak value of the sine-wave graph. More specifically, the peak value is exactly equal to the radius of the circle.

Up to this point we have been interested in explaining the nature of the sine-wave shape, the phase relationships between two, or more, sine waves which do not peak or pass through zero at the same time, and the numerical significance of peak (maximum) values, average values, and effective values. There is still another way of examining the sine wave that will be especially useful to us later. We propose to introduce the idea now and to develop it further in Chapter 19.

We know that the sine of an angle has a single numerical value. For instance, $\sin 30° = 0.5$. Any positive or negative choice of angle is possible, so that the graph is simply a curve passing through all of these numerical values plotted vertically against the numerical values of the angles plotted horizontally as illustrated in Fig. 11-30. However, the angles depend on the moment of time, since the angles must be those of the wheel rotation. Therefore, we are able to attach another horizontal scale to the graph of Fig. 11-31. Since one complete wheel rotation corresponds exactly to one complete cycle of sine-wave values, it is seen that the time scale can be expressed in terms of the period, T.

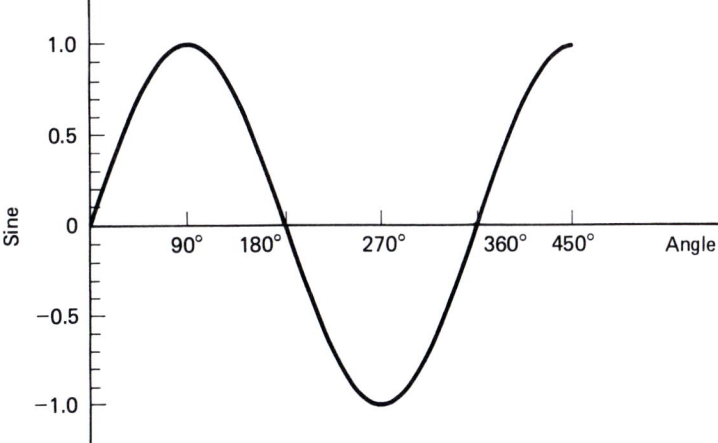

FIGURE 11-30 Graph of sine as function of angle.

The next logical step in this approach is to express the angle mathematically as a function of time t. One should make a clear distinction between t and T. The letter t is the symbol for time, which is assumed to change in the same way as is indicated by a clock. In contrast, T represents a fixed or constant quantity which is numerically related to the frequency as expressed by

$$T = \frac{1}{f} \qquad (11\text{-}1\text{-}2)$$

The angle is proportional to the value of t, being exactly 360° or 2π radians when $t = T$. We may then write

$$\text{angle in degrees} = \frac{360t}{T} = 360\,ft \qquad (11\text{-}6\text{-}1)$$

$$\text{angle in radians} = \frac{2\pi t}{T} = 2\pi ft \qquad (11\text{-}6\text{-}2)$$

The angle is commonly expressed in radians conforming to Eq. (11-6-2).

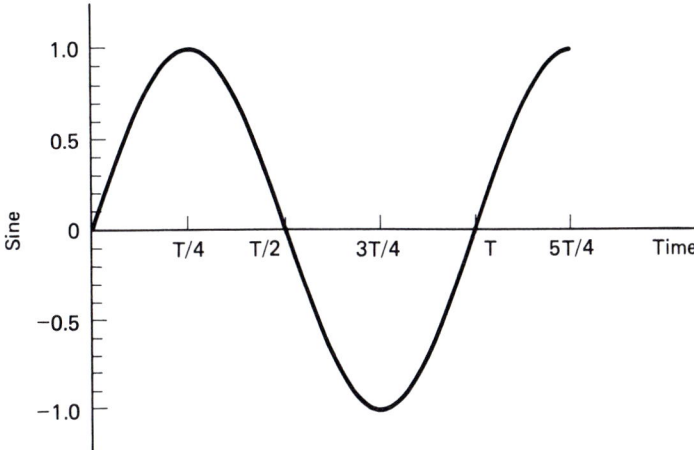

FIGURE 11-31 Graph of sine as a function of time.

Sec. 11-6 Mathematical Representation of Sine Wave

Since the value of the sine of an angle, using either Eq. (11-6-1) or (11-6-2), must lie between -1 and $+1$, the desired equation must be adjusted for the amplitude (peak value) of the sine-wave voltage, or current, being expressed mathematically. This is done through the multiplication by the constant amplitude of the sine wave. The resulting form of the mathematical function for a sine-wave voltage appears as follows:

$$v = V_{max} \sin 2\pi ft \qquad (11\text{-}6\text{-}3)$$

It should be recognized that v will have the same voltage unit as is assigned to V_{max}. A similar form is used for a current:

$$i = I_{max} \sin 2\pi ft \qquad (11\text{-}6\text{-}4)$$

A further modification of Eqs. (11-6-3) and (11-6-4) is necessary to convey information about the phase angle. In the general situation, the several sine-wave voltages and currents in a given circuit will not reach the positive peaks at the same instant of time. With this in mind the equations for the mathematical functions should be rewritten to incorporate the phase angle ϕ:

$$v = V_{max} \sin(2\pi ft + \phi) = V_{max} \sin(\omega t + \phi) \qquad (11\text{-}6\text{-}5)$$
$$i = I_{max} \sin(2\pi ft + \phi) = I_{max} \sin(\omega t + \phi) \qquad (11\text{-}6\text{-}6)$$

It will be seen that ω (Greek lowercase omega) has been substituted in these equations to simplify and shorten the more cumbersome form with f. The quantity ω is referred to as the angular velocity or angular frequency and is most often given in radians per second.

Chapter Summary

11.1. An alternating current reverses direction of flow periodically. The period T is the reciprocal of the frequency.

$$T = \frac{1}{f} \qquad (11\text{-}1\text{-}2)$$

Similarly,

$$f = \frac{1}{T} \qquad (11\text{-}1\text{-}1)$$

T is measured in seconds and f in hertz (cycles per second). All unit prefixes may be used.

11.2. Ohm's law applies for instantaneous values of voltage v and current i,

$$i = \frac{v}{R}$$

and for peak values V_p of voltage and current I_p for sine waves,

$$I_p = \frac{V_p}{R} \qquad (11\text{-}2\text{-}2)$$

11.3. The power dissipated in a resistor is determined using

$$P = IV \qquad (11\text{-}3\text{-}5)$$

$$P = I^2 R \qquad (11\text{-}3\text{-}6)$$

$$P = \frac{V^2}{R} \qquad (11\text{-}3\text{-}7)$$

where I and V are effective values of current and voltage, respectively, and are related to peak values for sine waves by

$$I = 0.707 I_p \qquad (11\text{-}3\text{-}3)$$

$$V = 0.707 V_p \qquad (11\text{-}3\text{-}4)$$

Instantaneous values for current or voltage may also be used with the power equations, giving rise to the instantaneous value of fluctuating power. A special case is that of peak current I_p and peak voltage V_p.

11.4. A changing current in an inductor causes an induced EMF whose value at any moment is proportional to current rate of change. Viewing the rate of change when current is graphed,

$$v_L = L \times \text{slope of current}$$

with v_L in volts and the slope in A/s.

11.5. With alternating current, the induced EMF in an inductor causes an opposition to current flow called inductive reactance.

$$X_L = \frac{V_p}{I_p} = \frac{V}{I}$$

In these equations V_p and I_p represent peak values as before, while V and I are effective values. With basic units of current and voltage, the unit of X_L is the ohm. All prefixes may be judiciously used.

11.6. Inductive reactance, X_L, is related to inductance L and frequency f.

$$X_L = 2\pi f L \qquad (11\text{-}4\text{-}1)$$

$$L = \frac{X_L}{2\pi f} \qquad (11\text{-}4\text{-}2)$$

$$f = \frac{X_L}{2\pi L} \qquad (11\text{-}4\text{-}3)$$

with X_L in ohms, L in henrys, and f in hertz.

11.7. Sine-wave ac quantities can be considerd as generated by a rotating wheel. The vertical distance between a point on the rim and the level of the axis of rotation is a sine function of the angle of rotation. Two or more separated points on the rim represent sine waves whose peak values do not coincide. They are said to have phase-angle differences.

11.8. Directed line segments (arrows) can represent ac quantities. The length of each line segment, called a phasor, represents the amount of current (or voltage). The angular difference between them is a measure of the phase-angle difference.

11.9. For a resistor, the phasors for I and V have the same direction. I and V are in phase.

For an inductor, I and V are at right angles (perpendicular). There is a phase difference of 90°. V is said to lead I by 90°. It is equally correct to say that I lags V by 90°.

11.10 For a capacitor, I and V are at right angles (perpendicular). There is a phase difference of 90°. V is said to lag I by 90°. It is equally correct to say that I leads V by 90°.

Self-Review

True–False Questions

1. **T. F.** The period of the ac wave is the reciprocal of the frequency.
2. **T. F.** If the period of an alternating current is doubled, the frequency must also double.
3. **T. F.** In an inductor the curent leads the inductor voltage by 90°.
4. **T. F.** A simple ac generator with two poles (one N and one S) will create one cycle of sine-wave voltage for each revolution of the rotating coil.
5. **T. F.** In a resistor the sine-wave voltage and sine-wave current will have peak values at the same instant of time.
6. **T. F.** For the same inductor the inductance will be twice as large if the frequency is doubled.
7. **T. F.** When the instantaneous current through a resistor is zero, the power dissipated will also be zero.
8. **T. F.** The average power for a 10 A (rms) ac resistive circuit is the same as that for 10-A dc current in the same circuit.
9. **T. F.** If the slope of the current graph is doubled, the voltage across the inductor will be halved.
10. **T. F.** If the peak value of alternating voltage across a resistor is doubled, the power will be increased by a factor of 4.
11. **T. F.** In an alternating-current circuit, the direction of current flow reverses once each cycle.
12. **T. F.** A wave that repeats after a fixed interval and multiples of that interval is said to be "periodic."
13. **T. F.** For an alternating current, the name of the unit indicating cycles per second is "hertz."
14. **T. F.** The period of a 60-Hz alternating voltage is constant.
15. **T. F.** If the waveform of a square wave is inverted, the frequency will be doubled.
16. **T. F.** The current passing through a load resistor will have the same waveform as that of the applied voltage.
17. **T. F.** In a simple ac generator, the induced voltage (EMF) will be greatest at the moment that the flux lines passing through the loop are maximum.
18. **T. F.** An instantaneous peak power of 100 W resulting from alternating current passing through a resistor is equivalent to an average power level of 70.7 W.
19. **T. F.** The average power of an inductive circuit with sine-wave alternating current is 50% of the peak instantaneous power.
20. **T. F.** Inductive reactance is directly proportional to the current passing through the inductor.

Multiple-Choice Questions

21. An alternating current changes from -10 mA to $+10$ mA at $t = 0$ and remains constant until $t = 0.005$ s. It then reverses polarity, going from $+10$ mA to -10 mA at $t = 0.005$ s, retains the same value until $t = 0.01$ s, and then reverses direction a second time as occurred at $t = 0$. The frequency, f, is
    (a) 200 Hz (b) 0.01 Hz (c) 20 Hz (d) 100 Hz (e) 10 Hz

22. For the periodic wave of Question 21, the current at $t = 0.017$ s will be
    (a) 0   (b) -10 mA   (c) $+0.005$ A   (d) $+10$ mA
    (e) insufficient information to solve
23. A circuit is connected with a dc source to cause an alternating current in a resistor which reverses direction every 5 s. The period is
    (a) 5 s   (b) 10 s   (c) 15 s   (d) 2.5 s
    (e) solution impossible
24. A 50-kHz ac signal has a period of
    (a) 0.005 s   (b) 100 ms   (c) 200 ms   (d) 20 μs   (e) 0.02 s
25. A sine-wave current having a peak value of 3 mA flows through a 1.2-kΩ resistor. The voltage across the resistor will peak at
    (a) 0.4 V   (b) 2.5 V   (c) 2.5 kV   (d) 3.6 V   (e) 3.0 mV
26. A sine wave of voltage or current is represented by
    (a) a phasor   (b) a resistance
    (c) an opposition   (d) a phase angle
    (e) a resistance diagram
27. In a resistor carrying sine-wave alternating current the power varies from zero to 250 W as the current changes. The average power is
    (a) zero   (b) 125 W   (c) 250 W   (d) 500 W   (e) 1000 W
28. If $X_C = 10$ kΩ and $C = 0.1$ μF, what is the frequency, f?
    (a) 6.3 kHz   (b) 159 Hz   (c) 6.3 Hz   (d) 10^7 Hz
    (e) no correct choice
29. In a circuit consisting of an ac source and a capacitor, the current will lead the voltage by
    (a) 90°   (b) 180°   (c) 270°   (d) $-90°$
    (e) no correct choice
30. If the voltage across a capacitor is represented by a horizontal phasor, the current phasor will point in what direction?
    (a) down   (b) left   (c) up   (d) right   (e) 45°
31. A square wave has a peak value of 24 V. When applied to a resistor of 2 kΩ, the current will be
    (a) $+12$ mA   (b) 0   (c) -12 mA
    (d) (a), (b), and (c) are correct at particular times in the cycle
    (e) no correct choice
32. A current-reversing circuit is constructed in accordance with Fig. 11-1. The DPDT switch is reversed once every millisecond, which causes
    (a) the frequency to be 1 Hz   (b) the period to be 1 ms
    (c) current reversal once per microsecond
    (d) an average power of zero   (e) no correct choice
33. A square-wave voltage with 10 V (peak-to-peak) is applied to a resistive load of 100 Ω. The power is
    (a) 1 W   (b) -1 W   (c) 0.25 W   (d) $-\frac{1}{4}$ W
    (e) both (c) and (d)
34. An alternating sine-wave voltage changes polarity at 5.0 μs, and 10 μs. The frequency is
    (a) 0 Hz   (b) 100 kHz   (c) 200 kHz   (d) 5 MHz   (e) 10 MHz
35. A 25-kHz sine wave has an instantaneous value of -7 mA at $t = 25$ ms. Determine the time at which the current will next be known to be the same.
    (a) 0 μs   (b) 40 μs   (c) 25 ms   (d) 25.04 ms   (e) 50 ms
36. The instantaneous power dissipated in a 1-kΩ resistor varies between 0 and 25 mW.
    (a) The peak current is equal to 5 mA.
    (b) The peak voltage is equal to 25 mV.
    (c) The average power is equal to zero.
    (d) The effective current is equal to 5 mA.
    (e) The effective voltage is equal to 5 V.
37. A sine-wave voltage is applied to a half-wave rectifier and provides exactly 12 V average dc level at the output. Assuming no losses, the effective value of the original voltage is
    (a) 6 V   (b) 12 V   (c) 13.3 V   (d) 18.9 V   (e) 26.7 V
38. The current passing through an inductor is 25 mA at $f = 750$ Hz with 4.5 V applied.

The inductance L is equal to
(a) 6.08 μH (b) 19.1 mH (c) 38.2 mH (d) 60 mH (e) 180 mH

39. The voltage across a capacitor has a peak value of 30 V for which the current peak is 2.75 mA. If $C = 0.01$ μF, the frequency is
(a) 1460 Hz (b) 2920 Hz (c) 9.17 kHz (d) 14.6 kHz (e) 29.2 kHz

40. A phasor diagram representing the voltage and current for an inductor has the current shown as directed horizontally to the right. The voltage will then be pointed
(a) to the right   (b) up
(c) to the left   (d) down
(e) at a 45° angle from the horizontal

Practice Problems

41. A voltage source is connected in series with a switch so that the current can be reversed every 200 ms. If the current delivered by the source is 100 mA at a particular time, how much will it be 400 ms later?

42. In Fig. 11-1, $V = 12$ V and $R = 6$ kΩ. The switch is first thrown to the right, so that the current from A to B is $+2$ mA. What will be the current when the switch is thrown to the left?

43. The period of an alternating current wave is 500 μs. What is the frequency?

44. Find the power dissipated in a 500-Ω resistor when the peak alternating sine-wave current is 50 mA.

45. Find the peak-to-peak voltage across a 2.2-kΩ resistor when 10 mW is dissipated.

46. A sine-wave voltage across a resistor has a peak value of 15 V and the peak value of the current in the resistor is 8.8 mA. Find the resistance.

47. Calculate the inductive reactance of a 0.1-H inductor at 200 Hz.

48. At a frequency of 60 Hz the inductive reactance of an inductor is 40 Ω. What is the inductance?

49. What frequency will result in an inductive reactance of 10 kΩ for a 100-μH inductor?

50. Calculate the capacitive reactance for a capacitor having 0.002 μF of capacitance at a frequency of 0.1 MHz.

51. A square-wave current has an amplitude of 25 mA and a period of 40 μs. Calculate the frequency.

52. The square-wave current of Problem 51 is passed through a 1000-Ω resistor. Calculate (a) the power for the positive half-cycle and (b) the power for the negative half-cycle.

53. A sine-wave current crosses the time axis in the upward direction at $t = 0.1$ ms and reaches a positive peak at $t = 0.2$ ms. What is the frequency?

54. A sine-wave voltage is applied to a resistive circuit such that the maximum positive current is 500 mA and the corresponding peak instantaneous power is 10 W. Determine the resistance of the circuit.

55. For the circuit of Problem 54, find the instantaneous power at the moment that the current is 500 mA in the reverse direction, that is, -500 mA.

56. The power delivered by a 115-V sine-wave generator to a resistive load is 50 W. What current will be measured by an ammeter placed in series with the load?

57. Calculate the inductance required so that the current is equal to 100 mA for an 8-V 400-Hz sine-wave voltage applied to the terminals of the inductor.

58. After rectification of a 12-V sine wave with a half-wave rectifier, its application to a 100-Ω resistive load will cause an average (dc) current to flow. Calculate that current assuming no voltage loss in the rectifier.

59. In looking at a phasor diagram, the current of 50 mA is shown directed to the right. The voltage phasor is 18 V and is directed upward. Describe the single circuit element.

60. A known capacitor of 0.1 μF is placed in series with an unknown capacitor value to allow a current of 100 μA with a total 1000-Hz voltage of 15 V. Find the C of the second capacitor.

Advanced Problems

61. A square-wave alternating voltage is -10 V at $t = 0$, rises to $+10$ V at $t = 50$ μs, falls to -10 V at $t = 150$ μs, and rises again to $+10$ V at $t = 250$ μs. Calculate the frequency.

62. An alternating sine wave of current passes through a 1.8-kΩ resistor. If the peak instantaneous power is 100 mW, find the average power and the effective value of the current.

63. Three resistors having values of 470 Ω, 560 Ω, and 1200 Ω are connected in parallel across a 120-V, sine-wave source. Determine the minimum and maximum values of the instantaneous power.

64. A sine-wave alternating current of 75 mA (peak) passes through a 20-mH inductor, so that the alternating voltage has an instantaneous minimum value of -40 V and a maximum value of $+40$ V. What is the frequency of the alternating current?

65. When a 5-V 10-kHz signal generator is connected to an inductor the measured current flow is 15 mA. Calculate L.

66. An inductor is connected cross the terminals of an alternating-voltage source and a current of 50 mA is measured. Without changing the source voltage, the frequency is increased by 1 kHz and a current of 40 mA is measured. At what frequencies were the measurements made?

67. A sine-wave alternating current flows through a 10-mH inductor. If the maximum instantaneous time rate of current change is 10,000 A/s, what is the effective value of voltage?

68. A sine-wave alternating current has a period of 100 μs. What is the capacitive reactance for a 0.1-μF capacitor with this current?

69. The application of 36 V to a 0.01-μF capacitor causes 5 mA to flow. At what frequency did this occur?

70. A 1.8-kΩ resistor is connected in series with a 0.5-μF capacitor. Application of a 400-Hz voltage to the combination causes power dissipation of 5 W. What is the capacitor voltage?

71. A square wave has a voltage level of $+50$ V for a time interval of 50 μs and a level of -50 V for 50 μs. Calculate the frequency.

72. At a particular time, the level of a 500-kHz square wave is -150 V. Determine the voltage level 1 μs later.

73. The sine-wave voltage applied to the terminals of an inductor reaches its positive peak at $t = 10$ μs. The current passing through the inductor is at its maximum at $t = 20$ μs. Calculate the frequency.

74. A sine-wave voltage is created in the turns of a conducting loop that is rotating in a magnetic field. Calculate the instantaneous induced voltage after the loop has turned 60° beyond the point of the maximum 50 V.

75. The induced loop voltage of Problem 74 is applied to a resistive load such that the total series resistance is 20 Ω. Find (a) the maximum amount of current flow at any one time and (b) the least amount of current flow at any one time.

76. A sine-wave voltage is measured using a cathode ray oscilloscope to be 17.5 V (peak to peak). Calculate the average level after this voltage has been full-wave rectified.

77. Two resistors are placed in parallel across an 18-V source. One resistor is rated at 3.3 kΩ and the second at 5.6 kΩ. Determine the peak value of the current flowing through the larger resistor.

78. The resistors of Problem 77 are rearranged to be in series across the same ac source. Determine the peak-to-peak voltage across the larger resistor.

79. The peak instantaneous power is 250 mW for an ac voltage of 12 V and a resistive load. Calculate the current in the circuit as would be read by a standard ammeter.

80. A 400-Hz ac source supplies 48 V to a 10-μF capacitor. Determine the maximum charge of the capacitor at any moment.

Self-Review

12

SEE NOTES

Series and Parallel AC Circuits

The objectives of this chapter are as follows:

1. To determine by calculations using phasors the current in ac series circuits with resistance and inductance.
2. To provide a working definition of impedance, methods for its calculation, and practical applications in series *RL* circuits.
3. To determine by calculations using phasors, the current and component voltages in series *RC* circuits.
4. To determine by calculations using phasors the branch currents and total current in parallel circuits for which one path contains *R* and the other either *L* or *C*.
5. To establish the meaning of and applications using the total equivalent impedances for a parallel circuit having *R* in one branch and *L* or *C* in the other.
6. To extend the analysis of series ac circuits to include resistance, inductance, and capacitance.
7. For series *RLC* circuits, to construct phasor diagrams relating amplitudes of ac currents, component voltages, and total applied voltage with the identification of phase relationships.
8. To extend the concept of impedance and the impedance diagram for series *RLC* circuits whereby numerical calculations with phasor amplitude ratios and phase relationships are expressed numerically.
9. To expand the analysis of series ac circuits so that more than one each of *R*, *L*, and *C* elements can be included with emphasis on measurements of current and voltage components.
10. To extend the concept of parallel *R*, *L*, and *C* circuits to include a third branch so that each of the three branches has resistance, inductance, or capacitance.
11. To emphasize the description of lagging and leading currents in the general parallel *RLC* circuit together with the impedance of each branch as it influences the phasor diagram relating applied voltage and total current.
12. To expand the concept of total equivalent impedance for the general parallel *RLC* circuit and to detail exact methods for its calculation.

12-0
Introduction

Practical circuits in which alternating currents flow are made up of various circuit elements in different combinations. The first circuit to be analyzed consists of a resistor in series with an inductor. Phasors are employed to combine the voltages across the resistor and inductor. A similar procedure is used for a resistor in series with a capacitor. The method is broadened to include more than one resistor in series with more than one inductor or capacitor. Parallel RL and RC circuits are investigated next with the phasor analysis techniques. Finally, combinations of resistors, inductors, and capacitors in series and in parallel are analyzed.

12-1
Series Circuit with Resistance and Inductive Reactance

We have learned that the sine-wave current through a resistor and the sine-wave voltage across the resistor both cross the time axis at the same point and in the same direction. Moreover, the two peak together and reach the negative maxima at the same time. For a resistor the voltage and the current are in phase.

We also know that the current passing through an inductor will lag the inductor voltage by 90°. There is a 90° phase difference between the two.

It is now time to consider a series circuit containing R and X_L and an ac sine-wave voltage source V_T (see Fig. 12-1). Let the voltage across the resistor be V_R and the voltage across the inductive reactance be V_L. We will now combine our knowledge about phase angles in a single phasor diagram, shown in Fig. 12-2. The construction of this diagram proceeded in the following way:

1. As a point of departure, V_L was oriented in the upward direction.
2. Current I lags V_L by 90°.
3. V_R is in phase with I. Note that V_R should be drawn to the same scale as V_L.

One step remains to be done. We should be able to find V_T on the phasor diagram. Kirchhoff's voltage law still applies. It is possible to add the instantaneous sine-wave values of V_R and of V_L point by point and find V_T. Since V_R and V_L have a phase difference of 90°, the peak value found will not be the arithmetic sum of the individual peak values. We can only say that the peak value of V_T will be larger than the peak values of V_R or V_L.

However, it is possible to sum the two voltages in order to find V_T by adding the phasors. This is done in Fig. 12-3 by moving V_L (see Fig. 12-2) so that the end without the arrowhead (tail end) is placed in contact with the

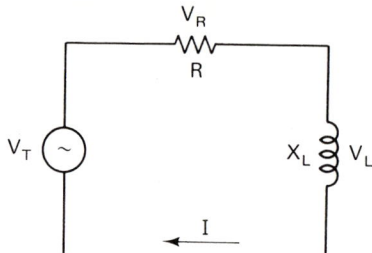

FIGURE 12-1 Resistance and inductive reactance in series.

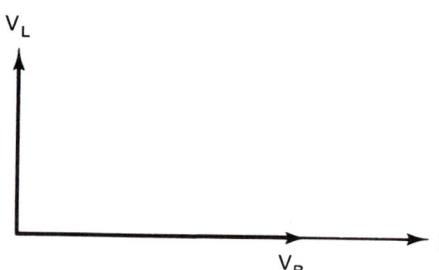

FIGURE 12-2 Phasor diagram for series RL circuit.

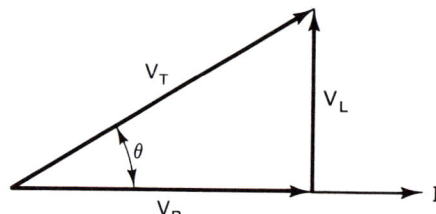

FIGURE 12-3 Phasor addition.

arrowhead end (head end) of V_R. The resultant voltage V_T has the length and direction of the phasor in Fig. 12-3. Note that the tail end of V_T coincides with the tail end of V_R. Similarly, the head end of V_T coincides with the head end of V_L. Also note that a right triangle is formed having legs of length V_L and V_R and a hypotenuse of length V_T. For a right triangle the square of the hypotenuse is equal to the sum of the squares of the sides.

$$V_T^2 = V_R^2 + V_L^2$$

Taking the square root of both sides gives

$$V_T = \sqrt{V_R^2 + V_L^2} \qquad (12\text{-}1\text{-}1)$$

EXAMPLE 12-1 A series circuit consists of a resistor and an inductive reactance. Find the peak value of the total voltage when the peak values of the two are as follows: $V_R = 3$ V, $V_L = 4$ V.

Solution

$$V_T = \sqrt{V_R^2 + V_L^2} \qquad (12\text{-}1\text{-}1)$$

$$= \sqrt{3^2 + 4^2} = 5 \text{ V} \quad \text{(peak value)}$$

EXAMPLE 12-2 A 10-Ω resistor is connected in series with a 15-Ω inductive reactance. Find the total voltage required to produce a current having a 4-A peak value.

Solution

$$V_R = IR = 4 \times 10 = 40 \text{ V}$$

$$V_L = IX_L = 4 \times 15 = 60 \text{ V}$$

$$V_T = \sqrt{V_R^2 + V_L^2} = \sqrt{40^2 + 60^2} = 72.1 \text{ V} \quad \text{(peak)}$$

We see in Fig. 12-3 phase angles other than 90°. To determine the phase angles, we will need to apply some trigonometric functions. To find the angle between V_R and V_T, use

$$\tan \theta = \frac{V_L}{V_R}$$

We can also use the inverse tangent.

$$\theta = \tan^{-1} \frac{V_L}{V_R} \qquad (12\text{-}1\text{-}2)$$

Tables and slide rules were used formerly in finding the tangent of a given angle or the angle whose tangent is given. Since trigonometric functions are available on inexpensive pocket calculators, it is assumed that each student owns one. The trick is to use it properly. For example, let us assume that $V_L = 11.6$ V and $V_R = 7.4$ V. The procedures for finding V_T and θ are as follows: To find the value of V_T:

Operation			Display	
11.6	x^2	+	134.56	(V_L^2)
7.4	x^2		54.56	(V_R^2)
	=		189.32	($V_L^2 + V_R^2$)
	$\sqrt{x}$		13.76	(V_T)

The value of V_T is therefore 13.76 V.

Sec. 12-1 Series Circuit with Resistance and Inductive Reactance

To find the value of θ:

Operation	Display
$\boxed{\div}$	
$\boxed{=}$	
$\boxed{INV}$ $\boxed{TAN}$	

The value of θ is 57.5°, rounded off.

In some calculators the $\boxed{INV}$, $\boxed{TAN}$ combination is replaced by either a $\boxed{TAN^{-1}}$ key or an $\boxed{ARC\ TAN}$ key.

EXAMPLE 12-3 Find the angle between V_L and V_T for the data of Example 12-1.

Solution

$$\theta = \tan^{-1} \frac{V_L}{V_R} \qquad (12\text{-}1\text{-}2)$$

$$= \tan^{-1} \frac{4}{3} = \tan^{-1} 1.333 \ldots$$

$$= 53.1°$$

EXAMPLE 12-4 Find the angle between V_T and I for the problem of Example 12-2.

Solution Be careful. The question asks for the angle between V_T and I and we know how to find the angle between V_R and V_T. Since V_R and I are in phase, the phase difference between V_T and I is the same as the angle between V_R and V_T.

$$\theta = \tan^{-1} \frac{V_L}{V_R} \qquad (12\text{-}1\text{-}2)$$

$$= \tan^{-1} \frac{60}{40} = 56.3°$$

We have seen how it is possible to combine two sine-wave voltages when the angle between them is 90°. We will use this knowledge to arrive at a measure of the total opposition to current flow. This total opposition, known as *impedance*, uses the symbol Z. Recall that in the series circuit

$$V_T^2 = V_R^2 + V_L^2$$

However by Ohm's law,

$$V_R = IR$$
$$V_L = IX_L$$
$$V_T = IZ$$
$$(IZ)^2 = (IR)^2 + (IX_L)^2$$
$$I^2Z^2 = I^2R^2 + I^2X_L^2$$

Each term has a common factor I^2 which can be divided out and made to disappear.

$$Z^2 = R^2 + X_L^2$$

$$Z = \sqrt{R^2 + X_L^2} \qquad (12\text{-}1\text{-}3)$$

$$\theta = \tan^{-1}\frac{X_L}{R} \qquad (12\text{-}1\text{-}4)$$

Remember that the square of the hypotenuse is equal to the sum of the squares of the sides. Z is then the length of the hypotenuse and the legs of the right triangle have lengths of R and X_L, as indicated in Fig. 12-4. We are also interested in the angle θ (theta) shown in Figs. 12-3 and 12-4. Angle θ is the angle between V_T and the lagging current I.

The preceding development suggests that an impedance requires two pieces of information. In many cases we either know R and X_L or know Z and θ. When Z and θ are known, we write the impedance as $Z\angle\theta$. This is a very convenient shorthand notation.

We feel it may be useful to summarize the ideas leading up to impedance in a way that uses the least amount of mathematics. Resistance comes most easily. The resistance applies to ac or dc. Inductive reactance represents the opposition to current flow for an inductor. It is measured in ohms, as is resistance. When resistance and inductive reactance are placed in series, a different form of opposition to current flow is created. We call this form "impedance." The calculation of impedance for such a series circuit requires a strange process for combining the two. They cannot be added directly as we intuitively add the costs of items in a market basket. Series resistance and inductive reactance must be combined by viewing the two as legs of a right triangle whose hypotenuse is the impedance. With this in mind, you might want to review the preceding pages leading to the concept of impedance.

The performance of calculations involving resistance and reactance is much easier using an electronic calculator having coordinate conversion keys. An electronics student should choose a scientific calculator having such features. When R and X_L are known, we refer to rectangular coordinates. The same information expressed as $Z\angle\theta$ is in polar coordinate form. In either coordinate form, a right triangle is completely described. Throughout most of the remaining lessons on ac you will be solving right triangles. You will be converting

FIGURE 12-4 Impedance diagram.

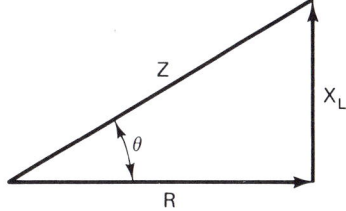

from rectangular coordinates to polar coordinates or from polar coordinates to rectangular coordinates. It will sometimes be necessary to convert back and forth during the solution of a single problem solution. For this reason a method of conversion using an electronic calculator is of great benefit.

At this time we recommend that you refer to the booklet most manufacturers provide to the calculator buyer. It describes the procedure for performing the basic functions, such as addition, subtraction, multiplication, and division. It would seem that all calculators would require the same procedures for these simple operations, but they do not. The booklet (or manual) will also explain how the values of different mathematical functions can be obtained using the calculator. Included are the trigonometric functions, sine, cosine, and tangent, and their inverses, arcsine, arccosine, and arctangent. In addition, we find logarithms, powers, and so on. You are advised to scan the booklet now to learn the capabilities. At the time they are needed, read the booklet carefully to learn and master the procedures. If the calculator does not have the scientific functions, it should be replaced whenever the occasion arises for evaluating scientific functions.

Now look for the directions for coordinate conversions. Assuming that your calculator has coordinate conversion capabilities, the booklet should describe the keys used, the procedures, and the method through use of examples. Do not rush through the examples. Do each one repeatedly until it can be accomplished without the use of the booklet. It is especially beneficial in the beginning to perform the conversions both ways for each case. If the example refers to conversion from rectangular to polar form, then upon completion, convert back from polar to rectangular form.

Thereafter, when following the examples in this book, use your own procedure. You should obtain the same answers in every case. You may also choose to do the calculations by the methods shown in the book. However, you will quickly learn that the coordinate conversion features of your calculator are far more convenient and trustworthy if you persist and practice.

EXAMPLE 12-5 In a series circuit $R = 20\ \Omega$ and $X_L = 20\ \Omega$. Find $Z\angle\theta$.

Solution Use Eqs. (12-1-3) and (12-1-4).

$$Z = \sqrt{20^2 + 20^2} = \sqrt{400 + 400} = \sqrt{800}$$
$$= 28.3\ \Omega$$

$$\theta = \tan^{-1}\frac{20}{20} = \tan^{-1} 1 = 45°$$

$$Z\angle\theta = 28.3\ \angle 45°\ \Omega$$

EXAMPLE 12-6 In a series circuit $R = 1\ \text{k}\Omega$ and $X_L = 10\ \text{k}\Omega$. Find $Z\angle\theta$.

Solution

$$Z = \sqrt{1^2 + 10^2} = \sqrt{1 + 100} = \sqrt{101} = 10.05\ \text{k}\Omega$$

$$\theta = \tan^{-1}\frac{10}{1} = \tan^{-1} 10 = 84.3°$$

$$Z\angle\theta = 10.05\ \angle 84.3°\ \text{k}\Omega$$

EXAMPLE 12-7 In a series circuit $R = 1.8$ MΩ and $X_L = 200$ kΩ. Find $Z\angle\theta$.

Solution Be careful. Different prefixes are used. It is best to immediately convert and use only one prefix. Let $R = 1800$ kΩ.

$$Z = \sqrt{1800^2 + 200^2} = \sqrt{32,400,000 + 40,000}$$
$$= 1811 \text{ k}\Omega$$
$$\theta = \tan^{-1}\frac{200}{1800} = 6.34°$$
$$Z\angle\theta = 1811 \angle 6.34° \text{ k}\Omega$$

In Examples 12-6 and 12-7 we see that the hypotenuse is nearly equal to the longer side when one side is much greater than the other. Also, when X_L is much greater than R, the angle approaches 90°. However, when R is much larger than X_L, the angle approaches 0°. In Example 12-5, $R = X_L$, tan $\theta = 1$, and $\theta = 45°$.

If we combine Ohm's law with our concept of impedance, it will be possible to analyze circuits in which sufficient information in various forms is given. These analyses will be demonstrated using examples.

EXAMPLE 12-8 A current of 3 A flows through a series circuit containing 4 Ω of resistance and 5 Ω of inductive reactance. Find the voltages in the circuit.

Solution Since the given current is not specified as a peak value, we must assume that effective or rms current is intended. Using Ohm's law, we obtain

$$V_R = IR = 3 \times 4 = 12 \text{ V}$$
$$V_L = IX_L = 3 \times 5 = 15 \text{ V} \qquad (12\text{-}1\text{-}1)$$
$$V_T = \sqrt{V_R^2 + V_L^2}$$
$$= \sqrt{12^2 + 15^2} = 19.2 \text{ V}$$

Alternatively,

$$Z = \sqrt{R^2 + X_L^2} \qquad (12\text{-}1\text{-}3)$$
$$= \sqrt{4^2 + 5^2} = 6.40 \text{ }\Omega$$
$$V_T = IZ = 3 \times 6.40 = 19.2 \text{ V}$$

EXAMPLE 12-9 In Example 12-8, what are the phase relations for the electrical current and voltages?

Solution The impedance can be expressed in the form $Z\angle\theta$, where

$$\theta = \tan^{-1}\frac{X_L}{R} \qquad (12\text{-}1\text{-}4)$$

$$= \tan^{-1}\frac{5}{4} = 51.3°$$

Therefore, I lags V_T by 51.3°, V_L will lead I by 90°, and V_R is in phase with I.

EXAMPLE 12-10 The voltage across the 15-kΩ resistor in a series circuit is 25 V. Determine all voltages in the circuit for $X_L = 10$ kΩ.

Solution

$$Z = \sqrt{R^2 + X_L^2} \qquad (12\text{-}1\text{-}3)$$
$$= \sqrt{15^2 + 10^2} = 18.0 \text{ k}\Omega$$

By Ohm's law,

$$I = \frac{V}{R} = \frac{25}{15} = 1.67 \text{ mA}$$

Applying Ohm's law again, we obtain

$$V_T = IZ = 1.67 \times 18.0 = 30.0 \text{ V}$$
$$V_L = IX_L = 1.67 \times 10 = 16.7 \text{ V}$$

In the preceding examples the voltage and current peak values were used. The same laws and formulas apply when effective or rms current and voltage values are specified or calculated. Normally, sine-wave voltage and current values are stated as effective values unless deliberately labeled as peak or peak-to-peak. The trick, of course, is never to mix effective values, peak values, or peak-to-peak values in the same calculation. Since the overwhelming majority of voltage and current statements are of one kind and seldom mixed, the dangers we are warning of are somewhat exaggerated. Reasonable caution will see you through.

12-2

Series Circuit with Resistance and Capacitive Reactance

In a circuit containing a capacitor and a voltage source, the phase relationship is as summarized in Fig. 12-5. Note that the current in the phasor diagram is oriented to point to the right and is considered to be the reference. Current I leads V_C by 90°. Both current and voltage peak values are represented by the individual phasors in this diagram.

A series RC circuit appears in Fig. 12-6. In it V_R is the resistor voltage, V_C is the capacitor voltage, and V_T is the total (source) voltage. Current I is

FIGURE 12-5 Phasor diagram for a capacitor.

FIGURE 12-6 Series RC circuit.

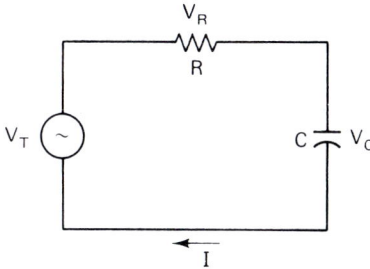

FIGURE 12-7 Phasor diagram for series RC circuit.

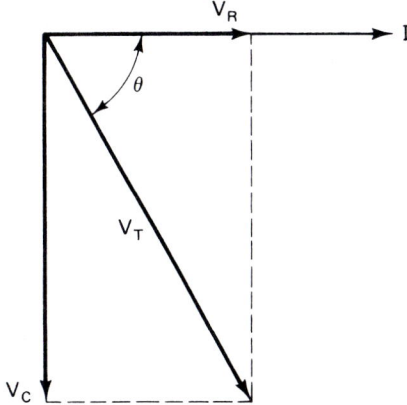

common to R, C, and the voltage source. Figure 12-7 is the phasor diagram for the series circuit in which I has been taken as the reference. Of course, I leads V_C by 90° and V_R must be in phase with I.

This phasor diagram emphasizes that the current leads the applied voltage by an angle which is less than 90°.

The angle between the total circuit voltage, V_T, and current, I, appears in Fig. 12-7. It is labeled θ. This angle is considered to be negative when the current leads the voltage.

When V_R and V_C are known, or can be found from the given data or measurements, V_T and θ are easily calculated.

$$V_T = \sqrt{V_R^2 + V_C^2} \qquad (12\text{-}2\text{-}1)$$

$$\theta = -\tan^{-1}\frac{V_C}{V_R} \qquad (12\text{-}2\text{-}2)$$

EXAMPLE 12-11 In a series RC circuit the peak voltage across the resistor is 25 V and the peak voltage across the capacitive reactance is 50 V. Find V_T and θ.

Solution

$$V_T = \sqrt{V_R^2 + V_C^2} \quad (12\text{-}2\text{-}1)$$

$$= \sqrt{25^2 + 50^2} = 55.9 \text{ V} \quad \text{(peak)}$$

$$\theta = -\tan^{-1} \frac{V_C}{V_R} \quad (12\text{-}2\text{-}2)$$

$$= -\tan^{-1} \frac{50}{25} = -63.4°$$

Using our shorthand notation, the answer is

$$55.9 \angle -63.4° \text{ V} \quad \text{(peak)}$$

which says that V_T has a peak value of 55.9 V and lags I by 63.4°. The angle θ is expressed as a negative quantity in Eq. (12-2-2) to indicate that the total voltage lags the current.

EXAMPLE 12-12 In a series circuit the peak sine-wave current is 5 A with $R = 23 \, \Omega$ and $X_C = 15 \, \Omega$. Find V_T and θ.

Solution It is first necessary to find V_R and V_C before completing the solution.

$$V_R = IR = 5 \times 23 = 115 \text{ V} \quad \text{(peak)}$$
$$V_C = IX_C = 5 \times 15 = 75 \text{ V} \quad \text{(peak)}$$

Use Eqs. (12-2-1) and (12-2-2).

$$V_T = \sqrt{V_R^2 + V_C^2} = \sqrt{115^2 + 75^2} = 137 \text{ V} \quad \text{(peak)}$$

$$\theta = -\tan \frac{V_C}{V_R} = -\tan \frac{75}{115} = -33.1°$$

The total voltage complete with angle is

$$137 \angle -33.1° \text{ V} \quad \text{(peak)}$$

In some types of analyses of RC series circuits an impedance must be found directly from R and X_C. To do this an impedance diagram is useful. An impedance diagram for a series RC circuit is shown in Fig. 12-8. It is quite similar in appearance to the impedance diagram for the RL circuit. The dif-

FIGURE 12-8 Impedance diagram for RC circuit.

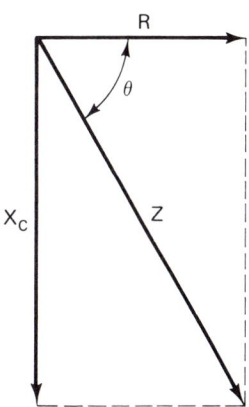

ference between the two is a downward direction for X_C and an upward direction for X_L. For the RC circuit

$$Z = \sqrt{R^2 + X_C^2} \qquad (12\text{-}2\text{-}3)$$

$$\theta = -\tan \frac{X_C}{R} \qquad (12\text{-}2\text{-}4)$$

In some cases the negative sign for the angle can be disregarded. This negative sign means that the circuit current leads the applied voltage. A positive sign found for a series RL circuit means that the current lags the applied voltage. Most people find it easier to remember that current leads in a capacitive circuit and lags in an inductive circuit than to remember the sign of θ.

In the examples that follow, the series resistance and capacitive reactance are combined to obtain the magnitude of the impedance as given by Eq. (12-2-3) and the angle provided by Eq. (12-2-4). Before plunging into these calculations, it might be well to reconsider the idea of impedance. Impedance is the total opposition to current flow in an ac circuit. This opposition exists for any series, parallel, and series-parallel combination of resistance and reactance. Of course, the impedance value (ohms) depends on the values of the resistances and reactances. It also depends on which kind of reactance—inductive or capacitive—is present. Moreover, the impedance depends on the manner in which the resistances and reactances are connected. At this particular point we are interested only in a single resistor connected in series with a single capacitor. Other configurations are considered later.

EXAMPLE 12-13 A series circuit consists of 50 kΩ of resistance in series with 30 kΩ of capacitive reactance. Find Z and the angle of current lead.

Solution Substitute into Eqs. (12-2-3) and (12-2-4).

$$Z = \sqrt{R^2 + X_C^2} = \sqrt{50^2 + 30^2} = 58.3 \text{ k}\Omega$$

$$\theta = -\tan \frac{X_C}{R} = -\tan^{-1} \frac{30}{50} = -31.0°$$

The current leads the applied voltage by 31.0°.

EXAMPLE 12-14 A 4.7-kΩ resistor is placed in series with a 0.0265-μF capacitor at 400 Hz. Find Z and the angle of current lead.

Solution Use Eqs. (11-5-5), (12-2-3), and (12-2-4).

$$X_C = \frac{1}{2\pi fC} = \frac{1}{2\pi \times 400 \times 2.65 \times 10^{-8}}$$
$$= 1.50 \times 10^4 \, \Omega = 15 \text{ k}\Omega$$

$$Z = \sqrt{R^2 + X_C^2} = \sqrt{4.7^2 + 15^2} = 15.7 \text{ k}\Omega$$

$$\theta = -\tan^{-1}\frac{X_C}{R} = -\tan\frac{15}{4.7} = -72.6°$$

The current leads the voltage by 72.6°.

The current in a series ac circuit can be found knowing the source voltage when there is sufficient information to determine the impedance. Both the magnitude of the impedance and the associated angle are necessary to find the current and the phase difference. Knowing Z and angle θ, we have a version of Ohm's law,

$$I = \frac{V_T}{Z} \quad (12\text{-}2\text{-}5)$$

and I will lead V_T by the amount of θ, without regard for sign, in an RC circuit. If a negative sign is retained for θ, this is the indication that the current is leading.

EXAMPLE 12-15 A 75 Ω resistor is connected in series with a 50-Ω capacitive reactance and a 250-V (peak) source. Calculate the current and the phase angle.

Solution Use Eqs. (12-2-3), (12-2-4), and (12-2-5).

$$Z = \sqrt{75^2 + 50^2} = 90.1 \, \Omega$$

$$\theta = -\tan^{-1}\frac{X_C}{R} = -\tan^{-1}\frac{50}{75} = -33.7°$$

$$I = \frac{V_T}{Z} = \frac{250}{90.1} = 2.77 \text{ A} \quad (\text{peak})$$

This current leads the voltage by 33.7°.

EXAMPLE 12-16 When $V_T = 50$ V (peak) and a leading current of 2 mA (peak) flows, what is the impedance and what kind of circuit elements will be found in the series circuit?

Solution

$$I = \frac{V_T}{Z} \tag{12-2-5}$$

$$Z = \frac{V_T}{I} = \frac{50}{2} = 25 \text{ k}\Omega$$

Since the current is leading, X_C must be connected in series. We do not have enough information to find the values of R and X_C.

EXAMPLE 12-17 A current of 15 mA (peak) passes through $R = 3$ kΩ and $X_C = 7$ kΩ connected in series. Calculate the total applied voltage and state the phase relationship between the total applied voltage and the current.

Solution

$$Z = \sqrt{R^2 + X_C^2} = \sqrt{3^2 + 7^2} = 7.62 \text{ k}\Omega \tag{12-2-3}$$

$$\theta = -\tan^{-1}\frac{X_C}{R} = -\tan^{-1}\frac{7}{3} = -66.8°$$

$$V_T = IZ \tag{12-2-5}$$

$$= 15 \times 7.62 = 114.2 \text{ V} \quad \text{(peak)}$$

The current will lead the total voltage by 66.8° since the circuit is capacitive.

12-3 Parallel RL and RC Circuits

Series circuits can be identified by having two or more elements connected end to end. With this connection arrangement the current in each element must be the same. If a resistor is connected in series with an inductor, the same current must pass through the resistor as well as the inductor. In Fig. 12-9 we have a series RL circuit. Ammeters A_1 and A_2 will indicate the same effective currents. Moreover, if the currents are observed using an oscilloscope, the current in A_1 will be seen to be in phase with the current in A_2. A third current meter A_3 will indicate that the current is the same as that measured by A_1 and

FIGURE 12-9 Series RL circuit.

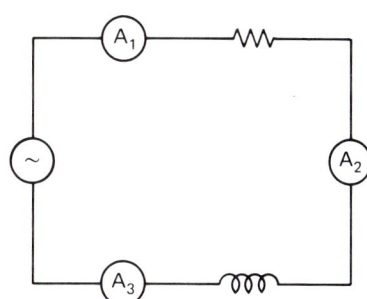

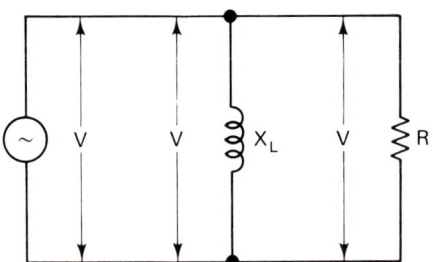

FIGURE 12-10 Parallel *RL* circuit.

A_2. The common current is the unique characteristic of a series circuit. Only a series circuit has the same electrical charges flowing through each element.

Parallel circuits also have a particular feature. Each element in a parallel circuit has the same voltage. As seen in Fig. 12-10, voltage *V* appears across the voltage source and also across *R* and across *L*. In the laboratory we connect one terminal of the source, one terminal of the resistor, and one terminal of the inductor together. Then the remaining free terminals are connected together. That is all there is to it.

Let us now look into the analysis of parallel circuits. You will be surprised to find out that no new information is needed to do this. We only need to know how Ohm's law applies to a single resistor, a single inductor, and a single capacitor. Kirchhoff's current law will also be used. If you are a bit hazy about the current law, it would be wise to go back and spend a few minutes reviewing Kirchhoff's current law as applied to dc circuits. As we proceed with parallel circuits, the same law applies. However, when the currents flowing to the point (node) are summed (added), they must be combined by using a phasor diagram. They cannot be added arithmetically or even algebraically. They must be added as phasors with proper consideration of the angles. Let us see how this works. We will first run through the steps of finding the source

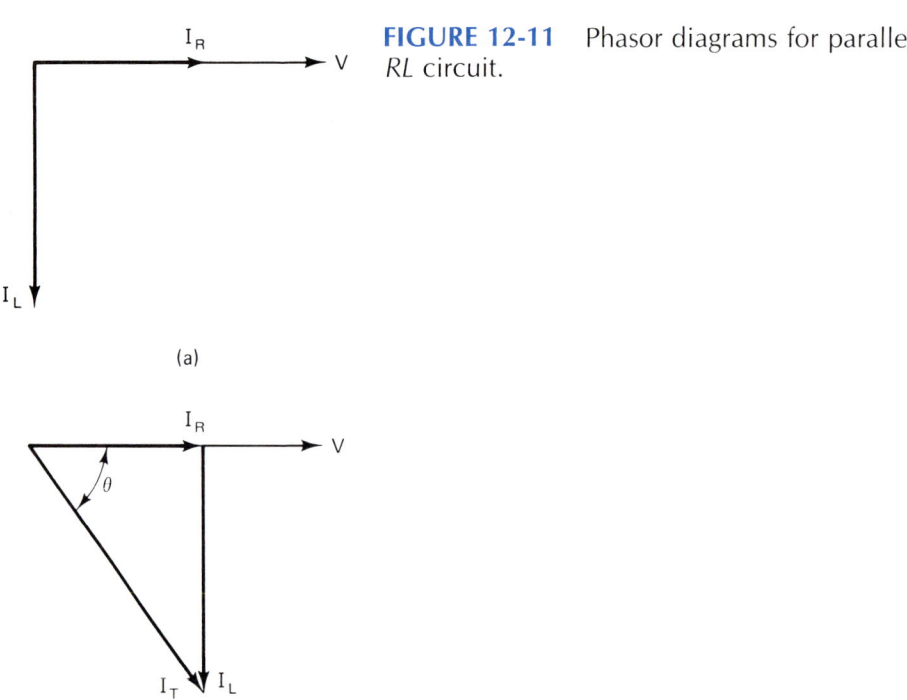

FIGURE 12-11 Phasor diagrams for parallel *RL* circuit.

494 Series and Parallel AC Circuits Ch. 12

current when V, R, and X_L are known (see Fig. 12-10). We will then look at a few examples using numerical values for V, R, and X_L.

When V, R, and X_L are known, Ohm's law is first applied.

$$I_R = \frac{V}{R} \qquad (12\text{-}3\text{-}1)$$

$$I_L = \frac{V}{X_L} \qquad (12\text{-}3\text{-}2)$$

Next, the quantities are shown on a phasor diagram [Fig. 12-11(a)]. Remember that I_R is in phase with V, here taken as the reference. I_L lags V by 90°.

In Fig. 12-11(b), I_L is also shown in the position which allows the two currents to be combined, giving I_T. With I_R and I_L at right angles (90° angle difference), the total current, I_T, is the hypotenuse of a right triangle, so that

$$I_T = \sqrt{I_R^2 + I_L^2} \qquad (12\text{-}3\text{-}3)$$

To find the phase angle, θ, use

$$\theta = \tan^{-1} \frac{I_L}{I_R} \qquad (12\text{-}3\text{-}4)$$

Notice that θ is the angle between V and the lagging current I_L and is considered to be positive for an inductive circuit.

EXAMPLE 12-18 In a parallel circuit the source voltage is 120 V (effective), the resistance branch has 40 Ω, and the inductive reactance branch has 30 Ω. Find the total current and the phase angle.

Solution Apply Eqs. (12-3-1) and (12-3-2).

$$I_R = \frac{120}{40} = 3 \text{ A}$$

$$I_L = \frac{120}{30} = 4 \text{ A}$$

Substitute into Eqs. (12-3-3) and (12-3-4).

$$I_T = \sqrt{I_R^2 + I_L^2} = \sqrt{3^2 + 4^2} = 5 \text{ A}$$

$$\theta = \tan^{-1} \frac{I_L}{I_R} = \tan^{-1} \frac{4}{3} = 53.1°$$

The current is 5 A (effective) and it lags the voltage by 53.1°. Expressed another way,

$$V = 120 \angle 0° \text{ V} \quad \text{(effective)}$$

$$I = 5 \angle -53.1° \text{ A} \quad \text{(effective)}$$

Sec. 12-3 **Parallel *RL* and *RC* Circuits**

EXAMPLE 12-19 Find I_T for the circuit of Fig. 12-12.

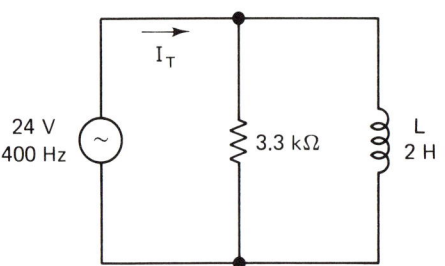

FIGURE 12-12 Circuit for Example 12-19.

Solution Wait a minute! Do we have all the necessary information? In comparing this problem with Example 12-18, it seems that X_L is missing. The total current, I_T, cannot be found until we can obtain both I_R and I_L. The current through the resistance branch is easy.

$$I_R = \frac{V}{R} = \frac{24}{3.3} = 7.27 \text{ mA} \quad \text{(effective)}$$

We are now stuck unless X_L is given. But wait! Even though X_L does not appear in Fig. 12-12, we do find that $L = 2$ H and $f = 400$ Hz. This is all we need to find X_L and then to proceed with the solution.

$$X_L = 2\pi f L = 2\pi \times 400 \times 2 = 5027 \text{ }\Omega$$

$$I_L = \frac{V}{X_L} = \frac{24}{5027} = 0.00477 \text{ A} = 4.77 \text{ mA}$$

$$I_T = \sqrt{I_R^2 + I_L^2} = \sqrt{7.27^2 + 4.77^2} = 8.70 \text{ mA}$$

$$\theta = \tan^{-1}\frac{I_L}{I_R} = \tan^{-1}\frac{4.77}{8.70} = 33.3°$$

The total current, I_T, lags the supply by 33.3°. When the supply voltage is expressed as $24\angle 0°$V (effective), the total current can be written as $8.70\angle -33.3°$ mA (effective).

Figure 12-13 is the schematic diagram for a parallel RC circuit. It is very much like the RL circuit we have been studying. Both the parallel RC and

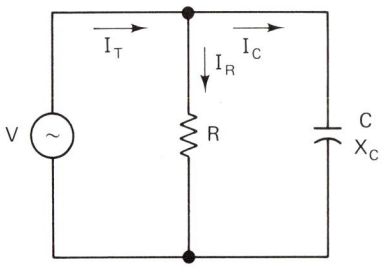

FIGURE 12-13 Parallel RC circuit.

RL circuits have a voltage source and one resistive branch. However, the reactive branch of the parallel RC circuit has a capacitor.

The analysis of an RC circuit is very similar to that of the RL circuit we have been solving. Suppose that we know the supply voltage, V, the resistance, R, and the capacitive reactance, X_C, in Fig. 12-13. This information will allow the branch currents to be calculated using Ohm's law.

$$I_R = \frac{V}{R}$$

$$I_C = \frac{V}{X_C}$$

These currents can now be combined, but not added directly, to find I_T. Kirchhoff's current law still applies. The method of applying it requires the use of a right triangle. The right triangle we need appears in Fig. 12-14(a). In it, the common voltage, V, is the reference. The current passing through the resistive branch, I_R, is in phase with V. Note that the current I_C leads V by 90°. This is because V is not only the supply voltage but is also the voltage across the capacitive branch.

By rearranging the phasors in the phasor diagram as shown in Fig. 12-14(b), the method of combining the two currents is shown. It is now recognized as the hypotenuse of the right triangle whose sides are I_R and I_C. Therefore,

$$I_T = \sqrt{I_R^2 + I_C^2} \qquad (12\text{-}3\text{-}5)$$

Moreover, the phase angle, θ, can be found.

$$\theta = -\tan^{-1} \frac{I_C}{I_R} \qquad (12\text{-}3\text{-}6)$$

The negative value for θ indicates that I_T leads V. When I_C is large compared to I_R, the angle approaches $-90°$. When I_C is small compared to I_R, the angle is small. However, I_T always leads V in a parallel RC circuit.

FIGURE 12-14 Phasor diagram for parallel RC circuit.

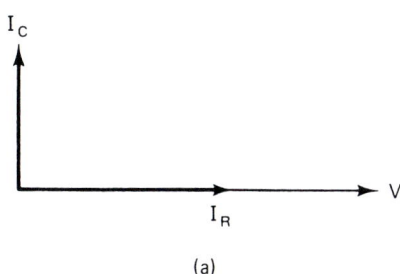

(a)

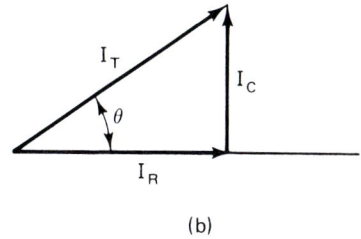

(b)

Sec. 12-3 Parallel RL and RC Circuits

EXAMPLE 12-20 Two kilohms of resistance are connected in parallel with 3 kΩ of capacitive reactance across a 75-V source. Calculate the total current and the phase angle.

Solution First apply Ohm's law for each branch.

$$I_R = \frac{V}{R} = \frac{75}{2} = 37.5 \text{ mA}$$

$$I_C = \frac{V}{X_C} = \frac{75}{3} = 25 \text{ mA}$$

Then combine the two currents and determine the phase angle. Apply Eqs. (12-3-5) and (12-3-6).

$$I_T = \sqrt{I_R^2 + I_C^2} = \sqrt{37.5^2 + 25^2}$$
$$= 45.1 \text{ mA}$$

$$\theta = -\tan^{-1}\frac{I_C}{I_R} = -\tan^{-1}\frac{25}{37.5} = -33.7°$$

The total current, supply current, or source current leads the voltage by 33.7°. The voltage can be written as 75∠0° V and the currents as follows:

$$I_R = 37.5 \angle 0° \text{ mA}$$
$$I_C = 25 \angle 90° \text{ mA}$$
$$I_T = 45.1 \angle 33.7° \text{ mA}$$

EXAMPLE 12-21 Find the total current and the phase angle for the circuit of Fig. 12-15.

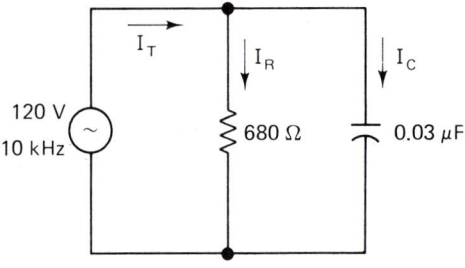

FIGURE 12-15 Circuit for Example 12-21.

Solution In Example 12-20, the first step consisted of calculating the branch currents using V, R, and X_C. But we do not have X_C given. It will be necessary to find X_C first using f and C.

$$X_C = \frac{1}{2\pi fC} = \frac{1}{2\pi \times 10^4 \times 0.03 \times 10^{-6}}$$
$$= 531 \text{ Ω}$$

Now, we can find both branch currents.

$$I_R = \frac{V}{R} = \frac{120}{680} = 0.176 \text{ A}$$

$$I_C = \frac{V}{X_C} = \frac{120}{531} = 0.226 \text{ A}$$

$$I_T = \sqrt{I_R^2 + I_C^2} = \sqrt{0.176^2 + 0.226^2} = 0.287 \text{ A}$$

$$\theta = -\tan^{-1}\frac{I_C}{I_R} = -\tan^{-1}\frac{0.226}{0.176} = -52.0°$$

The total current is 0.287 A and it leads the circuit voltage by 52.0°. We can also say that

$$V = 120\angle 0° \text{ V}$$
$$I_R = 0.176\angle 0° \text{ A}$$
$$I_C = 0.226\angle 90° \text{ A}$$
$$I_T = 0.287\angle 52.0° \text{ A}$$

We now move on to the total impedance of a parallel circuit. We recall that the total resistance of parallel dc circuits can be calculated using reciprocals. The same general rules can be used with parallel ac circuits. We wish to avoid that procedure now. The mathematics is too difficult to manage with the techniques already learned.

There is another simple approach to finding the total circuit impedance of parallel ac circuits. It requires the calculation of the total current. This procedure is exactly like the one we used for the parallel *RC* circuit and the parallel *RL* circuit.

After the total current is calculated, it is then possible to find the total impedance using Ohm's law. So you see that we have already learned the more difficult part of finding the total parallel circuit impedance. The next part is easy. We find the impedance using the known values of V and I_T.

$$Z = \frac{V}{I_T}$$

Also, the angle of Z is the phase angle, being positive for an inductive circuit and negative for a capacitive circuit.

EXAMPLE 12-22 A 22-kΩ resistor is connected in parallel with 15 kΩ of inductive reactance in a circuit with a 12-V voltage source. Determine the impedance.

Solution First find the individual branch currents and the total current.

$$I_R = \frac{V}{R} = \frac{12}{22} = 0.545 \text{ mA}$$

$$I_L = \frac{V}{X_L} = \frac{12}{15} = 0.800 \text{ mA}$$

$$I_T = \sqrt{I_R^2 + I_L^2} = \sqrt{0.545^2 + 0.8^2} = 0.968 \text{ mA}$$

$$\theta = \tan^{-1}\frac{I_L}{I_R} = \tan^{-1}\frac{0.800}{0.545} = 55.7°$$

Next, apply Ohm's law to find Z.

$$Z = \frac{V}{I_T} = \frac{12}{0.968} = 12.4 \text{ k}\Omega$$

Since the phase angle is 55.7°, the impedance can be written as $12.4\underline{/55.7°}$ kΩ.

Alternatively, the values of the impedance Z_T and the angle θ may be found from the formulas

$$Z_T = \frac{R \times X_L}{\sqrt{R^2 + X_L^2}} \quad \Omega$$

and

$$\theta = \tan^{-1}\frac{R}{X_L} \quad \left(not \; \tan^{-1}\frac{X_L}{R}\right)$$

The expression for Z_T looks very similar to the "product-over-sum" formula for two resistors in parallel [Eq. (3-3-3)]. However, the denominator in the Z_T formula is involved with the "Pythagorean" sum as opposed to simple addition. The calculator sequences are as follows: To find the value of Z_T:

Operation				Display	
22				22	
15				330	($R \times X_L$)
[(]	[(]	22	[x^2]	484	(R^2)
[+]		[x^2]	[)]	709	($R^2 + X^2$)
[$\sqrt{x}$]	[)]			26.63	($\sqrt{R^2 + X^2}$)
[=]				12.39	(Z_T)

The value of Z_T is 12.4 kΩ, rounded off.

To find the value of θ:

Operation		Display	
22	$\div$	22	(R)
15	$=$	1.467	$\left(\dfrac{R}{X_L}\right)$
	INV TAN	55.7°	(θ)

The value of θ is $+55.7°$.

EXAMPLE 12-23 A 48-V ac voltage is connected in parallel with 2200 Ω of resistance and 5700 Ω of capacitive reactance. What is the impedance of the circuit?

Solution

$$I_R = \frac{V}{R} = \frac{48}{2.2} = 21.8 \text{ mA}$$

$$I_C = \frac{V}{X_C} = \frac{48}{5.7} = 8.42 \text{ mA}$$

$$I_T = \sqrt{I_R^2 + I_C^2} = \sqrt{21.8^2 + 8.42^2} = 23.4 \text{ mA}$$

$$Z = \frac{V}{I} = \frac{48}{23.4} = 2.05 \text{ k}\Omega$$

$$\theta = -\tan^{-1}\frac{I_C}{I_R} = -\tan\frac{8.42}{21.8} = -21.1°$$

The current *leads* the supply voltage by 21.1° because the circuit is capacitive.

A convenient way of expressing the phase angle of the circuit is to include the angle with the magnitude of the impedance.

$$2.05\angle{-21.1°} \text{ k}\Omega$$

12-4
RLC Circuits

We will now consider a series circuit containing X_L and X_C. Once again the basic electrical laws for dc are good, but it is necessary to use slightly different methods in their application. Suppose that we have a series circuit consisting of X_L and X_C for which the current is known. It is now possible to find the supply voltage V_T. To do this we must first find V_L and V_C. When properly combined, V_L and V_C will give us V_T. Using Ohm's law, we have

$$V_L = IX_L$$
$$V_C = IX_C$$

These voltages and the current are shown in Fig. 12-16. Note that I lags V_L by 90° and leads V_C by 90°. V_L and V_C lie along a vertical line but point in opposite directions. This means that they can be combined by simply taking the difference. And this difference is V_T.

$$V_T = V_L - V_C$$

Since V_L is larger than V_C, V_T will be in the same direction as V_L and will lead I by 90°. We say that the circuit is inductive.

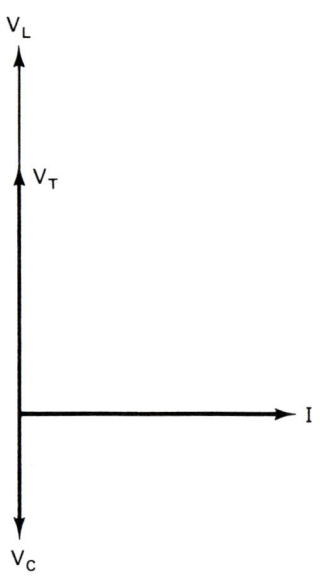

FIGURE 12-16 Phasor diagram for LC circuit.

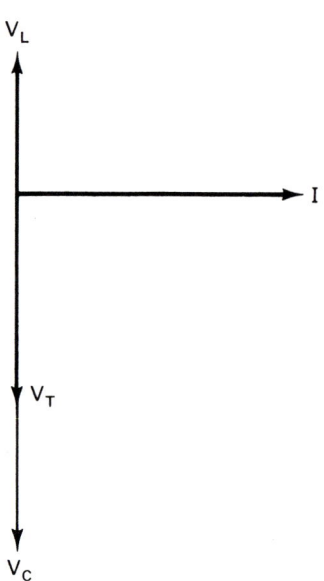

FIGURE 12-17 Phasor diagram for LC circuit.

If V_C should be larger than V_L, as seen in Fig. 12-17, the difference still gives us V_T.

$$V_T = V_C - V_L$$

In this case V_T is in phase with the larger of the two voltages. V_T now lags I by 90°. The circuit is capacitive.

EXAMPLE 12-24 In Fig. 12-16, $V_L = 150$ V and $V_C = 80$ V. What is V_T?

Solution First notice that V_L is larger than V_C.

$$V_T = V_L - V_C = 150 - 80 = 70 \text{ V}$$

and V_T leads I by 90°. There appears to be something wrong with the circuit. Both V_L and V_C are larger than V_T. This cannot happen in a dc or ac circuit containing only resistors. However, in a series ac circuit containing both X_L and X_C, V_L and V_C tend to cancel each other, so that only the difference survives as the total. However, if you place a voltmeter in the circuit, it will indicate 150 V across L, 80 V across C, and 70 V for V_T.

EXAMPLE 12-25 In Fig. 12-17, $V_L = 25$ V and $V_C = 80$ V. Find V_T.

Solution Since V_C is greater than V_L, the difference is

$$V_T = V_C - V_L = 80 - 25 = 55 \text{ V}$$

and the phasor V_T is in phase with V_C and lags I by 90°.

It is now possible to show how the circuit impedance can be found for X_L in series with X_C. For X_L greater than X_C,

$$V_T = V_L - V_C$$

Written another way,

$$IZ = IX_L - IX_C$$

so that

$$Z = X_L - X_C \qquad (12\text{-}4\text{-}1)$$

and the angle associated with Z is 90° because the circuit is inductive.
    For X_C greater than X_L,

$$V_T = V_C - V_L$$

and

$$IZ = IX_C - IX_L$$

Sec. 12-4 *RLC Circuits*

so that

$$Z = X_C - X_L \quad (12\text{-}4\text{-}2)$$

and the angle for Z is $-90°$ because the circuit is capacitive.

EXAMPLE 12-26 A series circuit consists of 15 kΩ of inductive reactance and 10 kΩ of capacitive reactance. Find the value of Z and its angle.

Solution Note that X_L is larger than X_C. Therefore,

$$Z = X_L - X_C = 15 - 10 = 5 \text{ k}\Omega$$

The angle for Z is 90°.

EXAMPLE 12-27 In a series circuit there is 900 Ω of capacitive reactance and 200 Ω of inductive reactance. Find Z and its angle.

Solution We first find the net reactance, which is the value of Z.

$$Z = X_C - X_L = 900 - 200 = 700 \text{ }\Omega$$

Since X_C is the larger, the circuit is capacitive and the angle is $-90°$.

In addition to having inductive reactance and capacitive reactance, a series circuit can also have resistance [see Fig. 12-18(a)]. The resistor and reactances can be rearranged as shown in Fig. 12-18(b) or (c) without changing the total circuit impedance.

Suppose that R, X_L, and X_C are known and we wish to find the magnitude of Z and the angle. The method of finding the magnitude of Z consists of first combining the two reactances by taking the difference. That is, for X_L greater than X_C,

$$X_T = X_L - X_C \quad (12\text{-}4\text{-}3)$$

For X_C greater than X_L,

$$X_T = X_C - X_L \quad (12\text{-}4\text{-}4)$$

Notice that the reactance found by taking the difference is not called Z, but rather is written as X_T for total reactance or net reactance. This is because the Z depends also on the value of R in the circuit. A second step is now necessary to combine R and X_T. This step is exactly like the one we have used before when combining R and X_L or R and X_C for a single reactance in

FIGURE 12-18 Series *RLC* circuits.

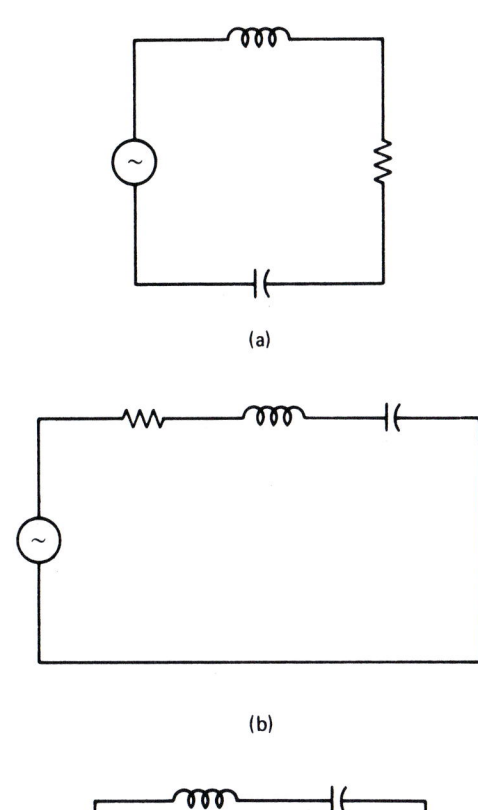

the circuit. With both X_L and X_C in the circuit, R must be combined with X_T using the same method as before.

$$Z = \sqrt{R^2 + X_T^2} \qquad (12\text{-}4\text{-}5)$$

Moreover, the angle can also be found in the same way that was learned previously. If X_T is inductive,

$$\theta = \tan^{-1} \frac{X_T}{R} \qquad (12\text{-}4\text{-}6)$$

If X_T is capacitive,

$$\theta = -\tan^{-1} \frac{X_T}{R} \qquad (12\text{-}4\text{-}7)$$

Sec. 12-4 *RLC* Circuits

EXAMPLE 12-28 A series circuit consists of $R = 10\ \Omega$, $X_L = 75\ \Omega$, and $X_C = 60\ \Omega$. Find the magnitude of Z and the angle.

Solution We first note that X_L is greater than X_C, so that

$$X_T = X_L - X_C \qquad (12\text{-}4\text{-}3)$$
$$= 75 - 60 = 15\ \Omega$$
$$Z = \sqrt{R^2 + X_T^2} \qquad (12\text{-}4\text{-}5)$$
$$= \sqrt{10^2 + 15^2} = 18.0\ \Omega$$
$$\theta = \tan^{-1}\frac{X_T}{R} \qquad (12\text{-}4\text{-}6)$$
$$= \tan^{-1}\frac{15}{10} = 56.3°$$

If a voltage is applied to this circuit, the current will lag the voltage by 56.3°. The circuit behaves like one consisting of 10 Ω of resistance in series with 15 Ω of inductive reactance. The 60 Ω of capacitive reactance has been balanced out by an equal 60 Ω of inductive reactance, leaving a net or total inductive reactance of 15 Ω.

EXAMPLE 12-29 A series circuit has 10 kΩ of resistance, 7 kΩ of capacitive reactance, and 4 kΩ of inductive reactance. Find the impedance and the angle of the impedance.

Solution We first notice that the capacitive reactance is greater than the inductive reactance. This means that

$$X_T = X_C - X_L \qquad (12\text{-}4\text{-}4)$$
$$= 7 - 4 = 3\ \text{k}\Omega$$

and also that

$$\theta = -\tan^{-1}\frac{X_T}{R} \qquad (12\text{-}4\text{-}7)$$
$$= -\tan\frac{3}{10} = -16.7°$$

We now need to find Z.

$$Z = \sqrt{R^2 + X_T^2} \qquad (12\text{-}4\text{-}5)$$
$$= \sqrt{10^2 + 3^2} = 10.4\ \text{k}\Omega$$

If this series circuit is connected across a voltage source, the current will lead the voltage by 16.7°.

Once the impedance magnitude and the angle are known, the current in a series RLC circuit can easily be determined. Moreover, the phase angle between the applied voltage and the current can be found using the angle of the impedance.

$$I = \frac{V_T}{Z}$$

$$\text{phase angle} = \theta = \tan^{-1}\frac{X_L - X_C}{R}$$

(12-4-8)

Remember that θ is positive for an inductive circuit and negative for a capacitive circuit. This means a lagging current for an inductive circuit. Similarly, a negative θ reminds us that the current is leading.

EXAMPLE 12-30 Referring to Fig. 12-18(a), $R = 1500\ \Omega$, $X_L = 3.5\ \text{k}\Omega$, and $X_C = 2.0\ \text{k}\Omega$. Find I and the phase angle when $V_T = 110\ \text{V}$ (effective).

Solution

$$X_T = X_L - X_C = 3.5 - 2.0 = 1.5\ \text{k}\Omega$$

$$Z = \sqrt{R^2 + X_T^2} = \sqrt{1.5^2 + 1.5^2} = 2.12\ \text{k}\Omega$$

$$I = \frac{V_T}{Z} = \frac{110}{2.12} = 51.9\ \text{mA}\ \text{(effective)}$$

$$\theta = \tan^{-1}\frac{X_T}{R} = \tan^{-1}\frac{1.5}{1.5} = 45.0°$$

Current I lags V_T by 45.0°.

After the current in a series circuit has been established, it is then possible to apply Ohm's law for each circuit element. By this means we are able to calculate the voltages across V_R, X_L, and X_C using

$$V_R = IR$$
$$V_L = IX_L$$
$$V_C = IX_C$$

A phasor diagram for a series RLC circuit is shown in Fig. 12-19(a). The current I is used as a reference in this phasor diagram. We see that V_R is in phase with I, V_L leads I by 90°, and V_C lags I by 90°. Furthermore,

$$V_T = \sqrt{V_R^2 + (V_L - V_C)^2}$$

(12-4-9)

and for the case shown,

$$\theta = \tan^{-1}\frac{V_L - V_C}{V_R}$$

(12-4-10)

Sec. 12-4 *RLC* Circuits

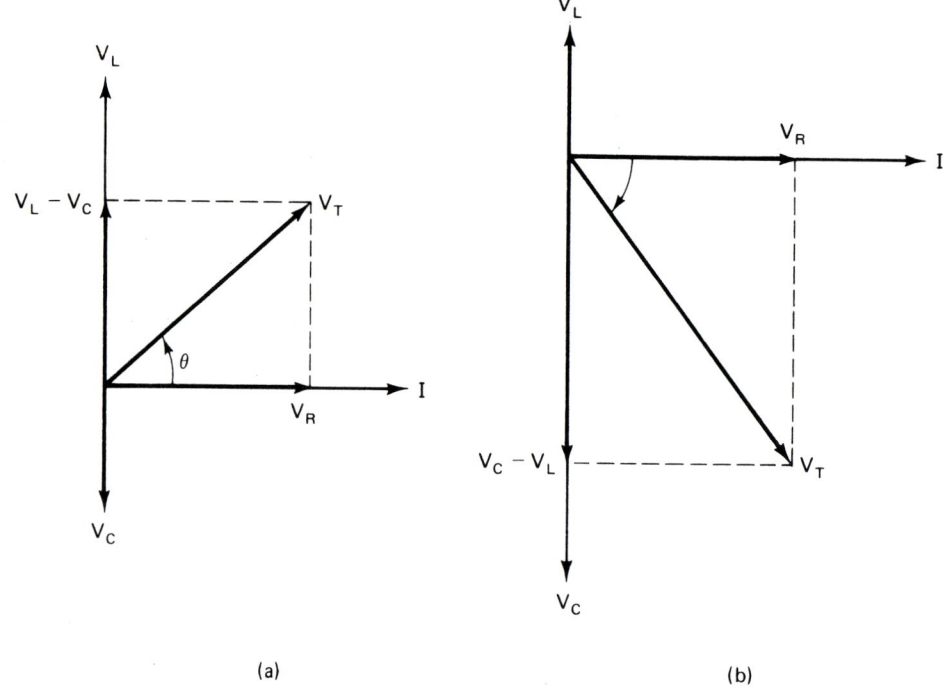

FIGURE 12-19 Phasor diagram for series *RLC* circuit.

because V_L is greater than V_C. (We also know that X_L is greater than X_C, but we do not mix an impedance diagram and a phasor diagram on the same figure.) Refer now to Fig. 12-19(b), in which V_C is greater than V_L. Therefore,

$$V_T = \sqrt{V_R^2 + (V_C - V_L)^2} \qquad (12\text{-}4\text{-}11)$$

$$\theta = -\tan^{-1}\frac{V_C - V_L}{V_R} \qquad (12\text{-}4\text{-}12)$$

EXAMPLE 12-31 In a series *RLC* circuit the resistance is 3.3 kΩ, the inductive reactance is 5.1 kΩ, and the capacitive reactance is 2.2 kΩ. Find the voltages V_R, V_L, V_C, and V_T when $I = 15$ mA.

Solution

$$V_R = IR = 15 \times 3.3 = 49.5 \text{ V}$$

$$V_L = IX_L = 15 \times 5.1 = 76.5 \text{ V}$$

$$V_C = IX_C = 15 \times 2.2 = 33.0 \text{ V}$$

$$V_T = \sqrt{V_R^2 + (V_L - V_C)^2} \qquad (12\text{-}4\text{-}9)$$

$$= \sqrt{49.5^2 + (76.5 - 33.0)^2} = 65.9 \text{ V}$$

$$\theta = \tan^{-1}\frac{V_L - V_C}{R} \qquad (12\text{-}4\text{-}10)$$

$$= \tan^{-1}\frac{76.5 - 33.0}{49.5} = 41.3°$$

EXAMPLE 12-32 In a series circuit $R = 560\ \Omega$, $X_L = 1500\ \Omega$, and $X_C = 1800\ \Omega$. (a) Find all voltages and the phase angle for a current of 200 mA. (b) First find the impedance of the circuit and then the total voltage and phase angle.

Solution

(a) $V_R = IR = 200 \times 0.56 = 112\ \text{V}$

$V_C = IX_C = 200 \times 1.8 = 360\ \text{V}$

$V_L = IX_L = 200 \times 1.5 = 300\ \text{V}$

$V_T = \sqrt{V_R^2 + (V_C - V_L)^2}$ (12-4-11)

$= \sqrt{112^2 + 60^2} = 127\ \text{V}$

$\theta = -\tan^{-1}\dfrac{V_C - V_L}{V_R}$ (12-4-12)

$= -\tan^{-1}\dfrac{60}{112} = -28.2°$

The current leads V_T by 28.2°.

(b) $Z = \sqrt{R^2 + (X_C - X_L)^2}$

$= \sqrt{560^2 + (1800 - 1500)^2} = 635\ \Omega$

$\theta = -\tan^{-1}\dfrac{X_C - X_L}{R}$

$= -\tan^{-1}\dfrac{1800 - 1500}{560} = -28.2°$

$V_T = IZ = 0.2 \times 635 = 127\ \text{V}$

We see that the same result was obtained for V_T and the angle with either approach. However, with the first, the individual voltages, which can be measured in the circuit, were calculated. With the second approach, only V_T was found.

When obtaining the values of Z and θ by using a scientific calculator, the procedures are as follows: To find the value of Z:

Operation						Display	
560	x^2	+				313600	(R^2)
	(	1800	−				
1500			)	x^2		90000	$(X_C - X_L)^2$
		=				403600	$R^2 + (X_C - X_L)^2$
		$\sqrt{x}$				635.3	(Z)

The value of Z is 635 Ω, rounded off.

To find the value of θ:

Operation		Display	
1800	−	1800	(X_C)
1500	÷	300	$(X_C - X_L)$
560	=	5.357	$\left(\dfrac{X_C - X_L}{R}\right)$
	INV TAN	28.18°	(θ)

The value of θ is $-28.2°$, rounded off.

In general, a series *RLC* circuit may have more than one resistor, more than one inductive reactance, and more than one capacitive reactance. Finding the impedance for the entire circuit is almost exactly the same as for one resistor, one inductive reactance, and one capacitive reactance in series. The method differs in one respect only. First, the individual resistors must be combined by adding their resistance values directly. Also, the inductive reactances can be combined by adding them directly and the capacitive reactances combined by adding them directly. From that point the individual combined resistances and reactances are handled as though there is one of each. Consider the circuit of Fig. 12-20, containing two resistors, R_1 and R_2, three inductive reactances, X_{L1}, X_{L2}, and X_{L3}, and two capacitive reactances, X_{C1} and X_{C2}. There is no significance in the number of each kind of element. We merely want to have more than one of each but not necessarily two of each. First, we combine the elements of like kind.

$$R = R_1 + R_2$$
$$X_L = X_{L1} + X_{L2} + X_{L3}$$
$$X_C = X_{C1} + X_{C2}$$

From this point on X_L and X_C are combined by finding the difference using Eq. (12-4-3) or (12-4-4) and then Z is determined using Eq. (12-4-5). The angle is then calculated using Eq. (12-4-6) or (12-4-7).

Whenever the current in the series circuit is known, the individual voltages can be found using Ohm's law. More specifically, for the circuit of Fig. 12-20, the voltages across the individual circuit elements for a known current I are

$$V_{R1} = IR_1$$
$$V_{R2} = IR_2$$
$$V_{L1} = IX_{L1}$$
$$V_{L2} = IX_{L2}$$
$$V_{L3} = IX_{L3}$$
$$V_{C1} = IX_{C1}$$
$$V_{C2} = IX_{C2}$$

Note that V_{R1} and V_{R2} are in phase with current I. V_{L1}, V_{L2}, and V_{L3} lead current I by 90°. V_{C1} and V_{C2} lag I by 90°.

FIGURE 12-20 Series *RLC* circuit with multiple elements.

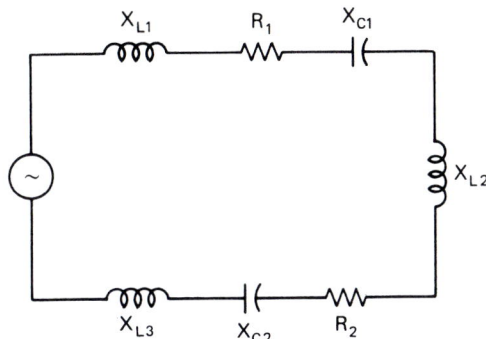

EXAMPLE 12-33 Refer to Fig. 12-20. $R_1 = 2.2$ kΩ, $R_2 = 3.3$ kΩ, $X_{L1} = 1.0$ kΩ, $X_{L2} = 1.5$ kΩ, $X_{L3} = 1.8$ kΩ, $X_{C1} = 2.5$ kΩ, and $X_{C2} = 5.8$ kΩ. Find Z for the circuit, I, and all the element voltages for $V_T = 220$ V.

Solution Since we have more pieces of information to deal with in this problem than in previous problems, it is wise to pause a moment to review the data. We should especially try to understand whether the data are complete or incomplete for each possible course of action. Very often the sequence of steps to be followed is the only sequence possible. This is true of the present problem. Let us reason together as follows:

1. The required voltages can be calculated using Ohm's law and requires the common current I and the individual resistance or reactance values. By reviewing the data we see that we have everything needed except I.
2. To find I we need V_T, which is given, and Z, which can be obtained. Since Z is also requested, we should first concentrate on finding Z and then apply Ohm's law to find I.
3. In order to calculate Z, R_1 and R_2 must be combined by adding R_1 and R_2 to get R, X_{L1}, X_{L2}, and X_{L3} combined by adding to get X_L, X_{C1}, and X_{C2} combined by adding to get X_C.
4. Then X_L and X_C are combined by taking the difference to obtain X_T. R and X_T are finally combined as phasors to get Z.
5. By applying Ohm's law, $I = V_T/Z$ and the I can be used again to obtain each resistor or reactor voltage.

Here we go.

$$R = R_1 + R_2 = 2.2 + 3.3 = 5.5 \text{ k}\Omega$$

$$X_L = X_{L1} + X_{L2} + X_{L3} = 1.0 + 1.5 + 1.8 = 4.3 \text{ k}\Omega$$

$$X_C = X_{C1} + X_{C2} = 2.5 + 5.8 = 8.3 \text{ k}\Omega$$

$$Z = \sqrt{R^2 + (X_C - X_L)^2}$$

$$= \sqrt{5.5^2 + 4.0^2} = 6.80 \text{ k}\Omega$$

$$\theta = -\tan^{-1} \frac{X_C - X_L}{R}$$

$$= -\tan^{-1} \frac{4.0}{5.5} = -36.0°$$

Sec. 12-4 *RLC* Circuits

Notice that the phase angle is not required in the solution.

$$I = \frac{V_T}{Z} = \frac{220}{6.80} = 32.35 \text{ mA}$$

$$V_{R1} = IR_1 = 32.35 \times 2.2 = 71.2 \text{ V}$$

$$V_{R2} = IR_2 = 32.35 \times 3.3 = 106.8 \text{ V}$$

$$V_{L1} = IX_{L1} = 32.35 \times 1.0 = 32.4 \text{ V}$$

$$V_{L2} = IX_{L2} = 32.35 \times 1.5 = 48.5 \text{ V}$$

$$V_{L3} = IX_{L3} = 32.35 \times 1.8 = 58.2 \text{ V}$$

$$V_{C1} = IX_{C1} = 32.35 \times 2.5 = 80.9 \text{ V}$$

$$V_{C2} = IX_{C2} = 32.35 \times 5.8 = 187.6 \text{ V}$$

This completes the solution.

This chapter concludes with a study of parallel circuits containing R, L, or C in any number of branches. We start with only two branches and add branches as the techniques for analysis are developed. In the first circuit one parallel branch contains X_L, a second parallel branch has X_C, and both branches are connected across a voltage source V. The source voltage, V, is found across both X_L and X_C because that is the meaning of a parallel circuit. Applying Ohm's law allows us to find each of the two reactive branch currents.

$$I_L = \frac{V}{X_L}$$

$$I_C = \frac{V}{X_C}$$

Assuming that X_L is smaller than X_C, then I_L is larger than I_C, as indicated in Fig. 12-21. In this phasor diagram the common voltage, V, is taken as the reference. Of course, I_C leads V by 90° and I_L lags V by 90°. It is equally correct to say that V lags I_C by 90° and leads I_L by 90°.

Since the currents of Fig. 12-21 are in opposite directions along the same line, they can be combined to give total circuit current in a very simple way. To combine the two purely reactive branch currents it is only necessary to subtract the smaller from the larger. In our particular case, I_L is larger than I_C. The difference is given by

$$I = I_L - I_C$$

and is in the direction of I_L.

In Fig. 12-22 is seen the phasor diagram, in which I_C is greater than I_L. The difference

$$I = I_C - I_L$$

is in phase with I_C, so that I leads V by 90°.

FIGURE 12-21 Phasor diagram for L and C parallel branches.

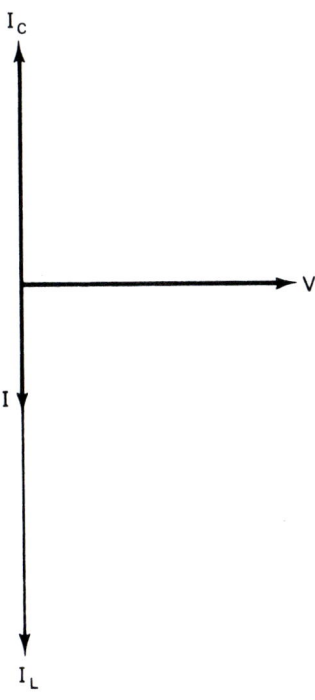

FIGURE 12-22 Phasor diagram for L and C parallel branches.

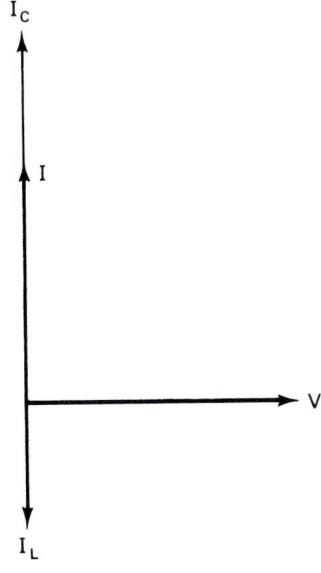

For parallel circuits with only inductive reactance and capacitive reactance in the two parallel branches, the net current or difference current may be either lagging or leading the supply voltage. That branch having the smaller reactance will dominate.

EXAMPLE 12-34 An inductive reactance of 100 Ω is placed in parallel with a capacitive reactance of 200 Ω across a 1000-V ac source. Find the total current I.

Sec. 12-4 *RLC* Circuits

Solution We must first find the two branch currents and then obtain the difference.

$$I_L = \frac{V}{X_L} = \frac{1000}{100} = 10 \text{ A}$$

$$I_C = \frac{V}{X_C} = \frac{1000}{200} = 5 \text{ A}$$

$$I = I_L - I_C = 10 - 5 = 5 \text{ A}$$

The total current, I, is 5 A and is lagging the source voltage by 90° (see Fig. 12-21).

EXAMPLE 12-35 In a parallel circuit, the inductive branch reactance is 45 kΩ, the capacitive branch reactance is 20 kΩ, and the supply voltage is 90 V. Find I.

Solution

$$I_L = \frac{90}{45} = 2 \text{ mA}$$

$$I_C = \frac{90}{20} = 4.5 \text{ mA}$$

$$I = I_C - I_L = 4.5 - 2 = 2.5 \text{ mA}$$

Total current is 2.5 mA, which leads the source voltage by 90° (see Fig. 12-22).

Sometimes it is necessary to find the equivalent reactance when two parallel branches contain X_L and X_C. When I and V are known, the reactance of the circuit as seen by the source is easily found.

$$X_T = \frac{V}{I}$$

The circuit will be inductive if I is a lagging current and capacitive if I is a leading current.

EXAMPLE 12-36 Find X_T for the data of Example 12-34.

Solution In Example 12-34 the total current was found to be 5 A lagging the 1000-V source voltage by 90°. Therefore,

$$X_T = \frac{V}{I} = \frac{1000}{5} = 200 \text{ }\Omega$$

and X_T is an inductive reactance because I is lagging by 90°.

EXAMPLE 12-37 Find the total reactance, X_T, for the parallel circuit of Example 12-35.

Solution In Example 12-35 the total current was found to be 2.5 mA leading the 90-V source voltage. We then find that

$$X_T = \frac{V}{I} = \frac{90}{2.5} = 36 \text{ k}\Omega$$

and X_T is a capacitive reactance because I is leading by 90°.

Let us now consider a parallel ac circuit having three branches. The first branch contains a resistor, the second branch an inductive reactance, and the third branch a capacitive reactance (see Fig. 12-23). It will be remembered that the mark of a parallel circuit is one common voltage for all the branches. This means that the branch currents can be found using Ohm's law when the voltage and branch resistance or reactances are known. One branch current does not depend on that flowing in the other branches.

$$I_R = \frac{V}{R}$$

$$I_L = \frac{V}{X_L}$$

$$I_C = \frac{V}{X_C}$$

Knowing the branch currents allows us to combine them to get the total current, the source current, I_T. Notice that the word is "combine" and not "add." In general, currents in parallel branches must be joined in ways that are slightly more difficult than straight arithmetic addition. As we have seen, phasors can always be combined by attaching the tail of the second to the head of the first. The resultant phasor consists of the straight line drawn directly from the tail of the first to the head of the second. We will see this principle applied again in the next few pages.

In particular cases where the phasors are along the same line, the process of combining is greatly simplified. As we have seen earlier in this section, the currents through an inductor and a parallel capacitor are represented by phasors which lie along a straight line. The capacitor branch current leads the voltage by 90° and the inductor branch current lags the same phasor voltage by 90°. To combine them, the smaller is subtracted from the larger. The resultant phasor has the same direction as the larger of the two. This is the same step that was taken when an L branch and a C branch alone were placed in parallel.

FIGURE 12-23 RLC parallel circuit.

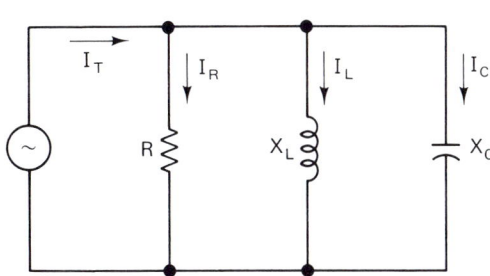

It is the first step to be taken when a third, resistive branch also exists. The resultant of this step is the net reactive current, which may be inductive or capacitive, depending on whether it lags or leads the voltage.

The next step is to combine the net reactive current with the resistive branch current to obtain the total circuit current. We use exactly the same method as that used previously when the two phasors are at right angles. There is nothing new to be learned here, so the next two examples are really in the nature of review.

EXAMPLE 12-38 In Fig. 12-23, $R = 240\ \Omega$, $X_L = 300\ \Omega$, and $X_C = 1200\ \Omega$. Find I_T for $V = 48$ V.

Solution First find the reactive branch currents and combine them by finding the difference.

$$I_L = \frac{V}{X_L} = \frac{48}{300} = 0.16\ \text{A}$$

$$I_C = \frac{V}{X_C} = \frac{48}{1200} = 0.04\ \text{A}$$

$$I_L - I_C = 0.16 - 0.04 = 0.12\ \text{A}$$

Next, combine the resistive branch current with the net reactive current as suggested in Fig. 12-24.

$$I_R = \frac{V}{R} = \frac{48}{240} = 0.2\ \text{A}$$

$$I_T = \sqrt{I_R^2 + (I_L - I_C)^2} = \sqrt{0.2^2 + 0.12^2} = 0.23\ \text{A}$$

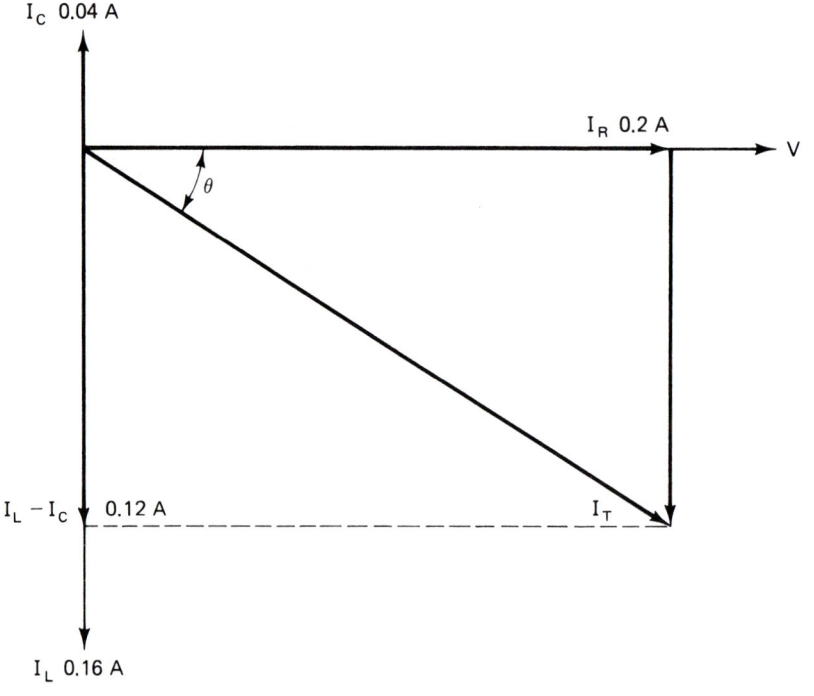

FIGURE 12-24 Phasor diagram for *RLC* parallel circuit.

EXAMPLE 12-39 A circuit is connected as illustrated in Fig. 12-23. In it the resistance is 1.8 kΩ, the inductive reactance is 3 kΩ, and the capacitive reactance is 900 Ω. With 24 V applied, the total circuit current will be how much?

Solution

$$I_C = \frac{V}{X_C} = \frac{24}{0.9} = 26.7 \text{ mA}$$

$$I_L = \frac{V}{X_L} = \frac{24}{3} = 8 \text{ mA}$$

$$I_C - I_L = 26.7 - 8 = 18.7 \text{ mA}$$

$$I_R = \frac{V}{R} = \frac{24}{1.8} = 13.3 \text{ mA}$$

$$I_T = \sqrt{13.3^2 + 18.7^2} = 22.9 \text{ mA} \quad \text{(Fig. 12-25)}$$

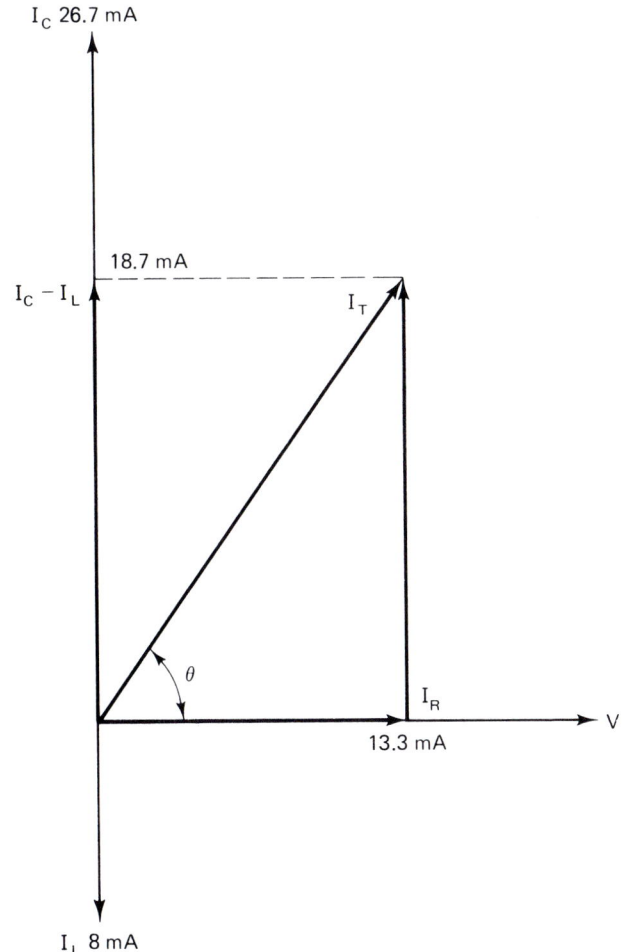

FIGURE 12-25 Phasor diagram for Example 12-39.

We will now take a moment to review the angle relations for the parallel circuit. Our primary interest is in the phase angle, the angle of total current as referred to the supply voltage. In Fig. 12-24 the phase angle is referred to as

$$\theta = \tan^{-1} \frac{I_L - I_C}{I_R}$$

Similarly, in Fig. 12-25 the angle θ is

$$\theta = -\tan \frac{I_C - I_L}{I_R}$$

Here we state the angle as being negative because the voltage lags. The phase angle is positive for an inductive circuit and negative for a capacitive circuit. That is all that one must know. Let us again illustrate these principles by means of examples.

EXAMPLE 12-40 Use the results of Example 12-38 and find the phase angle.

Solution It was found that $I_L = 0.16$ A, $I_C = 0.04$ A, and $I_R = 0.2$ A.

$$\theta = \tan^{-1} \frac{0.16 - 0.04}{0.2} = 31.0°$$

EXAMPLE 12-41 Use the results of Example 12-39 and find the phase angle.

Solution From the example $I_C = 26.7$ mA, $I_L = 8$ mA, and $I_R = 13.3$ mA.

$$\theta = -\tan^{-1} \frac{I_C - I_L}{I_R} = -54.5°$$

When the total current and voltage of a parallel circuit are known, the impedance can be found through the application of Ohm's law.

EXAMPLE 12-42 An *RLC* parallel circuit has a total current of 500 µA for an applied ac potential of 58 V. This current lags the applied potential by 23°. Find Z and θ.

Solution We must first recognize that potential is the electrical pressure forcing current flow through the branches. Since potential is measured in volts, we commonly substitute the word "voltage" for "potential." With this reminder out of the way, it is possible to complete the solution.

$$Z = \frac{V}{I} = \frac{58}{0.5} = 116 \text{ k}\Omega$$

$$\theta = 23°$$

A positive value of θ corresponds to a current that lags the voltage (potential). In other words, the circuit is inductive because the current through the inductive branch is greater than the current through the capacitive branch.

EXAMPLE 12-43 One branch of a parallel circuit has a current of 800 mA which is in phase with the 12-V applied ac potential. The second branch has a current of 2500 mA leading the source potential by 90°. A third branch has 2000 mA of current (lagging). Find the impedance and the associated angle.

Solution Before beginning the solution it is necessary to understand the information given in the problem statement. This information has been presented in several forms which are commonly used but seldom mixed as done for the present example. In the first place we have a 12-V supply or source and this voltage or potential is common to all three branches ($V = 12$ V). In the first branch mentioned there is an in-phase current of 800 mA ($I_R = 800$ mA). Next, there is a capacitive branch through which the leading current is 2500 mA ($I_C = 2500$ mA). Finally, there is an inductive branch with a lagging current of 2000 mA ($I_L = 2000$ mA).

It is also necessary to understand exactly what is being requested. The problem statement asks for the magnitude, the number of ohms of the impedance. And the problem statement requests the associated angle, which is the phase angle. In order to find Z the branch currents will be combined to determine I_T and Ohm's law applied using both V and I_T to find Z. Let us proceed.

$$I_T = \sqrt{I_R^2 + (I_C - I_L)^2}$$

$$= \sqrt{800^2 + (2500 - 2000)^2} = 943 \text{ mA}$$

$$Z = \frac{V}{I_T} = \frac{12}{0.943} = 12.7 \ \Omega$$

To find θ it will be necessary to use the three branch currents and the inverse tangent. Note that the net reactive current $I_C - I_L$ is leading, so that θ is negative.

$$\theta = -\tan^{-1} \frac{2500 - 2000}{800} = -32.0°$$

EXAMPLE 12-44 In the circuit of Fig. 12-23, $V = 75$ V, $f = 25$ kHz, $R = 3.3$ kΩ, $L = 10$ mH, and $C = 3000$ pF. Calculate the magnitude of Z and the phase angle θ.

Solution The steps of the solution are as follows:

Step 1. Calculate X_L and X_C.
Step 2. Find the three branch currents using Ohm's law.
Step 3. Combine the three branch currents to find I_T.
Step 4. Apply Ohm's law to get the magnitude of Z.
Step 5. Use the branch currents to determine θ using the inverse tangent.

Each step will now be demonstrated in order.

Step 1

$$X_L = 2\pi fL = 2\pi \times 25 \times 10^3 \times 10 \times 10^{-3} = 1570 \, \Omega = 1.57 \, k\Omega$$

$$X_C = \frac{1}{2\pi fC} = \frac{1}{2\pi \times 25 \times 10^3 \times 3000 \times 10^{-12}} = 2120 \, \Omega = 2.12 \, k\Omega$$

Step 2

$$I_R = \frac{V}{R} = \frac{75}{3.3} = 22.73 \text{ mA}$$

$$I_L = \frac{V}{X_L} = \frac{75}{1.57} = 47.75 \text{ mA}$$

$$I_C = \frac{V}{X_C} = \frac{75}{2.12} = 35.34 \text{ mA}$$

Step 3

$$I_T = \sqrt{I_R^2 + (I_L - I_C)^2}$$

$$= \sqrt{22.73^2 + (47.75 - 35.34)^2} = 25.89 \text{ mA}$$

Step 4

$$Z = \frac{V}{I} = \frac{75}{25.89} = 2.90 \, k\Omega$$

Step 5

$$\theta = \tan^{-1} \frac{I_L - I_C}{I_R} = \tan^{-1} \frac{47.75 - 35.34}{22.73} = 28.6°$$

Looking at Examples 12-42 to 12-44 it is seen that the second solution required one more step than the first. Moreover, the third solution required one more step than the first. Notice that the initial step was different for the three examples. This is the result of the form in which the problem information is provided.

The question arises: "What should be the order in which calculations are made?" The answer is surprisingly simple. *Just make the calculations in any order in which they can be made.* In Example 12-44, step 4 and step 5 can be exchanged. However, neither can be accomplished before step 2. Also, step 3 must precede step 4 but need not precede step 5. It is not necessary to memorize the step order. One should recognize when a step can or cannot be completed. When a step cannot be completed, it is necessary to identify the missing step or steps and to find the step procedure that can be accomplished with the known data.

We turn now to a very similar, but slightly different method of calculating the impedance of a parallel circuit. It is similar in that the steps are exactly the same as in the preceding three examples. The example that applies depends solely on the specific information known at the start.

Suppose that all the information is given, as in the preceding examples, and the problem statement is exactly the same with one exception. In each example you are to find Z and θ, but you are not given the value of V. Is it possible to find Z and θ? We will attempt to answer this question by using still another numerical example. Notice that it will be identical to Example 12-44 except that a new supply voltage of 150 V is specified.

EXAMPLE 12-45 In the circuit of Fig. 12-23, $V = 150$ V, $f = 25$ kHz, $R = 3.3$ kΩ, $L = 10$ mH, and $C = 3000$ pF. Calculate the magnitude of Z and the angle θ.

Solution
Step 1
The same as in Example 12-44.
Step 2

$$I_R = \frac{V}{R} = \frac{150}{3.3} = 45.45 \text{ mA}$$

$$I_L = \frac{V}{X_L} = \frac{150}{1.57} = 95.49 \text{ mA}$$

$$I_C = \frac{V}{X_C} = \frac{150}{2.12} = 70.69 \text{ mA}$$

Step 3

$$I_T = \sqrt{I_R^2 + (I_L - I_C)^2}$$
$$= \sqrt{45.45^2 + (95.49 - 70.69)^2} = 51.78 \text{ mA}$$

Step 4

$$Z = \frac{V}{I} = \frac{150}{51.78} = 2.90 \text{ k}\Omega$$

Step 5

$$\theta = \tan^{-1} \frac{95.49 - 70.69}{45.45} = 28.6°$$

Now let us compare the results of Example 12-45 with those for Example 12-44. In Example 12-45 all three currents are twice as great as the same currents of Example 12-44. This is caused, of course, by having a supply voltage of 150 V as contrasted with 75 V in Example 12-44.

So, when we get to step 4, both the numerator, V, and the denominator, I, have been doubled, as compared with Example 12-44. It should come as no surprise to learn that $Z = 2.90$ kΩ in both examples. Also, since the numerator and denominator in step 5 are both doubled, the quotient is unchanged and θ in Example 12-45 is exactly the same as was found for Example 12-44.

Sec. 12-4 *RLC Circuits*

As a matter of fact, Z and θ are independent of the applied voltage. If this was not true, Ohm's law would not be true and we would be forced to deal with extremely complicated solutions. Let us be thankful for Ohm's law.

We know that the impedance of a parallel circuit does not depend on the applied voltage. But how can we find the impedance and the phase angle when no applied voltage is given? The answer: Assume any value, especially some convenient value for V. Let us try it out using one more numerical example.

EXAMPLE 12-46 The three branches of a parallel circuit have the following elements: $R = 3\ \Omega$, $X_L = 12\ \Omega$, and $X_C = 3\ \Omega$. Find the magnitude of Z and the phase angle θ.

Solution We may choose any supply voltage and still obtain the same results. However, if we choose $V = 12$ V, the calculations are somewhat simplified. Let $V = 12$ V.

$$I_R = \frac{V}{R} = \frac{12}{3} = 4\text{ A}$$

$$I_L = \frac{V}{X_L} = \frac{12}{12} = 1\text{ A}$$

$$I_C = \frac{V}{X_C} = \frac{12}{3} = 4\text{ A}$$

$$I_T = \sqrt{I_R^2 + (I_C - I_L)^2} = \sqrt{4^2 + (4-1)^2} = 5\text{ A}$$

$$Z = \frac{12}{5} = 2.4\ \Omega$$

$$\theta = -\tan^{-1}\frac{I_C - I_L}{I_R} = -\tan^{-1}\frac{4-1}{4} = -36.9°$$

Alternatively, the magnitude of the total impedance, Z_T, may be found by first obtaining the total reactance, X, from the formula

$$X = \frac{X_L \times X_C}{X_L - X_C}\ \Omega \qquad (12\text{-}4\text{-}13)$$

The expression for X looks very similar to the "product-over-sum" formula for two resistors in parallel [Eq. (3-3-3)]. However, the denominator consists of a negative rather than a positive sign because the branch curents, I_L and I_C, are 180° out of phase. After calculating the value of X, the magnitude of the total impedance may be obtained from the formula

$$Z_T = \frac{R \times X}{\sqrt{R^2 + X^2}}\ \Omega \qquad (12\text{-}4\text{-}14)$$

The magnitude of the phase angle, θ, is given by

$$\theta = \tan^{-1}\frac{R}{X}$$

The expressions for Z_T and θ are the same as those that appeared in Example 12-22.

The calculator procedures are as follows: To find the magnitude of the impedance, Z_T:

Operation	Display	
12 ×	12	
3 ÷ (	36	$(X_L \times X_C)$
12 −		
3)	9	$(X_L - X_C)$
= ×	4	(X)
3 ÷ ((	12	$(R \times X)$
3 x^2 +		
4 x^2) x)	5	$(\sqrt{R^2 + X^2})$
=	2.4	(Z_T)

The magnitude of the impedance, Z_T, is 2.4 Ω.

To find the magnitude of θ:

Operation	Display	
3 ÷	3	(R)
4 =	0.75	$\left(\dfrac{R}{X}\right)$
INV TAN	36.87°	(θ)

The value of θ is $-36.9°$, rounded off.

Chapter Summary

12.1. In a series RL circuit where V_T is the total circuit voltage

$$Z = \sqrt{R^2 + X_L^2}$$
$$V_T = \sqrt{V_R^2 + V_L^2}$$

V_R is the voltage across the resistor, and V_L is the voltage across the inductor.

12.2. An angle exists between V_T and V_R which depends upon the relative values of V_L and V_R.

$$\text{angle, } \theta = \tan^{-1}\frac{V_L}{V_R}$$

where $\tan^{-1}$ is the inverse tangent or arctangent, meaning the angle whose tangent is V_L divided by V_R. The same angle exists between V_T and I in the series RL circuits since V_R and I are in phase.

12.3. Opposition to current flow also appears for a capacitor.

$$X_C = \frac{V_p}{I_p} = \frac{V_C}{I}$$

The capacitive reactance, X_C, has the basic unit of the ohm. It is the ratio of peak voltage to peak current (V_p/I_p) or the same ratio with effective values.

$$X_C = \frac{1}{2\pi f C}$$

In this equation the frequency is in hertz and C is in farads for X_C in ohms. All variations of this equation may be needed to solve for one quantity when the other two are known.

12.4. For the series RC circuit,

$$R = \sqrt{R^2 + X_C^2}$$
$$V_T = \sqrt{V_R^2 + V_C^2}$$
$$\theta = -\tan^{-1}\frac{V_C}{R}$$

It will be noted that a negative value is assigned to the angle to indicate that the current leads the applied voltage. Stated differently, the applied voltage lags the current.

12.5. For a parallel RL circuit,

$$I_T = \sqrt{I_R^2 + I_L^2}$$
$$\theta = \tan^{-1}\frac{I_L}{I_R}$$

Here θ is the angle between V and the lagging current I_T.

12.6. For a parallel RC circuit,

$$I_T = \sqrt{I_R^2 + I_C^2}$$
$$\theta = -\tan^{-1}\frac{I_C}{I_R}$$

Here θ is negative to indicate a leading total current I_T.

12.7. The impedance of a parallel circuit can be calculated when the applied voltage V is known together with the total current, I_T.

$$Z = \frac{V}{I_T}$$

12.8. In RLC series circuits,

$$Z = \sqrt{R^2 + X_T^2}$$

where

$$X_T = X_L - X_C$$

or

$$X_T = X_C - X_L$$

so arranged that the difference is positive. Also, the phase angle difference is

$$\theta = \tan^{-1} \frac{X_T}{R}$$

for X_T inductive, and

$$\theta = -\tan^{-1} \frac{X_T}{R}$$

for X_T capacitive. X_T is inductive when X_L is greater than X_C. X_T is capacitive when X_C is greater than X_L. For such a circuit with $R = 0$,

$$Z = X_T$$
$$\theta = 90° \text{ for inductive } X_T$$
$$\theta = -90° \text{ for capacitive } X_T$$

12.9. In an RLC series circuit,

$$Z = \sqrt{R^2 + (X_L - X_C)^2}$$

$$I = \frac{V_T}{Z}$$

$$\theta = \tan^{-1} \frac{X_L - X_C}{R}$$

$$V_R = IR$$
$$V_L = IX_L$$
$$V_C = IX_C$$
$$V_T = \sqrt{V_R^2 + (V_L - V_C)^2}$$

$$\theta = \tan^{-1} \frac{V_L - V_C}{R}$$

12.10. In an RLC series circuit containing more than one of each kind of circuit element, the same equations apply except that the resistances are added in series to obtain R, the inductive reactances are added in series to obtain X_L, and the capacitive reactances are added to obtain X_C. Then the individual voltages become

$$V_{R1} = IR_1$$
$$V_{R2} = IR_2$$
$$V_{L1} = IX_{L1}$$
$$V_{L2} = IX_{L2}$$
$$V_{C1} = IX_{C1} \quad \text{etc.}$$

which apply for any number of resistances, inductive reactances, and capacitive reactances in series.

Chapter Summary

12.11. For a parallel *RLC* circuit,

$$I_R = \frac{V}{R}$$

$$I_L = \frac{V}{X_L}$$

$$I_C = \frac{V}{X_C}$$

in which V is the common voltage, R is the resistance of one branch, X_L is the inductive reactance of another branch, and X_C is the capacitive reactance of still another branch. Note that the phase-angle difference between V and I_R is 0°, between V and I_L is 90° (I_L lagging), and between V and I_C is 90° (I_C leading).

12.12. Continuing, with a parallel *RLC* circuit,

$$I_T = \sqrt{I_R^2 + (I_L - I_C)^2}$$

$$\theta = \tan^{-1} \frac{I_L - I_C}{I_R}$$

$$\theta = -\tan^{-1} \frac{I_C - I_L}{I_R}$$

in which the first equation holds for any parallel circuit and in which the angle of the total current lags by angle θ for I_L greater than I_C (second equation) and leads for I_C greater than I_L (third equation).

Self-Review

True–False Questions

1. T. F. Resistance and inductive reactance can be combined to determine the impedance of a series circuit.
2. T. F. In a series circuit with V_R = 35 V (peak), V_L = 50 V (peak), and I = 2 A (peak), the reactance is approximately 50 Ω.
3. T. F. In a series *RC* circuit the applied voltage will lag the current by an angle less than 90°.
4. T. F. In a series *RL* circuit the three voltage phasors can be made to form a right triangle.
5. T. F. In a series *RC* circuit the voltage will lead by an angle greater than 0° and less than 90°.
6. T. F. A parallel circuit consists of inductive reactance in one branch and resistance in the second branch. The total current is the arithmetic sum of the individual branch currents.
7. T. F. The total current in a parallel circuit leads the applied voltage by 60°. If one branch contains *R*, the other will have *C*.
8. T. F. The frequency of an *RC* parallel circuit with equal currents in two branches is doubled. Therefore, the total current is doubled.
9. T. F. No voltage in a series *RLC* circuit can be greater than the supply voltage.
10. T. F. An impedance diagram for a series *RLC* circuit will show X_L and X_C at right angles to each other.
11. T. F. In a series *RL* circuit, the current passing through the inductor leads the current passing through the resistor by 90°.

12. **T. F.** The maximum angle between R and Z in the impedance diagram for the series RL circuit is 90°.
13. **T. F.** The hypotenuse of the impedance triangle cannot be longer than the numerical value of R or X, whichever is larger.
14. **T. F.** A positive angle for $Z\angle\theta$ indicates that the current lags the applied voltage for a series circuit.
15. **T. F.** In calculations involving V, I, and Z, peak values of the alternating quantities must be used, but effective or rms values cannot be used.
16. **T. F.** In the phasor diagram for a series RC circuit, V_R and I are collinear.
17. **T. F.** In a parallel circuit each element has the same current as every other element.
18. **T. F.** In a parallel RL circuit the current in the resistive branch is equal to V divided by R.
19. **T. F.** In a series RLC circuit the current will always lag the applied voltage.
20. **T. F.** In general, the addition of a new branch, having R, L, or C, in parallel with existing branches with R, L, or C will cause the total current to increase.

Multiple-Choice Questions

21. In a particular series circuit the values of resistance and inductive reactance are the same. Therefore, the angle between the current and the applied voltage is
    (a) 45°   (b) zero   (c) 90°   (d) 180°   (e) 2π radians
22. A 15-kΩ resistor is connected in series with 10 kΩ of capacitive reactance. The impedance of the circuit is
    (a) 25 kΩ (b) 18 kΩ (c) 5 kΩ (d) 17.5 kΩ (e) 19.7 kΩ
23. In a series RL circuit the peak voltage across the resistor is 9 V and the peak voltage across the inductor is 12 V. Therefore, the peak value of the total voltage is
    (a) 3 V   (b) 9 V   (c) 12 V   (d) 15 V   (e) 21 V
24. In a circuit consisting of an ac source and a capacitor, the current will lead the voltage by
    (a) 90°   (b) 180°   (c) 270°   (d) $-90°$
    (e) no correct choice
25. Four ohms of capacitive reactance is placed in series with 3 Ω of resistance. The impedance is
    (a) 3 Ω (b) 4 Ω (c) 5 Ω (d) 7 Ω (e) 37 Ω
26. In a series RL circuit the effective voltage across the resistor is 40 V and the effective voltage across the inductor is 30 V. If the effective value of current is 70 mA, the power is
    (a) 3.5 W   (b) 2100 mW   (c) 2800 mW   (d) 1200 mW   (e) 8.4 W
27. A resistor and a capacitor are connected in parallel. Each carries 45 mA. The total circuit current is
    (a) 90 mA   (b) 57 mA   (c) 45 mA   (d) 29 mA   (e) 64 mA
28. Connecting a 5-kΩ inductive reactance in parallel with 10 kΩ of capacitive reactance across a 50-V source will cause
    (a) a source current leading the voltage by 90°
    (b) a source current of 15 mA
    (c) a source current of 10 mA
    (d) a source current of 5 mA
    (e) zero source current
29. Adding a 20-kΩ resistor in parallel with the reactances in Question 28 will result in a new source current of
    (a) 20.2 mA   (b) 15.2 mA   (c) 10.3 mA   (d) 5.59 mA   (e) 2.5 mA
30. The impedance of the parallel combination for Question 29 is how much?
    (a) 2.48 kΩ (b) 3.29 kΩ (c) 4.85 kΩ (d) 8.94 kΩ (e) 20 kΩ
31. In a series RL circuit
    (a) V_T is in phase with V_R (b) V_T is in phase with I
    (c) V_R is in phase with I (d) V_L is in phase with I
    (e) V_R is in phase with R

32. A particular series RL circuit has a $Z = 5000\ \Omega$. Select the correct combination for R and L.
    (a) $R = 1000\ \Omega;\ X_L = 4000\ \Omega$ (b) $R = 4000\ \Omega;\ X_L = 1000\ \Omega$
    (c) $R = 2000\ \Omega;\ X_L = 3000\ \Omega$ (d) $R = 3000\ \Omega;\ X_L = 4000\ \Omega$
    (e) $R = 5000\ \Omega;\ X_L = 5000\ \Omega$
33. Consider a series circuit consisting of resistance R in series with inductance L and having impedance $Z\angle\theta$. If the frequency of the voltage source in the circuit is gradually reduced to zero, which of the following statements will be correct?
    (a) $Z = R$ (b) $\theta = 2\pi fL$ (c) $\theta = 0$ (d) $Z = L$
    (e) both (a) and (c)
34. An ac source of 48 V, 400 Hz is connected across a series combination consisting of a resistive element and an inductive element, both having the same number of ohms. The current must then
    (a) lead the voltage by 90°
    (b) lead the voltage by 45°
    (c) lag the voltage by 90°
    (d) lag the voltage by 45°
    (e) be in exact phase with the voltage
35. To increase the current flow in a series RC circuit, one should
    (a) decrease the frequency
    (b) increase the R
    (c) disconnect the voltage source
    (d) increase the frequency
    (e) both (a) and (b)
36. In a series RC circuit
    (a) R is in phase with I
    (b) V_C leads I by 90°
    (c) V_R leads V_C by 90°
    (d) V_T and I are in phase
    (e) no correct choice
37. In a parallel RL circuit
    (a) V_R must lead V_L by 90°
    (b) V_R must lag V_L by 90°
    (c) I_R must lag V_L by 90°
    (d) I_R must lead V_L by 90°
    (e) no correct choice
38. To find the total current in a parallel RL circuit
    (a) add I_R and I_L directly
    (b) add I_R and I_L algebraically
    (c) first find $V_T = V_R + V_L$
    (d) add I_R and I_L as phasors
    (e) write mesh equations using Kirchhoff's laws
39. In a series RLC circuit
    (a) the same current flows through each of the three elements
    (b) V_R is in phase with I
    (c) V_L and V_C are 180° out of phase
    (d) the total current cannot be larger than V/R
    (e) all choices are correct
40. A parallel circuit has three parallel branches, one having R, the second with L, and the third with C. In order that the total current be in phase with the applied voltage
    (a) $R = X_L$ (b) $R = X_C$ (c) $R = 0.707X_L$
    (d) $X_L = X_C$ (e) $R = 0.707X_C$

Practice Problems

41. A series circuit has peak values of 67 V across R and 23 V across X_L. Find the peak value of V_T and the phase angle (angle between V_R and V_T).
42. A 40-kΩ resistor is connected in series with a 30-kΩ inductive reactance. Calculate the magnitude of the impedance and the angle.
43. In a series RC circuit the voltage across the resistor is 57 V (peak) and the voltage across the capacitor is 41 V (peak). What is the total applied voltage?
44. With a capacitive reactance of 37 kΩ and a resistance of 63 kΩ in a series circuit, by how much will the current lead the applied voltage?
45. In a series RC circuit a current of 50 mA (peak value) flows through a 2.2-kΩ resistor and a capacitor with 3.75 kΩ of capacitive reactance. Find V_R, V_C, V_T, Z, and θ.

46. A parallel circuit has current of 200 mA in one branch in phase with the applied voltage. The 100-mA current in the second branch lags the applied voltage by 90°. Calculate the total current and the phase angle.
47. In a series RLC circuit, $R = 150$ kΩ, $X_L = 250$ kΩ, and $X_C = 150$ kΩ. Determine the impedance magnitude and the associated angle.
48. An inductive reactance of 1000 Ω is placed in parallel with a capacitive reactance of 2000 Ω across a 500-V source. Find the total current.
49. Three kilohms of resistance and 8 kΩ of capacitive reactance are connected in parallel. Calculate the total current and the phase angle for a 240-V source.
50. In a series LC circuit, $X_L = 5.1$ kΩ and $X_C = 3.4$ kΩ. Find the magnitude of Z and the phase angle. Express the impedance in a form with both the magnitude and angle.
51. A series circuit consists of 40 Ω of resistance in series with 30 Ω of inductive reactance. For a current of 200 mA, calculate V_T.
52. The voltages of a series RL circuit are measured to give $V_L = 13$ V and $V_R = 2$ V. Calculate the angle between V_R and V_T.
53. An 8.2-kΩ resistor is placed in series with a 30-mH inductor. Calculate the numerical values and express the impedance in the form $Z \angle \theta$ at $f = 25$ kHz.
54. Consider a phasor diagram for which the current is 2 A and lags the applied 24-V source voltage by 60°. Express the impedance in the form $Z \angle \theta$.
55. At a frequency of 100 kHz, the reactance of a capacitor is 5000 Ω. Calculate the frequency that will cause the reactance of the same capacitor to be 10,000 Ω.
56. R has the value of 50 kΩ and $X_C = 60$ kΩ in a series RC circuit. If $V = 24$ V, calculate the I, V_R, and V_C. Show the voltages and current on a single phasor diagram.
57. Let the frequency of the source in Problem 56 be doubled with no change in the values of R and C. Calculate I, V_R, and V_C for the new condition.
58. A parallel RC circuit has 1200 Ω of resistance and 800 Ω of capacitive reactance. If $I_R = 10$ mA, find the total current.
59. A 25-V ac source is connected across a parallel circuit consisting of $R = 1500$ Ω, $X_L = 750$ Ω, and $X_C = 1500$ Ω. Calculate the total current.
60. In Problem 59, calculate the total current if the frequency is doubled, with no other changes.

Advanced Problems

61. A 1-mH inductor and a 120-Ω resistor are placed in series. At what frequency will the impedance be 200 Ω?
62. At what frequency will the circuit of Problem 61 have an impedance angle of 60°?
63. In a series circuit, $V_R = 8$ V and $V_L = 12$ V with $I = 50$ mA. Find the impedance and its angle.
64. A capacitor is placed in series with a 2.2-kΩ resistor so that a current of 5 mA is measured with a total applied voltage of 15 V. For a frequency of 5 kHz, what value of capacitance is required?
65. A circuit has $V_R = 15$ V and $V_C = 11.5$ V with a 1-kHz signal applied to the series combination. At what frequency will $V_R = 10$ V with the same applied voltage?
66. A 3.3-kΩ resistor and a 0.5-mH inductor are placed in parallel across a 10-V source. Calculate the frequency for which the total current is 5 mA.
67. A 1.2-MΩ resistor is placed in parallel with a capacitor across the terminals of a 40-V, 50-kHz voltage source. Determine the value of the capacitance to make the total current equal to 40 μA.
68. Find the frequency for which the total current of an RC parallel circuit leads the voltage by 45°. For this problem $R = 1.8$ kΩ and $C = 0.01$ μF.
69. An RLC circuit contains a 3.3-kΩ resistor in series with 2.5 kΩ of inductive reactance. Determine the amount of capacitive reactance needed in series to obtain a total impedance of 5 kΩ.

70. A parallel RLC circuit is to be designed to have the total current in phase with the applied voltage. $V_T = 500$ mV, $I_T = 50$ mA, $L = 10$ mH, and $C = 0.01$ μF. Find R and f.

71. When 120 V is applied to a series RL circuit, the current is 60 mA. If the resistor is rated at 1.5 kΩ, determine the amount of inductive reactance.

72. The current in a series circuit lags the applied voltage by 30°. In this circuit $R = 2.2$ kΩ and $L = 500$ μH. At what frequency is it operating?

73. The impedance of a series RL circuit $Z = 3.5\angle 27.3°$ kΩ at $f = 10$ kHz. Express the impedance in the same basic form when $f = 20$ kHz.

74. The impedance of a series RL circuit is given by $1000\angle 30°$ Ω at $f = 500$ kHz. The frequency is now increased until the impedance angle becomes 45°. What is the new frequency for this condition?

75. Calculate the total voltage for a series RC circuit having $R = 15$ kΩ and $X_C = 20$ kΩ through which a current of $3\angle 0°$ mA flows. Express the answer in the form $V\angle \theta$ V in which θ is to have the appropriate sign.

76. A series circuit has a resistance value of 15 kΩ and current of 100 μA for 5 V, 1600 Hz applied. Determine the frequency for which the current will be doubled to 200 μA with no other circuit changes.

77. A 1-kΩ resistor is placed in parallel with a 100-mH inductor and 10 V is applied to the parallel combination. With the source adjusted to 1 MHz, the total curent is 10 mA. Estimate, with reasonable accuracy, a new frequency for which the currents in the two branches would be measured to be the same.

78. In a parallel RC circuit the total current is 50 mA and the currents in both branches are equal, though not in phase. Determine the total current if the resistor burns out (opens) with no other change in the circuit.

79. Three elements consisting of $R = 50$ kΩ, $L = 10$ μH, and $C = 100$ pF are connected in parallel. At a particular frequency $X_L = X_C$. Calculate the total current at this frequency for an applied voltage of 12 V.

80. Another resistor of 50 kΩ is now placed in parallel with the elements of Question 79. Determine the new current at the frequency for which $X_L = X_C$.

13

Power Circuits

The objectives of this chapter are as follows:

1. To define the meaning of power as it relates to resistance and the conversion of energy from the electrical form to heat.
2. To introduce and describe the concepts of apparent power and reactive power as they relate to ac circuits with both resistive and reactive elements.
3. To specify simple formulas for the calculation of power, apparent and reactive power, and their representation as compared to the phasor diagram.
4. To define the power factor and several methods for its calculation, including the use of the phasor diagram and phase angles.
5. To provide the equation for efficiency and to evaluate the significance of efficiency in power systems.
6. To describe the meaning and significance of power transfer and maximum power transfer.
7. To specify the conditions for maximum power transfer from a voltage source, with internal resistance, to a resistor load.
8. To specify the conditions for maximum power transfer from a voltage source with internal impedance (resistive and reactive components) to a load having both resistive and reactive components.
9. To demonstrate the application of maximum power transfer by means of examples.

13-0 Introduction

Electronic equipment often receives its power from alternating-current power sources. These sources are usually part of the public electrical utility network or the military equivalent for military electronic equipment. A brief study of power in this context is necessary. The power, reactive power, and apparent power are defined together with the power factor.

A second major aspect of power circuits is concerned with the transfer of power at both high and low levels between elements of communications and other systems. The basic question is associated directly with maximum power transfer between one element, the source, and a second element designated as the load. Whenever the source can be represented as a constant ac voltage in series with a resistance, it is seen that the load must also be resistive for proper matching. In the more general situation in which the source cannot be represented as having a series resistance, but rather a series impedance with reactance, the matching is more involved.

13-1 Power, Reactive Power, and Apparent Power

It will be recalled that in a series circuit containing both resistance and reactance, the voltage and currents can be represented as phasors. In Fig. 13-1, for example, it is seen that V_R is in phase with current I and V_L is oriented in the upward direction. V_T is the phasor sum of V_R and V_L. We also recall from Chapter 11 the ac methods of calculating the power supply for the power dissipated in a resistor.

$$P = IV \qquad (11\text{-}3\text{-}5)$$

$$P = I^2 R \qquad (11\text{-}3\text{-}6)$$

$$P = \frac{V^2}{R} \qquad (11\text{-}3\text{-}7)$$

in which V is the applied voltage, I is the current, and P is the power for a circuit containing resistance only.

The preceding equations apply to any series or parallel circuit or combination series-parallel circuit provided that the voltage is that across the resistor and the current is that flowing through the resistor. To emphasize this point, the equations will be rewritten with the subscript R.

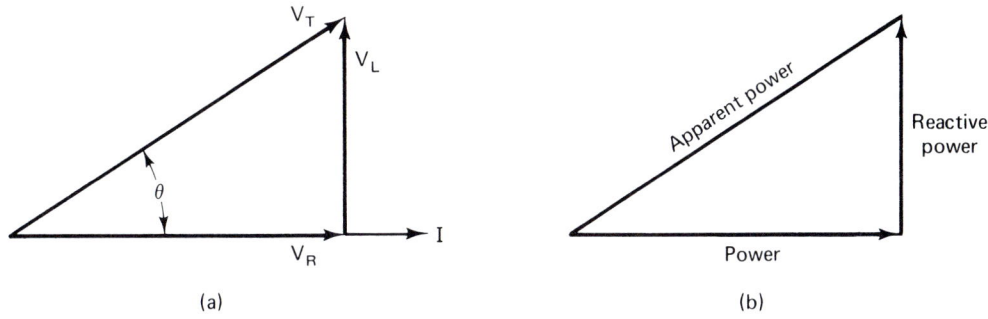

FIGURE 13-1 (a) Phasor diagram for series circuit; (b) power relationships.

$$P = I_R V_R \qquad (13\text{-}1\text{-}1)$$

$$P = I_R^2 R \qquad (13\text{-}1\text{-}2)$$

$$P = \frac{V_R^2}{R} \qquad (13\text{-}1\text{-}3)$$

EXAMPLE 13-1 A current of 5 A flows in a series circuit containing 4 Ω of resistance and 3 Ω of inductive reactance. Use Eqs. (13-1-1), (13-1-2), and (13-1-3) to find the power.

Solution

$$V_R = I_R R = 5 \times 4 = 20 \text{ V}$$

$$P = I_R V_R \qquad (13\text{-}1\text{-}1)$$
$$= 5 \times 20 = 100 \text{ W}$$

$$P = I_R^2 R \qquad (13\text{-}1\text{-}2)$$
$$= 5^2 \times 4 = 100 \text{ W}$$

$$P = \frac{V_R^2}{R} \qquad (13\text{-}1\text{-}3)$$
$$= \frac{20^2}{4} = 100 \text{ W}$$

If Eqs. (13-1-1), (13-1-2), and (13-1-3) are modified to apply to the reactance, the following equation forms have significance:

$$\text{reactive power} = I_X V_X \quad \text{var} \qquad (13\text{-}1\text{-}4)$$

$$\text{reactive power} = I_X^2 X \quad \text{var} \qquad (13\text{-}1\text{-}5)$$

$$\text{reactive power} = \frac{V_X^2}{X} \quad \text{var} \qquad (13\text{-}1\text{-}6)$$

Var signifies volt-amperes-reactive.

EXAMPLE 13-2 Find the reactive power for the circumstances of Example 13-1 in three different ways.

Solution

$$V_X = I_X X = 5 \times 3 = 15 \text{ V}$$

$$\text{reactive power} = I_X V_X \tag{13-1-4}$$
$$= 5 \times 15 = 75 \text{ var}$$
$$\text{reactive power} = I_X^2 X \tag{13-1-5}$$
$$= 5^2 \times 3 = 75 \text{ var}$$
$$\text{reactive power} = \frac{V_X^2}{X} \tag{13-1-6}$$
$$= \frac{15^2}{3} = 75 \text{ var}$$

Finally, when the product of the total voltage and total current of any series, parallel, or series-parallel combination circuit is taken, the result is called *apparent power*. Three forms are available for the calculation of apparent power.

$$\text{apparent power} = I_T V_T \tag{13-1-7}$$
$$\text{apparent power} = I_T^2 Z_T \tag{13-1-8}$$
$$\text{apparent power} = \frac{V_T^2}{Z_T} \tag{13-1-9}$$

The unit is the VA, signifying volt-ampere.

It should not be too surprising to learn that power, reactive power, and apparent power are related in the form of a right triangle. In Fig. 13-1(a) we see that the legs of the triangle are labeled "power" and "reactive power." The hypotenuse corresponds to the "apparent power." We suggest that you remember this triangle as a means of independently checking the results of power calculations. Remember, apparent power is usually larger than either the power or reactive power. Under no circumstances can the apparent power be less than the larger of the other two.

EXAMPLE 13-3 Find the apparent power for the circumstances of Example 13-1.

Solution Only one method will be used, as it should be apparent from the three forms of the equations demonstrated for power and reactive power that the three forms will give the same answer for apparent power.

$$Z = \sqrt{R^2 + X^2}$$
$$= \sqrt{4^2 + 3^2} = 5 \, \Omega$$
$$V_T = IZ$$
$$= 5 \times 5 = 25 \text{ V}$$
$$\text{apparent power} = \frac{V_T^2}{Z_T} \quad \quad (13\text{-}1\text{-}9)$$
$$= \frac{25^2}{5} = 125 \text{ VA}$$

The student is reminded that volts and amperes must be used to obtain VA for apparent power and var for reactive power.

13-2 Power Factor

The ratio of power to apparent power is commonly used and is designated as the power factor (PF), so that

$$P = \text{apparent power} \times \text{PF} \quad \quad (13\text{-}2\text{-}1)$$

EXAMPLE 13-4 The voltage in a circuit is 115 V and the current is 3 A for a load having a power factor of 0.8. What is the power?

Solution

$$\text{apparent power} = I_T V_T \quad \quad (13\text{-}1\text{-}7)$$
$$= 3 \times 115 = 345 \text{ VA}$$
$$P = \text{apparent power} \times \text{PF} \quad \quad (13\text{-}2\text{-}1)$$
$$= 345 \times 0.8 = 276 \text{ W}$$

The power factor can be identified as a trigonometric function of θ, the angle between the total circuit voltage and total circuit current.

$$\text{power factor (PF)} = \cos \theta \quad \quad (13\text{-}2\text{-}2)$$

EXAMPLE 13-5 In a particular load the current and voltage are 4 A and 220 V. The current lags the voltage by 15°. Calculate the total power.

Solution

$$\text{PF} = \cos\theta \qquad (13\text{-}2\text{-}2)$$
$$= \cos 15° = 0.966$$
$$\text{apparent power} = I_T V_T \qquad (13\text{-}1\text{-}7)$$
$$= 4 \times 220 = 880 \text{ VA}$$
$$P = \text{apparent power} \times \text{PF} \qquad (13\text{-}2\text{-}1)$$
$$= 880 \times 0.966 = 850 \text{ W}$$

13-3 Maximum Power Transfer with Resistive Load

Considerations of load power arise under several different circumstances. Electrical power for use in homes, commercial establishments, and factories, to cite a few examples, usually originates in large rotating machines called alternators. The prime source may be water power or steam power, and these may be further subdivided into specific types of prime movers and the type of fuel (if any) used. A major consideration is the generation and distribution of the greatest amount of electrical energy at the least cost. Clearly, the cost of fuel is only one factor in the total cost since the capital invested in power plants, transmission networks, and distribution grids will most often be even more significant in selecting from among alternative proposed new systems or modifications of existing systems.

A number of additional factors must be recognized as being essential to the basic nature of large-scale electrical utilities. Reliability is one of these. Also, nearly all electrical loads are intended to function normally over a relatively small range of terminal voltage. For example, an increase of only a few percent above a nominal rating of 120 V for the common incandescent lamp greatly increases the light produced, which is desirable, but significantly decreases the operating life, which is not desirable. Similarly, a decrease of a few percent reduces the amount of light produced while extending the operating life. One should recognize that the design of incandescent lamps represents a compromise that is acceptable to the mass market.

In addition, nearly all equipment, such as electrical motors composing utility loads, are also rated for a single voltage. The requirement for a nominal applied voltage means that the flow of rated current in the generators, transformers, and power lines causes voltage drops, due to the existence of series impedance, which are usually a small percentage of the rated voltage. We must also remember that the current supplied by generators (alternators) and other power system equipment is limited by the power dissipation within the equipment and the subsequent temperature rise resulting from the power dissipation.

The preceding discussion corresponds, in part, to the concept of circuit efficiency that was introduced in Chapter 5. Mathematically, circuit efficiency is defined as follows:

$$\text{circuit efficiency} = P_L \times \frac{100}{P_T}$$

$$= R_L \times \frac{100}{R_i + R_L} \qquad (5\text{-}4\text{-}6)$$

The second formulation of Eq. (5-4-6) was derived for a dc circuit for which the source internal resistance is R_i and load resistance is R_L.

Exactly the same circuit analysis applies for an ac circuit having an internal source resistance, R_i, and load resistance, R_L. In particular, one should notice that the greatest circuit efficiency is achieved for $P_L = 0$, as shown in Table 5-2. Circuit efficiency has meaning but little practical significance in practical power circuits. Clearly, we can increase circuit efficiency by reducing R_i, but the increase in power rating of the alternator through the reduction of R_i, because of the decrease of internal heating, is a more compelling reason.

We now change perspective by focusing attention on maximum power transfer. Maximum power transfer is analyzed for a dc circuit in Chapter 5. It is achieved by adjusting the load resistance, R_L, to match the internal generator resistance, R_i. This matching condition is expressed briefly as

$$R_L = R_i \tag{5-4-3}$$

and results in maximum load power $P_{L\text{max}}$, given by

$$P_{L\text{max}} = \frac{E^2}{4R_L} \tag{5-4-4}$$

in which E is the no-load dc source effective voltage. It is interesting to observe that the circuit efficiency is exactly 50%, as illustrated in Table 5-2, when maximum power transfer is attained.

The condition for maximum power transfer given by Eq. (5-4-3) is valid for an ac circuit having source impedance which has no reactive component and only resistance R_i. Similarly, the load impedance, R_L, is resistive only. To summarize, there is no reactive element in the circuit.

Moreover, Eq. (5-4-4) is valid for the same ac circuit when E is the effective value of the source no-load voltage, E.

EXAMPLE 13-6 A 1-MHz signal generator has a no-load (open-circuit) voltage of 10 V and an internal impedance of 600 Ω (resistive). Find the load impedance for maximum power transfer and the load power for this condition.

Solution For maximum power transfer,

$$R_L = R_i \tag{5-4-3}$$
$$= 600 \ \Omega$$

$$P_{L\text{max}} = \frac{E^2}{4R_L} \tag{5-4-4}$$

$$= \frac{10^2}{4(600)} = 0.0417 \ \text{W}$$

One must be aware that R_i includes the entire amount of series resistance between the open-circuit (no-load) voltage and the load terminals.

EXAMPLE 13-7 The signal generator of Example 13-6 is connected to a remote load by a pair of insulated wires, each having 100 Ω of resistance. Calculate the load power for (a) the 600-Ω resistive load and (b) a properly matched load.

Solution

$$R_i = 600 + 100 + 100 = 800 \ \Omega$$

(a) load power, $P_L = \dfrac{E^2 R_L}{(R_i + R_L)^2}$ \hfill (5-4-1)

$$= \dfrac{10^2(600)}{(800 + 600)^2}$$

$$= 0.03061 \ W$$

(b) load power, $P_L = \dfrac{E^2}{4R_L}$ \hfill (5-4-4)

$$= \dfrac{10^2}{4(800)} = 0.03125 \ W$$

13-4

Maximum Power Transfer with Reactive Circuit Components

We will now examine the extremely important situation in which the source impedance is not R_i but rather an impedance having resistive component R in series with reactive component X, as exemplified by Fig. 13-2. In this figure the circuit to the left of the load terminals cannot be changed by the designer.

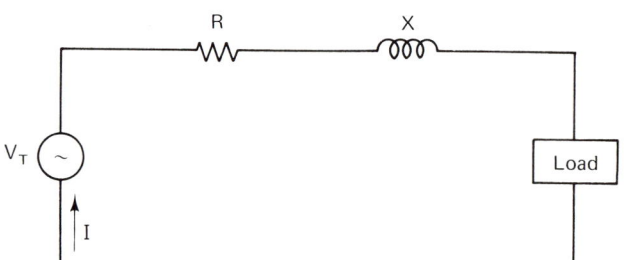

FIGURE 13-2 Series circuit with load.

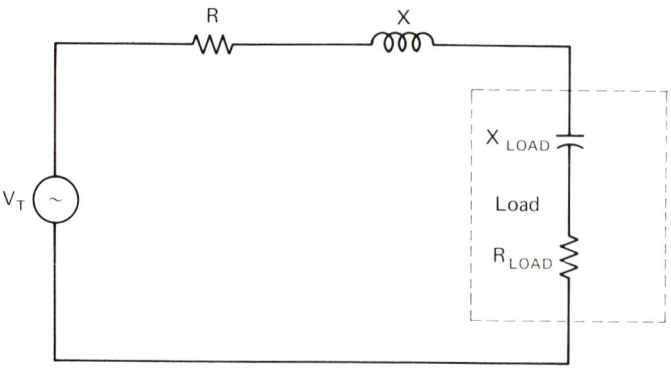

FIGURE 13-3 Maximum power transfer.

The designer must live with that part of the circuit. What we as designers can do is to select the proper load so as to maximize the load power. This requires that the load have two features.

1. The load resistance R_{LOAD} must be equal to the source resistance R, as is the case for maximum power transfer in a dc circuit.
2. There must be cancellation of the effects of source reactance X on circuit current. This can be achieved by introducing a series reactance X_{LOAD} whose value is equal to X but of the opposite kind.

The second feature requires further explanation. Figure 13-3 expands the circuit of Fig. 13-2 to illustrate the conditions for maximum power transfer. In both figures the source resistance is R and the source reactance is X. It is seen that X is pictured as an inductive reactance, as is most often true for practical electronic circuits. Therefore, X_{LOAD} must be a capacitive reactance with the same number of ohms. This means that

$$\frac{1}{2\pi f C_{LOAD}} = 2\pi f L \qquad (13\text{-}4\text{-}1)$$

where C_{LOAD} is the load capacitance whose reactance is X_{LOAD} and L is the source inductance whose reactance is X. A phasor diagram as seen in Fig. 13-4(a) shows the effect of having X_{LOAD} slightly less than X, so that the total circuit impedance is not the least amount possible. Compare Fig. 13-4(a) with (b) in which the latter has X_{LOAD} with the same magnitude as X. With the reactances perfectly balanced,

$$Z = R + R_{LOAD} = 2R = 2R_{LOAD} \qquad (13\text{-}4\text{-}2)$$

We will now consider a numerical example to confirm the conditions for maximum power transfer. A constant source voltage V_T will be given along with source resistance R and inductive reactance X. The load will have a fixed $R_{LOAD} = R$, but X_{LOAD} will be varied. For each value of X_{LOAD}, the magnitude of Z and the phase angle will be determined. Finally, I and the load power P_{LOAD} will be calculated.

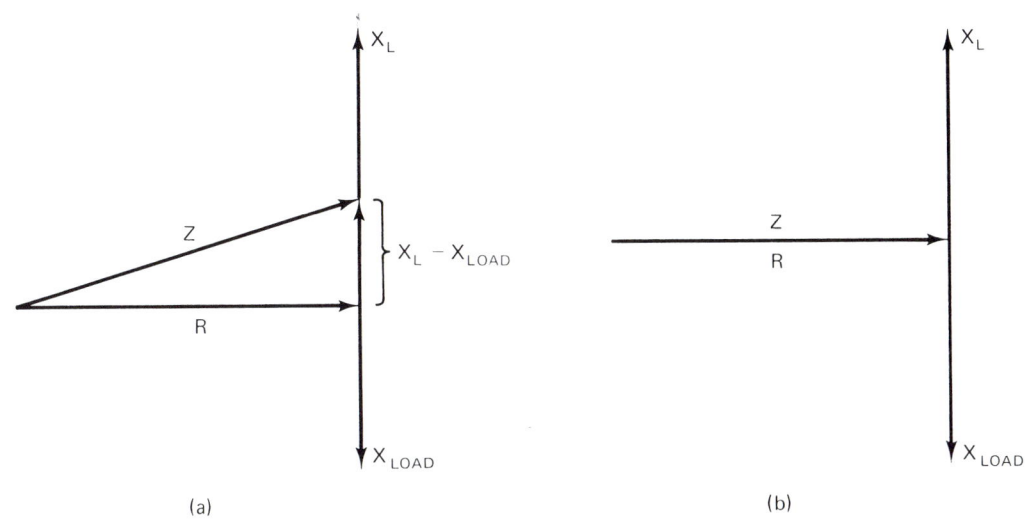

FIGURE 13-4 Impedance diagram illustrating maximum power transfer.

Sec. 13-4 Maximum Power Transfer with Reactive Circuit Components

EXAMPLE 13-8 A 100-V (effective) source has a series resistance $R = 10\ \Omega$ and reactance $X = 50\ \Omega$ (inductive). Select the proper R_{LOAD} for maximum power transfer and calculate Z, θ, I, and P_{LOAD} as X_{LOAD} (capacitive) is varied from 0 to 100 Ω in 10-Ω steps. Show the results in tabular form.

Solution A typical set of calculations will be shown here for $X_{LOAD} = 30\ \Omega$. Calculations for other values of X_{LOAD} will be identical in form; only the numerical quantities will differ. We first find Z, then θ, I, and P_{TOTAL}.

$$R_{TOTAL} = R + R_{LOAD} = 10 + 10 = 20\ \Omega$$

$$X_{TOTAL} = X - X_{LOAD} = 50 - 30 = 20\ \Omega$$

$$Z = \sqrt{R_{TOTAL}^2 + X_{TOTAL}^2} = \sqrt{20^2 + 20^2}$$

$$= 28.3\ \Omega$$

$$\theta = \tan^{-1}\frac{X_{TOTAL}}{R_{TOTAL}} = \tan^{-1}\frac{20}{20} = 45°$$

$$I = \frac{V_T}{Z} = \frac{100}{28.3} = 3.54\ A$$

$$P_{LOAD} = I^2 R_{LOAD} = (3.54)^2 \times 10$$

$$= 125\ W$$

The tabulated results appear in Table 13-1. It is seen that P_{LOAD} increases as X_{LOAD} is increased and reaches a maximum value of 250 W at $X_{LOAD} = 50\ \Omega$ (capacitive). At this point $\theta = 0°$, so that the current is in phase with the applied voltage.

As X_{LOAD} (capacitive) is increased beyond 50 Ω, the load power decreases. This proves that maximum power transfer occurred with $X_{LOAD} = 50\ \Omega$ (capacitive).

TABLE 13-1 Variation of Load Power with Load Reactance

X_{LOAD} (Ω)	Z (Ω)	θ (deg)	I (A)	P_{LOAD} (W)
0	53.9	68.2	1.85	34.5
10	44.7	63.4	2.24	50.0
20	36.1	56.3	2.77	76.9
30	28.3	45.0	3.54	125
40	22.4	26.6	4.47	200
50	20.0	0	5.00	250
60	22.4	−26.6	4.47	200
70	28.3	−45.0	3.54	125
80	36.1	−56.3	2.77	76.9
90	44.7	−63.4	2.24	50.0
100	53.9	−68.2	1.85	34.5

In Example 13-8 we discovered that the maximum load power was 250 W. This is to be expected, because the impedance for maximum power transfer is a resistance equal to the sum of the source resistance and the load resistance. Each being 10 Ω, the total power can be expressed as

$$\frac{V_T^2}{R_T} = \frac{V_T^2}{2R} = \frac{100^2}{20} = 500 \text{ W}$$

We must be very careful in calculating power in this way. In the first place the net circuit reactance must be zero, and the resistance used must be the total circuit resistance. The fact that half of the total power is dissipated in each of the two equal resistances suggests that

$$P_{\text{LOAD(MAX)}} = \frac{V_T^2}{4R} \qquad (13\text{-}4\text{-}3)$$

where R is the source resistance. Let us try it out for $V_T = 100$ V and $R = 10 \, \Omega$.

$$P_{\text{LOAD(MAX)}} = \frac{100^2}{4 \times 10}$$
$$= 250 \text{ W}$$

This result is the same as that found in the numerical example.

Up to this point we have considered only one possibility for the source reactance, that of inductive reactance. To achieve maximum power transfer with source inductive reactance, an equal magnitude capacitive reactance is required for the load. Of course, the load resistance must also be equal to the source resistance.

Suppose that the source has a capacitive reactance. Matching the source impedance will then require that the load reactance be inductive.

EXAMPLE 13-9 An ac source has an internal impedance consisting of 1000 Ω of resistance in series with 3000 Ω of capacitive reactance. The no-load source voltage is 48 V (effective). (a) Determine the load power when the load impedance is matched for maximum power transfer, and (b) state the amount of resistance and reactance (as well as kind of reactance) for the load giving maximum load power.

Solution Note that the no-load source voltage is the source voltage we have referred to as V_T. $R = 1000 \, \Omega = 1 \text{ k}\Omega$. Apply Eq. (13-4-3).

(a) $P_{\text{LOAD(MAX)}} = \dfrac{V_T^2}{4R} = \dfrac{48^2}{4 \times 1} = 576 \text{ mW}$

(b) $R_{\text{LOAD}} = 1000 \, \Omega \qquad X_{\text{LOAD}} = 3000 \, \Omega$ (inductive)

Chapter Summary

13.1. Three forms of power are distinguished: real power, reactive power, and apparent power. Only real power expresses an average rate of energy flow between the source and the load. Reactive power is the rate of energy flow between the inductive and capacitive elements. Apparent power is the product of voltage and current without consideration of the power factor.

13.2. The power factor (PF) is the ratio of real power to apparent power:

$$P = \text{apparent power} \times \text{PF}$$

The power factor is also cos θ, in which θ is the angle between the total circuit voltage and the total circuit current.

13.3. In the absence of reactance in the source, maximum power transfer requires that

$$\text{load resistance} = \text{source resistance}$$

$$P_{L\max} = \frac{E^2}{4R_L}$$

13.4. Additionally, when there is a reactive component associated with the source impedance, the load reactance is given by

$$\text{load reactance} = \text{source reactance (opposite)}$$

meaning that an inductive reactance for the source requires a capacitive reactance for the load, and vice versa.

Self-Review

True–False Questions

1. T. F. The unit of power in an ac circuit is the joule or volt-ampere.
2. T. F. In an ac circuit the apparent power will be equal to the reactive power when the circuit resistance is zero.
3. T. F. In a series RL circuit the reactive power is equal to the product of the total effective voltage and the effective circuit current.
4. T. F. In a series RL circuit the total applied voltage is doubled. Therefore, the reactive power will be quadrupled (factor of 4).
5. T. F. Maximum load power is obtained when the load resistance is made equal to the source resistance for a dc circuit.
6. T. F. For maximum power transfer the load reactance must be identical to the source reactance.
7. T. F. In a series circuit having the load adjusted for maximum power tranfer, the current will be in phase with both the source total voltage and the load voltage.
8. T. F. In a large public utility system, the load is deliberately adjusted for maximum power transfer.
9. T. F. When adjusted for maximum power transfer, the circuit efficiency of an ac circuit is 50%.
10. T. F. To obtain maximum load power in an ac circuit, the load reactance must always be zero.
11. T. F. The apparent power in an ac circuit is always greater than the power.
12. T. F. The reactive power in a parallel ac circuit may be greater than the power.
13. T. F. The power factor cannot be less than zero or greater than 1.
14. T. F. The name given to a generator designed for the production of ac power is the "alternator."
15. T. F. Reactive power is expressed in units of watts.

16. **T. F.** Power, reactive power, and apparent power can be represented by a right triangle with the hypotenuse being power.
17. **T. F.** If, for a parallel circuit consisting of a resistive branch and an inductive branch, the apparent power is found to be equal to the power, the inductive branch is "open."
18. **T. F.** The power factor of a series RLC circuit having $X_L = X_C$ is zero.
19. **T. F.** When a voltage source with internal resistance only is properly matched for maximum power transfer, the circuit efficiency is also maximum.
20. **T F.** When matched for maximum power transfer, the load voltage will be one-half the no-load source voltage if the internal impedance of the source includes R and X in series.

Multiple-Choice Questions

21. The correct expression for the power in watts in an ac circuit is
    (a) IV (b) I^2X (c) V_R^2/R
    (d) both (b) and (c) (e) all choices are correct
22. In calculating reactive power a correct equation to use is
    (a) $I_X V_X$ (b) $I_X^2 X$ (c) V_X^2/X
    (d) choices (a), (b), and (c) are correct
    (e) no correct choice
23. In comparing power, reactive power, and apparent power, the one that can be the largest is
    (a) $I_T^2 Z_T$ (b) V_X^2/X (c) $I_R^2 X$ (d) $I_R^2 R$
    (e) choices (a) and (b) are both correct
24. An ac source has a no-load voltage of 2.5 V and an internal impedance that is resistive and equal to 50 Ω. For maximum power transfer the load should be
    (a) 2.5 Ω (resistive) (b) 2.5 Ω (inductive)
    (c) 50 Ω (resistive) (d) 50 Ω (inductive)
    (e) 50 Ω (capacitive reactance)
25. The circuit of Question 24 is altered by doubling both the voltage and impedance. This will result in $P_{L\max}$ being multiplied by a factor of
    (a) 1 (b) 2 (c) 3 (d) 4 (e) 8
26. A source includes an internal impedance consisting of a resistance in series with an inductance. For maximum power transfer the load might properly be constructed of
    (a) two resistors in parallel
    (b) a resistor in parallel with a capacitor
    (c) an inductor in parallel with a capacitor
    (d) an inductor in series with a resistor
    (e) two capacitors in series
27. The electrical power that can be delivered by an alternator is limited by
    (a) a 50% circuit efficiency
    (b) the power in a matched load
    (c) the internal heating of the alternator
    (d) the internal capacitive reactance of the alternator
    (e) the internal inductive reactance of the alternator
28. The power factor of a practical load
    (a) is never greater than 1 (b) has a value between −1 and 1
    (c) is 0 for a purely resistive load (d) is 1 for a purely reactive load
    (e) choices (c) and (d) are both correct
29. After the load impedance is adjusted to properly match the source impedance to achieve the maximum load power, the circuit efficiency will be
    (a) 0% (b) 10% (c) 25% (d) 50% (e) 90%
30. The unit designation for reactive power is the
    (a) joule (b) watt (c) va (d) var (e) coulomb
31. The power in an ac circuit having resistance only is given by
    (a) IV (b) I^2R (c) V^2/R

(d) all preceding choices are correct
(e) at least one choice is incorrect

32. When the power, apparent power, and reactive power are shown as a right triangle, the hypotenuse will be the
    (a) power   (b) apparent power
    (c) reactance   (d) impedance power
    (e) resistive power

33. In a particular single-phase ac circuit the power = 10 W and the reactive power = 12 vars. Therefore, the power factor is
    (a) 0   (b) 0.64   (c) 0.768   (d) 0.83   (e) 1.00

34. The reactive power in an ac circuit can be calculated using
    (a) $I_X V_X$ (b) $I_X^2 X$ (c) V_X^2/X (d) $IV \cos \theta$
    (e) choices (a), (b), and (c) are all correct

35. In a particular circuit the net reactive power is zero. The apparent power is then equal to
    (a) $I^2 X$ (b) IV_X (c) V^2/X (d) power   (e) 0

36. In a simple circuit a resistor is placed in series with an inductor. At a particular frequency $R = X_L$, so that the power factor is then equal to
    (a) 0   (b) 0.5   (c) 0.707   (d) 0.866   (e) 1.0

37. A parallel RLC circuit is constructed with $R = X_L = X_C$. If the power dissipated is 100 W,
    (a) reactive power = 100 vars   (b) apparent power = 100 VA
    (c) reactive power = 141 vars   (d) apparent power = 141 VA
    (e) choices (a) and (d) are both correct

38. When a voltage source with internal resistance only is matched for maximum power transfer to a resistive load, the circuit efficiency will be
    (a) 0%   (b) 25%   (c) 50%   (d) 75%   (e) 100%

39. An ac source has an internal impedance consisting of R and X_C in series. Specify the conditions for maximum power transfer for a load with R_L and X in series.
    (a) $R_L = X_C$ (b) $R_L = R$ (c) $X = R_L$ (d) $X = -X_C$
    (e) choices (b) and (c) are both correct

40. For the condition of maximum power transfer
    (a) the impedance of the load equals the source impedance
    (b) the load current is minimum
    (c) the load current is in phase with the no-load (Thévenin) source voltage, V_T
    (d) the load resistance is zero
    (e) the reactive power is maximum

Practice Problems

41. In a series RL circuit, $V_R = 120$ V (effective), $V_L = 240$ V (effective), and the current is 5 A (effective). Determine the real power, reactive power, and apparent power.

42. A current of 20 mA flows in a series RL circuit containing 3.3 kΩ of resistance and 1.8 kΩ of inductive reactance. Find the power, reactive power, and apparent power.

43. The voltage in a circuit is 230 V and the current is 1.5 A for a load with 0.75 power factor. Calculate the power.

44. An ac generator has a terminal voltage of 440 V. When connected to a particular load, the current is 10 A and the current lags the voltage by 12.5°. Find the power delivered to the load.

45. A circuit consists of 1 Ω of inductive reactance in series with 5 Ω of resistance. Find the apparent power and reactive power when the current is 10 A.

46. A load consists of 3.3 kΩ of resistance in series with 2.5 kΩ of capacitive reactance properly matched to the source. If the no-load voltage of the source is 120 V (rms), the power rating of the load must be at least how much?

47. A source consists of $V_T = 48$ V, $R = 560$ Ω, and $X = 1000$ Ω (inductive reactance) at $f = 2$ kHz. Find the value of C_{LOAD}, in series with an unspecified resistance, to achieve maximum power transfer in a matched load at $f = 3$ kHz.

48. An ac source has a Thévenin voltage of 18 V and a Thévenin impedance of 600 + j300 Ω. Calculate the maximum load power when the load is correctly matched.
49. What amount of capacitive reactance is required in series with the 600 Ω in Problem 48 for maximum power transfer?
50. Calculate the load power for Problem 48 if, following the addition of $X_C = 300$ Ω in series with $R_L = 600$ Ω, the frequency of the source is doubled.
51. The ac current in a series circuit with 0.6 Ω of resistance and 0.8 Ω of inductive reactance is 25 A. Calculate the power delivered to this circuit.
52. Measurements made for a particular machine indicate that $P = 10$ kW and the apparent power = 15 kVA. Determine the reactive power.
53. Assume R and L in parallel for the load with the conditions of Problem 52 and a single-phase line voltage of 220 V. Find the values of R and X_L.
54. A current of 500 mA passes through a series circuit with 100 Ω of resistance and 50 Ω of capacitive reactance. Calculate the power, reactive power, and apparent power.
55. A load resistor dissipating 50 mW of power is placed in parallel with an inductor whose reactance in ohms is equal to the resistance. Calculate the apparent power.
56. Calculate the power factor for the condition that $R = X_C$ for R in series with C.
57. If, for the circuit of Problem 56, the frequency is doubled, other factors remaining the same, what will be the new power factor?
58. An ac source having a no-load voltage = 24 V and internal resistance of 600 Ω is properly matched for maximum power transfer. Find the load power.
59. An ac source has an internal resistance of 600 Ω and is matched to a resistive load of 600 Ω. What no-load voltage is required of the source in order that the maximum load power be 1.5 W?
60. Repeat Problem 59 with all the same given data except that the internal resistance is to be 50 Ω.

Advanced Problems

61. In a series circuit $V_R = 55$ V, $V_L = 27$ V, and $I = 5$ A. Calculate the power, apparent power, and reactive power.
62. In a parallel circuit $I_R = 2$ A, $I_C = 1$ A, and $V = 100$ V. Calculate the power, apparent power, and reactive power.
63. A series circuit has $L = 20$ mH in series with $C = 0.1$ μF and $R = 500$ Ω. At what frequency will the power factor be 1?
64. In Problem 63, determine two frequencies for which the power factor = 0.5.
65. The equivalent circuit of an ac load is represented by 24 Ω of resistance in parallel with 96 Ω of inductive reactance. If the load voltage is 550 V, determine the power and reactive power of the load.
66. For Problem 65, calculate the amount of capacitance to be placed in parallel with the existing load so that the resulting power factor is 1. Use 60 Hz for the frequency.
67. A power amplifier produces a no-load output voltage of 125 V at 400 Hz. The output impedance of the amplifier is 50 + j6Ω. Calculate the impedance of a load that achieves maximum power transfer and determine the load power for this matching condition.
68. Returning to Problem 67, let the original load consist of an 8-Ω resistor that is to be correctly matched to the amplifier with a step-down transformer. What is the required turns ratio of the transformer?
69. In Problem 68, how much capacitance must be connected in series with the 8-Ω resistor for maximum power transfer?
70. The power factor of a certain circuit is 0.8, with the 25-A current lagging the 230-V load voltage. Determine the circuit current if the load reactance is changed so that the new power factor is 0.9, with the current lagging as before but no change in load power.
71. The power delivered to a series circuit is 375 W. If the applied voltage is 25 V and the apparent power is 625 VA, what is the circuit resistance?

72. A 440-V, 12-kW ac circuit consists of a resistive element in parallel with a capacitive element for which the total apparent power of the two is 15,000 VA. Determine the inductive reactance to be added in parallel to the circuit in order that the total reactive power for all three is zero.

73. Design a series circuit for a 120-V single-phase source having a power factor of 0.85 for which the power is 1300 W. Specify the values of R and X for the circuit.

74. A particular circuit has an impedance consisting of 1000 Ω of resistance in series with an inductor. If the power dissipated is 900 mW, with source no-load voltage = 45 V, calculate the apparent power.

75. In Problem 74, what is the inductance value when the frequency is 100 kHz?

76. A particular parallel arrangement of R and C is intended to dissipate 100 mW and to absorb an equal quantity of vars. Calculate R and X_C to accomplish this with 10 V applied.

77. A series combination of R and X_L requires 1414 vars, with equal quantities of power and reactive power. Calculate the apparent power if the same elements are reconnected to be in parallel across the same ac source.

78. When properly matched for maximum power transfer, the load consists of C = 100 pF in series with a 50-Ω resistor for which the power is 500 mW. Determine the no-load source voltage.

79. For Problem 78, maximum power transfer is achieved at a frequency of 1 MHz. Describe the source.

80. If the no-load voltage of Problem 78 is increased to 15 V, what will be the new maximum load power?

14

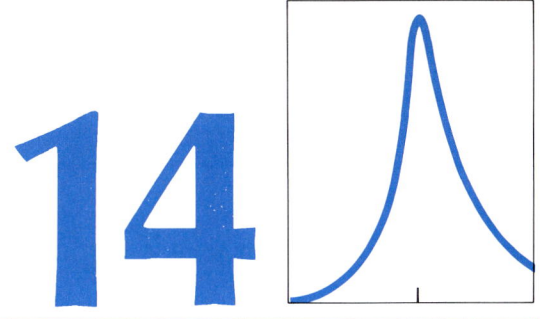

Resonant Circuits and Filters

SKIP ENTIRE CHAPTER

This chapter has the following objectives:

1. To establish the basis of series resonance as the condition for which the inductive reactance and the capacitive reactance are equal.
2. To examine the behavior of series *RLC* circuits at and near resonance.
3. To express the quantitative relationships among inductance *L*, capacitance *C*, and the resonant frequency f_r.
4. To define and employ the quality factor as a measure of the circuit selectivity in the vicinity of resonance.
5. To define the bandwidth of a resonant circuit and to relate it quantitatively to the resonant frequency f_r and quality factor *Q*.
6. To expand the resonance concept as it applies to parallel *RLC* circuits.
7. To determine practical, equivalent series and parallel resonant circuits.
8. To interpret resonant circuits as being particular examples of filters.
9. To introduce electrical filters as distinguishing between frequencies in different ranges: low, intermediate, and high.
10. To develop the characteristics of the low-pass, high-pass, and band-pass filters through direct calculation.
11. To introduce and develop the use of decibels for expressing electrical ratios as they are related to filter characteristics.
12. To introduce the concept of semilogarithmic graph paper and justify the employment of the Bode plot for filter characteristics.
13. To relate briefly the principles of *LC* filter sections and suggest more effective forms of passive filters.

14-0

Introduction

In this chapter we examine several characteristics of ac circuits which are especially important in that branch of electronics which is related to communications. Communications includes all forms of electromagnetic radiation and reception, such as AM radio, FM radio, CB radio, TV, public safety radio, microwave links, and so forth. The same basic principles apply to the equipment used in telephone systems, data transmission systems, and others that use solid electrical wires and cables for their interconnection.

14-1

Series Resonance

In Section 13-4 we achieved maximum power transfer by making the load resistance equal to the source resistance and by adjusting the load reactance to be the same as the source reactance but of the opposite kind (see Table 13-1).

We will here be concerned with a similar situation with RLC in series. We are not interested in a load but rather in the entire circuit. However,

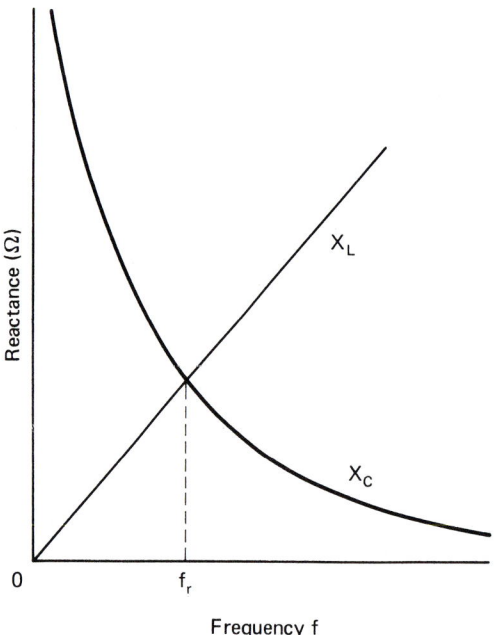

FIGURE 14-1 Conditions for series resonance.

instead of changing C while leaving L fixed or changing L while leaving C constant, the frequency, f, will be changed. Stated another way, one reactance alone will not be altered. Instead, by keeping the same L and C it will be possible to make their reactances have the same values at a particular frequency.

It will be recalled that

$$X_L = 2\pi f L$$

so that X_L plotted as a function of frequency is a straight line as drawn in Fig. 14-1. Inductive reactance is directly proportional to frequency.

Capacitive reactance, on the other hand, is inversely proportional to frequency. As the frequency is increased, X_C decreases. Capacitive reactance is shown as a decreasing-valued curved line in Fig. 14-1. Its curve grazes the vertical axis as the frequency becomes very small and the horizontal axis as the frequency becomes increasingly large.

The point at which the X_L line and X_C curve cross is of particular interest. At this point the two are equal and the series RLC circuit is said to be in resonance. This occurs for $f = f_r$, the resonant frequency.

We shall later look at several practical aspects of series resonance. However, it will be best to first apply our knowledge of RLC circuits and examine the behavior as the frequency is varied.

EXAMPLE 14-1 A series circuit consists of 8.9126768×10^{-5} H of inductance and $2.8420526 \times 10^{-10}$ F of capacitance connected in series with 56 Ω of resistance. Calculate X_L, X_C, Z magnitude, θ, and I with $V_T = 100$ V as the frequency is varied. Note that extremely accurate values are given for L and C. The purpose is to allow you an opportunity to duplicate a few calculations on your own.

Solution The calculations will not be shown here. However, the formulas to be used in the order of their use are listed.

$$X_L = 2\pi f L$$

$$X_C = \frac{1}{2\pi f C}$$

$$Z = \sqrt{R^2 + (X_L - X_C)^2} \quad \text{for } X_L > X_C$$

$$Z = \sqrt{R^2 + (X_C - X_L)^2} \quad \text{for } X_C > X_L$$

$$\theta = \tan^{-1}\frac{X_L - X_C}{R} \quad \text{for } X_L > X_C$$

$$\theta = -\tan^{-1}\frac{X_C - X_L}{R} \quad \text{for } X_C > X_L$$

$$I = \frac{V_T}{Z}$$

Table 14-1 lists all of the calculated results.

TABLE 14-1   Calculated Results for Example 14-1[a]

f (kHz)	X_L (Ω)	X_C (Ω)	Z (Ω)	θ (deg)	I (mA)
0	0	∞	∞	−90	0
100	56	5600	5600	−89.4	17.9
200	112	2800	2689	−88.8	37.2
300	168	1867	1700	−88.1	58.8
400	224	1400	1177	−87.3	84.9
500	280	1120	842	−86.2	118.8
600	336	933	600	−84.6	166.7
700	392	800	412	−82.2	242.8
800	448	700	258	−77.5	387.4
900	504	622	131	−64.7	764.4
950	532	589	80	−45.7	1246.2
1000	560	560	56	0	1785.7
1050	588	533	78	44.3	1277.8
1100	616	509	121	62.4	828.6
1200	672	467	470	83.2	212.8
1300	728	431	302	79.3	330.6
1400	783	400	388	81.7	257.7
1500	840	373	470	83.2	212.8
1600	896	350	549	84.1	182.2
1700	952	329	625	84.9	160.0
1800	1008	311	699	85.4	143.0
1900	1064	295	771	85.8	129.7
2000	1120	280	842	86.2	118.8

[a] $L = 8.9126768 \times 10^{-5}$ H; $C = 2.8420526 \times 10^{-10}$ F; $R = 56$ Ω; $f_r = 1000$ kHz; $V_T = 100$ V; $Q = 10$.

When data are tabulated as is done in Table 14-1 much can be learned by careful study. Here are some observations:

1. The frequency increases progressively line by line starting on the first line.
2. Inductive reactance X_L increases progressively line by line, starting on the first line.
3. Capacitive reactance X_C *decreases* progressively line by line, starting on the first line.
4. At a particular frequency (1000 kHz), $X_L = X_C = 560$ Ω.
5. The impedance magnitude decreases progressively line by line, starting on the top line, and reaches the lowest value of 56 Ω at 1000 kHz. As the frequency is increased, starting at 1000 kHz, the impedance increases progressively line by line.
6. Phase angle θ starts at −90° on the first line and approaches closer and closer to zero with each line until it reaches 0° on the line for which $f = 1000$ kHz. As the frequency increases beyond 1000 kHz, the phase angle becomes positive and becomes larger line by line. It is approaching, but never reaches, +90°.
7. The current starts at zero on the first line, reaches a maximum of 1785.7 mA at $f = 1000$ kHz, and then decreases as the frequency increases.

Figure 14-2 presents a graph of the circuit current plotted as a function of frequency. The peak value of current, the maximum current, occurs at 1000 kHz. We commonly say that the circuit is *resonant* at 1000 kHz and that 1000

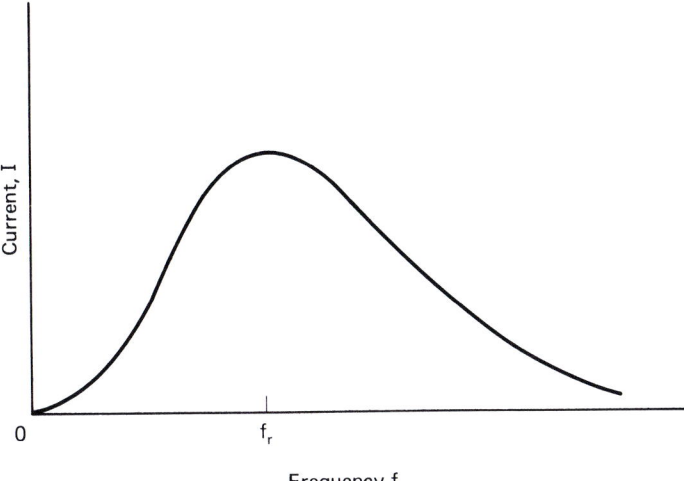

FIGURE 14-2 Current in series *RLC* circuit as frequency is varied for Example 14-1.

kHz is the *resonant frequency*. At the resonant frequency X_L and X_C cancel each other out and the current is limited solely by R. Below the resonant frequency X_C is dominant and the phase angle is negative. Above the resonant frequency X_L is dominant and the phase angle is positive. At the resonant frequency $X_L = X_C$ and the phase angle is zero because $Z = R$.

It is possible to find any one of the three quantities f, L, or C at resonance when the other two are known. The basic relationship comes from $X_{Lr} = X_{Cr}$ at resonance. That is,

$$2\pi f_r L = \frac{1}{2\pi f_r C}$$

at resonance. Solving for f_r gives

$$f_r = \frac{1}{2\pi\sqrt{LC}} \qquad (14\text{-}1\text{-}1)$$

where f_r is the resonance frequency. Solving for L yields

$$L = \frac{1}{4\pi^2 f_r^2 C} \qquad (14\text{-}1\text{-}2)$$

and for C,

$$C = \frac{1}{4\pi^2 f_r^2 L} \qquad (14\text{-}1\text{-}3)$$

Let us assume that $L = 150\ \mu\text{H} = 150 \times 10^{-6}$ H and $C = 250$ pF $= 250 \times 10^{-12}$ F. Using a scientific calculator, the procedure for obtaining the value of f_r is:

Operation					Display	
150	EE	+/−	6	×	1.5 − 4	(L)
250	EE	+/−	12	=	3.75 − 14	(LC)
	$\sqrt{x}$	×			1.936 − 7	($\sqrt{LC}$)
π	×	2		=	1.217 − 6	($2\pi\sqrt{LC}$)
	$\frac{1}{x}$				821873	(f_r)

The value of f_r is 821,873 Hz = 822 kHz, rounded off. In some scientific calculators the EE key is shown as an EXP key.

EXAMPLE 14-2 Three elements are connected in series. One is a resistor, the second is an inductor of 8.913×10^{-5} H, and a third is a capacitor of 2.842×10^{-10} F. Find the resonant frequency for the circuit.

Solution

$$f_r = \frac{1}{2\pi\sqrt{LC}} \qquad (14\text{-}1\text{-}1)$$

$$= \frac{1}{2\pi\sqrt{8.913 \times 10^{-5} \times 2.842 \times 10^{-10}}}$$

$$= 1{,}000{,}000 \text{ Hz} = 1000 \text{ kHz} = 1 \text{ MHz}$$

EXAMPLE 14-3 How much inductance is required to cause resonance at 150 kHz when C = 0.001 μF?

Solution

$$L = \frac{1}{4\pi^2 f_r^2 C} \qquad (14\text{-}1\text{-}2)$$

$$= \frac{1}{4\pi^2 (1.5 \times 10^5)^2 (10^{-9})}$$

$$= 1.13 \times 10^{-3} \text{ H}$$

$$= 1.13 \text{ mH}$$

EXAMPLE 14-4 Calculate the amount of capacitance required in a series circuit containing 10.1 μH of inductance if a resonant frequency of 5 MHz is required?

Solution

$$C = \frac{1}{4\pi^2 f_r^2 L} \qquad (14\text{-}1\text{-}3)$$

$$= \frac{1}{4\pi^2 (5 \times 10^6)^2 (10.1 \times 10^{-6})}$$

$$= 1.00 \times 10^{-10} \text{ F}$$

$$= 100 \text{ pF}$$

In Examples 14-1 through 14-4, no knowledge of the circuit resistance was used. That is, the resonant frequency does not depend on the circuit resistance for series circuits. There is one important aspect of series resonance that does depend on the amount of resistance in the circuit. We will examine this aspect using numerical examples.

EXAMPLE 14-5 Return to Example 14-1. Repeat the same calculations but use $R = 5.6 \: \Omega$ instead of $R = 56 \: \Omega$ as before. Also use $V_T = 10$ V instead of 100 V.

Solution All of the calculations are made using the same formulas as in Example 14-1. The results are given in Table 14-2.

TABLE 14-2 Calculated Results for Example 14-5[a]

f (kHz)	X_L (Ω)	X_C (Ω)	Z (Ω)	θ (deg)	I (mA)
0	0	∞	∞	−90	0
100	56	5600	5544.0	−89.9	1.8
500	280	1120	840.0	−89.6	11.9
600	336	933	597.4	−89.5	16.7
700	392	800	408.0	−89.2	24.5
800	448	700	252.1	−88.7	39.7
900	504	622	118.4	−87.3	84.5
950	532	589	57.7	−84.4	173.1
980	549	571	23.3	−76.1	429.0
990	554	566	12.6	−63.6	795.4
995	557	563	7.9	−45.1	1261.1
1000	560	560	5.6	0	1785.7
1005	563	557	7.9	44.9	1264.3
1010	566	554	12.5	63.3	801.8
1020	572	549	22.9	75.8	437.1
1050	588	533	55.0	84.2	182.0
1100	616	509	107.1	87.0	93.4
1200	672	467	205.4	88.4	48.7
1300	728	431	297.2	88.9	33.6
1500	840	373	466.7	89.3	21.4
2000	1120	280	840.0	89.6	11.9
5000	2800	112	2688.0	89.9	3.7

[a] $L = 8.9126768 \times 10^{-5}$ H; $C = 2.8420526 \times 10^{-10}$ F; $R = 5.6 \: \Omega$; $Q = 100$; $f_r = 1000$ kHz; BW = 10 kHz; $V_T = 10$ V.

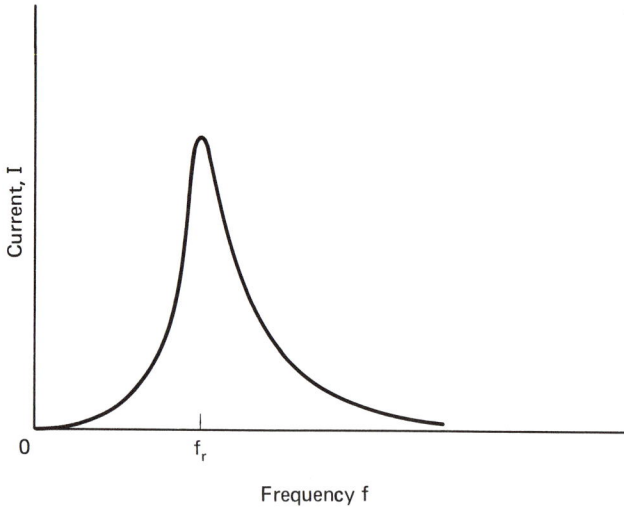

FIGURE 14-3 Current in series *RLC* circuit as frequency is varied for Example 14-5.

The entire discussion about Example 14-1 applies for Example 14-5. It might be well to read this discussion again while comparing the points made with Table 14-2.

To emphasize the difference between the two, let us consider the graph of *I* as a function of *f* for Example 14-5 (see Fig. 14-3). We see a graph with a sharp peak compared with the graph of Fig. 14-2. What is the significance of a sharp peak as contrasted with a rounded peak?

We are all aware that a large number of AM and FM radio stations, as well as TV, CB, public safety, and other types of stations, are transmitting simultaneously. Each is assigned to a different frequency band and every station in the band is required to transmit on a particular frequency. Table 14-3 gives an overview of frequency-band assignments for several radio and TV services. Table 14-4 lists several AM and FM stations in and around the city of Los Angeles to illustrate specific frequency assignments.

You should not be concerned with learning what frequency bands are assigned to any service. Nor is the knowledge of particular FM station assignments of much interest to anyone far removed from metropolitan Los Angeles. However, you might well be greatly interested in being able to tune to station KABC or KALI without receiving an interfering signal from station KBRT. You want the maximum response for the frequency to which your receiver is tuned. To avoid interference, the response for a neighbor (frequency-wise) should be very much less. The set is tuned so that the station broadcasting frequency is the same as the resonance frequency of the circuit in your receiver. Figure 14-4 puts the sharp and shallow responses of two

TABLE 14-3 Broadcast Radio and TV Bands

AM broadcast	535–1605 kHz
VHF TV	54–72 MHz
	76–88 MHz
	174–216 MHz
FM broadcast	88–108 MHz
UHF TV	470–890 MHz

TABLE 14-4 Broadcast Radio Stations in Los Angeles[a]

AM

KABC	790	KGRB	900	KRLA	1110
KALI	1430	KHJ	930	KTNQ	1020
KBRT	740	KIEV	870	KTYM	1460
KDAY	1580	KIIS	1150	KWIZ	1480
KEZY	1190	KKTT	1230	KWKW	1300
KFAC	1330	KLAC	570	KWOW	1600
KFI	640	KLIT	1220	KWRM	1370
KFRN	1280	KMPC	710	XEGM	950
KFWB	980	KNSE	1510	XPRS	1090
KGER	1390	KNX	1070	XTRA	690
KGIL	1260	KPOL	1540		
KGOE	850	KPPC	1240		

FM

KACE	103.9	KJOI	98.7	KPFK	90.7
KBIG	104.3	KKGO	105.1	KQLH	95.1
KBOB	98.3	KLON	88.1	KROQ	106.7
KCRW	89.9	KLOS	95.5	KRTH	101.1
KCSN	88.5	KLVE	107.5	KSAK	90.1
KDUO	97.5	KMAX	107.1	KSPC	88.7
KEZY	95.9	KMET	94.7	KSRF	103.1
KFAC	92.3	KNAC	105.5	KSUL	90.1
KFOX	93.5	KNJO	92.7	KUSC	91.5
KFSG	96.3	KNOB	97.9	KUTE	101.9
KGIL	94.3	KNTF	93.5	KWIZ	96.7
KHOF	99.5	KNX	93.1	KWST	105.9
KHTZ	97.1	KOCM	103.1	KWVE	107.9
KIIS	102.7	KORJ	94.3	KXLU	88.9
KIQQ	100.3	KOST	103.5	KYMS	106.3
KJLH	102.3	KPCS	89.3	KZLA	93.9

[a]AM frequencies in kHz; FM frequencies in MHz.

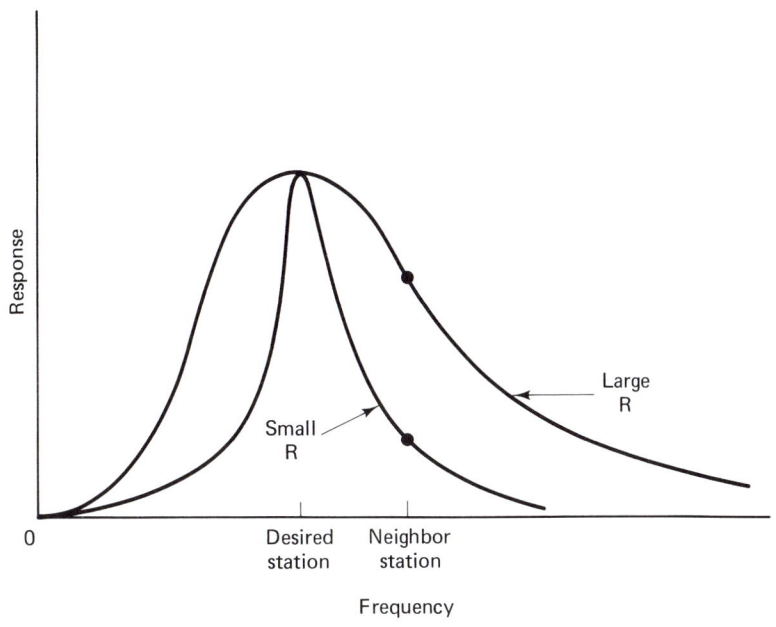

FIGURE 14-4 Frequency-response comparison.

circuits together. We can now compare the response for the neighboring station using the two curves. The upper curve has poor selectivity and allows the neighbor station to be received almost as well as the desired station. However, the sharp curve (low R) does not respond well to the neighbor station.

It is possible to illustrate the selectivity of series-resonant circuits using the data of Tables 14-1 and 14-2. In each, the current at the resonant frequency, 1000 kHz, is 1785.7 mA. In the circuit the current at 900 kHz is approximately one-tenth as much for the series circuit with the smaller resistor. This circuit is said to have a larger Q (Q stands for "quality factor") than the other. Q has the following definition:

$$Q = \frac{2\pi f_r L}{R} = \frac{X_{Lr}}{R} \qquad (14\text{-}1\text{-}4)$$

where f_r is the resonant frequency and X_{Lr} is the inductive reactance at the resonant frequency. Since $X_{Lr} = X_{Cr}$ for the series circuit,

$$Q = \frac{X_{Cr}}{R} \qquad (14\text{-}1\text{-}5)$$

EXAMPLE 14-6 Calculate the value of Q for the circuit of Example 14-1 for $f_r = 1000$ kHz.

Solution In Example 14-1, $R = 56\ \Omega$ and $X_{Lr} = X_{Cr} = 560\ \Omega$ at $f_r = 1000$ kHz (see Table 14-1). Therefore,

$$Q = \frac{X_{Lr}}{R} \qquad (14\text{-}1\text{-}4)$$
$$= \frac{560}{56} = 10$$

or

$$Q = \frac{X_{Cr}}{R} \qquad (14\text{-}1\text{-}5)$$
$$= \frac{560}{56} = 10$$

EXAMPLE 14-7 Calculate the value of Q for the circuit of Example 14-5.

Solution In this example $R = 5.6\ \Omega$ and $X_{Lr} = X_{Cr} = 560\ \Omega$ at 1000 kHz, the resonant frequency (see Table 14-2). Therefore,

$$Q = \frac{X_{Lr}}{R} = \frac{X_{Cr}}{R} = \frac{560}{5.6} = 100$$

We will now take a moment to explore one particular aspect of the series resonant circuit which is not emphasized in Table 14-2. At the resonant frequency of 1000 kHz both X_L and X_C are each 560 Ω, while $Z = R = 5.6$ Ω. Therefore, since the same current flows through all the circuit elements, 10 V appears across the resistor but 1000 V appears across X_L and X_C. Although this may seem to be impossible, we must remember that the two 1000-V phasors are 180° out of phase with each other and cancel one another. If V_T is the total applied voltage, at resonance

$$V_L = V_C = QV_T$$

The reader is advised to consult Table 14-1 and confirm this general rule for a different series resonant circuit.

A series *RLC* circuit is really an electrical filter. It allows an ac signal of one particular frequency to pass by having minimum impedance at that frequency. Other frequencies only slightly moved from the resonant frequency are also passed. Still other frequencies which are significantly different from the resonant frequency are passed weakly. This type of electrical filter is called a *bandpass* filter and the band of frequencies passed determines the *bandwidth*. From the numerical examples presented earlier we know that the bandwidth, abbreviated BW, is greater for the smaller Q and less for the larger Q.

There is a common way of describing the bandwidth of any bandpass device. This measure of bandwidth extends from the lower half-power point to the upper half-power point. Figure 14-5 shows a portion of the current versus frequency curve for an *RLC* circuit for which the maximum current is 100 mA at the resonant frequency. The lower half-power point occurs at A, for which the current is 70.7 mA. Also, the upper half-power point is at B, with the same 70.7 mA. Here the frequency span from A to B is the BW. It is not difficult to prove to yourself that I^2R for any resistance is one-half as much for 70.7 mA as for 100 mA.

It can be demonstrated that the half-power points occur at the frequencies for which the net reactance is exactly equal to the resistance. At the half-power frequency below the resonant frequency, $X_C - X_L = R$ at that half-

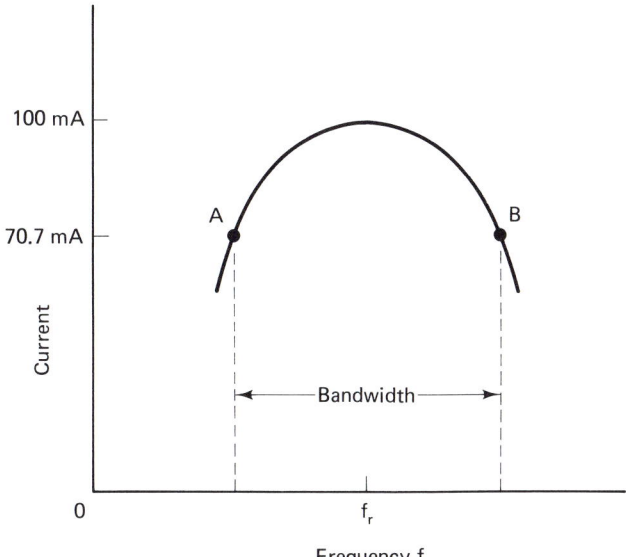

FIGURE 14-5 Demonstration of half-power points and bandwidth.

power frequency. At the half-power frequency above the resonant frequency, X_L is greater than X_C. Their difference $X_L - X_C = R$ at that half-power frequency. These facts help to explain why the half-power frequencies are so often used to specify the characteristics of resonant circuits. Still another reason is to follow.

The bandwidth is closely related to the resonant frequency and Q.

$$\text{BW} = \frac{f_r}{Q} \qquad (14\text{-}1\text{-}6)$$

The bandwidth is directly proportional to f_r and inversely proportional to Q.

EXAMPLE 14-8 A series RLC circuit is resonant at 1 MHz and has a Q of 200. What is the bandwidth?

Solution

$$\text{BW} = \frac{f_r}{Q} \qquad (14\text{-}1\text{-}6)$$

$$= \frac{10^6}{200} = 5 \times 10^3 \text{ Hz} = 5 \text{ kHz}$$

EXAMPLE 14-9 Calculate the Q required for a BW of 10 kHz for $f_r = 10$ MHz.

Solution

$$\text{BW} = \frac{f_r}{Q} \qquad (14\text{-}1\text{-}6)$$

$$Q = \frac{f_r}{\text{BW}} = \frac{10^7}{10^4} = 10^3 = 1000$$

EXAMPLE 14-10 A bandwidth of 200 Hz is achieved with a Q of 75. What is the resonant frequency?

Solution

$$\text{BW} = \frac{f_r}{Q} \qquad (14\text{-}1\text{-}6)$$

$$f_r = Q \times \text{BW}$$
$$= 75 \times 200$$
$$= 15{,}000 \text{ Hz} = 15 \text{ kHz}$$

14-2

Parallel Resonance

We have learned that the current and impedance of a parallel circuit can be found. If, for example, the values of V, R_p, X_L, and X_C are known for Fig. 14-6, the branch currents, total current, impedance magnitude, and phase angle can be calculated. One step in solving the circuit requires that the difference of the two reactive currents be found. The difference or net reactive current is then combined with the resistive branch current to obtain the total current. This method was demonstrated in the examples of Section 12-4. If the approach cannot be clearly recalled, a few minutes of browsing through that section will be invaluable at this point.

We turn now to the particular condition for which the two reactive currents are numerically equal. When this happens, the two reactive currents cancel and the total current is the same as the current through the resistive branch. This is the condition for parallel resonance. Parallel resonance exists for one particular frequency f_r when L and C are fixed. In order for the two reactive currents to be equal at the resonant frequency, the inductive reactance in one branch must equal the capacitive reactance in another branch. The requirement for parallel resonance is that $X_{Lr} = X_{Cr}$ in order that $I_L = I_C$. But wait. Doesn't this sound familiar? Is not this the same requirement that applies for series resonance? It is exactly the same. Therefore, the same formulas apply to a large extent. We need to learn very little that is new. That is, for resonance

$$2\pi f_r L = \frac{1}{2\pi f_r C} \tag{14-2-1}$$

$$f_r = \frac{1}{2\pi\sqrt{LC}} \tag{14-2-2}$$

We use Eq. (14-2-2) to obtain f_r when L and C are known. When L is expressed in henrys and C in farads, f_r will be in hertz.

There are other handy formulas obtained for series resonance which apply equally well for parallel resonance. Suppose that f_r and C are known. We need to find the required value of L.

$$L = \frac{1}{4\pi^2 f_r^2 C} \tag{14-2-3}$$

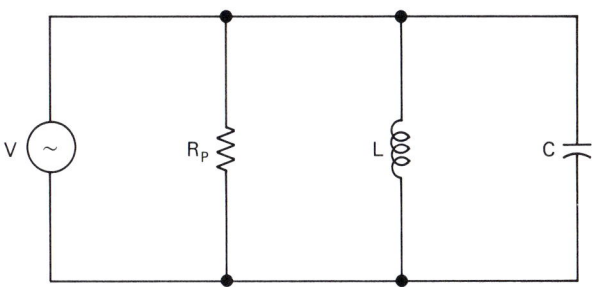

FIGURE 14-6 Parallel *RLC* circuit.

On the other hand, suppose that L is known and C is to be calculated.

$$C = \frac{1}{4\pi^2 f_r^2 L} \qquad (14\text{-}2\text{-}4)$$

Note that the basic units for all of these formulas are henrys, farads, and hertz.

Now for a few comments about this development for the student. If there is any element of confusion, some review may be necessary. Class discussion or discussion with your fellow students outside the laboratory or classroom may be useful. It requires much self-discipline at times to focus on the particular item or items that are not understood. One or two shaky points can endanger the whole structure of your understanding.

Another comment. Is this really necessary? Is it important? Yes, it is. The circuits and concepts of resonance are the backbone of communications in almost every form. To reinforce these ideas and to provide a basis for further development, we will take some time off to show how the parallel circuit behaves as frequency is varied. You will recall that this same approach was taken for series resonant circuits.

EXAMPLE 14-11 A parallel RLC circuit with $V = 100$ V is resonant at 1000 kHz. $R_p = 56$ kΩ. $X_L = X_C = 5.6$ kΩ at resonance. Calculate X_L, X_C, I_L, I_C, I_T, the magnitude of Z, and θ for a range of frequencies up to 2000 kHz. All data should be arranged in tabular form.

Solution Since X_L and X_C are known at 1000 kHz, it is possible to determine L and C using

$$L = \frac{X_{Lr}}{2\pi f_r}$$

and

$$C = \frac{1}{2\pi f_r X_{Cr}}$$

We can then find X_L and X_C at any other frequency using the following formulas:

$$X_L = 2\pi f L$$

$$X_C = \frac{1}{2\pi f C}$$

There is an easier way, especially when a calculator with sufficient memory is used. Recognize that if $X_L = 5.6$ kΩ at 1000 kHz, then $X_L = 0.56$ kΩ at 100 kHz. Then X_L at any other frequency is quickly found by multiplying 0.0056 by the number of kHz. Examination of Table 14-5 shows that X_L increases proportionally with the frequency.

Similarly, since $X_C = 5.6$ kΩ at 1000 kHz, it must be equal to 56 kΩ at 100 kHz. X_C can now be found by dividing 5600 by the number of kHz. In tabular form, X_C decreases from infinity (lazy 8) to 5.6 kΩ at 1000 kHz to 2.8 kΩ at 2000 kHz.

TABLE 14-5 Frequency Response of Parallel Circuit ($Q = 10$)[a]

f (kHz)	X_L (kΩ)	X_C (kΩ)	I_L (μA)	I_C (μA)	I_T (μA)	θ (deg)	Z (kΩ)
0	0	∞	∞	0	∞	90	0
100	0.5600	56.00	178,600	1,786	176,800	89.42	0.5656
200	1.120	28.00	89,290	3,571	85,730	88.81	1.166
300	1.680	18.67	59,520	5,357	54,200	88.11	1.845
400	2.240	14.00	44,640	7,143	37,540	87.27	2.664
500	2.800	11.20	35,710	8,929	26,850	86.19	3.725
600	3.360	9.333	29,762	10,710	19,130	84.64	5.227
700	3.920	8.000	25,510	12,500	13,130	82.18	7.613
800	4.480	7.000	22,320	14,290	8,232	77.47	12.15
900	5.040	6.222	19,840	16,070	4,171	64.65	23.97
950	5.320	5.895	18,800	16,960	2,559	45.74	39.08
1,000	5.600	5.600	17,860	17,860	1,786	0	56.00
1,050	5.880	5.333	17,010	18,750	2,495	−44.31	40.07
1,100	6.160	5.091	16,233	19,642	3,848	−62.35	25.98
1,200	6.720	4.667	14,880	21,430	6,787	−74.74	14.73
1,300	7.280	4.308	13,740	23,210	9,645	−79.33	10.37
1,400	7.840	4.000	12,760	25,000	12,370	−81.70	8.081
1,500	8.400	3.733	11,900	26,790	14,990	−83.16	6.672
1,600	8.960	3.500	11,160	28,570	17,500	−84.14	5.714
1,700	9.520	3.294	10,500	30,360	19,930	−84.86	5.017
1,800	10.08	3.111	9,921	32,140	22,290	−85.41	4.486
1,900	10.64	2.947	9,398	33,930	24,590	−85.84	4.066
2,000	11.20	2.800	8,929	35,710	26,850	−86.19	3.725

[a] $L = 8.9126768 \times 10^{-4}$ H; $C = 2.8420526 \times 10^{-11}$ F; $R_p = 56$ kΩ; $V = 100$ V; $I_R = 1785.7$ μA.

Depending on the particular capabilities of your calculator, different methods for effectively performing a large number of identical calculations are possible. If, for example, the calculator has coordinate conversions, and two or more memories, the best way is to do one line at a time. If the calculator does not have coordinate conversion keys, a good way consists of first calculating and tabulating X_L and X_C for all frequencies of interest. It will be assumed that the latter procedure is to be used. Numerical results will now be illustrated for $f = 800$ kHz.

$$I_R = \frac{V}{R} = \frac{100}{56} = 1.7857 \text{ mA} = 1785.7 \text{ }\mu\text{A}$$

$$I_L = \frac{V}{X_L} = \frac{100}{4.480} = 22.32 \text{ mA} = 22,320 \text{ }\mu\text{A}$$

$$I_C = \frac{V}{X_C} = \frac{100}{7.000} = 14.29 \text{ mA} = 14,290 \text{ }\mu\text{A}$$

$$I_T = \sqrt{I_R^2 + (I_L - I_C)^2}$$

$$= \sqrt{1785.7^2 + (22,320 - 14,290)^2} = 8232 \text{ }\mu\text{A}$$

$$Z = \frac{V}{I_T} = \frac{100}{8.232} = 12.15 \text{ k}\Omega$$

$$\theta = \tan^{-1} \frac{I_L - I_C}{R} = \frac{22,320 - 14,290}{1785.7} = 77.47°$$

The numerical results for all frequencies are tabulated in Table 14-5.

Let us now examine the shapes of the total current and impedance as plotted against frequency. Figure 14-7(a) shows that the impedance is greatest and the total current the least at resonance. Recall that the amount of impedance was least at resonance for the series resonant circuit. Also, in the series resonant circuit, the current was maximum at resonance. In a sense, the two resonant circuits are opposites.

When series resonant circuits were introduced, emphasis was placed on their ability to pass a single frequency while blocking others. This characteristic of a resonant circuit is used widely in the communications industry. However, it should be recognized that the parallel resonant circuit has the same properties. Indeed, the parallel resonant circuit finds even more applications in communications than does the series resonant circuit. The impedance levels required at the resonant frequency often govern which forms of resonant circuits are chosen.

In a parallel resonant circuit the value of Q is given by

$$Q = \frac{R_P}{X_{Lr}} \tag{14-2-5}$$

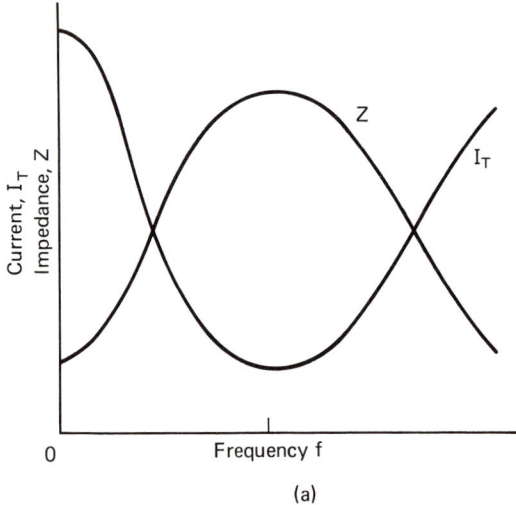
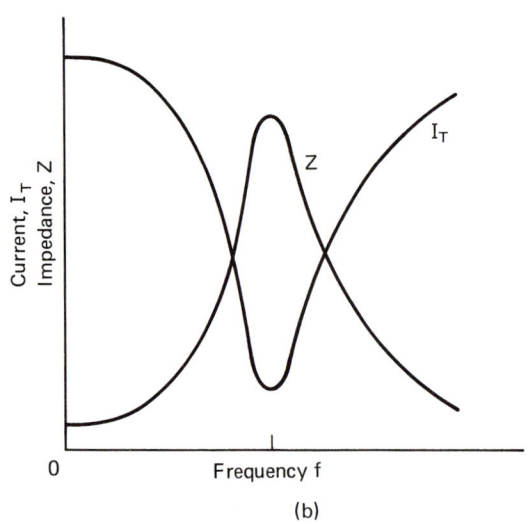

FIGURE 14-7 Frequency response for parallel resonance.

Be advised that this is not the same as Eq. (14-1-5), which applies to a series resonant circuit. In a series resonant circuit, the smaller the resistance, the larger the Q. In the parallel resonant circuit, the larger the parallel resistance R_P, the larger the Q. We will consider later the possibility of another location of the resistance.

The relationship between BW, Q and f_r still holds. For Example 14-11,

$$Q = \frac{R_P}{X_{Lr}} = \frac{56}{5.6} = 10$$

But what exactly is the bandwidth? In the parallel resonant circuit, the bandwidth is the span of frequency between two points on the impedance curve. These points occur when the impedance is 70.7% of the maximum value. Now if the bandwidth is 100 kHz, the lower end of the band should be about 50 kHz lower than 1000 kHz (for Example 14-11). It should be 950 kHz. Of course, the upper point should be approximately 50 kHz above 1000 kHz, which is 1050 kHz. Let us try it out. The maximum impedance occurs at 1000 kHz. It is 56 kΩ. At the lower and upper ends of the band, we have

$$56 \times 0.707 = 39.6 \text{ k}\Omega$$

However, from Table 14-5 we see that the impedance at 950 kHz is 39.08 kΩ and at 1050 kHz it is 40.07 kΩ. As stated earlier, the relationship is only approximate for lower values of Q. It is generally sufficiently accurate for most purposes when Q is greater than 10.

Here is another observation. It can be shown that the phase angle should be either 45° or −45° for the 70.7% points. Notice that the phase angle is 45.74° at 950 kHz and −44.31° at 1050 kHz. Once again, the values are reasonably close but not exact.

We now look at another numerical example for which $Q = 100$.

EXAMPLE 14-12 Repeat Example 14-11 for a parallel circuit which is identical except that $X_L = X_C = 0.56$ kΩ at resonance and $V = 10$ V.

Solution The tabulated results are presented in Table 14-6. Current and impedance are plotted as functions of frequency in Fig. 14-7(b).

We are now in a position to compare two circuits with different values of Q. In the last circuit

$$Q = \frac{R_P}{X_{Lr}} = \frac{56}{0.56} = 100$$

$$\text{BW} = \frac{f_r}{Q} = \frac{1000}{100} = 10 \text{ kHz}$$

This means that Z should be 70.7% of its maximum (resonant frequency) value at 995 kHz and 1005 kHz. As calculated for Example 14-11, $56 \times 0.707 = 39.6$ kΩ. From Table 14-6 we see 39.55 kΩ at 995 kHz and 39.65 kΩ at 1005 kHz. The two impedance values are very close.

In addition, we expect that the phase angle should be +45° at 995 kHz

Sec. 14-2 Parallel Resonance

TABLE 14-6 Frequency Response of Parallel Circuit ($Q = 100$)[a]

f (kHz)	X_L (kΩ)	X_C (kΩ)	I_L (μA)	I_C (μA)	I_T (μA)	θ (deg)	Z (kΩ)
100	0.0560	5.600	1,786,000	17,860	1,768,000	89.94	0.566
500	0.2800	1.120	357,100	89,290	267,900	89.61	0.3733
700	0.3920	0.800	255,100	125,000	130,100	89.21	0.7686
800	0.4480	0.700	223,200	142,900	80,380	88.73	1.244
900	0.5040	0.622	198,400	160,700	37,740	87.29	2.650
950	0.5320	0.590	188,000	169,600	18,410	84.43	5.431
980	0.5488	0.571	182,200	175,000	7,433	76.10	13.45
990	0.5544	0.566	180,400	176,800	4,009	63.55	24.94
995	0.5572	0.5628	179,500	177,700	1,790	45.07	39.55
1,000	0.5600	0.560	178,600	178,600	1,786	0	56.00
1,005	0.5628	0.5572	177,700	179,500	1,781	−44.93	39.65
1,010	0.5656	0.555	176,800	180,400	3,977	−63.32	25.14
1,020	0.5712	0.549	175,100	182,100	7,295	−75.83	13.71
1,050	0.5880	0.533	170,100	187,500	17,523	−84.15	5.707
1,100	0.6160	0.509	162,300	196,400	34,140	−87.00	2.929
1,200	0.6720	0.465	148,800	214,300	65,500	−88.44	1.527
1,300	0.7280	0.431	137,400	232,100	94,800	−88.92	1.055
1,500	0.8400	0.373	119,000	267,900	148,800	−89.31	0.6720
2,000	1.120	0.280	89,290	357,100	267,900	−89.62	0.3733
5,000	2.800	0.112	35,710	892,900	857,100	−89.88	0.1167

[a]$L = 8.9126768 \times 10^{-5}$ H; $C = 2.8420526 \times 10^{-10}$ F; $R_p = 56$ kΩ; $V = 100$ V; $I_R = 1786$ μA.

and $-45°$ at 1005 kHz. The angles at these frequencies in the table are very close to the expected values.

Inspection of Table 14-6 reveals an interesting aspect of the parallel resonant circuit when reactance values and impedance values are compared. At the resonant frequency, $X_L = X_C$, which is exactly what we expect to happen. Also, $Z = 56$ Ω with $\theta = 0°$. Therefore, the impedance is equal to the resistance of the one branch as though the inductive and capacitive branches did not exist. In other words, at the resonant frequency, the total impedance is the same as would appear when the inductor and capacitor are physically removed. Even so, because of the small values of equal inductive and capacitive reactances, the currents in these two branches are large compared with that in the resistive branch. In fact, $I_L = 178,600$ μA and $I_C = 178,600$ μA while $I_T = 1786$ μA. This may seem to be confusing, if not impossible. Remember that these two large reactive currents are represented by phasors pointing in opposite directions, and cancel each other when combining the three currents. If I_T represents the total current,

$$I_L = I_C = QI_T$$

at the resonant frequency.

The parallel resonant circuit with three branches has limited usefulness for one important reason. The reason is that any practical inductor has a significant amount of resistance. Representing an inductor as having only inductance in a branch simply ignores the practical side of inductor design, construction, and application. A realistic inductor can be represented as an L or X_L in series with a resistor R_S. If the inductor is constructed of a conducting wire coiled around a form or core, R_S is the resistance of the wire. Of course R_S can be measured. For relatively low frequency applications, R_S is the same as would be measured using an ohmmeter. At relatively high frequencies, the current tends to crowd to the outside of the conductor cross section. The result of this *skin effect* is to cause R_S to be larger than the measured dc resistance.

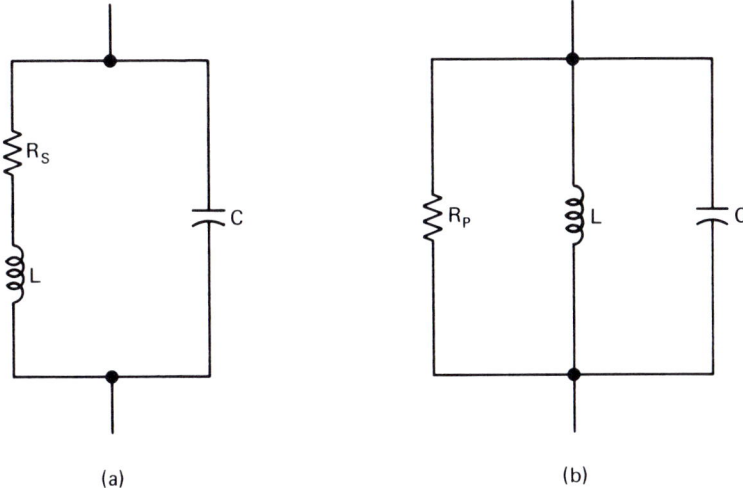

FIGURE 14-8 Practical parallel resonant circuit and its equivalent.

In Fig. 14-8(a) is shown a practical parallel resonant circuit. No parallel resistive branch appears in this figure. Figure 14-8(b) has the three-branch parallel circuit analyzed earlier. The first point to be made is that we can handle the calculations for the three-branch parallel circuit. However, we do not have the mathematical tools for the two-branch circuit. Now, if the Q for the one circuit is equal to that of the other and that Q is reasonably large, the two circuits have the same resonant frequency. They will also have the same bandwidth and the same total impedance.

In other words, it is possible to replace the practical circuit of Fig. 14-8(a) with the other. The practical "tank" circuit is found in so many types of communications equipment that it must be known to all people in the electronics field. Let us now nail down the essential features of the circuits of Fig. 14-8. First, the resonant frequency is true for both.

$$f_r = \frac{1}{2\pi\sqrt{LC}} \qquad (14\text{-}2\text{-}2)$$

Next, the resulting relationships between the circuit element and the resonant frequency for the series resonant circuit apply to the parallel resonant circuit in either form.

$$L = \frac{1}{4\pi^2 f_r^2 C} \qquad (14\text{-}2\text{-}3)$$

$$C = \frac{1}{4\pi^2 f_r^2 L} \qquad (14\text{-}2\text{-}4)$$

In addition,

$$\text{BW} = \frac{f_r}{Q} \qquad (14\text{-}1\text{-}6)$$

However, Q looks different.

$$Q = \frac{R_P}{X_{Lr}} \qquad (14\text{-}2\text{-}5)$$

$$Q = \frac{X_{Lr}}{R_S} \qquad (14\text{-}1\text{-}4)$$

Sec. 14-2 Parallel Resonance

where R_S and R_P are as shown in Fig. 14-8. In order that the two Q's be equal,

$$R_P = Q^2 R_S \qquad (14\text{-}2\text{-}6)$$

$$R_S = \frac{R_P}{Q^2} \qquad (14\text{-}2\text{-}7)$$

We quickly recognize that R_P must be relatively large for large Q. Similarly, R_S will be relatively small for large Q.

EXAMPLE 14-13 A three-branch parallel resonant circuit has $R_P = 100$ kΩ and $Q = 200$. What will be the R_S for a practical inductor so that the Q is the same?

Solution

$$R_S = \frac{R_P}{Q^2} \qquad (14\text{-}2\text{-}7)$$

$$= \frac{10^5}{4 \times 10^4} = 2.5 \ \Omega$$

EXAMPLE 14-14 Refer to the parallel circuit of Fig. 14-8(a). $R_S = 0.1 \ \Omega$, $L = 10 \ \mu\text{H}$, and $C = 1$ nF. Calculate f_r, BW, and the circuit elements for Fig. 14-8(b).

Solution There are several facts to be recognized and steps to be calculated. Since the C in each circuit is not changed by the presence of the other branch or branches, both the C and the L are the same in each. Next, let us find the resonant frequency

$$f_r = \frac{1}{2\pi\sqrt{LC}} \qquad (14\text{-}2\text{-}2)$$

$$= \frac{1}{2\pi\sqrt{10^{-5} \times 10^{-9}}} = 1.59 \times 10^6 = 1.59 \text{ MHz}$$

We will then need the value of Q.

$$Q = \frac{X_{Lr}}{R_S} = \frac{2\pi \times 1.59 \times 10^6 \times 10^{-5}}{0.1} = 1000$$

Finally

$$R_P = Q^2 R_S$$

$$= 1000^2 \times 0.1 = 100{,}000 \ \Omega = 100 \text{ k}\Omega$$

$$\text{BW} = \frac{f_r}{Q} = \frac{1.59 \times 10^6}{1000} = 1.59 \times 10^3 = 1.59 \text{ kHz}$$

In summary, the answers for a three-branch parallel circuit are:

$$R_P = 100 \text{ k}\Omega$$
$$L = 10 \text{ }\mu\text{H}$$
$$C = 1 \text{ nF}$$
$$BW = 1.59 \text{ kHz}$$

Note: The Q in this parallel circuit cannot be realized with practical values of R_S, L and C. Considerations of Q limitations, however important, will not be treated here.

14-3
Electrical Filter Concepts: Low-Pass Filters

An electrical filter is a circuit designed to "pass" ac signals in a specified frequency range. Of course, this also means that the filter will "block" ac signals having frequencies which do not fall within that range. We shall see shortly that it is both necessary and possible to be more specific about what is meant by passing or blocking ac signals. The term "signal" is commonly used in the field of electronic communications. We may think of a signal as a sine wave or a group of related sine waves whose purpose is to convey information.

To visualize a filter, let us first consider a sieve woven from thin metal strips with uniform spacing in the two perpendicular directions. The metal grid of the screen so formed has an array of identical square holes, as shown in Fig. 14-9. Such a sieve is commonly used to separate smaller particles of sand and gravel from the larger particles which cannot go through the holes. The larger particles are blocked, whereas the smaller ones pass through.

The action of the sieve is similar to that of an electrical low-pass filter which allows a lower-frequency sine wave to pass through while blocking other sine waves of higher frequencies. We will next consider how a simple electrical circuit can accomplish this. Such a circuit is shown in Fig. 14-10(a). In this schematic diagram a calibrated signal generator on the left provides a single sine-wave voltage of known effective value V_{IN} and frequency, f. A calibrated voltmeter is connected to measure the filter's V_{OUT}. The lower input and output terminals are common and are conveniently connected to the signal generator's ground output terminal.

The behavior of this circuit can be analyzed easily through its rearrangement, as shown in Fig. 14-10(b). Note that V_{IN} is applied to the impedance consisting of R and X_C connected in series; this assumes that the current drawn by the voltmeter is negligible. A modern electronic voltmeter justifies this assumption. V_{OUT} is the output voltage across the capacitor.

Extending the concept of the dc, unloaded voltage divider (Section 2-7), we are interested in the ratio of the output voltage to the input voltage, that is V_{OUT}/V_{IN}. This ratio describes the performance of the low-pass filter for any and all levels of input voltage. This ratio is equal to X_C divided by the magnitude of the impedance Z:

$$\frac{V_{OUT}}{V_{IN}} = \frac{X_C}{Z} = \frac{X_C}{\sqrt{R^2 + X_C^2}} \quad (14\text{-}3\text{-}1)$$

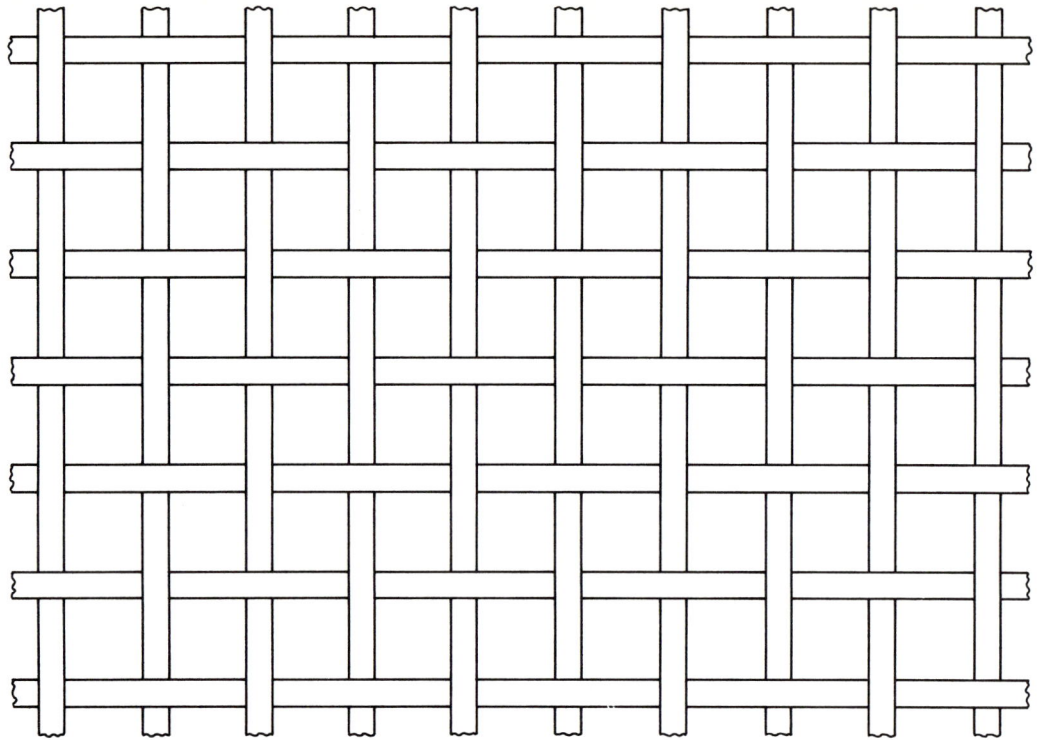

FIGURE 14-9 Sieve consisting of metal grid woven from metal strips.

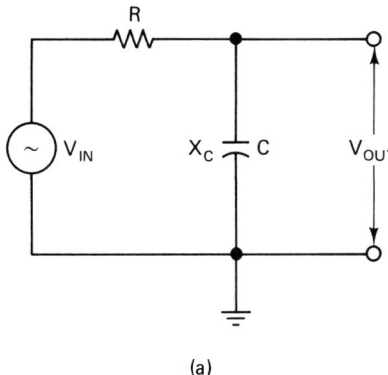

(a)

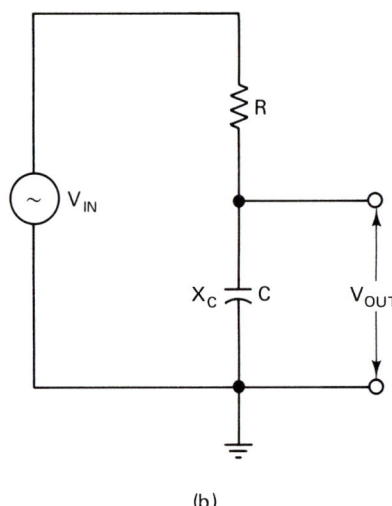

(b)

FIGURE 14-10 (a) Low-pass filter circuit; (b) low-pass filter circuit rearranged to resemble a voltage divider.

Resonant Circuits and Filters   Ch. 14

We should immediately recognize two important facts about Eq. (14-3-1). First, the form is more complicated than that of the dc voltage divider because R and X_C cannot be combined by direct addition to obtain the impedance, Z. Second, the ratio will not be the same at different frequencies because X_C is a function of the frequency. Recall that

$$X_C = \frac{1}{2\pi f C} \qquad (11\text{-}5\text{-}5)$$

However, there is no cause for alarm. The procedure for calculating the ratio for a particular circuit will be illustrated as an example. Perhaps more important, it is possible to understand the basic behavior of our low-pass filter without resorting to numerical calculations at all. The approach to doing this is useful in understanding a large variety of practical circuits.

Let us first imagine a low frequency such that X_C is large. Note that Eq. (11-5-5) tells us that when f, in the denominator, is permitted to be smaller and smaller, X_C becomes larger and larger. In particular, we wish to think of X_C being very large as compared with R. In the numerical example to be presented shortly, when $X_C = 10R$, the impedance, Z, is approximately equal to X_C, with the approximation differing from the exact value by less than 1%. This means that V_{OUT}/V_{IN}, as given in Eq. (14-3-1), is 1 to the same degree of accuracy. In other words, when the frequency is low, $V_{OUT}/V_{IN} = 1$ and we can say that low-frequency signals are passed by the filter.

We may employ similar reasoning in understanding what happens at high frequencies. Suppose that the frequency is sufficiently high to make X_C be one-tenth of R. In this case the impedance is approximately equal to R, with this approximation being less than the actual value by an error less than 1%. (For even higher frequencies the approximation is still more valid.) Therefore, at high frequencies, $V_{OUT}/V_{IN} = X_C/R$. We conclude that the output is less than 10% of the input voltage and are justified in saying that the low-pass filter blocks high-frequency signals. Note that the blocking action is even more effective, as the frequency is continually increased since X_C is inversely proportional to f.

EXAMPLE 14-15 Calculate the ratio V_{OUT}/V_{IN} for a low-pass RC filter having $R = 1592\ \Omega$ and $C = 0.1\ \mu\text{F}$ for $f = 2000$ Hz.

Solution

$$X_C = \frac{1}{2\pi f C} \qquad (11\text{-}5\text{-}5)$$

$$= \frac{1}{2\pi(2000)(10^{-7})}$$

$$= 795.8\ \Omega$$

$$\frac{V_{OUT}}{V_{IN}} = \frac{X_C}{\sqrt{R^2 + X_C^2}} \qquad (14\text{-}3\text{-}1)$$

$$= \frac{795.8}{\sqrt{1592^2 + 795.8^2}}$$

$$= 0.4472$$

To further illustrate the procedure, the calculations will be repeated and the results tabulated for selected low frequencies, high frequencies, and intermediate frequencies. "Intermediate" is the word chosen here to describe those frequencies that are neither low nor high, in accordance with the preceding discussion.

The calculations, whose results appear in Table 14-7, have been made with a scientific calculator. It will be noted that the tabulations appear to a numerical accuracy not justified with normal, or even with most precision, circuit elements. The reader can duplicate the calculations and check the numerical results with confidence but without the labor required before electronic calculators or computers were readily accessible. This low-pass filter has $C = 0.1\ \mu F = 10^{-7}$ F and $R = 10^4/2\pi = 1590\ \Omega$. This odd value of resistance has been chosen deliberately, for reasons that will be more apparent after Table 14-7 has been carefully examined.

In order that the full capability of the calculator is brought to bear in obtaining the tabulated numbers, the exact value of R should be stored in a memory and recalled each time it is needed. Proper use of memory for a multimemory calculator also requires that X_C be found one time only for each different frequency. The exact procedure to be used for the least effort, without sacrificing accuracy, depends on the particular calculator. A serious owner will invest some time in understanding effective techniques for applications of the calculating instrument.

We will now assess the performance of the low-pass filter whose schematic is shown in Fig. 14-10(a) and for which the response is tabulated in Table 14-7. A casual glance at the numbers in the column labeled V_{OUT}/V_{IN} indicates

TABLE 14-7   Low-Pass Filter Response[a]

f (Hz)	X_C (Ω)	Z (Ω)	V_{OUT}/V_{IN}	dB
10	159,200	159,200	1.000	−0.0004
20	79,580	79,590	0.9998	−0.00174
50	31,830	31,870	0.9988	−0.01084
100	15,920	15,990	0.9950	−0.04321
200	7,958	8,115	0.9806	−0.1703
300	5,305	5,539	0.9578	−0.3473
400	3,979	4,285	0.9285	−0.6446
500	3,183	3,559	0.8944	−0.9691
600	2,653	3,093	0.8575	−1.335
700	2,274	2,775	0.8192	−1.732
800	1,989	2,548	0.7809	−2.148
900	1,768	2,379	0.7433	−2.577
1,000	1,592	2,251	0.7071	−3.010
1,200	1,326	2,072	0.6402	−3.874
1,400	1,137	1,956	0.5812	−4.713
1,600	994.7	1,877	0.5300	−5.515
1,800	884.2	1,821	0.4856	−6.274
2,000	795.8	1,779	0.4472	−6.990
2,500	636.6	1,714	0.3714	−8.603
3,000	530.5	1,677	0.3162	−10.00
4,000	397.9	1,641	0.2425	−12.30
5,000	318.3	1,623	0.1961	−14.15
6,000	265.3	1,614	0.1644	−15.68
7,000	227.4	1,608	0.1414	−16.99
8,000	198.9	1,604	0.1240	−18.13
9,000	176.8	1,601	0.1104	−19.14
10,000	159.2	1,599	0.09950	−20.04
20,000	79.58	1,594	0.04994	−26.03
50,000	31.83	1,592	0.02000	−33.98
100,000	15.92	1,592	0.01000	−40.00

[a] $R = 10^4/2\pi\ \Omega$; $C = 0.1\ \mu F$; $f_C = 1000$ Hz.

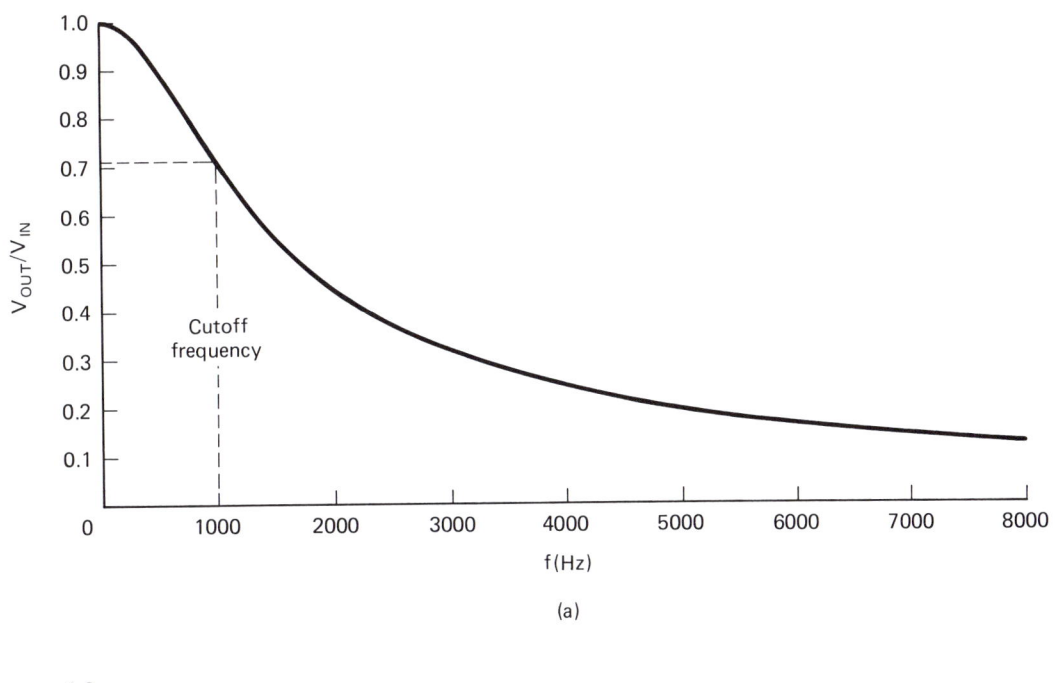

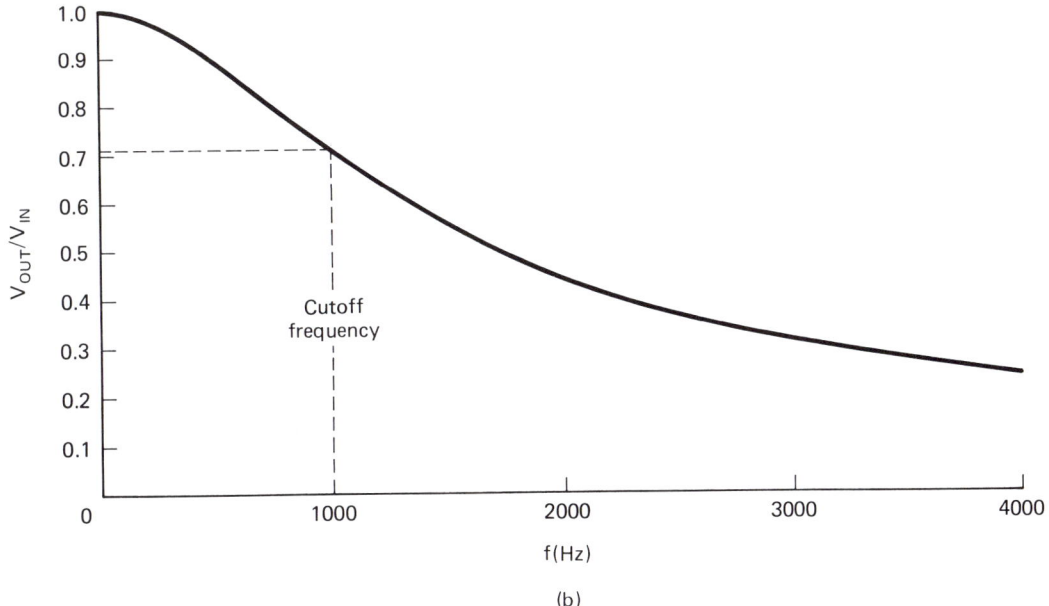

FIGURE 14-11 (a) Low-pass filter performance; (b) low-pass filter performance with expanded frequency scale.

that a sharp cutoff does not exist between the low frequencies and the high frequencies. Indeed, the rather smooth reduction in the ratio, as f is increased, contrasts with the sieve for which all gravel particles below a given size fall through; larger particles simply do not. The question regarding the low-pass electrical filter is as follows: What frequency represents the exact dividing line between the low frequencies and the high frequencies?

The answer for our filter is 1000 Hz, at which cutoff frequency, f_C, $X_C = R$, and $V_{OUT}/V_{IN} = 0.707$, referred to in the analysis of resonance as the half-power point. This particular selection for the cutoff frequency is not based on magical properties. However, it is convenient and enjoys international recognition. Figure 14-11(a) illustrates this cutoff frequency in a graph of the

Sec. 14-3 Electrical Filter Concepts: Low-Pass Filters

ratio, V_{OUT}/V_{IN} versus f. Figure 14-11(b) expands the frequency scale in the vicinity of f_C while necessarily sacrificing information about the filter response at frequencies significantly removed from f_C. A different, popular, and effective means of treating the frequency response graphically will be introduced later as the Bode plot.

14-4 High-Pass Filters

It should come as no surprise to learn that a high-pass filter can be constructed from R and C connected in series as an unloaded voltage divider. Figure 14-12(a) presents the circuit in the more conventional form in which the ac source provides V_{IN} on the left with V_{OUT} appearing on the right. Figure 14-12(b) represents a slight rearrangement to emphasize the voltage divider configuration.

At high frequencies the impedance of R and C in series is essentially R because X_C is relatively small. In other words, at high frequencies, there will be very little voltage across X_C, which means that $V_{OUT} = V_{IN}$. Indeed, this is approximately (within 1% error) true for any frequency such that X_C is less than 10% of R. Then the ratio $V_{OUT}/V_{IN} = 1$ and we consider that the filter passes the signal.

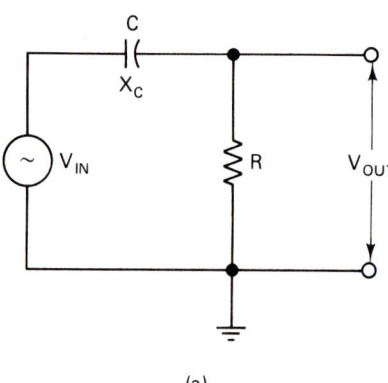

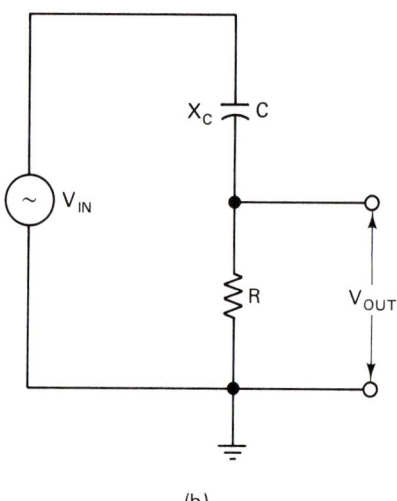

FIGURE 14-12 (a) High-pass filter circuit; (b) high-pass filter circuit rearranged to emphasize the voltage divider configuration.

TABLE 14-8 High-Pass Filter Response[a]

f (Hz)	X_C (Ω)	Z (Ω)	V_{OUT}/V_{IN}	dB
10	159,200	159,200	0.01000	−40.00
20	79,580	79,590	0.02000	−33.98
50	31,830	31,870	0.0499	−26.03
100	15,920	15,990	0.0995	−20.04
200	7,958	8,115	0.1961	−14.15
300	5,305	5,539	0.2874	−10.83
400	3,979	4,285	0.3714	−8.603
500	3,183	3,559	0.4472	−6.990
600	2,653	3,093	0.5145	−5.773
700	2,274	2,775	0.5735	−4.830
800	1,989	2,548	0.6247	−4.087
900	1,768	2,379	0.6990	−3.492
1,000	1,592	2,251	0.7071	−3.010
1,200	1,326	2,072	0.7682	−2.290
1,400	1,137	1,956	0.8137	−1.790
1,600	994.7	1,877	0.8480	−1.432
1,800	884.2	1,821	0.8472	−1.168
2,000	795.8	1,779	0.8944	−0.9691
2,500	636.6	1,714	0.9285	−0.3448
3,000	530.5	1,677	0.9487	−0.4576
4,000	397.9	1,641	0.9701	−0.2633
5,000	318.3	1,623	0.9806	−0.1703
6,000	265.3	1,614	0.9864	−0.1190
7,000	227.4	1,608	0.9900	−0.0877
8,000	198.9	1,604	0.9923	−0.0673
9,000	176.8	1,601	0.9939	−0.0533
10,000	159.2	1,599	0.9950	−0.0432
20,000	79.58	1,594	0.9988	−0.0108
50,000	31.83	1,592	0.9998	−0.0017
100,000	15.92	1,592	0.99995	−0.0004

[a] $R = 10^4/2\pi$ Ω; $C = 0.1$ μF; $f_C = 1000$ Hz.

At low frequencies the larger value of X_C predominates, so that essentially all of V_{IN} appears across X_C with none left for R, in which case V_{OUT} is very small and the signal is considered to be blocked. We may write the ratio as

$$\frac{V_{OUT}}{V_{IN}} = \frac{R}{\sqrt{R^2 + X_C^2}} \qquad (14\text{-}4\text{-}1)$$

in which the denominator $\sqrt{R^2 + X_C^2}$ is recognized as the magnitude of the total series impedance. At a low frequency for which $X_C = 10R$, $V_{OUT}/V_{IN} = R/X_C = 0.1$ with an error of less than 1%. The ratio R/X_C holds for all frequencies that are even less than the frequency just referred to.

It is informative to create a new table containing the calculated values of V_{OUT}/V_{IN} for the high-pass filter. Table 14-8 applies for $R = 10^4/2\pi$ Ω and $C = 0.1$ μF, which are the same values previously selected for the low-pass filter of Table 14-7. This selection automatically singles out 1000 Hz as the half-power frequency. The cutoff frequency, f_C, is commonly accepted to be 1000 Hz for this high-pass filter in the same way that it was for the low-pass filter having the same circuit values. If we set $X_C = R$ as the condition for cutoff, and then solve for the cutoff frequency, we find

$$f_C = \frac{1}{2\pi RC} \qquad (14\text{-}4\text{-}2)$$

Sec. 14-4 High-Pass Filters

Note that Eq. (14-4-2) also yields the cutoff frequency for the low-pass filter of Section 14-3.

EXAMPLE 14-16 Calculate the cutoff frequency using the numerical values of R and C given for Table 14-8.

Solution Using $R = 10^4/2\pi$ Ω and $C = 0.1$ μF $= 10^{-7}$ F,

$$f_C = \frac{1}{2\pi RC} \qquad (14\text{-}4\text{-}2)$$

$$= \frac{1}{(2\pi)(10^4/2\pi)(10^{-7})} = 1000 \text{ Hz}$$

Note that the strange choice of $R = 10^4/2\pi$ Ω was deliberately made to cause f_C to be exactly 1000 Hz.

Before moving on to filters that cannot properly be labeled as either low-pass or high-pass filters (like the two RC circuits that we have been studying), we should be aware that many different types of both low-pass and high-pass filters are possible. For example, we can construct either a low-pass or a high-pass filter having more than one capacitor and more than one resistor. These more complicated circuits are usually designed for sharper cutoff and to have other desirable features. Low-pass and high-pass filters are also commonly designed with capacitors and inductors. We have used the RC circuits to illustrate the basic ideas of filters with the least complexity.

The subject of filters is not only very important but also covers a lot of technical territory which is usually explored in more advanced courses. A brief overview of topics to be encountered in those future investigations will be presented before leaving the general subject of filters.

14-5 Band-Pass Filters

The resonant circuits of Sections 14-1 and 14-2 have provided an introduction to the more conventional band-pass filters which are widely used in practical electronic equipment. Almost all radio receivers have at least one resonant circuit which is tuned to match a transmitter frequency. Tuning is commonly accomplished by manually varying the capacitance which has been placed in series, or in parallel, with the fixed inductor. At this point the band-pass filter will be examined using circuits that do not require resonance. Emphasis will be placed on a simple design using the RC configurations of Sections 14-3 and 14-4 (low-pass filter and high-pass filter) rather than that of typical practical filters.

In the preceding explanations of RC low-pass and high-pass filters, an example of each, consisting of a single RC circuit combination with a cutoff frequency of 1000 Hz, was used. The frequency response of each was calculated and given in tabular form (see Tables 14-7 and 14-8).

We propose here to create a band-pass filter by connecting the output of the same low-pass filter to the input terminals of the high-pass filter already considered. The overall frequency response of this tandem combination will be that of a band-pass filter having a peak value for V_{OUT}/V_{IN} at 1000 Hz.

TABLE 14-9 Band-Pass Filter Response[a]

f, Hz	LOW-PASS SECTION		HIGH-PASS SECTION		BAND-PASS FILTER	
	V_{OUT}/V_{IN}	dB	V_{OUT}/V_{IN}	dB	V_{OUT}/V_{IN}	dB
10	1.000	−0.0004	0.01000	−40.00	0.01000	−40.00
20	0.9998	−0.00174	0.02000	−33.98	0.01999	−33.90
50	0.9988	−0.01084	0.04994	−26.03	0.04988	−26.04
100	0.9950	−0.04321	0.09950	−20.04	0.09901	−20.09
200	0.9806	−0.1703	0.1961	−14.15	0.1923	−14.32
300	0.9578	−0.3743	0.2874	−10.83	0.2752	−11.21
400	0.9285	−0.6446	0.3714	−8.603	0.3448	−9.248
500	0.8944	−0.9691	0.4472	−6.990	0.4000	−7.959
600	0.8575	−1.335	0.5145	−5.773	0.4412	−7.108
700	0.8192	−1.732	0.5735	−4.830	0.4698	−6.562
800	0.7809	−2.148	0.6247	−4.087	0.4878	−6.235
900	0.7433	−2.577	0.6690	−3.492	0.4972	−6.069
1000	0.7071	−3.010	0.7071	−3.010	0.5000	−6.021
1200	0.6402	−3.874	0.7682	−2.290	0.4918	−6.164
1400	0.5812	−4.713	0.8137	−1.790	0.4730	−6.503
1600	0.5300	−5.515	0.8480	−1.432	0.4494	−6.947
1800	0.4856	−6.274	0.8742	−1.168	0.4245	−7.442
2000	0.4472	−6.990	0.8944	−0.9691	0.4000	−7.959
2500	0.3714	−8.603	0.9285	−0.6444	0.3448	−9.248
3000	0.3162	−10.00	0.9487	−0.4576	0.3000	−10.46
4000	0.2425	−12.30	0.9701	−0.2633	0.2353	−12.57
5000	0.1961	−14.15	0.9806	−0.1703	0.1923	−14.32
6000	0.1644	−15.68	0.9864	−0.1190	0.1622	−15.80
7000	0.1414	−16.99	0.9900	−0.0877	0.1414	−17.08
8000	0.1240	−18.13	0.9923	−0.0673	0.1231	−18.20
9000	0.1140	−19.14	0.9939	−0.0533	0.1098	−19.19
10,000	0.09950	−20.04	0.9950	−0.0432	0.09901	−20.09
20,000	0.04994	−26.03	0.9988	−0.0108	0.04988	−26.04
50,000	0.02000	−33.98	0.9998	−0.0017	0.01999	−33.98
100,000	0.01000	−40.00	0.99995	−0.0004	0.01000	−40.00

[a] $R = 10^4/2\pi \ \Omega$; $C = 0.1 \ \mu F$.

Indeed, the frequency response of the band-pass filter will be found, point by point, as the product of V_{OUT}/V_{IN} for the individual, component filter circuits. One advantage in this approach lies in the present availability of the calculated data recorded in Tables 14-7 and 14-8.

Consider the form of Table 14-9. As before, selected frequencies are recorded in the leftmost column. In the next column appear the values of V_{OUT}/V_{IN} for the low-pass filter extracted from Table 14-7. Similarly, the fourth column contains the values of V_{OUT}/V_{IN} from Table 14-8. The number in the sixth and next-to-last column is the product, at any particular frequency, of the quantities from the other two columns.

EXAMPLE 14-17 Calculate the value of V_{OUT}/V_{IN} at the frequency of 700 Hz in Table 14-9.

Solution

from Table 14-7, $V_{OUT}/V_{IN} = 0.8192$

from Table 14-8, $V_{OUT}/V_{IN} = 0.5735$

from Table 14-9, $V_{OUT}/V_{IN} = 0.8192 \times 0.5735 = 0.4698$

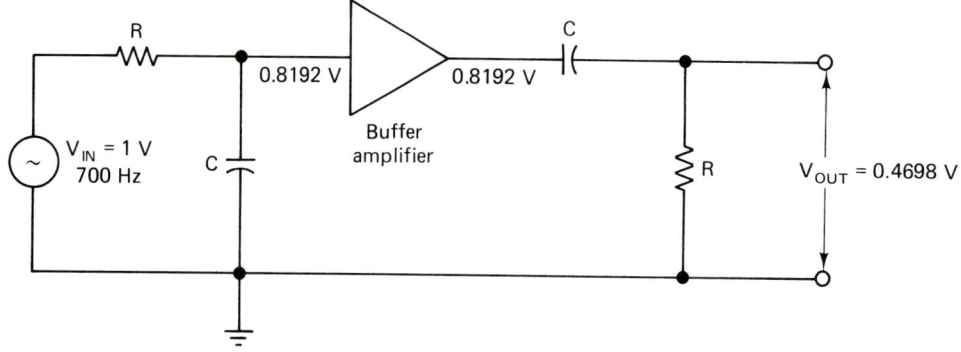

FIGURE 14-13 Schematic representation of band-pass filter consisting of low-pass and high-pass filters in tandem.

This calculation for 700 Hz is summarized in the schematic representation of the filter shown in Fig. 14-13 by assuming that $V_{IN} = 1$ V. A triangular symbol labeled BUFFER AMPLIFIER appears unannounced in this figure. One might ask: What is a buffer amplifier, why is it here, and how did it get in this book?

The buffer amplifier is included in Fig. 14-13 because, without it, the low-pass filter output will be loaded by the presence of the high-pass filter. Recall that the original low-pass and high-pass filters were analyzed as unloaded voltage dividers. The connection of the buffer amplifier to the output of the first filter section preserves the no-load condition because the high input impedance of the buffer amplifier does not disturb the low-pass filter circuit. This is similar to the manner in which an electronic voltmeter usually has very little effect in loading the circuit, as in the application of Fig. 14-11. Moreover, the buffer amplifier output voltage is identical to its input voltage. It is not necessary to know any more about this particular type of amplifier. However, this buffer amplifier is closely related to the other amplifiers (op amps), which are widely used in implementing a class of electrical filters known as "active filters."

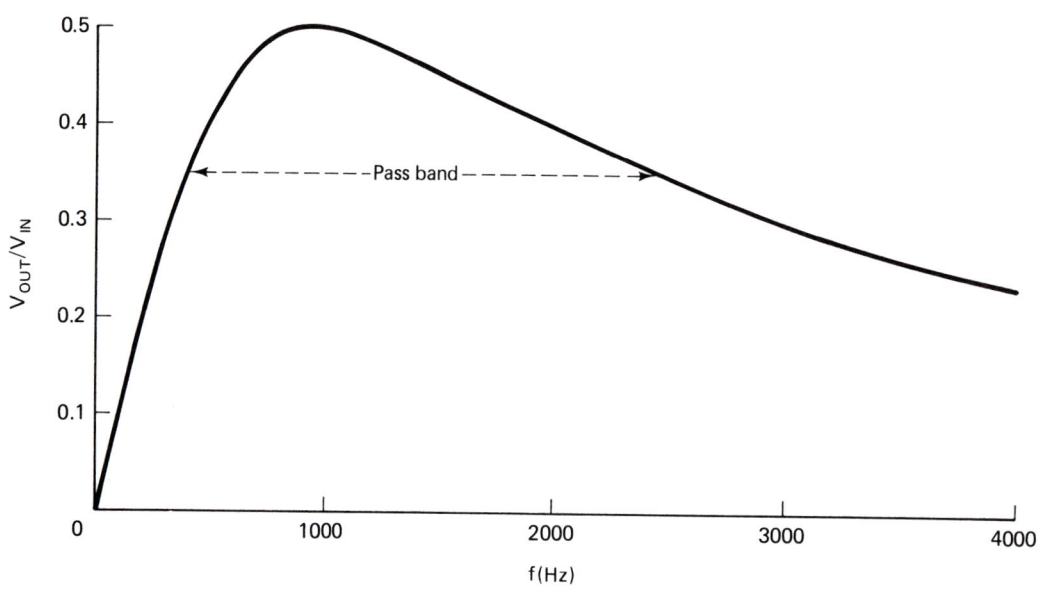

FIGURE 14-14 Performance of band-pass filter.

Figure 14-14 provides a graphical summary of the process detailed in Table 14-9. One can argue that the band-pass filter constructed by placing the other two in tandem is inefficient because the peak value of $V_{OUT}/V_{IN} = 0.5$ at the cutoff frequency. Such an argument is valid, but the present purpose is to demonstrate the principle of a band-pass filter in a simple way. Numerous other possibilities exist through modification of the present form or the adoption of the many other possible forms that practical band-pass filters take.

A band-pass filter is said to have a "bandwidth" which is the frequency difference between the upper and lower half-power points. The half-power points are those two frequencies, one below and one above the peak response, for which the numerical value of V_{OUT}/V_{IN} is 0.707 that of the peak frequency response.

EXAMPLE 14-18 Determine the bandwidth of the band-pass filter of Table 14-9 using the graph of Fig. 14-14.

Solution Referring to Fig. 14-14, the peak response is seen to be $V_{OUT}/V_{IN} = 0.5$. Therefore, the lower and upper half-power points will have $V_{OUT}/V_{IN} = 0.5 \times 0.707 = 0.3535$. Again from Fig. 14-14, $V_{OUT}/V_{IN} = 0.35$ approximately at $f = 400$ Hz and $f = 2450$ Hz. The bandwidth is then

$$2450 - 400 = 2050 \text{ Hz}$$

14-6 Band-Elimination (Notch) Filters

There are occasions when it is desirable to discriminate against a narrow band of frequencies. The frequency response graph should then show a notch covering those frequencies, as is indicated in a general way by Fig. 14-15. A radio

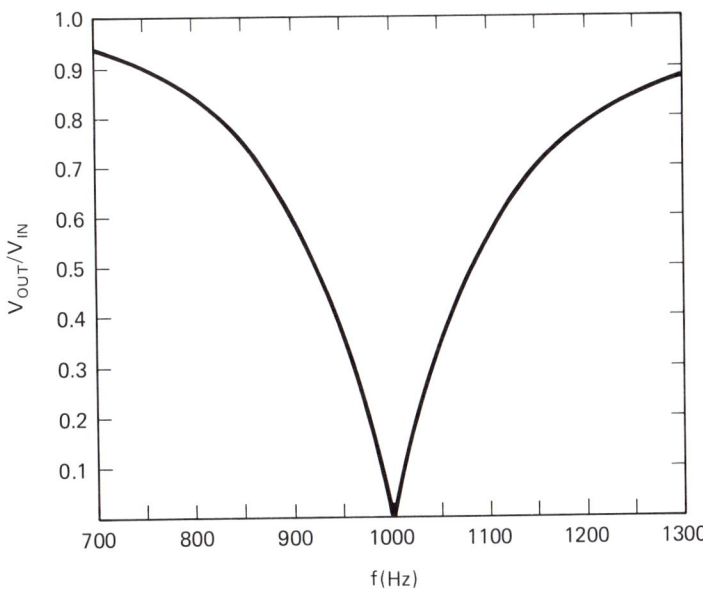

FIGURE 14-15 Characteristic of band-elimination filter.

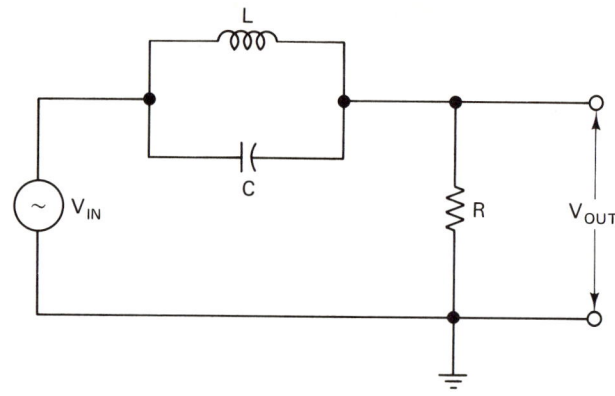

FIGURE 14-16 Simple band-elimination (notch) filter.

receiver operating in the vicinity of a powerful radio transmitter might perform more effectively if a band-elimination filter is used to greatly reduce the amplitude of the received signal at that transmitter frequency. For example, a commonly used circuit is shown in Fig. 14-16. The parallel LC combination is tuned to resonance at the frequency to be eliminated. At that frequency, the L and C in parallel present a high impedance. Following the line of reasoning used previously for the unloaded voltage divider, V_{OUT}, which appears across R, will be small. At frequencies off resonance, the parallel LC combination will yield a small impedance so that V_{OUT}/V_{IN} will be only slightly less than 1.

14-7 Decibels

Electronic equipment incorporating electrical filters often accommodates ac sine waves which are relatively large, as in the case of a powerful radio transmitter. Other equipment, such as a sensitive radio receiver, must handle extremely weak signals. More important, the overall range of signal levels, from weak to strong, in a system may be extremely great. For communications, the transmitter could have an output of several thousand volts. A distant receiver will have a sensitivity of several microvolts. The ratio of the stronger to the weaker signal levels is of the order of 1 billion (10^9). For frequencies well outside the passband of filters, the level of input, as compared to the output, may be similar or even much greater in magnitude.

Extremely large or small signal levels, and their ratios, may be expressed conveniently using a logarithmic scale and something called the decibel. It works like this: If two voltage levels in the same circuit are different by a factor of 10, the power levels differ by a factor of $10^2 = 100$. The exponent "2" is the key, as this is the number of bels for the power ratio. However, in electronics we commonly prefer the "decibel," so that this same power ratio is expressed as $2 \times 10 = 20$ decibels, abbreviated 20 dB, rather than 2 bels. The formula for finding the number of decibels with a known voltage ratio is

$$\text{number of decibels} = 20 \log_{10} (\text{voltage ratio}) \qquad (14\text{-}7\text{-}1)$$

where $\log_{10}$, or simply log, is the common, base 10, logarithm of that voltage ratio. The common logarithm is the power to which the base 10 must be raised to equal the number that is the voltage ratio in Eq. (14-7-1). There is usually no great difficulty in understanding and working with logarithms in certain cases. Thus $\log_{10}100 = 2$, $\log_{10}1000 = 3$, $\log_{10}10{,}000 = 4$, $\log_{10}0.1 = -1$, $\log_{10}0.01 = -2$, $\log_{10}0.001 = -3$, and so on. Finding the common logarithm for other numbers usually requires the use of tables, a slide rule, a scientific calculator, a computer, and so on. This task is so convenient with the scientific calculator that we need go no further in searching for an easy way. Also, subsequently multiplying the common logarithm value (in the calculator) by 20, in accordance with Eq. (14-7-1), requires the least number of keyboard entries.

EXAMPLE 14-19 The input to a certain filter at 15 kHz is 3 mV and the output is 1.5 mV. Calculate V_{OUT}/V_{IN} and express this ratio as the number of decibels.

Solution

$$\text{number of decibels} = 20 \log_{10} \frac{(V_{OUT})}{(V_{IN})}$$

$$= 20 \log_{10} \frac{(1.5)}{(3)} \qquad (14\text{-}7\text{-}1)$$

$$= 20(-0.30103) = -6.02 \text{ dB}$$

In order that the use of decibels and the calculation procedure be demonstrated further, the low-pass filter response table (Table 14-7) was repeated in Table 14-9. An additional column containing the number of decibels has also been added in Table 14-9. Corresponding to the value of V_{OUT}/V_{IN} for any frequency, the number of decibels appears in the third column. For example, selecting the row for which $f = 5000$ Hz, the value of $V_{OUT}/V_{IN} = 0.1961$ and $20 \log_{10} 0.1961 = -14.15$ dB.

Examination of Table 14-9 reveals that the frequency response for the low-pass filter, as expressed in decibels, is always negative. This means that the output voltage is less than the input voltage. We say that the circuit causes attenuation and that an applied voltage is attenuated. Of course, the amount of attenuation depends on the frequency. At low frequencies, the number of decibels in numerically small (and negative). Low-frequency signals are passed with little attenuation by the low-pass filter.

EXAMPLE 14-20 What is the attenuation of the low-pass filter in Fig. 14-10(a) at 100 Hz?

Solution At 100 Hz the ratio of $V_{OUT}/V_{IN} = 0.9950$.

$$\text{number of decibels} = 20 \log_{10} 0.9950 = -0.04$$

The amount of attenuation is 0.04 dB.

In direct contrast, high-frequency signals suffer much attenuation. This larger amount of attenuation is indicated by the numerically large, and negative, decibels in the table.

EXAMPLE 14-21 What is the attenuation of the low-pass filter in Fig. 14-10(a) at 10,000 Hz?

Solution At 10,000 Hz the ratio $V_{OUT}/V_{IN} = 0.09950$. $20 \log_{10} 0.09950$ is approximately -20 dB. The amount of attenuation is 20 dB.

A vigilant reader may notice that the conversion of V_{OUT}/V_{IN} into decibels may give answers that are not precisely the same as shown in Table 14-9. Start with $V_{OUT}/V_{IN} = 0.906$ and obtain $20 \log_{10} 0.9806 = -0.1702$. The slight difference occurs because 0.9086 is rounded as entered in the table, whereas -0.1702 was obtained using a more accurate result, that is, one with more digits calculated from scratch with the original data.

Table 14-8 for the high-pass filter has also been repeated, with the addition of a decibel column, as seen in Table 14-9. A procedure, identical to the one just described for the low-pass filter, was employed in calculating the response in decibels at any given frequency. We again see that all decibel entries are negative, so that the output voltage is always less than the input voltage. However, for the high-pass filter, there is very little attenuation of high-frequency sine waves and great attenuation of low-frequency sine waves. That is not surprising, as the purpose of the filter is to attenuate low frequencies while passing high frequencies with little attenuation.

Observe that the number of decibels is approximately -3 for the high-pass filter at the cutoff frequency, $f_C = 1000$ Hz. This is also true for the low-pass filter at 1000 Hz as is seen in Table 14-9. For both filters f_C was established through the selection of the same R and C values, in accordance with Eq. (14-4-2).

Table 14-9 for the bandpass filter applies to the schematic diagram of Fig. 14-13, for which the frequency response is presented in Fig. 14-14. We will now proceed to examine this table and the method used for its construction.

The leftmost column in Table 14-9 lists the frequency of V_{IN}, and hence V_{OUT}. These are exactly the same frequencies that are contained in the frequency response tables for the same low-pass filter and the same high-pass filter analyzed previously. Next, it will be noted that the remainder of the band-pass filter table (Table 14-9) contains the new information for the band-pass filter obtained by a procedure to be explained now.

For the band-pass filter, the ratio V_{OUT}/V_{IN} is obtained as the product of the individual values of V_{OUT}/V_{IN} for the component filters of which it is composed. This is a valid procedure because the two filters are separated by the buffer amplifier, so that each filter functions independently of the other.

EXAMPLE 14-22 Find the response for the band-pass filter at $f = 800$ Hz.

Solution From the table the two ratios are 0.7809 and 0.6247. Their product is $0.7809 \times 0.6247 = 0.4878$ as listed in the next-to-last column on the right side of Table 14-9.

It is also possible to determine the frequency response of the band-pass filter in decibels using the number of decibels for each of the two filter sections (low-pass and high-pass). This is done simply by adding the number of decibels algebraically.

EXAMPLE 14-23 Find the response for the band-pass filter at $f = 800$ Hz using the decibel method.

Solution Adding -2.148 and -4.087 gives -6.235 dB.

If this procedure is to be trusted, it should be possible to convert the V_{OUT}/V_{IN} value of 0.4878 to decibels and obtain the same answer as was obtained in Example 14-23. It works since $20 \log_{10} 0.4878 = -6.235$ dB. This illustrates one practical advantage of describing the performance of individual system elements in terms of decibels. Thinking ahead to communications systems composed of many stages having gain, such as amplifiers, or attenuation, the overall system gain in decibels can be found by addition and subtraction. The full significance of this can only be appreciated in a later phase of electronics studies.

14-8 Bode Plot

The Bode plot provides a convenient way for representing the frequency response of filters in graphical form. Figure 14-17(a) shows the basic arrangement of the vertical and horizontal scale lines employed for the Bode plot. The horizontal lines are evenly spaced and customarily represent the output level compared to the input level as a ratio expressed in decibels. However, the vertical lines are unevenly spaced, the distance separating them being less as one proceeds from left to right. These vertical lines are marked with the numerical values and frequency units in creating coordinate scales. The arrangement of scales and the graph paper are commonly referred to as "semilog" to indicate that the nonlinear, logarithmic spacing exists in one direction only. One cycle on the nonlinear scale, such as exists in Fig. 14-17(a), can accommodate a single-frequency decade so that the highest frequency is a power of 10, and at the same time, exactly 10 times greater than the lowest frequency. A number of cycles, corresponding to several decades of frequency range, are usually required to present the response of practical filters. Figure 14-17(b) shows the appearance of a two-cycle semilog graph paper sheet prior to the marking of the scales.

Figure 14-18 shows the Bode plot for the low-pass filter with $f_C = 1000$ Hz (see Table 14-7). Here the frequency scale includes one decade between 10 and 100 Hz, a second decade between 100 and 1000 Hz, a third decade between 1000 and 10,000 Hz, and a fourth decade between 10,000 and 100,000 Hz. For this filter the response is closely approximated by two straight-line segments, except in the immediate vicinity of the cutoff frequency, $f_C = 1000$ Hz. The straight-line approximations of the Bode plot are of considerable advantage in constructing the graph and visualizing the frequency response of the filter. It should also be noted that the intersection of the straight-line segments occurs at f_C, for which the true graph passes through a point 3 dB

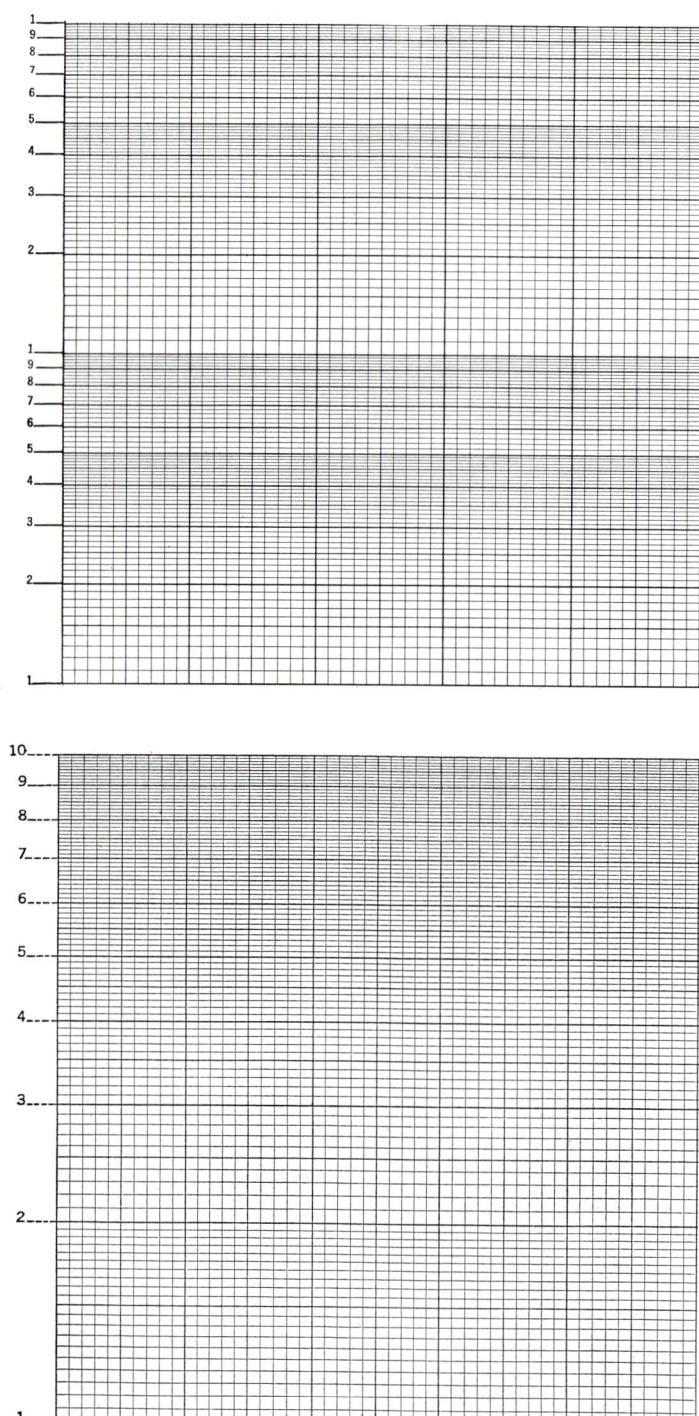

FIGURE 14-17 (a) One-cycle semilog graph paper; (b) two-cycle semilog graph paper.

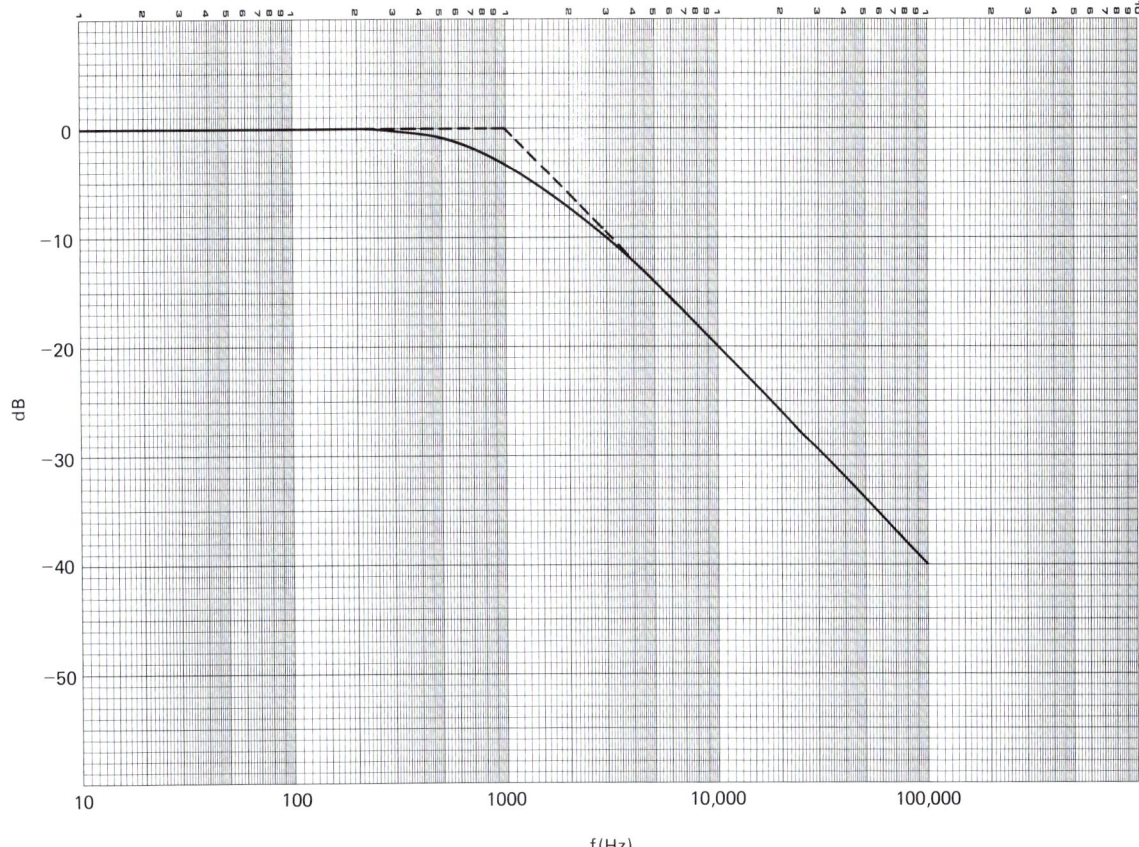

FIGURE 14-18 Bode plot for low-pass filter.

(the half-power point) below the intersection. Finally, the segment for frequencies above f_C has a slope of -20 dB per decade which continues indefinitely beyond the highest frequency (100,000 Hz) for the graph shown in the figure.

Figure 14-19 contains the Bode plot for the high-pass filter having $f_C = 1000$ Hz graphed for the data contained in Table 14-8. Again, the Bode plot demonstrates a simple appearance for which the response is two straight lines except in the neighborhood of f_C. The straight lines, when extended, intersect at f_C and the true graph passes 3 dB below the point of intersection. At frequencies well below f_C, the straight-line response has a constant slope of $+20$ dB per decade.

14-9 LC Filter Sections

As compared with the low-pass filter composed of R and C (Fig. 14-10), a more effective low-pass filter is shown in Fig. 14-20. If we regard this LC combination as an unloaded voltage divider, similar to that shown in Fig. 14-10(b), the ratio V_{OUT}/V_{IN} will decrease as the frequency increases for two reasons.

1. X_C decreases with increase of frequency.
2. X_L increases with increase of frequency.

Therefore, at higher frequencies a larger proportion of V_{IN} appears across X_L and a smaller proportion (V_{OUT}) appears across X_C. It should be noted

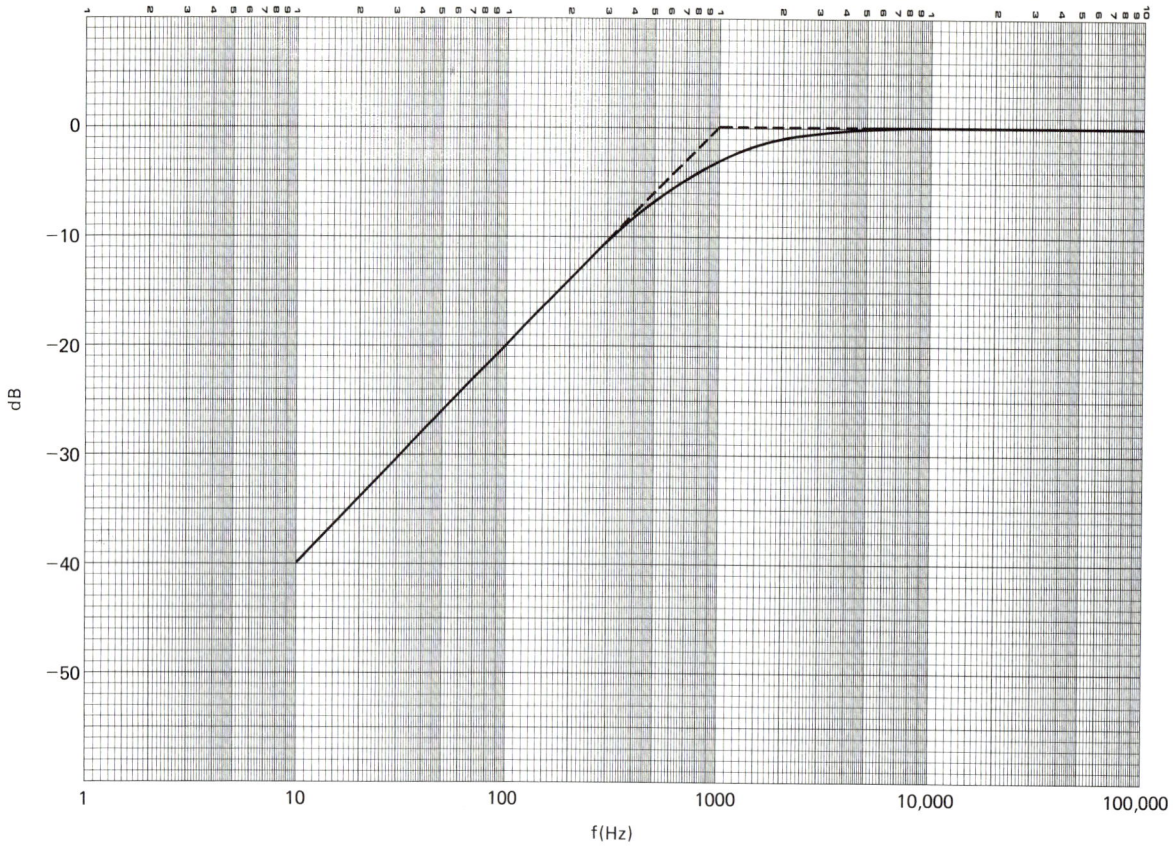

FIGURE 14-19 Bode plot for high-pass filter.

that normal applications of such filter sections, designated L-sections because of the resemblance to the letter L in the schematic diagram, the load cannot be disregarded. This means that the analysis of its behavior is made considerably more complicated and will not be pursued further here other than to present the general results in graphical form. Some additional facts will be reviewed before doing this.

Let us first consider what to expect when the positions of inductance L and capacitance C are interchanged in the circuit of Fig. 14-20. A little thought reveals that the response of the filter so obtained (i.e., V_{OUT}/V_{IN}) should rise with increase of frequency. This is the response of a high-pass filter. The loading of such a filter is such that its analysis as an unloaded voltage divider is not valid.

Filters composed of L-sections can assume different forms, as summarized in Fig. 14-21 for both the low-pass and high-pass configurations. These forms are commonly named the Π section and the T-section because of their resemblance to these symbols. Two L-sections (half-sections) properly placed in tandem compose a single Π or T section, with either combination regarded as a full section.

Figure 14-21 also summarizes the circuits composed of L and C having bandpass filter properties. The series elements are selected for series resonance at the center frequency of the passband. This selection yields the minimum series impedance at that frequency. The paralleled capacitance and inductance elements are selected for parallel resonance at the same center

FIGURE 14-20 (a) Low-pass filter with L and C elements; (b) low-pass filter represented as voltage divider.

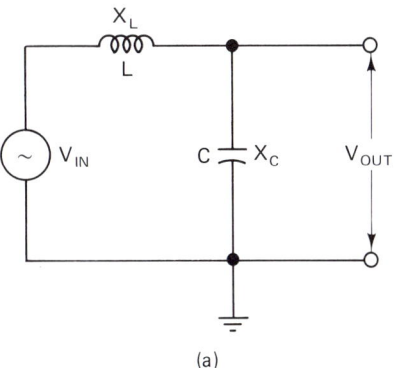

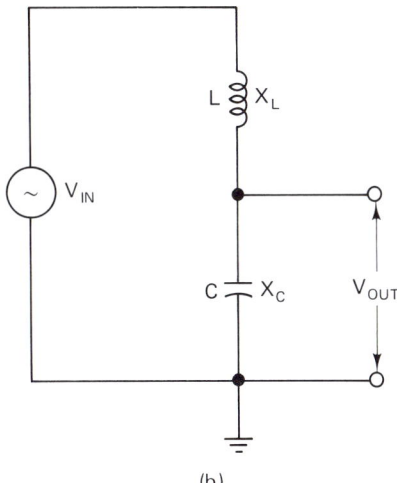

frequency. Therefore, the shunt impedance is greatest at that frequency and is relatively small at lower and higher frequencies. The combination of minimum series impedance and maximum shunt impedance is the key to the least attenuation in the passband.

Finally, a comparison of the band-elimination (notch filter) in Fig. 14-21 with the bandpass filter just described shows that the series elements in one are replaced with the paralleled elements of the other. The object is to achieve the greatest series impedance and the least shunt impedance at the center frequency for the band-elimination filter. Both series and parallel resonance at the center frequency meet these requirements.

A general form of electrical filter consists of cascaded identical sections such as those already introduced and is designated the constant-k filter. A half-section is shown in Fig. 14-22(a) with $Z_1/2$ being the impedance of a series reactive element (or elements) and $2Z_2$ being the impedance of a shunt reactive element (or elements). Properly joining such half-sections results in the T-section of Fig. 14-22(b) or Π-section of Fig. 14-22(c). When terminated with an "image" impedance Z_0, the input impedance of either the T or Π filter sections is also Z_0, for which the following relationship holds:

$$\frac{Z_0^2}{Z_1 Z_2} = 1 + \frac{Z_1}{4Z_2} \qquad (14\text{-}9\text{-}1)$$

Sec. 14-9 *LC Filter Sections*

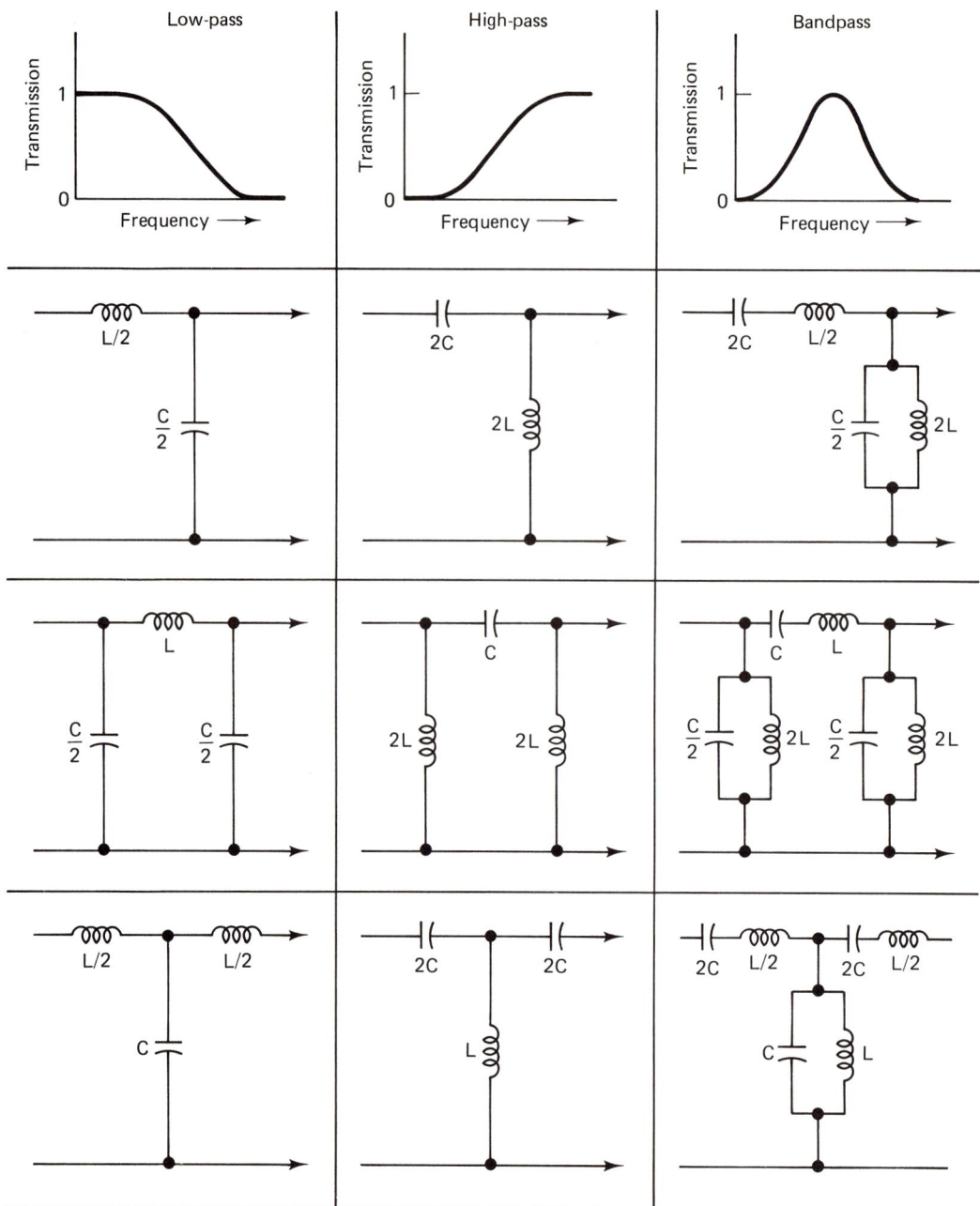

FIGURE 14-21 L, T, and π-section filter configurations.

With these terminations, any number of sections may be cascaded without changing the input impedance. The application of Eq. (14-9-1) requires that the impedances be handled mathematically as complex numbers using procedures such as those presented in Chapter 16.

As noted previously, filters can be constructed with Z_1 being inductive and Z_2 capacitive. For high-pass filters the reactive elements of the preceding arrangement are interchanged. Band-pass filters may have series-resonant

FIGURE 14-22 Filter sections: (a) half-section (L); (b) T-section; (c) π-section.

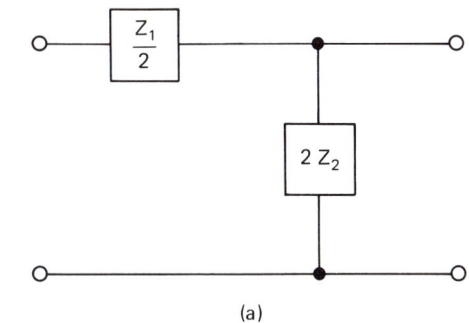

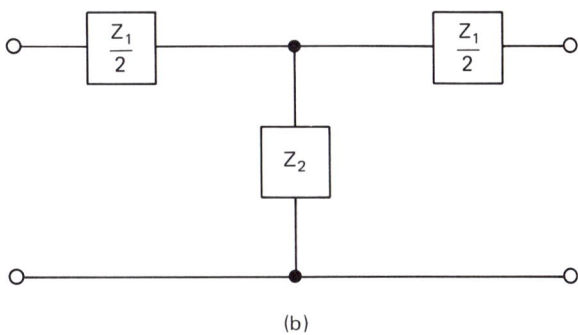

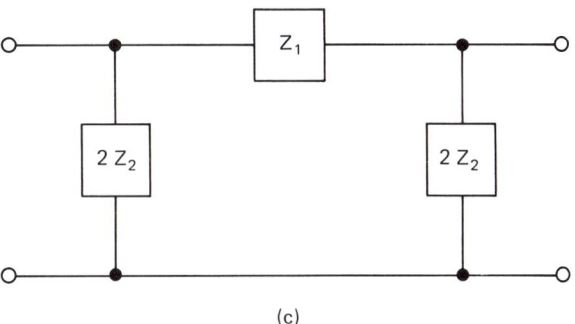

elements for Z_1 and parallel-resonant elements for Z_2. If the roles are reversed, a band-stop filter is the result. Meeting the termination requirements for Z_0 described in the preceding section causes the product of the two "image impedances" to be a constant (i.e., $k = \sqrt{Z_1 Z_2}$) giving rise to the constant-k filter name. When Z_1 and Z_2 are purely reactive, the product $Z_1 Z_2$ is independent of frequency. For example, if Z_1 is inductive and Z_2 is capacitive, $k = \sqrt{L/C}$. The attenuation-frequency curve for this T-section, low-pass filter appears in Fig. 14-23. Here the cutoff frequency is $f_C = 1$ unit.

One basic limitation of the constant-k filter is the limited slope of the skirts near the cutoff frequency. For this reason these filters are referred to as prototypes of a new line of filters, the m-derived filters, having more desirable attenuation characteristics. This class of filter will receive no further consideration in this book.

Sec. 14-9 *LC Filter Sections*

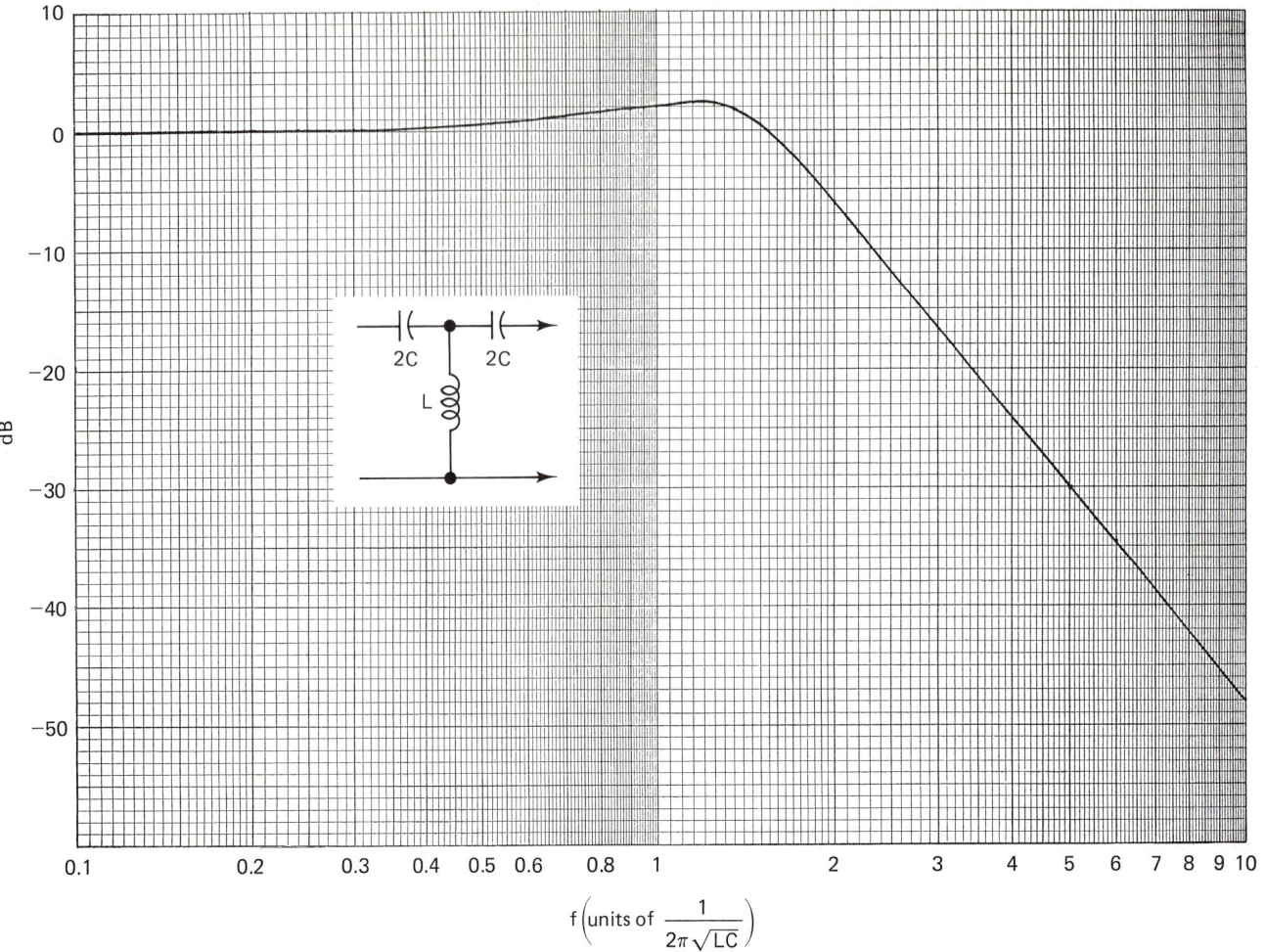

FIGURE 14-23 Characteristic curve (V_{OUT}/V_{IN}) for T-section low-pass filter.

Chapter Summary

14.1. For series resonance, the total impedance is minimum and equal to R_S at the resonant frequency.

$$f_r = \frac{1}{2\pi\sqrt{LC}}$$

$$L = \frac{1}{4\pi^2 f_r^2 C}$$

$$C = \frac{1}{4\pi^2 f_r^2 L}$$

14.2. For a series circuit the quality factor is

$$Q = \frac{2\pi f_r L}{R_S} = \frac{1}{2\pi f_r C R_S}$$

The bandwidth is related to Q:

$$\text{BW} = \frac{f_r}{Q}$$

14.3. For parallel resonance, the total impedance is maximum at resonance and equal to R_P at the resonant frequency.

$$f_r = \frac{1}{2\pi\sqrt{LC}}$$

$$L = \frac{1}{4\pi^2 f_r^2 C}$$

$$C = \frac{1}{4\pi^2 f_r^2 L}$$

14.4. For a parallel circuit the quality factor is

$$Q = \frac{R_P}{2\pi f L} = 2\pi f C R_P$$

$$\text{BW} = \frac{f_r}{Q}$$

14.5. In comparing the practical tank circuit with the three-branch parallel resonant circuit,

$$R_P = Q^2 R_S$$

14.6. For a series circuit at the resonant frequency,

$$V_L = V_C = QV_T$$

14.7. For a parallel circuit at the resonant frequency,

$$I_L = I_C = QI_T$$

14.8. For an RC low-pass filter,

$$\frac{V_{\text{OUT}}}{V_{\text{IN}}} = \frac{X_C}{\sqrt{R^2 + X_C^2}}$$

$$f_C = \frac{1}{2\pi RC}$$

14.9. For an RC high-pass filter,

$$\frac{V_{\text{OUT}}}{V_{\text{IN}}} = \frac{R}{\sqrt{R^2 + X_C^2}}$$

$$f_C = \frac{1}{2\pi RC}$$

14.10. For a known voltage ratio,

$$\text{number of decibels} = 20 \log_{10} (\text{voltage ratio})$$

14.11. The Bode plot of frequency response uses a logarithmic scale for frequency and a linear scale for the voltage ratio converted to decibels. The graph paper is referred to as semilogarithmic.

14.12. *LC* sections consist of two L-sections in either a Π or a T configuration. Each arm consists of single reactive elements, or *LC* elements in series or in parallel, depending on the purpose of the filter.

Self-Review

True-False Questions

1. T. F. In a series circuit, resonance occurs when $X_L = X_C$.
2. T. F. The Q of a series resonant circuit is decreased by increasing R.
3. T. F. The bandwidth of a resonant circuit is the frequency span between the lower and upper half-power points.
4. T. F. When a parallel circuit is resonant, the currents in the reactive branches will have the same effective values.
5. T. F. A low-pass filter will block the passage of low-frequency sine waves.
6. T. F. At any frequency above the cutoff frequency for a low-pass filter, the output voltage will be zero.
7. T. F. A high-pass filter may be constructed by placing a resistor and a capacitor in series with the output voltage appearing across the capacitor.
8. T. F. The graph of output voltage plotted against frequency for a band-pass filter will have exactly two half-power points.
9. T. F. A voltage ratio of 10 corresponds to exactly 2 dB.
10. T. F. In the vicinity of f_C the Bode plot consists only of straight-line segments.
11. T. F. If the capacitor value in a series *RCL* circuit is increased from 100 pF to 400 pF, the resonant frequency will be doubled.
12. T. F. At the half-power point of a series resonant circuit the net reactance is equal to the resistance.
13. T. F. Placing a larger value resistor in parallel with an inductor and a capacitor will decrease the Q.
14. T. F. A band-pass filter will "pass" low- and high-frequency signals and "block" signals within a limited frequency band.
15. T. F. The presence of a resistive load on the output of a simple *RC* high-pass filter will cause the cutoff frequency to increase.
16. T. F. A "band-eliminator" filter is also referred to as a "notch" filter.
17. T. F. A parallel combination of *L* and *C* placed in series with the antenna leads of a receiver will perform as a notch filter.
18. T. F. In the Bode plot the frequency scale is logarithmic.
19. T. F. The straight-line sections in the Bode plot for a low-pass filter, when extended, intersect at f_C.
20. T. F. A band-pass filter composed of *L* and *C* will have parallel resonance in the series arm and series resonance in the parallel arm at the center frequency.

Multiple-Choice Questions

21. When the graphs of *I* plotted against frequency are compared, those with sharper peaks indicate larger values of
    (a) V_T (b) Q (c) X_C (d) R (e) Z
22. If the resistance of a series resonant circuit only is doubled, the Q will be changed by what factor?
    (a) one-half   (b) one-fourth   (c) one-eighth
    (d) two   (e) will be no change

23. In a parallel resonant circuit the R_P will be
  (a) in series with X_L
  (b) in series with X_C
  (c) in series with R_S
  (d) in parallel with X_L and X_C
  (e) in series with the parallel combination of X_L and X_C

24. A parallel circuit is tuned to a particular resonant frequency. The bandwidth
  (a) is the frequency span between the lower and upper half-power frequencies
  (b) is inversely proportional to Q
  (c) is smaller with small values of R_S
  (d) is smaller with large values of R_P
  (e) all the preceding choices

25. An ac voltage of constant effective value is applied to a series RC combination. As the frequency is increased,
  (a) the voltage across R will increase
  (b) the voltage across C will increase
  (c) the current in the circuit will be constant
  (d) both (a) and (b)
  (e) all choices are correct

26. In order that the cutoff frequency of a simple RC low-pass filter be reduced, one should
  (a) decrease R (b) decrease C
  (c) increase the product RC (d) reduce the signal frequency
  (e) change from a 10% tolerance to a 5% tolerance resistor

27. In a high-pass filter the frequencies above the cutoff frequency
  (a) are greatly attenuated   (b) are passed relatively freely
  (c) are completely blocked   (d) both (a) and (c)
  (e) no correct choice

28. The cutoff frequency of a RC high-pass filter can be increased by
  (a) increasing R (b) decreasing R (c) increasing C
  (d) either (a), (c), or both (a) and (c)
  (e) no correct choice

29. A voltage of 10 V (peak to peak) is applied to the input terminals of a series RC high-pass filter. If the frequency of the source is f_C, the output voltage, as measured by a standard VOM (voltohmmeter) or DMM (digital multimeter) will be
  (a) 1.25 V   (b) 2.5 V   (c) 5.0 V   (d) 7.07 V   (e) 10 V

30. A band-pass filter will cause
  (a) low-frequency sine waves to be attenuated
  (b) sine waves of very high and very low frequencies to be severely attenuated
  (c) high-frequency sine waves to be attenuated
  (d) all choices are correct
  (e) attenuation will occur at one frequency only

31. A series circuit is resonant at 1 MHz for an inductor having L = 89.1 μH, R = 10 Ω and a capacitor with C = 284 pF. Find the resonant frequency if a second identical inductor and a second identical capacitor are each placed in series with the original combination.
  (a) 250 kHz   (b) 500 kHz   (c) 1 MHz   (d) 2 MHz   (e) 4 MHz

32. Two inductors have the same values of L. However, the effective resistance of the second inductor is twice that of the first. If the Q of the first inductor is 100 at a given frequency, the Q of the second at the same frequency will be
  (a) 50   (b) 70.7   (c) 100   (d) 141   (e) 200

33. A resonant circuit has an equivalent circuit consisting of L, C, or R_P, respectively, in three parallel branches. With R_P = 50 kΩ the Q of the circuit is 100. When a new value of R_P = 100 kΩ is placed in parallel with the L and C branches, the new Q will be
  (a) 50   (b) 70.7   (c) 100   (d) 141   (e) 200

34. A low-pass filter is constructed in the form of a voltage divider with R and C in series across V_{IN}. At the cutoff frequency
  (a) $R = C$ (b) $R = 1/C$ (c) $R = X_C$
  (d) $R = 0.707 X_C$ (e) $X_C = 0.707 R$

35. To reduce the effects of "loading" when constructing a filter using passive elements such as R and C, it is possible to use an active element known as the
    (a) negative capacitor
    (b) high-Q resistor
    (c) buffer amplifier
    (d) potentiometer
    (e) parabolic capacitor
36. The "bandwidth" of a band-pass filter is the
    (a) upper half-power point
    (b) middle half-power point
    (c) lower half-power point
    (d) the difference between (a) and (c)
    (e) choice (b) divided by $\sqrt{2}$
37. Interference is suffered in the operation of a radio communications receiver in the presence of a powerful nearby transmitter. Improvement might be obtained using a
    (a) low-pass filter
    (b) high-pass filter
    (c) band-pass filter
    (d) notch filter
    (e) all choices are equally correct
38. The vertical scale in the Bode plot customarily is marked in
    (a) volts
    (b) amperes
    (c) watts
    (d) decibels
    (e) joules
39. An LC filter section arranged in the configuration for a low-pass filter will have
    (a) L in the series arm
    (b) C in the series arm
    (c) L in the shunt arm
    (d) C in the shunt arm
    (e) both (a) and (d)
40. A band-pass filter composed of inductance and capacitance elements may have at the center frequency
    (a) series resonance in the series arm
    (b) parallel resonance in the shunt arm
    (c) series resonance in the shunt arm
    (d) parallel resonance in the series arm
    (e) both (a) and (b)

Practice Problems

41. At what frequency will a series circuit be resonant if $C = 0.001$ mF $= 1$ μF and $L = 1$ μH?
42. Determine the necessary capacitance to cause series resonance with a 100-mH inductor at 5 kHz.
43. Find the inductance for a series resonant circuit with $f_r = 25$ kHz and $C = 0.01$ mF.
44. Calculate the quality factor of a series resonant circuit for which $X_{Lr} = 5$ kΩ and $X_{Cr} = 5$ kΩ while $R = 500$ Ω.
45. A resistor of 1.8 kΩ is placed in series with a capacitor of 0.001 μF to form a low-pass filter. At what frequency will the capacitive reactance be equal to the resistance of the resistor?
46. A low-pass filter consists of a 3.3-kΩ resistor placed in series with a 0.001-μF capacitor in the arrangement of Fig. 14-10. Calculate the ratio of the output voltage to the input voltage at $f = 500$ Hz.
47. Determine the cutoff frequency, f_C, for a low-pass RC filter having $R = 3.3$ kΩ and $C = 0.001$ μF.
48. Making use of Tables 14-7 and 14-8, confirm the entry in Table 14-9 for a frequency of 1200 Hz.
49. For a high-pass filter consisting of a series RC arrangement, find the proper value of R to obtain a cutoff frequency of 5 kHz using a capacitor for which $C = 0.01$ μF.
50. The input to a particular filter at 100 Hz is 250 mV and the output voltage is 25 mV. Determine the ratio V_{IN}/V_{OUT} and convert the answer to decibels.
51. Calculate the amount of capacitance that must be placed in series with a 200-μH inductor to obtain a resonant frequency of 800 kHz.
52. A coil has an inductance of 0.2 mH and 15 Ω of resistance. When placed in parallel with a 0.001-μF capacitor, a resonant circuit is formed. Calculate the bandwidth.

53. A 10-V ac voltage source is placed across the parallel combination of Problem 52. Calculate the total current at the resonant frequency.
54. A 0.1-μF capacitor and a 10-kΩ resistor are used to construct a low-pass filter. Calculate the cutoff frequency.
55. A band-pass filter consists of a low-pass filter with 55 kHz cutoff frequency and a high-pass filter with a 45 kHz cutoff frequency placed in cascade. Estimate the center frequency and the bandwidth of this combination.
56. The input signal for an amplifier is 5 mV and the output is 2.5 V. Calculate the amplifier gain and express it in decibels.
57. An attenuator yields an output level of 10 mV for an input of 200 mV. Calculate the attenuation and express it in decibels.
58. In Fig. 14-16, $L = 150$ μH and $C = 100$ pF. Determine the frequency for which the attenuation will be greatest.
59. Refer to the semilog paper shown in Fig. 14-17. The span between two neighboring vertical lines, each marked with a "1," is one cycle. If a point on the vertical line marked "3" in one cycle represents 300 Hz, what is the interpretation of a point on the vertical line marked "3" two cycles to the right?
60. Refer to Fig. 14-18. Determine the frequency for which the Bode plot shows -3 dB.

Advanced Problems

61. A series circuit has $X_L = 50$ kΩ, $X_C = 50$ kΩ, and $R = 1$ kΩ at $f = 1$ MHz. What is the bandwidth?
62. The series circuit of Problem 61 is to be replaced by a parallel circuit that is resonant at the same frequency and having the same Q. Calculate the resistance of the resistive branch.
63. A parallel circuit is resonant at 5 MHz, has a Q of 100, and has a resistive branch consisting of 500 kΩ. Find the values of L and C in the other two branches.
64. In Problem 63, determine the additional amount of resistance to be placed in parallel with the circuit to reduce the Q to 50.
65. Convert the circuit resulting from the solution of Problem 64 to an equivalent series circuit having the same resonant frequency and the same $Q = 50$.
66. A 1-kΩ resistor is placed in series with a capacitor in constructing a low-pass filter. Determine the capacitive reactance at the frequency for which $V_{OUT}/V_{IN} = 0.1$.
67. A low-pass filter with a 1.8-kΩ resistor in series with a capacitor has a cutoff frequency, f_C, equal to 25 kHz. Calculate the new cutoff frequency when a 3.3-kΩ resistor is substituted for the other without further changes in the circuit.
68. A high-pass RC filter is to be constructed having a cutoff frequency of approximately 4.8 kHz. The available components include a 560-Ω resistor and a 820-Ω resistor together with two capacitors rated at 0.012 μF. What is the correct way to connect the resistors and capacitors?
69. The ratio V_{OUT}/V_{IN} for a filter at a particular frequency is 0.1. Express the ratio in decibels.
70. Refer to Table 14-7 for the low-pass filter. Determine the change in V_{OUT}/V_{IN} as the frequency is increased from 10 kHz to 20 kHz. Express this change in decibels and as a change in the numerical ratios.
71. An inductor has 3 Ω of resistance and unknown inductance L. When placed in series with C, resonance is achieved at 2.8 MHz. Determine L and C if the bandwidth is 50 kHz.
72. The inductor and capacitor of Problem 71 are now placed in parallel. Determine the equivalent circuit having L, C and R_P all in parallel. In this equivalent configuration the inductive branch is assumed to have no resistance.
73. A practical "tank" circuit has a 100-μH inductor placed in parallel with a variable capacitor. The capacitor is adjusted for a resonant frequency of 2 MHz, at which frequency the Q of the inductor is 125. Determine the amount of additional resistance, R_X, to be placed in parallel with the tank to obtain a new Q of 25.

74. In an emergency situation a technician must construct a high-pass filter with resistance and capacitance elements only. He has one 5.6-kΩ resistor and one 10.2-kΩ resistor, one 0.001-μF capacitor and one 0.002-μF capacitor. What will be the greatest cutoff frequency attainable using these available components?
75. An unloaded voltage divider consists of a 50-Ω resistor tapped at 10 Ω above the grounded terminal for the output. Assume an input voltage level of 500 mV and find the attenuation as expressed in decibels.
76. The distance between the vertical line marked "1" and that marked "2" in Fig. 14-17(a) is approximately 3.8 cm. Calculate the expected distance between "1" and "5."
77. Refer to Fig. 14-18. Determine the frequencies for which the response is -20 dB and -40 dB. Interpret the meaning of these readings.
78. Refer to Fig. 14-19, which is the Bode plot for a high-pass filter. The reading for 10 Hz is -40 dB. Assuming that the response rises 20 dB for each decade (factor of 10), predict the reading for 100 Hz.
79. In Fig. 14-19, what response can be predicted for $f = 0.1$ Hz?
80. In Fig. 14-23, determine the frequency for -30 dB assuming $L = 10$ μH and $C = 50$ pF.

15

Mutually Coupled Circuits

> *This chapter has the following objectives:*
>
> 1. To establish the fundamental relationships between the primary and secondary properties of ideal transformers.
> 2. To relate input (primary) and output (secondary) voltage, current, and power.
> 3. To consider the more common applications of transformers in power transmission and distribution systems and in electronic power supplies.
> 4. To investigate the role of transformers in impedance matching for maximum power transfer.
> 5. To establish the equations for transformer turns ratios required in impedance matching.
> 6. To explore the conventions commonly met in marking transformer primary and secondary windings to account for voltage phase angles.
> 7. To consider losses and their origin in practical transformers.
> 8. To explore efficiency and the consequences of transformer copper and core losses.
> 9. To describe the procedures used in troubleshooting a transformer.
> 10. To define inductances and the mathematical expressions relating self-inductances, mutual inductance, and the coefficient of coupling.
> 11. To construct equations based on Kirchhoff's laws for circuits having common physical current paths in two inductively coupled windings.
> 12. To present the equations for total equivalent inductance for mutually coupled windings connected in series and in parallel.

15-0
Introduction

The present chapter is concerned with circuits having common magnetic field linkage. Magnetic flux caused by the flow of current in one circuit links with the conductors of a second circuit. When the current through the first is changing, an electromotive force (EMF) is induced in the second. With alternating current in the first, there will be an alternating EMF in the second circuit. These circuits are said to be mutually coupled.

As a result of mutual coupling between two circuits, an ac source in one can cause current flow in another with no electrical conducting path between the two. There are practical ac circuits of this nature, while others may be electrically connected as well as magnetically coupled.

Transformers are introduced first because of the basic simplicity of their construction and analysis as treated here. This is followed by a more general approach to mutual inductance. It should be understood at the outset that mutually coupled circuits, including transformers, have mutual inductance.

Two transformers are shown in Fig. 15-1. One is considerably larger than the other and the reader should be made aware of even larger and smaller transformers. The number of leads depends on the particular application.

15-1
Transformers

In Chapter 11 we learned that self-inductance opposes the flow of alternating current through an inductor. This opposition is more directly associated with current change, so that the current in an inductor lags the applied voltage by

FIGURE 15-1 Two ac transformers.

90°. The opposition to current flow, the inductive reactance, is increased by winding more turns around a core of magnetic material such as iron. It is said that the changing current in the winding causes a change in the magnetic flux in the core which induces a voltage to oppose the change. The polarity of the induced voltage is described by Lenz's law.

There is a slightly different way of looking at the induced voltage. For instance, let us consider a coil with 1200 turns wrapped around a ferromagnetic core. If 120 V, 60 Hz is applied to the terminals of the coil, the matching induced voltage will be 120 V, 60 Hz. This statement assumes that the *IR* drop in the coil is negligible, that is, too small to be considered. This 120 V is shared equally by every single turn on the coil. In our particular case the 120 V is divided into 1200 parts because there are 1200 turns. The amount of voltage induced in each turn is 0.1 V. As a matter of fact, if a voltmeter is connected across one turn, it will indicate 0.1 V. Scraping the insulation off the wire at two points is a messy way of finding out the voltage for a single turn. A better way consists of making a separate single turn of wire and measuring the 0.1 V. In other words, the number of volts for each turn is really dictated by the amount of alternating magnetic flux inside the core and the frequency. Both are the same whether or not the single turn is one of 1200 turns or is a separate single turn.

Well, now. What if the voltmeter will not give accurate readings for voltages in the neighborhood of 0.1 V? Suppose that the lowest range for accurate readings is the 1-V range. Why not, then, make a 10-turn winding for which the induced voltage will be 1 V? We can go even further. A second winding of 100 turns will produce 10 V and a third winding of 1200 turns will produce 120 V. The device we have invented is called the *transformer*. Incidentally, the frequency need not be 60 Hz for all the preceding discussion to be true. There are some practical limitations on how much the frequency can be changed without changing the number of volts for each turn of the second winding, but we will avoid these limitations at the present time, as they are imposed by electrical laws that we have not yet studied.

The second winding of a transformer is named the *secondary*. Naturally, the first winding is called the *primary* winding. The primary winding is always connected to the ac source and the secondary winding is always connected to a load. In our discussion the load is considered to be small, as the current in the secondary is extremely small. This is the result of attaching a voltmeter of high resistance to the terminals of the secondary. The only load is the voltmeter. We will consider large loads later.

Since the total voltage for either winding is proportional to the number of turns, we can say that

$$\frac{E_p}{N_p} = \frac{E_s}{N_s} \qquad (15\text{-}1\text{-}1)$$

where E_p and E_s are the primary and secondary voltages and N_p and N_s are the number of turns in the windings. Now, a simple word of caution. We cannot double E_p by doubling N_p. In doubling N_p, we simply spread E_p over twice as many turns. This equation simply says that the voltage per turn of the primary is numerically the same as the voltage per turn of the secondary. An alternative form is as follows:

$$\frac{E_p}{E_s} = \frac{N_p}{N_s} \qquad (15\text{-}1\text{-}2)$$

Sec. 15-1 Transformers

or

$$\frac{E_s}{E_p} = \frac{N_s}{N_p} \qquad (15\text{-}1\text{-}3)$$

EXAMPLE 15-1 A transformer has $E_p = 120$ V, $N_p = 600$ turns, $E_s = 40$ V, and $N_s = 200$ turns. What is the voltage per turn in the primary and the secondary?

Solution We can use Eq. (15-1-1) to find both answers.

$$\frac{E_p}{N_p} = \frac{E_s}{N_s} \qquad (15\text{-}1\text{-}1)$$

$$\frac{120}{600} = \frac{40}{200} = 0.2$$

Incidentally, this is an unusual problem statement. We have four pieces of information given. Normally, the instructor would give you three and ask that the fourth be calculated. As this example problem is stated, only two pieces of information are necessary, but the two must be properly matched. We demonstrate the use of two pieces of information in the next example.

EXAMPLE 15-2 A transformer has $E_s = 40$ V and $N_s = 200$ turns. What is the voltage per turn in the primary and in the secondary?

Solution

$$\frac{E_p}{N_p} = \frac{E_s}{N_s} = \frac{40}{200} = 0.2 \qquad (15\text{-}1\text{-}1)$$

As you can see, the question can be answered if only E_s and N_s are known. It can also be answered if both E_p and N_p are known. Note how the two pieces of information must be properly matched.

EXAMPLE 15-3 The number of secondary turns is 1500 and the number of primary turns is 500. Find the secondary voltage when the applied primary voltage is 12 V.

Solution

$$\frac{E_s}{E_p} = \frac{N_s}{N_p} \qquad (15\text{-}1\text{-}3)$$

$$E_s = \frac{N_s}{N_p} \times E_p$$

$$= \frac{1500}{500} \times 12 = 36 \text{ V}$$

It is seen that the secondary voltage can be greater than the primary voltage. Such a transformer is called a *step-up transformer*. Very often a transformer will have a lower secondary voltage and is referred to as a *step-down transformer*.

A modern power transformer is very efficient. Essentially all of the input power is transferred to the load. If we neglect the relatively small amount of power dissipated as heat in the transformer, the power supplied by the source in the primary circuit (P_p) is equal to the secondary load power (P_s). For a purely resistive load

$$P_s = P_p$$

$$E_s I_s = E_p I_p \tag{15-1-4}$$

We can now show that

$$\frac{I_s}{I_p} = \frac{N_p}{N_s} \tag{15-1-5}$$

EXAMPLE 15-4 In the transformer of Example 15-3, the primary current is 9 A. What will the current in the secondary be for a resistive load?

Solution

$$\frac{I_s}{I_p} = \frac{N_p}{N_s} \tag{15-1-5}$$

$$I_s = I_p \frac{N_p}{N_s} = 9 \times \frac{500}{1500} = 3 \text{ A}$$

The solution of Example 15-4 is easily checked under the assumption that the power supplied by the source at the primary is equal to the power supplied to the load by the secondary. For a resistive load the secondary current will be in phase with the secondary voltage. Similarly, the primary current is approximately in phase with the primary voltage. Therefore, the load (secondary) power is

$$P_L = P_s = E_s I_s = 36 \times 3 = 108 \text{ W}$$

and the primary power is

$$P_p = E_p I_p = 12 \times 9 = 108 \text{ W}$$

The loads in the secondary circuits are most often not resistive. A typical load can be represented as a resistor in series with an inductor. We then know that the product of secondary voltage and secondary current is only an apparent power. Apparent power is greater than the number of watts of real power. Nevertheless, the equating of products in Eq. (15-1-4) is still valid. That is, the apparent power at the primary input is equal to the apparent power of the secondary load.

Sec. 15-1 Transformers

EXAMPLE 15-5 In the secondary of a transformer the current is 500 mA with 100 V across the load. For this condition the primary current is 2 A. Find the primary voltage.

Solution

$$E_s I_s = E_p I_p \qquad (15\text{-}1\text{-}4)$$

Solve for E_p.

$$E_p = \frac{E_s I_s}{I_p} = \frac{100 \times 0.5}{2} = 25 \text{ V}$$

Note that we cannot claim that $100 \times 0.5 = 50$ W in the load because we lack information to find more than the apparent power.

Transformers are widely used in public power utilities to change the voltage and current levels for the safest and most efficient operation of the system. Power is generated at moderately high voltage levels by large generators. These, for the most part, are driven by steam turbines or are hydroelectric. The voltage levels are too low for transmission over long distances to the area where the loads exist. Step-up transformers are used for this purpose. Levels of 500,000 V (effective) and more are commonly used for power transmission. At the receiving end the voltage level is stepped down for distribution and at least one more time near the consumer's location to give relatively safe levels such as 120 V (effective), as used in the home.

Electronics technicians are not often involved in planning and designing any aspect of a power generation, transmission, and distribution system. However, the persons doing this employ the same basic electrical principles which the electronics technician must know. The power expert will have additional training on the specialized equipment, measurement apparatus, and practices for the field. In the same way you must have both training and experience in your field of specialization.

However, before excusing yourself from further interest in the power transformer, do not forget the power supplies which are essential to the operation of most electronics devices. Most power supplies have transformers with characteristics similar to those of their big brothers in the power industry. For this reason, transformers are very much the concern of the technician. There is still another reason why the transformer is important in electronics. In Section 5-4 we learned that maximum load power can be achieved by matching the load resistance and the source resistance. This means that the load resistance must be adjusted to be equal to the source resistance. The same principle applies for ac as for dc. We assume that the source resistance is fixed in either case. What, then, can be done in matching when the load resistance is also fixed? This may sound like a senseless question, but the transformer may provide some assistance for ac circuits. In Fig. 15-2(a) a particular step-down transformer is shown together with the turns in the primary winding and the secondary winding, both voltages and both currents. The load is a resistance of 25 Ω. Suppose that we now calculate the load that the source experiences. Before continuing, let us explain and amplify the last statement. We all know that the source is an electronic device of some type. This device "sees" only by providing a current at some voltage. If another source is

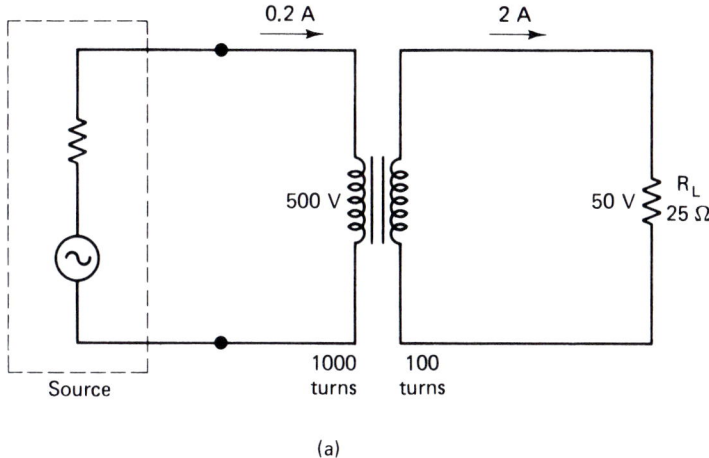

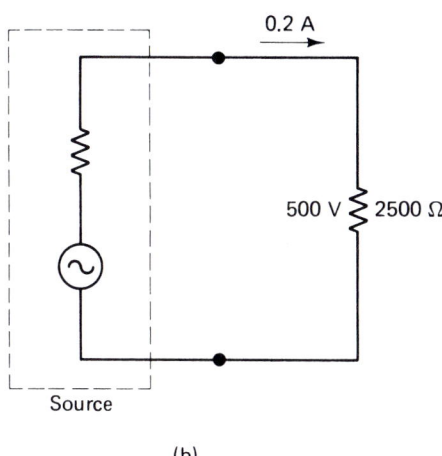

FIGURE 15-2 Transformation of load resistance using a transformer.

attached, it may provide a different current and a different voltage. However, the resistance seen by each source must be the same. It is the ratio of the primary voltage to the primary current.

$$R_p = \frac{E_p}{I_p} = \frac{500}{0.2} = 2500 \ \Omega$$

We have an interesting situation. There is a load of 25 Ω in the secondary circuit. Yet it is seen as a load of 2500 Ω by the source. Because of the limited senses of the source, the voltage source can only "see" current when a load is applied. By this means, the source can distinguish between a low-resistance load and a high-resistance load. Because of the known values of 500 V and 0.2 A, the source is exposed to a load of 2500 Ω. The source cannot distinguish between the actual circuit and the equivalent circuit shown in Fig. 15-2(b).

There is a formula to express the resistance R_p as seen in the primary circuit due to the load resistance R_s present in the secondary.

$$\frac{R_p}{R_s} = \left(\frac{N_p}{N_s}\right)^2 \qquad (15\text{-}1\text{-}6)$$

Sec. 15-1 Transformers

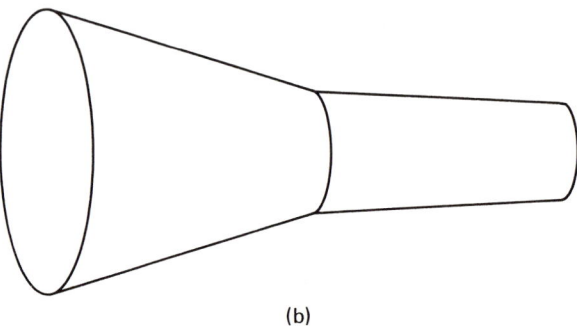

FIGURE 15-3 Illustrating load resistance transformation.

Using the transformer circuit of Fig. 15-2(a), we obtain

$$\frac{R_p}{25} = \left(\frac{1000}{100}\right)^2 = 10^2$$

$$R_p = 25 \times 10^2 = 2500\ \Omega$$

Figure 15-3(a) gives a better idea of the two windings of the 10:1 ratio step-down transformer. Without deliberately counting the number of turns in each winding, the intent is to show graphically that the primary has more turns. Figure 15-3(b) is a common funnel with the large opening on the left and the small opening on the right. If we associate the large end of the funnel with the larger number of transformer windings, a memory aid can be developed. The larger end of the funnel, larger number of turns for the transformer, larger voltage, and larger apparent resistance all go together. Similarly, the smaller end of the funnel, the smaller number of transformer winding turns, the smaller voltage, and the smaller resistance go together. It is not necessary that the large end be the primary end of the funnel, as will be demonstrated next.

EXAMPLE 15-6 A transformer has 1100 turns for the primary winding and 2200 turns for the secondary winding. Calculate the primary (source) current when the secondary load resistance is 50 Ω and the primary voltage is 40 V.

Solution The most direct method of solution is to find the resistance as seen by the voltage source. Then the primary current can be calculated using Ohm's law.

$$\frac{R_p}{R_s} = \left(\frac{N_p}{N_s}\right)^2 \tag{15-1-6}$$

$$\frac{R_p}{50} = \left(\frac{1100}{2200}\right)^2 = \left(\frac{1}{2}\right)^2$$

$$R_p = 50 \times \left(\frac{1}{2}\right)^2 = 12.5 \ \Omega$$

$$I_p = \frac{E_p}{R_p} = \frac{40}{12.5} = 3.2 \ \text{A}$$

Check
Always keep in mind that any numerical answer can usually be checked in one or more ways. The ability to check the results of a calculation quickly requires a deeper understanding of the whole picture. Let us now look at one way of checking the answer. One very elementary way of doing this compares the powers in the primary and secondary circuits.

Use Eq. (15-1-3).

$$\frac{E_s}{E_p} = \frac{N_s}{N_p}$$

$$\frac{E_s}{40} = \frac{2200}{1100} = 2$$

$$E_s = 2 \times 40 = 80 \ \text{V}$$

Now, applying Ohm's law in the secondary circuit,

$$I_s = \frac{E_s}{R_s} = \frac{80}{50} = 1.6 \ \text{A}$$

In the primary circuit, by Eq. (15-1-5),

$$\frac{I_p}{I_s} = \frac{N_s}{N_p}$$

$$\frac{I_p}{1.6} = \frac{2200}{1100} = 2$$

$$I_p = 2 \times 1.6 = 3.2 \ \text{A}$$

Finally,

$$P_p = E_p I_p = 40 \times 3.2 = 128 \ \text{W}$$

$$P_s = E_s I_s = 80 \times 1.6 = 128 \ \text{W}$$

Since the primary input power to this 1:2 ratio set-up transformer is equal to the secondary load power, we have a check of the numbers. Checking a calculation cannot be a guarantee. The potential error can appear in the check. This is more unlikely when the checking calculation is done in the way differing the most in every respect from the original calculation.

In a transformer the secondary voltage will either be in phase with the primary voltage or 180° out of phase. Figure 15-4(a) represents a transformer with the primary and secondary windings mounted side by side on one leg of a common core. The windings are shown in such a way that the directions in which the turns are wound is clear. There can be no doubt that the primary and secondary coils are wound in the same direction. Also, one end of both windings is connected to ground. With this arrangement the voltage between ground and the free terminal of the secondary are in phase with the applied primary voltage. Such an arrangement is shown schematically in Fig. 15-4(b). The two spots indicate that the primary and secondary voltages with respect to ground, are in phase.

In Fig. 15-5(a) the secondary winding has been reversed. Therefore, the secondary voltage will be 180° out of phase with the primary voltage.

In the two schematic diagrams, the relative primary and secondary polarities are indicated by the dots near the ends of the coil symbol. The ends with the dots will have a positive polarity at the same time, relative to the other ends of the windings. Of course, this is true of negative voltage polarities as well.

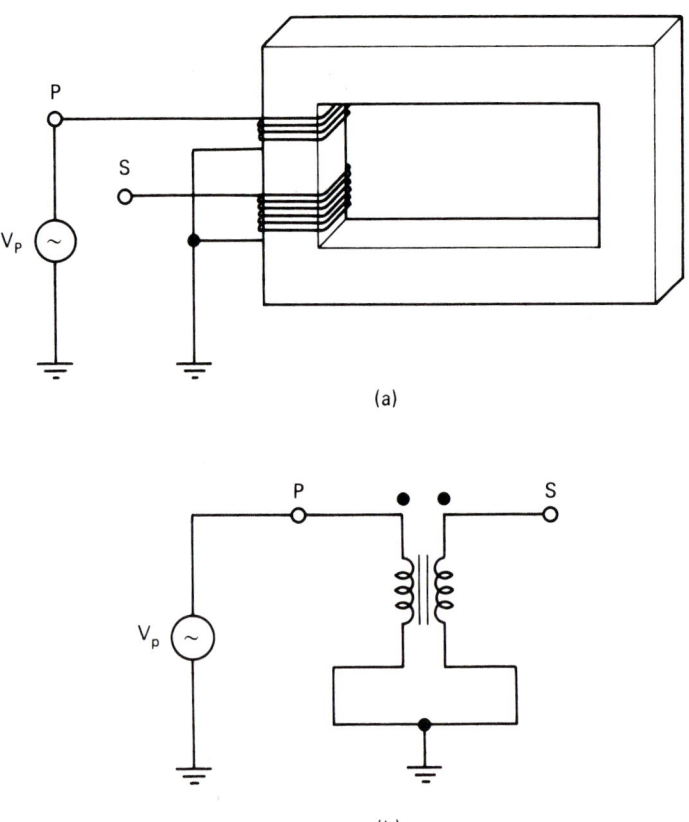

FIGURE 15-4 Transformer with primary and secondary voltages in phase.

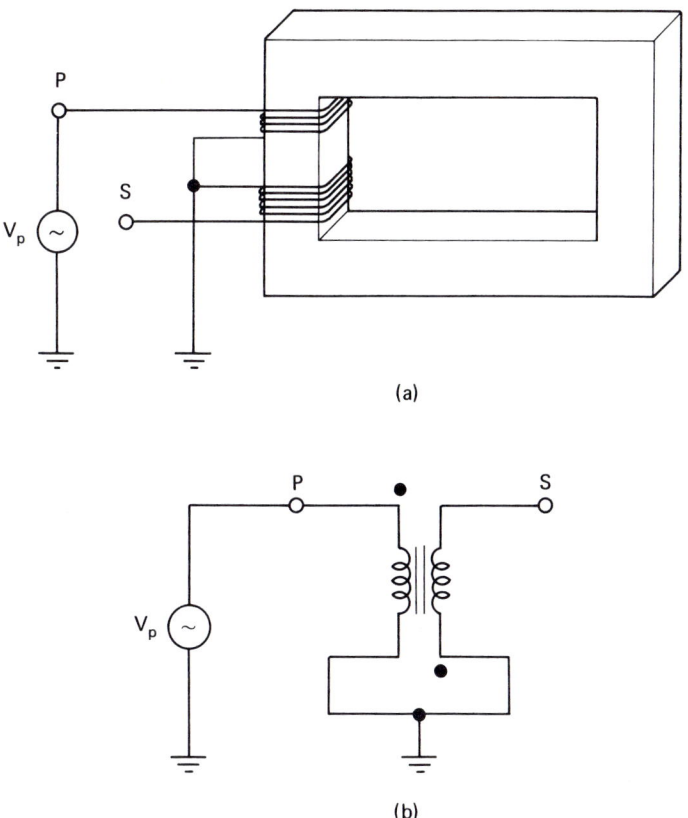

FIGURE 15-5 Transformer with 180° phase difference between the primary and secondary windings.

EXAMPLE 15-7 In Fig. 15-4(a) the number of primary turns is 10 times that of the secondary winding. What is the secondary voltage when the primary voltage is 120 V for this 10:1 step-down transformer? Use the primary voltage as the reference.

Solution

$$\frac{E_s}{E_p} = \frac{N_s}{N_p} \tag{15-1-3}$$

$$E_s = E_p \times \frac{N_s}{N_p} = 120 \times \frac{1}{10} = 12 \text{ V}$$

Since E_p is the reference, the phasor will be horizontal with the arrow pointed to the right. This corresponds to an angle of 0°. Therefore, the angle of the secondary voltage will also be 0°. Written another way, for the primary the voltage, with angle, is $120 \angle 0°$ V and for the secondary $12 \angle 0°$ V.

EXAMPLE 15-8 For Fig. 15-5 the number of turns in the primary winding is 1000 and the number of turns in the secondary winding is 3000. With 120 V applied to the primary, find the phasor representing the secondary voltage.

Sec. 15-1 Transformers

Solution

$$\frac{E_s}{E_p} = \frac{N_s}{N_p} \qquad (15\text{-}1\text{-}3)$$

$$\frac{E_s}{120} = \frac{3000}{1000}$$

$$E_s = 120 \times \frac{3000}{1000} = 360 \text{ V}$$

Assume the primary voltage to be the reference. For the primary the phasor is $120 \angle 0°$ V and for the secondary it is $360 \angle 180°$ V.

In the preceding development of transformers, the existence of power losses was ignored. However, we know that a working transformer will become warm to the touch after a period of time. This is especially true for a transformer having significant amounts of current in the windings. A warm transformer is dissipating heat and therefore must have power losses. Next, we briefly examine the causes of these losses and their consequences.

One loss is caused by the resistance of the windings. This *copper loss* or I^2R *loss* should be no stranger, because it occurs in all conductors having resistance.

Another loss results from the flow of *eddy currents* in the core material itself. If the core is solid, for example, current paths exist in the core material at right angles to the direction of the alternating magnetic flux. Currents flow in these paths because of the induced, alternating electromagnetic force (EMF) in much the same way that currents flow in the windings proper due to the same alternating flux. Magnetic cores for transformers are commonly laminated, with the adjacent laminations insulated from one another, to reduce this loss. The amount of flux in a lamination is less, with a corresponding decrease in the EMF for any current path.

A third loss occurs due to the alternations of the magnetic domains (regions) in the core material. We may think of this as a form of friction loss, but it is called *hysteresis loss* (Section 7-6). Hysteresis loss can be reduced by the proper choice of core materials and transformer design.

The existence of losses in the transformer means that the output power from the secondary winding (P_S) must be less than the primary input power (P_P). The difference is the power loss (P_L), where

$$\boxed{P_L = P_p - P_s} \qquad (15\text{-}1\text{-}7)$$

Efficiency for any machine and for the transformer is defined as follows:

$$\text{Eff} = \frac{P_{\text{OUT}}}{P_{\text{IN}}}$$

where P_{OUT} is the output power and P_{IN} is the input power. It is usually expressed as a percentage. For the transformer

$$\boxed{\text{Eff} = \frac{P_s}{P_p} \times 100} \qquad (15\text{-}1\text{-}8)$$

EXAMPLE 15-9 The output of a certain transformer is 4890 W and the input power is 5430 W. Calculate (a) the power loss and (b) the efficiency.

Solution

$$\text{(a) } P_L = P_p - P_s \quad (15\text{-}1\text{-}7)$$

$$= 5430 - 4890 = 540 \text{ W}$$

$$\text{(b) Eff} = \frac{4890}{5430} \times 100 = 90\% \quad (15\text{-}1\text{-}8)$$

EXAMPLE 15-10 A 10-kW transformer has an efficiency of 92%. Find (a) the input power when delivering its rated power and (b) the losses under those conditions.

Solution

$$\text{(a) Eff} = \frac{P_s}{P_p} \times 100 \quad (15\text{-}1\text{-}8)$$

$$P_p = \frac{P_s \times 100}{\text{Eff}}$$

$$= \frac{10 \times 100}{92} = 10.87 \text{ kW}$$

$$\text{(b) } P_L = P_p - P_s \quad (15\text{-}1\text{-}7)$$

$$= 10.87 - 10 = 0.87 \text{ kW} = 870 \text{ W}$$

15-2
Troubleshooting for Faults in Transformers

Troubleshooting procedures for transformers are similar to those for inductors (Section 8-6). However, instead of a single coil there will be at least one primary winding and one secondary winding; there may also be two or more secondary windings. If the transformer is of the voltage step-down variety, the primary has more turns than the secondary, which carries the greater current and therefore requires thicker wire. Consequently, the primary winding has the higher resistance, so that an ohmmeter is used to test the continuity and also to measure the resistance of each winding. The resistance readings are then compared with the rated values as established by the manufacturer; this comparison should reveal whether the transformer has one or more faults.

After the terminals of the transformer have been identified correctly and the resistance readings taken, there are a number of possibilities to consider:

1. The resistance reading of a winding is infinite ohms. Clearly, the winding has an open circuit. Hopefully, the open is the result of a break between the winding and one of the terminal leads; such a fault can be repaired simply by resoldering the terminals back to the winding.

However, if the open is somewhere inside the winding, the transformer will probably have to be thrown away.

2. Finite resistance between primary and secondary windings. The two windings are insulated from each other and therefore an ohmmeter that is connected between the primary and secondary terminals should read infinite ohms (this may only be true provided that the transformer has been totally removed from the circuitry). However, if the reading is not infinite, there exists some form of breakdown in the insulation between the windings and the transformer may have to be discarded.

3. The resistance of one winding is much higher than its rated value. If the measured resistance is high but not infinite, there may exist a bad solder connection between the end of the winding and the terminal lead. You should also inspect the connections between the terminals and the circuit in which the transformer is installed.

4. The resistance of one winding is much lower than its rated value. This indicates that a number of turns in the winding have been shorted out. Alternatively, the winding at one or more points has become shorted to the transformer's frame.

15-3 Mutual Inductance

In the transformer the alternating current in the primary sets up alternating magnetic flux in the core. This alternating flux causes alternating electromotive forces which oppose the alternating current change. We call this opposition *inductive reactance*. The winding has a property known as *inductance*. More properly it is called *self-inductance*, because the induced voltage and current are in the same physical circuit. In the primary circuit it is possible to trace electron current flow from a starting point around the entire closed circuit and back again to the starting point [see Fig. 15-6(a)].

With the secondary circuit open so that no secondary current can flow, we know that

$$X_p = \frac{E_p}{I_p}$$

where

$$X_p = 2\pi f L_p$$

Recall that X_p is the inductive reactance (of the primary winding), L_p is the inductance of the primary winding, and f is the frequency. When L_p is expressed in henrys and f is in hertz, the reactance X_p unit is the ohm. We already know that there will also be a measurable secondary voltage in the secondary winding which we will call E_s. This voltage is related to the primary voltage by the turns ratio for the transformer. There is, however, a slightly different way of looking at what is happening. As you will see later, this way has more general application. Suppose that we think of another form of inductance, that existing between two circuits. This inductance is called *mutual inductance*. It is defined in much the same way as is self-inductance.

$$e_p = L_p \times \text{rate of change of primary current}$$

$$e_s = M \times \text{rate of change of primary current}$$

FIGURE 15-6 Circuits illustrating self- and mutual inductances.

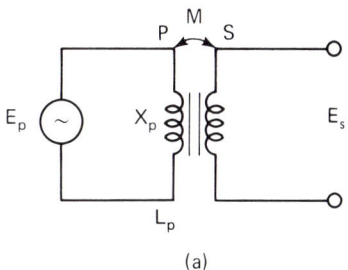

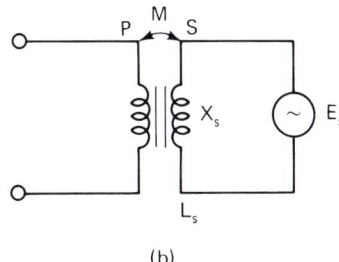

when e_p and e_s are the instantaneous voltages in the primary and secondary circuits and M is the mutual inductance. When dealing with alternating currents the effective voltages are

$$E_p = I_p X_p \qquad (15\text{-}3\text{-}1)$$

$$E_s = I_p X_M \qquad (15\text{-}3\text{-}2)$$

where I_p is the effective value of alternating current and

$$X_M = 2\pi f M \qquad (15\text{-}3\text{-}3)$$

The same units apply for X_p and X_M (ohms) as well as L_p and M (henry).

Suppose that we now apply the voltage source to the secondary winding and leave the primary winding open-circuited [see Fig. 15-6(b)].

$$E_s = I_s X_s \qquad (15\text{-}3\text{-}4)$$

$$E_p = I_s X_M \qquad (15\text{-}3\text{-}5)$$

Perhaps the most remarkable single feature of mutual inductance is the fact that it is the same in both directions. The X_M of Eq. (15-3-5) is the same as the X_M of Eqs. (15-3-2) and (15-3-3).

We now use some examples to illustrate the nature of M and X_M.

EXAMPLE 15-11 Calculate the secondary open-circuit voltage for a transformer having 200 Ω of mutual reactance and 500 mA of primary current.

Solution

$$E_s = I_p X_M \qquad (15\text{-}3\text{-}2)$$

$$= 0.5 \times 200 = 100 \text{ V}$$

Sec. 15-3 Mutual Inductance

EXAMPLE 15-12 Find the primary current when a voltage of 10 V appears across the secondary winding and the mutual reactance is 50 Ω.

Solution

$$E_s = I_p X_M \tag{15-3-2}$$

$$I_p = \frac{E_s}{X_M} = \frac{10}{50} = \frac{1}{5} \text{ A} = 200 \text{ mA}$$

EXAMPLE 15-13 The following apply for a transformer:

$$X_p = 40 \text{ Ω}$$
$$X_M = 30 \text{ Ω}$$
$$I_s = 0.65 \text{ A}$$

Find the value of E_p.

Solution

$$E_p = I_s X_M \tag{15-3-5}$$
$$= 0.65 \times 30 = 19.5 \text{ V}$$

It can be shown using Eqs. (15-3-1), (15-3-2), (15-3-4), and (15-3-5) that the self-inductances and the mutual inductance of a power transformer are related in a most unusual and useful way.

$$X_M^2 = X_p X_s$$

Therefore, the inductances have a similar relationship.

$$M^2 = L_p L_s$$

Changing the subscripts and allowing "1" to refer to one winding and "2" to the other,

$$M^2 = L_1 L_2 \tag{15-3-6}$$

This equation says that the product of the self-inductances is equal to the square of the mutual inductance. If the square root is taken for each member (side) of the equation, we have a more popular form:

$$M = \sqrt{L_1 L_2} \tag{15-3-7}$$

Please notice that the subscripts have been changed for a reason. This is because we wish to back away from the power transformer concept and deal with mutually coupled circuits, for which there may not be a clearly identifiable primary or secondary winding. However, before passing on, let us pause to

FIGURE 15-7 Comparison of ideal transformer and other magnetically coupled circuits.

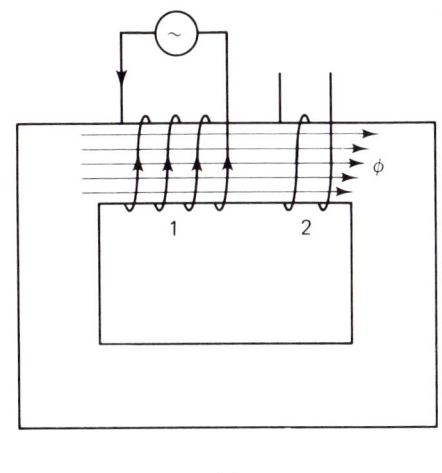

(a)

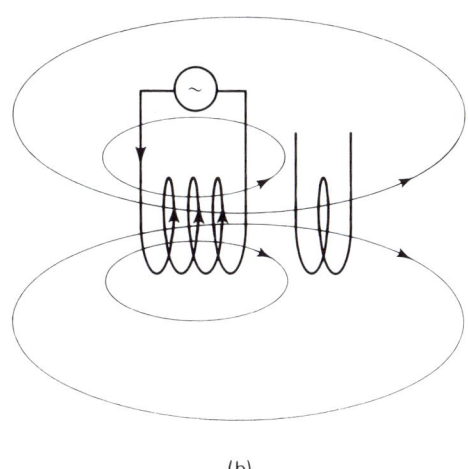

(b)

review the particular characteristics of the ideal (power) transformer that were the subjects of Sections 15-1 and 15-2 and a good part of Section 15-3.

1. The magnetic flux in the core links all of the turns of the windings [see Fig. 15-7(a)].
2. There are no appreciable power losses in the core or in the conductors making up the windings.
3. The following equations, and their reciprocal relationships where fractions are involved, apply.

$$\frac{E_p}{E_s} = \frac{N_p}{N_s} \qquad (15\text{-}1\text{-}2)$$

$$\frac{I_s}{I_p} = \frac{N_p}{N_s} \qquad (15\text{-}1\text{-}5)$$

$$E_s I_s = E_p I_p \qquad (15\text{-}1\text{-}4)$$

$$\frac{R_p}{R_s} = \left(\frac{N_p}{N_s}\right)^2 \qquad (15\text{-}1\text{-}6)$$

Sec. 15-3 Mutual Inductance

$$M^2 = L_1 L_2 \tag{15-3-6}$$

$$M = \sqrt{L_1 L_2} \tag{15-3-7}$$

The last two equations will now be demonstrated by means of examples.

EXAMPLE 15-14 A transformer has two windings. One has a self-inductance of 300 mH and the second has a self-inductance of 1500 mH. Calculate the mutual inductance.

Solution

$$M = \sqrt{L_1 L_2} \tag{15-3-7}$$

$$= \sqrt{300 \times 10^{-3} \times 1500 \times 10^{-3}}$$

$$= \sqrt{300 \times 1500} \times 10^{-3}$$

$$= 671 \times 10^{-3} \text{ H} = 671 \text{ mH}$$

EXAMPLE 15-15 For a transformer $L_1 = 750$ mH and $M = 450$ mH. Find L_2.

Solution

$$M^2 = L_1 L_2 \tag{15-3-6}$$

$$L_2 = \frac{M^2}{L_1} = \frac{(450 \times 10^{-3})^2}{750 \times 10^{-3}} = 270 \times 10^{-3} \text{ H}$$

$$= 270 \text{ mH}$$

A power transformer is designed to approach the ideal characteristics we have been concerned with thus far. In general, circuits having self-inductances and mutual inductance do not have the same ideal characteristics. For example, if the ferromagnetic core of Fig. 15-7(a) is eliminated, the self-inductances and mutual inductance of two windings will be significantly reduced. This will occur for two reasons. First, there will be much less magnetic flux produced by a given amount of winding current. Second, not all of the flux produced by the current in one winding will link all of the turns of the other. This is illustrated in Fig. 15-7(b). The flux produced in winding 2 has been dramatically reduced.

Under these circumstances the relationship between the self-inductances and the mutual inductance must take a modified form [see Eq. (15-3-6) for comparison].

$$M^2 = k^2 L_1 L_2 \tag{15-3-8}$$

where k is the *coefficient of coupling*. With the ideal power transformer, $k = 1$ and it was omitted from Eq. (15-3-6) and Eq. (15-3-7). Any practical transformer will have a value of k less than 1. In practical coupled circuits which are not intended to be like the windings of a power transformer, the k can be relatively small. In the limiting case, $k = 0$ simply means that no magnetic coupling exists.

EXAMPLE 15-16 Two coils have 10 mH and 30 mH of self-inductance, respectively. Find the mutual inductance if the coefficient of coupling $k = 0.2$.

Solution

$$M = k\sqrt{L_1 L_2} \qquad (15\text{-}3\text{-}8)$$

$$= 0.2\sqrt{10 \times 30} = 3.46 \text{ mH}$$

Alternatively,

$$M = 0.2\sqrt{10 \times 10^{-3} \times 30 \times 10^{-3}}$$

$$= 0.2 \times 10^{-3}\sqrt{300}$$

$$= 3.46 \times 10^{-3} \text{ H} = 3.46 \text{ mH}$$

EXAMPLE 15-17 Find L_2 when $L_1 = 215$ mH, $M = 140$ mH, and $k = 0.6$.

Solution

$$M^2 = k^2 L_1 L_2 \qquad (15\text{-}3\text{-}8)$$

$$L_2 = \frac{M^2}{k^2 L_1} = \frac{140^2}{0.6^2 \times 215} = 253 \text{ mH}$$

The amount of voltage induced in one winding by the flow of ac current in another was given in Eqs. (15-3-2) and (15-3-5) for a transformer. These equations still apply for two coils which are not intended to be primary or secondary transformer windings.

$$E_2 = I_1 X_M \qquad (15\text{-}3\text{-}9)$$

$$E_1 = I_2 X_M \qquad (15\text{-}3\text{-}10)$$

where X_M is given by

$$X_M = 2\pi f M \qquad (15\text{-}3\text{-}11)$$

EXAMPLE 15-18 Find the voltage induced in coil 2 for $X_M = 3\ k\Omega$ and $I_1 = 5\ mA$.

Solution

$$E_2 = I_1 X_M \quad (15\text{-}3\text{-}9)$$
$$= 5 \times 3 = 15\ V$$

The phase angle of the induced voltage in one winding with respect to that applied to the other may be of concern in some applications. For this reason dots are attached to the schematic diagrams as shown in Fig. 15-8. This means that the alternating voltage at the ungrounded terminal of winding 1 is in phase with that of the ungrounded terminal of winding 2.

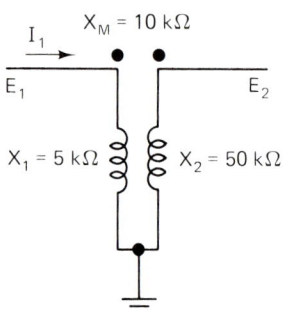

FIGURE 15-8 Ungrounded terminal voltages in-phase.

EXAMPLE 15-19 In Fig. 15-8, the current in coil 1 is 20 mA. Find E_1 and E_2 and their phase angles using I_1 as the reference.

Solution

$$E_1 = I_1 X_1$$
$$= 20 \times 5 = 100\ V$$
$$E_2 = I_1 X_M$$
$$= 20 \times 10 = 200\ V$$

Recall that E_1 must lead I_1 by 90°. Since E_2 must be in phase with E_1, its phase angle must also be 90°.

In all of the coupled circuits examined so far, the current in one winding has been isolated from the other. In other words, the electrons passing through one coil cannot reach the other. However, this need not be the case. Consider the circuit of Fig. 15-9(a), in which the two inductors are actually connected

FIGURE 15-9 Mutual inductances aiding and opposing.

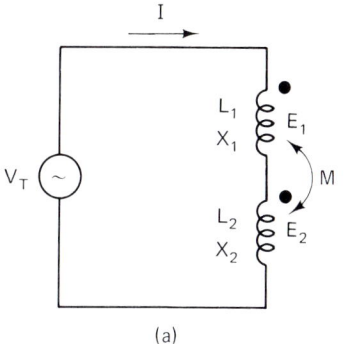

(a)

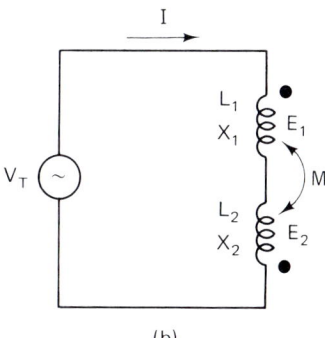

(b)

in series. We know that $V_T = E_1 + E_2$. Also, when there is no mutual inductance, $E_1 = IX_1$ and $E_2 = IX_2$. This is a series circuit for which the current I is the same in both inductors.

With mutual inductance the values of E_1 and E_2 cannot be calculated so easily. For example, there is an additional component due to the mutual inductance and the current through X_2 when writing the expression for E_1.

$$E_1 = IX_1 + IX_M$$

Similarly, E_2 has still another component because of the mutual inductance and the current through X_1.

$$E_2 = IX_2 + IX_M$$

Adding these two voltages together and factoring the common current I gives

$$V_T = E_1 + E_2 = IX_1 + IX_M + IX_2 + IX_M$$
$$= I(X_1 + X_2 + 2X_M)$$

The equivalent inductance for the two coupled inductors in series is

$$X_s = X_1 + X_2 + 2X_M \qquad (15\text{-}3\text{-}12)$$

We can reason then that

$$L_s = L_1 + L_2 + 2M \qquad (15\text{-}3\text{-}13)$$

Sec. 15-3 Mutual Inductance

EXAMPLE 15-20 Two inductors are connected in series as in Fig. 15-9(a). If $L_1 = 0.03$ H, $L_2 = 0.07$ H, and $M = 0.01$ H, what will be the total series inductance of the combination?

Solution

$$L_s = L_1 + L_2 + 2M \qquad (15\text{-}3\text{-}13)$$

$$= 0.03 + 0.07 + 2(0.01) = 0.12 \text{ H}$$

EXAMPLE 15-21 Calculate the total inductive reactance for the inductors of the last example for a frequency of 60 Hz.

Solution

$$X_s = 2\pi f L_s = 2\pi \times 60 \times 0.12 = 45.2 \text{ }\Omega$$

We have seen that the two inductors can be connected so that the total inductance is greater than the sum of the individual self inductances. This requires that the dot ends of the inductors be observed in a similar way as the positive terminals of batteries connected in series aiding. This is shown as a memory aid in Fig. 15-10(a). Note that there is no characteristic of series-aiding batteries comparable to the mutual inductance.

When the leads of one inductor are reversed as shown in Fig. 15-10(b), the sign of the $2M$ term in Eq. (15-3-13) is changed to give

$$L_s = L_1 + L_2 - 2M \qquad (15\text{-}3\text{-}14)$$

Once again the polarity is emphasized in Fig. 15-10(b).

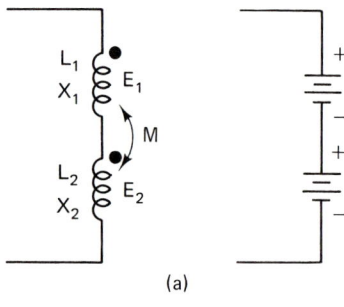

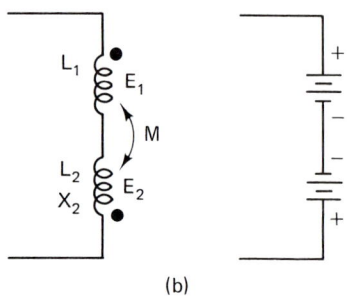

FIGURE 15-10 Aiding and opposing concepts.

EXAMPLE 15-22 In Fig. 15-10(a), $L_1 = 300$ mH, $L_2 = 500$ mH, and $M = 350$ mH. Calculate the equivalent total inductance.

Solution

$$L_s = L_1 + L_2 + 2M \qquad (15\text{-}3\text{-}13)$$

$$= 300 + 500 + 2 \times 350 = 1500 \text{ mH}$$

EXAMPLE 15-23 Two inductors are connected in series as shown in Fig. 15-10(b). The self-inductances are 50 μH and 400 μH and the coefficient of coupling is 0.1. What is the equivalent inductance of the series arrangement?

Solution

$$M = k\sqrt{L_1 L_2} \qquad (15\text{-}3\text{-}8)$$

$$= 0.1\sqrt{50 \times 400} = 14.14 \text{ μH}$$

$$L_s = L_1 + L_2 - 2M \qquad (15\text{-}3\text{-}14)$$

$$= 50 + 400 - 2 \times 14.14 = 422 \text{ μH}$$

It is also possible to connect two inductors in parallel [see Fig. 15-11(a)]. We are already familiar with this situation when there is no mutual inductance. The reactances are combined by adding the reciprocals to get the reciprocal of the equivalent reactance. (The method for combining parallel reactances is the same as that used for combining parallel resistances.) However, when there is a mutual inductance, the same method cannot be used. The equivalent inductance for two inductors in parallel, as for Fig. 15-11(a), is

$$L_p = \frac{L_1 L_2 - M^2}{L_1 + L_2 - 2M} \qquad (15\text{-}3\text{-}15)$$

This formula applies when the polarity dots for the inductors are connected to the same point. Of course, the other terminals of the inductors must be

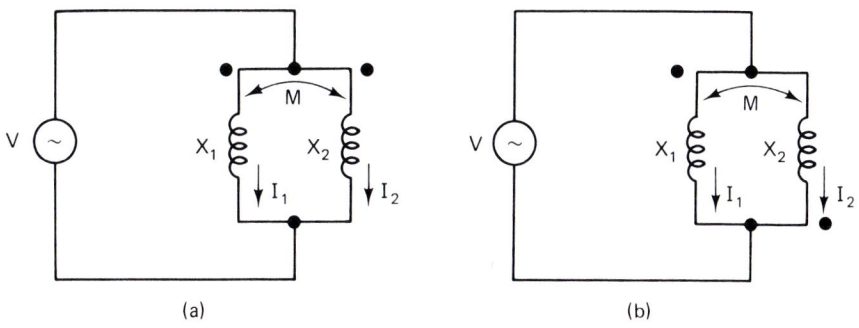

FIGURE 15-11 Mutually coupled inductors in parallel.

connected to a second common point. The equation for the equivalent reactance is

$$X_p = \frac{X_1 X_2 - X_M^2}{X_1 + X_2 - 2X_M} \qquad (15\text{-}3\text{-}16)$$

The inductors can also be connected in parallel, as shown in Fig. 15-11(b). With this arrangement, the equivalent inductance L_p is given by

$$L_p = \frac{L_1 L_2 - M^2}{L_1 + L_2 + 2M} \qquad (15\text{-}3\text{-}17)$$

and the equivalent inductive reactance is

$$X_p = \frac{X_1 X_2 - X_M^2}{X_1 + X_2 + 2X_M} \qquad (15\text{-}3\text{-}18)$$

EXAMPLE 15-24 In Fig. 15-11(a), $X_1 = 10\ \Omega$, $X_2 = 20\ \Omega$, and $X_M = 12\ \Omega$. Calculate the equivalent reactance.

Solution

$$X_p = \frac{X_1 X_2 - X_M^2}{X_1 + X_2 - 2X_M} \qquad (15\text{-}3\text{-}16)$$

$$= \frac{10 \times 20 - 12^2}{10 + 20 - 2 \times 12} = 9.33\ \Omega$$

EXAMPLE 15-25 In Fig. 15-11(b), $X_1 = 10\ \Omega$, $X_2 = 20\ \Omega$, and $X_M = 12\ \Omega$. Calculate the equivalent reactance.

Solution

$$X_p = \frac{X_1 X_2 - X_M^2}{X_1 + X_2 + 2X_M} \qquad (15\text{-}3\text{-}18)$$

$$= \frac{10 \times 20 - 12^2}{10 + 20 + 2 \times 12} = 1.04\ \Omega$$

Chapter Summary

15.1. For a transformer the primary and secondary voltages are related to the number of turns in each as follows:

$$\frac{E_p}{N_p} = \frac{E_s}{N_s} \qquad \frac{E_p}{E_s} = \frac{N_p}{N_s} \qquad \frac{E_s}{E_p} = \frac{N_s}{N_p}$$

Note that these are quite accurate for a lightly loaded transformer and provide a reasonable approximation for practical transformers under normal usage conditions.

15.2. For a resistive load, neglecting transformer losses,

$$P_s = P_p$$
$$E_s I_s = E_p I_p$$
$$\frac{I_s}{I_p} = \frac{N_p}{N_s}$$

15.3. Load resistances in the secondary circuit of a transformer are seen by the source in the primary circuit as having been transformed as follows:

$$\frac{R_p}{R_s} = \left(\frac{N_p}{N_s}\right)^2$$

15.4. Changing currents in magnetically coupled circuits cause induced voltages. Thinking in terms of a transformer, the instantaneous values are

$$e_p = L_p \times \text{rate of change of primary current}$$
$$e_s = M \times \text{rate of change of primary current}$$

where L_p represents the primary self-inductance and M the mutual inductance. In addition,

$$e_s = L_s \times \text{rate of change of secondary current}$$
$$e_p = M \times \text{rate of change of secondary current}$$

15.5. Corresponding to the instantaneous values just given, the effective voltages and currents are used.

$$E_p = I_p X_p$$
$$E_s = I_p X_M$$

or

$$E_s = I_s X_s$$
$$E_p = I_s X_M$$

where

$$\text{mutual inductive reactance, } X_M = 2\pi f M$$

15.6. For the ideal transformer in which all the flux links all the turns of the windings,

$$X_M^2 = X_p X_s$$
$$M^2 = L_p L_s$$

With incomplete flux linkage for magnetically coupled circuits,

$$M^2 = k^2 L_1 L_2$$
$$M = k\sqrt{L_1 L_2}$$

in which L_1 and L_2 are the self-inductances and k is the coefficient of coupling.

Chapter Summary

15.7. For a series connection of two magnetically coupled inductors, the total inductance L_s is given by

$$L_s = L_1 + L_2 + 2M$$
$$L_s = L_1 + L_2 - 2M$$

in which the sign $(+)$ or $(-)$ depends on whether the mutual coupling aids or opposes the establishment of magnetic flux by the current.

15.8. Magnetically coupled inductors may also be placed in parallel. When the polarity dots are connected to the same point, the total inductance is

$$L_p = \frac{L_1 L_2 - M^2}{L_1 + L_2 - 2M}$$

Otherwise,

$$L_p = \frac{L_1 L_2 - M^2}{L_1 + L_2 + 2M}$$

Self-Review

True-False Questions

1. T. F. It is possible to double the primary voltage of a transformer by simply doubling the number of turns in the primary winding.
2. T. F. It is possible to double the secondary voltage of a transformer by simply doubling the number of turns in the secondary winding.
3. T. F. Transformer loads always consist of a resistor connected across the secondary winding terminals.
4. T. F. The resistance "seen" by a source is smaller than the load resistance for a step-down transformer.
5. T. F. The mutual inductance between the windings of a power transformer is zero.
6. T. F. Because of the mutual reactance, the primary and secondary currents must be equal.
7. T. F. The basic SI unit for mutual inductance is the ohm.
8. T. F. If the self-inductance of one transformer winding is increased by changing the number of turns while that of the other remains constant, the mutual inductance must decrease.
9. T. F. In the case of actual coupled coils, the flux produced by the current in one coil will not link all the turns of the other.
10. T. F. When the coefficient of coupling is 1, no magnetic coupling exists between the coils.
11. T. F. When the current flow in one circuit results in its magnetic field linking with the conductors of another circuit, mutual inductance exists.
12. T. F. Self-inductance opposes the constant flow of dc current in an inductor.
13. T. F. The winding of a transformer connected to the ac source is referred to as the "primary winding."
14. T. F. The number of volts per turn of the primary winding is greater than the number of volts per turn of the secondary winding of a step-down transformer.
15. T. F. In a power system the voltage from the alternator at the power station is usually stepped up by transformers before transmission to distant loads.
16. T. F. Core losses in a transformer result in heat dissipation in the core material.

17. **T. F.** A particular feature of self-inductance is that the current whose change causes the induced EMF and the EMF itself are both in the same physical circuit.
18. **T. F.** Because of mutual inductance, the current whose change causes the induced EMF and the EMF itself must be in physically separated circuits, that is, have no conducting linkages.
19. **T. F.** The efficiency of a transformer can be expressed as a decimal fraction less than 1 or the equivalent percentage which is less than 100.
20. **T. F.** The coefficient of coupling has no units or dimensions, but mutual inductance is expressed in henrys.

Multiple-Choice Questions

21. The name of the law that allows the polarity of induced voltage to be found is
    - (a) Ohm's law
    - (b) Kirchhoff's law
    - (c) Lenz's law
    - (d) inductance law
    - (e) induced voltage law
22. In a transformer there is one quantity that is the same for the primary and secondary windings.
    - (a) apparent power
    - (b) voltage
    - (c) current
    - (d) volts per turn
    - (e) both (a) and (d)
23. When the secondary voltage is less than the primary voltage, the transformer is referred to as a
    - (a) step-down transformer
    - (b) power transformer
    - (c) step-up transformer
    - (d) power supply transformer
    - (e) all choices are correct
24. If $N_p/N_s = 0.5$, the source in the primary circuit "sees" what amount of resistance for a 1-kΩ load resistance?
    - (a) 4000 Ω
    - (b) 2000 Ω
    - (c) 1000 Ω
    - (d) 500 Ω
    - (e) 250 Ω
25. The mutual inductance between the windings of a transformer is 3 mH. If the current in the primary is changing at the rate of 50,000 A/s, the instantaneous secondary voltage will be
    - (a) 16,670 V
    - (b) 150,000 V
    - (c) 60 MV
    - (d) 0.6 MV
    - (e) 150 V
26. At 60 Hz the value of X_M for a pair of magnetically coupled coils is 250 Ω. At 180 Hz the new value of X_M will be
    - (a) 180 Ω
    - (b) 250 Ω
    - (c) 750 Ω
    - (d) 83.3 Ω
    - (e) 60 Ω
27. Two coils are wound on a common core. If an ac source is applied to one coil where the other is open-circuited, which of the following is true?
    - (a) Current will flow in the first coil.
    - (b) Current will flow in the second coil.
    - (c) Voltage will appear across the second coil.
    - (d) All of the previous choices are true.
    - (e) Choices (a) and (c) are true.
28. An ideal power transformer will have
    - (a) identical primary and secondary self-inductances
    - (b) primary and secondary winding resistances proportional to the square of the turns ratio
    - (c) both (a) and (b)
    - (d) zero power losses
    - (e) all choices are correct
29. The coefficient of coupling for two coils is limited to
    - (a) an upper limit of ∞
    - (b) an average of 0.5

(c) a lower limit of zero
(d) the mutual inductance divided by the product of the self-inductances
(e) the product of the self-inductances divided by the mutual inductance

30. The phase angles of the voltages across two coupled coils are indicated by
    (a) + and − signs on the schematic
    (b) 0° and 180° angles stamped on the metal case
    (c) dots on the ends of the coils in the schematic
    (d) color coding of the coil winding insulations
    (e) graffiti marks on the coils

31. The induced EMF in one circuit resulting from current change in another occurs because of
    (a) parasitic capacitance   (b) mutual inductance
    (c) self inductance   (d) hysteresis
    (e) capacitive coupling

32. The name given to the property of a coil or winding whereby a changing current induces electromotive force is
    (a) reluctance   (b) momentum   (c) resistance
    (d) capacitance   (e) inductance

33. The load power delivered by a transformer is extracted from the
    (a) primary winding   (b) secondary winding   (c) magnetic core
    (d) steel laminations (e) capacitance existing between the primary and secondary windings

34. An ideal transformer will have no power losses. For such a transformer
    (a) $P_S = P_P$ (b) $E_S I_S = E_P I_P$ (c) $N_P V_S = N_S V_P$
    (d) $N_P I_P = N_S I_S$ (e) all choices are correct

35. A certain amount of power is transmitted a short distance at 110 V ac, for which the power loss is 500 W. If transformers are now introduced so that the same amount of power is transmitted over the same line at 220 V, the power loss will be
    (a) 125 W   (b) 250 W   (c) 500 W   (d) 1000 W   (e) 2000 W

36. Power losses in a transformer are classified as
    (a) friction   (b) hysteresis   (c) copper
    (d) eddy current   (e) only one choice is incorrect

37. In ac circuits the changing current induces an EMF that is accounted for as a voltage drop caused by
    (a) resistance   (b) inductive reactance   (c) capacitive reactance
    (d) conductance   (e) Kirchhoff's voltage

38. The symbol for mutual inductance is
    (a) L (b) M (c) I (d) C (e) E

39. The reactance appearing in Kirchhoff's voltage law equation because of mutual reactance has the form
    (a) L (b) M (c) ωL (d) ωM (e) $j\omega L$

40. Two windings having self-inductances L_1 and L_2 will have mutual inductance $M = \sqrt{L_1 L_2}$ if
    (a) the magnetic flux in the core links all the turns of the windings
    (b) $E_1 = I_2 X_M$ (c) $E_2 = I_1 X_M$ (d) $k = 1$ (e) both (a) and (d)

Practice Problems

41. A transformer has 1000 turns in the primary winding and 250 turns in the secondary winding. How many volts per turn will there be when the primary voltage is 120 V?

42. What is the secondary voltage for the transformer of Problem 41?

43. Find the primary current for the transformer of Problem 41 with a 10-Ω load connected across the secondary winding terminals.

44. A step-up transformer has an output of 500 mA at 480 V. Determine the primary current if power is to be supplied to the transformer at 120 V.

45. A 20,000-Ω electronic amplifier is to be matched to an 8-Ω speaker to achieve maximum power transfer. Find N_p/N_s.

46. Calculate the secondary open-circuit voltage for a transformer having 20 kΩ of mutual reactance and a primary current of 0.43 mA.
47. At 13.7 kHz the value of X_M is measured to be 149 Ω. Determine the mutual inductance.
48. An ideal transformer has two windings with self-inductance of 0.3 H and 0.7 H, respectively. Determine the mutual inductance.
49. Calculate the mutual inductance for two coils having 300 mH and 100 mH of self-inductance, respectively. The coefficient of coupling is 0.1.
50. Determine L_1 when $L_2 = 2000$ µH, $M = 200$ µH, and $k = 0.05$.
51. A 1200-turn coil wound on a magnetic core has 120 V ac applied to its terminals. Expess the induced voltage in the winding in a way that is independent of the exact total voltage and number of turns.
52. A transformer delivers 12.6 V with 120 V applied to the primary terminals. If the number of secondary turns is 126, how many turns must the primary winding have?
53. A transformer is rated at 120 V for the primary winding and 24 V for the secondary winding. Calculate the primary current with a 750-W load under rated voltage conditions.
54. A transformer has a rating of 1 kW. Under normal load, the power losses are 100 W total. What is the efficiency?
55. The mutual inductance between two windings is 0.1 H. Calculate the value of X_M at $f = 1000$ Hz.
56. Two coils having self-inductances $L_1 = 0.1$ mH and $L_2 = 1.6$ mH are wound on the same core so as to have mutual inductance $M = 0.3$ mH. Find the coefficient of coupling, k.
57. Two inductors, each having self-inductances of 0.2 H, are placed on the same core and connected in series. If the total inductance of the series combination is measured to be 0.5 H, what is the value of the mutual inductance?
58. A single coil or winding on a core is tapped at the center so that $L_1 = L_2 = 5$ H. Calculate the total inductance, as would be measured between the ends of the coil.
59. A small, practical transformer delivers 20 kW and dissipates 2 kW in the windings and core. Determine the efficiency of the transformer and express it to two significant digits.
60. A large transformer has an efficiency of 98% while delivering 100 kW of power. How much power is converted into heat in the transformer?

Advanced Problems

61. Two inductors having self-inductances of 3 mH and 5 mH also have a mutual inductance of 1 mH. If the two are connected in series to obtain maximum total inductance, what will be the effective inductance of the series combination?
62. The leads of one inductor in the electrical circuit of Problem 61 are reversed. Calculate the value of effective inductance for the new series combination.
63. The leads of both inductors in the electrical circuit of Problem 61 are reversed. What will be the total value of effective inductance for the series combination?
64. A 10,000-Ω electronic amplifier is to be matched to a 16 Ω load. How many turns will be in the primary winding when the secondary winding has 100 turns?
65. A 100 V ac source was connected to the terminals of an inductor and a current of 50 mA was measured in the circuit. At the same time a voltage of 12.5 V was measured across the terminals of the second inductor which is mutually coupled to the first. A high impedance electronic voltmeter was used for this measurement so that current flow in the second inductor was negligible. Find the primary self-inductive reactance and the mutual inductive reactance.
66. If the self-inductance of the first inductor in Problem 65 is 50 mH, what is the mutual inductance?
67. The ratio of primary turns to secondary turns is 10 for a step-down transformer (10:1 step-down). If 500 mW is being supplied to a 250 Ω load resistance, what is the primary current?

68. Determine the number of volts per turn for the transformer of Problem 67 if there are 2000 turns in the primary.
69. Two inductors are mutually coupled. $L_1 = 300$ µH, $L_2 = 800$ µH, and $M = 200$ µH. They are now connected in parallel to achieve the maximum inductance for this combination. Find the maximum effective inductance for that parallel combination.
70. The leads of one inductor in Problem 69 are reversed. Find the effective inductance for the new parallel combination.
71. A transformer is to be designed to deliver approximately 6.3 V to an ac load with a nominal 125 V applied to the primary winding. Use 0.1 V/turn and specify the number of turns in each winding and the exact output voltage to be expected.
72. The ratio of primary to secondary turns of a transformer is 10. Calculate the currents in both windings if the transformer delivers 50 W to a resistive load with 115 V on the primary.
73. A load requires 15 A, which is to be supplied from a 120-V ac source by means of a transformer. Calculate the current in the primary winding if the load resistance is 10 Ω.
74. The current in a winding rises linearly from 500 mA to 600 mA in 10 ms. Calculate the induced voltage in the winding if $L = 50$ mH.
75. An ideal transformer has the following parameters:

$$L_P = 1 \text{ H}$$

$$L_S = 40 \text{ mH}$$

$$M = 200 \text{ mH}$$

Find the current drawn from the 120-V 60-Hz ac source with the secondary open.

76. Two separate windings (coils) are placed on the same core so that the coefficient of coupling $k = 0.5$. The self-inductances of the coils are 10 mH and 20 mH, respectively. The two coils are now connected to form a continuous series circuit. Calculate the maximum possible equivalent inductance of the series combination.
77. In Problem 76, let the two coils be reconnected to obtain the least inductance for the series combination. Calculate the minimum inductance obtained.
78. The coils of Problem 76 are now connected to form a parallel arrangement of self-inductance with mutual inductance between the branches (see Fig. 15-11). Determine the maximum equivalent total inductance possible with this arrangement.
79. Repeat Problem 78 except that the minimum equivalent total inductance for the parallel combination is to be determined.
80. Review the results of Problems 76 through 79 and list all the equivalent total inductances available through switching the coils in the various series and parallel patterns.

16

Complex Algebra

This chapter has the following objectives:

1. To introduce the *j*-operator and its definition as an imaginary number.
2. To present the *j*-operator as a means of rotating phasors through 90° in the counterclockwise direction with successive products giving rotations of 180°, 270°, and so on.
3. To define complex numbers as consisting of real numbers in combination with imaginary numbers.
4. To express a complex number in rectangular form as the sum of a real number and an imaginary number.
5. To express complex numbers in the polar form for which an absolute value (length of the hypotenuse) and angle are identified and in which no *j*-operator need be written.
6. To establish procedures for converting complex numbers expressed in polar form to equivalent complex numbers in rectangular form.
7. To establish procedures for conversion from rectangular to polar form.
8. To state the rules and procedures for adding or subtracting phasors expressed in rectangular form.
9. To state the rules and procedures for multiplying or dividing phasors expressed in either rectangular or polar form.
10. To state the rules and procedures for performing any sequence of chain calculations involving both multiplication and division.
11. To employ complex numbers in the expression of impedances having (real) resistive components and (imaginary) reactive components.
12. To demonstrate the application of complex numbers in the solution of simple series ac circuits and parallel ac circuits.

16-0
Introduction

We have seen that the solution of ac problems is often associated with the solution of right triangles. It is necessary to find various unknown sides and angles of right triangles when other sides and angles are known. The sides (legs and hypotenuse) of the right triangle are called *phasors* when they represent alternating voltages and currents. It is customary to indicate a direction for each phasor using arrowheads. The angles between these phasors are the phase angle differences of the ac quantities. (This is the basis for the name "phasors.")

One particular step in analyzing ac circuits consists of adding voltages or currents that are not in phase. When two ac voltages are 90° out of phase, the addition is performed by finding the hypotenuse of the right triangle whose legs are the voltages. The same method is used when adding currents that are 90° out of phase. Whenever the voltages (or currents) to be added are not 90° out of phase, the method must be modified. However, after modification, the solution still requires that right triangles be solved.

The preceding discussion emphasizes the importance of phasors that are perpendicular to each other. These phasors are best represented in a phasor diagram as horizontal phasors and vertical phasors. This chapter introduces a very consistent and useful way of representing such phasors.

16-1
The *j* Operator

We begin by considering the phasors shown in Fig. 16-1(a), which are 2, 3, and 5 units in length. All three phasors are horizontal and point to the right. They are drawn to the same scale since the unit of length for each occupies the same space on the paper. The number appearing above each denotes the phasor length. Equally important is the positive sign of these numbers, which indicates the positive direction.

In Fig. 16-1(b) the three original phasors have been rotated 180° so that they all now point to the left, which is the negative horizontal direction. The numbers associated with these rotated phasors are -2, -3, and -5. For our purpose it is advisable to think of the minus sign $(-)$ as an operator. For example, the -3 means that a $+3$ phasor [see Fig. 16-1(a)] has been rotated 180° in the counterclockwise direction. (We use this particular direction to be consistent with what is to follow shortly.)

In Fig. 16-1, the phasors are shown at different levels so that they can be seen more clearly. It is, of course, possible to place them at the same level, as shown in Fig. 16-2. When this is done, the result resembles the number line which is often used in elementary mathematics to introduce the concept

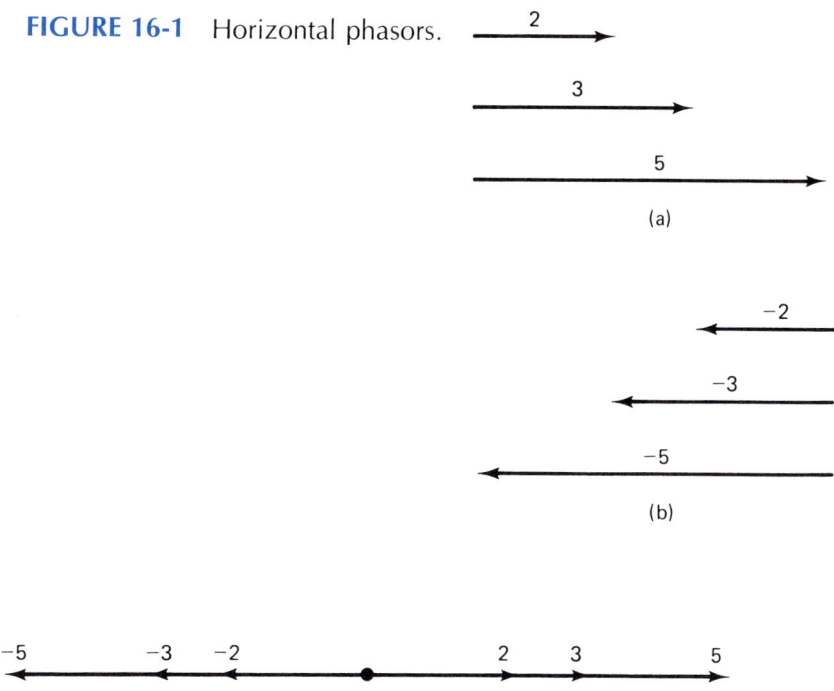

FIGURE 16-1 Horizontal phasors.

FIGURE 16-2 Superimposed positive and negative horizontal phasors.

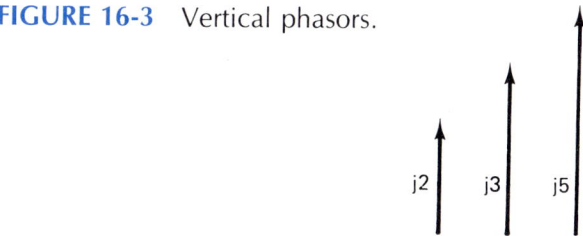

FIGURE 16-3 Vertical phasors.

of negative numbers. Although our approach is perfectly consistent with the ideas of positive and negative numbers, the emphasis is different. We want to think of a positive number as a phasor pointing to the right. A negative number indicates that the phasor has been rotated 180° counterclockwise so that it now points to the left.

So far, so good. It may not seem that much progress has been made in this chapter so far. However, believe it or not, we have come a long way in looking at complex algebra. (Please do not substitute the word "complicated" for "complex," as this implies a difficult subject.) We now know how to express the phasor as a number preceded by a + or − sign, as long as the phasor is horizontal. What if the phasor is vertical? Can we use an operator to tag a vertical phasor? As you may guess, the answer is yes. Otherwise, the authors would be embarrassed at this point and the chapter would be extremely short. Figure 16-3 shows our old friends, the phasors of Fig. 16-1, but pointing up.

There is another difference, too. Notice that the phasors are represented numerically by $j2$, $j3$, and $j5$. What is this j? Where did it come from? The "j" for the phasors is a shorthand way of stating that each has a counterpart in Fig. 16-1(a) that was rotated 90° counterclockwise to obtain $j2$, $j3$, and $j5$. Perhaps a few examples will serve to nail down these points before continuing.

Sec. 16-1 **The j Operator**

EXAMPLE 16-1 For the phasor shown, how is it expressed as a number with an operator to indicate the direction?

$$\xrightarrow{\qquad 13.7 \qquad}$$

Solution The length of the phasor is 13.7 units and the direction is to the right, the positive horizontal direction. Therefore, the expression for the phasor is

$$+13.7 = 13.7$$

where the + sign is implied when it is not specifically written.

EXAMPLE 16-2 A phasor is 5.6 units in length and is pointed to the left, the negative horizontal direction. How is it expressed as a number with an operator to indicate the direction?

Solution The number 5.6 must be preceded by the 180° operator, which is the minus sign ($-$). Therefore, the phasor is

$$-5.6$$

EXAMPLE 16-3 A phasor is 161.3 units in length and is pointed directly up. How is it expressed as a number with an operator to indicate the direction?

Solution The proper operator is the j. The answer is then

$$+j161.3 = j161.3$$

It is now time to examine the j operator more closely. So far it has only one meaning. When it precedes a number, the result is a phasor in the positive vertical direction. That is, if the original phasor is 5, otherwise written as $+5$, then $j5$ has been rotated 90° counterclockwise, and points up. Suppose that we now rotate the phasor another 90°. In order to illustrate the sequence of events, look at Fig. 16-4(a), which has the original unrotated phasor (5), then Fig. 16-4(b), showing the results of the 90° counterclockwise rotation, and finally Fig. 16-4(c), which shows what happens after a second rotation of 90°. The initial rotation yielded $j5$, which may be interpreted as the product of operator "j" and phasor "5." The second rotation has a similar interpretation as the product of operator "j" and phasor "$j5$." From Fig. 16-4(c),

$$j \times j5 = j^2 5 = -5$$

Since two 90° rotations are equivalent to a single 180° rotation,

$$j^2 = -1$$

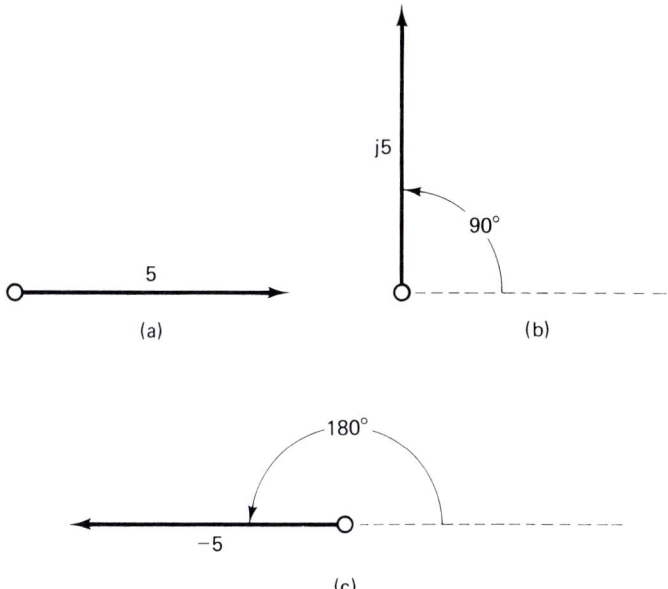

FIGURE 16-4 Phasor rotation.

and taking the square root of each side,

$$j = \sqrt{-1} \qquad (16\text{-}1\text{-}1)$$

Big deal! We have identified the mathematical definition of j. We can believe that a mathematician will be very happy with this, but does it do us any good? Once again the answer is "yes." Without changing any of the common rules for algebra, we can now do addition, subtraction, multiplication, and division with phasors! We can use all of the electrical laws learned for dc in solving ac without change, with one exception: the standard arithmetic processes must be completed using phasors in place of plain, ordinary numbers. At first glance Eq. (16-1-1) does not tell us much. However, it then becomes clear that j is not $+1$ and it is not -1 because

$$(+1)^2 = 1$$

and

$$(-1)^2 = 1$$

In fact, there is no real number that can be squared to give -1. For this reason we call j an *imaginary number*. This does not mean that j cannot exist in the normal pattern of a number system. Fortunately, it is not necessary to go more deeply into the philosophical nature of the matter, for we have a valuable working tool as matters stand.

Let us now try for additional rotations of 90°. Starting with

$$+5$$

rotate 90°:

$$j5$$

and another 90°:

$$j \times j5 = j^25 = -5$$

and another 90°:

$$j \times j \times j5 = j^35 = (-1)j5 = -j5$$

and still another 90°:

$$j \times j \times j \times j5 = -j^25 = (-1) \times (-1) \times 5 = 5$$

Now it is all coming together. The mathematics works out fine when $j = \sqrt{-1}$ and we do not even worry about the use of the term "imaginary" for j.

In Section 12-1 a different notation for the representation of phasors was introduced. Thus, if a current of 25 mA leads the applied voltage of 110 V in a circuit by 45°, the current can be written as

$$25 \underline{/45°} \text{ mA}$$

with the understanding that the voltage is the reference which has an angle of 0° as follows:

$$110 \underline{/0°} \text{ V}$$

Adopting this notation, called the *polar notation*, gives us an alternative and useful way of writing the j operator and successive applications of the j operator.

$$j = 1\underline{/90°}$$
$$j^2 = -1 = (1\underline{/90°})(1\underline{/90°}) = 1\underline{/180°}$$
$$j^3 = -j = (1\underline{/90°})(1\underline{/90°})(1\underline{/90°}) = 1\underline{/270°} = 1\underline{/-90°}$$
$$j^4 = 1 = (1\underline{/90°})(1\underline{/90°})(1\underline{/90°})(1\underline{/90°}) = 1\underline{/360°} = 1\underline{/0°}$$

The multiplication of two or more j operators or their polar notation equivalents really amounts to adding the angles so that the resulting angle is a multiple of 90°. A few additional examples will aid us in grasping these points.

EXAMPLE 16-4 A current of 1 A leads the circuit applied voltage by 90°. Express the current in two different ways, assuming the voltage to be the reference.

Solution By attaching the appropriate unit to Eq. (16-1-1), we have the answer.

$$I = j1 = 1\underline{/90°} \text{ A}$$

EXAMPLE 16-5 The current in a circuit lags the circuit applied voltage by 90°. If this current is measured to be 250 mA, express it as a phasor in two different ways, assuming the voltage to be the reference.

Solution When the phasor is expressed as 250 mA, the angle is assumed to be zero. Thus

$$250 \text{ mA} = 250 \underline{/0°} \text{ mA}$$

It is now necessary to rotate the current phasor by 270° counterclockwise or 90° clockwise.

$$I = j^3 \times 250 = -j250 = 250 \underline{/-90°} \text{ mA}$$

16-2
Rectangular and Polar Conversions

The j operation can also be used in representing phasors for angles other than 90° and multiples of 90°. Consider the parallel circuit in Fig. 16-5(a). We immediately know that I_1 must be in phase with the voltage, V, and I_2 must lead V by 90°. Assuming V to be the reference, the phasor diagram for the branch currents and common voltage, V, is given in Fig. 16-5(b). To be specific, let the current as measured by an ammeter in the first branch be 4 A and that in the capacitive branch be 3 A. We can immediately put our knowledge of the j operator to work.

$$I_1 = 4 \text{ A}$$
$$I_2 = j3 \text{ A}$$

Moreover, the total current becomes

$$I_T = I_1 + I_2$$
$$= 4 + j3 \text{ A}$$

FIGURE 16-5 Parallel *RC* circuit.

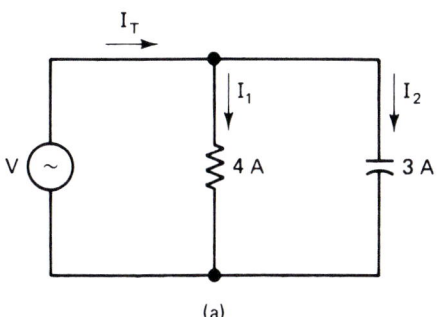

(a)

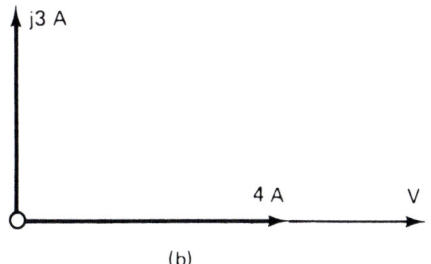

(b)

Sec. 16-2 Rectangular and Polar Conversions

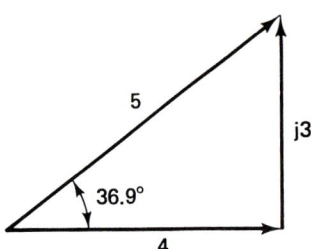

FIGURE 16-6 Phasor addition of currents for Fig. 16-5.

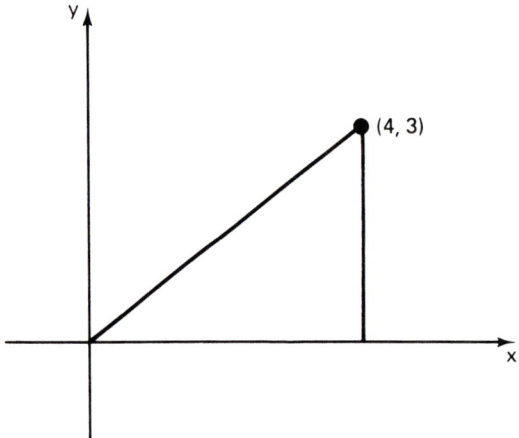

FIGURE 16-7 Phasor of Fig. 16-6 superimposed on rectangular coordinates.

Note that we have added the currents correctly to obtain a complex number consisting of a "real" part (4) and an "imaginary" part ($j3$). To add $4 + j3 = 7$ will provide a meaningless result.

Referring now to the phasor diagram for the currents as pictured in Fig. 16-6, the right triangle can be solved to find the hypotenuse and the angle.

$$I_T = \sqrt{I_R^2 + I_C^2} \qquad (16\text{-}2\text{-}1)$$

$$-\theta = \tan^{-1}\frac{I_C}{I_R} \qquad (16\text{-}2\text{-}2)$$

$$= \tan^{-1}\tfrac{3}{4} = 36.9°$$

However, you will recall that the phase angle difference of a lagging current is considered to be positive. Therefore, the phase angle difference of a leading current is negative. In this particular example we may express the total in two different ways.

$$I_T = 4 + j3 = 5\underline{/36.9°}$$

The first form is called *rectangular* because the two quantities conform to the x and y of conventional rectangular coordinates. The phasor diagram of Fig. 16-6 has been superimposed on the axes of a rectangular coordinate graph in Fig. 16-7 to illustrate this. The tip of the arrow is the point $x = 4$, $y = 3$, usually written (4, 3) in parentheses. We can also represent the total current phasor in polar coordinates as shown in Fig. 16-8. For the moment we restrict our attention to converting phasors written in rectangular form to phasors written in polar form.

FIGURE 16-8 Phasor in polar coordinates.

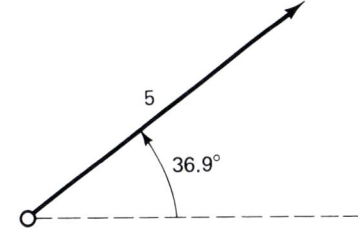

EXAMPLE 16-6 The current in a parallel RC circuit is $4 + j1$ A. Express this current in polar form.

Solution

$$I_T = \sqrt{I_R^2 + I_C^2} \qquad (16\text{-}2\text{-}1)$$

$$= \sqrt{4^2 + 1^2} = 4.12 \text{ A}$$

$$-\theta = \tan^{-1}\frac{I_C}{I_R} \qquad (16\text{-}2\text{-}2)$$

$$= \tan^{-1}\tfrac{1}{4} = 14.0°$$

Therefore,

$$I_T = 4.12\underline{/14.0°} \text{ A}$$

It is also possible to express lagging currents in rectangular form. In Fig. 16-9 are shown two parallel branches, one of which is resistive and the other inductive. The phasor diagram for the branch currents appears in Fig. 16-10. To simplify this discussion, the current in the first branch is given the value 4 A. It is easy to estimate that the total current will be larger than 4 A and that

FIGURE 16-9 Parallel inductive circuit.

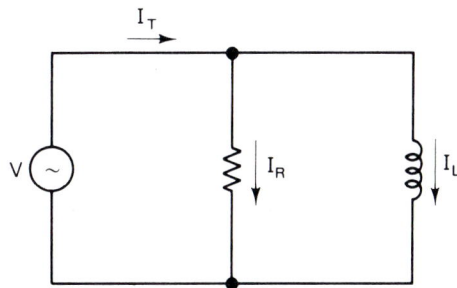

FIGURE 16-10 Phasor diagram for parallel RL circuit.

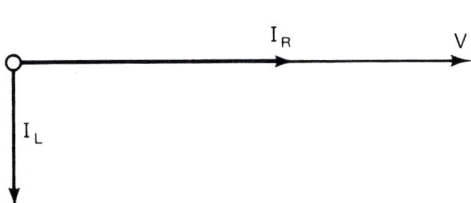

Sec. 16-2 Rectangular and Polar Conversions

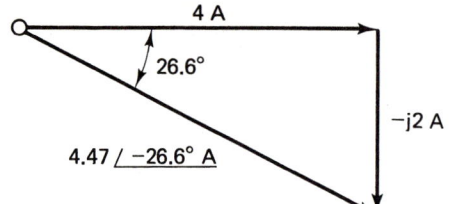

FIGURE 16-11 Phasor diagram of parallel RL circuit.

the total current will lag the voltage by an angle less than 45°. However, it is not necessary to continue guessing, since

$$I_T = \sqrt{I_R^2 + I_L^2} \qquad (16\text{-}2\text{-}3)$$

$$= \sqrt{4^2 + 2^2} = 4.47 \text{ A}$$

$$\theta = \tan^{-1}\frac{I_L}{I_R} \qquad (16\text{-}2\text{-}4)$$

$$= \tan^{-1}\tfrac{2}{4} = 26.6°$$

In rectangular form the total current is

$$4 - j2 \text{ A}$$

and in polar form it is

$$4.47\angle{-26.6°} \text{ A}$$

Bear in mind that the angle is 26.6° clockwise from the phasor (see Fig. 16-11). At the same time the phase-angle difference is 26.6°. It is proper to say that

$$I_T = 4 - j2 = 4.47\angle{-26.6°} \text{ A}$$

EXAMPLE 16-7 In a parallel circuit the current through one branch is 3.5 A in phase with the applied voltage. The current in the second branch is 6.3 A and lags the applied voltage by 90°. Find the total current.

Solution

$$I_T = \sqrt{I_R^2 + I_L^2} \qquad (16\text{-}2\text{-}3)$$

$$= \sqrt{3.5^2 + 6.3^2} = 7.21 \text{ A}$$

$$\theta = \tan^{-1}\frac{I_L}{I_R} \qquad (16\text{-}2\text{-}4)$$

$$= \tan^{-1}\frac{6.3}{3.5} = 61.0°$$

$$I_T = 3.5 - j6.3 = 7.21\angle{-61.0°} \text{ A}$$

FIGURE 16-12 Conversion from polar to rectangular form.

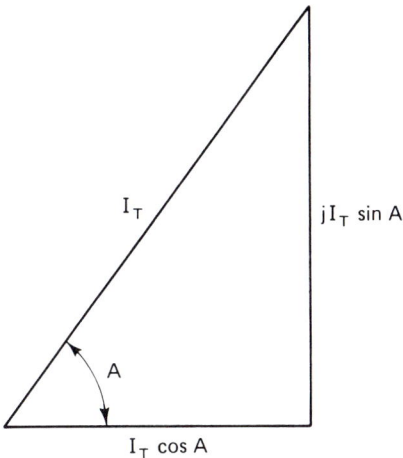

Thus far we have been concerned with conversions from rectangular to polar form. It is equally practical to convert from polar form to rectangular form. Suppose that we have a circuit consisting of one resistor in parallel with a capacitor. [Refer to Fig. 16-5(a) as an illustration of the circuit and to Fig. 16-5(b) phasor diagram with the branch currents.] However, in this particular case the total current and phase angle are known. We now wish to find the branch currents. To accomplish this we will use some additional trigonometric relationships shown in Fig. 16-12. In this phasor diagram the angle is designated A. The horizontal leg of the right triangle is

$$I_R = I_T \cos A \qquad (16\text{-}2\text{-}5)$$

and the vertical leg is

$$jI_C = jI_T \sin A \qquad (16\text{-}2\text{-}6)$$

For any given angle A, the values of the function $\sin A$ and $\cos A$ can be found using tables, slide rules, or electronic calculators.

EXAMPLE 16-8 In Fig. 16-12, $I_T = 10$ A and $A = 50°$. Find the currents through the resistive branch and the capacitive reactive branch.

Solution

$$I_R = I_T \cos A \qquad (16\text{-}2\text{-}5)$$
$$= 10 \cos 50° = 6.43 \text{ A}$$
$$jI_C = jI_T \sin A \qquad (16\text{-}2\text{-}6)$$
$$= j10 \sin 50° = j7.66 \text{ A}$$

EXAMPLE 16-9 An inductor is placed in parallel with a resistor. The total current is 53.6 mA and lags the applied voltage by 15.5°. Find the current through each branch.

Sec. 16-2 Rectangular and Polar Conversions

Solution We have another angle sign to be concerned with. A leading current corresponds to a positive A and a lagging current to a negative A.

Therefore,

$$I_R = I_T \cos A \qquad (16\text{-}2\text{-}5)$$
$$= 53.6 \cos(-15.5°) = 51.7 \text{ mA}$$

$$jI_L = jI_T \sin A \qquad (16\text{-}2\text{-}7)$$
$$= j53.6 \sin(-15.5°) = -j14.3 \text{ mA}$$

The preceding examples suggest that a major difficulty in making coordinate conversions is the sign of the result. One method of coping with this difficulty accounts for all signs in writing the equations. A better method for most technicians and engineers attaches signs on the basis of the phasor diagram. If one can sketch a crude diagram, the sign can usually be attached without error after a calculation has been made.

We must recognize that all the coordinate conversions made to this point apply for phasors with angles between $-90°$ and $+90°$. It is possible to extend the same methods to any angle. In Fig. 16-13, angle A is greater than 90° and is less than 180°. Note that the supplementary angle $(180° - A)$ is used in obtaining the real rectangular component

$$I_X = -I_T \cos(180° - A) \qquad (16\text{-}2\text{-}8)$$

and the imaginary rectangular component

$$jI_Y = jI_T \sin(180° - A) \qquad (16\text{-}2\text{-}9)$$

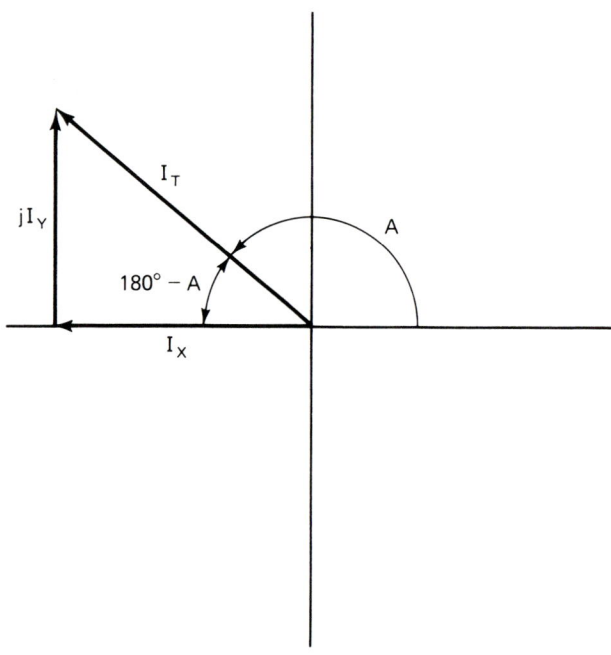

FIGURE 16-13 Illustration of phasor angle greater than 90°.

In these equations I_X represents the horizontal component of the phasor and jI_Y represents the vertical component. It is not always proper to consider these two as branch currents in a parallel circuit as was done before. Also, we should be aware that the phasor need not be a current, but could be a voltage. In general, all of the equations introduced in this chapter apply to any phasor. The ideas will later be expanded to impedances and admittances.

When the rectangular coordinates are known, the polar coordinates can be found.

$$I_T = \sqrt{I_X^2 + I_Y^2} \qquad (16\text{-}2\text{-}10)$$

$$180° - A = \tan^{-1} \frac{|I_Y|}{|I_X|} \qquad (16\text{-}2\text{-}11)$$

EXAMPLE 16-10 A phasor has a length of 75 units and an angle of 160°. Find the rectangular component and express the phasor in both polar and rectangular forms.

Solution

$$I_X = -I_T \cos(180° - A) \qquad (16\text{-}2\text{-}8)$$
$$= -75 \cos(180° - 160°) = -70.5$$
$$jI_Y = jI_T \sin(180° - A) \qquad (16\text{-}2\text{-}9)$$
$$= j75 \sin(180° - 160°) = j25.7$$

Therefore,

$$I = 75\angle 160° = -70.5 + j25.7$$

EXAMPLE 16-11 A phasor is expressed in rectangular form as $-10.6 + j31.9$. Determine the polar form for this phasor.

Solution

$$I_T = \sqrt{I_X^2 + I_Y^2} \qquad (16\text{-}2\text{-}10)$$

$$= \sqrt{(-10.5)^2 + 31.9^2} = 33.6$$

$$180° - A = \tan^{-1} \frac{|I_Y|}{|I_X|} \qquad (16\text{-}2\text{-}11)$$

$$= \tan \frac{31.9}{10.6} = 71.6°$$

$$A = 180° - 71.6° = 108°$$

Therefore,

$$\text{phasor } I_T = 33.6\angle 108°$$

Sec. 16-2 Rectangular and Polar Conversions

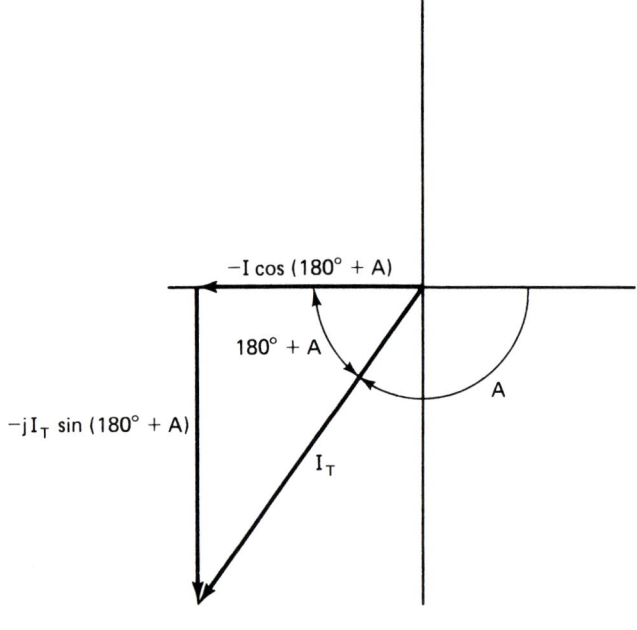

FIGURE 16-14 Phasor diagram with phasor in third quadrant.

Notice that positive values are used for the rectangular components in Eq. (16-2-11). This is indicated by the lines on either side of I_Y and I_X in the equation. It is possible that the phasor be in another quarter (quadrant), as shown in Fig. 16-14. Here A is negative since it represents a rotation of I in the clockwise (negative) direction from the reference direction (to the right). The angle $180° + A$ will be positive and less than $90°$. Both $\sin(180° + A)$ and $\cos(180° + A)$ are positive, so that the negative signs for the rectangular coordinates must be injected into the equations.

$$I_X = -I_T \cos(180° + A) \qquad (16\text{-}2\text{-}12)$$

$$jI_Y = -jI_T \sin(180° + A) \qquad (16\text{-}2\text{-}13)$$

Also,

$$I_T = \sqrt{I_X^2 + I_Y^2} \qquad (16\text{-}2\text{-}10)$$

$$180° + A = \tan^{-1} \frac{|I_Y|}{|I_X|} \qquad (16\text{-}2\text{-}14)$$

EXAMPLE 16-12 Convert $150 \angle -100°$ to rectangular form.

Solution

$$I_X = -I_T \cos(180° + A) \qquad (16\text{-}2\text{-}12)$$
$$= -150 \cos 80° = -26.0$$
$$jI_Y = -jI_T \sin(180° + A) \qquad (16\text{-}2\text{-}13)$$
$$= -j150 \sin 80° = -j148$$

Therefore,

$$150 \angle -100° = -26.0 - j148$$

EXAMPLE 16-13 Convert the phasor $-27.1 - j33.5$ to polar form.

Solution

$$I_T = \sqrt{I_X^2 + I_Y^2} \qquad (16\text{-}2\text{-}10)$$

$$= \sqrt{(-27.1)^2 + (-33.5)^2} = 43.1$$

$$180° + A = \tan^{-1} \frac{|I_Y|}{|I_X|} \qquad (16\text{-}2\text{-}14)$$

$$= \tan^{-1} \frac{|-33.5|}{|-27.1|} = 51.0°$$

$$A = 51.0° - 180° = -129°$$

Therefore,

$$-27.1 - j33.5 = 43.1 \angle -129°$$

16-3
Addition and Subtraction of Phasors

The application of Kirchhoff's Laws to ac circuits requires that phasors be added and subtracted. This does not present a severe difficulty when the angle is 90° or some multiple of 90°. However, when this is not the case, additional steps are necessary before addition and subtraction can be achieved numerically. Thus when two phasors are to be added analytically, these phasors must be in the rectangular form initially, or must be converted from the polar to the rectangular form before addition can take place.

In this section we first investigate the addition and subtraction of phasors as expressed in rectangular form. We will then add and subtract phasors in which at least one phasor is in polar form and for which the result may be needed in polar form. It will be seen that the steps required for adding and subtracting phasors under these circumstances are not difficult. Usually, a sequence of steps is required so that particular care must be taken to avoid errors in any one step. The situation is like the strength of a chain, which is no greater than that of the weakest link. When one step in phasor addition or subtraction is in error, the results are always in error. The good news is that no new basic operations need to be mastered. We will use simple algebra coupled with the coordinate conversion methods already studied. Consider now the addition of two phasors $a + jb$ and $c + jd$, where a, b, c, and d are real numbers.

$$(a + jb) + (c + jd) = a + c + j(b + d) \qquad (16\text{-}3\text{-}1)$$

This seems simple enough. All we do is add the real parts of the two complex numbers (rectangular form) to obtain the real part of the sum. Then we add the imaginary parts of the complex numbers to obtain the imaginary part of the sum. The same method is used for the sum of any number of phasors. Just handle the real and imaginary parts separately, as dictated by the rules of algebra.

For subtraction,

$$(a + jb) - (c + jd) = (a - c) + j(b - d) \qquad (16\text{-}3\text{-}2)$$

Like addition, subtraction follows the rules of algebra.

EXAMPLE 16-14 Add $2 + j5$ to $3 + j1$.

Solution

$$\begin{array}{r} 2 + j5 \\ (+)\ 3 + j1 \\ \hline 5 + j6 \end{array}$$

If we recognize that $a = 2$, $b = 5$, $c = 3$, and $d = 1$, Eq. (16-3-1) may be applied as

$$(2 + j5) + (3 + j1) = (2 + 3) + j(5 + 1)$$
$$= 5 + j6$$

As is easily seen, the sum is found very quickly by aligning the digits vertically in the conventional method for adding numbers.

EXAMPLE 16-15 Add $2 - j5$ to $-1 + j4$.

Solution

$$\begin{array}{r} 2 - j5 \\ (+)\ -1 + j4 \\ \hline 1 - j1 \end{array}$$

Ah, ha! This is not quite as easy. The original statement about how to add complex numbers should refer to adding the real parts "algebraically" and the imaginary parts "algebraically."

EXAMPLE 16-16 Subtract $2 + j3$ from $5 + j10$.

Solution

$$\begin{array}{r} 5 + j10 \\ -\ (2 + j3) \\ \hline 3 + j7 \end{array}$$

EXAMPLE 16-17 Subtract $-2 + j3$ from $5 - j10$.

Solution The particular difficulty arises from the signs and the changing of signs. A good rule to follow is to change the signs for the phasor being subtracted, then add the real and imaginary parts algebraically. Thus

$$(5 - j10) - (-2 + j3) = (5 - j10) + (2 - j3) = 7 - j13$$

As mentioned earlier, when one or more phasors to be added or subtracted are in polar form, conversions to rectangular form are necessary first. In the next few examples, the conversion results will be indicated but without showing every individual step.

EXAMPLE 16-18 Add $7.5 \angle -16.4°$ to $8.93 \angle 25.7°$.

Solution

$$7.59 \angle -16.4° = 7.28 - j2.14$$
$$8.93 \angle 25.7° = 8.05 + j3.87$$
$$(7.28 - j2.14) + (8.05 + j3.87) = 15.3 + j1.73 = 15.4 \angle 6.44°$$

Note: Since neither polar nor rectangular form has been specified for the answer, both have been provided.

EXAMPLE 16-19 Add $158 \angle -135°$ to $761 \angle 143°$.

Solution

$$158 \angle -135° = -112 - j112$$
$$761 \angle 143° = -608 + j458$$
$$(-112 - j112) + (-608 + j458) = -719 + j346 = 798 \angle 154°$$

EXAMPLE 16-20 Subtract $98.7 \angle -58.9°$ from $43.7 \angle 161°$.

Solution

$$43.7 \angle 161° = -41.3 + j14.2$$
$$98.7 \angle -58.9° = 51.0 - j84.5$$
$$(-41.3 + j14.2) - (51.0 - j84.5) = (-41.3 + j14.2) + (-51.0 + j84.5)$$
$$= -92.3 + j98.7 = 135 \angle 133°$$

EXAMPLE 16-21 Subtract $143\angle-107°$ from $88.9\angle 195°$.

Solution

$$88.9\angle 195° = -85.9 - j23.0$$
$$143\angle -107° = -41.8 - j136.8$$
$$(-85.9 - j23.0) - (-41.8 - j136.8) = (-85.9 - j23.0) + (41.8 + j136.8)$$
$$= -44.1 + j114 = 122\angle 111°$$

16-4

Multiplication and Division of Phasors

When one phasor is multiplied by a second, the product has a magnitude and an angle given by the following rules:

1. The magnitude of the product is equal to the product of the two phasor magnitudes.
2. The angle of the product is equal to the algebraic sum of the two phasor angles.

Let two phasors be represented in polar form as $M_1\angle A_1$ and $M_2\angle A_2$. The product is then

$$(M_1\angle A_1)(M_2\angle A_2) = M_1 M_2 \angle A_1 + A_2 \qquad (16\text{-}4\text{-}1)$$

EXAMPLE 16-22 Find the product $12.5\angle 39.5° \times 3.71\angle 21.4°$.

Solution

$$12.5\angle 39.5° \times 3.71\angle 21.4° = 12.5 \times 3.71\angle 39.5° + 21.4° = 46.4\angle 60.9°$$

EXAMPLE 16-23 Find the product $0.0513\angle -125° \times 721\angle 44.3°$.

Solution

$$0.0513\angle -125° \times 721\angle 44.3° = 0.0513 \times 721\angle -125° + 44.3°$$
$$= 37.0\angle -80.7°$$

The same rule can be applied when finding the product of any number of phasors $M_1\angle A_1$, $M_2\angle A_2$, $M_3\angle A_3$, and so forth. This is because the product of the first two is a phasor found using Eq. (16-4-1), that product can be multiplied by the third phasor using Eq. (16-4-1) again to obtain another product, and the process repeated to multiply by the fourth phasor and the fifth and so on, until the product of all given phasors is obtained.

EXAMPLE 16-24 Find the product

$$0.123 \underline{/-75.6°} \times 41.5 \underline{/-127°} \times 1.59 \underline{/218°}.$$

Solution

$$0.123 \underline{/-75.6°} \times 41.5 \underline{/-127°} \times 1.59 \underline{/218°}$$
$$= (0.123 \underline{/-75.6°} \times 41.5 \underline{/-127°}) \times 1.59 \underline{/218°}$$
$$= (0.123 \times 41.5 \underline{/-75.6° - 127°}) \times 1.59 \underline{/218°}$$
$$= (5.10 \underline{/-203°}) \times 1.59 \underline{/218°}$$
$$= 5.10 \times 1.59 \underline{/-203° + 218°}$$
$$= 8.12 \underline{/15°}$$

We should be aware that the steps need not be in the order shown. To illustrate a different way of obtaining the same result, the multiplication of the three magnitudes will be performed to obtain the magnitude of the final phasor product. Also, the angles will be summed to obtain the final algebraic sum without showing intermediate steps. Thus

$$0.123 \underline{/-75.6°} \times 41.5 \underline{/-127°} \times 1.59 \underline{/218°}$$
$$= 0.123 \times 41.5 \times 1.59 \underline{/-75.6° - 127° + 218°}$$
$$= 8.12 \underline{/15.4°}$$

EXAMPLE 16-25 Find the product

$$4.78 \underline{/-23.7°} \times 0.0257 \underline{/-117°} \times 0.516 \underline{/37.6°} \times 77.2 \underline{/126°}.$$

Solution

$$4.78 \underline{/-23.7°} \times 0.0257 \underline{/-117°} \times 0.516 \underline{/37.6°} \times 77.2 \underline{/126°}$$
$$= 4.78 \times 0.0257 \times 0.516 \times 77.2 \underline{/-23.7° - 117° + 37.6° + 126°}$$
$$= 4.89 \underline{/22.9°}$$

Both magnitudes and angles are involved in division. Division of one phasor by another is accomplished using the following rules:

1. The magnitude of the first is divided by that of the second.
2. The angle of the second is *subtracted* algebraically from that of the first.

In mathematical form,

$$\frac{M_1 \angle A_1}{M_2 \angle A_2} = \frac{M_1}{M_2} \angle A_1 - A_2 \qquad (16\text{-}4\text{-}2)$$

Instead of thinking in terms of subtracting the angle of the phasor in the denominator from the angle of the phasor in the numerator, it is sometimes easier to change the sign and add. Thus the sign of the denominator angle is changed and then the angle is added algebraically to the angle of the phasor in the numerator.

EXAMPLE 16-26 Find the quotient

$$\frac{77.3 \angle 81.4°}{2.49 \angle 17.5°}$$

Solution

$$\frac{77.3 \angle 81.4°}{2.49 \angle 17.5°} = \frac{77.3}{2.49} \angle 81.4° - 17.5°$$

$$= 31.0 \angle 63.9°$$

EXAMPLE 16-27 Find the quotient

$$\frac{0.105 \angle -115°}{12.6 \angle -27.3°}$$

Solution

$$\frac{0.105 \angle -115°}{12.6 \angle -27.3°} = \frac{0.105}{12.6} \angle -115° + 27.3°$$

$$= 0.00833 \angle -88°$$

Phasors expressed in rectangular form can be multiplied and divided directly using the rules of common algebra with a few tricks thrown in. This approach is demonstrated later. A particularly straightforward method for multiplication and division consists of first converting these phasors to polar form and then applying either Eq. (16-4-1) or (16-4-2).

EXAMPLE 16-28 Find the quotient

$$\frac{3 - j4}{-5 + j6}$$

Solution

$$3 - j4 = 5.00 \angle -53.1°$$

$$-5 + j6 = 7.81 \angle 130°$$

$$\frac{3 - j4}{-5 + j6} = \frac{5.00 \angle -53.1°}{7.81 \angle 130°} = 0.640 \angle -183° = 0.640 \angle 177°$$

$$= -0.639 + j0.0328$$

On occasion there is a need to perform chain calculations with phasors. By this is meant more than one factor in the numerator or in the denominator or in both the numerator and denominator. Equations (16-4-1) and (16-4-2) are still valid and are applied in any convenient order. Two examples will illustrate the principle.

EXAMPLE 16-29 Find the phasor given by

$$\frac{2.53 \angle -27.6° \times 37.1 \angle 74.7°}{0.149 \angle 21.4°}$$

Solution Three equally valid solutions will be shown. First multiply the numerical factors.

$$\frac{2.53 \angle -27.6° \times 37.1 \angle 74.7°}{0.149 \angle 21.4°} = \frac{93.9 \angle 47.1°}{0.149 \angle 21.4°}$$

$$= 630 \angle 25.7°$$

In the second method one numerator phasor is first divided by the denominator phasor.

$$\frac{2.53 \angle -27.6° \times 37.1 \angle 74.7°}{0.149 \angle 21.4°} = 17.0 \angle -49.0° \times 37.1 \angle 74.7°$$

$$= 630 \angle 25.7°$$

In the third method the other numerator phasor is first divided by the denominator phasor.

$$\frac{2.53 \angle -27.6° \times 37.1 \angle 74.7°}{0.149 \angle 21.4°} = 2.53 \angle -27.6° \times 249 \angle 53.3°$$

$$= 630 \angle 25.7°$$

EXAMPLE 16-30 Find the phasor given by

$$\frac{7.65 \angle 21.6°}{0.0198 \angle -87.2° \times 0.715 \angle 51.2°}$$

Sec. 16-4 Multiplication and Division of Phasors

Solution Three different step orders are demonstrated in this solution.

$$\frac{7.65 \,\underline{/21.6°}}{0.0198 \,\underline{/-87.2°} \times 0.715 \,\underline{/51.2°}} = \frac{7.65 \,\underline{/21.6°}}{0.0142 \,\underline{/-36.0°}}$$

$$= 540 \,\underline{/57.6°}$$

$$\frac{7.65 \,\underline{/21.6°}}{0.0198 \,\underline{/-87.2°} \times 0.715 \,\underline{/51.2°}} = \frac{386 \,\underline{/108.8°}}{0.715 \,\underline{/51.2°}}$$

$$= 540 \,\underline{/57.6°}$$

$$\frac{7.65 \,\underline{/21.6°}}{0.0198 \,\underline{/-87.2°} \times 0.715 \,\underline{/51.2°}} = \frac{10.7 \,\underline{/-29.6°}}{0.0198 \,\underline{/-87.2°}}$$

$$= 540 \,\underline{/57.6°}$$

As was stated earlier, phasors expressed in rectangular form can be multiplied directly. As will be seen, multiplication of such phasors is more effective when the real and imaginary parts are simple whole numbers. Otherwise, the mechanics of multiplication become more involved. Let the two phasors be expressed as $a + jb$ and $c + jd$. When the two are multiplied, like terms grouped and the j^2 factor recognized as being -1, the product is

$$(a + jb)(c + jd) = (ac - bd) + j(ad + bc) \qquad (16\text{-}4\text{-}3)$$

It is seen that four multiplications, one addition, and one subtraction are required. Of particular interest is the rectangular form of the result. This may be very advantageous for some applications.

EXAMPLE 16-31 Multiply $3 + j4$ and $5 + j2$.

Solution To apply Eq. (16-4-3), it is necessary to identify the given numerical data with the a, b, c, and d symbols of that equation. Let us say that $a = 3$, $b = 4$, $c = 5$, and $d = 2$. Substituting into the equation, we obtain

$$(3 + j4)(5 + j2) = (3 \times 5 - 4 \times 2) + j(3 \times 2 + 4 \times 5)$$
$$= (15 - 8) + j(6 + 20)$$
$$= 7 + j26$$

We should recognize that Eq. (16-4-3) is still valid when some or all of the quantities a, b, c, and d are negative. It is only necessary to remember that the product of a positive quantity and a negative quantity is negative. Moreover, the product of two negative quantities is positive.

EXAMPLE 16-32 Multiply $-3 + j4$ and $5 - j2$.

Solution Let us assign the data as follows: $a = -3$, $b = 4$, $c = 5$, and $d = -2$. Substitution into Eq. (16-4-3) gives

$$(-3 + j4)(5 - j2) = [(-3)(5) - (4)(-2)] + j[(-3)(-2) + (4)(5)]$$
$$= [-15 - (-8)] + j[6 + 20]$$
$$= [-15 + 8] + j26$$
$$= -7 + j26$$

EXAMPLE 16-33 Multiply $3 - j4$ and $5 - j2$.

Solution Use $a = 3$, $b = -4$, $c = 5$, and $d = -2$. Substitution into Eq. (16-4-3) yields

$$(3 - j4)(5 - j2) = [(3)(5) - (-4)(-2)] + j[(3)(-2) + (-4)(5)]$$
$$= [15 - 8] + j[-6 + (-20)]$$
$$= 7 - j26$$

The division of phasors in rectangular form can also be accomplished without changing to polar coordinates. The key step in the process eliminates the imaginary term in the denominator by multiplying both the numerator and the denominator by the complex conjugate of the denominator. To illustrate this, consider the indicated division,

$$\frac{a + jb}{c + jd}$$

The complex conjugate of the denominator is $c - jd$ and it can be shown that

$$(c + jd)(c - jd) = c^2 + d^2$$

Therefore,

$$\frac{a + jb}{c + jd} = \frac{(a + jb)(c - jd)}{(c + jd)(c - jd)} = \frac{(ac + bd) + j(bc - ad)}{c^2 + d^2}$$
$$= \frac{ac + bd}{c^2 + d^2} + j\frac{bc - ad}{c^2 + d^2} \quad (16\text{-}4\text{-}4)$$

EXAMPLE 16-34 Complete the indicated division.

$$\frac{9 + j30}{3 + j2}$$

Solution $a = 9$, $b = 30$, $c = 3$, $d = 2$. Substitute into Eq. (16-4-4).

$$\frac{9 + j30}{3 + j2} = \frac{(9)(3) + (30)(2)}{3^2 + 2^2} + j\frac{(30)(3) - (9)(2)}{3^2 + 2^2}$$

$$= \frac{27 + 60}{9 + 4} + j\frac{90 - 18}{9 + 4}$$

$$= \frac{87}{13} + j\frac{72}{13}$$

$$= 6.69 + j5.54$$

One particular phasor division is of special interest in converting a series circuit into an equivalent parallel circuit. Consider the practical "tank" circuit of Fig. 16-15. We first convert the impedance of the inductive branch into an admittance by finding the reciprocal using Eq. (16-4-4).

$$\frac{1}{R + jX_L} = \frac{R}{R^2 + X_L^2} - j\frac{X_L}{R^2 + X_L^2}$$

Note that in applying the equation $a = 1$, $b = 0$, $c = R$, and $d = X_L$. It is now possible to replace the circuit of Fig. 16-15 with one having three parallel branches, as shown in Fig. 16-16. Whereas in the preceding equation the two terms have the units of siemens, in the figure all three branches have the units of ohms. This is because the two terms representing a conductance in parallel with a susceptance in the preceding equation have each been inverted to obtain a resistance and a reactance. It will be recalled that the quality factor, Q, of an inductor is expressed by Eq. (14-1-4).

$$Q = \frac{X_L}{R}$$

Since Q is usually much larger than 10, the R^2 in $R^2 + X_L^2$ can usually be ignored. This greatly simplifies the expressions for two of the branches of Fig. 16-16.

$$\frac{R^2 + X_L^2}{X_L} \approx X_L \qquad (16\text{-}4\text{-}5)$$

$$\frac{R^2 + X_L^2}{R} \approx \frac{X_L^2}{R} = \frac{X_L^2}{R^2} \times R = Q^2 R \qquad (16\text{-}4\text{-}6)$$

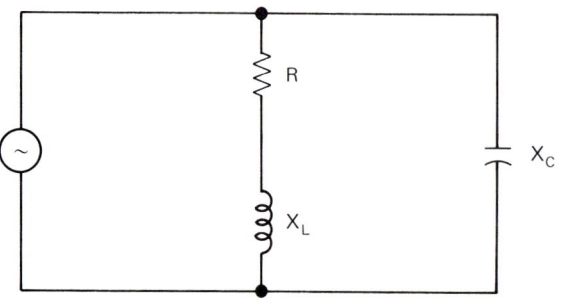

FIGURE 16-15 Practical "tank" circuit.

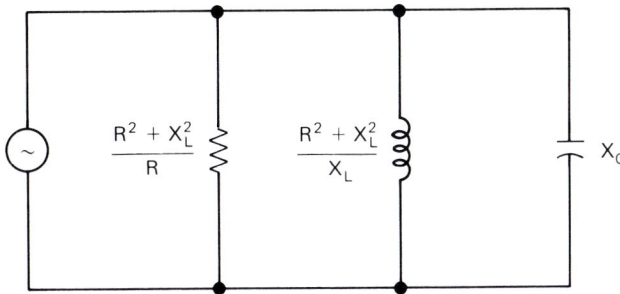

FIGURE 16-16 Circuit with three parallel branches.

when "≈" is read "is approximately equal to." We interpret the results of Eqs. (16-4-5) and (16-4-6) as follows. A practical coil having resistance R and quality factor Q can be replaced by two parallel branches. When Q is large, as is often the case, one parallel branch has an inductive reactance approximately equal to X_L. Under these circumstances, the other parallel branch has a resistance equal to Q^2R.

EXAMPLE 16-35 A coil has a resistance of 20 Ω and $Q = 100$. Find the resistance of one branch for the equivalent parallel circuit.

Solution Apply Eq. (16-4-6).

$$Q^2R = 100^2 \times 20 = 200{,}000 \text{ Ω}$$

EXAMPLE 16-36 The resistance of a coil is 5 Ω and the inductive reactance is 50 Ω at a given frequency. Calculate the percentage errors for the equivalent parallel circuit branches in using the approximations of Eqs. (16-4-5) and (16-4-6).

Solution The true inductive reactance for the parallel reactive branch is

$$\frac{R^2 + X_L^2}{X_L} = \frac{5^2 + 50^2}{50} = \frac{2525}{50} = 50.5 \text{ Ω}$$

and the approximate value is 50 Ω. Therefore, the approximation is 0.5 Ω low, which is about 1% of the true value. In round numbers the error is 1%. Continuing,

$$\frac{R^2 + X_L^2}{R} = \frac{2525}{5} = 505$$

$$Q = \frac{X_L}{R} = \frac{50}{5} = 10$$

$$Q^2R = 10^2 \times 5 = 500 \text{ Ω}$$

The true value is 505 Ω, whereas the approximation is 500 Ω, so that the approximate value is 5 Ω too low. The error is again about 1%.

Sec. 16-4 Multiplication and Division of Phasors

16-5 Series-Parallel Circuits

The application of complex algebra allows the solution of circuits containing R, L, and C, or various combinations of these elements, in series-parallel-connected branches. More generally, all the methods developed in earlier chapters for the analysis of dc circuits can be applied for the analysis of ac circuits having the same configuration. In doing so, two general rules must apply:

1. Each branch must contain an impedance expressed in (or convertible to) complex form.
2. The manipulation of complex numbers will be required in the use of Ohm's law, Kirchhoff's laws, and the other techniques developed for dc circuits.

To demonstrate the general method of circuit analysis using complex algebra with all practical circuit configurations so far encountered in this book, it would be necessary to add, one by one, most of the previous dc examples and problems in ac form. Fortunately, we need not show every conceivable circuit variation using Z where only R was needed before. Instead of showing hundreds of examples and comparable numbers of problems, only *two* examples will be given. The generality, hence general usefulness, of these techniques is discussed later.

EXAMPLE 16-37 Find the currents and voltages for the series-parallel circuit of Fig. 16-17.

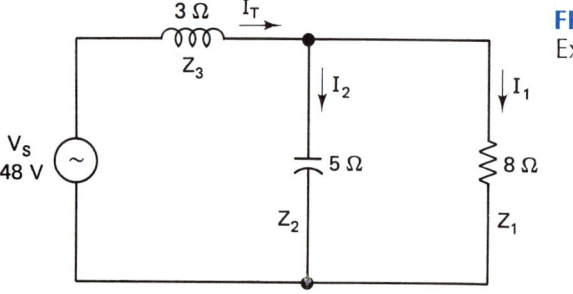

FIGURE 16-17 Circuit for Example 16-37.

Solution We first notice that the total applied voltage source of 48 V is given. One parallel branch has 8 Ω of resistance and the second parallel branch has 5 Ω of capacitive reactance. Our basic strategy is the same as previously used for dc circuits of the same form. Here we reduce the parallel branches to a single impedance, which must then be combined with the series 3-Ω inductive reactance. Let us accomplish these steps before explaining the subsequent steps. The first application of a formula is here referenced to its initial appearance for dc. We start with the parallel RC combination.

$$\frac{1}{Z_P} = \frac{1}{Z_1} + \frac{1}{Z_2} \qquad (3\text{-}3\text{-}1)$$

$$= \frac{1}{8 + j0} + \frac{1}{0 - j5}$$

$$= \frac{1}{8\angle 0} + \frac{1}{5\angle -90°}$$

$$= 0.125\angle 0 + 0.2\angle 90°$$

$$= 0.125 + j0 + 0 + j0.2$$

$$= 0.125 + j0.2 = 0.236\angle +58.0°$$

$$Z_P = \frac{1}{0.236\angle 58.0°} = 4.24\angle -58.0°$$

$$= 2.25 - j3.60 \ \Omega$$

Pausing for a moment, we see that Eq. (3-3-1), which applies for resistors in parallel, has been rewritten for impedances in parallel. Next we see that the impedances are initially written in rectangular form, then in polar form preparatory to finding the reciprocal (dividing into 1). Next, the reciprocal impedances, formally named admittances with the units of siemens, are resolved into rectangular coordinates and combined. After that the admittance of the parallel branches is converted to polar form and inverted (reciprocal found) to yield the impedance of the parallel branches. This last impedance is then converted to rectangular form in anticipation of further calculations.

The final steps in finding the total circuit impedance will now be made.

$$Z_T = Z_S + Z_P$$

$$= 0 + j3 + 2.25 - j3.60$$

$$= 2.25 - j0.6$$

$$= 2.32\angle -14.8° \ \Omega$$

Using the applied voltage as the reference, the source current can be found. It is then possible to find V_3, V_P, I_1, and I_2 to complete the solution.

$$I_T = \frac{V_S}{Z_T} = \frac{48\angle 0°}{2.32\angle -14.8°} = 20.6\angle 14.8° \ \text{A}$$

$$V_3 = I_T Z_3 = (20.6\angle 14.8°)(3\angle 90°) = 61.9\angle 104.8° \ \text{V}$$

$$V_P = I_T Z_P = (20.6\angle 14.8°)(4.24\angle -58.0°) = 87.5\angle -43.2° \ \text{A}$$

$$I_1 = \frac{V_P}{Z_1} = \frac{87.5\angle -43.2°}{8\angle 0°} = 10.9\angle -43.2° \ \text{A}$$

$$I_2 = \frac{V_P}{Z_2} = \frac{87.5\angle -43.2°}{5\angle -90°} = 17.5\angle 46.8° \ \text{A}$$

Although the circuit of Example 16-37 was simplified by having only one element in each branch, the procedure for solution is perfectly general. When more than one element exists in a branch, it is first necessary to combine them in series before taking the steps demonstrated in this example. This aspect of the analysis of series–parallel circuits using complex algebra is shown in the next example.

EXAMPLE 16-38 Find the currents for the circuit of Fig. 16-18.

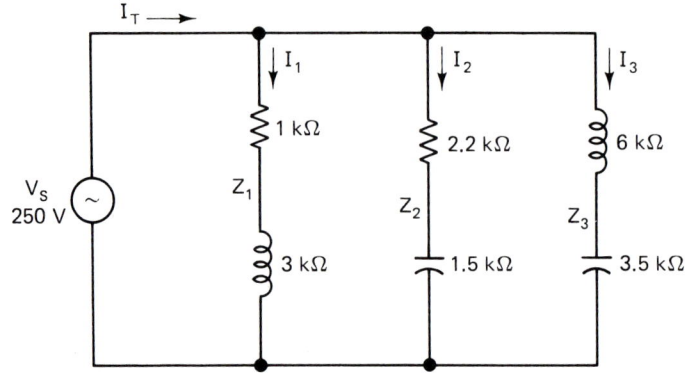

FIGURE 16-18 Circuit for Example 16-38.

Solution

$$Z_1 = 1 + j3 = 3.16\underline{/71.6°}\text{ k}\Omega$$

$$Z_2 = 2.2 - j1.5 = 2.66\underline{/-34.3°}\text{ k}\Omega$$

$$Z_3 = 0 + j(6 - 3.5) = j2.5 = 2.5\underline{/90°}\text{ k}\Omega$$

$$\frac{1}{Z_T} = \frac{1}{Z_1} + \frac{1}{Z_2} + \frac{1}{Z_3}$$

$$= \frac{1}{3.16\underline{/71.6°}} + \frac{1}{2.66\underline{/-34.3°}} + \frac{1}{2.5\underline{/90°}}$$

$$= 0.316\underline{/-71.6°} + 0.376\underline{/34.3°} + 0.4\underline{/-90°}$$

$$= 0.1 - j0.3 + 0.310 + j0.212 - j0.4$$

$$= 0.410 - j0.488 = 0.638\underline{/-50.0°}\text{ mS}$$

$$Z_T = \frac{1}{0.638\underline{/-50.0°}} = 1.57\underline{/50.0°}\text{ k}\Omega$$

$$I_T = \frac{V_S}{Z_T} = \frac{250\underline{/0°}}{1.57\underline{/50.0°}} = 159\underline{/-50.0°}\text{ mA}$$

$$I_1 = \frac{V_S}{Z_1} = \frac{250\underline{/0°}}{3.16\underline{/71.6°}} = 79.1\underline{/-71.6°}\text{ mA}$$

$$I_2 = \frac{V_S}{Z_2} = \frac{250\underline{/0°}}{2.66\underline{/-34.3°}} = 93.9\underline{/34.3°}\text{ mA}$$

$$I_3 = \frac{V_S}{Z_3} = \frac{250\underline{/0°}}{2.5\underline{/90.0°}} = 100\underline{/-90.0°}\text{ mA}$$

In Example 16-38, as in Example 16-37, the given source voltage was taken to be the reference phasor. Therefore, all other voltages and the currents have phase angles relative to their respective reference. Selecting the reference is only a matter of convenience during problem solution. Once the circuit is solved, any voltage or current can be made the reference by rotating the phasor diagram until that particular phasor is directed to the right (has zero angle). One must not forget to add or subtract the same angle from all the phasors, in polar form, when altering the reference.

The variety of series–parallel configurations is too great to be explored exhaustively here. Using complex algebra, substantial labor (as contrasted with dc), and great care, the student should be able to solve any ac circuit that he or she is able to solve in the same dc configuration with arithmetic. This statement can be expanded to include other circuits, such as those requiring delta–wye transformations. Following the two examples should make the student keenly aware of the need for effective polar–rectangular and rectangular–polar transformations. Whatever the calculation tool selected, and a scientific calculator having dedicated keys for these transformations is recommended, success is dependent on two factors:

1. An overall understanding of the strategy to be employed as based on the given information and the results desired.
2. A complete knowledge of the individual step calculations and the knowledge of how they are to be performed. In addition, consistent format and the neat recording of results are almost mandatory.

Chapter Summary

16.1. The j operator preceding a phasor, or other directed quantity, results in a 90° counterclockwise (ccw) rotation. Repeated application of the j operator causes repeated rotations of 90° ccw. In particular,

$$j \times j = -1$$
$$j \times j \times j \times j = 1$$

16.2. In polar notation the j operator appears as

$$j = 1 \angle 90°$$

16.3. Combining phasor elements expressed in rectangular form is simply expressed

$$I_T = \sqrt{I_X^2 + I_Y^2}$$

$$A = \tan^{-1} \frac{I_Y}{I_X}$$

in which I_X is parallel to the x-axis and I_Y is parallel to the y-axis.

16.4. The preceding method is applicable to impedances as well as phasors. It can be extended to all four quadrants with proper care of signs.

16.5. Multiplication and division are most easily accomplished with the polar form.

$$(M_1 \angle A_1)(M_2 \angle A_2) = M_1 M_2 \angle A_1 + A_2$$

$$\frac{M_1 \angle A_1}{M_2 \angle A_2} = \frac{M_1}{M_2} \angle A_1 - A_2$$

16.6. Conversion from the polar form to the rectangular form is accomplished by use of

$$I_X = I_T \cos A$$
$$I_Y = I_T \sin A$$

This method can be extended to all quadrants with proper care for signs.

16.7. Addition and subtraction are most easily accomplished using the rectangular form.

$$(a + jb) + (c + jd) = (a + c) + j(b + d)$$
$$(a + jb) - (c + jd) = (a - c) + j(b - d)$$

16.8. When the phasors are in the rectangular form, multiplication and division are possible without prior conversion to polar form.

$$(a + jb)(c + jd) = (ac - bd) + j(ad + bc)$$
$$\frac{a + jb}{c + jd} = \frac{ac + bd}{c^2 + d^2} + j\frac{bc - ad}{c^2 + d^2}$$

Self-Review

True–False Questions

1. **T. F.** The addition of two phasors that differ in phase by 90° involves finding the hypotenuse of a right triangle.
2. **T. F.** When phasors are drawn to the same scale, the unit of length for each occupies the same space on the paper.
3. **T. F.** Phasors pointing to the right and others pointing to the left cannot properly be shown on the same phasor diagram.
4. **T. F.** It is correct to say that $3.5 = +3.5$ because in the absence of the sign, a + is implied.
5. **T. F.** A current expressed as a phasor $250\angle -270°$ mA is the same when expressed as $250\angle -90°$ mA.
6. **T. F.** A circuit consists of a resistive branch placed in parallel with a capacitive branch. The total current supplied by the source will lag the voltage by an angle of less than 90°.
7. **T. F.** In addition to converting from rectangular to polar form, it is possible to convert from the polar to the rectangular form.
8. **T. F.** In converting a phasor in polar form to one in rectangular coordinates, a sketch of the phasor diagram is very useful in attaching the proper signs.
9. **T. F.** A phasor could possibly lie in any quadrant of a circle.
10. **T. F.** Two phasors are more easily added if they are in rectangular form.
11. **T. F.** Phasors may be used in representing alternating voltages and currents in the solution of ac circuit problems.
12. **T. F.** In designating the direction of a phasor, $+j$ means a horizontal direction to the right.
13. **T. F.** The length of a phasor is proportional to the numerical value of the physical quantity represented.
14. **T. F.** Multiplying any given phasor by j results in a 90° clockwise rotation.
15. **T. F.** The total current for a parallel circuit must always measure more than that in any one branch.
16. **T. F.** The total current for two or more parallel branches, each consisting of R, L, or C, will always have an angle between $-90°$ and $+90°$.

17. T. F. Phasors can either be in rectangular form or in polar form, but not mixed, in performing multiplication.
18. T. F. Series–parallel ac circuits can be solved using complex numbers for impedances and phasors.
19. T. F. The complex conjugate can be obtained by changing the sign of the complex number's imaginary part.
20. T. F. A practical tank circuit has an equivalent circuit consisting of an inductor in one parallel branch and a capacitor in series with a resistor in the other.

Multiple-Choice Questions

21. A horizontal phasor and a vertical phasor are said to be
    - (a) perpendicular
    - (b) in phase
    - (c) parallel
    - (d) right angles
    - (e) collinear
22. A horizontal phasor pointed to the left is represented by
    - (a) a negative number
    - (b) the square root of -1
    - (c) the j operator and a number
    - (d) a positive number
    - (e) the negative j operator and a number
23. The distinction between a phasor represented by a positive number and one represented by a negative number is
    - (a) twice the difference
    - (b) They are based on completely different ideas.
    - (c) The negative number is imaginary.
    - (d) Only the positive number is rational.
    - (e) a rotation of 180°
24. When the j operator is applied twice to a horizontal phasor of length seven pointing to the right, the result is $j \cdot j7 = j^2 \cdot 7 = -7$, so that the total rotation has been
    - (a) 0°
    - (b) 90°
    - (c) 270°
    - (d) 360°
    - (e) 180°
25. A current of 10 mA leads the applied voltage by 90° in a capacitor branch. The voltage is $115 \angle 0°$ V in polar form. Express the current in polar form.
    - (a) $10 \angle -90°$ mA
    - (b) $10 \angle 0°$ mA
    - (c) $115 \angle 90°$ mA
    - (d) $11.5 \angle -90°$ mA
    - (e) $10 \angle 90°$ mA
26. A parallel circuit consists of a resistor of 1 kΩ in parallel with a capacitor having a capacitor reactance of 1 kΩ both connected across an ac voltage source. Try to choose the correct phasor for the current with the voltage as the reference without solving the circuit numerically.
    - (a) $16.4 - j16.4$ mA
    - (b) $10.6 \angle -45°$ A
    - (c) $27.3 \angle -90°$ mA
    - (d) $14.4 \angle 90°$ mA
    - (e) $4.35 + j4.35$ mA
27. If the total current I_T and angle A are known for a parallel circuit, the current through the resistive branch is given by
    - (a) $I_T \sin A$
    - (b) $I_T \tan A$
    - (c) $-Q = \tan^{-1} I_C/I_R$
    - (d) $I_T \cos A$
    - (e) $\theta = \tan^{-1} I_L/I_R$
28. When the angle for a phasor is between 90° and 180°, the horizontal component is
    - (a) zero
    - (b) 1
    - (c) positive
    - (d) negative
    - (e) infinite
29. A phasor in rectangular form is to be added to a phasor in polar form. The first step in the addition is to
    - (a) align the digits vertically
    - (b) convert the second phasor to rectangular form
    - (c) convert both phasors before aligning the digits vertically
    - (d) convert the first phasor to polar form
    - (e) call the instructor for help
30. When two phasors are multiplied,
    - (a) the real and imaginary parts are inverted
    - (b) the magnitudes are multiplied
    - (c) the complex conjugate is used

(d) the angles are added
  (e) choices (b) and (d) are both correct
31. A right triangle has
   (a) no 90° angles
   (b) one 90° angle
   (c) two 90° angles
   (d) three 90° angles
   (e) No correct choice
32. Using the j-notation a phasor representing three units directed horizontally to the left might be written as
   (a) $+3$
   (b) $+j3$
   (c) $+j^2 3$
   (d) $+j^3 3$
   (e) $+j^4 3$
33. In carrying the idea of the j operator to a ridiculous extreme, the phasor $j^7 5$ is equivalent to
   (a) $-j5$
   (b) $+j5$
   (c) -5^2
   (d) $+j^2 5$
   (e) $-j^3 5$
34. Which statement regarding phasors using the j-notation and polar notation is correct?
   (a) $j6 = 6\angle 90°$
   (b) $j^2 3^2 = 9\angle -180°$
   (c) $-j^3 15 = -15\angle 90°$
   (d) all choices are correct
   (e) at least one choice is incorrect
35. A number consisting of a "real" part and an "imaginary" part is called a
   (a) complicated number
   (b) complex number
   (c) j-number
   (d) funny number
   (e) unreal number
36. One parallel branch of a circuit consists of a resistor and the other an inductor. The current in the inductive branch will lead that in the resistive branch by
   (a) 0°
   (b) 45°
   (c) 90°
   (d) 180°
   (e) 270°
37. When converted to rectangular coordinate form the phasor $100\angle -260°$ will have a real part equal to
   (a) -98.5
   (b) -17.4
   (c) 0
   (d) 17.4
   (e) 98.5
38. The complex conjugate of $-9 + j12$ is
   (a) $15\angle -126.9°$
   (b) $15\angle -36.9°$
   (c) $15\angle 36.9°$
   (d) $15\angle 126.9°$
   (e) $9 - j12$
39. The reciprocal of $0.1234\angle 56.78°$ is
   (a) $0.1234\angle -56.78°$
   (b) $0.5678\angle +12.34°$
   (c) $4.440 - j6.779$
   (d) $-4.440 + j6.779$
   (e) $4.440 + j6.779$
40. The reciprocal of $0.3 - j0.4$ is
   (a) $2\angle -53.1°$
   (b) $2\angle +53.1°$
   (c) $-0.3 + j0.4$
   (d) $3.33 - j2.5$
   (e) $3.33 + j2.5$

Practice Problems

41. Five phasors are represented by the positive numbers 2, 9.6, 7/2, and A. Write expressions for these phasors after they have been rotated 90° in a counterclockwise (ccw) direction.
42. Start with a phasor of length 154 oriented to the right ($+154$) and show the results of four successive applications of the j operator.
43. A phasor has a length of 1000 units and an angle of 113°. Express the phasor both in polar and rectangular forms.
44. Add $5 + j2$ to $1 + j3$.
45. Add $-2 + j7$ to $13 - j14$.
46. Subtract $-10 + j8$ from $27 - j13$.
47. Multiply $20.1\angle 13.7°$ and $1.23\angle 41.6°$.
48. Multiply $755\angle -71.9°$ and $24.6\angle 154.3°$.
49. Divide $4350\angle 180°$ by $27.3\angle 75.0°$.
50. Divide $47.6\angle -17.3°$ by $418\angle -89.7°$.
51. Express the complex number $12 + j9$ using polar notation.
52. A capacitor and a resistor are placed in parallel so that $I_C = 200$ mA and $I_R = 100$ mA. Calculate the total current.

53. Two phasors are expressed as 14 − j20 and −36 + j44. Find the sum of the two and express the answer in polar notation.
54. Multiply −3 + j5 by 6 − j3.
55. Divide −2 + j5 by 3 − j2 and express the quotient both in rectangular and polar forms.
56. Find the product.

$$(0.0158\angle 315.9°)(115.3\angle -21.7°)(0.167\angle 66.5°)$$

57. Find the product.

$$(2 + j3)(0.707\angle -46.2°)(-6 - j1.5)$$

58. Find the product.

$$(0.0158\angle 315.9°)(115.3\angle -21.7°)(0.167\angle 66.5°)$$

59. Perform the operation indicated.

$$(2 - j3)(-6 + j7) + 15 - j19$$

60. Calculate the equivalent impedance of $Z_1 = 15 + j10$ kΩ in parallel with $Z_2 = 10 - j5$ kΩ.

Advanced Problems

61. A parallel circuit has two branches, one having a resistor with 7.5 mA of current and the other having a capacitor with 3.7 mA of current. Express the total current in rectangular and in polar form.
62. In a parallel circuit there is 150 mA in one branch that is in phase with the applied voltage. In the second branch the current is 450 mA and lags the applied voltage by 90°. Express the total current in both rectangular and polar forms.
63. In a parallel circuit the total current is 50 mA and the current leads the applied voltage by 25°. Find the two branch currents assuming that one branch is resistive and the other branch has capacitive reactance only.
64. Two branches are connected in parallel. One is composed of a resistor and the other an inductor. The total current is 300 mA and lags the applied voltage by 32.7°. Calculate the current in each branch.
65. In polar form a phasor is expressed as $261\angle -131.8°$. Find I_X and I_Y.
66. Convert the phasor $-335 - j271$ to polar form.
67. Add $29.5\angle 37.4°$ to $36.8\angle -71.5°$.
68. Subtract $315\angle -64.7°$ from $688\angle 169°$.
69. Perform the indicated operations.

$$\frac{66.3\angle -218.5° \times 0.0711\angle 21.7°}{0.000145\angle -79.8°}$$

70. Complete the division without converting either phasor to polar form.

$$\frac{3 - j4}{4 - j3}$$

71. The total current in a parallel circuit is 25 mA. Calculate the current in the inductive branch if the current in the other, resistive branch is 20 mA.

Sec.   Self-Review

72. The current through a 1-kΩ resistor is 50 mA. A second branch consisting only of inductance is added and the total current becomes 75 mA. For a frequency of 1000 Hz, determine the amount of inductance in this branch.

73. The impedance of a circuit is expressed as $555 \angle -75°$ Ω and the current passing through this impedance is $20 + j33$ mA. Calculate the applied voltage.

74. The effective Q of a 100-μH inductor is 60 at 1 MHz. Find the L and parallel R_P for the equivalent circuit at that frequency.

75. In Problem 74, calculate the parallel capacitance required to cancel the inductive reactance of the 100-μH inductor and describe the equivalent circuit.

76. Perform the operations indicated:

$$\frac{(3.869 \angle -79.21°)(0.1593 \angle +256.77°)}{0.04333 \angle -21.66°}$$

77. Perform the operations indicated:

$$\frac{(16.1 - j21.7)(243 \angle +315°)}{47.6 + j89.9}$$

78. Refer to Fig. 16-17 and change Z_3 to $3 + j4$ Ω. Find the total Z_T of the circuit.

79. Referring to Problem 78, find the total current and the voltage across the parallel combination for a source voltage $V_S = 12 \angle 0°$ V.

80. Referring to Problems 78 and 79, determine the current through each parallel branch.

17

Kirchhoff's Laws and the Network Theorems in AC Circuits

This chapter has the following objectives:

1. To restate and demonstrate the application of Kirchhoff's laws to ac circuits in which mathematical expressions for phasors appear in the equations.
2. To develop procedures for the solution of simultaneous equations for which the unknown quantities (voltages or currents) are phasors and the constants (source voltages and impedances) are expressed as complex numbers.
3. To apply mesh analysis as a basic procedure for writing simultaneous equations for ac networks.
4. To apply Millman's theorem as a procedure for simplifying ac networks for which sources exist in parallel.
5. To apply nodal analysis in establishing simultaneous equations containing node voltages in ac networks.
6. To restate and demonstrate the application of the superposition theorem to ac networks for which mathematical expressions for phasors appear in the equations.
7. To restate and demonstrate the application of Thévenin's theorem to ac networks for which mathematical expressions for phasors appear in the equations.
8. To restate and demonstrate the application of Norton's theorem to ac networks for which mathematical expressions for phasors appear in the equations.
9. To establish the basis and application formulas for delta–wye and wye–delta conversions.
10. To restate and demonstrate the application of the reciprocity theorem to ac networks for which mathematical expressions for phasors appear in the equations.

17-0 Introduction

In Chapter 6 we discussed the application of Kirchhoff's laws and the network theorems to the analysis of dc circuits. However the same analytical methods may be applied to the sine-wave ac circuits provided that we take into account the concepts of reactance, impedance, and phase angles, all of which have no meaning in dc circuits. This will be achieved by using the j operator and the complex rules of algebra (Chapter 16); if necessary, a network theorem will be restated in terms of its ac version. In addition, the values of the constant voltage and current sources (for example, $9 \angle 90°$ V and $4 \angle -20°$ A) will be expressed in polar form with the phase angle related to some reference sine wave. In a particular ac circuit all voltages and currents will be assumed to possess the same frequency.

17-1 Kirchhoff's Laws

Restating Kirchhoff's laws as they appeared on pages 229 and 231,

> *Kirchhoff's voltage law (KVL)*. The algebraic sum of the constant voltage EMFs and the voltage drops around any closed electrical loop is always zero.
>
> *Kirchhoff's current law (KCL)*. The algebraic sum of the currents existing at any junction point is zero.

In applying these laws to ac circuits we have a problem as far as the algebraic sum is concerned. Clearly, we cannot give a fixed polarity to an alternating voltage or a constant direction to an alternating current. It is therefore necessary to assign an instantaneous polarity to a source voltage and show the associated direction of the (electron) current flow. In addition, we must refer to the phasor sums of the voltages and the currents. These considerations appear in the circuit that is illustrated in Fig. 17-1.

You must recognize the importance of the indicated polarity; for example, if the terminals of the E_2 source were reversed, the polarity would have to be changed from "−" to "+."

To analyze the current of Fig. 17-1, we will follow the method described in Section 6-1. Apply KVL to the loop *PNGP* (starting at *P*):

$$V_1 + V_2 - E_1 = 0 \qquad (17\text{-}1\text{-}1)$$
$$I_1(3 - j2) + I_3(1 + j4) - 14\angle 60° = 0$$

For the loop $QNGQ$ (starting at Q):

$$V_3 + V_2 - E_2 = 0$$
$$I_2 \times 2 + I_3(1 + j4) - 10\angle 30° = 0 \qquad (17\text{-}1\text{-}2)$$

The currents I_1, I_2, and I_3 are assumed to be measured in milliamperes.
Use KCL at point N:

$$I_1 + I_2 = I_3 \quad \text{or} \quad I_1 = I_3 - I_2 \qquad (17\text{-}1\text{-}3)$$

Substitute for I_3 in Eqs. (17-1-1) and (17-1-2):

$$I_1(3 - j2) + (I_1 + I_2)(1 + j4) = 14\angle 60°$$
$$I_1(4 + j2) + I_2(1 + j4) = 14\angle 60° \qquad (17\text{-}1\text{-}4)$$
$$I_2 \times 2 + (I_1 + I_2)(1 + j4) = 10\angle 30°$$
$$I_1(1 + j4) + I_2(3 + j4) = 10\angle 30° \qquad (17\text{-}1\text{-}5)$$

To eliminate I_2 from these equations and to solve for I_1 multiply Eq. (17-1-4) by $(3 + j4)$ and Eq. (17-1-5) by $(1 + j4)$:

$$I_1(4 + j2)(3 + j4) + I_2(1 + j4)(3 + j4) = 14\angle 60° \times (3 + j4)$$
$$I_1(4 + j22) + I_2(1 + j4)(3 + j4) = 14\angle 60° \times (3 + j4) \qquad (17\text{-}1\text{-}6)$$
$$I_1(1 + j4)(1 + j4) + I_2(3 + j4)(1 + j4) = 10\angle 30° \times (1 + j4)$$
$$I_1(-15 + j8) + I_2(3 + j4)(1 + j4) = 10\angle 30°(1 + j4) \qquad (17\text{-}1\text{-}7)$$

Subtract Eq. (17-1-7) from Eq. (17-1-6):

$$I_1(19 + j14) = 14\angle 60° \times (3 + j4) - 10\angle 30° \times (1 + j4)$$
$$= (7 + j12.12)(3 + j4) - (8.66 + j5)(1 + j4)$$
$$= (-27.48 + j64.36) - (-11.34 + j39.64)$$
$$= -16.14 + j24.72$$

This yields

$$\text{current, } I_1 = \frac{-16.14 + j24.72}{19 + j14} = \frac{29.51\angle 123.11°}{23.6\angle 36.38°}$$
$$= 1.25\angle 86.7° \text{ mA}$$

Substitute $I_1 = I_3 - I_2$ in Eq. (17-1-1):

$$(I_3 - I_2)(3 - j2) + I_3(1 + j4) = 14\angle 60°$$
$$-I_2(3 - j2) + I_3(4 + j2) = 14\angle 60° \qquad (17\text{-}1\text{-}8)$$

Sec. 17-1 Kirchhoff's Laws

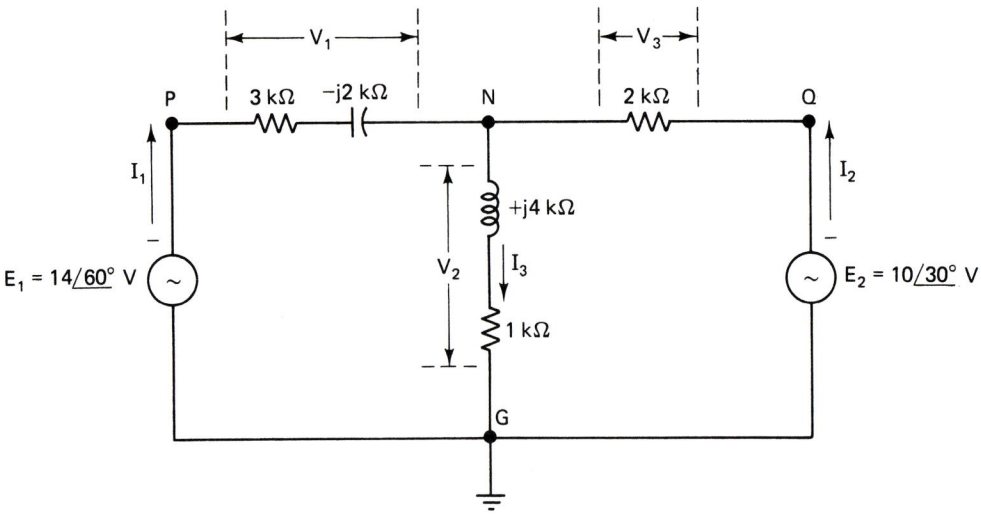

FIGURE 17-1 Circuit illustrating Kirchhoff's laws.

Repeat Eq. (17-1-2):

$$I_2 \times 2 + I_3(1 + j4) = 10\angle 30° \quad (17\text{-}1\text{-}2)$$

Multiply Eq. (17-1-8) by 2 and Eq. (17-1-2) by $(3 - j2)$:

$$-I_2 \times 2(3 - j2) + 2I_3(4 + j2) = 28\angle 60° \quad (17\text{-}1\text{-}9)$$

$$I_2 \times 2(3 - j2) + I_3(1 + j4)(3 - j2) = 10\angle 30° \times (3 - j2) \quad (17\text{-}1\text{-}10)$$

Add Eqs. (17-1-9) and (17-1-10):

$$2I_3(4 + j2) + I_3(1 + j4)(3 - j2) = 28\angle 60° + 10\angle 30° \times (3 - j2)$$

$$= 14 + j24.25 + (8.66 + j5)(3 - j2)$$

$$I_3(19 + j14) = 14 + j24.25 + 35.98 - j2.32$$

$$= 49.98 + j21.93$$

This yields

$$\text{current, } I_3 = \frac{49.98 + j21.93}{19 + j14} = \frac{54.58\angle 23.70°}{23.6\angle 36.38°} = 2.31\angle -12.7° \text{ mA}$$

From Eq. (17-1-3),

$$\text{current, } I_2 = I_3 - I_1 = \frac{49.98 + j21.93 - (-16.12 + j24.72)}{19 + j14}$$

$$= \frac{66.10 - j2.79}{19 + j14} = \frac{66.16\angle -2.42°}{23.6\angle 36.38°}$$

$$= 2.80\angle -38.8° \text{ mA}$$

Voltage Checks

1. From Eq. (17-1-1):

$$V_1 + V_2 = I_1 \times (3 - j2) + I_3 \times (1 + j4)$$

$$= \frac{(-16.12 + j24.72)(3 - j2) + (49.98 + 21.93)(1 + j4)}{19 + j14}$$

$$= \frac{-48.36 + 49.44 + j106.5 + 49.98 - 87.72 + j199.92 + j21.93}{19 + j14}$$

$$= \frac{-36.66 + j328.35}{19 + j14} = \frac{330 \angle 96.37°}{23.6 \angle 36.38°} \approx 14 \angle 60° = E_1$$

2. From Eq. (17-1-2):

$$V_3 + V_2 = I_2 \times 2 + I_3 \times (1 + j4)$$

$$= \frac{2 \times (66.10 - j2.79) + (49.98 + j21.93)(1 + j4)}{19 + j14}$$

$$= \frac{132.2 - j5.58 + 49.98 - 87.52 + j199.92 + j21.93}{19 + j14}$$

$$= \frac{94.46 + j216.27}{19 + j14} = \frac{236.0 \angle 66.40°}{23.6 \angle 36.38°} \approx 10 \angle 30° = E_2$$

EXAMPLE 17-1 In Fig. 17-2, calculate the values of the currents I_1, I_2, and I_3.

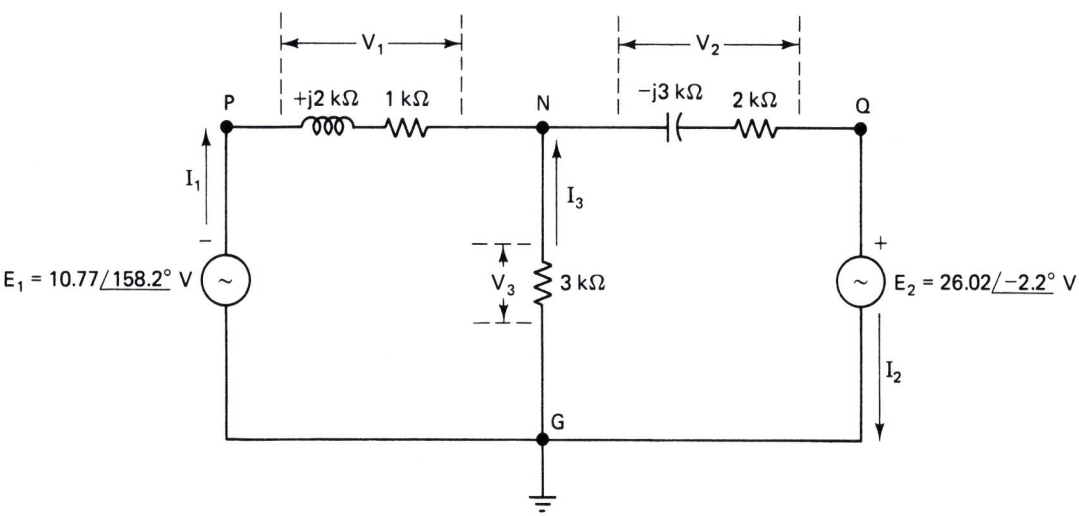

FIGURE 17-2 Circuit for Example 17-1.

Solution With the branch currents as indicated, V_1 and V_3 are opposing voltages but the voltages V_2 and V_3 are aiding. Use KVL for the loop $PNGP$ (starting at P):

$$V_1 - V_3 - E_1 = 0$$

$$I_1(1 + j2) - 3I_3 = 10.77 \angle 158.2° = -10 + j4$$

Sec. 17-1 Kirchhoff's Laws

Use KVL for the loop $GNQG$ (starting at G):

$$V_3 + V_2 - E_2 = 0$$

$$3I_3 + I_2(2 - j3) = 26.02\angle -2.2° = 26 - j1$$

For the junction point, N, KCL yields

$$I_1 + I_3 = I_2 \quad \text{or} \quad I_3 = I_2 - I_1$$

Substitute $I_3 = I_2 - I_1$ in the two loop equations:

$$I_1(1 + j2) - 3(I_2 - I_1) = -10 + j4$$

or

$$I_1(4 + j2) - 3I_2 = -10 + j4 \tag{17-1-1-1}$$

$$I_2(2 - j3) + 3(I_2 - I_1) = 26 - j1$$

or

$$-3I_1 + I_2(5 - j3) = 26 - j1 \tag{17-1-1-2}$$

Multiply Eq. (17-1-1-1) by $(5 - j3)$ and Eq. (17-1-1-2) by 3:

$$I_1(4 + j2)(5 - j3) - 3I_2(5 - j3) = (-10 + j4)(5 - j3) \tag{17-1-1-3}$$

$$-9I_1 + 3I_2(5 - j3) = 3 \times (26 - j1) \tag{17-1-1-4}$$

Adding Eqs. (17-1-1-3) and (17-1-1-4):

$$I_1(17 - j2) = -50 + 12 + j50 + 78 - j3 = 40 + j47$$

This yields

$$\text{current, } I_1 = \frac{40 + j47}{17 - j2} = 2 + j3 = 3.6\angle 56.3° \text{ mA}$$

Since $V_3 = V_1 - E_1$,

$$\text{current, } I_3 = \frac{V_3}{3 \text{ k}\Omega} = \frac{I_1 \times (1 + j2) - (-10 + j4)}{3}$$

$$= \frac{(2 + j3)(1 + j2) - (-10 + j4)}{3}$$

$$= \frac{-4 + j7 - (-10 + j4)}{3}$$

$$= \frac{6 + j3}{3} = 2 + j1 = 2.24\angle 26.6° \text{ mA}$$

Since $I_2 = I_3 + I_1$,

$$\text{current, } I_2 = (2 + j1) + (2 + j3)$$

$$= 4 + j4 = 5.66\angle 45.0° \text{ mA}$$

17-2

Mesh Analysis

In Section 6-2 we discussed the advantages of mesh analysis compared with the use of Kirchhoff's laws. Referring back to Fig. 17-1 we will replace the three branch currents with the two mesh currents I_1 and I_2 (Fig. 17-3). As far as the polarities of the voltage sources are concerned, the normal KVL convention will be used. The mesh equations can then be derived by inspection.

In Fig. 17-3 the mesh equation for the loop $PNGP$ is

$$I_1(3 - j2 + 1 + j4) - I_2(1 + j4) = 14\underline{/60°} \qquad (17\text{-}2\text{-}1)$$

For the loop $NQGN$ the mesh equation is

$$-I_1(1 + j4) + I_2(1 + j4 + 2) = -10\underline{/30°} \qquad (17\text{-}2\text{-}2)$$

where the currents I_1 and I_2 are measured in milliamperes. These equations yield

$$I_1(4 + j2) - I_2(1 + j4) = 14\underline{/60°} = 7 + j12.12 \qquad (17\text{-}2\text{-}3)$$

$$-I_1(1 + j4) + I_2(3 + j4) = -10\underline{/30°} = -8.66 - j5 \qquad (17\text{-}2\text{-}4)$$

Multiply Eq. (17-2-3) by $(3 + j4)$ and Eq. (17-2-4) by $(1 + j4)$:

$$I_1(4 + j2)(3 + j4) - I_2(1 + j4)(3 + j4) = (7 + j12.12)(3 + j4) \qquad (17\text{-}2\text{-}5)$$

$$-I_1(1 + j4)(1 + j4) + I_2(3 + j4)(1 + j4) = (-8.66 - j5)(1 + j4) \qquad (17\text{-}2\text{-}6)$$

The addition of Eqs. (17-2-5) and (17-2-6) gives

$$I_1(4 + j22 + 15 - j8) = 21 - 48.48 + j64.36 + 20 - 8.66 - j39.64$$

$$I_1(19 + j14) = -16.14 + j24.72$$

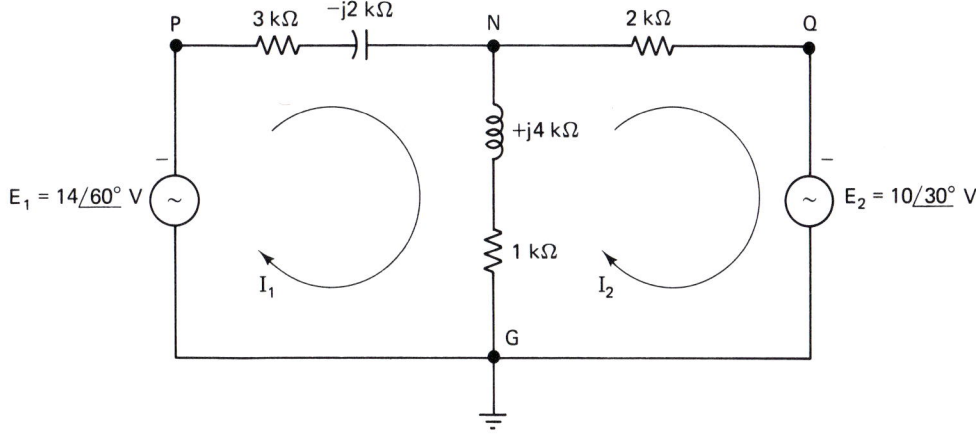

FIGURE 17-3 Circuit illustrating mesh analysis.

Therefore,

$$\text{current, } I_1 = \frac{-16.14 + j24.72}{19 + j14} = 1.25\underline{/86.7°} \text{ mA}$$

This result is precisely the same as we obtained from the Kirchhoff's laws analysis of Section 17-1.

EXAMPLE 17-2 In Figure 17-4 find the value of the current from the point N to the point G.

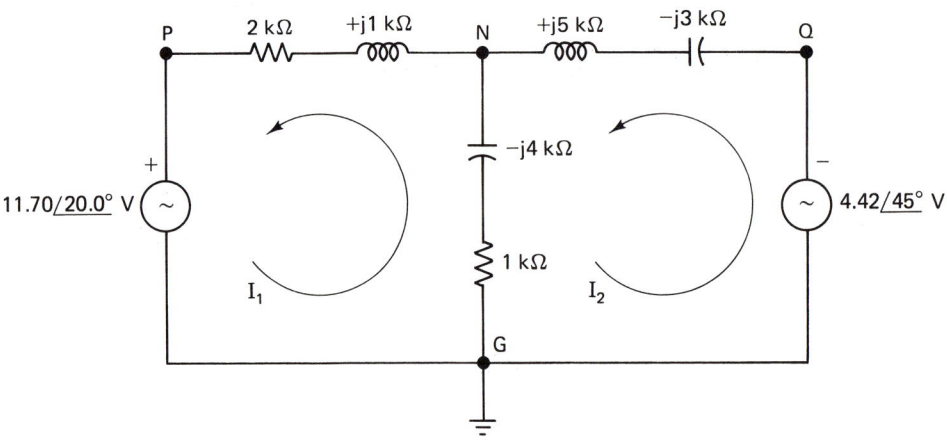

FIGURE 17-4 Circuit for Example 17-2.

Solution The mesh equation for the loop *PNGP* is

$$I_1(2 + j1 + 1 - j4) - I_2(1 - j4) = -11.7\underline{/20°} = -11 - j4$$

or

$$I_1(3 - j3) - I_2(1 - j4) = -11 - j4$$

The mesh equation for the loop *NQGN* is

$$-I_1(1 - j4) + I_2(1 - j4 + j5 - j3) = -4.24\underline{/45°}$$

or

$$-I_1(1 - j4) + I_2(1 - j2) = -4.24\underline{/45°} = -3 - j3$$

Multiply the first mesh equation by $(1 - j2)$ and the second mesh equation by $(1 - j4)$. This yields

$$I_1(3 - j3)(1 - j2) - I_2(1 - j4)(1 - j2) = (-11 - j4)(1 - j2)$$

or

$$I_1(-3 - j9) - I_2(1 - j4)(1 - j2) = -19 + j18 \qquad (17\text{-}2\text{-}1\text{-}1)$$

$$-I_1(1 - j4)(1 - j4) + I_2(1 - j2)(1 - j4) = (-3 - j3)(1 - j4)$$

or

$$-I_1(-15 - j8) + I_2(1 - j2)(1 - j4) = -15 + j9 \qquad (17\text{-}2\text{-}1\text{-}2)$$

The addition of Eqs. (17-2-1-1) and (17-2-1-2) yields

$$I_1(12 - j1) = -34 + j27$$

Therefore,

$$\text{mesh current, } I_1 = \frac{-34 + j27}{12 - j1} = -3 + j2 = 3.6 \underline{/146.3°} \text{ mA}$$

in the direction P to N.

Substitute for I_1 in the mesh equation for the loop $NQGN$:

$$I_2(1 - j2) = -3 - j3 + (-3 + j2)(1 - j4) = 2 + j11$$

Therefore,

$$\text{mesh current, } I_2 = \frac{2 + j11}{1 - j2} = -4 + j3 = 5 \underline{/143.1°} \text{ mA}$$

in the direction of N to Q.

It follows that the current in the direction N to G is

$$I_1 - I_2 = (-3 + j2) - (-4 + j3) = 1 - j1 = 1.41 \underline{/-45°} \text{ mA}$$

17-3 Millman's Theorem

Referring to Section 6-4, we can restate Millman's theorem as it relates to general ac circuits. The new statement will be as follows:

> Any number of constant alternating-current sources that are *directly* connected in parallel can be connected into a single alternating current source whose total current is the phasor sum of the individual source currents and whose total internal impedance is the result of combining the individual source impedances in parallel.

Let us apply this theorem to the example illustrated in Fig. 17-1. Our object will be to determine the current I_3 so that we will regard the internal impedances of E_1, E_2 to be $3 - j2$ kΩ and 2 kΩ, respectively. After converting to the constant current generators, the equivalent circuit is illustrated in Fig. 17-5.

Notice that I_1 and I_2 are driving in the *same* direction. Therefore, the value, I_T, of the total equivalent current generator is

$$I_T = I_1 + I_2 = \frac{14 \underline{/60°}}{3 - j2} + \frac{10 \underline{/30°}}{2}$$

$$= \frac{7 + j12.12}{3 - j2} + 4.33 + j2.5$$

$$= \frac{7 + j12.12 + (4.33 + j2.5)(3 - j2)}{3 - j2}$$

$$= \frac{24.99 + j10.96}{3 - j2} \text{ mA}$$

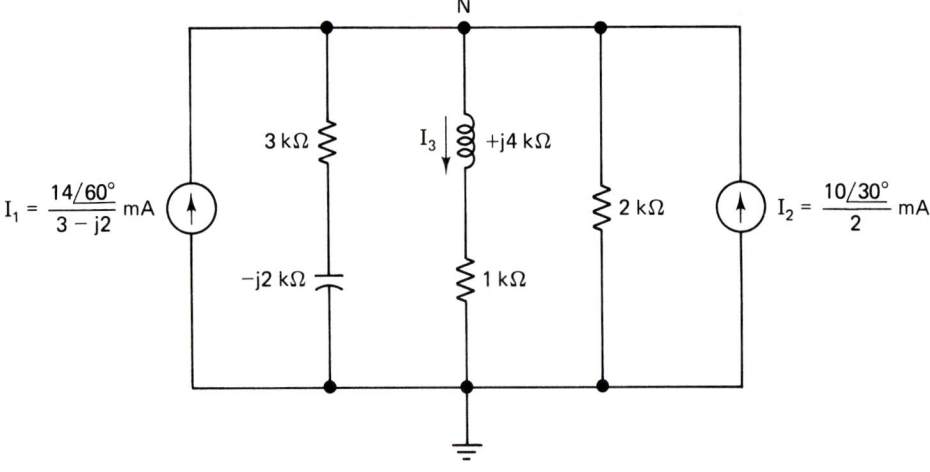

FIGURE 17-5 Circuit illustrating Millman's theorem.

The total equivalent impedance, Z_T, of $(3 - j2) \| (1 + j4) \| 2$ kΩ is given by

$$\frac{1}{Z_T} = \frac{1}{3 - j2} + \frac{1}{1 + j4} + \frac{1}{2} = \frac{2 + j8 + 6 - j4 + (3 - j2)(1 + j4)}{2(3 - j2)(1 + j4)}$$

$$= \frac{8 + j4 + 11 + j10}{2(3 - j2)(1 + j4)}$$

$$= \frac{19 + j14}{2(3 - j2)(1 + j4)}$$

By the current division rule (CDR),

$$\text{current, } I_3 = I_T \times \frac{Z_T}{Z_3}$$

$$= \frac{24.99 + j10.96}{3 - j2} \times \frac{2(3 - j2)(1 + j4)}{(19 + j14)(1 + j4)}$$

$$= \frac{49.98 + j21.92}{19 + j14} = 2.31 \angle -12.7° \text{ mA}$$

This is the same result as we obtained in the Kirchhoff's law analysis of Section 17-1.

Although the determination of I_3 by Millman's theorem was obviously quicker than using either Kirchhoff's laws or mesh analysis, it should be emphasized that the currents flowing through the internal impedances (Fig. 17-5) are *not* the same as the branch currents and the mesh currents (I_1 and I_2).

EXAMPLE 17-3 In Fig. 17-2, use Millman's theorem to calculate the value of the current, I_3.

Solution By converting to constant current generators, the circuit reduces to that of Fig. 17-6. Notice that I_2 and I_1 are driving in opposite directions.

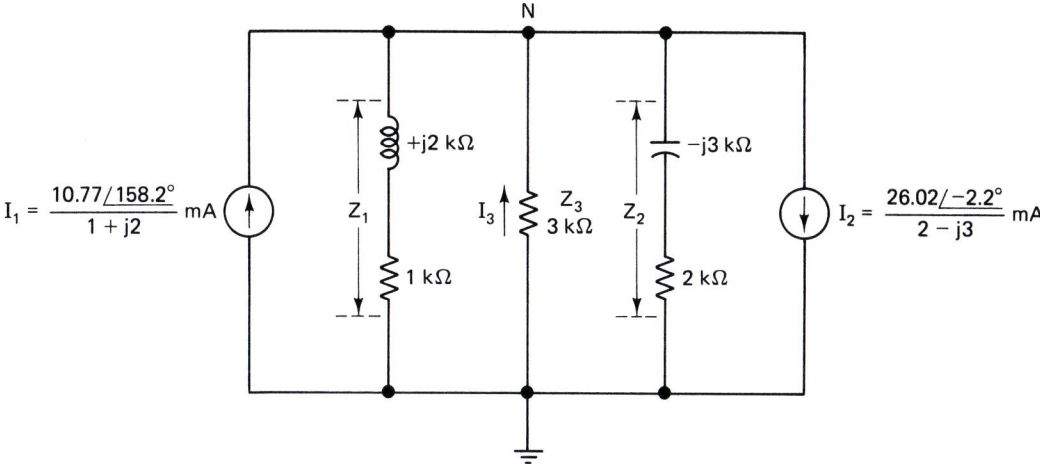

FIGURE 17-6 Circuit for Example 17-3.

Therefore, the value, I_T, of the total equivalent current generator is

$$I_T = I_2 - I_1 = \frac{26.02\angle -2.2°}{2 - j3} - \frac{10.77\angle 158.2°}{1 + j2}$$

$$= \frac{26 - j1}{2 - j3} - \frac{(-10 + j4)}{1 + j2}$$

$$= \frac{(26 - j1)(1 + j2) - (-10 + j4)(2 - j3)}{(1 + j2)(2 - j3)}$$

$$= \frac{36 + j13}{(1 + j2)(2 - j3)} \text{ mA}$$

The total equivalent impedance, Z_T, of $(1 - j2)\|3\|(2 - j3)$ kΩ is given by

$$\frac{1}{Z_T} = \frac{1}{1 + j2} + \frac{1}{3} + \frac{1}{2 - j3}$$

$$= \frac{6 - j9 + (1 + j2)(2 - j3) + 3 + j6}{3(1 + j2)(2 - j3)}$$

$$= \frac{17 - j2}{3(1 + j2)(2 - j3)}$$

From the current division rule (CDR),

$$\text{current, } I_3 = I_T \times \frac{Z_T}{Z_3} = \frac{(36 + j13)}{(1 + j2)(2 - j3)} \times \frac{3(1 + j2)(2 - j3)}{(17 - j2) \times 3}$$

$$= \frac{36 + j13}{17 + j2} = 2 + j1 = 2.24\angle 26.6° \text{ mA}$$

The same result appeared in the solution of Example 17-1.

Sec. 17-3 Millman's Theorem

17-4 Nodal Analysis

In Section 6-3 we made the following statement regarding nodal analysis:

> At a node the algebraic sum of the source currents entering = the sum of the currents leaving through the resistors.

For ac circuits the statement will have to be modified to take into account the various phase relationships between the currents:

> At a node the phasor sum of the source currents entering = the phasor sum of the currents leaving through the impedances.

As an example we will use nodal analysis to determine the voltage, V_N, at the node N in Fig. 17-1. The first step will be to convert the constant voltage sources, E_1 and E_2, into their constant current equivalents (Fig. 17-7).

The currents I_1 and I_2 are both driving into the node while the currents I_3, I_4, and I_5 are all leaving the point N. Therefore, the single nodal equation is

$$I_1 + I_2 = \frac{V_N}{3 - j2} + \frac{V_N}{1 + j4} + \frac{V_N}{2} \qquad (17\text{-}4\text{-}1)$$

where I_1 and I_2 are measured in milliamperes.

Referring to the solution to the same problem in Section 17-3 gives

$$I_1 + I_2 = \frac{14\angle 60°}{3 - j2} + \frac{10\angle 30°}{2} = \frac{24.99 + j10.96}{3 - j2} \text{ mA} \qquad (17\text{-}4\text{-}2)$$

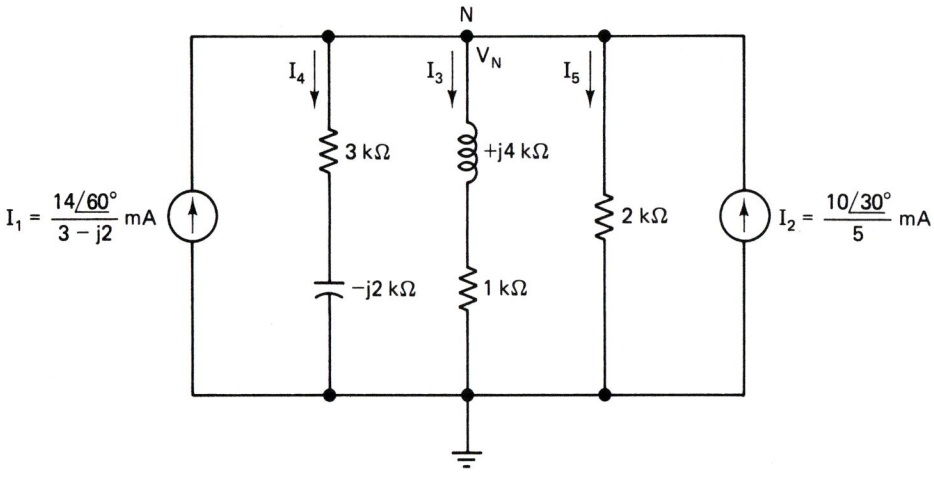

FIGURE 17-7 Circuit illustrating nodal analysis.

and

$$\frac{V_N}{3-j2} + \frac{V_N}{1+j4} + \frac{V_N}{2} = \frac{(19+j14)V_N}{2(3-j2)(1+j4)} \qquad (17\text{-}4\text{-}3)$$

Therefore,

$$\frac{(19+j14)V_N}{2(3-j2)(1+j4)} = \frac{24.99 + j10.96}{3-j2}$$

This yields

$$V_N = \frac{2(24.99 + j10.96)(1+j4)}{19+j14}$$

$$= \frac{(49.98 + j21.92)(1+j4)}{19+j14}$$

$$= \frac{54.58 \angle 23.70° \times 4.12 \angle 75.96°}{23.6 \angle 36.38°} = 9.53 \angle 63.28° \text{ V}$$

This value may be checked by finding the value of the current, I_3:

$$\text{current, } I_3 = \frac{V_N}{1+j4} = \frac{9.53 \angle 63.28°}{4.12 \angle 75.96°} = 2.31 \angle -12.68° \text{ mA}$$

This is the same result as we obtained from the Kirchhoff's law analysis of Section 17-1 and the Millman's theorem analysis of Section 17-3.

EXAMPLE 17-4 Use nodal analysis to find the voltage at the node N in Fig. 17-4.

Solution The first step is to convert the constant voltage sources into their equivalent constant current generators. This equivalent circuit is shown in Fig. 17-8. The current I_2 drives into the node N while the current I_1 flows

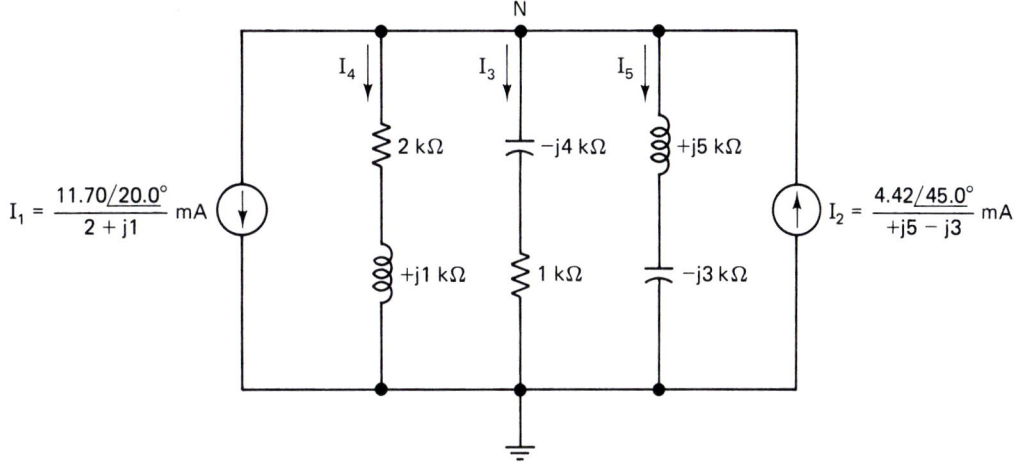

FIGURE 17-8 Circuit for Example 17-4.

away from the node. Therefore, the nodal equation is

$$I_2 - I_1 = \frac{V_N}{1 - j4} + \frac{V_N}{2 + j1} + \frac{V_N}{j5 - j3}$$

where V_N is the alternating voltage at the node N. Then

$$I_2 - I_1 = \frac{4.42 \angle 45.0°}{j5 - j3} - \frac{11.70 \angle 20.0°}{2 + j1}$$

$$= \frac{3 + j3}{j5 - j3} - \frac{11 + j4}{2 + j1} = \frac{3 + j3}{j2} - \frac{11 + j4}{2 + j1}$$

$$= (3 + j3)(-j0.5) - \frac{(11 + j4)(2 - j1)}{2^2 + 1^2}$$

$$= -j1.5 + 1.5 - \frac{26 - j3}{5}$$

$$= -j1.5 + 1.5 - 5.2 + j0.6$$

$$= -3.7 - j0.9 = 3.81 \angle -166.33° \text{ mA}$$

and

$$\frac{V_N}{1 - j4} + \frac{V_N}{2 + j1} + \frac{V_N}{j5 - j3} = \frac{V_N(1 + j4)}{1^2 + 4^2} + \frac{V_N(2 - j1)}{2^2 + 1^2} + V_N \times (-j0.5)$$

$$= V_N \times (0.0588 + j0.235)$$
$$+ V_N \times (0.4 - j0.2) + V_N \times (-j0.5)$$
$$= V_N \times (0.4588 - j0.465)$$
$$= V_N \times 0.653 \angle -45.38°$$

Therefore,

$$V_N \times (0.653 \angle -45.38°) = 3.81 \angle -166.33°$$

voltage at the node N, $V_N = \dfrac{3.81 \angle -166.33°}{0.653 \angle -45.38°} = 5.83 \angle 120.95° = -3 - j5 V$

To check this result we can calculate the value of the current, I_3:

$$\text{current, } I_3 = \frac{V_N}{1 - j4} = \frac{-3 - j5}{1 - j4} = 1 - j1 = 1.414 \angle -45° \text{ mA}$$

This agrees with the value obtained in the solution of Example 17-2.

Superposition Theorem

17-5

This theorem was applied to dc circuits in Section 6-5. For ac circuits the theorem could be stated as follows:

> If a network of *linear* impedances contains more than one alternating voltage source, the current flowing at any point is the phasor sum of the currents which would flow at that point if each alternating source were considered separately with all other alternating sources replaced by impedances equal to their internal impedances.

We will now use the superposition theorem to find the current in the direction N to G for the circuit of Fig. 17-4.

STEP 1 If the $11.70 \angle +20.0°$ V source is removed, the remaining circuit is illustrated in Fig. 17-9(a). The total impedance presented to the $4.24 \angle 45°$ V

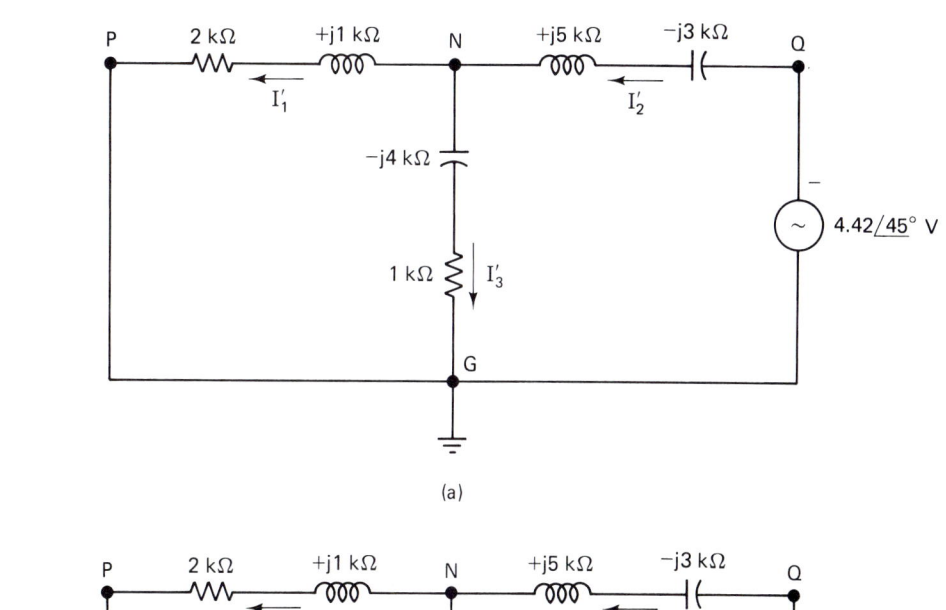

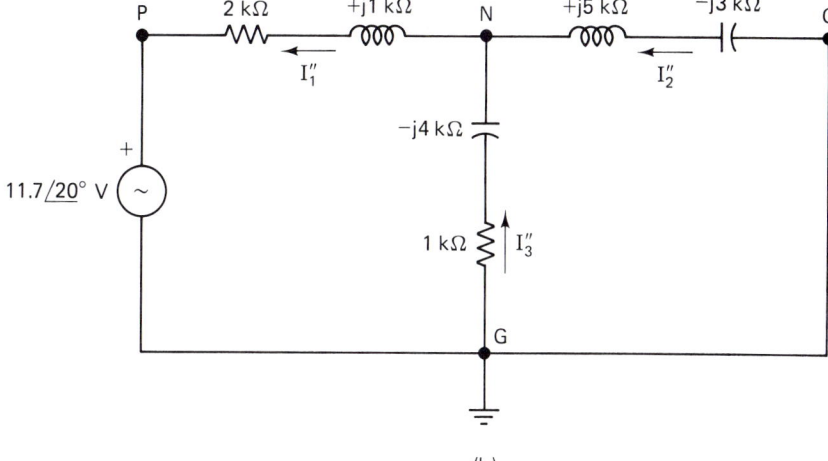

FIGURE 17-9 Circuit illustrating the superposition theorem.

Sec. 17-5 Superposition Theorem

source is

$$Z'_T = \frac{(1-j4)(2+j1)}{1-j4+2+j1} + j5 - j3 = \frac{6-j7}{3-j3} + j2 = \frac{(6-j7)(3+j3)}{18} + j2$$

$$= \frac{39-j3}{18} + j2$$

$$= 2.167 + j1.833$$

$$= 2.838\underline{/40.23°}\,k\Omega$$

$$\text{current,}\ I'_2 = \frac{4.24\underline{/45°}}{2.838\underline{/40.23°}} = 1.494\underline{/4.77°}\ \text{mA}$$

By the current division rule, the current in the direction N to G is

$$\text{current,}\ I'_3 = 1.494\underline{/4.77°} \times \frac{2+j1}{2+j1+1-j4}$$

$$= 1.494\underline{/4.77°} \times \frac{2.236\underline{/26.56°}}{4.242\underline{/-45°}}$$

$$= 0.787\underline{/76.33°}\ \text{mA}$$

STEP 2 When the $4.24\underline{/45°}$ V source is removed, the remaining circuit is shown in Fig. 17-9(b). The total impedance presented to the $11.70\underline{/20°}$ V source is

$$Z''_T = \frac{(1-j4)(j5-j3)}{1-j4+j5-j3} + 2 + j1$$

$$= \frac{(1-j4)(j2)}{1-j2} + 2 + j1$$

$$= \frac{8+j2}{1-j2} + 2 + j1$$

$$= \frac{(8+j2)(1+j2)}{5} + 2 + j1$$

$$= 2.8 + j4.6 = 5.385\underline{/58.67°}\ k\Omega$$

$$\text{current,}\ I''_1 = \frac{11.7\underline{/+20.0°}\ \text{V}}{5.385\underline{/58.67°}\ k\Omega} = 2.172\underline{/-38.67°}\ \text{mA}$$

Using the CDR, the current in the direction G to N is

$$\text{current,}\ I''_3 = 2.172\underline{/-38.67°} \times \frac{j5-j3}{1-j4+j5-j3}$$

$$= 2.172\underline{/-38.67°} \times \frac{j2}{1-j2}$$

$$= 2.172\underline{/-38.67°} \times \frac{2\underline{/90°}}{2.236\underline{/-63.43°}}$$

$$= 1.943\underline{/114.76°}\ \text{mA}$$

STEP 3 By the principle of superposition the current in the direction *N to G* is

$$\text{current, } I_3 = I'_3 - I''_3 = 0.787 \underline{/76.33°} - 1.943 \underline{/114.76°}$$
$$= 0.186 + j0.765 + 0.814 - j1.764$$
$$= 1 - j1 \text{ mA} \quad \text{(rounded off)}$$

This is the same result as we obtained in Example 17-2.

EXAMPLE 17-5 In the circuit of Fig. 17-10(a) use the superposition theorem to determine the value of the circuit I_2.

Solution

STEP 1 Remove the source E_1 so that the remaining circuit is as shown in Fig. 17-10(b). Total impedance, Z'_T, presented to the E_2 source is

$$Z'_T = 3 - j2 + \frac{j1 \times (2 - j3)}{j1 - j3 + 2}$$
$$= \frac{5 - j8}{2 - j2} \text{ k}\Omega$$

$$\text{current, } I'_2 = \frac{E_2}{Z'_T} = \frac{10 \underline{/-53.1°} \text{ V} \times (2 - j2)}{(5 - j8) \text{ k}\Omega}$$
$$= \frac{(6 - j8) \times (2 - j2)}{5 - j8}$$
$$= \frac{-4 - j28}{5 - j8} \text{ mA}$$

STEP 2 Remove the source E_2 so that the remaining circuit is as shown in Fig. 17-10(c). Total impedance, Z''_T, presented to the E_1 source is

$$Z''_T = 2 - j3 + \frac{j1 \times (3 - j2)}{3 - j2 + j1}$$
$$= 2 - j3 + \frac{2 + j3}{3 - j1}$$
$$= \frac{(2 - j3)(3 - j1) + (2 + j3)}{3 - j1}$$
$$= \frac{5 - j8}{3 - j1} \text{ k}\Omega$$

$$\text{current, } I''_1 = 9.22 \underline{/-40.6°} \times \frac{3 - j1}{5 - j8}$$
$$= \frac{(7 - j6) \times (3 - j1)}{5 - j8} \text{ mA}$$

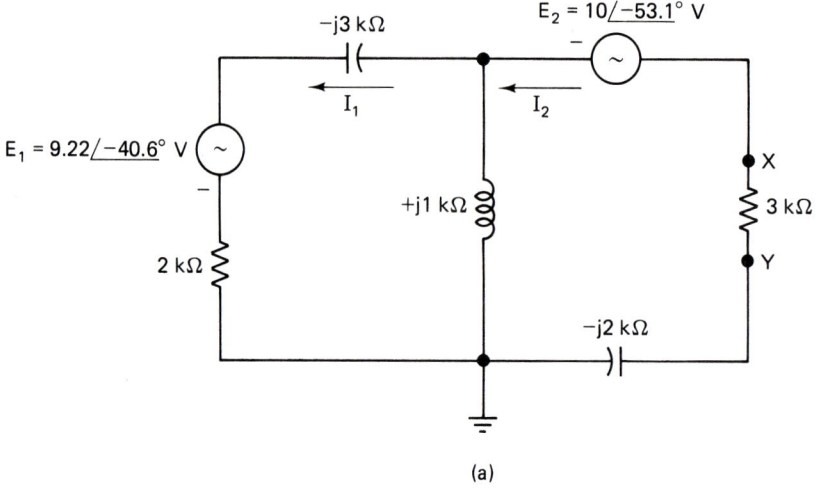

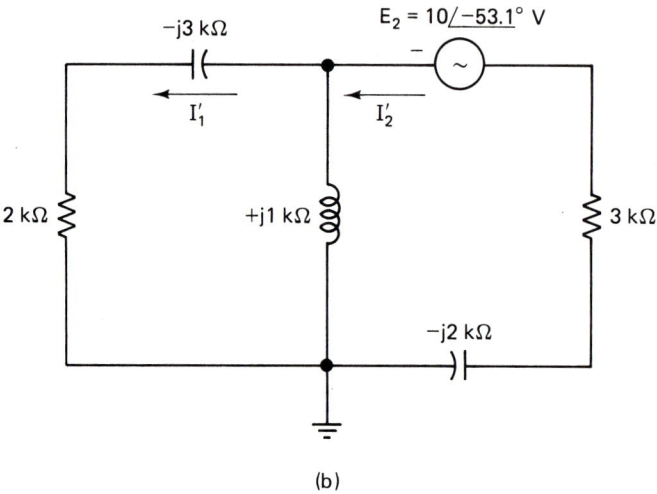

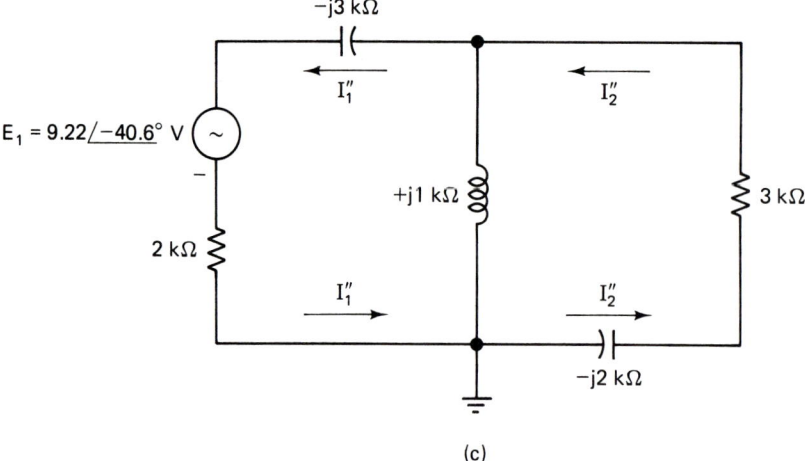

FIGURE 17-10 Circuit for Example 17-5.

By the current division rule,

$$\text{current, } I_2'' = (7 - j6) \times \frac{3 - j1}{5 - j8} \times \frac{j1}{3 - j1}$$

$$= \frac{(7 - j6) \times j1}{5 - j8}$$

$$= \frac{6 + j7}{5 - j8} \text{ mA}$$

STEP 3 By the principle of superposition,

$$\text{current, } I_2 = I_2' + I_2''$$

$$= \frac{-4 - j28 + 6 + j7}{5 - j8}$$

$$= \frac{2 - j21}{5 - j8}$$

$$= 2 - j1 = 2.24 \angle -26.6° \text{ mA}$$

17-6
Thévenin's Theorem

The steps that we take in Thévenizing an ac circuit are the same as those which we discussed for dc circuits in Section 6-6. However, in the case of ac circuits the theorem is restated as follows:

> The current in a load impedance connected between two terminals X and Y of a network of impedances and alternating sources is the same as if that load impedance were connected to a single alternating source whose output is the open-circuit voltage as measured between X and Y and whose internal impedance is the impedance of the network looking back into terminals X and Y with all alternating sources replaced by impedances equal to their internal impedances.

To illustrate this theorem, let us Thévenize the circuit of Example 17-1; this circuit has been redrawn in Fig. 17-11(a). The load Z_L is represented by the series combination of the 2-kΩ resistance and the $-j3$-kΩ capacitive reactance. The circuit will be Thévenized to determine the value of I_2 which is the same as the load current.

STEP 1 Remove the load Z_L so that the remaining circuit appears as in Fig. 17-11(b). By the voltage division rule, the voltage, V_1 between N and G is

$$\text{voltage, } V_1 = 10.77 \angle 158.2° \text{ V} \times \frac{3 \text{ k}\Omega}{3 + (1 + j2) \text{ k}\Omega}$$

$$= (-10 + j4) \times \frac{3}{4 + j2} = \frac{-15 + j6}{2 + j1} \text{ V}$$

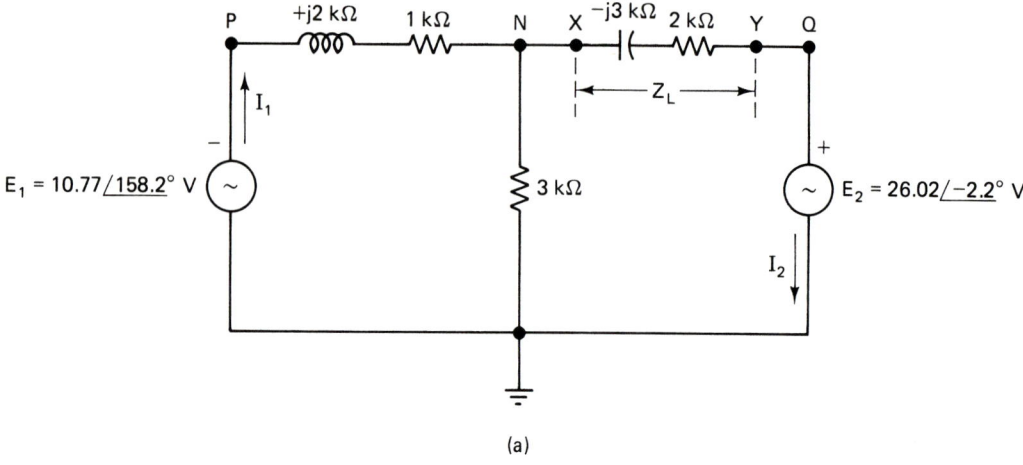

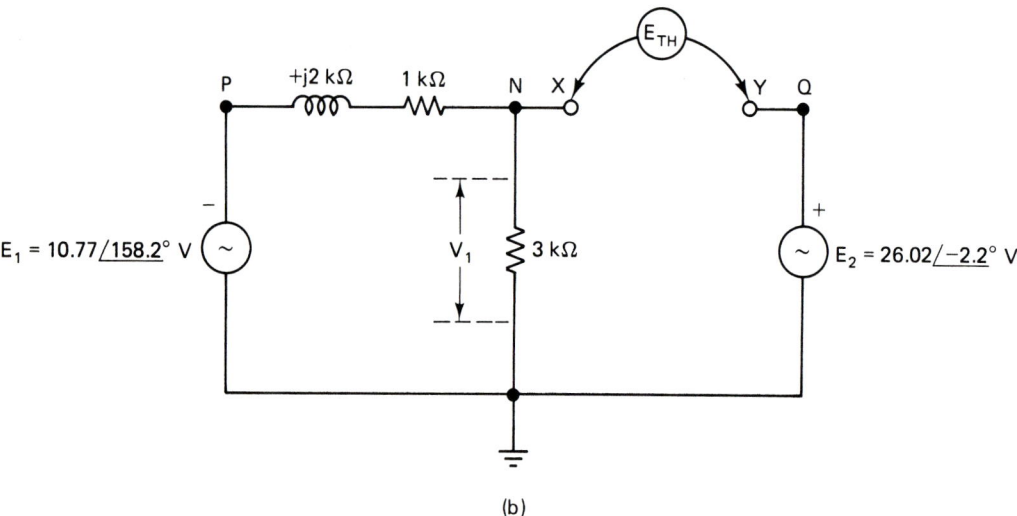

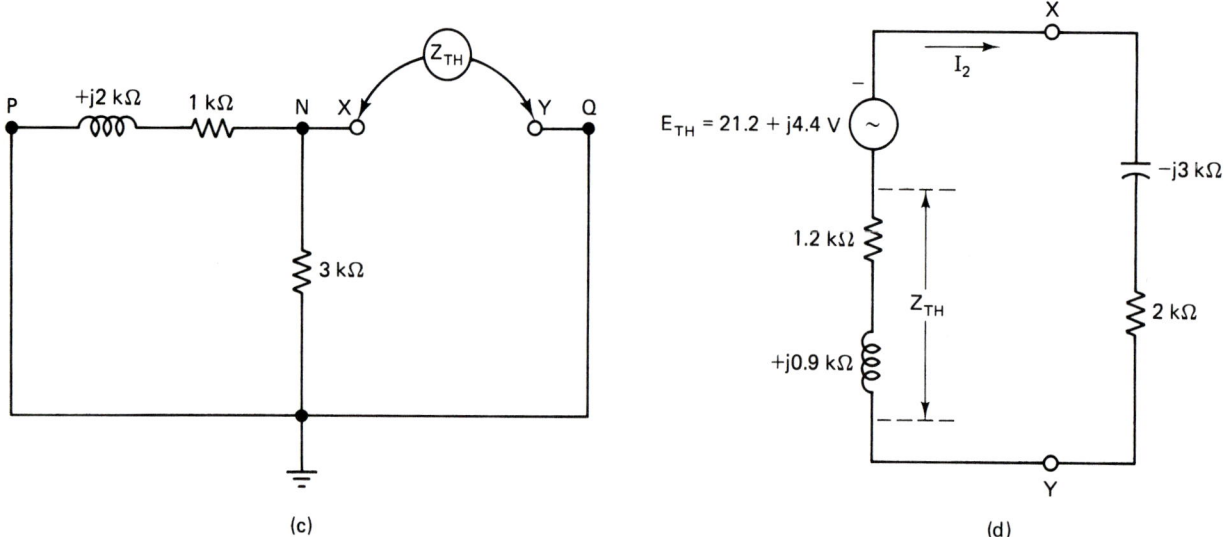

FIGURE 17-11 Circuit illustrating Thévenin's theorem.

open-circuit voltage between X and Y, $E_{TH} = V_1 + E_2$

$$= \frac{-15 + j6}{2 + j1} + 26 - j1$$

$$= \frac{38 + j30}{2 + j1}$$

$$= 21.2 + j4.4 \text{ V}$$

STEP 2 Remove the sources E_1 and E_2 and replace them with short circuits [Fig. 17-11(c)]. The Thévenin impedance between terminals X and Y is given by

$$\text{Thévenin impedance, } Z_{TH} = (1 + j2) \| 3 \text{ k}\Omega$$

$$= \frac{(1 + j2) \times 3}{(1 + j2) + 3}$$

$$= \frac{3 + j6}{4 + j2}$$

$$= 1.2 + j0.9 \text{ k}\Omega$$

STEP 3 Construct the equivalent Thévenin alternator and replace the load between terminals X and Y [Fig. 17-11(d)]. Then

$$\text{current, } I_2 = \frac{21.2 + j4.4}{3.2 - j2.1}$$

$$= \frac{21.65 \angle +11.73°}{3.83 \angle -33.27°}$$

$$= 5.65 \angle 45° \text{ mA}$$

This result agrees with the solution of Example 17-1.

EXAMPLE 17-6 In Fig. 17-12(a), derive the Thévenin equivalent voltage source between terminals X and Y and obtain the value of the load current, I_L.

Solution

STEP 1 Remove the 5-kΩ load and obtain the open-circuit voltage between X and Y [Fig. 17-12(b)]. Using the voltage division rule (VDR),

$$\text{Thévenin source voltage, } E_{TH} = 10\angle 30° \text{ V} \times \frac{j4 \text{ k}\Omega}{(1 - j3) + j4 \text{ k}\Omega}$$

$$= \frac{10\angle 30° \times 4\angle 90°}{1 + j1}$$

$$= \frac{10\angle 30° \times 4\angle 90°}{1.414\angle 45°}$$

$$= 28.3 \angle 75° \text{ V}$$

Sec. 17-6 Thévenin's Theorem

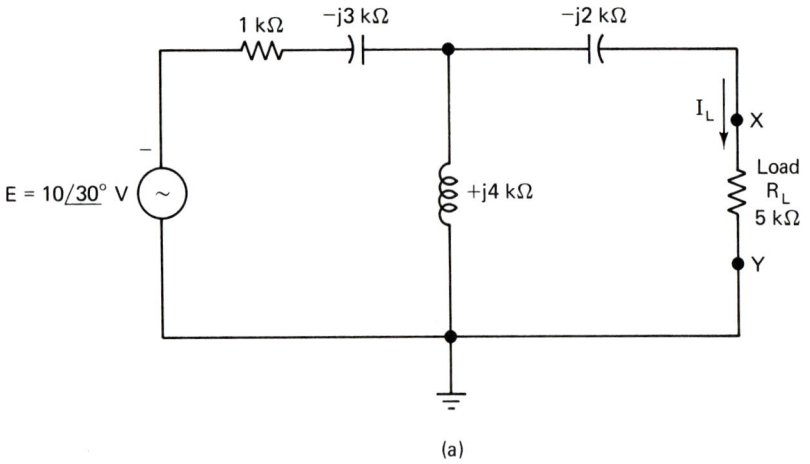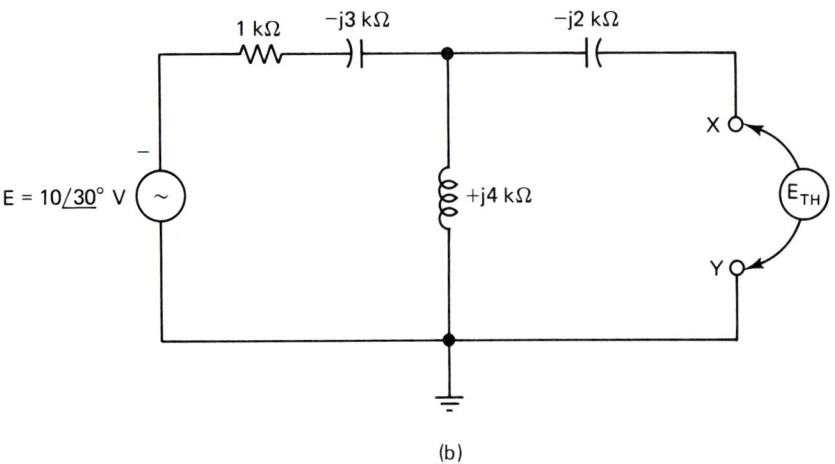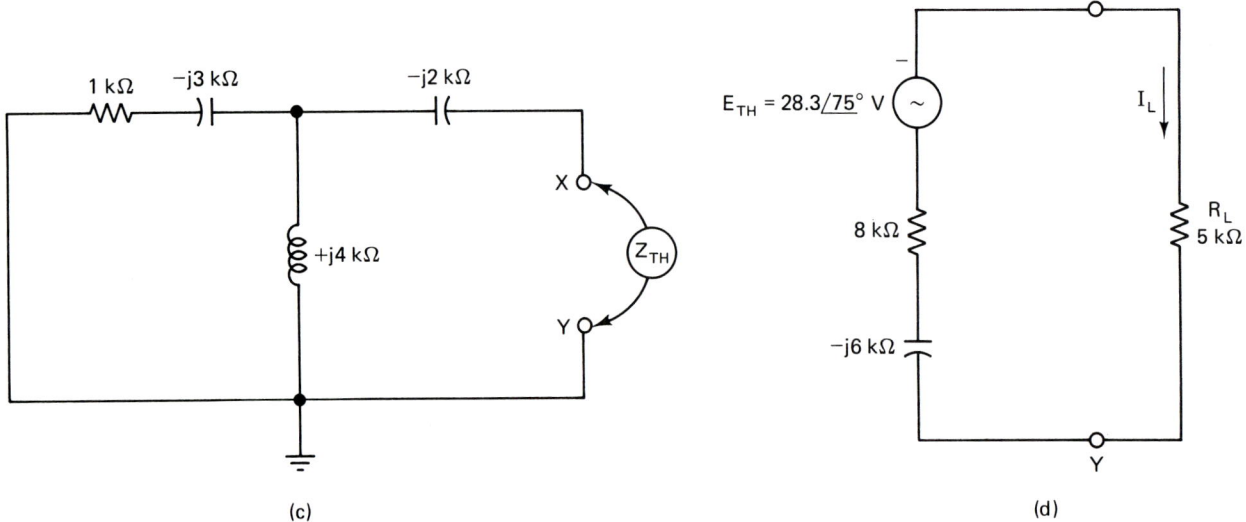

FIGURE 17-12 Circuit for Example 17-6.

STEP 2 Replace the $10\angle 30°$ V source by a short circuit and find the value of the impedance phasor between X and Y. Then

$$\text{impedance phasor, } Z_{TH} = -j2 + \frac{j4 \times (1 - j3)}{j4 + (1 - j3)}$$

$$= -j2 + \frac{(j4 + 12)(1 - j1)}{2}$$

$$= -j2 + 8 - j4 = 8 - j6 \text{ k}\Omega$$

STEP 3 Draw the circuit of the equivalent Thévenin voltage source and replace the load between terminals X and Y [Fig. 17-12(d)]. Then the load current is

$$\text{load current, } I_L = \frac{E_{TH}}{Z_{TH} + R_L} = \frac{28.3 \angle 75° \text{ V}}{(8 - j6) + 5 \text{ k}\Omega} = \frac{28.3 \angle 75°}{13 - j6}$$

$$= \frac{28.3 \angle 75°}{14.3 \angle -24.78°}$$

$$= 1.98 \angle 99.78° \text{ mA}$$

17-7 Norton's Theorem

Norton's theorem was discussed originally in Section 6-7. The principles of this theorem apply equally well to both dc and ac circuits and there is no difference in the steps required to Nortonize a complex network of alternating sources and impedances. Once again the equivalent Norton alternating source will be of the constant current type with its associated impedance in parallel.

In relation to ac circuits, the formal statement of Norton's theorem is:

> The alternating current in a load connected between two output terminals, X and Y, of a complex network containing alternating sources and impedances is the same as if that load were connected to a constant alternating source whose current, I_N, is equal to the short-circuit current measured between X and Y. This constant alternating current source, I_N, is placed in parallel with an impedance, Z_N, which is equal to the impedance of the network looking back into the terminals X and Y with all sources replaced by impedances equal in value to their internal impedances.

This last step would theoretically involve short-circuiting the constant alternating voltage sources and open-circuiting the constant alternating current sources.

As an example, let us Nortonize the circuit of Fig. 17-13(a) by regarding the 3-kΩ resistor as the load between terminals X and Y.

STEP 1 Remove the load and place a short circuit between X and Y [Fig. 17-13(a)].

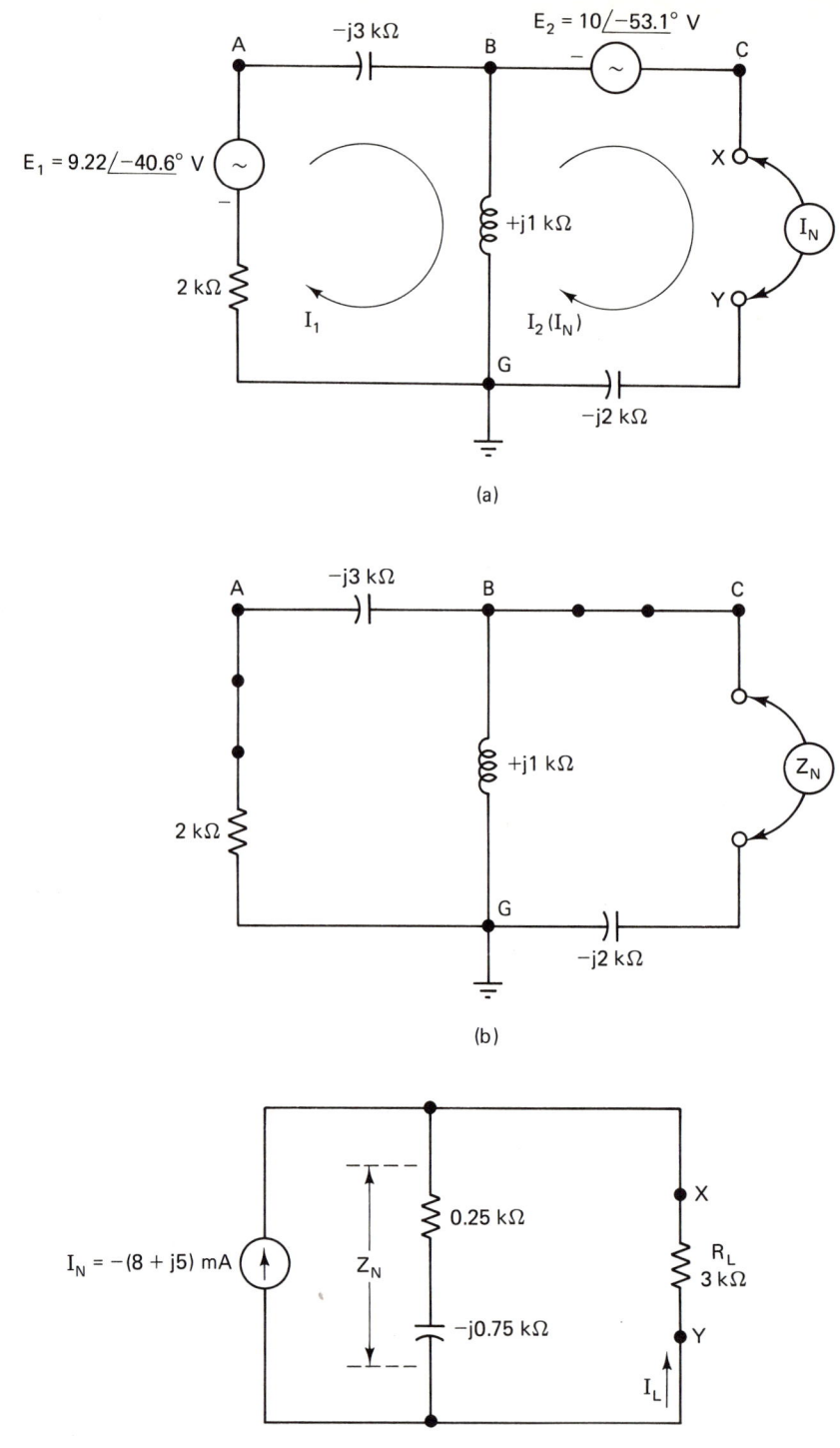

FIGURE 17-13 Circuit illustrating Norton's theorem.

We may conveniently solve for the Norton current, I_N, by using mesh analysis. For the mesh $ABGA$ the equation is

$$I_1(-j3 + j1 + 2) - I_2 \times j1 = -9.22\underline{/-40.6°} = -7 + j6$$

$$I_1(2 - j2) - I_2 \times j1 = -7 + j6 \qquad (17\text{-}7\text{-}1)$$

For the mesh $BCGB$ the equation is

$$-I_1 \times j1 + I_2(-j2 + j1) = -10\angle-53.1° = -6 + j8$$

$$\boxed{-I_1 \times j1 - I_2 \times j1 = -6 + j8} \qquad (17\text{-}7\text{-}2)$$

or

$$\boxed{I_1 + I_2 = \frac{+6 - j8}{+j1} = -8 - j6} \qquad (17\text{-}7\text{-}3)$$

Subtract Eq. (17-7-2) from Eq. (17-7-1):

$$I_1(2 - j2 + j1) = -7 + j6 - j8 + 6 = -1 - j2$$

$$\text{current, } I_1 = \frac{-1 - j2}{2 - j1} = -j1 \text{ mA}$$

Substitute the value of I_1 in Eq. (17-7-3):

$$\text{Norton current, } I_N = I_2 = -8 - j6 + j1 = -8 - j5 \text{ mA}$$

Note that in this particular problem it is more difficult to obtain the Norton current than the Thévenin voltage.

STEP 2 Replace the voltage sources E_1 and E_2 by short circuits [Fig. 17-13(b)]. The Norton impedance between X and Y is given by

$$\text{Norton impedance, } Z_N = -j2 + \frac{(2 - j3) \times j1}{(2 - j3) + j1}$$

$$= -j2 + \frac{3 + j2}{2 - j2}$$

$$= \frac{-j2(2 - j2) + 3 + j2}{2 - j2}$$

$$= \frac{-1 - j2}{2 - j2} = \frac{1 - j3}{4} \text{ k}\Omega$$

The equivalent Norton current source is shown in Fig. 17-13(c). When the load is replaced between terminals X and Y, the current, I_L, is found by the current division rule.

$$\text{current, } I_L = I_N \times \frac{Z_N}{Z_N + Z_L} = (-8 - j5) \times \frac{1 - j3}{4\left(\dfrac{1 - j3}{4} + 3\right)}$$

$$= \frac{-23 + j19}{13 - j3}$$

$$= -(2 - j1) \text{ mA (in the direction of } X \text{ to } Y)$$

This result is identical to the value of I_2 that we obtained by using the superposition theorem in Fig. 17-10.

Sec. 17-7 Norton's Theorem

EXAMPLE 17-7 In Fig. 17-12(a), derive the Norton equivalent current source between terminals X and Y and then obtain the value of the load current I_L.

Solution

STEP 1 Remove the 5-kΩ load and obtain the short-circuit current between X and Y [Fig. 17-14(a)]. The total impedance, Z_T, presented to the voltage source is

$$\text{total impedance, } Z_T = 1 - j3 + \frac{j4 \times (-j2)}{j4 + (-j2)}$$

$$= 1 - j3 + \frac{8}{j2}$$

$$= 1 - j3 - j4 = 1 - j7 \text{ k}\Omega$$

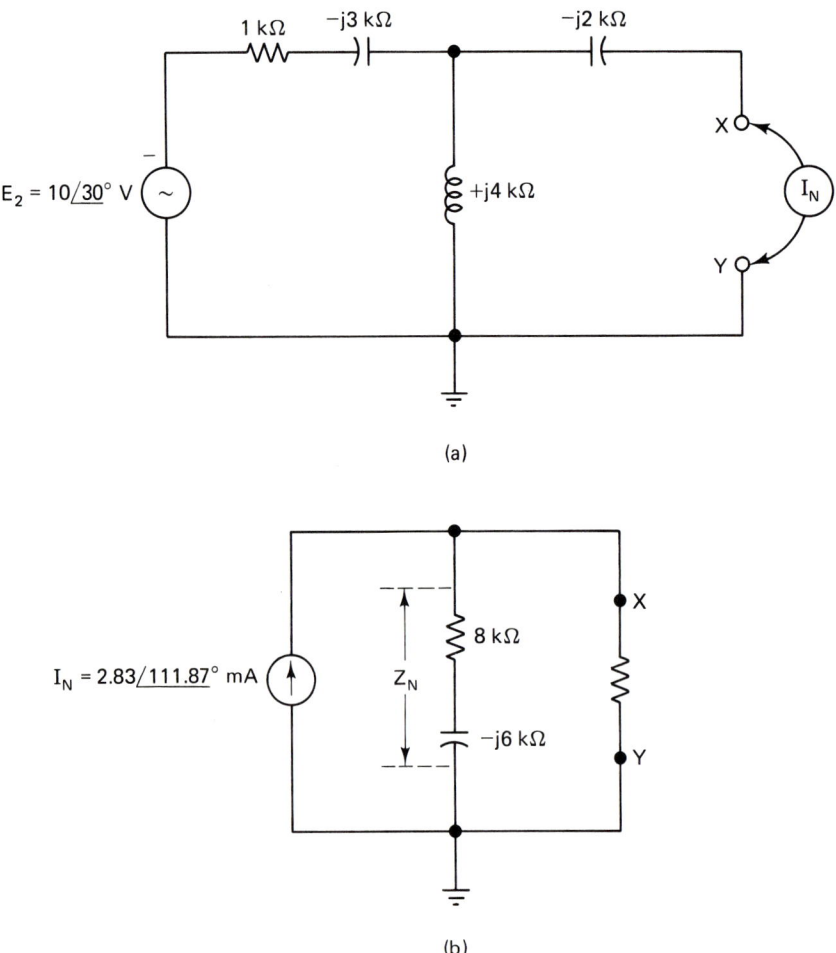

FIGURE 17-14 Circuit for Example 17-7.

The current, I_T, drawn from the voltage source is

$$\text{current, } I_T = \frac{10\angle 30°}{1 - j7} \text{ mA}$$

Using the current division rule (CDR), the short-circuit current, I_N, between terminals X and Y is

$$\text{current, } I_N = \frac{10\angle 30°}{1 - j7} \times \frac{j4}{j4 + (-j2)}$$

$$= \frac{20\angle 30°}{1 - j7}$$

$$= \frac{20\angle 30°}{7.07\angle -81.87°} = 2.83\angle 111.87° \text{ mA}$$

STEP 2 The Norton impedance, Z_N, is the same as the Thévenin impedance, Z_{TH}, and therefore $Z_N = 8 - j6$ kΩ (Example 17-6).

STEP 3 Remove the short circuit and replace the load in the Norton equivalent alternator circuit of Fig. 17-14(b). By the current division rule,

$$\text{load current, } I_L = 2.83\angle 111.87 \times \frac{(8 - j6)}{(8 - j6) + 5}$$

$$= \frac{2.83\angle 111.87° \times 10\angle -36.87°}{14.32\angle -24.78°}$$

$$= 1.98\angle 99.78° \text{ mA}$$

This is the same result as we obtained by Thévenin's theorem in Example 17-6.

17-8
Wye–Delta and Delta–Wye Transformations

In Section 6-8 we derived the equations for delta–wye transformations that involved resistors. The same equations can be applied to ac networks provided that we replace the resistors with the necessary impedance phasors.
  Referring to Fig. 17-15, the equations are
Wye-to-delta transformation

$$Z_{AB} = \frac{Z_A Z_B + Z_B Z_C + Z_C Z_A}{Z_C} \quad (17\text{-}8\text{-}1)$$

$$Z_{BC} = \frac{Z_A Z_B + Z_B Z_C + Z_C Z_A}{Z_A} \quad (17\text{-}8\text{-}2)$$

$$Z_{CA} = \frac{Z_A Z_B + Z_B Z_C + Z_C Z_A}{Z_B} \quad (17\text{-}8\text{-}3)$$

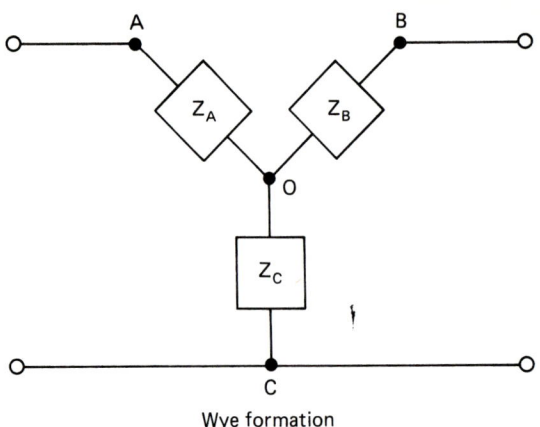

FIGURE 17-15 Wye and delta formations.

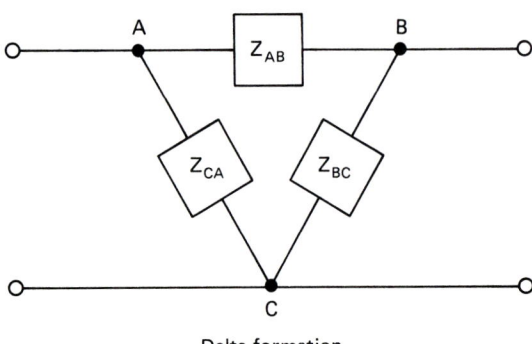

Delta-to-wye transformation

$$Z_A = \frac{Z_{AB}Z_{CA}}{Z_{AB} + Z_{BC} + Z_{CA}} \qquad (17\text{-}8\text{-}4)$$

$$Z_B = \frac{Z_{BC}Z_{AB}}{Z_{AB} + Z_{BC} + Z_{CA}} \qquad (17\text{-}8\text{-}5)$$

$$Z_C = \frac{Z_{CA}Z_{BC}}{Z_{AB} + Z_{BC} + Z_{CA}} \qquad (17\text{-}8\text{-}6)$$

We will use the delta–wye transformation to obtain the current I_T in the bridge circuit of Fig. 17-16.

Referring to the A, B, C delta formation,

$$Z_{AB} = 2 \text{ k}\Omega \qquad Z_{BC} = 3 - j2 \text{ k}\Omega \qquad Z_{CA} = -j3 \text{ k}\Omega$$

Then

$$Z_A = \frac{2 \times (-j3)}{2 + 3 - j2 - j3}$$

$$= \frac{-j6}{5 - j5} = \frac{(-j6)(5 + j5)}{5^2 + 5^2} = \frac{30 - j30}{50} = 0.6 - j0.6 \text{ k}\Omega$$

$$Z_B = \frac{(3 - j2) \times (2)}{5 - j5} = \frac{(6 - j4)(5 + j5)}{50} = \frac{50 + j10}{50}$$

$$= 1 + j0.2 \text{ k}\Omega$$

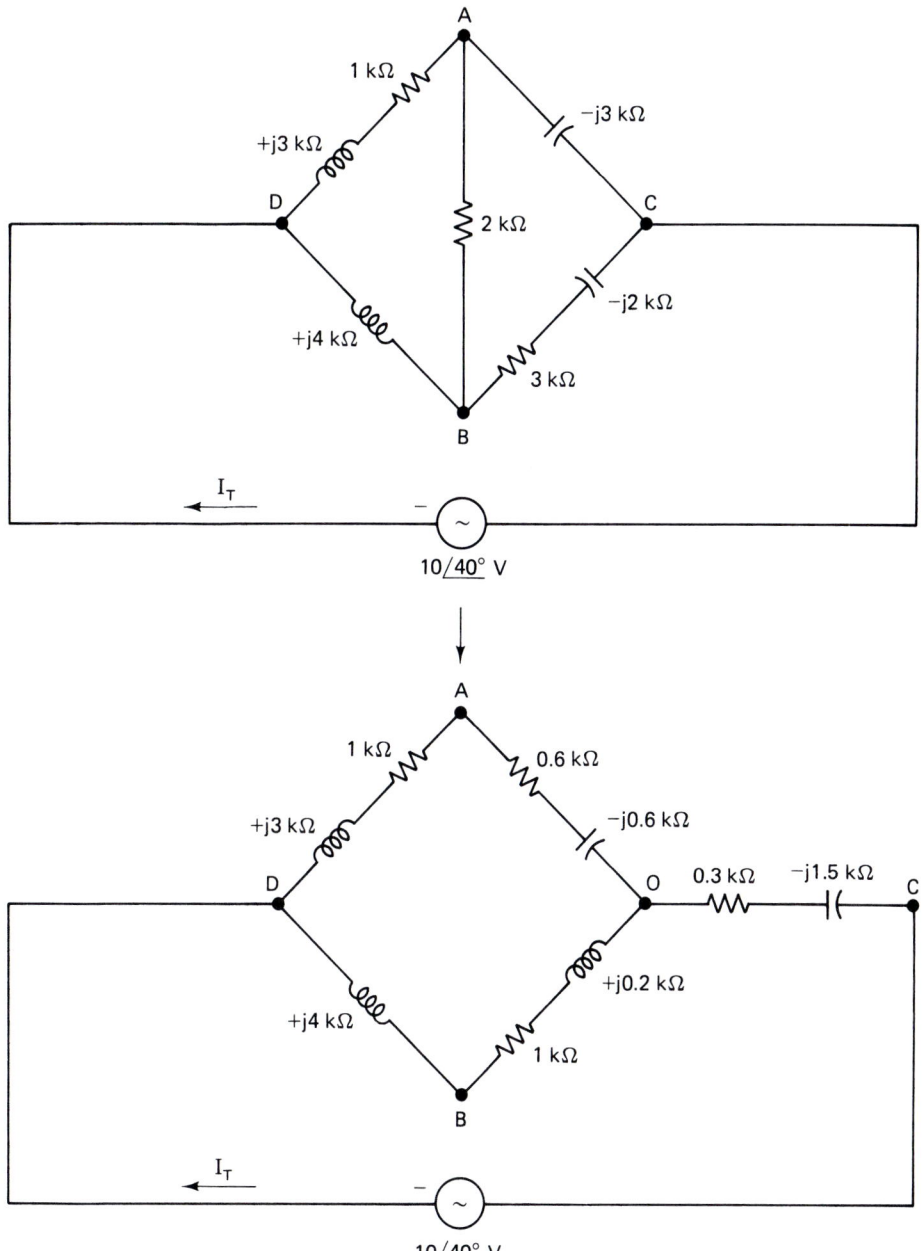

FIGURE 17-16 Delta to wye transformations.

$$Z_C = \frac{(-j3) \times (3 - j2)}{5 - j5}$$

$$= \frac{(-6 - j9)(5 + j5)}{50} = \frac{15 - j75}{50} = 0.3 - j1.5 \text{ k}\Omega$$

Total impedance, Z_T, presented to the $10\underline{/40°}$ V source is

$$\text{total impedance, } Z_T = \frac{(1.6 + j2.4)(1 + j4.2)}{2.6 + j6.6} + 0.3 - j1.5$$

$$= \frac{2.88\underline{/56.3°} \times 4.32\underline{/76.6°}}{7.09\underline{/68.5°}} + 0.3 - j1.5$$

Sec. 17-8 Wye–Delta and Delta–Wye Transformations

$$= 1.755\underline{/64.4°} + 0.3 - j1.5$$
$$= 0.758 + j1.583 + 0.3 - j1.5$$
$$= 1.058 + j0.083 = 1.06\underline{/4.5°}\ k\Omega$$

$$\text{current,}\ I_T = \frac{10\underline{/40°}\ V}{1.06\underline{/4.5°}\ k\Omega} = 9.4\underline{/35.5°}\ mA$$

EXAMPLE 17-8 In Fig. 17-16, convert the wye formation *DBCA* (*A* is the wye point) into its delta equivalent and then calculate the value of the total impedance presented to the $10\underline{/40°}$ V source.

Solution The result of the wye–delta transformation is shown in Fig. 17-17.

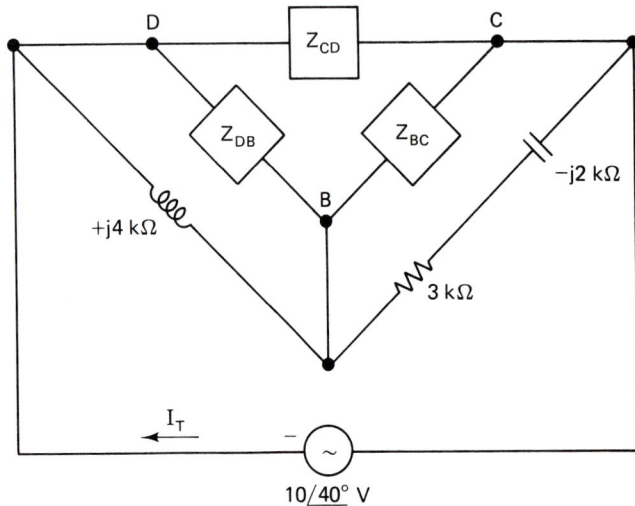

FIGURE 17-17 Circuit for Example 17-8.

From Eqs. (17-8-1), (17-8-2), and (17-8-3)

$$Z_{DB} = \frac{Z_D Z_B + Z_B Z_C + Z_C Z_D}{Z_C}$$

$$= \frac{(1 + j3)(2) + (2)(-j3) + (-j3)(1 + j3)}{-j3}$$

$$= \frac{2 + j6 - j6 - j3 + 9}{-j3}$$

$$= \frac{11 - j3}{-j3} = 1 + j3.67\ k\Omega$$

$$Z_{BC} = \frac{11 - j3}{1 + j3} = \frac{(11 - j3)(1 - j3)}{10} = 0.2 - j3.6\ k\Omega$$

$$Z_{CD} = \frac{11 - j3}{2} = 5.5 - j1.5\ k\Omega$$

$$\text{impedance of } Z_{DB} \| j4 \text{ k}\Omega = \frac{(1 + j3.67)j4}{1 + j7.67}$$

$$= \frac{(-14.68 + j4)(1 - j7.67)}{1^2 + 7.67^2}$$

$$= \frac{16 + j116.6}{59.8}$$

$$= 0.267 + j1.95 \text{ k}\Omega$$

$$\text{impedance of } Z_{BC} \| (3 - j2) \text{ k}\Omega = \frac{(0.2 - j3.6)(3 - j2)}{3.2 - j5.6}$$

$$= \frac{(-6.6 - j11.2)(3.2 + j5.5)}{3.2^2 + 5.6^2}$$

$$= \frac{41.6 - j72.8}{41.6}$$

$$= 1.0 - j1.75 \text{ k}\Omega$$

$$\text{total impedance } Z_T = [(0.267 + j1.95) + (1.0 - j1.75)] \| (5.5 - j1.5)$$

$$= \frac{(1.267 + j0.2) \times (5.5 - j1.5)}{6.767 - j1.3}$$

$$= \frac{(7.2685 - j0.8)(6.767 + j1.3)}{6.767^2 + 1.3^2}$$

$$= \frac{50.22 + j4.035}{47.48}$$

$$= 1.057 + j0.085 \text{ k}\Omega$$

This is the same result as we obtained with the delta–wye transformation.

17-9 Reciprocity Theorem

The reciprocity theorem is equally true for dc and ac networks that contain only linear passive elements. However, this theorem is applied primarily to alternating circuits and was therefore not covered in Chapter 6.

The reciprocity theorem may be stated as follows:

> An ac network of impedances is supplied with current from an alternating voltage source which is connected between two points, X and Y. The current through a particular element is measured by an ac ammeter that is inserted between two points, A and B. If the positions of the alternating source and the ammeter are interchanged (in other words, the source is connected between the points A and B while the ammeter is inserted between points X and Y), the reading of the ammeter is unchanged. If the ammeter possesses an appreciable internal impedance, this impedance remains between points A and B and is regarded as being in series with the voltage source.

This result is independent of the complexity of the network. For example, it allows you to deduce that the directional properties of a receiving antenna are the same as those of the identical antenna when it is used for transmission purposes.

We will use the circuit of Fig. 17-18(a) to verify the reciprocity theorem. In Fig. 17-18(a) the total impedance, Z_T, presented to the alternating source is given by

$$Z_T = +j1 + \frac{4 \times (-j3)}{4 + (-j3)}$$

$$= +j1 - \frac{j12}{4 - j3}$$

$$= \frac{3 - j8}{4 - j3} \text{ k}\Omega$$

current, $I_T = 20\underline{/-20°} \times \frac{4 - j3}{3 - j8}$ mA

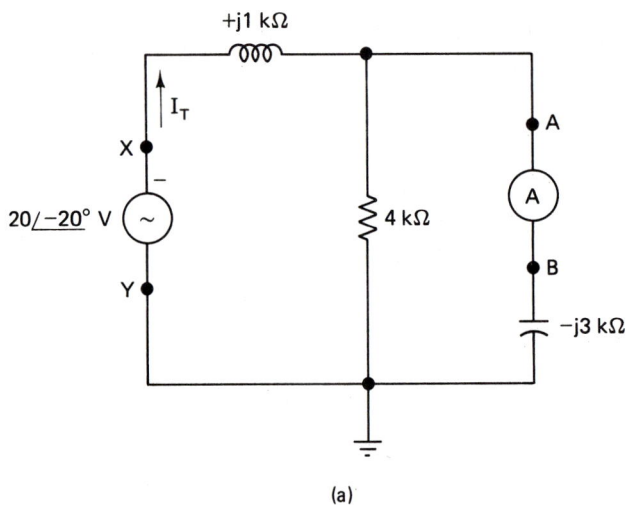

(a)

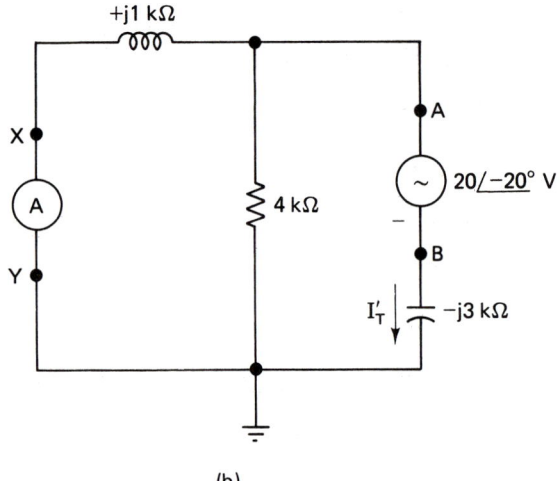

(b)

FIGURE 17-18 Circuit illustrating the reciprocity theorem.

By the current division rule, the reading of the ammeter, A, is

$$20\angle-20° \times \frac{4-j3}{3-j8} \times \frac{4}{4-j3} = \frac{80\angle-20°}{3-j8}$$

$$= \frac{80\angle-20°}{8.544\angle-69.4°} = 9.36\angle 49.4° \text{ mA}$$

In Fig. 17-18(b) the source and the ammeter have been interchanged. The new total impedance, Z_T, presented to the alternating source is

$$Z_T = -j3 + \frac{4 \times j1}{4 + j1}$$

$$= \frac{3-j8}{4+j1} \text{ k}\Omega$$

$$\text{current, } I'_T = 20\angle-20° \times \frac{4+j1}{3-j8} \text{ mA}$$

By the current division rule, the reading of the ammeter, A, is

$$20\angle-20° \times \frac{4+j1}{3-j8} \times \frac{4}{4+j1} = \frac{80\angle-20°}{3-j8} = 9.36\angle 49.4° \text{ mA}$$

The reading of the ammeter is therefore the same in each of the circuits in Fig. 17-18.

Chapter Summary

17.1. Kirchhoff's voltage law (KVL):
The phasor sum of the alternating voltage sources and the voltage drops around any closed electrical loop is zero.

17.2. Kirchhoff's current law (KCL):
The phasor sum of the currents existing at any junction point is zero.

17.3. Nodal analysis:
The phasor sum of the alternating source currents entering the node is equal to the phasor sum of the currents leaving through the impedances.

17.4. Millman's theorem:
Any number of alternating-current sources which are directly connected in parallel can be converted into a single alternating current source whose total alternator current is the phasor sum of the individual source currents and whose internal impedance is the result of combining the individual source impedances in parallel.

17.5. Superposition theorem:
In a network of *linear* impedances containing more than one alternating source, the current flowing at any point is the phasor sum of the currents which would flow at that point if each alternating source is considered separately with all other sources replaced by their equivalent internal impedances.

17.6. Thévenin's theorem:
The alternating current in an impedance load connected between two output terminals X and Y of a complex network of impedances and alternating sources, is the same as if that load were connected across a simple alternator whose voltage, e_{TH}, is the *open-circuit* value measured between X and Y and whose

series internal impedance, z_{TH}, is the impedances of the network looking back into terminals X and Y with all sources replaced by impedances equal to their internal impedances.

17.7. Norton's theorem:

The alternating current in a load connected between two output terminals X and Y of a complex network containing alternating sources and impedances is the same as if that load were connected to an alternating source whose alternator current, I_N, is equal to the *short-circuit* value measured between X and Y. This current alternator is placed in *parallel* with an impedance, Z_N, which is equal to the network's impedance, looking back into terminals X and Y with all sources replaced by impedances equal to their internal impedances.

Relationships between Thévenin and Norton alternators:

$$E_{TH} = I_N \times Z_N; \quad I_N = \frac{E_{TH}}{Z_{TH}}$$

$$Z_{TH} = Z_N$$

17.8. Delta–wye transformation:

$$Z_A = \frac{Z_{AB} Z_{CA}}{Z_{AB} + Z_{BC} + Z_{CA}} \quad \text{etc.}$$

17.9. Wye–delta transformation:

$$Z_{AB} = \frac{Z_A Z_B + Z_B Z_C + Z_C Z_A}{Z_C} \quad \text{etc.}$$

17.10. Reciprocity theorem:

An ac network of impedances is supplied with current from one alternating source that is connected between two points X and Y. The current through a particular element is measured by an ac ammeter that is inserted between two points A and B. If the positions of the alternating source and the ammeter are interchanged (in other words, the source is connected between points A and B while the ammeter is inserted between points X and Y), the reading of the ammeter is unchanged. If the ammeter possesses an appreciable internal impedance, this impedance remains between points A and B and is regarded as being in series with the voltage source.

Self-Review

True-False Questions

1. **T. F.** Kirchhoff's current law for ac circuits states that the phasor sum of the currents around any closed electrical loop is zero.
2. **T. F.** Kirchhoff's voltage law for ac circuits states that the phasor sum of the potentials at a particular junction point is zero.
3. **T. F.** To use Millman's theorem, every alternating-current source must be converted into its voltage equivalent.
4. **T. F.** The superposition theorem may be used only in cases that involve linear impedances.
5. **T. F.** When using the reciprocity theorem, the current meter's internal impedance must be transferred with the meter.
6. **T. F.** With Norton's theorem, I_N is found by removing the load and then calculating the phasor value of the short-circuit current between the load terminals.

7. **T. F.** When comparing the equivalent Norton and Thévenin alternating sources, $Z_{TH} = Z_N$.
8. **T. F.** The equation

$$Z_{AB} = \frac{Z_A Z_B + Z_B Z_C + Z_C Z_A}{Z_C}$$

is used in the delta-to-wye transformation.

9. **T. F.** With Thévenin's theorem, E_{TH}, is found by removing the load and then calculating the phasor value of the open-circuit voltage between the load terminals.
10. **T. F.** To use nodal analysis, every alternating voltage source must be converted into its current equivalent.
11. **T. F.** Inductive reactance has no meaning for ac circuits.
12. **T. F.** Kirchhoff's voltage law applies both to dc and ac circuits.
13. **T. F.** When writing equations for voltages around a loop, the phasors must be expressed in polar form.
14. **T. F.** In performing mesh analysis, simultaneous equations are written for which each branch current is separately identified.
15. **T. F.** The principle of superposition states essentially that the total effect of all sources is the superposition of the individual source effects.
16. **T. F.** In obtaining the Thévenin equivalent of an ac network, one dc voltage source and series resistance replaces all existing ac sources and internal impedances.
17. **T. F.** In applying Thévenin's theorem using experimental procedures, all active internal sources of the network must be suppressed, that is, replaced by their individual internal impedances.
18. **T. F.** Application of Norton's theorem to a network yields a current source in series with a single impedance.
19. **T. F.** One terminal of the wye configuration, or its equivalent delta configuration, must be common to the input and output terminals.
20. **T. F.** In comparing the wye configuration and its equivalent delta configuration, it is clear that $Z_{AB} = Z_A + Z_B$.

Multiple-Choice Questions

21. Using Kirchhoff's laws in the circuit of Fig. 17-19, the current flowing through the inductor is
    (a) $3 + j1$ mA
    (b) $2 + j3$ mA
    (c) $5 + j4$ mA
    (d) $1 - j2$ mA
    (e) $2 + j1$ mA

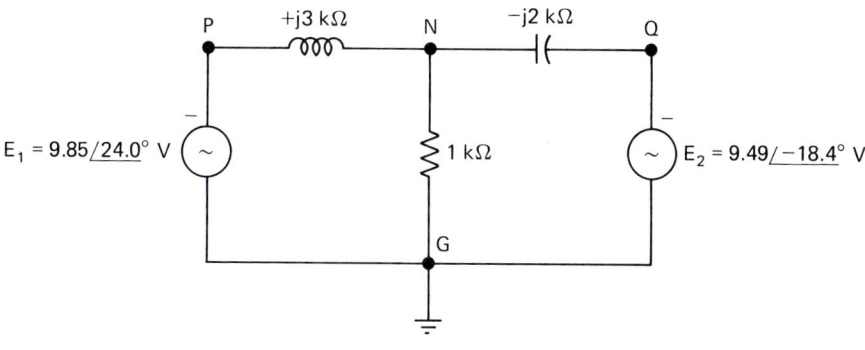

FIGURE 17-19 Circuit for Questions 21 through 30.

22. Using mesh analysis in the circuit of Fig. 17-19, the current associated with the capacitor is
    (a) $3 - j2$ mA
    (b) $1 - j2$ mA
    (c) $2 + j3$ mA
    (d) $-3 - j1$ mA
    (e) $2 + j1$ mA

23. Using Millman's theorem in the circuit of Fig. 17-19, the current flowing through the resistor is
    (a) $3 + j1$ mA
    (b) $2 - j3$ mA
    (c) $4 + j1$ mA
    (d) $-3 - j1$ mA
    (e) $2 - j1$ mA

24. Using nodal analysis in the circuit of Fig. 17-19, the voltage at the node N is
    (a) $1 - j3$ V
    (b) $3 - j1$ V
    (c) $3 - j2$ V
    (d) $-3 - j1$ V
    (e) $2 + j3$ V

25. In Fig. 17-19 use the superposition theorem to determine the alternating current drawn from the E_1 source. The value of this current is
    (a) $1 - j2$ mA
    (b) $2 + j1$ mA
    (c) $3 + j1$ mA
    (d) $2 + j3$ mA
    (e) $1 + j3$ mA

26. In the circuit of Fig. 17-19, Thévenize the circuit between the points N and G. The voltage of the Thévenin alternator is
    (a) $3 + j1$ V
    (b) $17 + j9$ V
    (c) $4 - j13$ V
    (d) $13 - j4$ V
    (e) $-9 + j17$ V

27. In Question 26, the value of the Thévenin impedance is
    (a) $j1$ kΩ
    (b) $-j6$ kΩ
    (c) $-j5$ kΩ
    (d) 5 kΩ
    (e) $+j6$ kΩ

28. In the circuit of Fig. 17-19, Nortonize the circuit between points N and G. The value of the Norton current is
    (a) $2.83 + j1.5$ mA
    (b) $1.5 - j2.83$ mA
    (c) $1.5 + j2.83$ mA
    (d) $2.83 - j1.5$ mA
    (e) $3.2 \angle 17.9°$ mA

29. In the circuit of Fig. 17-19, Nortonize the circuit between points N and Q. The value of the Norton impedance is
    (a) $3 + j4$ kΩ
    (b) $0.3 - j0.9$ kΩ
    (c) $0.3 + j0.9$ kΩ
    (d) $0.9 + j0.3$ kΩ
    (e) $0.9 - j0.3$ kΩ

30. In Fig. 17-19 convert the wye connection between points P, Q, and G into its delta equivalent. The value of the delta impedance between points P and G is
    (a) $-0.5 + j3$ kΩ
    (b) $6 + j1$ kΩ
    (c) $1 - j6$ kΩ
    (d) $2 + j0.33$ kΩ
    (e) $0.33 - j2$ kΩ

31. Indicate the circuit parameter that has meaning in both dc and ac circuits.
    (a) resistance
    (b) inductive reactance
    (c) capacitive reactance
    (d) impedance
    (e) all choices are correct

32. The rule, or law, that describes current behavior at a node or junction bears the name
    (a) conservation of momentum
    (b) Einstein's law of relativity
    (c) Kirchhoff's current law
    (d) Kirchhoff's junction and node law
    (e) all choices are correct

33. Regardless of the number of voltage and current sources, the application of one particular theorem usually avoids the writing and solutions of simultaneous equations. Its name is
    (a) Thévenin's theorem
    (b) Norton's theorem
    (c) Millman's theorem
    (d) the superposition theorem
    (e) Fermat's last theorem

34. This theorem permits all internal sources in a network to be replaced by a single voltage source. Its name is
    (a) Thévenin's theorem
    (b) Norton's theorem
    (c) Millman's theorem
    (d) the superposition theorem
    (e) Fermat's last theorem

35. After two points in the network have been selected, what measurements must be taken to find the Thévenin equivalent?
   (a) the open-circuit voltage across the points with internal sources active
   (b) the impedance across the points with internal sources active
   (c) the impedance across the points with the internal sources suppressed
   (d) the short-circuit current with sources active
   (e) choices (a) and (c) are both correct
36. The Norton equivalent circuit is described as having
   (a) a current source in one branch   (b) a voltage source in one branch
   (c) an impedance in one branch   (d) choices (a) and (c) are both correct
   (e) choices (b) and (c) are both correct
37. In the process of performing measurements to find either the Thévenin or Norton equivalents,
   (a) the ac voltage sources must be set with all voltages in phase
   (b) all internal impedances must be adjusted to be identical
   (c) the ac current sources must be set with all currents in phase
   (d) all choices are correct
   (e) no correct choice
38. A delta configuration has only resistance and capacitance elements. The impedances of the equivalent wye configuration are
   (a) either resistances alone or capacitive reactances alone
   (b) combinations of resistance and capacitive reactance
   (c) either resistances alone or inductive reactances alone
   (d) combinations of resistance and inductive reactance
   (e) combinations of R, L, and C
39. Interchange of ac source and current measurement locations is a feature of
   (a) the superposition theorem   (b) Thévenin's theorem
   (c) Norton's theorem   (d) Millman's theorem
   (e) the reciprocity theorem
40. Combining ac sources in parallel is facilitated by the application of
   (a) the superposition theorem   (b) Thévenin's theorem
   (c) Norton's theorem   (d) Millman's theorem
   (e) the reciprocity theorem

Practice Problems

41. In the circuit of Fig. 17-20, determine the value of the alternating current in the direction of point P to point N.

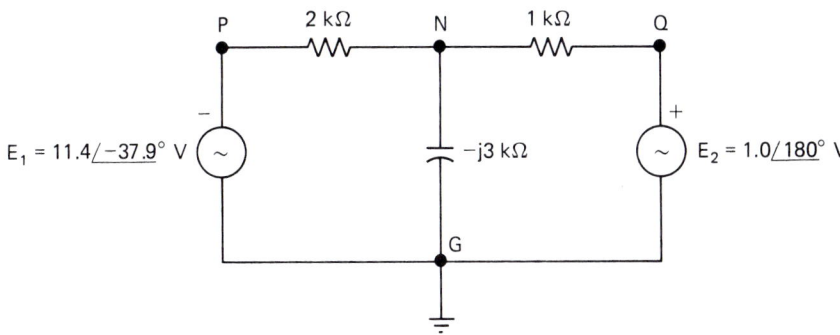

FIGURE 17-20   Circuit for Problems 41 through 50.

42. In Fig. 17-20, use mesh analysis to obtain the value of the alternating current in the direction of point N to point Q.
43. Use Millman's theorem in the circuit of Fig. 17-20 to derive the value of the alternating current from point N to point G.

44. Use nodal analysis in the circuit of Fig. 17-20 to calculate the voltage at point N.
45. In Fig. 17-20, use the superposition theorem to obtain the value of the alternating current drawn from the E_2 source.
46. In the circuit of Fig. 17-20, Thévenize the circuit between points N and G. What is the voltage, E_{TH}, of the Thévenin alternator?
47. In Problem 46, what is the value of the Thévenin impedance, Z_{TH}?
48. In the circuit of Fig. 17-20, Nortonize the circuit between points N and G. What is the value of the Norton current, I_N?
49. In Problem 48, what is the value of the Norton impedance, Z_N?
50. In Fig. 17-20 convert the wye connection between points P, Q, and G into its delta equivalent. What is the value of the delta impedance between points P and Q?
51. Determine the mesh currents I_1 and I_2 for the circuit of Fig. 17-21.

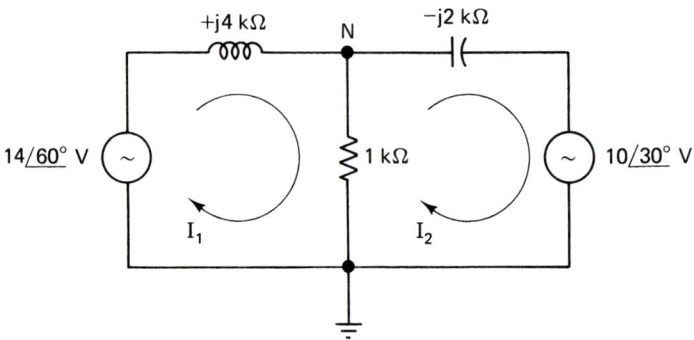

FIGURE 17-21 Circuit for Problem 51.

52. In Problem 51, find the voltage at node N with respect to ground.
53. Using nodal analysis, examine the circuit of Fig. 17-22 to find the voltage at node N.

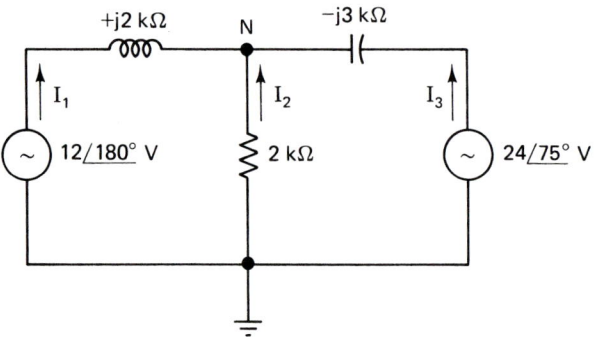

FIGURE 17-22 Circuit for Problem 53.

54. In Problem 53, find currents I_1, I_2, and I_3.
55. Apply Millman's theorem to find the voltage at node N for the circuit of Fig. 17-21. Construct the equivalent circuit schematic showing all values of voltages and currents.

56. In Fig. 17-23, find the values of Z_A, Z_B, and Z_C.

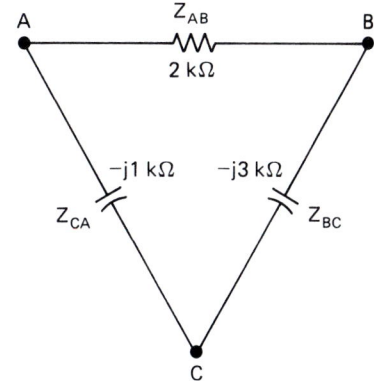

FIGURE 17-23   Circuit for Problem 56.

57. In Fig. 17-24, $Z_A = 1000 + j0\ \Omega$, $Z_B = 0 + j1000\ \Omega$, and $Z_C = 0 - j1000\ \Omega$. Find Z_{AB}, Z_{BC}, and Z_{CA} for the equivalent delta network.

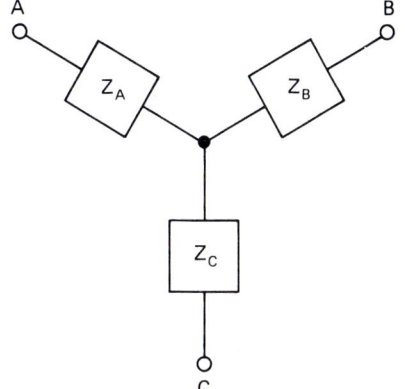

FIGURE 17-24   Circuit for Problem 57.

58. In Fig. 17-25, find the voltage across points A and B before the 20-kΩ load is connected.

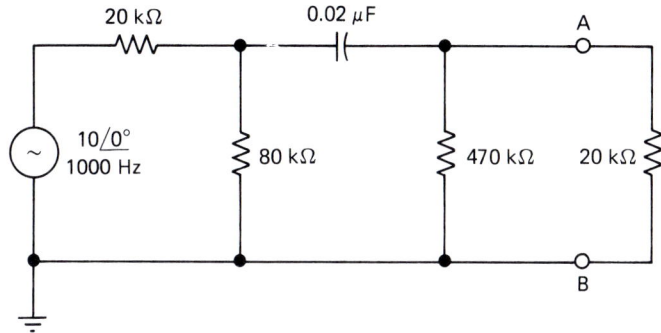

FIGURE 17-25   Circuit for Problems 58 and 59.

59. In Fig. 17-25, determine the voltage across points A and B after the 20-kΩ load is connected.

60. Find the equivalent T-network for the circuit of Fig. 17-26 at 400 Hz.

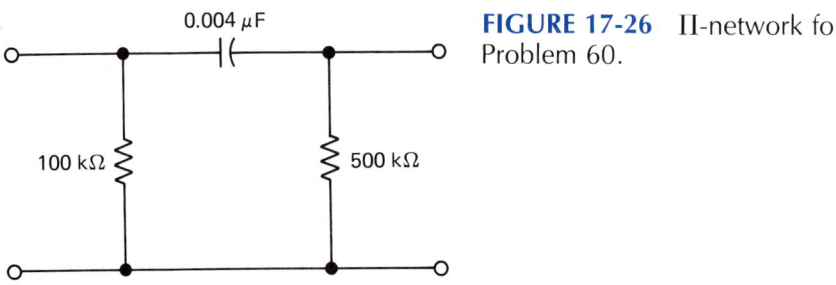

FIGURE 17-26 Π-network for Problem 60.

Advanced Problems

61. In Fig. 17-27, use Kirchhoff's laws to find the value of the alternating current in the direction of point N to point P.

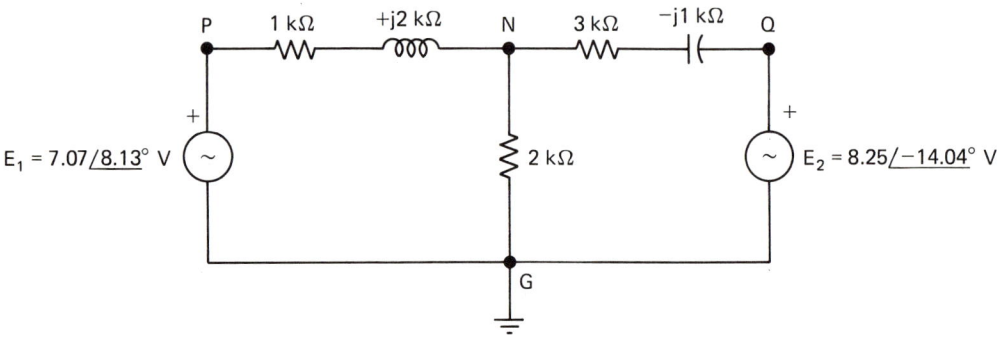

FIGURE 17-27 Circuit for Problems 61 through 64.

62. In Fig. 17-27, use nodal analysis to find the voltage at point N.

63. In Fig. 17-27 use the superposition theorem to find the value of the alternating current drawn from the voltage source E_2.

64. In Fig. 17-27, use Millman's theorem to find the alternating current in the direction of the point G to point N.

65. In Fig. 17-28, use the method of mesh analysis to determine the frequency for which V_o and V_i are 180° out of phase. At this frequency calculate the value of the ratio $V_i:V_o$. *Note:* This analysis illustrates the principle behind the RC phase shift oscillator.

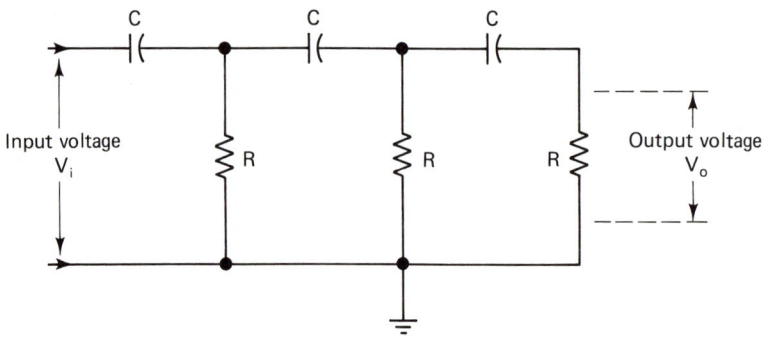

FIGURE 17-28 Circuit for Problem 65.

66. In Fig. 17-29, Thévenize the circuit between points X and Y. Calculate the values of E_{TH} and Z_{TH}.

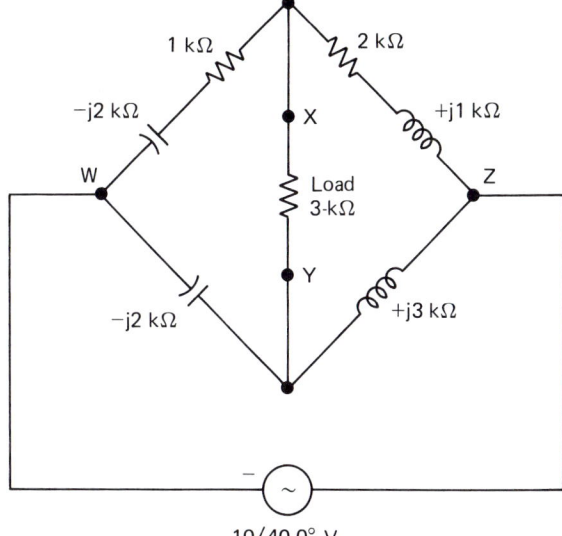

FIGURE 17-29 Circuit for Problems 66 through 70.

67. In Fig. 17-29, Nortonize the circuit between points X and Y. Calculate the value of i_N.
68. In Fig. 17-29, transform the delta network between points X, Y, and Z into its wye equivalent. Using the equivalent circuit, calculate the value of the total impedance presented to the source.
69. In Fig. 17-29, transform the wye network between points W, X, and Z into its delta equivalent. Using the equivalent circuit, calculate the value of the total impedance presented to the source.
70. Determine the value of the load current in the circuit of Fig. 17-29.
71. Find the current I_1 for the circuit of Fig. 17-30.

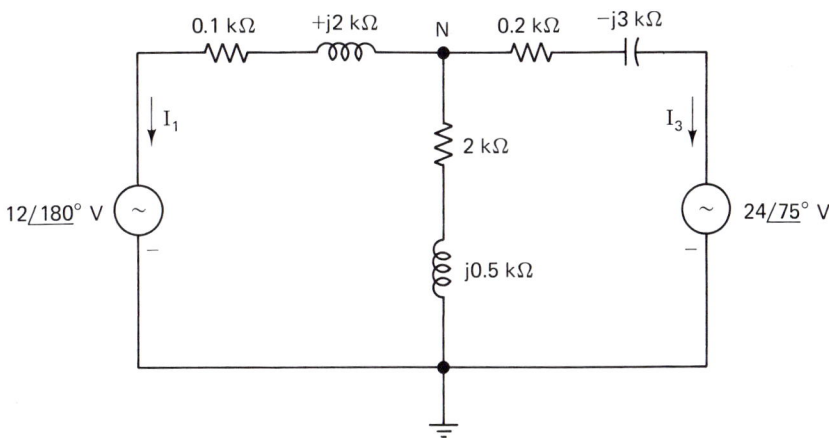

FIGURE 17-30 Circuit for Problems 71, 72, and 73.

72. Find current I_3 for the circuit of Fig. 17-30.
73. Using any convenient theorem, calculate the voltage at node N in Fig. 17-30.

74. Calculate I_1 and I_2 in Fig. 17-31 using the superposition theorem.

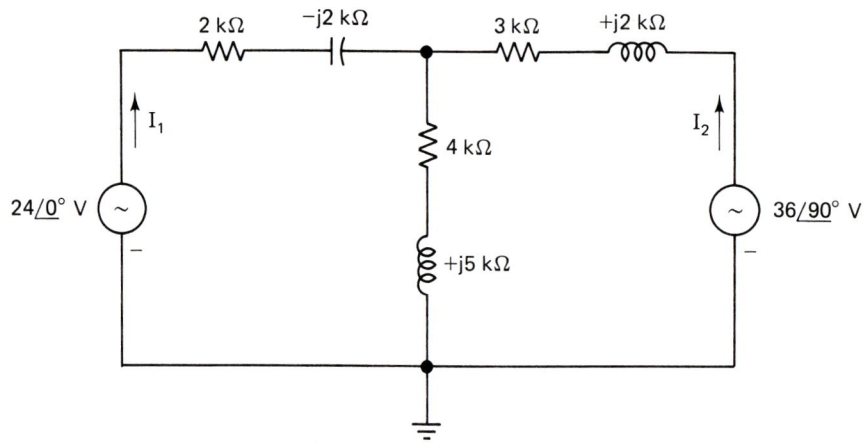

FIGURE 17-31 Circuit for Problems 74 and 75.

75. Verify the results of Problem 74 by choosing any convenient method.
76. The bridge of Fig. 17-32 has $Z_1 = 200 + j0\ \Omega$, $Z_2 = 100 + j0\ \Omega$, $Z_3 = 20 + j0\ \Omega$, and Z_4 consists of 10 Ω of resistance in series with 200 pF of capacitance and 126.7 μH of inductance. Assuming that the galvanometer G has a resistance of 50 Ω, calculate the current through G for a frequency of 1010 kHz.

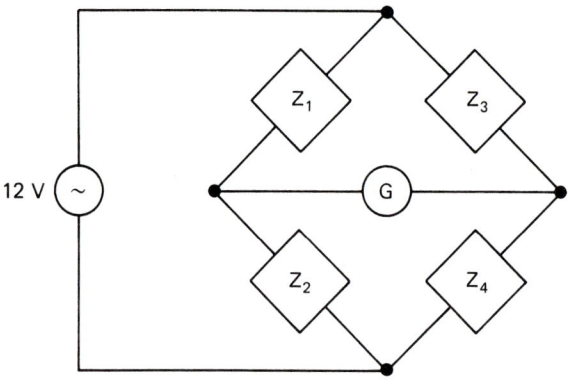

FIGURE 17-32 Bridge for Problem 76.

77. In the bridged-T network of Fig. 17-33, $Z_L = 5 + j5\ \text{k}\Omega$. Using the Thévenin theorem, determine the load current.

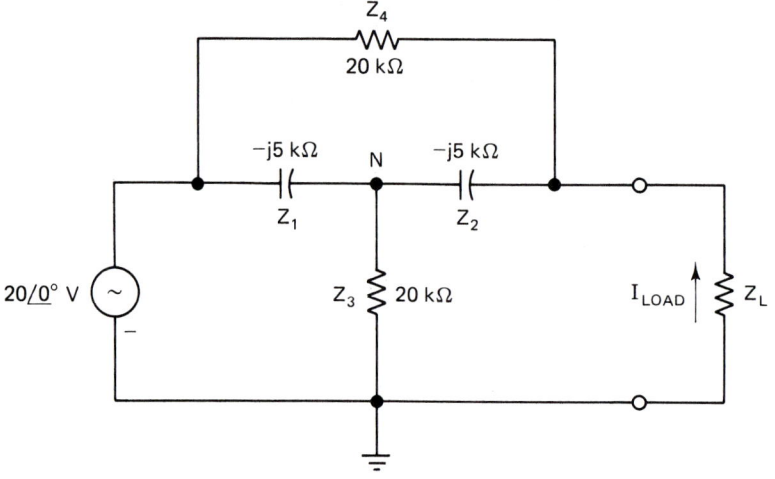

FIGURE 17-33 Bridged-T network for Problems 77 and 78.

78. In Fig. 17-33, the load Z_L is removed and replaced by another voltage source of $20\,\underline{/180°}$ V. Find the voltage at node N for this condition.
79. Determine Z_A, Z_B, and Z_C for the equivalent wye configuration of the delta configuration shown in Fig. 17-34(a).
80. Repeat Problem 79 for the unfortunate condition of the capacitor in branch BC being shorted.

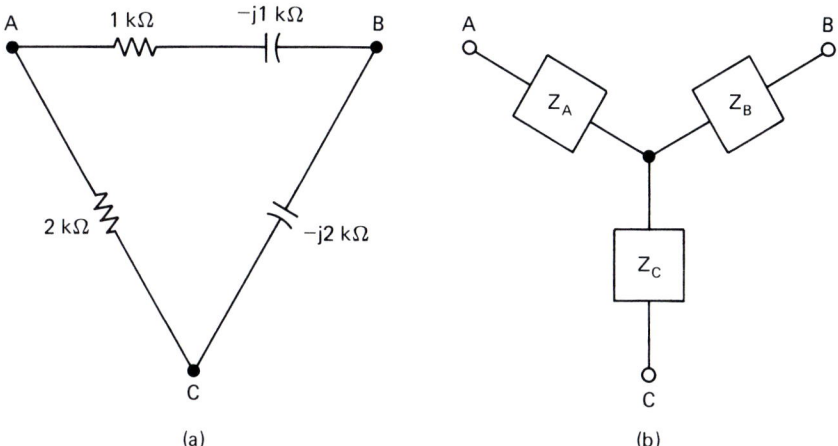

FIGURE 17-34 Delta–wye transformation for Problems 79 and 80.

18

Polyphase Systems

This chapter has the following objectives:

1. To state and demonstrate formulas for the generated voltages in terms of the rotational speed and number of pole pairs.
2. To provide practical descriptions of alternators constructed to produce two (two-phase) or three (three-phase) generated ac voltages having specific relative phase angles.
3. To describe the more significant design features of practical alternators, including the stator function and construction, rotor field function, and the generated voltages as affected by chord factor and angular displacement on the stator frame.
4. To analyze the electrical features of two-pole power systems, including line and phase voltages, currents, and power.
5. To analyze the electrical features of three-phase power systems, including line and phase voltages, currents, and power.
6. To make comparisons of the relative advantages of single-phase, two-phase, and three-phase power systems.
7. To analyze the line and phase relationships for the voltages and currents of the delta connection for three-phase alternators and loads.
8. To analyze the line and phase relationships for the voltages and currents of the wye connection for three-phase alternators and loads with consideration of the neutral conductor currents.
9. To determine the operating principles of the three-phase transformer and single-phase transformers connected for three-phase operation.
10. To determine transformer connections for the production of six-phase and twelve-phase voltages.

18-0
Introduction

In Chapters 11 through 17 we have regarded every alternating source as producing one sine wave of voltage or current; this is referred to as single-phase operation. By contrast an alternator using polyphase operation produces two or more sine-wave voltages which bear certain phase relationships to each other. For example, in the commonly used three-phase system, there are three equal alternating voltages which are separated in phase from each other by 120°.

A simple picture of a single-phase alternator contains several turns of insulated copper wire which are wound on a laminated soft-iron armature. By supplying mechanical energy to the armature shaft, the single coil is rotated to cut the flux of a stationary magnetic field (Fig. 18-1). The sine-wave voltage output from the coil is then taken through slip rings and stationary carbon brushes to the load. If the direct magnetic field is created by a single pair of poles, one complete rotation of the coil will generate one cycle of the alternating voltage. However, if four magnetic poles are involved, one complete rotation of the armature produces two cycles of the sinusoidal output. It follows that

$$\text{generated frequency}, f = \frac{Np}{60} \quad \text{hertz} \quad \text{(18-0-1)}$$

where N = armature speed (revolutions per minute or rpm)
p = number of *pairs* of poles

Consequently, if the armature of a four-pole alternator revolves at 1800 rpm, the generated frequency is

$$\frac{1800 \times (4/2)}{60} = 60 \text{ Hz}$$

As described, the simple revolving coil arrangement has a number of disadvantages:

1. The output voltage is developed between sliding contacts which are subject to sparking and to wear caused by friction.
2. The contacts are exposed and are therefore liable to arc-over at high voltages.
3. As the coil rotates at high speed, its many turns are subjected to centrifugal forces, which may cause a breakdown in the coil's insulation; as a result, there is a further danger of arc-over.

These difficulties are overcome by using a revolving field alternator. In this arrangement the conductors are stationary but the magnetic field is rotating.

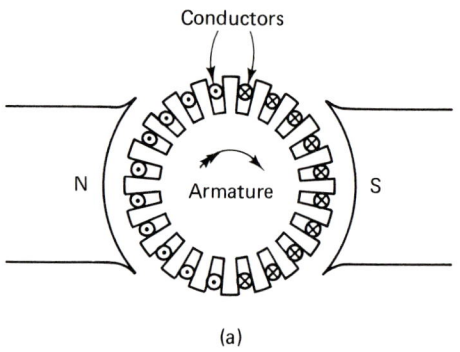

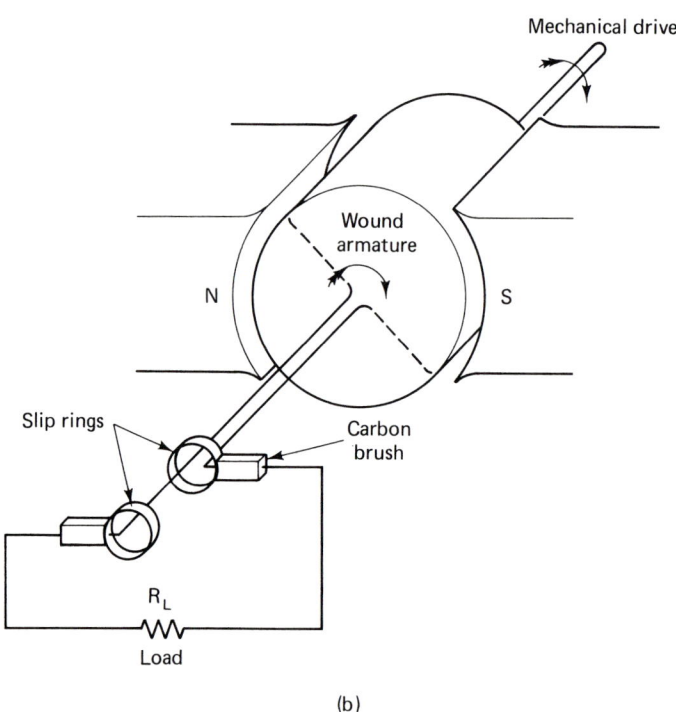

FIGURE 18-1 Single-phase alternator with a revolving armature.

Here you will recall that either the conductor or the field may be moving in the application of Faraday's law; it is only necessary that there be relative motion between the conductor and the magnetic flux. A discussion of the revolving field single-phase and three-phase alternators is covered in this chapter.

18-1

Revolving Field Single-Phase Alternator

In the revolving field alternator, a low-voltage dc source drives a direct current through the slip rings, brushes, and the winding on a soft-iron rotor that is driven around by mechanical means. This creates a rotating magnetic field which cuts the conductors embedded in the surrounding stator [Fig. 18-2(a)]. An alternating voltage then appears between the ends, S (start) and F (finish), of the stator winding. Since S and F are fixed terminals, there are no sliding

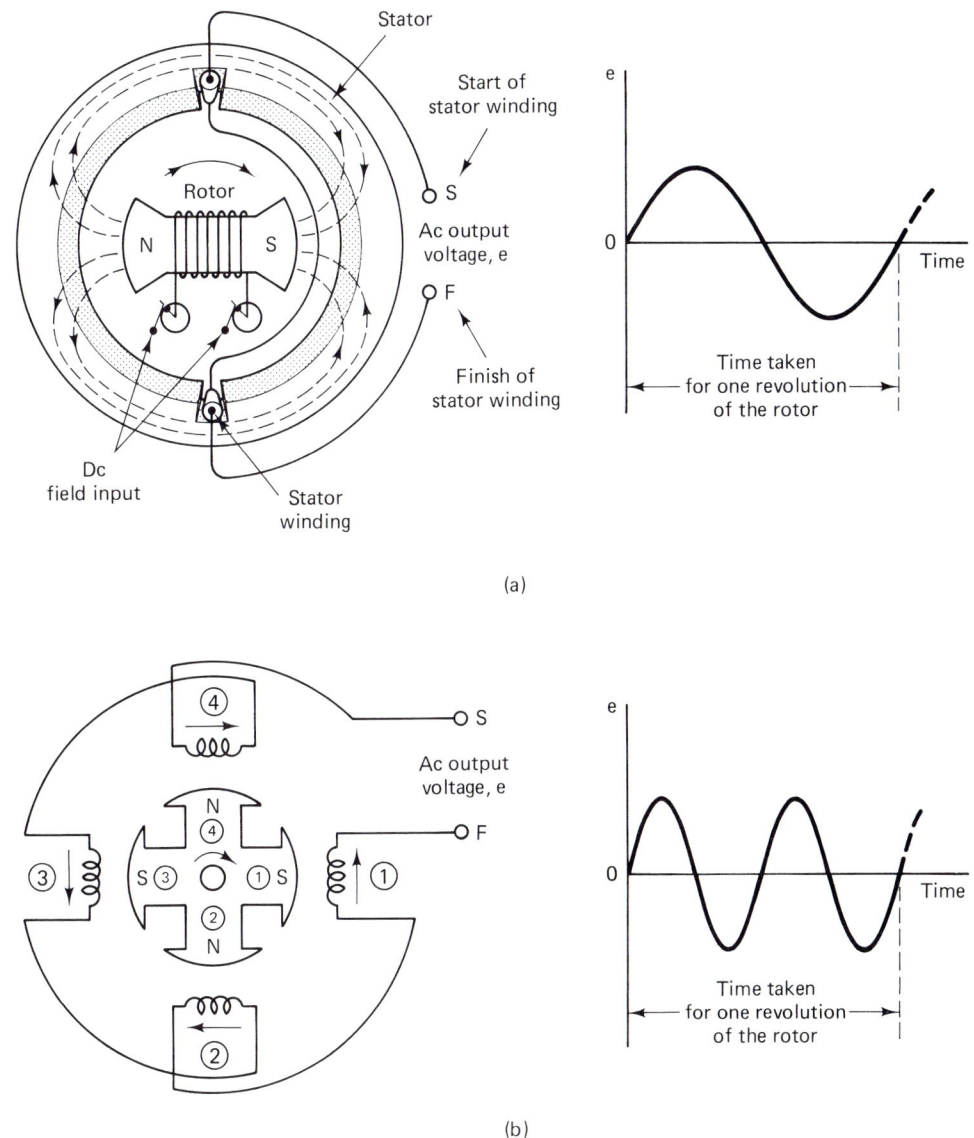

FIGURE 18-2 Principle of the revolving field alternator.

contacts and the whole of the stator winding may be continuously insulated. Moreover, the rotor winding has relatively few turns of thick copper wire, so that the effect of the centrifugal forces is considerably reduced.

There are three ways by which we can obtain a greater rms value for the output voltage:

1. By increasing the rotor speed, but this would also have the effect of changing the frequency.
2. By raising the current in the rotor winding to increase the magnetic density. There is a limit to this solution since the soft iron of the rotor will eventually saturate.
3. By winding a greater portion of the stator so that more conductors are cut by the revolving magnetic field.

If the stator is fully wound [Fig. 18-3(a)], we must consider the fact that the magnetic field is not cutting all conductors at the same time. For example,

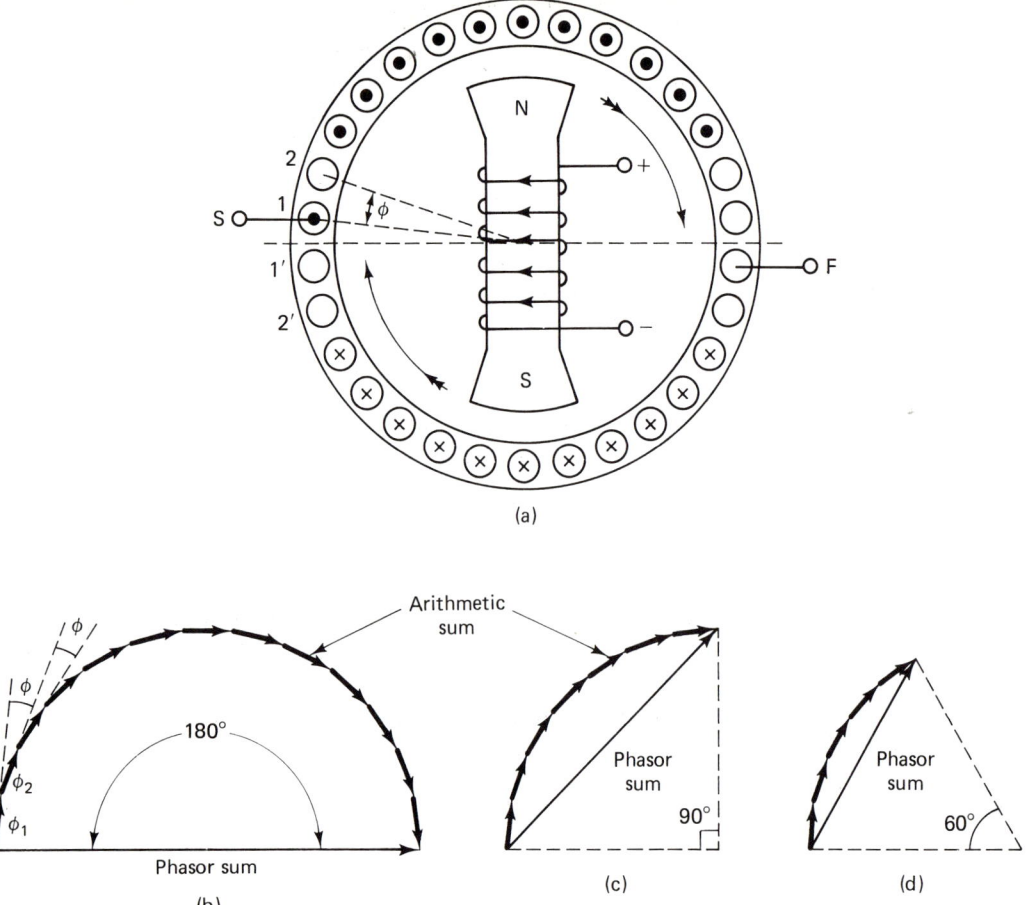

FIGURE 18-3 Chord distribution factor.

there is an angular separation of ϕ degrees between conductor 1 and conductor 2. If we assume that the magnetic field is revolving in the clockwise direction, the phasor voltage e_1 (induced in conductors 1 and 1′) will lead the phasor voltage e_2 (induced in conductors 2 and 2′) by the angle, ϕ. Consequently, the total effective output voltage will be the phasor (*not* the arithmetic) sum of the individual voltages.

If we assume that there are a large number of conductors embedded in the fully wound stator, the voltage phasor diagram resembles a semicircle [Fig. 18-3(b)]. Then the perimeter of the semicircle approximately equals the arithmetic sum of the individual voltages, while the diameter is their phasor sum. The relationship between the arithmetic and phasor sums is measured by the

$$\text{chord distribution factor (CDF)} = \frac{\text{phasor sum}}{\text{arithmetic sum}}$$
$$= \frac{\text{diameter}}{\pi \times \text{diameter}/2}$$
$$= 0.637$$

It is clear that the higher the value of the chord distribution factor, the more effectively is the stator being wound. For example, if only half the stator is wound [Fig. 18-3(c)], the value of the CDF rises to

$$\frac{\sqrt{2} \times \text{radius}}{(\pi/2) \times \text{radius}} = 0.907$$

In a three-phase alternator each winding occupies only one-third of the stator [Fig. 18-3(d)], for which the chord distribution factor is

$$\frac{\text{radius}}{(\pi/3) \times \text{radius}} = 0.935$$

For single-phase operation we have a further problem. We have already seen in Section 11-3 that the *instantaneous* power output fluctuates at twice the generated frequency. Consequently, the load on the mechanical source of energy, and therefore the necessary torque, will not be constant. The result will be some degree of vibration and wear on the bearings in which the alternator's shaft rotates.

To summarize, the revolving field alternator is superior to the rotating armature type as far as sparking and insulation problems are concerned. However, the single-phase alternator still suffers from a low chord distribution factor and an uneven torque requirement from the mechanical source of energy.

EXAMPLE 18-1 In a revolving field alternator a six-pole rotor is turning at 1000 rpm. What is the value of the generated frequency?

Solution

$$\text{generated frequency, } f = \frac{N \times p}{60} = \frac{1000 \times (6/2)}{60} = 50 \text{ Hz} \qquad (18\text{-}0\text{-}1)$$

18-2

Two-Phase Alternator

In this type of revolving field alternator, two identical coils are mounted on the stator with their axes separated by 90°. Figure 18-4(a) shows a simplified arrangement of the windings which are cut by the magnetic flux of a two-pole rotor. The EMFs e_1 and e_2 induced in the windings will therefore be equal in magnitude but 90° out of phase [Fig. 18-4(b)]. For each winding the chord distribution factor is 0.907.

The fact that e_2 leads e_1 by 90° is a result of the assumed clockwise movement of the rotor. This phase relationship is commonly represented by drawing the two stator windings 90° apart [Fig. 18-4(c)]. Each of the windings may be connected to separate loads, but it is more convenient to use a neutral line so that the number of lines required is reduced from four to three; this is illustrated in Fig. 18-4(d). Assuming identical (balanced) resistive loads across the windings, the total instantaneous power of the two-phase alternator is constant as opposed to the fluctuating power output of the single-phase machine. Note that this advantage is achieved only when the loads are balanced.

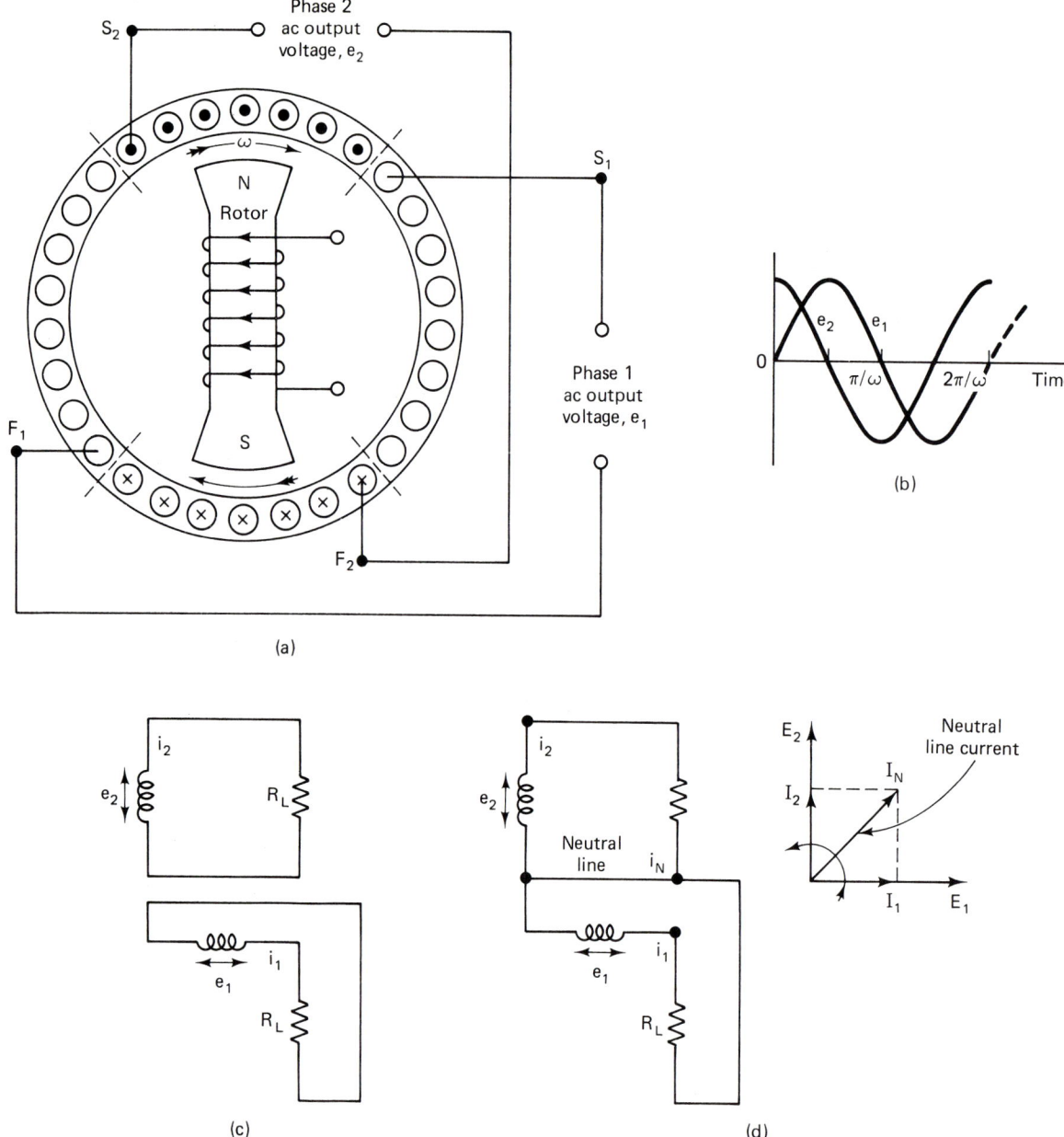

FIGURE 18-4 Two-phase alternator.

If the effective current of each load is I_{RMS}, the neutral line current is $1.414 I_{rms}$. The total load current in the three lines is therefore $(2 + \sqrt{2})I_{rms} = 3.414 I_{rms}$. Assuming that a neutral line is used, the two-phase alternator requires less copper than the comparable single-phase generator when supplying a given voltage to a particular value of total load resistance.

Compared with a single-phase alternator, a high-power two-phase generator requires a smaller stator and is more efficient. It is also less subject to vibration and wear on the bearings because the torque required from the mechanical source is constant rather than fluctuating at twice the generated frequency.

In Fig. 18-4(b),

$$\text{phase voltage, } e_1 = E_{max} \sin \omega t \text{ volts} \qquad (18\text{-}2\text{-}1)$$

$$\text{phase voltage, } e_2 = E_{max} \sin\left(\omega t + \frac{\pi}{2}\right)$$

$$= E_{max} \cos \omega t \text{ volts} \qquad (18\text{-}2\text{-}2)$$

where the angular frequency, $\omega = 2\pi f \text{(rad/s)}$

E_{max} = peak value of the phase voltage (V)

e_1, e_2 = instantaneous phase voltages (V)

Equations (18-2-1) and (18-2-2) show that e_2 leads e_1 by $\pi/2$ radians or 90°. Assuming balanced resistive loads, we have

$$\text{instantaneous power, } p_1 = \frac{e_1^2}{R_L} = \frac{E_{max}^2 \sin^2 \omega t}{R_L} \text{ watts} \qquad (18\text{-}2\text{-}3)$$

$$\text{instantaneous power, } p_2 = \frac{e_2^2}{R_L} = \frac{E_{max}^2 \cos^2 \omega t}{R_L} \text{ watts} \qquad (18\text{-}2\text{-}4)$$

$$\text{total instantaneous power, } p_1 + p_2 = \frac{E_{max}^2}{R_L} \text{ watts} \qquad (18\text{-}2\text{-}5)$$

since $\sin^2 \omega t + \cos^2 \omega t = 1$. The total instantaneous power is independent of the time and is therefore a constant.

In Fig. 18-4(d), assuming balanced resistive loads,

effective value of the neutral line current,

$$I_N = \sqrt{I_{rms}^2 + I_{rms}^2} \qquad (18\text{-}2\text{-}6)$$

$$= \sqrt{2}\, I_{rms} \quad \text{amperes}$$

where I_{rms} is the effective load current for each phase. As shown in Example 18-2, the amount of copper required for a two-phase transmission line is less than the amount needed for single-phase operation (assuming that the same amount of power is to be transmitted).

Although two-phase operation has some advantages with respect to single phase, it has virtually no advantages when compared with three phase; consequently, a two-phase system is comparatively rare.

EXAMPLE 18-2 A two-phase alternator has a four-pole rotor whose speed is 1800 rpm. What is the value of the generated frequency? The effective voltage of each phase is 240 V and the two resistive loads are balanced so that each has an equivalent resistance of 80 Ω. Calculate the current in the neutral line and the instantaneous power output of the alternator.

Solution

$$\text{generated frequency} = \frac{1800 \times 2}{60} = 60 \text{ Hz} \qquad (18\text{-}0\text{-}1)$$

$$\text{load current for each phase, } I_{\text{rms}} = \frac{240 \text{ V}}{80 \text{ }\Omega} = 3 \text{ A rms}$$

$$\text{neutral-line current} = \sqrt{2}\, I_{\text{rms}} \qquad (18\text{-}2\text{-}6)$$

$$= \sqrt{2} \times 3 = 4.242 \text{ A rms}$$

$$\text{total instantaneous power} = \frac{E_{\text{max}}^2}{R_L} \qquad (18\text{-}2\text{-}5)$$

$$= \frac{(\sqrt{2} \times 240 \text{ V})^2}{80}$$

$$= 1440 \text{ W}$$

The average power delivered over the cycle is also 1440 W. *Note:* For the corresponding single-phase alternator, the total load resistance is 80 Ω/2 = 40 Ω. The maximum instantaneous power is

$$\frac{(\sqrt{2} \times 240 \text{ V})^2}{40} = 2880 \text{ W}$$

so that the average power over the cycle is 2880 W/2 = 1440 W. The total of the currents in the two supply lines is

$$2 \times \frac{240 \text{ V}}{40 \text{ }\Omega} = 12 \text{ A rms.}$$

When we compare in this example the single-phase and two-phase systems for the same amount of total power delivered to an arrangement of balanced loads:

1. The single-phase system requires two lines each of which carries a current of 6 A.
2. The two-phase system needs three lines. Of these two must each carry 3 A while the third, neutral line has a return current of 4.24 A.

By consulting wire gage tables and considering the single-phase system as the basis of comparison, the two-phase system has a percentage saving of about 25% in the volume of copper required for the conducting lines. The value of the percentage depends to some degree on the type of insulation surrounding the individual lines.

18-3
Three-Phase Alternator

The three-phase alternator of Fig. 18-5(a) has three single-phase windings which are equally spaced on the stator so that the voltage induced in each winding is 120° ($2\pi/3$ radians) out of phase with the voltages induced in the other two windings [Fig. 18-5(b)]. These windings, each of which has a chord distribution factor of 0.935, are independent of each other and can be connected to three separate loads. Such connections would require a six-line system [Fig. 18-5(c)].

Assume that the three loads are balanced in the sense that they all have the same ohms value and the same phase angle. It then follows that they

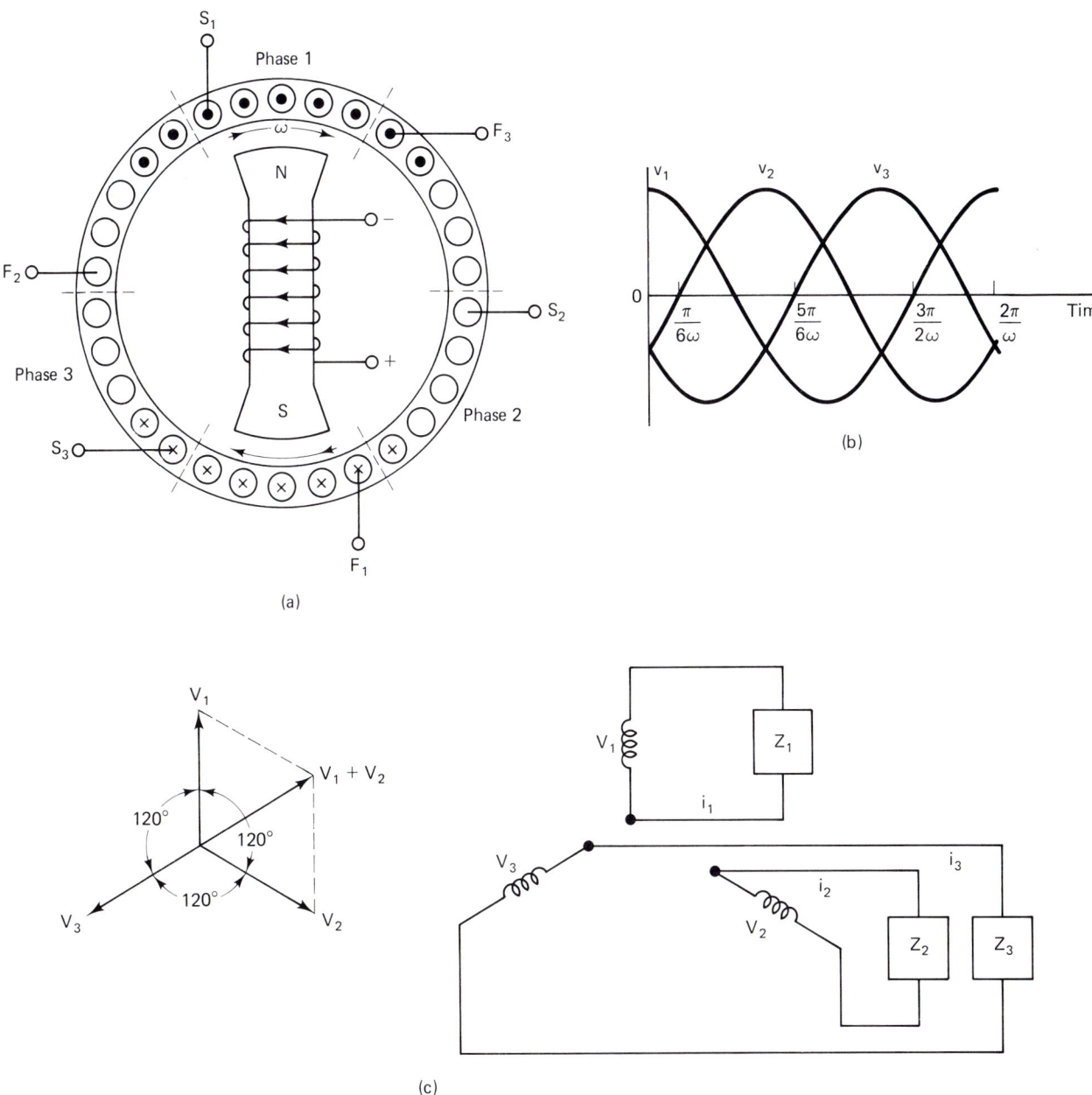

FIGURE 18-5 Three-phase alternator.

would all have identical power factors, which in our example are assumed to be lagging. The three load currents will also be equal in magnitude, but 120° out of phase with each other so that their phasor sum is zero [Fig. 18-6(a)]. It is then possible to join the corresponding ends of the three windings to a common point and to connect the other ends to separate terminals. A single neutral line is then attached to the common point to produce a four-wire wye (Y) system [Fig. 18-6(b)]. With this arrangement the individual alternator phase voltages are each applied to one of the three balanced load impedances.

Apart from the advantage of reducing the number of wires from six to four, the theoretical current flowing in the neutral line is zero. In a practical system the thin neutral line requires only a low current capacity since any current that it carries exists only as a result of an imbalance between the loads. Alternatively, the neutral line may be replaced by a ground return. The saving in the copper required by the supply lines is therefore greater for a balanced three-phase system than for a balanced two-phase system.

We will show that the total instantaneous power for the balanced three-phase system is constant and therefore independent of the time. Consequently, there is no change required in the instantaneous torque provided by the mechanical source of energy.

In Fig. 18-5(b), the phase voltages v_1, v_2, and v_3 are shown equally separated by $2\pi/3$ radians. If V_1 is used as the reference phasor and the rotor is assumed to be revolving in the clockwise direction,

$$\text{phase 1 voltage, } v_1 = E_{max} \sin \omega t \quad \text{volts} \tag{18-3-1}$$

$$\text{phase 2 voltage, } v_2 = E_{max} \sin\left(\omega t - \frac{2\pi}{3}\right)$$

$$= -\frac{E_{max} \sin \omega t}{2} - \frac{\sqrt{3}}{2} E_{max} \cos \omega t \quad \text{volts} \tag{18-3-2}$$

$$\text{phase 3 voltage, } v_3 = E_{max} \sin\left(\omega t - \frac{4\pi}{3}\right)$$

$$= -\frac{E_{max} \sin \omega t}{2} + \frac{\sqrt{3}}{2} E_{max} \cos \omega t \quad \text{volts} \tag{18-3-3}$$

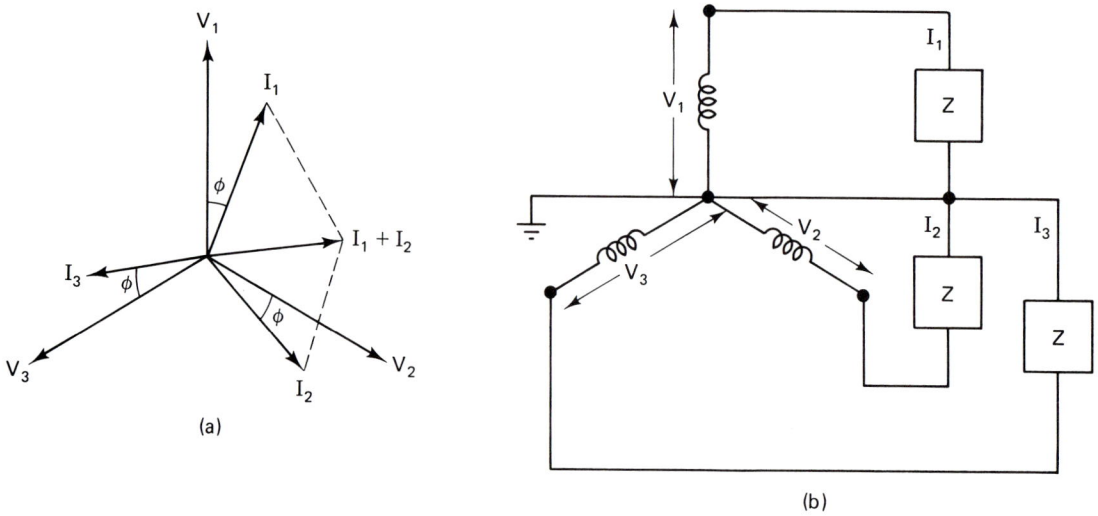

FIGURE 18-6 Four-wire system with balanced loads.

where E_{max} = peak value of the phase voltage (V)
ω = angular frequency (rad/s)

By adding Eqs. (18-3-1), (18-3-2), and (18-3-3), we obtain

$$v_1 + v_2 + v_3 = 0 \qquad (18\text{-}3\text{-}4)$$

For three balanced loads with the same lagging power factor, each load current will lag its corresponding phase voltage by the same phase angle, ϕ [Fig. 18-6(a)]. It follows that

$$i_1 + i_2 + i_3 = 0 \qquad (18\text{-}3\text{-}5)$$

If each of the balanced loads is equal to a resistance R_L, the instantaneous powers delivered by the three phases are:

$$\text{instantaneous power, } p_1 = \frac{E_{max}^2}{R_L}\sin^2\omega t \quad \text{watts} \qquad (18\text{-}3\text{-}6)$$

$$\text{instantaneous power, } p_2 = \frac{E_{max}^2}{R_L}\sin^2\left(\omega t - \frac{2\pi}{3}\right)$$

$$= \frac{E_{max}^2}{R_L}\left(\frac{-\sin\omega t - \sqrt{3}\cos\omega t}{2}\right)^2 \quad \text{watts} \qquad (18\text{-}3\text{-}7)$$

$$\text{instantaneous power, } p_3 = \frac{E_{max}^2}{R_L}\sin^2\left(\omega t + \frac{2\pi}{3}\right)$$

$$= \frac{E_{max}^2}{R_L}\left(\frac{-\sin\omega t + \sqrt{3}\cos\omega t}{2}\right)^2 \quad \text{watts} \qquad (18\text{-}3\text{-}8)$$

$$\text{total instantaneous power, } P_T = p_1 + p_2 + p_3$$

$$= \frac{E_{max}^2}{R_L}\left(\sin^2\omega t + \frac{\sin^2\omega t}{4} + \frac{\sin^2\omega t}{4} + \frac{3}{4}\cos^2\omega t + \frac{3}{4}\cos^2\omega t\right)$$

$$= \frac{3E_{max}^2}{2R_L} = \frac{3E_{rms}^2}{R_L} \quad \text{watts} \qquad (18\text{-}3\text{-}9)$$

The total instantaneous power, P_T, for a balanced three-phase system is therefore constant and is independent of the time. Just as in the two-phase system, there is no change required in the instantaneous torque provided by the mechanical source of energy.

In high-power communications transmitters, a three-phase system has a considerable advantage over single-phase operation when large amounts of dc power are required. Rectified three-phase voltages have less ripple than a rectified single-phase voltage, and furthermore, the three-phase ripple frequency is higher, so that the low-pass smoothing filters can be designed with smaller and cheaper components.

Sec. 18-3 Three-Phase Alternator

EXAMPLE 18-3 A wye-connected three-phase alternator has a four-pole rotor whose speed is 1500 rpm. What is the value of the generated frequency? The effective voltage of each phase is 120 V and the resistive loads, also connected in the wye configuration, are balanced so that each has an equivalent resistance of 60 Ω. Calculate the total of the line and neutral currents and determine the total instantaneous power output of the alternator.

Solution

$$\text{generated frequency, } f = \frac{1500 \times 2}{60} = 50 \text{ Hz} \quad (18\text{-}0\text{-}1)$$

$$\text{load current for each phase} = \frac{120 \text{ V}}{60 \text{ }\Omega} = 2 \text{ A}$$

Neutral line current is *zero* for a balanced load system.

$$\text{total of the load and neutral currents} = 3 \times 2 \text{ A} = 6 \text{ A}$$

$$\text{total instantaneous power} = \text{average power over the cycle}$$

$$= \frac{3 E_{\text{rms}}^2}{R} \quad (18\text{-}3\text{-}9)$$

$$= 3 \times \frac{(120 \text{ V})^2}{60 \text{ }\Omega} = 720 \text{ W}$$

Note: For the equivalent two-phase alternator the load on each phase is $2 \times 60 \text{ }\Omega/3 = 40 \text{ }\Omega$. The current for each phase is $120 \text{ V}/40 \text{ }\Omega = 3 \text{ A}$ and the neutral line current is $3 \times \sqrt{2} = 4.242 \text{ A}$.

$$\text{total instantaneous power} = \text{average power over the cycle}$$

$$= 2 \times \frac{(120 \text{ V})^2}{40 \text{ }\Omega} = 720 \text{ W}$$

With the comparable single-phase alternator the total load resistance would be $60 \text{ }\Omega/3 = 20 \text{ }\Omega$. The current in each of the supply lines is $120 \text{ V}/20 \text{ }\Omega = 6 \text{ A}$ and the average power over the cycle is $(120 \text{ V})^2/20 \text{ }\Omega = 720 \text{ W}$.

In this example the single-phase, two-phase, and three-phase systems are compared for the same total load power. The following results are obtained:

1. The single-phase system requires two lines, each of which carries 6 A.
2. The two-phase system requires three lines. Two lines each carry a current of 3 A. The third, neutral line has a return current of 4.24 A.
3. The balanced three-phase system requires four lines. Three lines each conduct a current of 2 A. The fourth, neutral line carries zero current, so long as the load is perfectly balanced. In practice the neutral current is usually very small, as compared with the line currents, so that a smaller conductor is adequate.

By consulting wire gage tables and considering the single-phase system as the basis of comparison, the two-phase system has a percentage saving of about 25% and the three-phase system has a percentage saving of about 50% in the volume of copper required for the conducting lines. The value of the percentage depends to some degree on the type of insulation surrounding the individual lines.

18-4 Wye Connection

The alternator EMF applied across one of the three balanced loads in Section 18-3 was the voltage that existed between ground (or the neutral line) and the line connected to the load. This EMF is normally called the phase voltage, V_p. However, with three-phase systems, it is normal practice to make use of the line voltage, V_L, as well as the phase voltage. As illustrated in Fig. 18-7(a), the line voltage is the alternating EMF that exists between points X and Y and is related to the voltages generated in the first and second phases of the stator windings.

To determine the relationship between V_L and V_p, it is important to realize that the correct stator terminals must be connected to the common point if the phase voltages are to be represented as 120° ($2\pi/3$ radians) apart; for example, if the two terminals of one phase winding are reversed, its phase voltage is shifted by 180°. The phase voltages V_x, V_y, and V_z are the alternating (phasor) voltages monitored at points X, Y, and Z, with respect to the common point O. Consequently, the (phasor) line voltage, V_{xy}, is the alternating voltage at point X with respect to point Y and is therefore the voltage difference between the two points [Fig. 18-7(b)]. The line voltages are

$$V_{xy} = V_x - V_y$$
$$V_{yz} = V_y - V_z$$
$$V_{zx} = V_z - V_x$$

By adding together these three equations, it follows that the phasor sum of the line voltages is zero. Mathematically, we shall also show that the phasors of the three line voltages are 120° apart and are shifted by 30° from the phase voltages.

When three balanced loads are connected between the lines, the phasors of the three line currents are also equally spaced by 120°. Since each line is

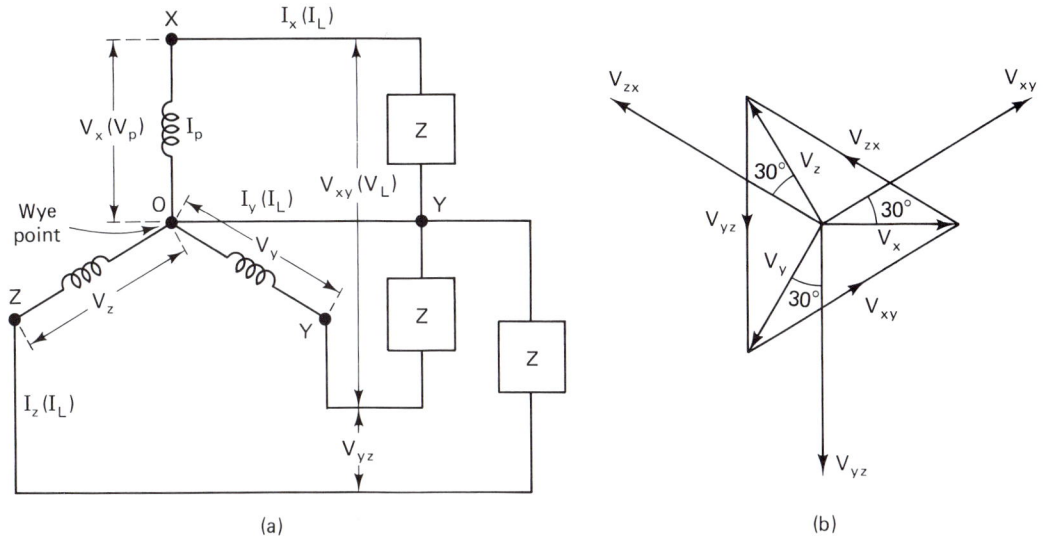

FIGURE 18-7 Wye connection.

directly connected to a terminal of a phase winding, the line current, I_L, must be equal to the phase current, I_p.

In Fig. 18-7(b), let V_x be the reference phase voltage, so that

$$\text{phase voltage } V_x = E\angle 0° \text{ V} \quad (18\text{-}4\text{-}1)$$

Therefore,

$$\text{phase voltage, } V_y = E\angle -2\pi/3 \text{ V} \quad (18\text{-}4\text{-}2)$$

$$\text{phase voltage, } V_z = E\angle +2\pi/3 \text{ V} \quad (18\text{-}4\text{-}3)$$

where E is the effective value of the phasor voltage. Then

$$\begin{aligned}
\text{line voltage, } V_{xy} &= V_x - V_y \\
&= E\angle 0 - E\angle -2\pi/3 \\
&= E - E\left[\left(-\frac{1}{2}\right) + j\left(-\frac{\sqrt{3}}{2}\right)\right] \quad (18\text{-}4\text{-}4) \\
&= E\left[\frac{3}{2} + j\left(\frac{\sqrt{3}}{2}\right)\right] \\
&= \sqrt{3}E\angle \pi/6 = \sqrt{3}\,E\angle 30° \text{ V}
\end{aligned}$$

This means that the line voltage, V_{xy}, has a magnitude of $\sqrt{3}\,E$ and leads V_x by 30°.

$$\begin{aligned}
\text{line voltage, } V_{yz} &= V_y - V_z \\
&= E\angle -2\pi/3 - E\angle +2\pi/3 \\
&= E\left[\left(-\frac{1}{2}\right) + j\left(-\frac{\sqrt{3}}{2}\right)\right] - E\left[\left(-\frac{1}{2}\right) + j\left(\frac{\sqrt{3}}{2}\right)\right] \quad (18\text{-}4\text{-}5) \\
&= \sqrt{3}\,E\angle -\pi/2 = \sqrt{3}\,E\angle -90° \text{ V}
\end{aligned}$$

The line voltage, V_{yz}, has a magnitude of $\sqrt{3}\,E$ volts and leads V_y by 30°.

$$\begin{aligned}
\text{line voltage, } V_{zx} &= V_z - V_x \\
&= E\angle +2\pi/3 - E\angle 0° \\
&= E\left[\left(-\frac{1}{2}\right) + j\left(+\frac{\sqrt{3}}{2}\right)\right] - E \quad (18\text{-}4\text{-}6) \\
&= E\left[\left(-\frac{3}{2}\right) + j\left(+\frac{\sqrt{3}}{2}\right)\right] \\
&= \sqrt{3}\,E\angle 5\pi/6 = \sqrt{3}\,E\angle 150° \text{ V}
\end{aligned}$$

The line voltage, V_{zx}, has a magnitude of $\sqrt{3}\ E$ volts and leads V_z by 30°. Equations (18-4-4), (18-4-5), and (18-4-6) all show that the line voltage is $\sqrt{3}\ \times$ the phase voltage or 1.732 × the phase voltage. In addition, the line phasors are 120° apart and each line voltage phasor is shifted by 30° from one of the phase voltages.

EXAMPLE 18-4 In a three-phase, wye-connected system the reference phase voltage is $120\angle 50°$ V, 60 Hz. Balanced loads, each equal to $15\angle 20°\ \Omega$, are connected between the supply lines and the common neutral point. Express the line voltages and the line currents in their polar forms.

Solution The phase voltages are $120\angle 50°$ V, $120\angle 50° - 120° = 120\angle -70°$ V, and $120\angle 50° + 120° = 120\angle +170°$ V.

The line voltages are

$$V_{xy} = \sqrt{3}\ E\angle 50° + 30° \qquad (18\text{-}4\text{-}4)$$
$$= 1.732 \times 120\angle 80°$$
$$= 209\angle 80°\ \text{V}$$

$$V_{yz} = \sqrt{3}\ E\angle -70° + 30° \qquad (18\text{-}4\text{-}5)$$
$$= 1.732 \times 120\angle -40°$$
$$= 209\angle -40°\ \text{V}$$

$$V_{zx} = \sqrt{3}\ E\angle +170° + 30° \qquad (18\text{-}4\text{-}6)$$
$$= 1.732 \times 120\angle 200°$$
$$= 209\angle 200°\ \text{V}$$

The phase currents are equal to the line currents, which are

$$\text{phase current, } I_x = \frac{120\angle 50°\ \text{V}}{15\angle 20°\ \Omega}$$
$$= 8\angle 30°\ \text{A}$$

$$\text{phase current, } I_y = \frac{120\angle -70°\ \text{V}}{15\angle 20°\ \Omega}$$
$$= 8\angle -90°\ \text{A}$$

$$\text{phase current, } I_z = \frac{120\angle 170°\ \text{V}}{15\angle 20°\ \Omega}$$
$$= 8\angle 150°\ \text{A}$$

The three-phase currents are also equally separated by 120°.

18-5 Delta Connection

Since the sum of the three instantaneous phase voltages is at all times zero, it is possible to connect the phase windings in a delta (Δ) formation. This is illustrated in Fig. 18-8(a) and provided that the correct connections are made, there will be zero circulating current in the delta loop. Three supply lines may then be connected to the corners of the delta formation, and these are joined to three loads, also arranged in delta. It is clear that the voltage applied across any one of the loads must be the same as the voltage generated in the corresponding phase winding across which that particular load is connected. Therefore, in the three-phase delta system, the line voltage, V_L, is equal to the phase voltage, V_p.

By contrast with the voltage relationship, a particular line current phasor, I_x, is associated with two phase currents, I_{yx} and I_{xz}. Each of the load currents will then be equal to its corresponding phase current. However, each line current will be equal to the phasor difference between two of the load currents, so that the related equations are

$$I_x = I_{xz} - I_{yx} \qquad (18\text{-}5\text{-}1)$$

$$I_y = I_{yx} - I_{zy} \qquad (18\text{-}5\text{-}2)$$

$$I_z = I_{zy} - I_{xz} \qquad (18\text{-}5\text{-}3)$$

The sum of the three line currents is therefore zero; this is true irrespective of whether the loads are balanced or not. However, if the loads are balanced, the currents in the three windings are separated from each other by a phase difference of 120°. The relationship between the line and phase currents of the delta system are then comparable with the equations relating the line and phase voltages in the wye arrangement (Section 18-4). Therefore, in the delta

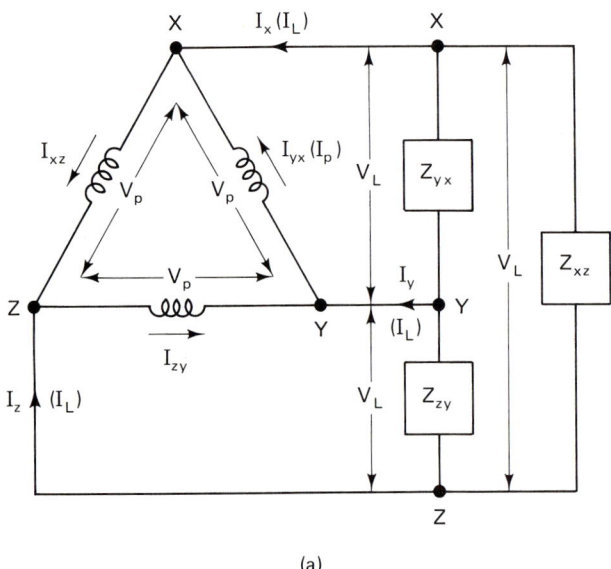
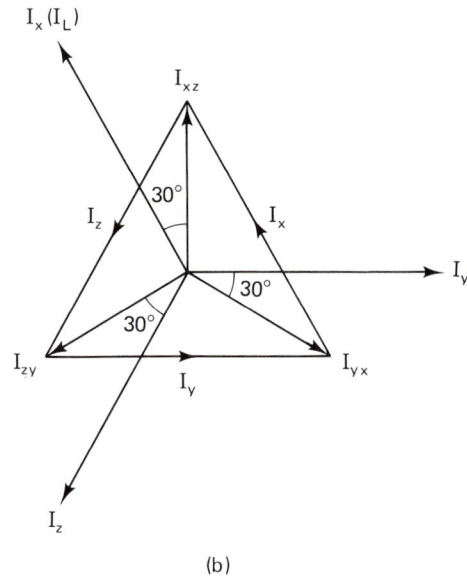

FIGURE 18-8 Delta connection.

system, the line current, I_L, is equal to $\sqrt{3} \times$ the phase current, I_p (or 1.732 $\times I_p$). In addition, each current is shifted by 30° from one of the phase currents [Fig. 18-8(b)].

The total power output from the alternator can be measured in terms of the line voltage and current values together with the power factor for each of the balanced loads. The expression for the total power is, in fact, the same for both wye and delta connections.

Three-Phase Power

In a balanced wye or delta system the power in each load is the same and therefore the total power is

$$\text{total power, } P_T = 3 \times E_{\text{rms}}(\text{load}) \times I_{\text{rms}}(\text{load}) \times \cos\phi \text{ watts} \quad \text{(18-5-4)}$$

where $\cos\phi$ is the power factor for each of the balanced loads.

For the wye connection,

$$E_{\text{rms}}(\text{line}) = \sqrt{3} \times E_{\text{rms}}(\text{load}) \text{ and } I_{\text{rms}}(\text{line}) = I_{\text{rms}}(\text{load}) \quad \text{(18-5-5)}$$

Combining Eqs. (18-5-4) and (18-5-5) gives

$$\text{total power, } P_T = 3 \times \frac{E_{\text{rms}}(\text{line})}{\sqrt{3}} \times I_{\text{rms}}(\text{line}) \times \cos\phi$$

$$= \sqrt{3} \times E_{\text{rms}}(\text{line}) \times I_{\text{rms}}(\text{line}) \times \cos\phi \text{ watts} \quad \text{(18-5-6)}$$

For the delta connection,

$$E_{\text{rms}}(\text{line}) = E_{\text{rms}}(\text{load}) \text{ and } I_{\text{rms}}(\text{line}) = \sqrt{3}\, I_{\text{rms}}(\text{load}) \quad \text{(18-5-7)}$$

Combining Eqs. (18-5-4) and (18-5-7) yields

$$\text{total power, } P_T = 3 \times E_{\text{rms}}(\text{line}) \times \frac{I_{\text{rms}}(\text{line})}{\sqrt{3}} \times \cos\phi$$

$$= \sqrt{3} \times E_{\text{rms}}(\text{line}) \times I_{\text{rms}}(\text{line}) \times \cos\phi \text{ watts} \quad \text{(18-5-8)}$$

The expressions for the total power are therefore the same for both the wye and delta connections.

EXAMPLE 18-5 In a three-phase delta-connected alternator the phase voltages are $110 \angle 0°$ V, $110 \angle 120°$ V, and $110 \angle -120°$ V. Each of the three balanced, delta-connected loads has an impedance of $5.5 \angle 25°$ Ω. Calculate the values of the line currents and the total power delivered from the three-phase alternator.

Solution The phase currents are

$$I_{xz} = \frac{110 \angle 0° \text{ V}}{5.5 \angle 25° \text{ } \Omega} = 20 \angle -25° \text{ A}$$

$$I_{zy} = \frac{110 \angle 120° \text{ V}}{5.5 \angle 25° \text{ } \Omega} = 20 \angle 95° \text{ A}$$

$$I_{yx} = \frac{110 \angle -120° \text{ V}}{5.5 \angle 25° \text{ } \Omega} = 20 \angle -145° \text{ A}$$

The line currents are

$$I_x = I_{xz} - I_{yx}$$
$$= 20 \angle -25° - 20 \angle -145°$$
$$= 18.1 - j8.45 + 16.4 + j11.47$$
$$= 34.5 + j3.02 = 34.6 \angle 5° \text{ A}$$

The same result may be derived from

$$I_x = 20 \times \sqrt{3} \angle (-25°) + (+30°) = 34.6 \angle 5° \text{ A}$$

Similarly,

$$I_y = 34.6 \angle (+95°) + (+30°)$$
$$= 34.6 \angle 125° \text{ A} = -19.84 + j28.35 \text{ A}$$

and

$$I_z = 34.6 \angle (-145°) + (+30°)$$
$$= 34.6 \angle -115° \text{ A} = -14.66 - j31.37 \text{ A}$$

The sum of the line currents is $(+34.5 + j3.02) + (-19.84 + j28.35) + (-14.66 - j31.37) = 0$ A. Each of the line currents has a magnitude of $20\sqrt{3} = 20 \times 1.732 = 34.6$ A and each line current is shifted by 30° from one of the phase currents.

The total power delivered from the alternator is

$$\text{total power, } P_T = \sqrt{3} \times E_{\text{rms}}(\text{line}) \times I_{\text{rms}}(\text{line}) \times \cos \phi \qquad (18\text{-}5\text{-}8)$$

$$= \sqrt{3} \times 110 \text{ V} \times 34.6 \text{ A} \times \cos 25°$$
$$= 5980 \text{ W}$$
$$= 5.98 \text{ kW}$$

EXAMPLE 18-6 In a three-phase delta-connected alternator the phase voltages are $160 \angle 80°$ V, $160 \angle -160°$ V, and $160 \angle -40°$ V. The corresponding three delta-connected, unbalanced loads are $20 \angle 20°$ Ω, $40 \angle -30°$ Ω, and $80 \angle 130°$ Ω. What are the values of the phase and line currents?

Solution The phase currents are

$$\frac{160\angle 80° \text{ V}}{20\angle 20°} = 8\angle 60° \text{ A}$$

$$\frac{160\angle -160° \text{ V}}{40\angle -30°} = 4\angle -130° \text{ A}$$

$$\frac{160\angle -40°}{80\angle 130°} = 2\angle -170° \text{ A}$$

The corresponding line currents are

$$8\angle 60° - 2\angle -170° = (4 + j6.928) - (-1.97 + j0.347)$$
$$= 5.97 + j7.27$$
$$= 9.41\angle 50.6° \text{ A}$$

$$2\angle -170° - 4\angle -130° = (-1.97 - j0.347) - (-2.57 - j3.064)$$
$$= 0.60 + j2.72$$
$$= 2.79\angle 77.6° \text{ A}$$

$$4\angle -130° - 8\angle 60° = (-2.57 - j3.064) - (4 + j6.928)$$
$$= -6.57 - j9.99$$
$$= 11.96\angle -123° \text{ A}$$

Although the loads are unbalanced, the total of the three line currents is

$$(5.97 + j7.27) + (0.60 + j2.72) + (-6.57 - j9.99) = 0$$

18-6 Three-Phase Transformers. Six-Phase and Twelve-Phase Voltages

Alternating power may be supplied through three-phase circuits which contain transformers with their primary and secondary windings in various wye and delta formations. We may either use separate single-phase transformers or one three-phase transformer with three limbs in its laminated core.

There are four possible combinations of the primary and secondary windings; these are:

1. Primary windings in the delta formation and secondary windings in the delta formation
2. Primary windings in the delta formation and secondary windings in the wye formation
3. Primary windings in the wye formation and secondary windings in the delta formation
4. Primary windings in the wye formation and secondary windings in the wye formation

The wye and delta relationships between (a) the line and phase voltages and (b) the line and phase currents have already been derived for the three-phase alternator. The same relationships apply to the primary and secondary formations of three-phase transformers.

Figure 18-9 shows three single-phase transformers with both the primary and the secondary windings in delta formations. As we emphasized in Section 18-5, it is essential that the correct terminals of the windings be connected in the delta formation. If any one of the three windings is wrongly reversed, the total voltage around the delta loop will no longer be zero but will be twice the

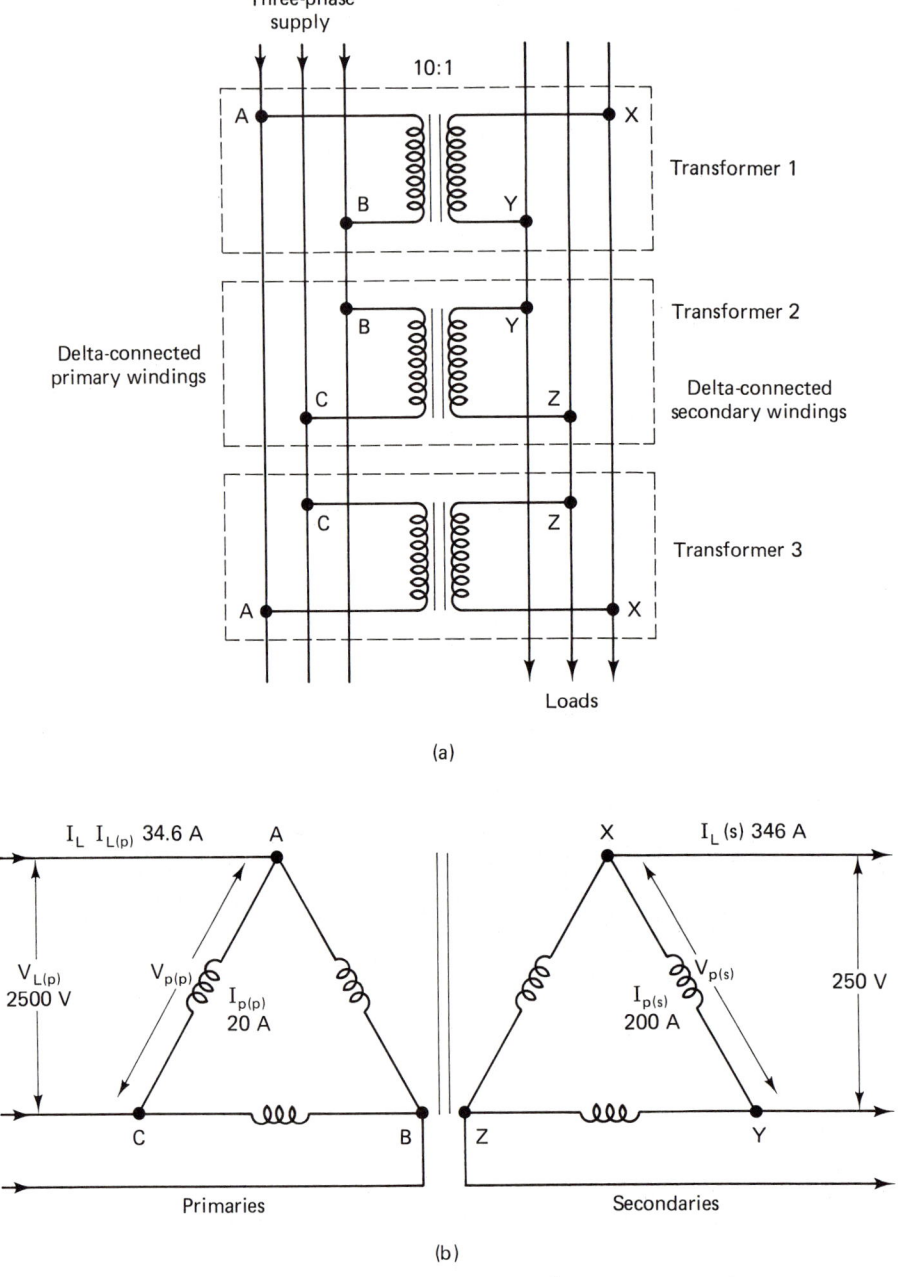

FIGURE 18-9 Delta-to-delta three-phase transformer connections.

value of the phase voltage. This will cause a large circulating current and consequent damage to the transformers' windings.

For the primary formation the line current, I_L, is $\sqrt{3} \times$ the phase current, I_p, while the line voltage, V_L, equals the phase voltage, V_p; the same equations apply to the secondary circuits. Each of the three transformers has the same turns ratio, which will determine the value of the secondary phase voltage.

As an example, let the secondary delta circuit supply be a three-phase balanced load that requires a total power of 150 kW. Each of the loads has a unity power factor and is supplied with 250 V. The turns ratio for each of the transformers is 10:1 step-down. We may then calculate the following:

$$\text{secondary power for each phase} = \frac{150}{3} = 50 \text{ kW}$$

$$\text{secondary phase current} = \frac{50 \text{ kW}}{250 \text{ V}} = 200 \text{ A}$$

$$\text{secondary line current} = \sqrt{3} \times 200 = 1.732 \times 200 = 346 \text{ A}$$

$$\text{value of each of the balanced resistive loads} = 250 \text{ V}/346 \text{ A} = 0.72 \text{ }\Omega$$

$$\text{primary phase voltage} = 10 \times 250 = 2500 \text{ V}$$

$$\text{primary phase current} = \frac{200}{10} = 20 \text{ A}$$

$$\text{primary line current} = \frac{346}{10} = 34.6 \text{ A}$$

The main advantage of the delta–delta connection is that the phase current in each winding is only $1/\sqrt{3} = 0.577$ of the corresponding line current.

Let us now consider the case in which the primary windings are still delta connected but the secondary windings are reconnected in a wye formation (Fig. 18-10). This arrangement allows us to increase the secondary output voltage so that it is often used to provide high voltages for the output stages of powerful transmitters. For example, if the secondary phase voltages are each 2500 V, the secondary line voltages (across which the three-phase balanced loads may be connected) are all $\sqrt{3} \times 2500 = 1.732 \times 2500 = 4330$ V.

Additional single-phase balanced loads can also be connected between the three lines and the neutral line so that each of these loads is supplied with 2500 V. To achieve these results it is essential that the proper connections are made to create secondary wye formations. For the secondary windings three corresponding leads must be joined to form the common or neutral line, while the other three leads are brought out to the lines that feed the balanced loads. If any one of the three windings is reversed, the voltages between the three lines become unbalanced and the loads are not supplied with their correct currents. In addition, the line currents will no longer be separated in phase by 120°.

Another possible arrangement with wye-connected primaries and delta-connected secondaries is illustrated in Fig. 18-11. As an example, this could be used to step down the line voltage of approximately 4000 V on the primary to the secondary output voltage of 115 V. The primary phase voltage should be $4000 \text{ V}/\sqrt{3} \approx 2400$ V, so that a 20:1 turns ratio (step-down) would be required. This provides a reduction in transmission losses since the primary line voltage is 73.2% higher than the phase voltage and the primary line current is correspondingly less.

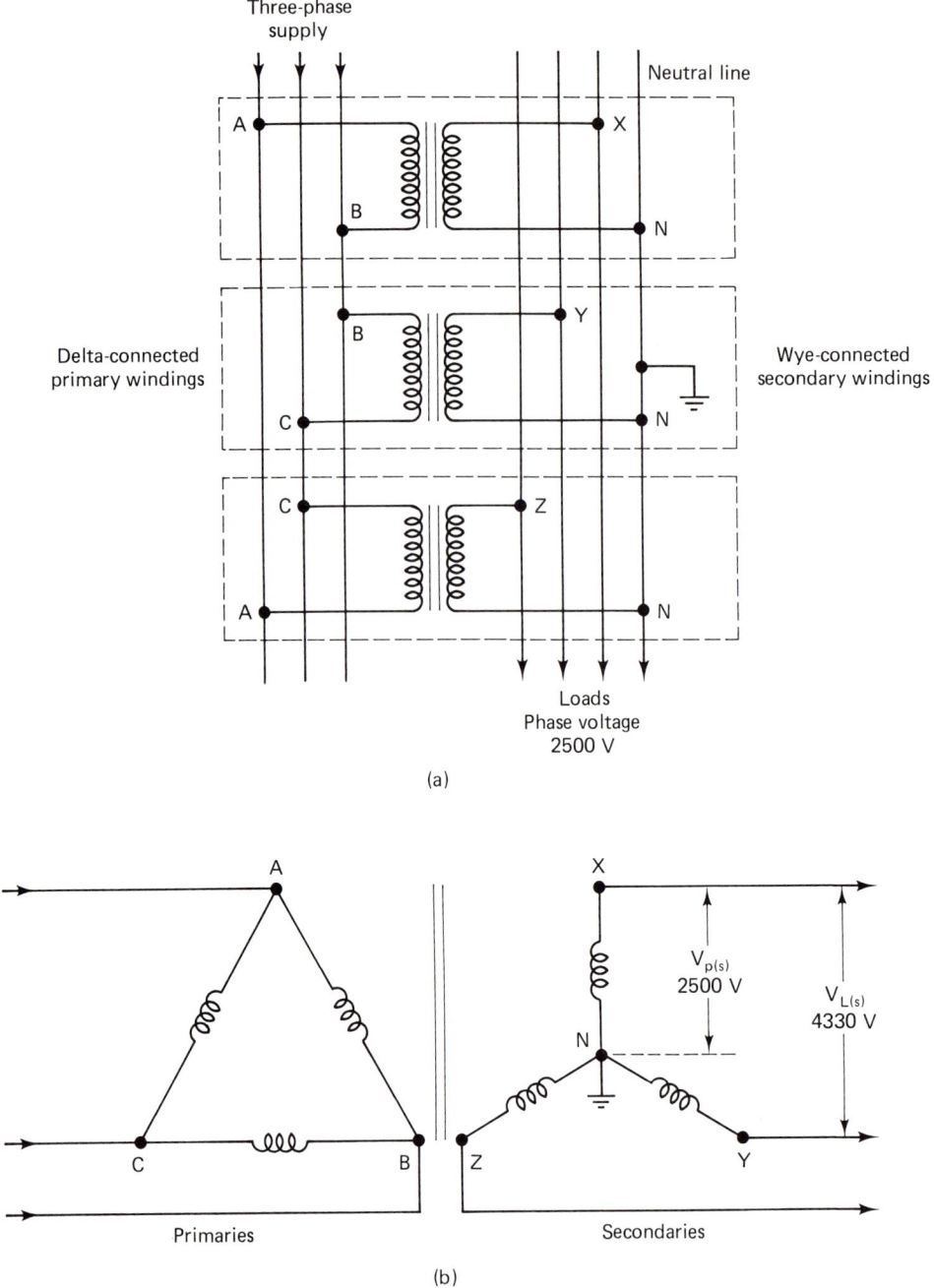

FIGURE 18-10 Delta-to-wye three-phase transformer connections.

Six-Phase and Twelve-Phase Voltages

In six-phase operation there will be six alternating voltages, each of whose phasors is separated from its neighbors by 60°. It is possible to create such a system by using a three-phase supply and three single-phase transformers [Fig. 18-12(a)]. The primary windings are connected in delta, but the centers of the three secondary windings are joined together to form the neutral point of a (double) wye formation [Fig. 18-12(b)]. Since, for example, the phase voltage, $V_{A'}$ is 180° out of phase with the phase voltage, V_A (with respect to the

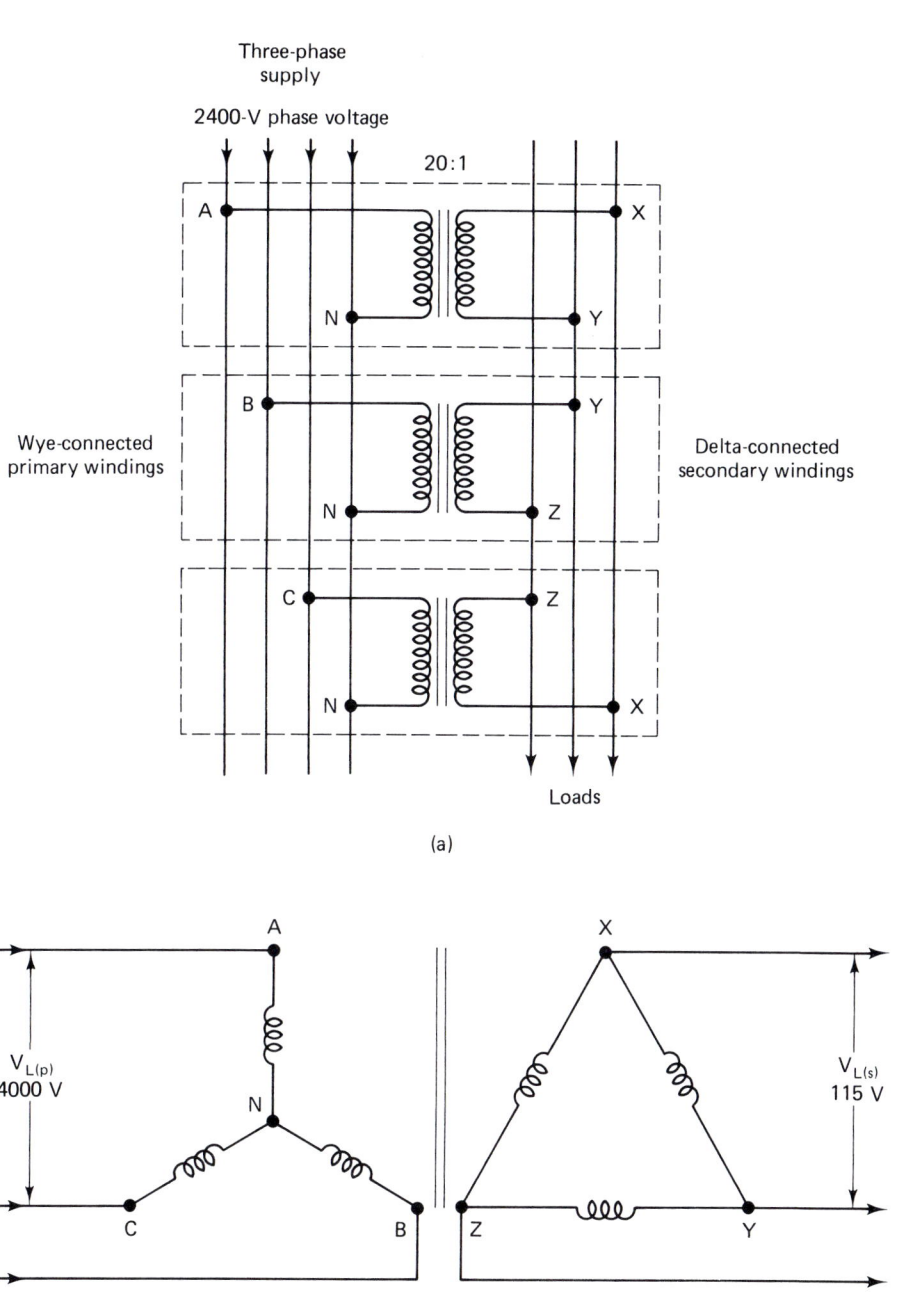

FIGURE 18-11 Wye-to-delta three-phase transformer connections.

grounded neutral point), the ends of the secondary windings will provide the required six-phase output.

For twelve-phase operation we need 12 alternating voltages each of whose phasors is separated from its neighbors by 30°. This can be achieved by starting off with a six-phase system and then introducing an additional six phases with the necessary 30° shift. In Fig. 18-13(a) there are six single-phase transformers with three of the primary windings connected in a wye formation while the other three are joined in a delta formation. The two formations are in parallel and the primary voltages associated with the wye formation are shifted by 30°

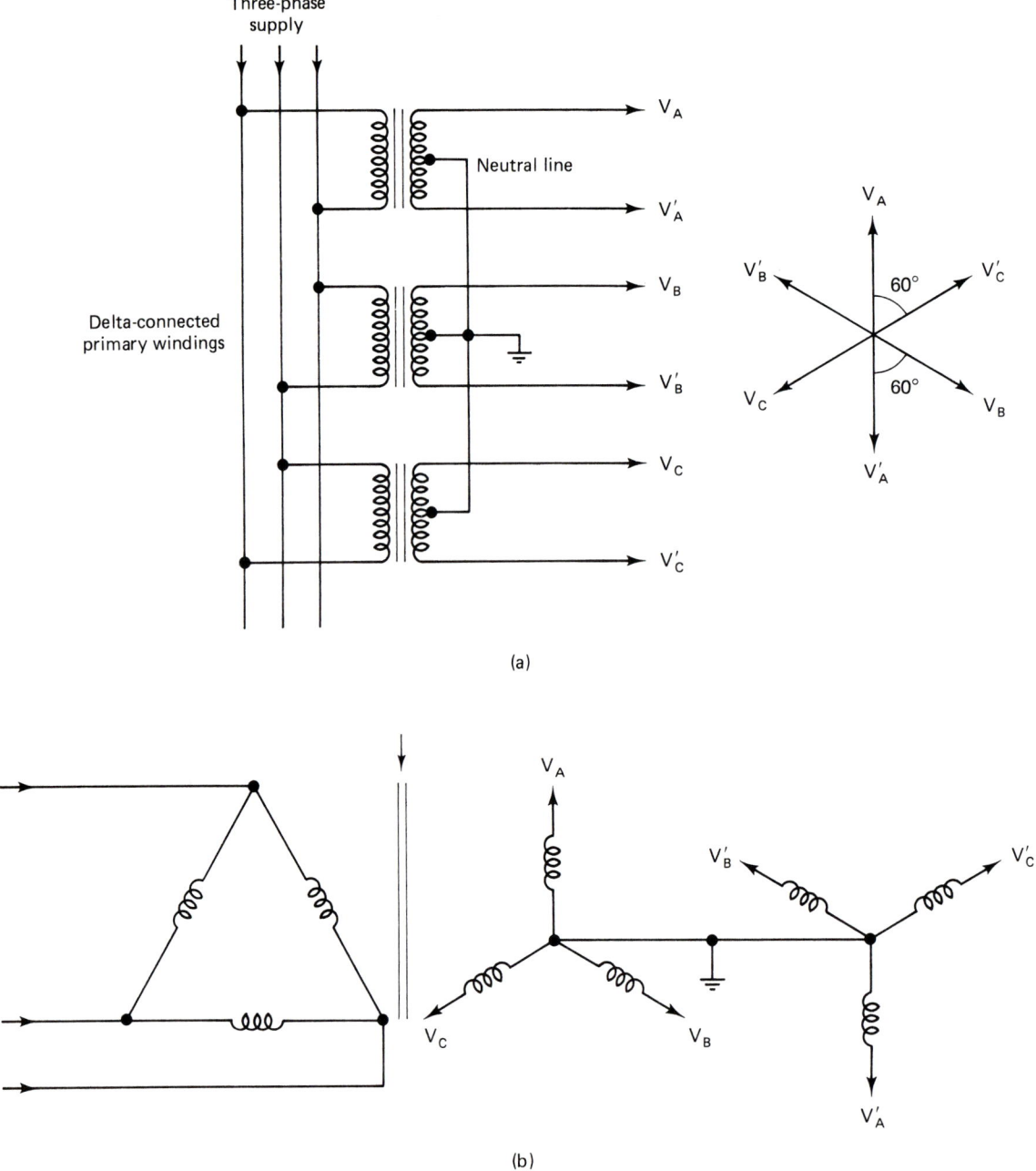

FIGURE 18-12 Six-phase voltage system.

from those of the delta formation (Section 18-4). The inequality between the wye and delta phase voltages (delta phase voltage = $\sqrt{3}$ × wye phase voltage) can be compensated by using different turns ratios for the two groups of transformers. All six secondary voltages will then be equal in magnitude.

The centers of the three secondary windings in group 1 are joined together to form the neutral point of a double-wye formation. The same arrangement is made in group 2 and the two neutral points are then connected. The ends of the windings will then provide the required twelve-phase output [Fig. 18-13(b)].

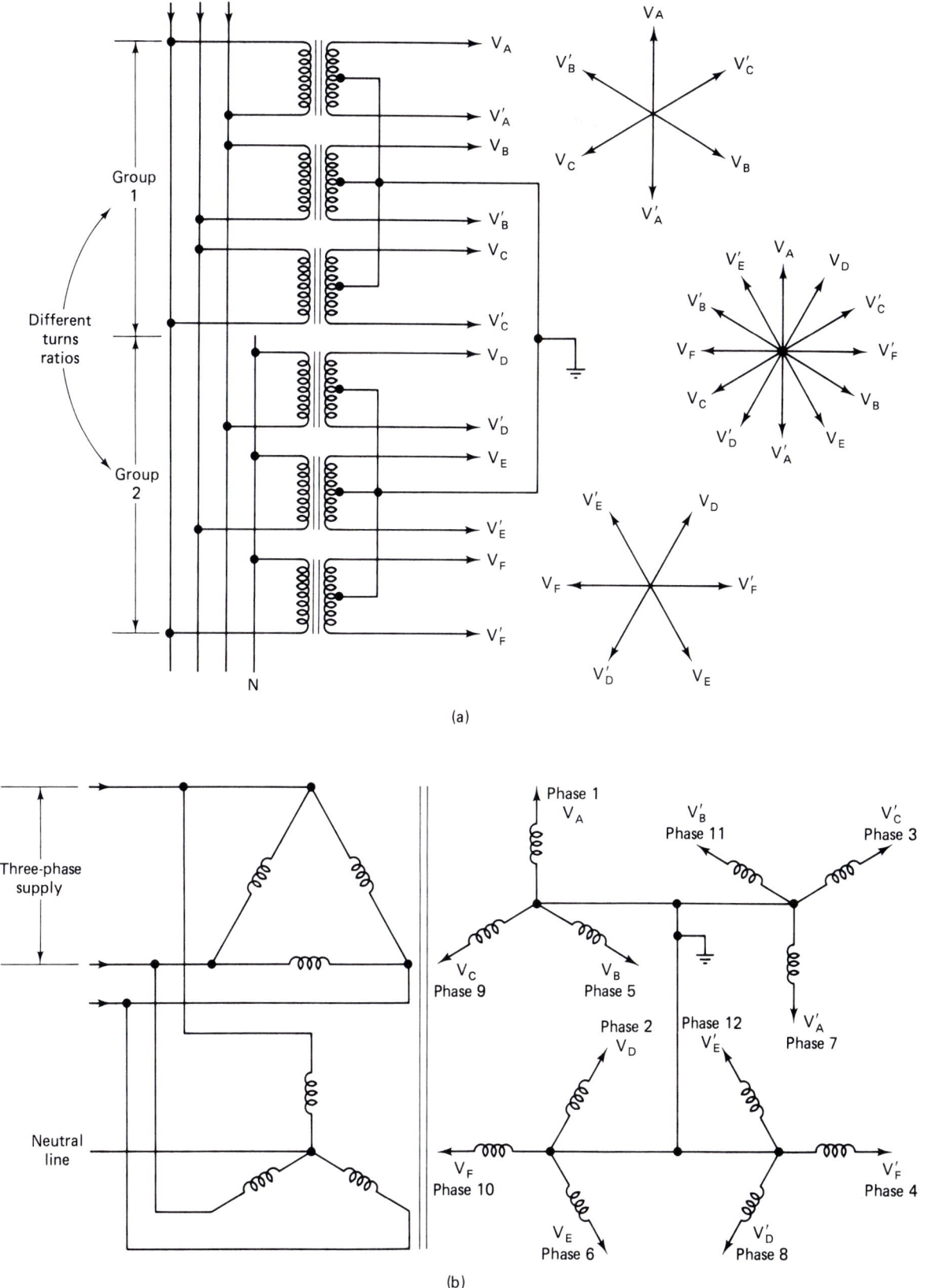

FIGURE 18-13 Three-phase voltage system.

Sec. 18-6 Three-Phase Transformers. Six-Phase and Twelve-Phase Voltages

Chapter Summary

18.1. Generated frequency = $Np/60$ hertz.

18.2. Two-phase alternator:
Two-phase voltages are 90° out of phase.
Neutral line current for balanced loads = $\sqrt{2}\, I_{rms}$ amperes.
Total instantaneous power is constant.

18.3. Three-phase alternator:
Three-phase voltages are 120° out of phase.
Neutral line current for balanced loads is zero.
Total instantaneous power is constant.

18.4. Wye three-phase connection:

$$\text{line current} = \text{phase current}$$
$$\text{line voltage} = \sqrt{3} \times \text{phase voltage}$$
$$= 1.732 \times \text{phase voltage}$$

18.5. Delta three-phase connection:

$$\text{line current} = \sqrt{3} \times \text{phase current}$$
$$= 1.732 \times \text{phase current}$$
$$\text{line voltage} = \text{phase voltage}$$

18.6. Three-phase power:
For both wye and delta connections,

$$\text{total power, } P_T = \sqrt{3} \times E_{rms}\,(\text{line}) \times I_{rms}\,(\text{line}) \times \cos\theta \text{ watts}$$

Self-Review

True-False Questions

1. **T. F.** If a two-pole alternator generates a frequency of 50 Hz, the armature's speed of rotation is 3000 rpm.
2. **T. F.** For a two-phase alternator the neutral line current is $\sqrt{2}\, I_{rms}$, where I_{rms} is the line current in each of the two balanced loads.
3. **T. F.** A three-phase generator has three separate windings, each with its own separate shaft.
4. **T. F.** Three-phase line voltages all reach their peak values simultaneously.
5. **T. F.** The angles between the generated voltages for an alternator depend solely on the relative number of turns of the windings.
6. **T. F.** The neutral line will carry some current if a delta-connected load is not balanced.
7. **T. F.** If no neutral current flows for wye-connected resistive loads, we can conclude that the resistive loads are equal in magnitude.
8. **T. F.** If a delta-connected load is rewired into a wye configuration with no change in impedance values, the line current will decrease.
9. **T. F.** A balanced load has identical impedance in each phase of the load.
10. **T. F.** At least three conductors are required to supply a delta-connected load and at least four if it is wye-connected.

11. T. F. The maximum speed for a conventional 60-Hz alternator is 3600 rpm.
12. T. F. The voltage of an alternator is generated in the windings of the stator, not the rotor.
13. T. F. The power delivered by a two-phase alternator supplying a balanced resistive load fluctuates at twice the basic frequency.
14. T. F. With a balanced two-phase load, the current in the common, neutral conductor is zero.
15. T. F. With a balanced three-phase load, the current in the common, neutral conductor is zero.
16. T. F. The equation for total power with a balanced three-phase load having line voltage, line current, and power factor is independent of whether wye connections or delta connections are used.
17. T. F. The neutral current of a delta-connected alternator may not be zero if the load is unbalanced.
18. T. F. Three-phase transformer banks can consist of single-phase transformers connected properly.
19. T. F. A six-phase system can be devised by making the proper connections with center-tapped primary windings.
20. T. F. A three-phase balanced wye-connected load is supplied by a neutral conductor and three line conductors. The neutral conductor current is equal to the line current divided by $\sqrt{3}$.

Multiple-Choice Questions

21. In a three-phase alternator the phase voltages are
    (a) always equal to the line voltage
    (b) generated in each of three windings
    (c) larger than the line voltage
    (d) 120° out of phase with the current
    (e) 120° out of phase with the line voltage
22. A wye-connected, three-phase alternator
    (a) has three windings and three or four terminals
    (b) has equal phase and line voltages with a balanced load
    (c) has phase and line currents related by 1.732
    (d) can function only with a precisely balanced load
    (e) will not work with a balanced, delta-connected load
23. Select one of the given choices as pertaining to a wye-connected load.
    (a) The line current and phase current are equal.
    (b) It cannot be connected to a delta-connected alternator.
    (c) No neutral current flows with a balanced load.
    (d) Choices (a) and (c) are both correct.
    (e) Choices (a) and (b) are both correct.
24. A delta-connected alternator
    (a) has equal line and phase voltages
    (b) cannot be connected to a wye-connected load
    (c) will have zero neutral current only if the load is balanced
    (d) requires a precisely balanced, delta-connected load
    (e) has unequal line and phase voltages
25. If one conductor feeding a delta-connected load is disconnected,
    (a) the neutral conductor will overheat
    (b) single-phase currents will continue to flow
    (c) all line currents will be zero
    (d) all phase currents will be zero
    (e) all choices are correct
26. In a balanced wye-connected load
    (a) the neutral current is zero   (b) the line currents are equal
    (c) the phase currents are equal   (d) the line voltages are equal
    (e) all choices are correct

27. A resistor, a capacitor, and an inductor are connected to form a three-phase load.
    (a) Three-phase resonance will be the result.
    (b) The situation described is impossible.
    (c) This is possible for a delta connection only.
    (d) The load is not balanced.
    (e) The load is balanced if the same number of ohms is in each phase.
28. Three resistors have equal values of resistance and are connected to form a three-phase load.
    (a) The power factor is 1.
    (b) If wye-connected, the neutral current is zero.
    (c) The phase voltage is in phase with the phase current.
    (d) The load is balanced.
    (e) All choices are correct.
29. Three identical capacitors are delta-connected to form a load.
    (a) The three-phase currents are separated by 90° on a phasor diagram.
    (b) The three line currents are separated by 90° on a phasor diagram.
    (c) The line currents and phase currents are zero.
    (d) The power factor is zero.
    (e) The arrangement has no practical value.
30. A delta-connected alternator supplies a wye-connected load.
    (a) There can be no neutral current.
    (b) The line voltages are equal to the phase voltages for both.
    (c) The line currents are equal to the phase currents for both.
    (d) The arrangement is not possible.
    (e) The load cannot be balanced.
31. An alternator that produces a frequency of 50 Hz has a rotational speed of
    (a) 375 rpm (b) 400 rpm
    (c) 514.29 rpm (d) 720 rpm
    (e) 900 rpm
32. The current passing to and from the rotor windings of an alternator is
    (a) direct current (b) half-wave rectified ac
    (c) full-wave rectified ac (d) a 60-Hz sine wave
    (e) primarily semiconductor holes
33. The number of turns in a three-phase alternator stator is
    (a) 1 turn (b) 2 turns (c) 3 turns (d) between 60 and 3600 turns
    (e) more information is required to answer this question sensibly
34. The physical separation of the stator windings for a two-phase, two-pole alternator is 90°. What will be the angle if the number of poles is four?
    (a) 22.5° (b) 45° (c) 90° (d) 135° (e) 180°
35. The minimum number of conductors required to connect loads to a three-phase alternator for a balanced system is
    (a) 1 (b) 2 (c) 3 (d) 4 (e) 6
36. The phase voltage of an alternator is 240 V and the phase voltage of the load is also 240 V. We must conclude that
    (a) the alternator is wye-connected and the load wye-connected
    (b) the alternator is wye-connected and the load delta-connected
    (c) the alternator is delta-connected and the load wye-connected
    (d) the alternator is delta-connected and the load delta-connected
    (e) exactly two choices are correct
37. A possible connection arrangement for the primary and secondary windings of a three-phase transformer bank is:
    (a) primaries are delta-connected and secondaries delta-connected
    (b) primaries are delta-connected and secondaries wye-connected
    (c) primaries are wye-connected and secondaries delta-connected
    (d) primaries are wye-connected and secondaries wye-connected
    (e) all choices are correct
38. Assuming a turns ratio of 1 for single-phase transformers connected to form three-phase transformer banks, the ratio of secondary to primary line voltages for the connections of Question 37 are respectively
    (a) 1 (b) $\sqrt{3}$ (c) $1/\sqrt{3}$ (d) $\sqrt{2}$
    (e) all choices are correct

39. Considering all six phase voltages of a six-phase system, the minimum angle separating any two voltages is
    (a) 30°   (b) 60°   (c) 90°   (d) 120°   (e) 180°
40. A balanced three-phase system requires less copper than
    (a) a two-phase system of the same power
    (b) a single-phase system of the same power
    (c) three single-phase systems with the same total power
    (d) an unbalanced three-phase system with neutral conductor
    (e) all are correct

Practice Problems

41. The primary windings for a three-phase transformer are delta connected and the secondary windings are joined in a wye formation. If each pair of windings has a step-up ratio of 1:5 and the primary phase voltage is 250 V, what is the value of the secondary line voltage?
42. Each stator winding of a two-phase alternator has a phase voltage of 350 V and is connected to 70-Ω resistive load. Calculate the value of the neutral current.
43. The phase voltage for a wye-connected, three-phase load is 127 V. What is the line voltage?
44. Find the phase voltage corresponding to a line voltage of 3300 V for a three-phase wye-connected alternator.
45. The line current for a three-phase, wye-connected load is 75 A. What is the phase current?
46. For a three-phase, wye-connected alternator, the current in the neutral conductor is zero. Briefly describe the known nature of the load.
47. When the line voltage and the phase voltage of a three-phase load are the same, what type of connection has been used?
48. A particular delta-connected load has a phase current of 3.7 A. Calculate the line current.
49. The line voltage of a three-phase system is 240 V. Determine the line current if three 200-Ω resistors form a wye-connected load.
50. Three 200-Ω resistors form a delta-connected load with a three-phase 240-V source. What is the line current?
51. A single-phase alternator has two poles and rotates at 3600 rpm. What is the frequency?
52. Calculate the voltage between the line terminals of a two-phase alternator generating 220 V per phase and having a common, neutral conductor.
53. Calculate the maximum total instantaneous power delivered by a three-phase alternator if one phase winding delivers an average of 5000 W. Assume a balanced, three-phase load.
54. The voltage generated in each winding of a three-phase alternator is 240 V. Calculate the winding current if the total load power is 10 kW and the power factor = 1. (The load behaves as resistance.)
55. A delta-connected alternator supplies 15 kW at 440 V to a 0.85 power factor load. Calculate the line current.
56. In Problem 55, find the phase current of the alternator.
57. A three-phase transformer is connected delta–delta to drop 13,800 V to 440 V. Calculate the currents in the primary lines for a load of 50 kW, 0.8 power factor.
58. In Problem 57, calculate the secondary current in each transformer.
59. Referring to Fig. 18-12, let the individual transformer voltages be 220 V. Calculate the minimum voltage to be measured between any two points in the secondary circuit.
60. A three-phase transformer bank is rated 15 kVA with a primary line voltage of 1200 V and a secondary line voltage of 240 V. Calculate the impedance for each leg of a three-phase, balanced, delta-connected load.

Advanced Problems

61. A three-phase alternator has an eight-pole rotor and generates a frequency of 440 Hz. Calculate the value of the rotor's speed in revolutions per minute.
62. In a three-phase transformer the secondary windings are connected in a wye formation. Balanced 50-Ω loads are joined between the three secondary lines. If the secondary phase voltage is 450 V, calculate the value of the total secondary power.
63. Given the choice of delta or wye connections for a three-phase alternator and a three-phase load, which arrangement will provide the greatest phase voltage for 10-Ω load resistors?
64. Given the choice of delta or wye connections for a three-phase alternator and three-phase load, which arrangement will provide the greatest phase load current for a specified total load power?
65. For a 5-kW delta-connected resistive load, the line voltage is 220 V. Calculate the apparent power, the real power, and the reactive power.
66. For a wye-connected resistive load, the line voltage is 220 V. Calculate the line current if the power is 5 kW.
67. The power factor of a 10-kW balanced delta-connected load is 0.8. If the line voltage is 440 V, what is the line current?
68. The power factor of a 10-kW balanced wye-connected load is 0.8. If the line voltage is 440 V, what is the line current?
69. A 200-Ω resistor and a 100-Ω inductive reactance are connected in series for each part of a three-phase delta-connected load. With a line voltage of 240 V, what power is taken by the load?
70. The phase loads of Problem 69 are now connected in a wye configuration with the same line voltages. Calculate the load power.
71. Calculate the exact frequency if mechanical difficulties cause the 60-Hz alternator to slow down slightly to 3540 rpm.
72. An alternator is designed to operate at a speed of 4000 rpm or less and to produce ac power at 400 Hz. Determine the least number of pole pairs to meet this requirement.
73. A balanced two-phase resistive load is connected to a two-phase alternator so that two lines and a common, neutral conductor carry current. If the phase voltage = 120 V, find the line and neutral currents for a 5-kW load.
74. One line conductor in Problem 73 is gnawed through by a rat, now deceased, thereby creating an open line. Calculate the resulting line and neutral conductor currents.
75. If, in Problem 74, the rat had bitten the neutral instead, he might well have survived. Determine the new line current if the neutral conductor had been severed.
76. A three-phase alternator produces power at 220 V (line) and delivers 20 A (line) to a 7.5-kW balanced load. Calculate the power factor.
77. Determine the turns ratio for a twelve-phase system as shown in Fig. 18-13(a). Assume that the primary line voltages are 440 V and that each phase voltage of the secondary is 50 V.
78. Three single-phase transformers are connected to supply three 1-Ω resistors in a delta configuration. Each transformer secondary provides 120 V. Calculate the secondary current of each transformer for wye connection of the secondary windings.
79. The primary-to-secondary voltage ratings of the transformers in Problem 78 is 13,800:120. Find the primary currents for the conditions of that problem.
80. Calculate the total power delivered when the load of Problem 78 is changed to three impedances of $0.8 + j0.6$ Ω instead of $1 + j0$ Ω, as originally given.

19

Nonsinusoidal Waveforms

This chapter has the following objectives:

1. To introduce the concept of periodic, nonsinusoidal waveforms as consisting of sinusoidal components.
2. To establish the meaning of fundamental and harmonic sine waves.
3. To formulate mathematical expressions for sine-wave functions and their combination to form nonsinusoidal functions.
4. To compare the representation of a sawtooth wave, as synthesized with a finite number of sine-wave components, with the actual function.
5. To compare the representation of a square wave, as synthesized with a finite number of sine-wave components, with the actual function.
6. To introduce the formal definition of a Fourier series having a period of 2π radians.
7. To explore the meaning of the words "finite," "single-valued," and "continuous" as they apply to functions that can be synthesized as Fourier series.
8. To present the Fourier series for the triangular waveform and that of the rectified half-wave ac sine wave.
9. To relate the amplitudes of the Fourier series components to parameter values of selected nonsinusoidal functions.
10. To establish the mean value of the Fourier series as the dc level of the nonsinusoidal function.
11. To express the effective (rms) value of the nonsinusoidal function in terms of the Fourier series component amplitudes.
12. To analyze a series ac circuit having nonsinusoidal sources through the employment of their Fourier components.
13. To analyze a series–parallel ac circuit having nonsinusoidal sources through the employment of their Fourier components.

19-0

Introduction

In Chapter 11 we used the square wave to lead to the concept of the sine wave and then in all succeeding chapters we were concerned only with sinusoidal voltages and currents. But what about other alternating waveforms such as the sawtooth and the triangle (Fig. 19-1)? Do we have to establish new methods of analysis for each type of alternating source? Fortunately, the answer is *no!* In 1780, a French mathematician, Baron Jean Baptiste Joseph Fourier (1768–1830), was trying to solve a problem involving the flow of heat. He developed a means of analysis by which any finite, continuous, and periodic waveform could be analyzed into a series of sine waves. The general series contained a fundamental sine wave whose frequency is the same as that of the periodic waveform; together with the fundamental wave were harmonic components so that the derivation of the Fourier series is sometimes known as harmonic analysis.

In electronics a practical example of a periodic waveform is the sawtooth voltage. Such a voltage is applied to a cathode ray tube (Section 20-4) so that the beam is deflected horizontally and causes the spot on the screen to move across from left to right with a constant velocity. This nonsinusoidal waveform is composed of a fundamental component and a large number of harmonics. If all these components are added together (a process called "synthesis"), the result, of course, is the original sawtooth voltage.

In this chapter we start our discussion by adding together (synthesizing) a number of sine waves and studying the summation. We then quote the Fourier series and the results for a number of nonsinusoidal waveforms. Finally, we apply nonsinusoidal voltages to various *LCR* circuits and obtain the corresponding currents.

19-1

Synthesis of Nonsinusoidal Waveforms

Let us consider a nonsinusoidal voltage that contains fundamental, second-harmonic, and third-harmonic components (Fig. 19-2). We assume that the amplitudes of the second and third harmonics are, respectively, one-half and one-third of the fundamental amplitude. Then if the fundamental sine wave is represented by $e_1 = E \sin \omega t$, the expressions for the second and third harmonics are $e_2 = (E/2) \sin 2\omega t$ and $e_3 = (E/3) \sin 3\omega t$. Therefore, the complete equation for the nonsinusoidal voltage is

$$e = E \sin \omega t + \frac{E}{2} \sin 2\omega t + \frac{E}{3} \sin 3\omega t$$

$$= E(+1.00 \sin \omega t + 0.500 \sin 2\omega t + 0.333 \sin 3\omega t)$$

$$= E(1.00 \sin 2\pi ft + 0.500 \sin 4\pi ft + 0.333 \sin 6\pi ft)$$

$$= E(1.00 \sin \theta + 0.500 \sin 2\theta + 0.333 \sin 3\theta) \quad (19\text{-}1\text{-}1)$$

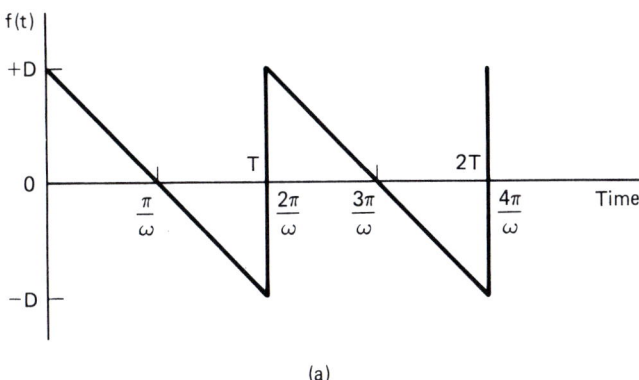

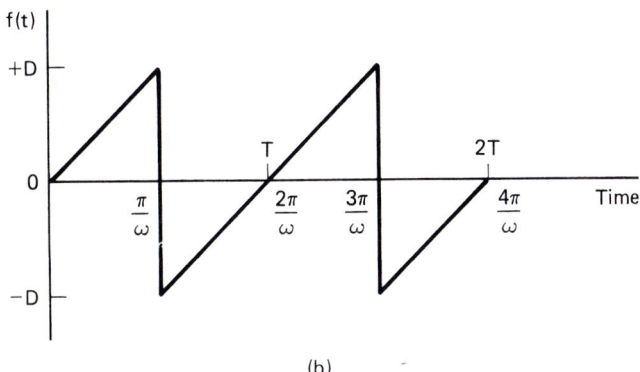

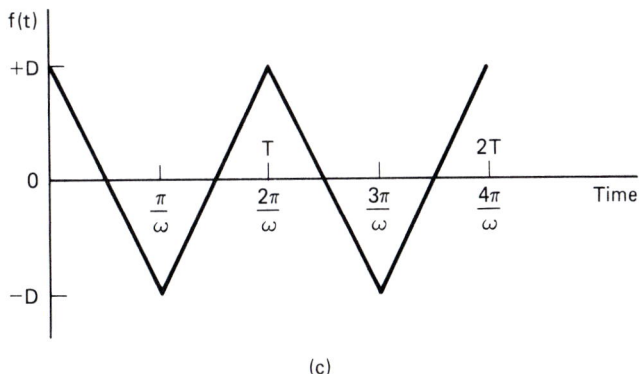

FIGURE 19-1 Examples of nonsinusoidal waveform: (a) negative sawtooth waveform; (b) positive sawtooth waveform; (c) triangular waveform; (d) symmetrical square waveform; (e) half-wave (HW) rectification; (f) full-wave (FW) rectification.

Sec. 19-1 Synthesis of Nonsinusoidal Waveforms

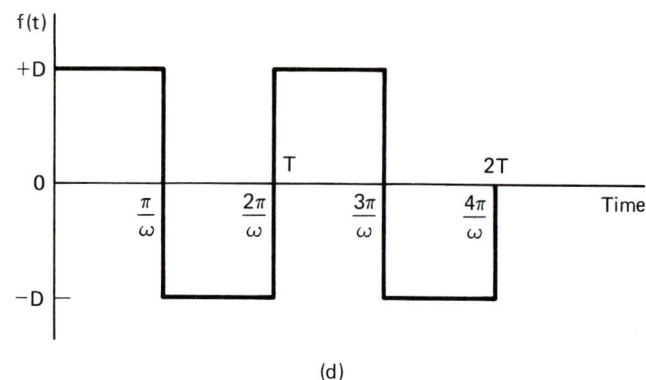

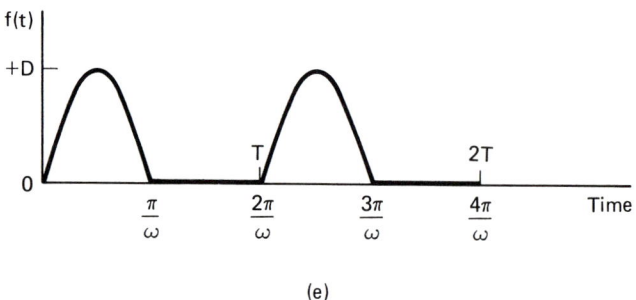

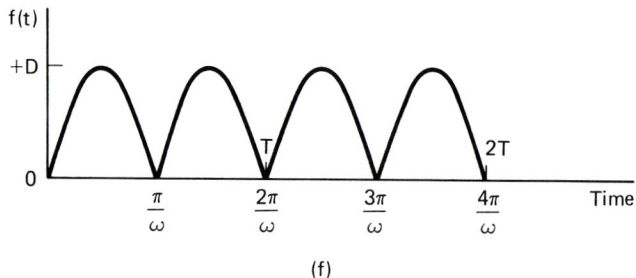

FIGURE 19-1 Continued

where e = instantaneous value of the nonsinusoidal voltage (V)
 E = amplitude of the fundamental sine wave (V)
 ω = angular frequency (rad/s)
 f = frequency (Hz)
 θ = angle (rad)
 t = time (s)

Notice that the three components (fundamental and harmonics) start off together at the beginnings of their positive half-cycles. This accounts for the positive signs in front of all the terms in Eq. (19-1-1).

To derive the nonsinusoidal wave we plot the fundamental and the two harmonics on a common graph by using a horizontal radian (or degree) scale. At a number of conveniently chosen points we then take the algebraic sum of the vertical distances associated with the instantaneous values of the three components.

When we study the resultant or e waveform, we are struck with its sim-

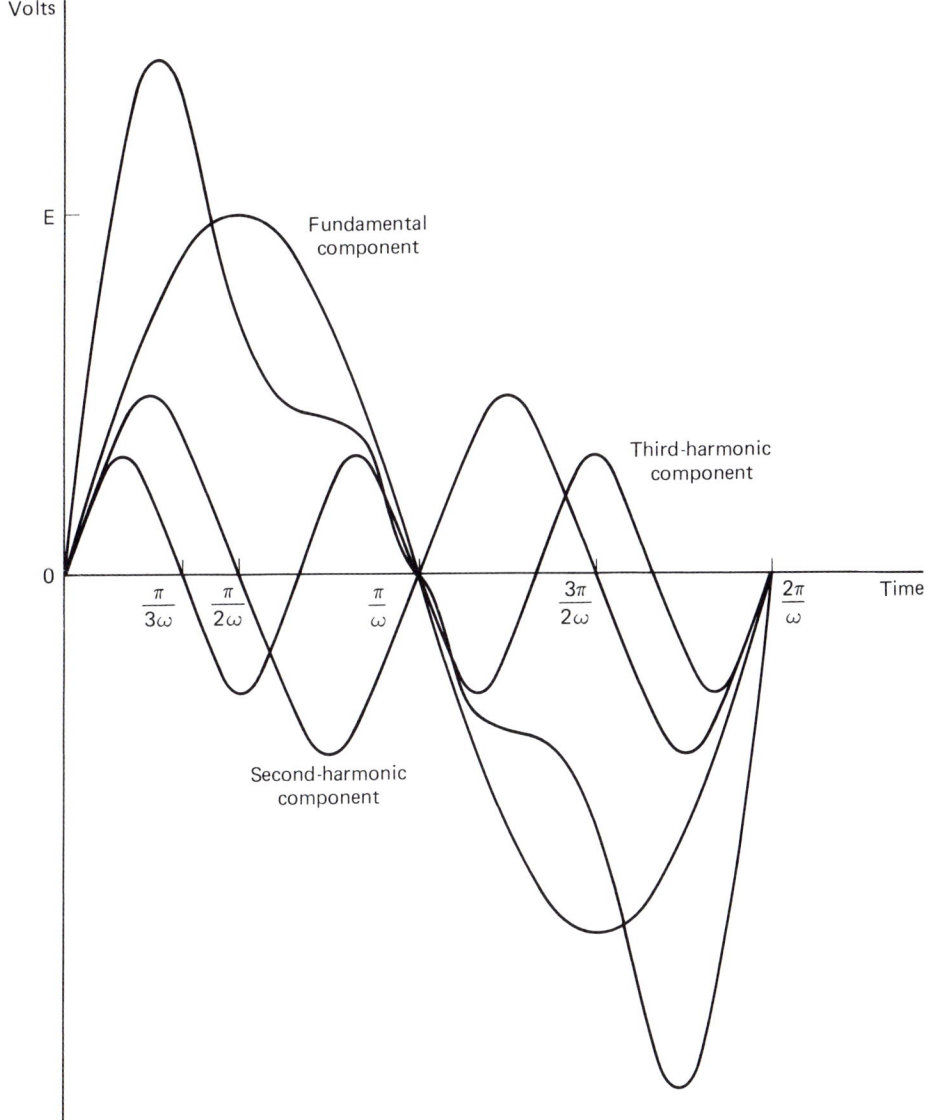

FIGURE 19-2 Synthesis of a nonsinusoidal voltage waveform.

ilarity to a sawtooth voltage with a negative slope [Fig. 19-1(a)]. In fact, if we add more and more harmonics (of the correct size), we obtain closer and closer approximations to the negative sawtooth waveform. This is well illustrated by the waveforms of Fig. 19-3.

The perfect sawtooth can only be obtained by including an infinite number of harmonics. This would lead you to question the value of Fourier analysis. However, we should note that the higher the order of the harmonic, the less is its amplitude. In our example of the negative sawtooth, the amplitude of the hundredth harmonic [$e_{100} = (E/100) \sin 100\omega t$] is only 1% of the fundamental amplitude. It follows that only a limited number of harmonics are required to obtain a "good" approximation to the negative sawtooth voltage. When sufficient harmonics are included, the peak value of the sawtooth waveform is approximately $1.57 \times E$ volts (quoted result).

As a second example we consider the symmetrical square-wave voltage of Fig. 19-1(d). The word "symmetrical" means that the time intervals marked

Sec. 19-1 Synthesis of Nonsinusoidal Waveforms

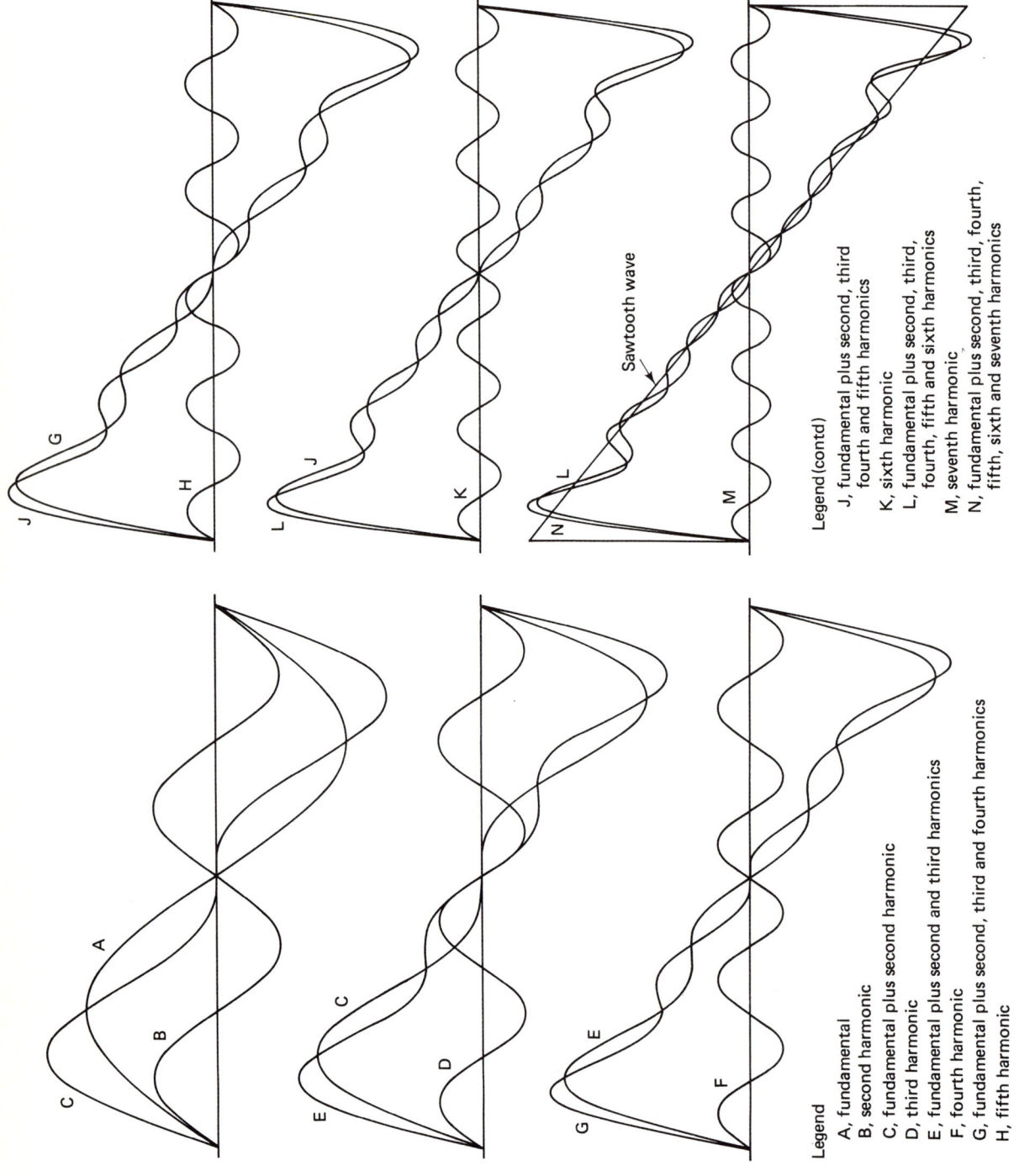

FIGURE 19-3 Synthesis of the negative sawtooth waveform.

T in Fig. 19-1(d), are equal in value. Up to and including the seventh harmonic, the equation of this nonsinusoidal waveform is

$$e = E \sin \omega t + \frac{E}{3} \sin 3\omega t + \frac{E}{5} \sin 5\omega t + \frac{E}{7} \sin 7\omega t$$
$$= E(+1.00 \sin \omega t + 0.333 \sin 3\omega t + 0.200 \sin 5\omega t + 0.143 \sin 7\omega t) \quad \text{(19-1-2)}$$
$$= E(+1.00 \sin \theta + 0.333 \sin 3\theta + 0.200 \sin 5\theta + 0.143 \sin 7\theta)$$

The meanings of E, ω, θ, and t are the same as in Eq. (19-1-1).

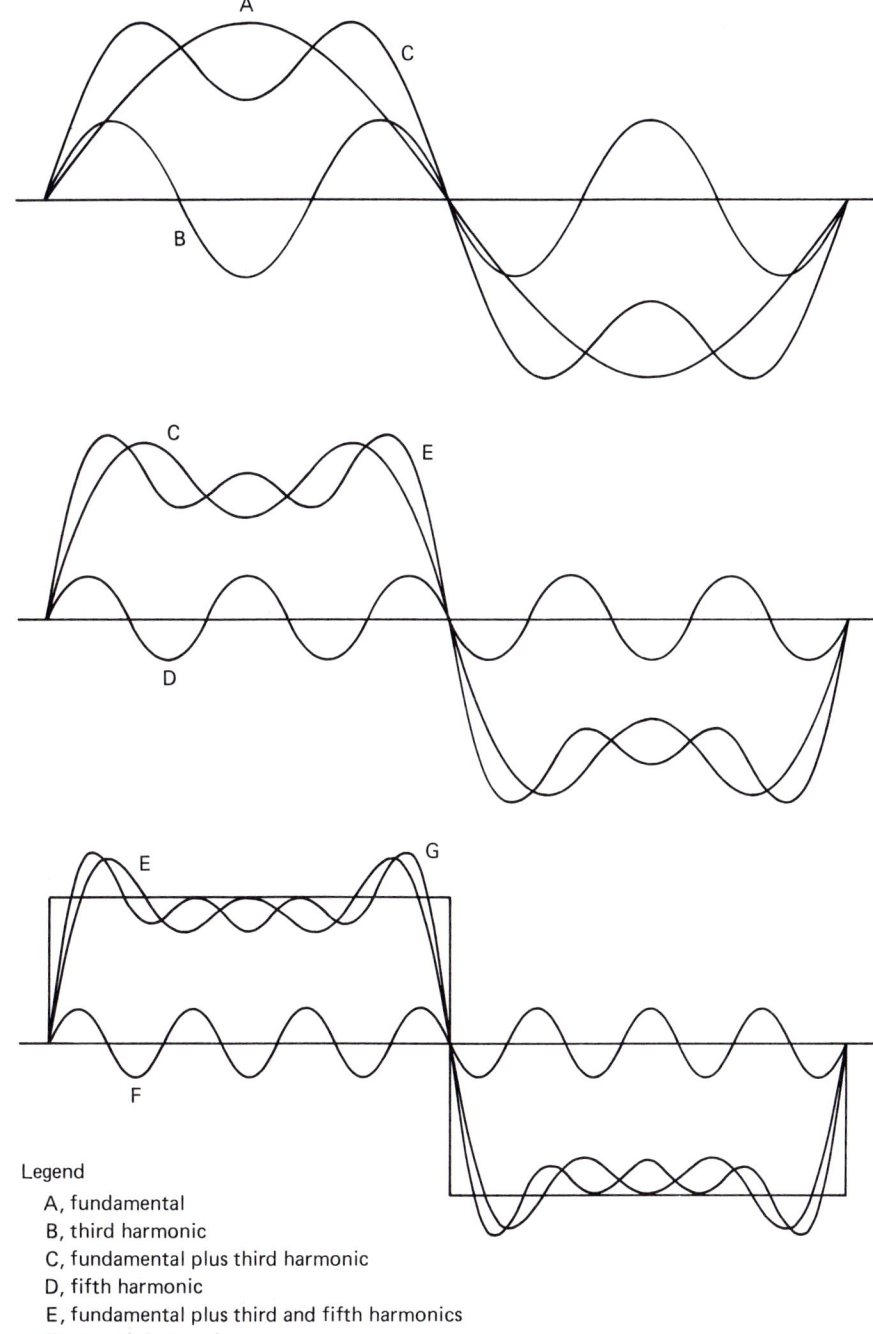

Legend
A, fundamental
B, third harmonic
C, fundamental plus third harmonic
D, fifth harmonic
E, fundamental plus third and fifth harmonics
F, seventh harmonic
G, fundamental plus third, fifth and seventh harmonics

FIGURE 19-4 Synthesis of the symmetrical square wave.

Sec. 19-1 Synthesis of Nonsinusoidal Waveforms

Notice that Eq. (19-1-2) contains only odd harmonics, whereas Eq. (19-1-1) included both odd and even harmonics. This is due to the different symmetries of the two waveforms.

The synthesis of the square wave is illustrated in Fig. 19-4; its peak value, D, is approximately $1.273\,E$ volts (quoted result).

Summarizing, we can obtain nonsinusoidal ac waveforms by combining fundamental and harmonic components. Theoretically, the Fourier series contains an infinite number of harmonics but in practice only a limited number are necessary in order to obtain a "good" approximation to the required waveform.

EXAMPLE 19-1 A symmetrical square wave has a top level of $+4$ V and a bottom level of -4 V. Calculate the amplitudes of the fundamental component and the third, fifth, and seventh harmonics.

Solution

$$\text{peak value of the square wave} = 4 \text{ V}$$

$$\text{fundamental amplitude, } E_1 = 0.785 \times 4 = 3.14 \text{ V}$$

$$\text{third-harmonic amplitude, } E_3 = \frac{3.14}{3} = 1.05 \text{ V} \qquad (19\text{-}1\text{-}2)$$

$$\text{fifth-harmonic amplitude, } E_5 = \frac{3.14}{5} = 0.63 \text{ V}$$

$$\text{seventh-harmonic amplitude, } E_7 = \frac{3.14}{7} = 0.45 \text{ V}$$

19-2

Fourier Series. Analysis of Nonsinusoidal Waveforms

The Fourier theorem may be formally stated as follows:

> Any finite, continuous, single-valued function $f(t)$ which has a period of 2π radians (360°) may be expressed by the following series:
>
> $$f(t) = a_0 + A_1 \sin(\omega t + \phi_1) + A_2 \sin(2\omega t + \phi_2)$$
> $$+ A_3 \sin(3\omega t + \phi_3) + \cdots \qquad (19\text{-}2\text{-}1)$$

It may be shown that

$$A \sin(\omega t + \phi) = A \sin \omega t \cos \phi + A \cos \omega t \sin \phi$$
$$= a \cos \omega t + b \sin \omega t$$

where $a = A \sin \phi$ and $b = A \cos \phi$.

Therefore, the series may be restated as

$$f(t) = a_0 + a_1 \cos \omega t + a_2 \cos 2\omega t + a_3 \cos 3\omega t + \cdots \qquad (19\text{-}2\text{-}2)$$
$$+ b_1 \sin \omega t + b_2 \sin 2\omega t + b_3 \sin 3\omega t + \cdots$$

For some waveforms this is an infinite series, but for other waveforms only a limited number of terms are required.

Now what does all this mean? In the first place, $f(t)$ is the mathematical expression for "a function of time," or, in other words, $f(t)$ is any quantity whose instantaneous value is dependent on the time. For example, $f(t) = \sin \omega t$ is a function of time because the instantaneous value of a sine curve changes from moment to moment.

The word "finite" indicates that $f(t)$ contains no "infinities," since the sum of a Fourier series can never tend to infinity. For example, $f(t) = 1/(t-1)$ could not be expressed by a Fourier series because as $t \to 1$, $f(t) \to \infty$.

"Single-valued" means that $f(t)$ cannot have more than one value for a particular time. This is perhaps obvious since the sum of a Fourier series would not have two different answers at the same time. The function $f(t) = \pm\sqrt{t}$ is double valued since each positive real number has both a positive square root and a negative root; such a function cannot be analyzed into a Fourier series.

"Continuous" shows that there are no discontinuities in the function. If, for example, $f(t) = +\sqrt{(t-1)(t-2)}$, $f(t)$ has no real value between $t = 1$ and $t = 2$ and clearly the sum of the Fourier series can never be an imaginary quantity.

Turning to the series itself, the term a_0 [Eq. (19-2-1)] represents the mean level of $f(t)$. If the positive and negative excursions of a nonsinusoidal waveform do not have a zero average value, the waveform may be regarded as composed of a dc value (a_0) together with the alternating components [$A_1 \sin(\omega t + \phi_1)$, $A_2 \sin(2\omega t + \phi_2)$, and so on]. Examples are the half- and full-wave rectified ac voltages of Fig. 19-1(e) and (f).

Notice that the fundamental and harmonic components are each associated with the phase angles ϕ_1, ϕ_2, and so on. Such phase angles were zero in our examples of the negative sawtooth and the symmetrical square wave, but in general the existence of these phase angles means that the fundamental and the harmonics do not start off together at the beginnings of their positive half-cycles.

In the alternative equation (19-2-2), the complete series is shown as composed of both sine and cosine terms (you will recall that a cosine wave leads a sine wave by 90°). Due to the symmetries of the waveforms, there were no cosine terms in the two series for the sawtooth and the symmetrical square waves. However, in the following series there are a number of cosine terms:

Modified sawtooth waveform (Fig. 19-5):

$$f(t) = \frac{D}{4} - \frac{2D}{\pi^2}\left(\cos \omega t + \frac{\cos 3\omega t}{3^2} + \frac{\cos 5\omega t}{5^2} + \cdots\right)$$
$$+ \frac{D}{\pi}\left(\sin \omega t - \frac{\sin 2\omega t}{2} + \frac{\sin 3\omega t}{3} + \cdots\right) \qquad (19\text{-}2\text{-}3)$$
$$= \frac{D}{4} - \frac{2D}{\pi^2}\left(\cos \omega t + \frac{\cos 3\omega t}{9} + \frac{\cos 5\omega t}{25} + \cdots\right)$$
$$+ \frac{D}{\pi}\left(\sin \omega t - \frac{\sin 2\omega t}{2} + \frac{\sin 3\omega t}{3} + \cdots\right)$$

Sec. 19-2 Fourier Series. Analysis of Nonsinusoidal Waveforms

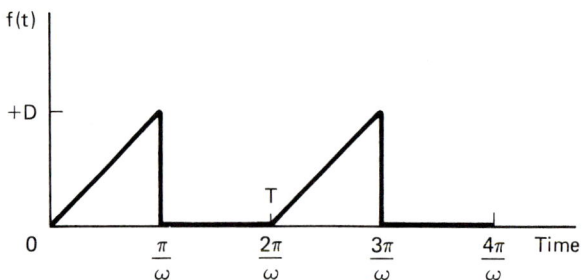

FIGURE 19-5 Modified sawtooth waveform.

Triangular waveform [Fig. 19-1(c)]:

$$f(t) = \frac{8D}{\pi^2}\left(\cos \omega t + \frac{\cos 3\omega t}{3^2} + \frac{\cos 5\omega t}{5^2} + \cdots\right)$$
$$= \frac{8D}{\pi^2}\left(\cos \omega t + \frac{\cos 3\omega t}{9} + \frac{\cos 5\omega t}{25} + \cdots\right) \qquad (19\text{-}2\text{-}4)$$

Half-wave rectified ac waveform [Fig. 19-1(e)]:

$$f(t) = \frac{2D}{\pi}\left(\frac{1}{2} + \frac{\pi}{4}\sin \omega t - \frac{1}{1 \times 3}\cos 2\omega t - \frac{1}{3 \times 5}\cos 4\omega t\right.$$
$$\left.- \frac{1}{5 \times 7}\cos 6\omega t - \cdots\right)$$
$$= \frac{2D}{\pi}\left(\frac{1}{2} + \frac{\pi}{4}\sin \omega t - \frac{1}{3}\cos 2\omega t - \frac{1}{15}\cos 4\omega t\right.$$
$$\left.- \frac{1}{35}\cos 6\omega t - \cdots\right) \qquad (19\text{-}2\text{-}5)$$

Full-wave rectified ac waveform [Fig. 19-1(f)]:

$$f(t) = \frac{4D}{\pi}\left(\frac{1}{2} - \frac{1}{1 \times 3}\cos 2\omega t - \frac{1}{3 \times 5}\cos 4\omega t - \frac{1}{5 \times 7}\cos 6\omega t - \cdots\right)$$
$$= \frac{4D}{\pi}\left(\frac{1}{2} - \frac{1}{3}\cos 2\omega t - \frac{1}{15}\cos 4\omega t - \frac{1}{35}\cos 6\omega t - \cdots\right) \qquad (19\text{-}2\text{-}6)$$

In each of the expressions, D is the peak value of the waveform.

The square-wave, sawtooth, and rectified ac waveforms are frequently encountered in electronics. The pulse waveform is also quite common, but its analysis is beyond the scope of this book.

To summarize, we have examined a number of periodic, finite, single-valued, and continuous waveforms whose Fourier expansions in general contain a constant value together with sine and cosine terms that represent the fundamental or harmonic components.

EXAMPLE 19-2 The half-wave rectified ac waveform of Fig. 19-1(e) has a peak value of 100 V. Calculate the amplitudes of the fundamental component and all the harmonics up to and including the sixth.

Solution

mean value or dc level, $a_0 = 100 \text{ V} \times \dfrac{1}{\pi} = 31.8 \text{ V}$ (19-2-5)

amplitude of the fundamental sine-wave component, b_1

$$= 100 \text{ V} \times \dfrac{1}{2} = 50 \text{ V}$$

amplitude of the second-harmonic component, a_2

$$= (-)100 \text{ V} \times \dfrac{2}{3\pi} = (-)21.2 \text{ V}$$

amplitude of the fourth-harmonic component, a_4

$$= (-)100 \text{ V} \times \dfrac{2}{15\pi} = (-)4.24 \text{ V}$$

amplitude of the sixth-harmonic component, a_6

$$= (-)100 \times \dfrac{2}{35\pi} = (-)1.82 \text{ V}$$

19-3 Effective Value of a Nonsinusoidal Wave

We have already seen that a nonsinusoidal wave may, in general, be expressed in terms of a dc level together with fundamental and harmonic components [Eq. (19-2-1)]. Let us assume that this wave is in a form of a source voltage which is connected across a resistor, R. Each component provides its own power dissipation and since the individual powers dissipated are additive, we can derive an equation for the effective voltage of the source. This value would be the reading of an ac voltmeter connected across the source provided that the instrument could respond to the frequencies of the higher harmonics.

total power dissipated, $P_T = \dfrac{V_{DC}^2}{R} + \dfrac{V_1^2}{R} + \dfrac{V_2^2}{R} + \cdots$ watts   (19-3-1)

where V_{dc} = dc level of the voltage source (V)
V_1 = effective value of the fundamental component (V)
V_2 = effective value of the second-harmonic component (V)

But

$$P_T = \dfrac{V_T^2}{R} \quad \text{watts} \quad (19\text{-}3\text{-}2)$$

where V_T is the effective value of the voltage source (V).
Comparing Eqs. (19-3-1) and (19-3-2) gives

$$V_T^2 = V_{dc}^2 + V_1^2 + V_2^2 + \cdots$$

or

$$V_T = \sqrt{V_{dc}^2 + V_1^2 + V_2^2 + \cdots} \quad \text{volts} \qquad (19\text{-}3\text{-}3)$$

A similar analysis reveals that

$$I_T = \sqrt{I_{dc}^2 + I_1^2 + I_2^2 + \cdots} \quad \text{amperes} \qquad (19\text{-}3\text{-}4)$$

where I_T is the effective value of a nonsinusoidal current.

EXAMPLE 19-3 Calculate the total effective value of the component voltages calculated in Example 19-2.

Solution

$$\text{total effective voltage} = \sqrt{31.8^2 + \frac{50^2}{2} + \frac{21.2^2}{2} + \frac{4.24^2}{2} + \frac{1.82^2}{2}}$$
$$\approx 50 \text{ V}$$

This shows that the rms value of the half sine wave is 0.5 of the peak value. By comparison, the rms value of the full sine wave is 0.707 of the peak value.

19-4 Nonsinusoidal Current in a Series AC Circuit

If a nonsinusoidal voltage source is applied across a series ac circuit, the resulting current will contain fundamental and harmonic components provided that all these components are contained in the voltage waveform.

When the series ac circuit contains both inductors and capacitors, their reactances must be calculated separately for the fundamental and harmonic frequencies in order to obtain the value of the corresponding current values.

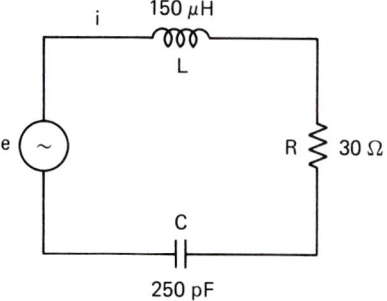

FIGURE 19-6 Nonsinusoidal voltage applied to a series *LCR* circuit.

In other words, for each frequency the current must be recalculated. When the values of all the current components are known, we can use Eq. (19-3-4) to obtain the effective value of the nonsinusoidal current drawn from the source. Let us illustrate these principles by the following example.

A nonsinusoidal source has a voltage that is represented by the equation $e = 100 \sin \omega t - 80 \sin (2\omega t + 40°) + 40 \sin (3\omega t - 20°)$ V, where the frequency $f = \omega/2\pi = 400$ kHz. This voltage is then applied across the series circuit of Fig. 19-6. We will now obtain the equation of the nonsinusoidal current and then calculate its effective value.

STEP 1: THE FUNDAMENTAL COMPONENT

$$\text{inductive reactance, } X_L = 2 \times \pi \times f \times L$$
$$= 2 \times \pi \times 400 \times 10^3 \times 150 \times 10^{-6}$$
$$= 377 \ \Omega$$

$$\text{capacitive reactance, } X_C = \frac{1}{2 \times \pi \times f \times C}$$
$$= \frac{1}{2 \times \pi \times 400 \times 10^3 \times 250 \times 10^{-12}}$$
$$= 1592 \ \Omega$$

$$\text{total impedance, } Z = 30 + j377 - j1592$$
$$= 30 - j1215$$
$$= 1215 \angle -88.6° \ \Omega$$

$$\text{peak value of the fundamental current, } I_1 = \frac{100 \angle 0° \text{ V}}{1215 \angle -88.6° \ \Omega}$$
$$= 0.082 \angle 88.6° \text{ A}$$

$$\text{effective value of the fundamental current} = 0.082 \times 0.707$$
$$= 0.058 \text{ A}$$

STEP 2: THE SECOND-HARMONIC COMPONENT

$$\text{inductive reactance, } X_L = 2 \times 377 = 754 \ \Omega$$

$$\text{capacitive reactance, } X_C = \frac{1592}{2} = 796 \ \Omega$$

$$\text{total impedance, } Z = 30 + j754 - j796$$
$$= 30 - j42$$
$$= 51.6 \angle -54.5° \ \Omega$$

$$\text{peak value of the second-harmonic current, } I_2 = \frac{-80 \angle 40° \text{ V}}{51.6 \angle -54.5° \ \Omega}$$
$$= -1.55 \angle 94.5°$$
$$= 1.55 \angle -85.5° \text{ A}$$

$$\text{effective second-harmonic current} = 0.707 \times 1.55 = 1.10 \text{ A}$$

Sec. 19-4 Nonsinusoidal Current in a Series AC Circuit

STEP 3: THE THIRD-HARMONIC COMPONENT

$$\text{inductive reactance, } X_L = 3 \times 377 = 1131 \ \Omega$$

$$\text{capacitive reactance, } X_C = \frac{1592}{3} = 531 \ \Omega$$

$$\text{total impedance, } Z = 30 + j1131 - j531$$
$$= 30 + j600$$
$$= 600.7 \underline{/87.1°} \ \Omega$$

$$\text{peak value of the third-harmonic current, } I_3 = \frac{40 \underline{/-20°} \ \text{V}}{600.7 \underline{/87.1°} \ \Omega}$$
$$= 0.067 \underline{/-107.1°} \ \text{A}$$

$$\text{effective value of the third-harmonic current} = 0.707 \times 0.067$$
$$= 0.047 \ \text{A}$$

$$\text{total nonsinusoidal current, } i = 0.082 \sin(\omega t + 88.6°) + 1.55 \sin(2\omega t - 85.50°) + 0.067 \sin(3\omega t - 107.1°) \ \text{A}$$

$$\text{effective nonsinusoidal current, } I = \sqrt{0.058^2 + 1.10^2 + 0.047^2}$$
$$= 1.10 \ \text{A} \qquad (19\text{-}3\text{-}4)$$

$$\text{total power dissipated} = I^2 R$$
$$= (1.10 \ \text{A})^2 \times 30 \ \Omega = 36.5 \ \text{W}$$

The analysis of a parallel LCR circuit is similar to that of the series circuit. The individual branch currents for the fundamental and harmonic components are first calculated. Using the rules of complex algebra, the branch currents are combined to create the supply current, which is drawn from the nonsinusoidal voltage source.

19-5

Nonsinusoidal Currents in a Series–Parallel Circuit

When a nonsinusoidal voltage source is applied across the series–parallel circuit of Fig. 19-7, we can determine the fundamental current components in each of the two branches. The fundamental line current is then the phasor sum of the branch currents. The procedure is repeated for each of the harmonic components. Finally, the various currents are combined [Eq. (19-3-4)] to obtain the effective value of the line current drawn from the source. These principles are illustrated in the following example.

In Fig. 19-7, the nonsinusoidal voltage source is expressed by the equation $e = 14.14 \sin(\omega t + 20°) + 3.535 \sin(3\omega t + 40°) - 1.414 \sin(5\omega t - 60°)$ V, where $\omega = 2\pi f = 1000$ rad/s. The reactance values shown correspond to the fundamental frequency. Obtain the equation of the instantaneous line current and then calculate its effective value.

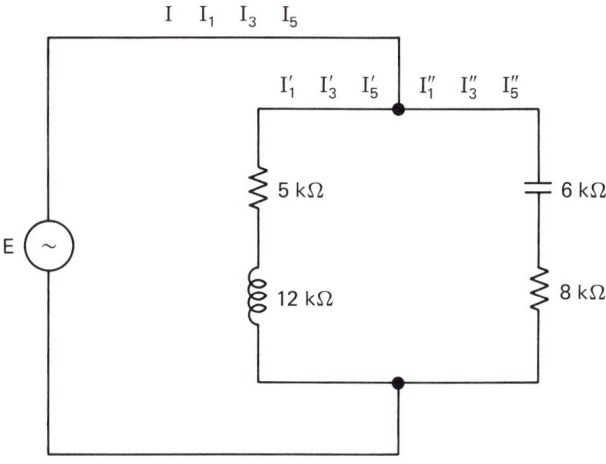

FIGURE 19-7 Nonsinusoidal voltage applied to a series–parallel ac circuit.

STEP 1: FUNDAMENTAL CURRENT

$$\text{fundamental current component, } I'_1 = \frac{14.14 \angle 20° \text{ V}}{5 + j12 \text{ k}\Omega}$$

$$= \frac{14.14 \angle 20° \, (5 - j12)}{5^2 + 12^2}$$

$$= \frac{14.14 \angle 20° \times 13 \angle -67.4°}{169}$$

$$= 1.09 \angle -47.4° \text{ mA}$$

$$\text{fundamental current component, } I''_1 = \frac{14.14 \angle 20° \text{ V}}{8 - j6 \text{ k}\Omega}$$

$$= \frac{14.14 \angle 20° \times (8 + j6)}{100}$$

$$= \frac{14.14 \angle 20° \times 10 \angle 36.8°}{100}$$

$$= 1.414 \angle 56.9° \text{ mA}$$

$$\text{fundamental line current, } I_1 = 1.09 \angle -47.4° + 1.414 \angle 56.9°$$

$$= 0.737 - j0.800 + 0.773 + j1.18$$

$$= 1.501 + j0.384 = 1.56 \angle 14.26° \text{ mA}$$

$$\text{effective value of the fundamental line current} = 0.707 \times 1.56$$

$$= 1.10 \text{ mA}$$

STEP 2: THIRD-HARMONIC CURRENT

At the third-harmonic frequency, the new reactance of the inductor is $3 \times 12 = 36$ kΩ and the new reactance of the capacitor is $6/3 = 2$ kΩ.

Sec. 19-5 Nonsinusoidal Currents in a Series–Parallel Circuit

third-harmonic current component, $I_3' = \dfrac{3.535 \angle 40° \times (5 - j36)}{5^2 + 36^2}$

$= \dfrac{3.535 \angle 40° \times 36.3 \angle -82.1°}{1321}$

$= 0.097 \angle -42.1°$ mA

third-harmonic current component, $I_3'' = \dfrac{3.535 \angle 40° \times (8 + j2)}{8^2 + 2^2}$

$= \dfrac{3.535 \angle 40° \times 8.25 \angle 14.0°}{68}$

$= 0.429 \angle 54.03°$ mA

third-harmonic line current, $I_3 = 0.097 \angle -42.1° + 0.429 \angle 54.03°$

$= 0.072 - j0.065 + 0.252 + j0.347$

$= 0.324 + j0.282$

$= 0.429 \angle 41.02°$ mA

effective value of the third-harmonic line current $= 0.707 \times 0.429$

$= 0.304$ mA

STEP 3: FIFTH-HARMONIC CURRENT At the fifth-harmonic frequency the inductive reactance is $5 \times 12 = 60$ kΩ and the capacitive reactance is $6/5 = 1.2$ kΩ.

fifth-harmonic current component, $I_5' = \dfrac{-1.414 \angle -60° \times (5 - j60)}{5^2 + 60^2}$

$= \dfrac{-1.414 \angle -60° \times 60.2 \angle -85.2°}{3625}$

$= -0.0235 \angle -145.2°$

$= 0.0235 \angle 34.8°$ mA

fifth-harmonic current component, $I_5'' = \dfrac{1.414 \angle -60° \times (8 + j1.2)}{8^2 + 1.2^2}$

$= \dfrac{-1.414 \angle -60° \times 8.09 \angle 8.53°}{65.44}$

$= -0.175 \angle -51.47°$

$= 0.175 \angle 128.5°$ mA

fifth-harmonic line current, $I_5 = 0.0235 \angle 34.8° + 0.175 \angle 128.5°$

$= 0.0193 + j0.0134 - 0.109 + j0.1367$

$= -0.0896 + j0.1501$

$= 0.175 \angle 120.8°$ mA

effective value of the fifth-harmonic line current $= 0.707 \times 0.175$

$= 0.124$ mA

STEP 4 The expression for the instantaneous line current is

$$i = i_1 + i_3 + i_5$$
$$= 1.56 \sin(\omega t + 14.26°) + 0.429 \sin(3\omega t + 41.02°)$$
$$+ 0.175 \sin(5\omega t + 120.8°) \text{ mA}$$

The effective line current is

$$I = \sqrt{1.10^2 + 0.304^2 + 0.124^2} = 1.15 \text{ mA}$$

Chapter Summary

19.1. Fourier's theorem:
Any finite, continuous, single-valued function $f(t)$ which has a period of 2π radians (360°) may be expressed by the following series:

$$f(t) = a_0 + A_1 \sin(\omega t + \phi_1) + A_2 \sin(2\omega t + \phi_2) + A_3 \sin(3\omega t + \phi_3)$$
$$= a_0 + a_1 \cos \omega t + a_2 \cos 2\omega t + a_3 \cos 3\omega t + \cdots$$
$$+ b_1 \sin \omega t + b_2 \sin 2\omega t + b_3 \sin 3\omega t + \cdots$$

19.2. Series equations for various nonsinusoidal waveforms:
Symmetrical square wave [Fig. 19-1(d)]:

$$f(t) = \frac{4D}{\pi}\left(\sin \omega t + \frac{1}{3}\sin 3\omega t + \frac{1}{5}\sin 5\omega t + \frac{1}{7}\sin 7\omega t + \cdots\right)$$

where D is the peak value of the symmetrical square wave.
Negative sawtooth waveform [Fig. 19-1(a)]:

$$f(t) = \frac{2D}{\pi}\left(\sin \omega t + \frac{1}{2}\sin 2\omega t + \frac{1}{3}\sin 3\omega t + \frac{1}{4}\sin 4\omega t + \cdots\right)$$

where D is the peak value of the negative sawtooth waveform.
Positive sawtooth waveform [Fig. 19-1(b)]:

$$f(t) = \frac{2D}{\pi}\left(\sin \omega t - \frac{1}{2}\sin 2\omega t + \frac{1}{3}\sin 3\omega t - \frac{1}{4}\sin 4\omega t + \cdots\right)$$

where D is the peak value of the positive sawtooth waveform.
Modified sawtooth waveform (Fig. 19-5):

$$f(t) = \frac{D}{4} - \frac{2D}{\pi^2}\left(\cos \omega t + \frac{\cos 3\omega t}{3^2} + \frac{\cos 5\omega t}{5^2} + \frac{\cos 7\omega t}{7^2} + \cdots\right)$$
$$+ \frac{D}{\pi}\left(\sin \omega t - \frac{\sin 2\omega t}{2} + \frac{\sin 3\omega t}{3} + \cdots\right)$$

where D is the peak value of the modified sawtooth waveform.

Triangular waveform [Fig. 19-1(c)]:

$$f(t) = \frac{8D}{\pi^2}\left(\cos \omega t + \frac{\cos 3\omega t}{3^2} + \frac{\cos 5\omega t}{5^2} + \frac{\cos 7\omega t}{7^2} + \cdots\right)$$

where D is the peak value of the triangular waveform.
Half-wave rectification [Fig. 19-1(e)]:

$$f(t) = \frac{2D}{\pi}\left(\frac{1}{2} + \frac{\pi}{4}\sin \omega t - \frac{1}{1 \times 3}\cos 2\omega t - \frac{1}{3 \times 5}\cos 4\omega t - \frac{1}{5 \times 7}\cos 6\omega t - \cdots\right)$$

where D is the peak value of the half-rectified sine wave.
Full-wave rectification:

$$f(t) = \frac{4D}{\pi}\left(\frac{1}{2} - \frac{1}{1 \times 3}\cos 2\omega t - \frac{1}{3 \times 5}\cos 4\omega t - \frac{1}{5 \times 7}\cos 6\omega t - \cdots\right)$$

where D is the peak value of the full-rectified sine wave.

19.3. Effective values of a nonsinusoidal wave:

$$\text{total effective current} = \sqrt{I_{dc}^2 + I_1^2 + I_2^2 + I_3^2 + \cdots}$$

$$\text{total effective voltage} = \sqrt{V_{dc}^2 + V_1^2 + V_2^2 + V_3^2 + \cdots}$$

19.4. Application of a nonsinusoidal voltage source to ac circuits:

The individual currents must be calculated separately for each of the fundamental and harmonic components contained in the waveform of the voltage source. These currents are then combined to obtain the equation of the instantaneous current at any point in the circuit. From a knowledge of the instantaneous equations, the effective values of the currents can be calculated.

Self-Review

True-False Questions

1. **T. F.** The Fourier series for the symmetrical square wave of Fig. 19-1(d) contains only sine terms.
2. **T. F.** The harmonic analysis of the triangular waveform in Fig. 19-1(c) contains both sine and cosine terms.
3. **T. F.** Fourier analysis may be applied to any periodic waveform that is finite, single-valued, and discontinuous.
4. **T. F.** The higher the order of a harmonic component in a Fourier series, the greater is its peak value.
5. **T. F.** In Fig. 19-1(e), the dc level of the rectified sine wave is $1/\pi \times$ peak value.
6. **T. F.** Referring to Fig. 19-1(f), the peak value of the fourth-harmonic component is one fifth of the second harmonic's peak value.
7. **T. F.** The analysis of the negative sawtooth waveform in Fig. 19-1(a) contains only cosine terms.
8. **T. F.** In the analysis of the modified sawtooth waveform (Fig. 19-5), the peak value of the eighth-harmonic cosine term is $2D/(8\pi)^2$.
9. **T. F.** In the analysis of the positive sawtooth waveform [Fig. 19-1(b)] the peak value of the second harmonic cosine term is zero.
10. **T. F.** The effective value of a nonsinusoidal current is found by arithmetically adding the rms values of the fundamental and harmonic components.

11. T. F. A periodic wave is one that is repeated at regular intervals.
12. T. F. Any waveform that is finite, continuous, and periodic may be decomposed into sine waves.
13. T. F. The peak value of the fundamental component for a square wave is automatically equal to the peak value of the square wave.
14. T. F. Theoretically, the Fourier series of a sawtooth wave requires an infinite number of harmonics.
15. T. F. The Fourier series for a sine wave, after full-wave rectification, has no fundamental component.
16. T. F. A nonsinusoidal wave has an effective value that is equal to the square of the sum of the component effective values squared.
17. T. F. The effective value of the Fourier series is not affected by the amount of dc component present.
18. T. F. The principle of superposition applies when determining the fundamental and harmonic voltages and currents in a linear network in which there exists one or more nonsinusoidal sources.
19. T. F. The higher the order of the harmonic, the greater will be the frequency of the sine wave component.
20. T. F. To be symmetrical, it is *only* necessary that the positive and negative peaks be numerically equal and differ in polarity.

Multiple-Choice Questions

21. The negative sawtooth waveform of Fig. 19-1(a) has a peak value of 15 V. The amplitude of its fourth-harmonic component is
    (a) 3.75 V   (b) 2.39 V   (c) 7.5 V   (d) 4.78 V
    (e) 1.19 V
22. The positive sawtooth waveform of Fig. 19-1(b) has a peak value of 8 V. The amplitude, with sign, of its second-harmonic component is
    (a) +4 V   (b) +2.55 V   (c) −4 V   (d) −2.55 V
    (e) −5.09 V
23. The triangular waveform of Fig. 19-1(c) has a peak value of 12 V. The amplitude of its fifth-harmonic component is
    (a) 0.39 V   (b) zero   (c) 1.22 V   (d) 1.95 V   (e) 2.44 V
24. The symmetrical square wave of Fig. 19-1(d) has a peak value of 18 V. The amplitude of its fundamental component is
    (a) 5.73 V   (b) 11.47 V   (c) 2.87 V   (d) 9 V   (e) 22.9 V
25. A half-wave rectified voltage [Fig. 19-1(e)] has a peak value of 7 V. The value of its mean dc level is
    (a) zero   (b) 2.23 V   (c) 4.46 V   (d) 3.5 V   (e) 1.12 V
26. A full-wave rectified voltage [Fig. 19-1(f)] has a peak value of 23 V. The amplitude of its fundamental component is
    (a) 14.6 V   (b) 7.3 V   (c) 9.76 V   (d) 19.52 V   (e) zero
27. The modified sawtooth waveform of Fig. 19-5 has a peak value of 6 V. The peak value of its second-harmonic component is
    (a) 0.636 V   (b) −0.955 V   (c) 1.22 V   (d) 1.5 V
    (e) 1.91 V
28. At a particular point in an ac circuit the instantaneous current consists of a fundamental component with a peak value of 3 A, together with a second-harmonic component whose peak value is 2 A. The effective value of the current is
    (a) 5 A   (b) 3.6 A   (c) 2.5 A   (d) 3.2 A   (e) 1.5 A
29. The instantaneous voltage across a 1-kΩ resistor may be expressed as $4 + 3 \sin \omega t$ V. The power dissipated in the resistor is
    (a) 8.42 mW   (b) 25 mW   (c) 19.5 mW   (d) 20.5 mW
    (e) 18.7 mW
30. The instantaneous value of a nonsinusoidal current is expressed by the equation $i = 7 - 4 \sin \omega t + 2 \sin 2\omega t$ A. Its effective value is
    (a) 6.56 A   (b) 11.24 A   (c) 9.31 A   (d) 8.44 A
    (e) 7.68 A

31. Identify a periodic function or functions.
    - (a) wallpaper patterns
    - (b) lightning stroke
    - (c) power supply ripple voltage
    - (d) automobile ignition sparks
    - (e) all choices except (b) are correct
32. The process of adding the fundamental and harmonics for a periodic function is called
    - (a) constructive addition
    - (b) Fourier analysis
    - (c) synthesis
    - (d) Fourier addition
    - (e) logical addition
33. The frequency of the fifth harmonic of a 4-kHz square wave is
    - (a) 1 kHz
    - (b) 3 kHz
    - (c) 4 kHz
    - (d) 5 kHz
    - (e) no correct choice
34. The period of the fifth harmonic for a 4-kHz square wave is
    - (a) 40 µs
    - (b) 50 µs
    - (c) 100 µs
    - (d) 200 µs
    - (e) 250 µs
35. A modified sawtooth wave, as shown in Fig. 19-5 has a period of 15 ms. In each cycle the wave voltage is zero during a time interval of
    - (a) π ms
    - (b) π/ω ms
    - (c) 5 ms
    - (d) 7.5 ms
    - (e) 15 ms
36. The time span between the peaks of the modified sawtooth wave in Fig. 19-5 is 25 µs. The lowest frequency of any harmonic is
    - (a) 20 kHz
    - (b) 40 kHz
    - (c) 60 kHz
    - (d) 120 kHz
    - (e) 200 kHz
37. The inductive reactance of a particular series circuit is 400 Ω at the fundamental frequency. The inductive reactance for the fifth harmonic is
    - (a) 80 Ω
    - (b) 179 Ω
    - (c) 400 Ω
    - (d) 566 Ω
    - (e) 2000 Ω
38. The capacitive reactance of a particular series circuit is 400 Ω at the fundamental frequency. The capacitive reactance for the fifth harmonic is
    - (a) 80 Ω
    - (b) 179 Ω
    - (c) 400 Ω
    - (d) 566 Ω
    - (e) 2000 Ω
39. Two particular harmonics of the Fourier series for a square wave have frequencies of 400 Hz and 720 Hz, respectively. The fundamental frequency is
    - (a) 50 Hz
    - (b) 80 Hz
    - (c) 90 Hz
    - (d) 100 Hz
    - (e) 360 Hz
40. A periodic wave has a dc component of 4 V, a fundamental frequency component of 10 V (peak), and a second harmonic component of 5 V (peak). The rms value is
    - (a) 19 V
    - (b) 41 V
    - (c) 8.86 V
    - (d) 125 V
    - (e) 141 V

Practice Problems

41. The negative sawtooth waveform of Fig. 19-1(a) has a peak value of 150 V. In its harmonic analysis, what is the peak value of its sixth-harmonic component?
42. The positive sawtooth waveform of Fig. 19-1(b) has a peak-to-peak value of 80 V. In its harmonic analysis, what is the peak value of its fifth-harmonic component?
43. The triangular waveform of Fig. 19-1(c) has a peak value of 35 V. Obtain its Fourier series up to and including the seventh-harmonic component.
44. The symmetrical square wave of Fig. 19-1(d) has a peak-to-peak value of 45 V. What is the peak value of its ninth-harmonic component?
45. The half-wave rectified voltage of Fig. 19-1(e) has a mean dc level of 32 V. What is the peak value of its fourth-harmonic component?
46. The full-wave rectified voltage of Fig. 19-1(f) has a sixth-harmonic component whose peak value is 3 V. What is the value of its mean dc level?
47. The modified sawtooth waveform of Fig. 19-5 has a dc level of 9 V. What is the peak value of its second-harmonic component?
48. The instantaneous voltage output of a nonsinusoidal source is expressed by the equation $e = 24 \sin \omega t + 12 \sin 2\omega t + 8 \sin 3\omega t$ V, where $\omega = 377$ rad/s. If this voltage is impressed across a 4-kΩ resistor, obtain the expression for the instantaneous current and calculate its effective value. What is the total amount of the power dissipated in the resistor?
49. The voltage of Problem 48 is applied across an 8-H inductor. Obtain the expression for the instantaneous current and calculate its effective value.
50. The voltage of Problem 48 is applied across a 20-µF capacitor. Obtain the expression for the instantaneous current and calculate its effective value.

51. The modified sawtooth waveform of Fig. 19-5 has a peak value of exactly 25 V. Calculate the amplitude of the fundamental component.
52. The triangular waveform of Fig. 19-1(c) is measured to have a peak-to-peak value of 20 V. Calculate the amplitude of the fundamental component.
53. A sine wave, after half-wave rectification, has a peak value of 100 V. Determine the dc component.
54. In Problem 53, what is the amplitude of the fundamental component?
55. A negative sawtooth waveform [Fig. 19-1(a)] has a peak value of 25 V. Calculate the dc component.
56. In Problem 55, what is the amplitude of the fundamental component?
57. A particular nonsinusoidal voltage waveform has a 5-V dc component, a 15-V fundamental, and a 1.5-V third-harmonic component. Determine the effective voltage.
58. The nonsinusoidal voltage wave of Problem 57 is applied to a 100-Ω resistor. Calculate the effective current and the power dissipated.
59. The nonsinusoidal voltage of Problem 57 is now applied to a series combination of C, having negligible reactance at 400 Hz, and L, whose reactance is 50 Ω at 400 Hz. Determine the effective current.
60. The nonsinusoidal voltage of Problem 57 is now applied to a single capacitor having a reactance of 50 Ω at 400 Hz. Determine the effective current.

Advanced Problems

61. A nonsinusoidal voltage is expressed by the equation $e = 23 + 17 \sin \omega t - 11 \sin 3\omega t + 8 \sin 5\omega t$ V. Calculate its effective value. If this voltage is applied across a 2-kΩ resistor, what is the total amount of the power dissipated?
62. The full-wave rectified sine wave of Fig. 19-1(f) has a peak value of 75 V. What is the effective value of its tenth-harmonic component?
63. In the harmonic analysis of a symmetrical square wave [Fig. 19-1(d)], the peak value of the fifth-harmonic component is 3.5 V. What are the peak and effective values of the square wave?
64. The modified sawtooth waveform of Fig. 19-5 has a peak value of 24 V. Calculate the peak value of its fundamental sine-wave component.
65. In the triangular waveform of Fig. 19-1(c), the effective value of the third-harmonic component is 7.5 V. What is the peak value of the triangular waveform?
66. The instantaneous voltage output of a nonsinusoidal source is expressed by the equation $e = 120 \sin \omega t + 40 \sin 2\omega t - 24 \sin 3\omega t$ V, where $\omega = 628$ rad/s. If this voltage is applied across a 5-kΩ resistor in series with a 3-H inductor, obtain the expression for the instantaneous current and calculate its effective value. What is the total amount of power dissipated in the resistor?
67. The source voltage of Problem 66 is applied across a 2-kΩ resistor in parallel with a 1-μF capacitor. Obtain the expression for the instantaneous source current and calculate its effective value. What is the total amount of power dissipated in the resistor?
68. The source voltage of Problem 66 is applied across a series LCR circuit in which $L = 0.6$ H, $C = 1$ μF, and $R = 1$ kΩ. Obtain the expression for the instantaneous current and calculate its effective value.
69. The source voltage of Problem 66 is applied across a parallel LCR circuit in which $L = 0.6$ H, $C = 1$ μF, and $R = 1$ kΩ. Obtain the expression for the instantaneous source current and calculate its effective value.
70. In the series–parallel circuit of Fig. 19-8 the source voltage is expressed by the equation $e = 100 \sin \omega t - 50 \sin 2\omega t + 35 \sin 3\omega t$ V, where $\omega = 1257$ rad/s. Obtain the equation of the instantaneous current, i, drawn from the source and calculate its effective value.
71. A 60-Hz sine wave provides power to a full-wave rectifier as shown in Fig. 19-1(f). Calculate the amplitude of the fundamental if the amplitude of the third harmonic is known to be 35 mV.

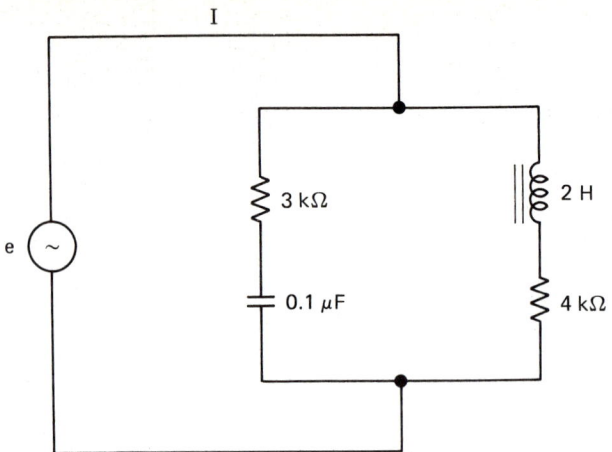

FIGURE 19-8 Circuit for Problem 70.

72. The fifth harmonic of a modified sawtooth wave (Fig. 19-5) can be written in equation form as $e = 5 \cos(4 \times 10^3)t$ V.
    Determine the frequency of the original modified sawtooth waveform.
73. The peak voltage for a modified sawtooth waveform (Fig. 19-5) is 75 V and the period is 100 μs. Write the equation for all components up to and including the third harmonic.
74. The peak-to-peak voltage for a triangular waveform [Fig. 19-1(c)] is 28 V and the frequency is 5 kHz. Write the equation for all components up to and including the third harmonic.
75. It can be shown that the effective value of a triangular wave with a peak value of 1 V (2 V peak to peak) is $1/\sqrt{3}$. Construct a table showing the effective value, adding harmonics, one at a time, up to and including the eleventh harmonic.
76. A simple low-pass filter appears in Fig. 19-9. Prepare a table of

$$\frac{V_o}{V_{IN}} = \frac{1}{1 + j\omega RC}$$

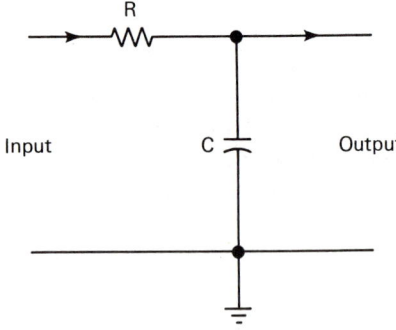

FIGURE 19-9 Low-pass filter.

for the fundamental component and each harmonic, up to and including the eleventh, assuming for the fundamental frequency that $\omega RC = 1$.

77. A simple high-pass filter appears in Fig. 19-10. Prepare a table of

$$\frac{V_o}{V_{IN}} = \frac{1}{1 - j/\omega RC}$$

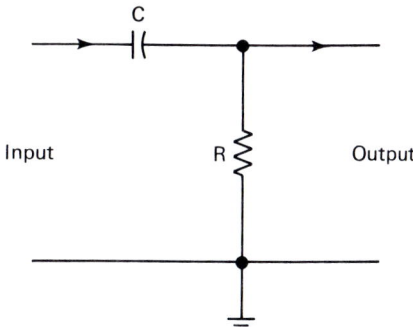

FIGURE 19-10 High-pass filter.

for the fundamental frequency component and each odd harmonic, up to and including the eleventh, assuming for the fundamental frequency that $\omega RC = 1$.

78. In Fig. 19-9, let $RC = 1$ to simplify the expression for V_o/V_{IN}. Also, a square wave voltage of peak value D is applied such that the fundamental frequency is significantly greater than the cutoff frequency so that

$$\frac{V_o}{V_{IN}} = \frac{1}{1 + j\omega} \approx \frac{1}{j\omega} = \frac{-j}{\omega}$$

$$V_{IN} = \frac{4D}{\pi}\left(\sin \omega t + \frac{\sin 3\omega t}{3} + \frac{\sin 5\omega t}{5} + \cdots\right) V$$

Find an expression for V_o.

79. Interpret the results of the solution for Problem 78 as representing a Fourier series for another waveform introduced in this chapter.
80. Calculate the effective values of V_{IN} and V_o for Problem 78 assuming $D = 1$ V.

20

AC Electrical Measurements

This chapter has the following objectives:

1. To indicate the methods for adapting the d'Arsonval meter movement, and other electromechanical meter movements, for ac measurements.
2. To analyze the design of rectifier circuits and their use in "peak responding" and "average responding" ac meters.
3. To relate scale markings of the basic meter movements as they must be modified for proper sine-wave ac calibration.
4. To consider thermocouple meter devices as providing true effective values for dc and all ac waveforms, relatively independent of frequency, with the same scale calibration.
5. To describe other meter movements for ac measurement, including the moving-iron and electrodynamometer instruments.
6. To indicate methods for ac power and energy measurement using the electrodynamometer and related principles.
7. To introduce and enumerate the advantages of electronic analog instruments using moving-coil meter movements.
8. To describe a range of specialized electronic meters named electrometers, microvoltmeters, nanometers, and picometers and vector instruments for ac applications.
9. To provide an abbreviated technical description of basic digital meters.
10. To indicate and analyze digital meter performance features as they are related to sensitivity, resolution, and accuracy.
11. To describe how digital meters can give true effective value readings.
12. To summarize ac measurements as made using the cathode ray oscilloscope (CRO).
13. To illustrate ac measurement procedures using the CRO.

20-0 Introduction

The subject of dc measurements was introduced and developed in Chapter 10. We are concerned here with extending the same underlying principles as they apply to ac electrical measurements. It will be discovered in comparing the techniques found here for ac measurements with those studied earlier for dc measurements that four areas can be distinguished:

1. The adaptation of the D'Arsonval meter movement and others covered in Chapter 10 to the measurement of ac voltage, current, power, and energy
2. Application of electronic amplifiers for the purpose of reducing loading effects and to improve sensitivity
3. The introduction of digital meters having greatly improved accuracy, sensitivity, and readability for both dc and ac instrumentation
4. Introduction of the basic cathode ray oscilloscope and its use in the direct viewing of waveforms, the measurement of dc components when they exist, and the direct measurement of ac quantities, including time delays and periods

The reader should be reminded that electronic amplifiers, digital devices, and the detailed internal operation of the cathode ray oscilloscope are beyond the technical level of this book. However, since this equipment is commonly encountered in the laboratory today, it is necessary that a working knowledge of their use be obtained.

20-1 Electromechanical Meters

As explained in Chapter 10, the permanent-magnet moving-coil movement, commonly known as the D'Arsonval movement, is the basis of dc voltmeters and ammeters. With this meter movement the torque produced by the interaction of the current in the coil and the permanent magnet's field causes the pointer to deflect an amount proportional to the current. When the direction of current flow is reversed, both the torque and direction of deflection are also reversed. However, the inertia of moving elements is such that the pointer cannot achieve full deflection whenever an ac current of frequency more than a few hertz passes through the winding, so that this meter is not suitable for ac measurements without some modification. We consider first the rectifier as the means of converting ac to dc to be one technique for making ac measurements with the D'Arsonval movement.

A rectifier is a device through which current passes freely in one direction only. Placing a rectifier such as a semiconductor diode in series with the moving coil of the D'Arsonval movement restricts deflection to one direction only when an ac voltage is applied. More specifically, the amount of deflection will depend on the average value of an ac alternation (half-cycle). Since this average value and the effective (rms) value, whose measurement is desired, are directly related for a sine wave, it is possible to modify the meter scale so that the effective value can be read directly for any pointer deflection.

The rectifier circuit just described is called a half-wave rectifier since every negative alternation is blocked, leaving only the positive half-waves as illustrated in Fig. 11-22. A different, more complicated circuit usually having two or possibly four diodes, yields an average current corresponding to both the positive alternation and the negative alternation inverted (see Fig. 11-23). With this full-wave rectifier arrangement the value averaged over a number of full cycles is

$$I_{av} = 0.637 I_p \tag{11-3-8}$$

in which I_{av} represents the average current value and I_p the peak current. Recall also that the effective value, I, of current is

$$I = 0.707 I_p \tag{11-3-3}$$

Solving these equations to find I in terms of I_{av}, we find that

$$I = 1.11 I_{av} \tag{20-1-1}$$

A similar set of equations exist for voltages:

$$V_{av} = 0.637 V_p \tag{11-3-9}$$

$$V = 0.707 V_p \tag{11-3-4}$$

Combining Eqs. (11-3-4) and (11-3-9), we can show that

$$V = 1.11 V_{av} \tag{20-1-2}$$

Equations of the same general form can be obtained for any periodic waveform. In general, the constants in these other equations will not be the same as those given here for the sine wave.

EXAMPLE 20-1 A sine-wave voltage has a peak value of 10 V. Determine (a) the deflection indication of a dc voltmeter when fitted with a full-wave rectifier, and (b) the effective voltage.

Solution

(a) $V_{av} = 0.637 V_p$
$= 0.637 \times 10 = 6.37$ V

(b) $V = 0.707 V_p$
$= 0.707 \times 10 = 7.07$ V

We should interpret this result to mean that the pointer should indicate 7.07 V (rms) on the new ac scale. This particular position of the pointer indicates a reading of 6.37 V on the existing dc scales. Needless to say, the last reading should be disregarded as erroneous for an ac measurement.

The arrangement just described, whereby a permanent-magnet moving-coil (D'Arsonval) movement and rectifier circuits are used for ac measurements, is referred to as "average responding." By adding a system of multipliers, shunts, and switches, its usefulness can be extended to cover the measurement of ac voltage and current in several ranges. The principles described in Chapter 10 for dc measurements and the basic plan for the VOM instrument also apply for ac measurements. However, as stated earlier, the meter face must also be fitted with ac scales to account for the difference between average value (as rectified) and the effective value.

Meter scales established for ac measurements with sine waves give erroneous results when the ac waveforms are not sinusoidal. A triangular wave, for instance, does not have average and effective values related to the peak values as given earlier in this section. Also, these equations do not apply to a square wave for which the peak, average, and effective values are identical.

EXAMPLE 20-2 A square wave of 5 V (peak) is applied to an average responding voltmeter calibrated to read the effective value of a sine wave. Determine the voltmeter reading and compare this reading to the correct effective value of voltage.

Solution Using Eq. (20-1-1) and recalling that the average value of the square wave is the same as the peak value, the meter indication will be

$$V = 1.11 V_{av} = 1.11 \times 5 = 5.55 \text{ V}$$

The correct effective value is exactly 5 V, so that the meter indication is 11% high.

An alternative circuit arrangement is possible for using rectifiers and the D'Arsonval movement in making ac measurements to obtain a "peak responding" meter. The details of the circuit will not be explored further here. However, it is important to recognize that the readings, as obtained on the meter's dc scales, will be the peak values of any ac waveform. In adapting the meter to obtain effective readings of sine waves directly, the ac scales must be superimposed on the meter face to match the 0.707 multiplication figure of Eqs. (11-3-3) and (11-3-4).

EXAMPLE 20-3 A sine wave of 5 V (peak) value is applied to a D'Arsonval meter movement equipped with rectifiers and other auxiliary circuits to be a "peak responding" ac instrument. Determine the readings on the calibrated dc and ac scales.

Sec. 20-1 Electromechanical Meters

Solution On the dc scale the reading will be the same as the peak value, that is, 5 V. On the ac scale the reading will be given correctly as

$$V = 0.707 V_p \quad (11\text{-}3\text{-}4)$$
$$= 0.707 \times 5 = 3.535 \text{ V}$$

EXAMPLE 20-4 A square wave of 5 V (peak) value is applied to the same meter as in Example 20-3. Determine the readings on the calibrated dc and ac scales and calculate the effective voltage error.

Solution On the dc scale the reading will be the same as the peak value, that is, 5 V. On the ac scale the reading will be given incorrectly to be 3.535 V, whereas the correct value is 5 V. The error is

$$\frac{3.535 - 5}{5} \times 100 = -29\%$$

The preceding consideration of rectifier circuits and the basic dc meter movement suggests that another approach to obtaining the rms or effective value might be desirable. Recalling that the rms value of an ac voltage (or current) is defined in terms of the heat generated leads to the idea that a thermocouple may be the answer. A thermocouple consisting of two dissimilar metals joined together supplies a dc voltage at the junction which is proportional to the temperature of the junction. In adapting the thermocouple to ac measurements, the junction is packaged with a heater element (wire) which carries the ac current that is, in turn, derived from the ac voltage or current to be measured by a system of multipliers and shunt resistors (see Fig. 20-1). Using this principle, a dc current of 100 mA, for example, or an ac current having an effective (rms) value of 100 mA will result in the same meter pointer de-

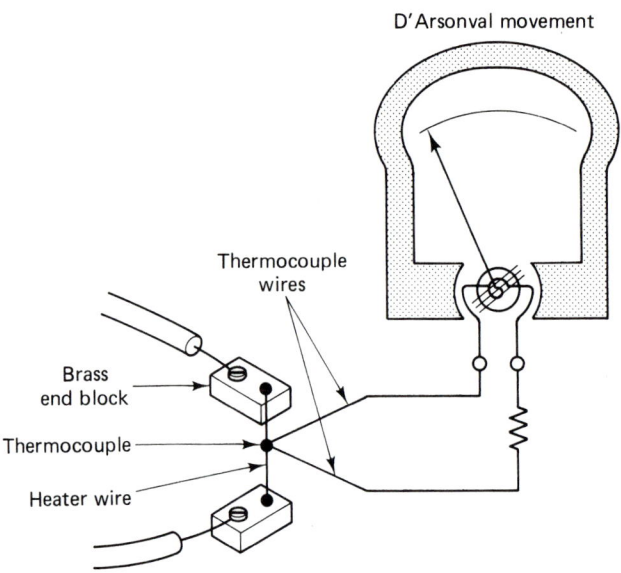

FIGURE 20-1 Thermocouple ac meter.

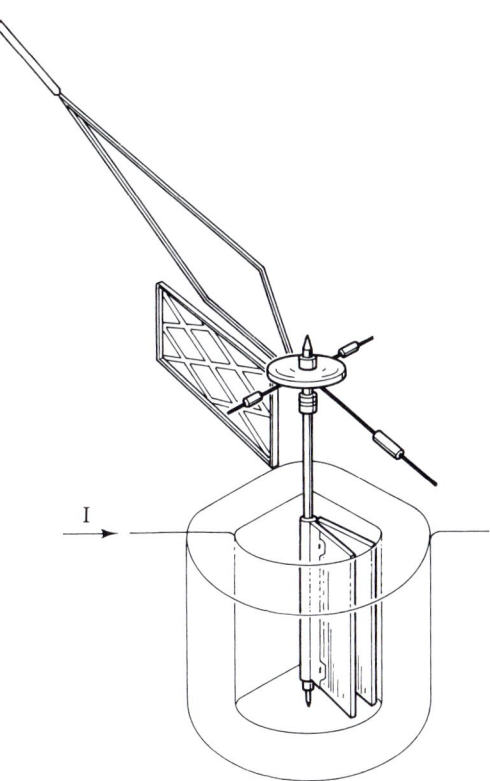

FIGURE 20-2 Radial moving-iron vane movement.

flection regardless of the ac waveform. Application of the thermocouple requires that additional circuitry be provided to compensate for the ambient operating temperature, which adds to the cost. One particular advantage of this plan for ac measurements is its relative independence from the ac frequency, an advantage not shared by the other ac measurement instruments to be introduced. The thermocouple meter arrangement is also discussed in Section 10-7.

One form of moving-iron instrument is shown in Fig. 20-2. Two soft-iron vanes are mounted inside a coil, one being stationary, as is the coil, the other fixed to the shaft upon which the pointer is mounted. When a dc current passes through the coil, the vanes are magnetized, with the corresponding ends being poles of the same kind, either both north or both south. The bar magnets, which the vanes have become, repel each other, causing the pointer to deflect. An alternating current will also magnetize the vanes except that the adjacent ends of the vanes will be magnetically alike at every moment, although alternating. The pointer will deflect an amount depending on the current. Through proper calibration, the scale will indicate the true rms current value. This instrument is useful for sine waves up to several hundred hertz. It is normally applied for power-line frequencies. The repulsion moving-iron meter is also discussed in Section 10-6.

The electrodynamometer meter movement contains a moving coil mounted on a shaft inside a coil, or coils, which produce the magnetic field. As seen in Fig. 20-3, the same ac current passes through both the fixed and movable coils. Since both the magnetic field and the movable coil interaction with it are each proportional to the current, the torque and resulting pointer movement are proportional to the square of the current. The scales of this meter are marked to indicate the square root of the pointer displacement, hence effective (rms) value of ac current. For example, on the 10 A range, full-scale deflection represents $10 = \sqrt{100}$ and the deflection at the midpoint represents $7.07 = \sqrt{50}$ A, the quarter point represents $5 = \sqrt{25}$ A, and so on. The result is a

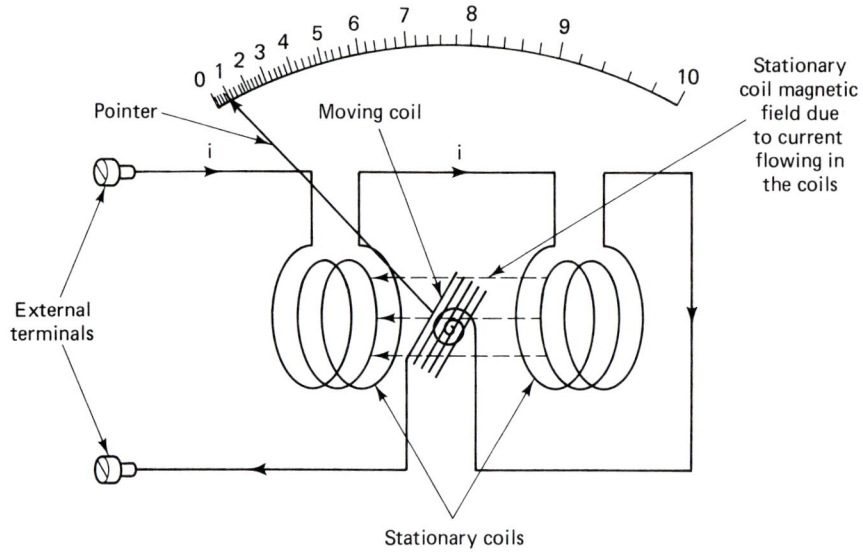

FIGURE 20-3 Electrodynamometer movement.

noticeably nonlinear scale with greater spacing between scale divisions on the upper end, as seen in Fig. 20-3.

A power meter, usually referred to as a wattmeter, can be constructed from the electrodynamometer meter movement as shown in Fig. 20-4. Here the stationary coils are connected in series with the load so that the alternating magnetic field established by these coils is proportional to, and in phase with, the load current. Moreover, the movable coil is connected in series with a large R, and this combination placed across the load so that its current is proportional to the load voltage. Assuming R to be much larger than the inductive reactance of the movable coil, the current through this coil will be essentially in phase with the load voltage. The torque produced by the interaction of the movable coil current and stationary coil current is proportional to the product of the two. When the load voltage and current are in phase, as is the case for a resistive load having a power factor of unity, the pointer deflection can indicate the power directly on a linear, calibrated scale. For frequencies above a few hertz, the shaft-mounted coil and pointer cannot follow the instantaneous power variations and will automatically indicate the average

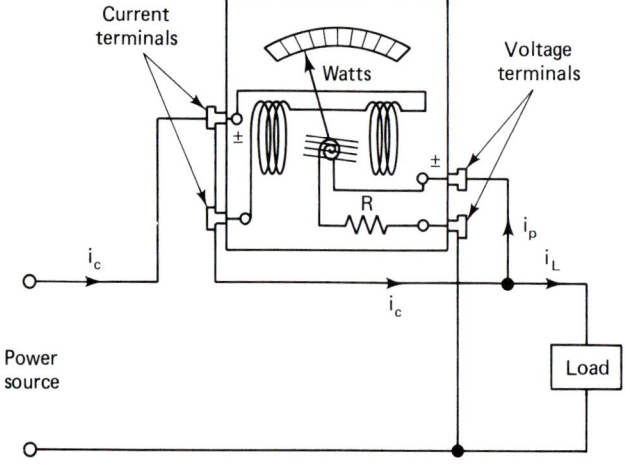

FIGURE 20-4 Dynanometer wattmeter.

762   AC Electrical Measurements Ch. 20

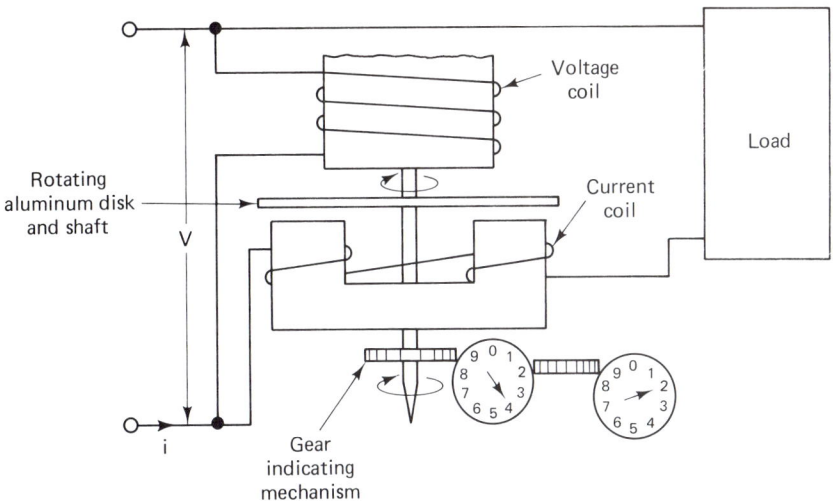

FIGURE 20-5 Watt-hour meter.

power. It can be shown that the torque produced is also proportional to the power factor, so that when the load voltage and current are not in phase, the pointer will still indicate the true power for other than a resistive load.

The watthour meter is commonly used to measure electrical energy supplied to customers by utilities. It is sufficiently accurate to be connected at the electrical entry point of every subscriber and provide the basis for payment for the energy consumed. A diagram illustrating the operation of the watthour meter is shown in Fig. 20-5. Two separate coils, a current coil through which the load current flows and a voltage coil placed across the load, are necessary. Eddy currents are set up by the transformer action in the aluminum disk when load current is flowing. These eddy currents interact with the current in the voltage coil so that a torque acts on the disk, causing a rotational speed proportional to the product of voltage, current, and power factor. The energy, being the time integral of power, is then determined by the number of disk revolutions, which are counted by a gear train and read out from needle positions on dial indicators.

20-2
Analog Electronic Meters

Analog electronic meters usually include an electronic amplifier connected to a D'Arsonval movement [see Fig. 20-6(a)]. Originally, vacuum-tube amplifiers were used and the instrument was known as a vacuum-tube voltmeter or VTVM. More recently, solid-state amplifiers have supplanted those driven by vacuum tubes and the names TVM (transistor voltmeter), FETVM (field-effect transistor voltmeter), or EVM (electronic voltmeter) are encountered. Sometimes VTVM is used as a generic reference for any electronic voltmeter, regardless of the presence of semiconductor amplifiers and the absence of vacuum tubes.

The electronic voltmeter often has ac voltage-measuring capability with a rectifier as shown in Fig. 20-6(b). The rectifier and amplifier circuits of most (but not all) EVMs is of the type that causes a response to the peak-to-peak value of the waveform. Some EVMs have true rms voltage response. There are also a variety of multipurpose analog electronic instruments having several ranges each of dc voltage, ac voltage, dc current, ac current, and resistance. Analog VTVMs first became popular because of the greatly reduced circuit

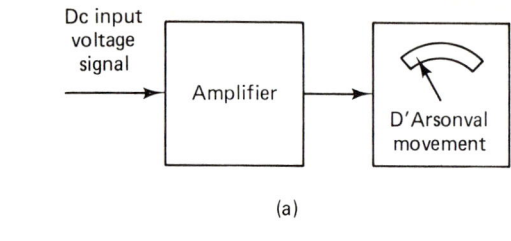

(a)

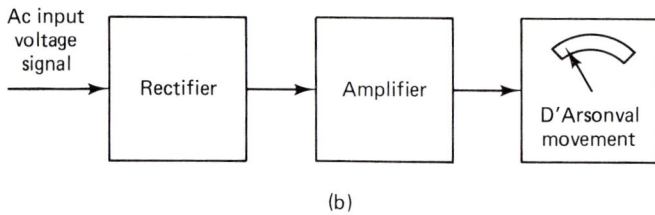

(b)

FIGURE 20-6 (a) Block diagram of a dc electronic analog meter; (b) block diagram of an ac electronic analog meter.

loading (see Section 10-4 for a discussion of dc voltmeter loading) and their utility for measurements at higher frequencies. In recent years sales of digital multimeters (DMM) have largely replaced those of the multipurpose analog electronic meters, but new electronic voltmeters are still available.

Measurement of ac voltages into the megahertz range, and into the hundreds of megaherts with special probes, emphasizes the input impedance, consisting of the input resistance in parallel with the inevitable distributed capacitances found in the amplifier circuits and the cables between the measurement points and the input terminals. Special probes must be used to minimize the effects of these capacitances, noise pickup, and other factors degrading the quality and dependability of the measurements.

A number of special-purpose electronic meters are available to meet measurement needs that cannot be satisfied with general-purpose meters. Although some are employed only for dc measurements, their effectiveness is based on the properties of electronic amplifiers.

Electrometers are dc voltmeters having high (10^{16} Ω or larger) input resistance. They are capable of detecting extremely small currents and quantities of charge.

Microvoltmeters are basically sensitive D'Arsonval movements and high-gain amplifiers. These are available for dc measurements or ac/dc measurements.

Nanometers and *picometers* are essentially microvoltmeters that measure the voltage drop across calibrated shunts through which the current flows.

Vector voltmeters are used to measure the phase difference between the voltages at two points in the circuit at the same time the voltage difference is measured. They are useful in establishing amplifier and filter phase shifts for frequencies into the hundreds of megahertz.

Vector impedance meters are designed to measure the magnitude and angle of an impedance for the frequency of interest. The element, or circuit, is connected to the input terminals of the instrument and the frequency is selected from an internal source. Readings are obtained from two panel meters. Some models can operate up to approximately 100 MHz.

20-3

Digital Multimeters

Digital multimeters are characterized by the display of numerical measurements as digits on the meter face. This, however, is only a small part of the difference between the newer generations of digital meters and the older generations of analog meters. The purpose of this section is to provide a brief description of digital multimeters with particular emphasis on general laboratory, shop, or field use.

The main feature of the digital voltmeter is the circuitry, which converts the voltage level to a digital form. Several methods for accomplishing this are possible. These may differ in time required for each sample point conversion, accuracy, and so on. For example, one method counts clock pulses to indicate the time interval required for a precision voltage ramp to match the input voltage level. The discrete count so generated is then displayed as three or more digits on the instrument case face. Since the input level may change with time, the measurement process must be repeated over and over again periodically to update the measurement indication.

Digital multimeters are available in a variety of packages, the smallest being that for the hand-held DMM, followed in size by the general-purpose bench DMM. The largest and most sophisticated is the microprocessor-controlled system model with interface circuitry for connection to completely automated data acquisition and test systems. None of these DMMs as we know them today would be practical without the existence of digital and linear (analog) semiconductor integrated circuits.

Digital meters are classified according to the number of digits. An overrange feature is often called a "one-half" or a "partial" digit since it cannot display all numbers through 9. Overranging greatly extends the DMM usefulness by maintaining resolution up to, and beyond, full scale. For example, if a voltage being measured changes from 9.99 V to 10.12 V, a three-digit DMM without overranging could measure the first voltage as 9.99 V and the second voltage as 10.1 V. With overranging the second reading will be displayed as 10.12 V with four digits. A three-digit DMM with 100% overranging would have a maximum display of 1999 without regard for the decimal point position and is referred to as a $3\frac{1}{2}$-digit meter.

Resolution is the ratio of the least number of counts that can be displayed to the maximum number of counts. Full-scale resolution of a three-digit DMM is 1 to 1000, or expressed in another way, 0.1%.

Sensitivity refers to the smallest incremental change that the DMM is able to detect. Mathematically, it is the lowest full-scale range multiplied by the resolution of the DMM.

$$\text{sensitivity} = \text{resolution} \times \text{lowest full-scale range} \qquad (20\text{-}3\text{-}1)$$

The voltage sensitivity of a five-digit DMM with a 100 mV lowest voltage range is 10^{-5} times 10^{-1} V $= 10^{-6}$ V $= 1$ μV. For a four-digit DMM having a low range of 100 mV, the sensitivity is $10^{-4} \times 10^{-1} = 10^{-5}$ V $= 10$ μV. A $4\frac{1}{2}$-digit DMM of equal sensitivity will have a low range of 200 mV and a corresponding resolution of 5×10^{-5}.

EXAMPLE 20-5 Calculate the sensitivity of a $3\frac{1}{2}$-digit DMM on the (lowest) 200 mV range.

Solution

$$\text{sensitivity} = \text{resolution} \times \text{lowest full-scale range} \quad (20\text{-}3\text{-}1)$$
$$= 5 \times 10^{-4} \times 0.2 = 10^{-4} \text{ V} = 100 \text{ }\mu\text{V}$$

Accuracy is the exactness to which a voltage, for example, can be determined relative to the accepted standard. To be meaningful, accuracy must be stated along with the conditions under which it will hold. Accuracy is usually expressed as a percent of the reading plus a percent of the range (or full scale).

EXAMPLE 20-6 The accuracy specification of a particular $4\frac{1}{2}$-digit DMM is (0.1% + 1 count). Determine the accuracy, expressed as the uncertainty of the reading, with an actual reading of 10 V on the 20-V range. Also state the result as a percentage of the reading.

Solution

$$0.1\% = 0.001 = 1 \times 10^{-3}$$
$$1 \times 10^{-3} \times 10^{1} = 1 \times 10^{-2} \text{ V} = 0.01 \text{ V}$$
$$1 \text{ count is equivalent to } 5 \times 10^{-5} \times 20 = 10^{-3} = 0.001 \text{ V}$$
$$\text{reading uncertainty} = 0.01 + 0.001 = 0.011 \text{ V}$$
$$\text{percent uncertainty} = 0.011 \times \frac{100}{10} = 0.11\%$$

EXAMPLE 20-7 The accuracy specification of a particular $3\frac{1}{2}$-digit DMM is (0.1% + 1 count). Determine the accuracy, expressed as the uncertainty of the reading, with an actual reading of 10 V on the 20 V range. Also state the result as a percentage of the reading.

Solution

$$0.1\% = 0.001 = 1 \times 10^{-3}$$
$$1 \times 10^{-3} \times 10^{1} = 1 \times 10^{-2} = 0.01 \text{ V}$$
$$1 \text{ count is equivalent to } 5 \times 10^{-4} \times 20 = 10^{-2} = 0.01 \text{ V}$$
$$\text{reading uncertainty} = 0.01 + 0.01 = 0.02 \text{ V}$$
$$\text{percent uncertainty} = 0.02 \times \frac{100}{10} = 0.2\%$$

Table 20-1 summarizes the reading uncertainty for readings of 1 V, 5 V, 10 V, and 19 V on the 20 V scale for the same meter specification as used in Examples 20-5 and 20-6. This same reading uncertainty information is incor-

TABLE 20-1 Reading uncertainty of different voltage using 20 V range of $3\frac{1}{2}$-digit and $4\frac{1}{2}$-digit DMMs

Voltage Level (V)	Percent of Full Scale	Percent Uncertainty	
		$3\frac{1}{2}$ Digits $\pm(0.1\% + 1\text{ count})$	$4\frac{1}{2}$ Digits $\pm(0.1\% + 1\text{ count})$
1	5	1.1	0.2
5	25	0.3	0.12
10	50	0.2	0.11
19	95	0.15	0.105

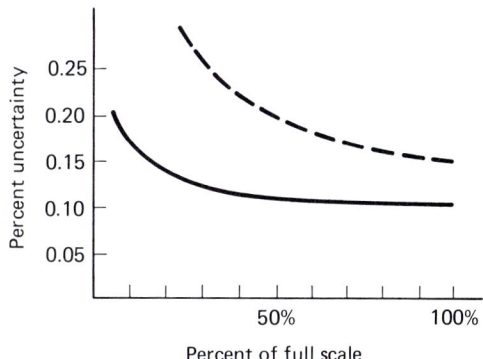

FIGURE 20-7 Reading uncertainty for $3\frac{1}{2}$- and $4\frac{1}{2}$-digit DMMs versus percent of full scale. Dashed line, $3\frac{1}{2}$ digits at $\pm(0.1\% + 1\text{ count})$; solid line, $4\frac{1}{2}$ digits at $\pm(0.1\% + 1\text{ count})$.

porated graphically in Fig. 20-7. It is valid for any voltage, current, or resistance range for any $3\frac{1}{2}$-digit or $4\frac{1}{2}$-digit DMM having the same specification.

EXAMPLE 20-8 A reading of 17152 is obtained on a $4\frac{1}{2}$-digit 20 kΩ range. The accuracy specification for the instrument is (0.1% + 1 count). Using Fig. 20-7, estimate the reading uncertainty.

Solution The reading of 17152 as a percentage of full scale is

$$17{,}152 \times \frac{100}{20{,}000} = 85.8\%$$

From Fig. 20-7, the percentage uncertainty is approximately 0.1%. Therefore, the reading uncertainty is

$$0.001 \times 17{,}152 = 17.2 \text{ Ω}$$

Digital multimeters may have the capability of directly measuring true rms values of alternating voltages and currents regardless of the waveform. The method of accomplishing this consists of squaring the voltage (or current), averaging the squared value, and extracting the square root. These operations correspond exactly to the rms (meaning root-mean-square) designation for effective value.

TABLE 20-2 Measurement values for common waveforms

DC-Coupled Input Waveform	DC Component Only	AC Component Only			Total
		Peak Responding	Average Responding	True RMS	True RMS
Sine	0.000	0.707P	0.707P	0.707P	0.707P
Rectified sine (full)	0.637P	0.707P	0.298P	0.308P	0.707P
Rectified sine (half)	0.318P	0.707P	0.390P	0.386P	0.500P
Square	0.000	0.707P	1.111P	1.000P	1.000P
Rectified square	0.500P	0.707P	0.555P	0.500P	0.707P
Pulse train ($D = W/T$)	DP	0.707P	$2.221(D - D^2)P$	$\sqrt{D - D^2}\,P$	$\sqrt{D}\,P$
Triangular	0.000	0.707P	0.555P	0.577P	0.577P

Table 20-2 summarizes the variety of measurement values possible for some of the more common waveforms. Whenever sine waves only are encountered, no particular care is necessary in making measurements provided that the frequency limitations and loading of the instrument are properly allowed for. The measured values will be directly observed regardless of the specific response (peak responding, average responding, or true rms) of the instrument. This is so because the DMM, analog electronic meter, or electromechanical meter is calibrated to indicate the effective value of a sine wave.

However, if the waveform differs in any significant manner from a sine wave with zero dc component, the meter readings or displayed digits cannot be relied on as being the effective (rms) value. Table 20-2 not only provides warnings regarding the potential dangers of improper measurement interpretations, but also suggests procedures for obtaining the information desired.

EXAMPLE 20-9 An average responding DMM is calibrated to display the effective value of a sine wave. A half-wave-rectified sine wave displays a reading of 3.472 V. What is the true rms value of the ac component?

Solution From Table 20-2, $3.472 = 0.390P$. Therefore,

$$P = 3.472/0.390$$

From Table 20-2, true rms = $0.386P$. Therefore,

$$\text{true rms} = 0.386 \times \frac{3.472}{0.390} = 3.436 \text{ V}$$

20-4
Cathode Ray Oscilloscopes

The key element of the CRO is the cathode ray tube, which is very similar to the "picture tube" of a black-and-white television set. A stream of electrons is produced by a heated cathode acting together with other metallic elements to form the electron gun. The stream of electrons is accelerated by high voltage toward the inside surface of the tube face. Light is produced when the electrons strike the fluorescent screen coating this inside surface.

Between the electron gun and the fluorescent screen are a number of electrodes held at positive potentials with respect to the cathode. Some of these serve the purpose of focusing the electron beam to a point at the fluorescent screen. A control on the front panel, to be described later, can be caused to improve the focus and affect the size and appearance of the spot seen by the operator. Other electrodes cause the electron beam to be deflected. One pair of electrodes, called plates, causes horizontal deflection proportional to the impressed voltage. Another pair causes vertical deflection in the same way.

In the most commonly used mode of operation, the horizontal deflection is linear. The beam, and spot that it produces, move equal distances in equal intervals of time. For slow movement, as caused by the varying plate voltages, the spot can be seen moving to the right, leaving a rapidly disappearing trail. Upon reaching the end of its travel on the right, the spot disappears and quickly reappears on the left to repeat its linear journey. With fast movements of the beam, required for most electronic measurements, the beam cannot be seen, but its trace, repeated hundreds, thousands, and even millions of times per second, appears as a solid horizontal line.

The vertical deflection of the beam is usually caused by an amplified signal to be observed and measured. The amount of amplification, hence deflection, is controlled by a knob on the CRO panel.

The combined horizontal sweep motion and vertical signal deflection give a graphical picture of the signal waveform. If the controls are set to their calibration positions, two possibilities for measurement are obtained.

1. The peak value of the waveform can be accurately measured.
2. The period of the waveform can be measured and the frequency calculated. Other time intervals can also be measured.

The remaining description of the CRO stresses the terminals and controls on the front panel. The present intention is to introduce the basic nature of the general capabilities of this class of instrument. The operations manual should be consulted for the particular make and model of oscilloscope you will be using before attempting to turn it on. By consulting the manual and examining the front panel you will observe several prominent features of a high-quality CRO. First, terminals and controls having associated functions are usually grouped in the same area on the panel (see Fig. 20-8). Second, control knobs, switches, and panel markings may be color coded. This feature is particularly helpful when two knobs, an inner knob and an outer knob, occupy the same position on the panel. Third, there may be one or more lights to inform you of the status of some functions. Many of the controls to be described are shown in Fig. 20-9.

POWER: a knob setting or switch that turns on the CRO. It may be marked on the panel as ON/OFF. It provides a connection of the ac power to the CRO power supplies. This control is often accompanied with an indicator light.

INTENSITY: controls the brightness of the spot, horizontal sweep trace, or waveform. The intensity setting should be high enough to view the waveform, but not too high. Care should be taken to obtain a low brightness when the electron beam is not moving. A high-intensity spot can burn the phosphorescent coating called the phosphor.

FOCUS: controls the size of the spot or width of line. The sharpest line is usually desired and is obtained when the line-generating spot is in focus.

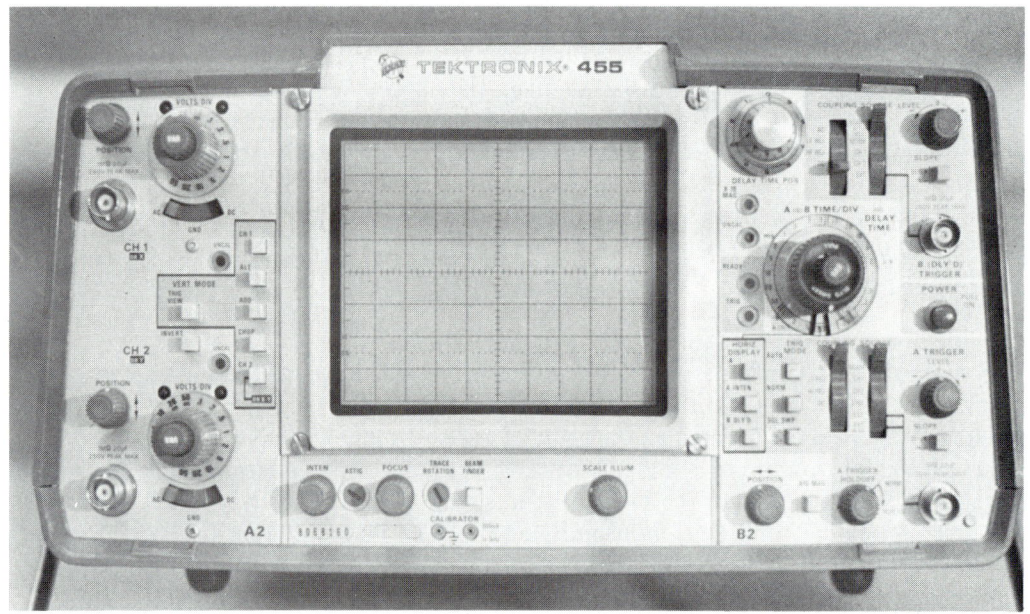

FIGURE 20-8 Dual-trace triggered-sweep oscilloscope.

FIGURE 20-9 Front panel of inexpensive cathode ray oscilloscope.

ASTIGMATISM: counteracts the tendency for the electron beam to defocus in some areas of the CRO tube face. This control, together with the FOCUS control, is used to obtain the sharpest line everywhere on the cathode ray tube face.

SCALE ILLUMINATOR: provides a variable level of illumination for the grid (called graticule) on the tube face.

HORIZONTAL POSITION: a control knob that moves the spot, sweep trace, or displayed waveform to the left or right horizontally.

VERTICAL POSITION: a control knob that moves the spot, sweep trace, or displayed waveform up or down vertically (lower right-hand knob of Fig. 20-10.

TIME/DIV: when set in the calibration position, the TIME/DIV control knob determines the time required by the sweep to travel the distance between two major division marks. On most modern oscilloscopes the major division marks are separated by 1 cm.

VOLTS/DIV: when set in the calibration position, the VOLTS/DIV control knob determines the voltage required to deflect the electron beam vertically one major division (usually 1 cm).

INPUT: the input terminal for the signal to be observed and measured. Test leads or a probe are attached to this terminal and to the circuit under test. Many scope probes have an attenuation factor of 10, which must be taken under consideration in using the vertical calibration of the oscilloscope to measure voltage levels. The proper connectors and/or adaptors must be used for the scope leads or probes.

INPUT SELECTOR: for the purpose of our experiments the dc selection is used to measure dc levels with the CRO. AC selection permits the measurement of ac signals without regard for any direct-current component that may be present. A GND selection places the input at ground level in preparation for centering or otherwise making VERTICAL POSITION adjustments.

Sec. 20-4 Cathode Ray Oscilloscopes

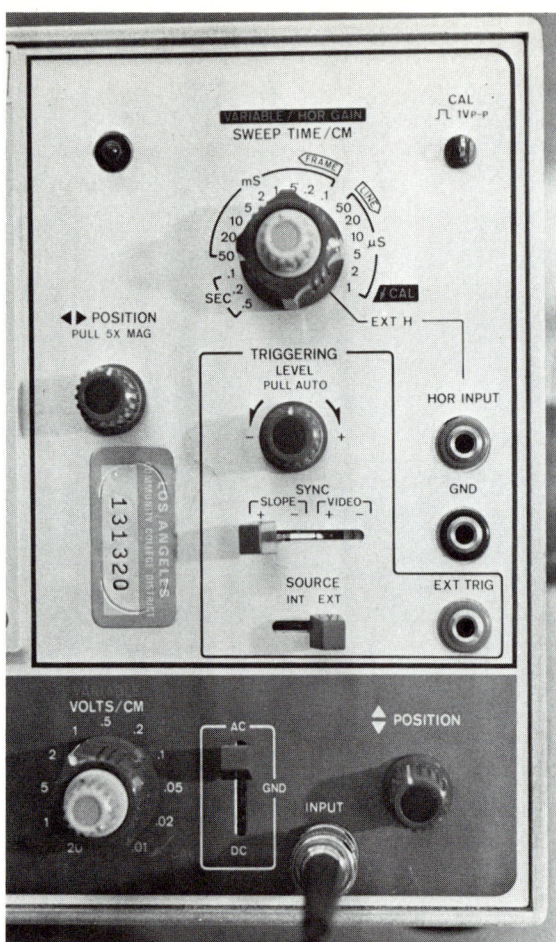

FIGURE 20-10 Control knobs and switches for cathode ray oscilloscope.

TRIGGER MODE: the INTERNAL trigger mode causes the horizontal sweep to be triggered by the input signal being observed and measured. After being started the sweep moves the spot from the left side of the cathode ray tube to the right. Note that the HORIZONTAL POSITION control determines exactly where the sweep trace begins and ends without changing its length. In the absence of an ac signal, the sweep will not trigger and no sweep trace will be seen when the INTERNAL trigger mode is selected.

AUTOMATIC TRIGGER: similar to INTERNAL. An important exception causes triggering of the sweep in the absence of any ac signal at the INPUT. This feature allows dc voltage level measurements. It also permits the trace and/or waveforms to be observed before other adjustments, not yet described, have been made.

EXTERNAL MODE SELECTION: results in the sweep being triggered by a signal other than that being observed on the CRO. The signal used for this purpose is connected to the EXTERNAL terminal of the CRO by means of a scope lead or probe. With the time of sweep triggering established by this means, phase comparisons can be made for all the voltages at points in a circuit.

TRIGGER LEVEL: establishes the voltage level on the waveform at which triggering occurs.

TRIGGER SLOPE: determines whether triggering occurs as the voltage is rising (+) or falling (−).

Next we consider the procedures for making simple measurements using the dual-trace, triggered-sweep oscilloscope.

EXAMPLE 20-10 A sine-wave display, as seen in Fig. 20-11, is obtained with the TIME/CM setting of 50 μs and a VOLTS/CM setting of 20, taking into consideration the normal use of a probe. It should be noted that the TRIGGER LEVEL and TRIGGER SLOPE controls have been adjusted so that the sine wave passes upward through the leftmost extreme end of the graticule center horizontal line. Determine the effective value and frequency of the sine wave.

FIGURE 20-11 Sine wave displayed by CRO for Example 20-10.

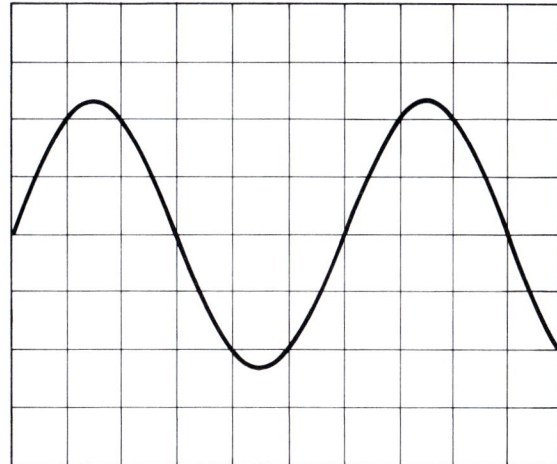

Solution From the figure the vertical distance from peak to peak is approximately 4.7 cm. Therefore, the peak-to-peak value is $4.7 \times 20 = 94$ V and the effective value is

$$V = 94 \times \frac{0.707}{2} = 32 \text{ V (rms)}$$

The period corresponds to 6 cm measured for one cycle on the horizontal scale or $6 \times 50 = 300$ μs.

$$f = \frac{1}{T} \qquad (11\text{-}1\text{-}1)$$

$$= \frac{1}{300} \times 10^6 = 3300 \text{ Hz}$$

EXAMPLE 20-11 Determine the effective value of the voltage displayed in Fig. 20-12 with the following control settings:

$$\text{VOLTS/CM} = 5 \text{ mV}$$

$$\text{TIME/CM} = 2 \text{ μs}$$

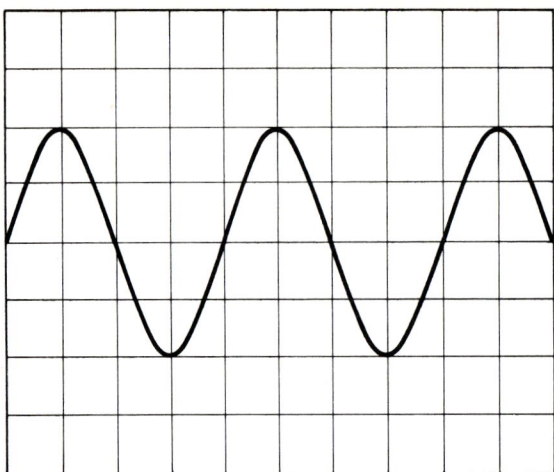

FIGURE 20-12 Sine wave displayed by CRO for Example 20-11.

Solution The peak-to-peak voltage is 4 × 5 = 20 mV. Therefore, the effective value is

$$20 \times \frac{0.707}{2} = 7.07 \text{ mV}$$

One cycle occupies a time span of approximately 4 cm, so that

$$f = \frac{1}{T} \tag{11-1-1}$$

$$= \frac{1}{4 \times 2 \times 10^{-6}} = 125{,}000 \text{ Hz}$$

Alternatively, $2\frac{1}{2}$ cycles occupy a span of approximately 9.9 cm, so that

$$f = \frac{2.5}{9.9 \times 2 \times 10^{-6}} = 126{,}000 \text{ Hz}$$

This second determination of the sine-wave period and frequency demonstrates the more accurate reading possible using the graticule when a single cycle is only a small fraction of the visible waveform.

EXAMPLE 20-12 In Fig. 20-13 the control settings are as follows:

$$\text{VOLTS/CM} = 0.2$$
$$\text{TIME/CM} = 1 \text{ ms}$$

Find the peak values, frequencies, and phase relationship of the two sine waves.

Solution It should first be understood that the two sine waves are of the same frequency and both appear stationary even though the sweep is triggered from the A waveform.

FIGURE 20-13 Sine waves displayed by CRO for Example 20-12.

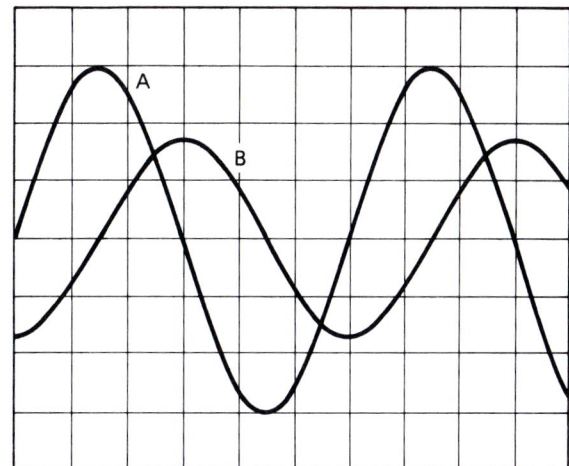

The peak-to-peak value of the A sine wave is

$$6 \times 0.2 = 1.2 \text{ V}$$

and the peak value is 0.6 V.

The peak-to-peak value of the B sine wave is approximately

$$3.6 \times 0.2 = 0.72 \text{ V}$$

and the peak value is 0.36 V. Using the A waveform, it is seen that one cycle spans approximately 6 cm or $6 \times 1 = 6$ ms. Therefore,

$$f = \frac{1}{6 \times 10^{-3}} = 167 \text{ Hz}$$

Waveform B peaks approximately 1.5 cm later than waveform A. Therefore,

$$\text{B lags A by } 1.5 \times \frac{360}{6} = 90°$$

Alternatively, A leads B by 90°.

EXAMPLE 20-13 The waveform seen in Fig. 20-14 was obtained with a VOLTS/CM setting of 20 mV and a TIME/CM setting of 1 μs with the INPUT SE-

FIGURE 20-14 Waveform with sine-wave and dc components for Example 20-13.

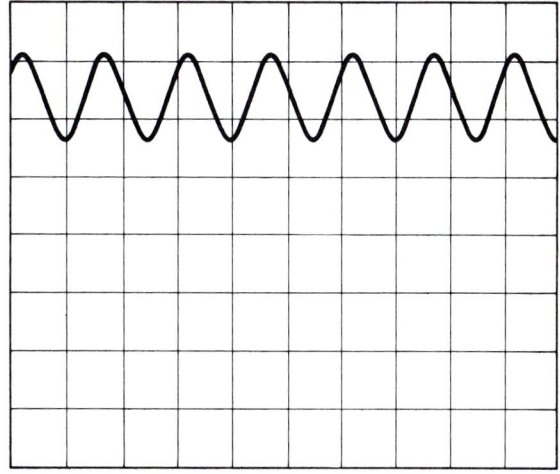

Sec. 20-4 Cathode Ray Oscilloscopes

LECTOR on dc after initially centering the trace with the AUTOMATIC trigger. Convert the information displayed to numerical values.

Solution The waveform displayed is that of a sine wave superimposed on a dc level. The sine-wave component has minimum points that are approximately 1.6 cm above the ground level and maximum points that are 3.1 cm above the ground level. The average input waveform level is

$$\frac{1.6 + 3.1}{2} = \frac{4.7}{2} = 2.35 \text{ cm}$$

and the peak-to-peak value is

$$3.1 - 1.6 = 1.5 \text{ cm}$$

Therefore, the sine-wave component is $20 \times 1.5 = 30$ mV peak to peak and the dc component is $20 \times 2.35 = 47$ mV. (One should be aware of the fact that a simple change of the INPUT SELECTOR control from dc to ac will cause the dc component to disappear from the oscilloscope face. The sine-wave component will then be centered.)

It is seen that two complete cycles span 3 cm in the horizontal direction. The frequency is then

$$f = \frac{1}{T} = \frac{2}{3 \times 1 \times 10^{-6}} = 667{,}000 \text{ Hz}$$

Chapter Summary

20.1 The D'Arsonval meter movement is adapted to ac voltage and current measurements by the addition of a half-wave or full-wave rectifier. The formulas to be remembered with a full-wave rectifier for the sine-wave voltage are

$$V_{av} = 0.637 V_P$$
$$V = 0.707 V_P$$
$$V = 1.11 V_{av}$$

in which V_{av} is the average value over one alternation, V_p is the peak value, and V is the effective value.

The formulas for current are the same and are found by substituting I when V appears in these same equations.

20.2. Some instruments are applicable for the direct measurement of dc and the effective value of ac voltage (or current). These include

Moving-iron meters
Electrodynamometer meters
Thermocouples with dc meter movement

Additionally, the electrodynamometer movement can be adapted to the direct measurement of power and energy consumed in ac circuits.

Self-Review

True-False Questions

1. T. F. The permanent-magnet moving-coil movement is also known as the D'Arsonval movement.
2. T. F. A permanent-magnet moving-coil movement is equally suitable for indicating the value of a dc current or an ac current passing through the coil.
3. T. F. An ac voltmeter using the D'Arsonval movement equipped with a rectifier circuit and meter scale calibrated to indicate the effective value of a sine wave will not provide accurate readings for a triangular or a square wave.
4. T. F. The average value of a single ac alternation is less than the effective value.
5. T. F. When the current reverses in the winding of a two-vane repulsion-type moving-iron meter, the magnetic flux lines will reverse direction in one (of two) vanes only.
6. T. F. The indication of an electrodynamometer meter movement connected as a wattmeter will be zero if the power factor of the load is unity.
7. T. F. A watthour meter indicates the consumption of energy.
8. T. F. The primary purpose of the ASTIGMATISM control on the CRO is to adjust the displayed waveforms so as to match the individual viewer's eyeglass corrections.
9. T. F. The addition of a rectifier circuit permits the permanent magnet (D'Arsonval) movement to be used in the measurement of ac voltage and current.
10. T. F. The connection of a conventional permanent magnet meter movement voltmeter, with rectifier, across two points of an ac circuit usually reduces the actual value of the voltage to be measured as a result of loading.
11. T. F. Application of ac to the D'Arsonval meter will cause needle deflection proportional to the effective (rms) value.
12. T. F. The simple equation $V_{EFF} = 1.11 V_{av}$ applies only for a sine wave.
13. T. F. A "peak responding" D'Arsonval meter movement circuit calibrated for sine-wave ac will give correct readings for triangular and sawtooth waves.
14. T. F. A thermocouple meter will give effective value readings for triangular-wave voltages.
15. T. F. The electrodynamometer meter will have its scale crowded at the lower end.
16. T. F. An electrodynamometer meter movement connected to measure power will have the current in one winding proportional to the load voltage.
17. T. F. The purpose of placing a rectifier in an ac instrument is to reverse the flow of current passing through a dc meter movement.
18. T. F. An "average responding" ac meter, when calibrated to indicate effective (rms) value for a sine wave, will give erroneous results when connected to a square wave.
19. T. F. The "resolution" of a DMM might be expressed in volts, millivolts, or microvolts, for example.
20. T. F. With the CRO, it is possible to measure both the peak value and period, hence frequency, of a sine wave.

Multiple-Choice Questions

21. A device which, when placed in series with a meter movement, will block flow in one direction while permitting flow in the opposite direction is called
    - (a) a rectifier
    - (b) a circuit breaker
    - (c) a resistor
    - (d) a metallic conductor
    - (e) no correct choice

22. The average value of a single alternation (one-half cycle) or of two alternations with the negative alternation inverted is
    (a) equal to the peak value
    (b) equal to the effective (rms) value
    (c) equal to 0.637 multiplied by the peak value
    (d) equal to 0.637 multiplied by the effective value
    (e) equal to 0.707 multiplied by the peak value
23. The electrodynamometer meter movement includes
    (a) a heating wire, thermocouple, and D'Arsonval meter movement
    (b) two magnetic vanes in an alternating field
    (c) two separate windings of which one is stationary
    (d) a full-wave rectifier and an electronic amplifier
    (e) a half-wave rectifier and a digital readout
24. An electronic voltmeter may have
    (a) an electronic amplifier   (b) a rectifier circuit
    (c) a D'Arsonval meter movement   (d) a high input impedance
    (e) all choices are correct
25. An electronic meter capable of measuring small electrical charges is named the
    (a) electrometer   (b) microvoltmeter
    (c) nanometer   (d) picometer
    (e) megameter
26. The fundamental process for obtaining the effective (rms) value of a voltage (or current) waveform in an electronic analog or digital meter consists of
    (a) squaring the input voltage
    (b) averaging the squared value
    (c) extracting the square root
    (d) all the preceding steps
    (e) at least one of the preceding steps is not necessary
27. A correct unit for the indicated measurement made by an electrodynamometer meter movement connected as a power meter is
    (a) volt   (b) ampere   (c) ohm   (d) watt   (e) joule
28. A correct unit for the indicated measurement of energy consumption is
    (a) volt   (b) ampere   (c) ohm   (d) watt   (e) joule
29. Select from among the following choices that one which does not have any function control effect on sweep triggering.
    (a) TRIGGER LEVEL   (b) TRIGGER SLOPE
    (c) FOCUS   (d) EXTERNAL mode selection
    (e) AUTOMATIC trigger
30. When the effective voltage applied to a fixed resistive load is *doubled*, the power indication of a correctly-connected electrodynamometer wattmeter will be changed by a factor of
    (a) 0.637   (b) 0.707   (c) 1 (no change)   (d) 2   (e) 4
31. The application of electronic amplifiers to electrical measurements can result in
    (a) fine tuning   (b) reduced loading
    (c) more reliability   (d) increased sensitivity
    (e) both (b) and (d)
32. A permanent-magnet moving-coil (D'Arsonval) movement with full-wave rectifier forms an "average-responding" meter that is calibrated to read effective (rms) voltages with sine waves. A 10-V (peak) square wave, when applied to this instrument, will cause an indication of
    (a) 5.55 V   (b) 6.36 V   (c) 7.07 V   (d) 10 V   (e) 11.1 V
33. A "peak-responding" ac meter, when calibrated for sine-wave ac voltage, is connected to a 20-V (peak-to-peak) square-wave voltage. The meter reading will be
    (a) 5 V   (b) 7.07 V   (c) 10 V   (d) 14.14 V   (e) 20 V
34. An electrodynamometer meter necessarily has a nonlinear scale. If the full-scale deflection on a particular range is 2.5 A, what will be the percentage scale deflection for a current of 1 A?
    (a) 2.5%   (b) 16%   (c) 25%   (d) 40%   (e) 62.5%
35. For a wattmeter, the torque produced by the interaction of the coil currents is

proportional to
- (a) effective load voltage
- (b) effective load current
- (c) power factor
- (d) eddy-current losses
- (e) (a), (b), and (c) are all correct

36. The primary purpose of adding an electronic amplifier to the analog meter in creating the EVM is to
    - (a) reduce instrument weight
    - (b) increase scale linearity
    - (c) reduce circuit loading
    - (d) lower costs
    - (e) all choices are correct

37. An instrument especially designed to have high input impedance and capable of detecting extremely small currents is named the
    - (a) electrometer
    - (b) dosimeter
    - (c) microvoltmeter
    - (d) vector voltmeter
    - (e) microminiature nanopicometer

38. A meter that has a maximum display of 19999 without regard for decimal-point position is rated as a
    - (a) $1\frac{1}{2}$-digit meter
    - (b) $2\frac{1}{2}$-digit meter
    - (c) $3\frac{1}{2}$-digit meter
    - (d) $4\frac{1}{2}$-digit meter
    - (e) $5\frac{1}{2}$-digit meter

39. The smallest incremental change that a DMM is able to detect is called the
    - (a) resolution
    - (b) sensitivity
    - (c) range
    - (d) accuracy
    - (e) calibration setting

40. The exactness to which a voltage, or other quantity, can be determined relative to the accepted standard is called the
    - (a) resolution
    - (b) sensitivity
    - (c) range
    - (d) accuracy
    - (e) calibration setting

Practice Problems

41. A sine-wave ac current has a peak value of 250 mA. When rectified using a full-wave rectifier, what will the average value be?

42. After full-wave rectification, the average value of a rectified sine wave is 8.07 V. Calculate the effective value of the original voltage sine wave.

43. A square wave of 50 V (peak) is applied to a peak-responding ac meter that is calibrated for the effective value of sine waves. Determine the percentage error in the indicated rms value of this square wave.

44. Calculate the sensitivity of a $5\frac{1}{2}$-digit laboratory multimeter on the (lowest) 20 mV range.

45. For the DMM and range given in Problem 44, what will be the largest number that can be displayed?

46. An average-responding DMM is calibrated to display the effective value of a sine wave. A full-wave-rectified sine wave displays a reading of 251.3 mV. What is the true rms value of the ac component only?

47. A 150-W incandescent lamp is turned on at 8:30 P.M. and off at 6:30 A.M. the next morning. Calculate the energy consumption as indicated by a watthour meter.

48. Refer to Fig. 20-11. Determine the peak value and the frequency of the sine wave for the following control settings:

$$\text{VOLTS/DIV} = 0.1$$

$$\text{TIME/DIV} = 1\ \mu\text{s}$$

49. A periodic waveform whose frequency is 47.5 kHz is to be observed with at least one whole cycle visible. Determine the setting of the TIME/DIV control to meet this requirement with the least number of cycles seen.

50. In Problem 49, the TIME/DIV control of the CRO is now adjusted to 0.2 ms/division. How many full cycles will be observed?

51. A D'Arsonval meter movement, calibrated for dc voltage, is combined with a full-wave rectifier for ac voltage measurement. Neglecting any voltage drop in the rectifier diodes, what will be the indicated voltage when 115 V sine-wave ac is being measured?

52. An "average-responding" D'Arsonval movement, with suitable rectifier, calibrated to give effective ac voltage readings with sine waves, is inadvertently connected across a 100-V dc source. Calculate the meter indication.

53. A thermocouple instrument correctly gives a reading of 12 V when a dc source of exactly 12 V is connected to its terminals. Calculate the expected reading when a 12 V (peak-to-peak) square wave is being measured with the same instrument.

54. The input impedance of an ac voltmeter with electronic amplifier is that of 10 MΩ and 30 pF in parallel. Calculate the high-frequency, half-power point for which the resistance and capacitive reactance are equal.

55. A reading of 1.5 kΩ is obtained using the 2 kΩ range of a $3\frac{1}{2}$-digit meter. Using Fig. 20-7 estimate the reading uncertainty.

56. A sine-wave display, as seen in Fig. 20-11, is obtained with a TIME/CM setting of 2 ms and a VOLTS/CM setting of 500 mV, taking into consideration the normal use of the probe. Determine the peak-to-peak value and the frequency.

57. Determine the effective value of the voltage displayed in Fig. 20-12 with the following control setting:

$$\text{VOLTS/CM} = 2 \text{ V}$$
$$\text{TIME/CM} = 5 \text{ } \mu\text{s}$$

58. In Fig. 20-13, the control settings are as follows:

$$\text{VOLTS/CM} = 500 \text{ mV}$$
$$\text{TIME/CM} = 20 \text{ } \mu\text{s}$$

Find the peak values of the two sine waves.

59. For the conditions of Problem 58, find the frequency and phase relationship of the waves.

60. The waveform in Fig. 20-14 was obtained with a VOLTS/CM setting of 2 V and a TIME/CM setting of 5 μs with the INPUT SELECTOR on dc after initially centering the trace using AUTOMATIC trigger. Convert the information displayed to numerical values.

Advanced Problems

61. A sine-wave alternating current is rectified (full-wave) so that the negative half-cycles (alternations) are inverted and then measured to be 400 mA using a dc current measuring instrument. Calculate the effective (rms) value of the original sine-wave current.

62. The average value of a rectified triangular wave is equal to one-half the peak value. Also, the effective value is 0.577 the peak value. Determine the percentage error when the measurement of the effective value of the wave is made with an instrument using the D'Arsonval dc movement with rectifier, the combination being calibrated for sine waves.

63. An electrodynamometer meter, when connected as an ac ammeter with both stationary and movable coils placed in series with the load, has full-scale deflection when the effective load current is 25 mA. Determine the total resistance, including that of the coils and multiplier resistor, such that the same meter can function as an ac voltmeter with a 250-V full-scale deflection.

64. A particular analog meter is arranged to have a linear decibel scale in which the range extends from -10 dBm on the left to $+2$ dBm on the right. A reference power level of 1 mW, with a resistance of 600 Ω, corresponds to 0 dBm. Calculate the voltage levels at the extreme ends of the decibel range.

65. Determine the energy consumption in kilowatthours for the following conditions:

Voltage (V)	Current (A)	Power Factor	Time (h)
120	3	0.9	3.5
120	2.9	0.85	1.7

66. The accuracy specification of a particular $5\frac{1}{2}$-digit true rms ac voltmeter is $+(0.3\% + 200$ counts). Determine the accuracy, expressed as the uncertainty of the reading, with an actual reading of 100 mV on the 200 mV range. Also state the result as a percentage of the reading.

67. Measurements are made on a pulse train as follows: dc component = 25 mV and total true rms voltage = 50 mV. Using the information given in Table 20-2, find the peak value P and the duty cycle D.

68. The waveforms visible on the dual-trace oscilloscope are shown in Fig. 20-15. Describe these waveforms for known control settings as follows:

$$\text{VOLTS/DIV} = 0.2 \text{ V}$$

$$\text{TIME/DIV} = 10 \text{ μs}$$

$$\text{INPUT SELECTOR: DC}$$

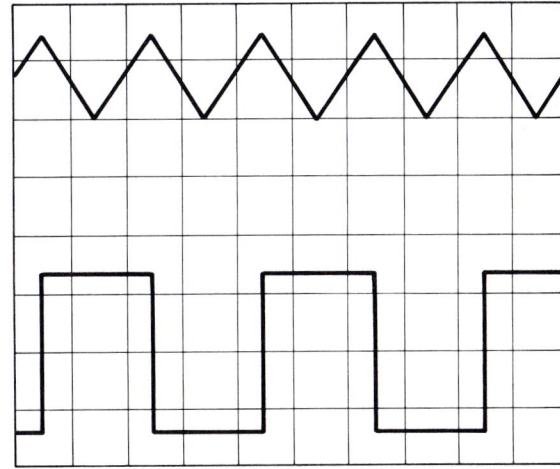

FIGURE 20-15 Waveforms displayed by CRO for Problem 68.

69. In Fig. 20-16, sine wave A is the voltage across a 10-kΩ resistor and sine wave B is the total voltage applied to the series circuit, which includes the 10-kΩ resistor and an unknown circuit element. Determine the nature and specific numerical value of the unknown reactive circuit element for oscilloscope control settings as follows:

$$\text{VOLTS/DIV} = 5 \text{ V}$$

$$\text{TIME/DIV} = 20 \text{ μs}$$

70. Rework Problem 69 for the conditions that sine wave A is the voltage across an unknown circuit element and sine wave B is the voltage across a 10-kΩ resistor in a series circuit.

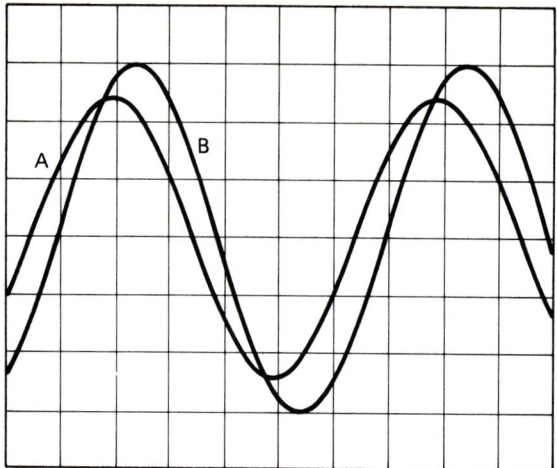

FIGURE 20-16 Waveforms displayed by CRO for Problems 69 and 70.

71. An "average-responding" ac meter is used to measure a triangular-wave voltage having a peak-to-peak value of 20 V. What will be the indicated voltage?

72. An "average-responding" ac meter is fabricated using a center-tapped transformer in the secondary and with two rectifier diodes. It is calibrated to read rms values for a sine wave. While reading 115 V ac, one diode fails and "opens" so that the current in the path is completely blocked. Determine the new reading of the meter.

73. A thermocouple meter indicates a reading of 48 V when a dc voltage of exactly 48 V is applied. Calculate the reading when a 48-V sine-wave ac voltage is applied after the negative alternations are eliminated by half-wave rectification.

74. When the windings of an electrodynamometer meter are connected in series for current measurements, full-scale deflection is achieved with 1 A. Determine the power that will cause full-scale deflection with the meter configured to measure power for which a 1000-Ω resistor is placed in series with the shunt winding.

75. Calculate the resolution of a $5\frac{1}{2}$-digit meter.

76. The lowest voltage range for the meter of Problem 75 is 200 mV. Determine the sensitivity.

77. The accuracy specification of a $3\frac{1}{2}$-digit DMM is (0.1% + 1 count). Determine the accuracy if a reading of 10 kΩ is taken on the 200 kΩ range. Also determine the resulting uncertainty as a percentage of the reading.

78. The reading of Problem 77 is repeated with the same DMM *except* that the reading is taken on the 20 kΩ range. Determine the accuracy. Also determine the resulting uncertainty as a percentage of the reading.

79. A full-wave rectified sine wave is measured with a "peak-responding" meter to obtain a reading of 7.12 V. What is the true rms value?

80. The reading obtained with an "average-responding" meter and a triangular wave is 12.34 V. What is the true rms value?

Appendices

A
SI and CGS Units. Conversion Factors.

Comparison of SI and CGS Units

Quantity	SI Units	Equivalent in CGS Units	
		Electromagnetic	Electrostatic
Length	1 meter	100 cm	100 cm
Mass	1 kilogram	1000 g	1000 g
Time	1 second	1 sec	1 sec
Force	1 newton	10^5 dynes	10^5 dynes
Work	1 joule	10^7 ergs	10^7 ergs
Power	1 watt	10^7 ergs/sec	10^7 ergs/sec
Current	1 ampere	0.1 abampere	3×10^9 statamperes
Charge	1 coulomb	0.1 abcoulomb	3×10^9 statcoulombs
EMF and PD	1 volt	10^8 abvolts	1/300 statvolt
Resistance	1 ohm	10^9 abohms	$1/(9 \times 10^{11})$ statohms
Inductance	1 henry	10^9 abhenrys	—
Capacitance	1 farad	—	9×10^{11} statfarads
Magnetomotive force	1 ampere-turn	0.4π gilberts	—
Magnetic field intensity	1 ampere-turn per meter	$4\pi \times 10^{-3}$ oersted	—
Magnetic flux	1 weber	10^8 maxwells	—
Magnetic flux density	1 tesla	10^4 gauss	—
Permeability of free space	$4\pi \times 10^{-7}$ henry per meter	1 unit	—
Electric field intensity	1 volt per meter	—	$\frac{1}{3} \times 10^{-4}$ unit/cm
Electric flux density	1 coulomb per meter^2	—	$12\pi \times 10^5$ units/cm^2
Permittivity of free space	8.85×10^{-12} farad per meter	—	1 unit

Conversion Factors

To Convert:	Multiply By:
radians to degrees	57.30
inches to meters	0.0254
yards to meters	0.9144
miles/hour to meters/s	0.447
pounds to kilograms	0.4536
pound weight (force) to newtons	4.45
kilogram weight (force) to newtons	9.81
horsepower to ft · lb/min	33,000
horsepower to ft · lb/s	550
horsepower to watts	746
ft · lb to joules	1.356
calories to joules	4.187
kWh to joules	3,600,000

Color Code for Low-Value Inductors

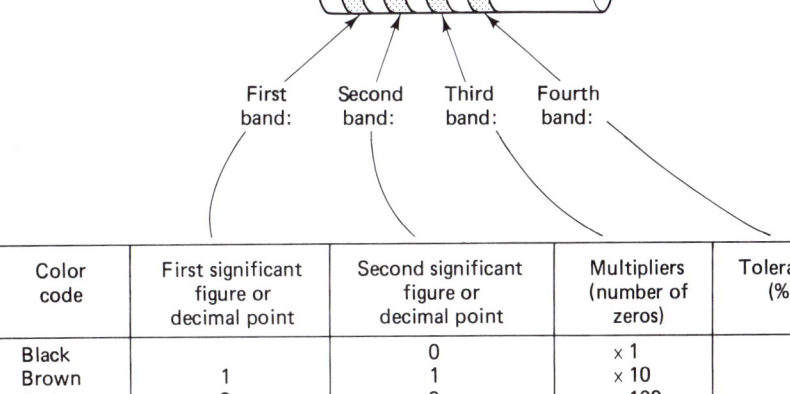

Color code	First significant figure or decimal point	Second significant figure or decimal point	Multipliers (number of zeros)	Tolerance (%)
Black		0	× 1	
Brown	1	1	× 10	
Red	2	2	× 100	
Orange	3	3		
Yellow	4	4		
Green	5	5		
Blue	6	6		
Violet	7	7		
Gray	8	8		
White	9	9		
Gold	Decimal point	Decimal point		±5%
Silver				±10%
No color				±20%

All inductor values indicated by the color code are measured in *microhenrys* (µH).

C

Capacitor Color Codes

The color coding of capacitors is the outcome of the composition resistor coding described in Section 1-8. However, as well as capacitance value and tolerance, a capacitor has a voltage rating and a temperature coefficient, so that its coding system is necessarily more complex. In addition, the color-coded capacitors are manufactured in a variety of types (paper, mica, ceramic) and shapes (rectangular, tubular, disk). Irrespective of type and shape, the color-coded value of the capacitance is always measured in picofarads.

The systems outlined in this appendix are based on EIA (Electronics Industries Association) and MIL (Military) standards.

1. Six-Band Code for Tubular Paper Capacitors

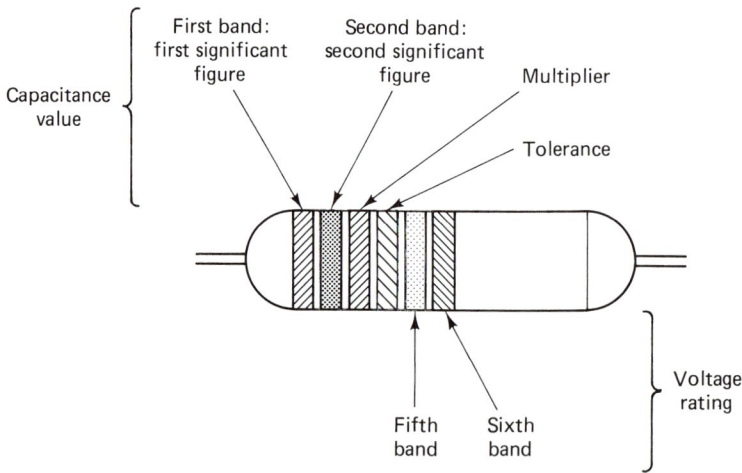

The first four bands are related to the capacitance value, while the fifth and sixth bands designate the voltage rating. If this rating is less than 900 V, there is no sixth band and the value of the fifth band is multiplied by 100. For example, a green fifth band would mean a voltage rating of $5 \times 100 = 500$ V. For ratings of 1000 V or more, the value indicated by the fifth and sixth bands is multiplied by 100; a 1000-V rating would be shown by brown (1) and black (0) bands, so that $10 \times 100 = 1000$ V.

2A. Six-Dot Code for Mica and Paper Molded Capacitors

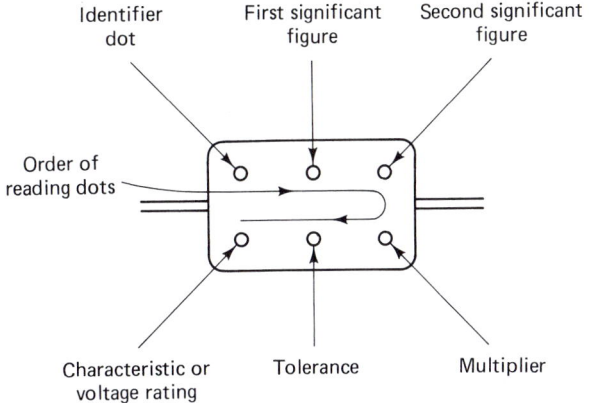

2B. Five-Dot Code for EIA Mica and Molded Paper Capacitors

In this system there is no identifier dot. Paper capacitors use a black case, while a dark brown or tan case means a mica dielectric; a higher-grade mica capacitor with silver surfaces has a red case.

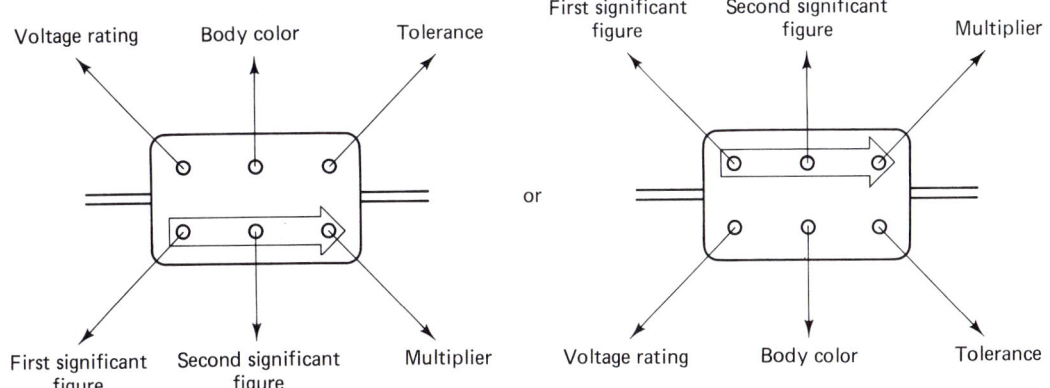

3. MIL Color Code for Ceramic Capacitors

A — Temperature coefficient
B — 1st digit
C — 2nd digit
D — Multiplier
E — Tolerance

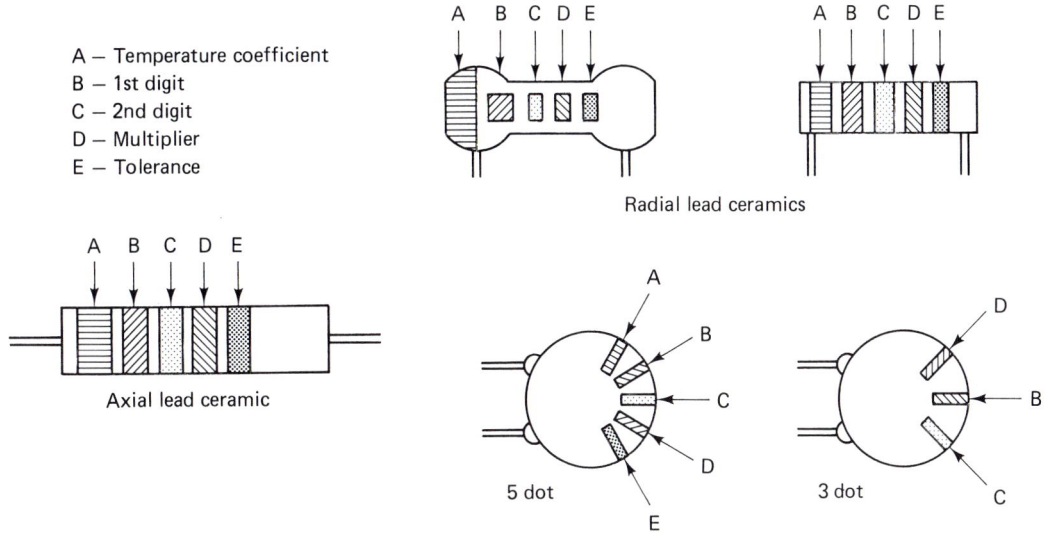

Radial lead ceramics

Axial lead ceramic

5 dot   3 dot

Ceramic disk capacitor marking

App. C Capacitor Color Codes

D Answers to End-of-Chapter Questions and Problems

Chapter 1 Self-Review

True–False Questions
1. F 2. T 3. T 4. T 5. T 6. T 7. T 8. F 9. T 10. F 11. T
12. F 13. T 14. F 15. T 16. F 17. F 18. F 19. F 20. F

Multiple-Choice Questions
21. (c) 23. (e) 25. (c) 27. (b) 29. (d) 31. (e) 33. (a) 35. (c)
37. (d) 39. (b)

Practice Problems
41. 1471.5 J 43. 0.833 A; 144 Ω; 0.3 kWh 45. 21.7 V 47. 3.9 Ω 49. 0.01779 kg
51. 2310 J 53. 53.3 mW 55. 9.825 Ω 57. 1.414 59. 1.15×10^{-6} m^2

Advanced Problems
61. 2.74 hp 63. 1.04 J 65. 101 V 67. 0.45 Ω 69. 108 W; 540 J
71. red, black, red 73. 69.8 m 75. +0.0036 Ω/Ω/°C 77. 13.6 °C 79. 1.293 mV

Chapter 2 Self-Review

True–False Questions
1. F 2. F 3. T 4. F 5. F 6. T 7. F 8. F 9. T 10. F 11. F
12. T 13. F 14. T 15. T 16. T 17. F 18. F 19. F 20. F

Multiple-Choice Questions
21. (d) 23. (a) 25. (c) 27. (a) 29. (a) 31. (e) 33. (a) 35. (a)
37. (c) 39. (c)

Practice Problems
41. 1876 Ω 43. 34 V 45. 2.8 W 47. 12 Ω 49. 3250 V 51. 25.6 mV
53. 33.3 W 55. 53.9 V 57. 4 W 59. 4.125 kΩ

Advanced Problems
61. 112 V 63. 12.96 Ω 65. +24 V 67. 40.9 mW
69. +10 V; +9.463 V; +8.806 V; +8 V 71. 1.5 W 73. +25.4 V; zero
75. (a) +1.6 V, +1.6 V, −7 V (b) +12 V, +12 V, −12 V 77. +20.8 V 79. −11.6 V

Chapter 3 Self-Review

True–False Questions
1. F 2. T 3. T 4. F 5. F 6. F 7. T 8. T 9. F 10. F 11. T
12. F 13. F 14. F 15. T 16. T 17. F 18. F 19. F 20. F

Multiple-Choice Questions
21. (d) 23. (a) 25. (b) 27. (c) 29. (d) 31. (c) 33. (d) 35. (a)
37. (d) 39. (e)

Practice Problems
41. 271 Ω; 3.7 mS 43. 1.11 mA; 83.0 mW 45. 40 W 47. 316 V 49. 73 mW
51. 34 kΩ 53. 2500 Ω 55. 12.9 mS 57. 972 Ω 59. 9.2 W

Advanced Problems
61. 4.73 mA; 10.4 mA; 232 mW 63. 286 Ω; 183 Ω 65. 16.5 mA 67. 7.4 W
69. 59.0 mA; 42.3 V 71. 15 W 73. 46.5 Ω 75. 1.43 kΩ 77. 3.22 kΩ 79. 7.2 V

Chapter 4 Self-Review

True–False Questions
1. F 2. T 3. T 4. F 5. T 6. T 7. T 8. F 9. T 10. F 11. F
12. F 13. T 14. F 15. F 16. F 17. F 18. F 19. F 20. F

Multiple-Choice Questions
21. (a) **23.** (d) **25.** (b) **27.** (a) **29.** (b) **31.** (a) **33.** (c) **35.** (b)
37. (b) **39.** (e)

Practice Problems
41. 18 mA **43.** 1.03 Ω; 870 W **45.** 6 V; 36 mW **47.** 333 Ω **49.** −70 V; −60 V
51. 4.5 kΩ **53.** 0.19 mA **55.** 5.45 mA **57.** 1.93 W **59.** +195.8 V

Advanced Problems
61. (a) 2.41 kΩ (b) 2.35 kΩ
63. 15 V; 225 mW; 22.5 V; 337.5 mW; 20.78 V; 196.2 mW; 10.01 V; 55.6 mW; 10.77 V; 24.7 V; 24.7 mW; 58.3 V
65. −10.77 V; −33.27 V; +10.01 V; +25.01 V **67.** zero; 1.37 mA **69.** 7.9 mA **71.** 100 Ω; 400 Ω; 400 Ω; 100 Ω **73.** 8 kΩ; 2.5 kΩ **75.** 10.7 V; 0.41 mA **77.** 65.7 V **79.** 438 Ω

Chapter 5 Self-Review

True–False Questions
1. F **2.** T **3.** T **4.** F **5.** T **6.** F **7.** F **8.** F **9.** T **10.** F **11.** F
12. T **13.** F **14.** T **15.** T **16.** T **17.** F **18.** F **19.** T **20.** T

Multiple-Choice Questions
21. (c) **23.** (b) **25.** (a) **27.** (c) **29.** (a) **31.** (a) **33.** (e) **35.** (d)
37. (b) **39.** (a)

Practice Problems
41. 172800 C **43.** 0.5 Ω; 5.5 Ω; 48 A; 0.5 Ω **45.** Voltage at point X is negative

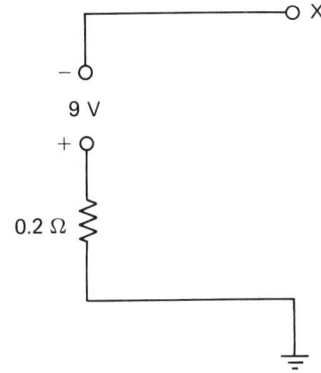

47. 4.8 Ω; 4.2 W
49.

R_L ohms	I_L amps	V_L volts	P_L watts
0	6	0	0
2	4	8	32
4	3	12	36
8	2	16	32
20	1	20	20
∞	0	24	0

51. 31.5 V **53.**

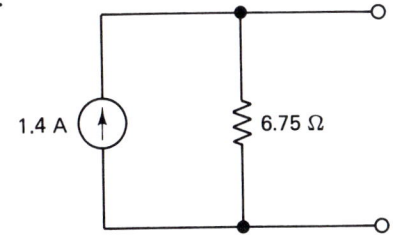

55. 13.2 V **57.** 0.48 Ω; 3.84 V. **59.** 20%

Advanced Problems
61. 40 Ω; 312.5 W **63.** +48 V **65.** 15.6 V; 19% **67.** 2.22 mA **69.** 2.12 mA; 18.1 mW; 71% **71.** 29.14 Ω; 85.4%; 0.858 Ω; 14.6% **73.** 0.091 A **75.** 72; 93.8 mA **77.** 39.2 W **79.** 7.2 mA

Chapter 6 Self-Review

True–False Questions
1. F **2.** T **3.** F **4.** T **5.** T **6.** T **7.** F **8.** F **9.** F **10.** T **11.** F
12. T **13.** T **14.** T **15.** F **16.** F **17.** T **18.** F **19.** F **20.** F

Multiple-Choice Questions
21. (a) **23.** (e) **25.** (a) **27.** (a) **29.** (a) **31.** (a) **33.** (c) **35.** (e)
37. (b) **39.** (e)

Practice Problems
41. +12 V **43.** −4.0 V **45.** 2 mA **47.** −9 V **49.** 1.26 Ω; 1.89 Ω; 2.84 Ω

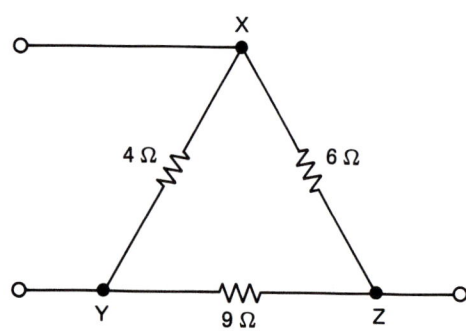
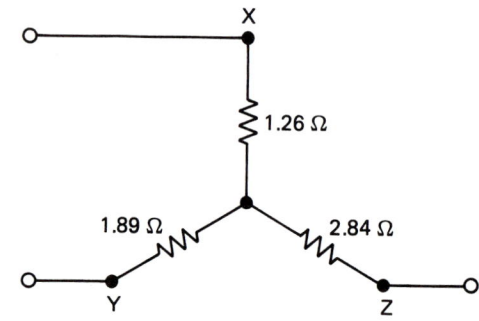

51. 3 mA; 6 mA **53.** 7.27 V **55.** 1.67 mA **57.** $I_N = 0.61$ mA; $R_N = 1052$ Ω

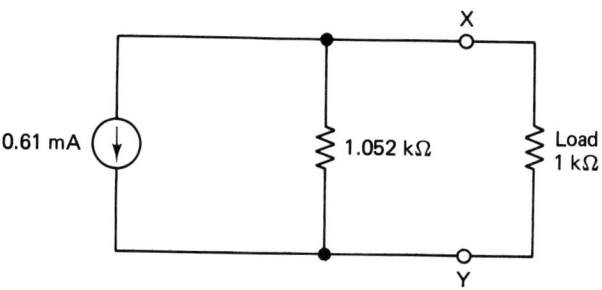

59. 6.67 kΩ; 4.44 kΩ; 13.33 kΩ

Advanced Problems
61. 83.3 Ω **63.**

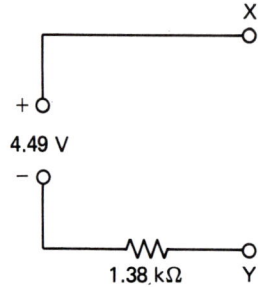

65. 7.01 V; 1.79 kΩ

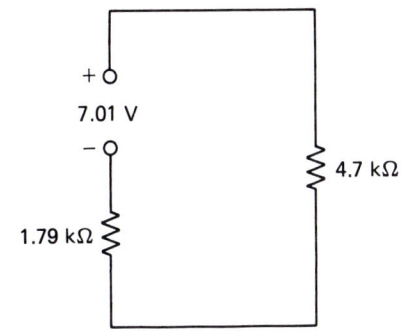

67. 1.11 mA; 2.03 kΩ

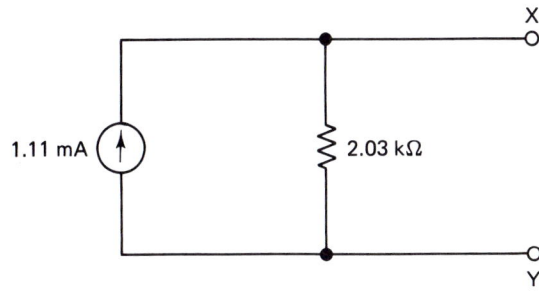

69. 2.64 kΩ   **71.** 0.75 mA
73.

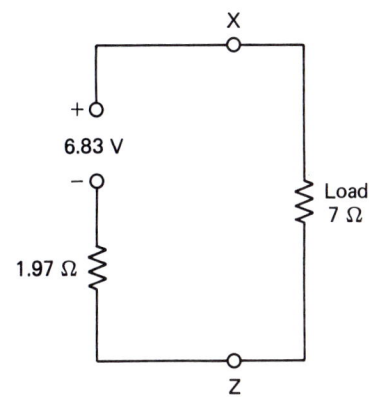

75. −3.1 mA   **77.** 2.59 mA
79.

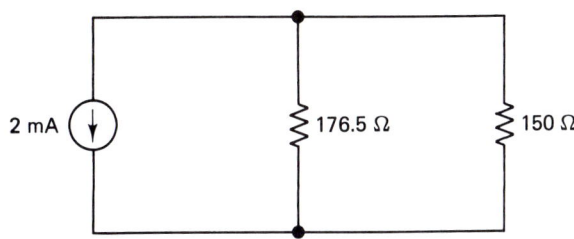

Chapter 7 Self-Review

True–False Questions
1. F 2. F 3. T 4. F 5. F 6. T 7. F 8. T 9. T 10. T 11. F
12. F 13. T 14. F 15. T 16. F 17. F 18. F 19. F 20. F

Multiple-Choice Questions
21. (d)   **23.** (b)   **25.** (a)   **27.** (c)   **29.** (a)   **31.** (a)   **33.** (a)   **35.** (e)   **37.** (d)   **39.** (b)

Practice Problems
41. 1.59×10^6 SI units   **43.** 4.97×10^5 SI units   **45.** 7.5×10^{-4} SI units   **47.** 36 V   **49.** 11.6 A   **51.** 1.15 A   **53.** 694 SI units   **55.** 0.24 Wb   **57.** 2514   **59.** 117 µWb

Advanced Problems
61. 3737 AT   **63.** 3.2 A   **65.** 6667 AT/m; 0.00838 T; 2.1 µWb   **67.** 1.2 N.m   **69.** 0.71 A   **71.** 25 cm   **73.** 0.0014 T   **75.** 71 turns   **77.** 213 A.T   **79.** 4.96 A

Chapter 8   Self-Review

True–False Questions
1. F   **2.** T   **3.** F   **4.** F   **5.** T   **6.** F   **7.** T   **8.** T   **9.** F   **10.** T   **11.** F   **12.** F   **13.** T   **14.** T   **15.** T   **16.** F   **17.** T   **18.** F   **19.** T   **20.** T

Multiple-Choice Questions
21. (e)   **23.** (b)   **25.** (a)   **27.** (c)   **29.** (e)   **31.** (a)   **33.** (b)   **35.** (a)   **37.** (e)   **39.** (b)

Practice Problems
41. 50 V   **43.** 15 H   **45.** 6 J   **47.** 8 V   **49.** 0.5 A   **51.** 22.5 mWb   **53.** 4 kΩ   **55.** 9630 turns   **57.** 22 mH

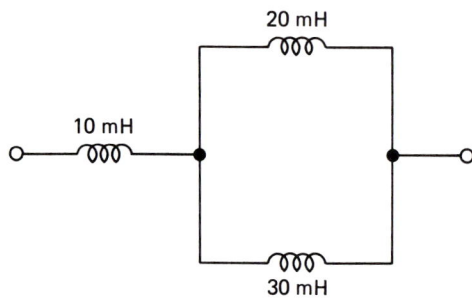

59. 1.6 H

Advanced Problems
61. 3.33 H   **63.** 20 mH; 10 V   **65.** 3.08 H; 2.30 mH   **67.** 0.126 A; 0.158 A; 0.196 A; 0.2 A   **69.** 404 A/s   **71.** 0.06 s   **73.** 4 mH   **75.** $I_2 = I_T = 6$ mA, $I_1 = 0$ mA

Chapter 9   Self-Review

True–False Questions
1. F   **2.** T   **3.** F   **4.** T   **5.** T   **6.** F   **7.** F   **8.** T   **9.** F   **10.** T   **11.** F   **12.** T   **13.** F   **14.** F   **15.** F   **16.** T   **17.** F   **18.** F   **19.** T   **20.** T

Multiple-Choice Questions
21. (d)   **23.** (b)   **25.** (b)   **27.** (e)   **29.** (c)   **31.** (d)   **33.** (d)   **35.** (c)   **37.** (d)   **39.** (e)

Practice Problems
41. 71 pF   **43.** 160000 V/m; 1.4175 µC/m²   **45.** 1080 µC; 540 µC; 97200 µJ; 48600 µJ   **47.** +100 V; zero V; +20 V; 0.4 ms   **49.** 4 µF   **51.** 0.0633 µF   **53.** 158 µs   **55.** 235 V; 78 V; 47 V   **57.** 100 V   **59.** 500 pF

Advanced Problems
61. 0.19824 µC; 8.0×10^5 V/m; 24.78×10^{-6} C/m²   **63.** 230.5 V; 1101250 µJ; 1089055 µJ   **65.** 36000 µJ   **67.** Initial: 15.23 mA; 6.28 mA; 8.95 mA; Final: 8.69 mA; zero mA; 163.6 µC   **69.** 34.3 V   **71.** 39.6 V; 438 ms   **73.** 16.4 V; 14.4 V; 30.8 V   **75.** 0.1 µF   **77.** 1.416 mm   **79.** 1000 times per second; 100000 times per second

Chapter 10 Self-Review

True–False Questions
1. F 2. F 3. T 4. T 5. T 6. F 7. T 8. T 9. F 10. F 11. T
12. F 13. F 14. T 15. T 16. F 17. F 18. T 19. T 20. T

Multiple-Choice Questions
21. (c) 23. (c) 25. (d) 27. (b) 29. (e) 31. (a) 33. (d) 35. (e)
37. (c) 39. (b)

Practice Problems
41. $0.505\ \Omega$; $0.5\ \Omega$ 43. 40 V 45. 2.75×10^{-5} N.m 47. $900\ \text{k}\Omega$ 49. 40 W
51. $3\ \text{k}\Omega$ 53. $1071.2\ \Omega$ 55. $120\ \text{k}\Omega$ 57. $350\ \text{k}\Omega$ 59. 16 A

Advanced Problems
61. $1.6875\ \text{k}\Omega$ 63. 13.3 V; 40 V 65. $1.11\ \Omega$; $10\ \Omega$ 67. 1.04 V; 4.8 V 69. 0.51%
71. 25.6 mA 73. $2.4\ \text{M}\Omega$ 75. $40000\ \Omega/\text{V}$ 77. $533.33\ \Omega$; $122.22\ \Omega$; $11.11\ \Omega$
79. 8 mA to 12 mA; ±20%

Chapter 11 Self-Review

True–False Questions
1. T 2. F 3. F 4. T 5. T 6. F 7. T 8. T 9. F 10. T 11. F
12. T 13. T 14. T 15. F 16. T 17. F 18. F 19. F 20. F

Multiple-Choice Questions
21. (d) 23. (b) 25. (d) 27. (b) 29. (a) 31. (d) 33. (c) 35. (d)
37. (e) 39. (a)

Practice Problems
41. 100 mA 43. 2 kHz 45. 13.3 V 47. $126\ \Omega$ 49. 15.9 MHz 51. 25 kHz
53. 2500 Hz 55. 10 W 57. 0.0318 H 59. $360\ \Omega\ (X_L)$

Advanced Problems
61. 5 kHz 63. 137 W (peak); 0 (min.) 65. 5.3 mH 67. 70.7 V 69. 2210 Hz
71. 10^4 Hz 73. 25 kHz 75. 2.5 A; 0 77. 4.55 mA 79. 10.4 mA

Chapter 12 Self-Review

True–False Questions
1. T 2. F 3. T 4. T 5. F 6. F 7. T 8. F 9. F 10. F 11. F
12. T 13. F 14. T 15. F 16. T 17. F 18. T 19. F 20. F

Multiple-Choice Questions
21. (a) 23. (d) 25. (c) 27. (e) 29. (d) 31. (c) 33. (e) 35. (d)
37. (e) 39. (e)

Practice Problems
41. 70.8 V (peak); 19.0° 43. 70.2 V (peak)
45. 110 V (peak); 187.5 V (peak); 217 V (peak); −59.6°; $4.35\ \text{k}\Omega$ 47. $180\ \text{k}\Omega$; 33.7°
49. 85.4 mA; −20.6°
51. 10 V 53. $9460\angle 29.9°\ \Omega$ 55. 50 kHz
57. 0.412 mA; 20.6 V; 12.3 V 59. 23.6 mA

Advanced Problems
61. 25,500 Hz 63. $288\ \Omega$; 56.3° 65. 717 Hz 67. 1.76 pF 69. $6.26\ \text{k}\Omega$
71. $1.32\ \text{k}\Omega$ 73. $4.46\angle 45.9°\ \text{k}\Omega$ 75. $75\angle -53.1°$ V 77. 1590 Hz 79. 0.24 mA

Chapter 13 Self-Review

True–False Questions
1. F 2. T 3. F 4. T 5. T 6. F 7. F 8. F 9. T 10. F 11. F
12. T 13. T 14. T 15. F 16. F 17. T 18. F 19. F 20. F

Multiple-Choice Questions
21. (c)   **23.** (a)   **25.** (b)   **27.** (c)   **29.** (d)   **31.** (d)   **33.** (b)   **35.** (d)
37. (b)   **39.** (b)

Practice Problems
41. 600 W; 1200 vars; 1340 VA   **43.** 259 W   **45.** 510 VA; 100 vars   **47.** 3.54×10^{-8} F
49. 300 Ω   **51.** 375 W   **53.** 4.84 Ω; 4.33 Ω   **55.** 0.707 VA   **57.** 0.894   **59.** 60 V

Advanced Problems
61. 275 W; 306 VA; 135 vars   **63.** 3560 Hz   **65.** 12,600 W; 3150 vars
67. $50 - j6$ Ω; 78.1 W   **69.** 4.14×10^{-4} F   **71.** 0.6 Ω   **73.** 8 Ω; 6.16 Ω
75. 1.44 mH   **77.** 2828 VA   **79.** 10 V; 253 μH; 50 Ω

Chapter 14 Self-Review

True–False Questions
1. T   **2.** T   **3.** T   **4.** T   **5.** F   **6.** F   **7.** F   **8.** T   **9.** F   **10.** F   **11.** F
12. T   **13.** F   **14.** F   **15.** T   **16.** T   **17.** T   **18.** T   **19.** T   **20.** F

Multiple-Choice Questions
21. (b)   **23.** (d)   **25.** (a)   **27.** (b)   **29.** (b)   **31.** (c)   **33.** (e)   **35.** (c)
37. (d)   **39.** (e)

Practice Problems
41. 159 kHz   **43.** 4.05 mH   **45.** 88.4 kHz   **47.** 48.2 kHz   **49.** 3180 Ω   **51.** 198 pF
53. 0.75 mA   **55.** 50 kHz; 10 kHz   **57.** 26 dB   **59.** 30 kHz

Advanced Problems
61. 20 kHz   **63.** 159×10^{-12} H; 6.37×10^{-12} F   **65.** $R_s = 100$ Ω   **67.** 13.6 kHz
69. -20 dB   **71.** 9.55×10^{-6} H; 3.38×10^{-10} F   **73.** 39.3 kΩ   **75.** $+14.0$ dB
77. 10 kHz and 100 kHz or 1 decade for 20 dB change.   **79.** -80 dB

Chapter 15 Self-Review

True–False Questions
1. F   **2.** T   **3.** F   **4.** F   **5.** F   **6.** F   **7.** F   **8.** F   **9.** T   **10.** F   **11.** T
12. F   **13.** T   **14.** F   **15.** T   **16.** T   **17.** T   **18.** F   **19.** T   **20.** T

Multiple-Choice Questions
21. (c)   **23.** (a)   **25.** (e)   **27.** (e)   **29.** (c)   **31.** (b)   **33.** (b)   **35.** (a)
37. (b)   **39.** (d)

Practice Problems
41. 0.12 V/turn   **43.** 0.75 A   **45.** 50   **47.** 1.73 mH   **49.** 17.3 mH   **51.** 0.1 V/turn
53. 6.25 A   **55.** 628 Ω   **57.** 0.05 H   **59.** 91%

Advanced Problems
61. 10 mH   **63.** 10 mH   **65.** 2000 Ω; 250 Ω   **67.** 4.47 mA   **69.** 286 μH
71. 1250 turns; 625 turns; 6.25 V   **73.** 18.75 A   **75.** 0.318 A   **77.** 15.9 mH
79. 3.40 mH

Chapter 16 Self-Review

True–False Questions
1. T   **2.** T   **3.** F   **4.** T   **5.** T   **6.** F   **7.** T   **8.** T   **9.** T   **10.** T   **11.** T
12. F   **13.** T   **14.** F   **15.** F   **16.** F   **17.** T   **18.** T   **19.** T   **20.** F

Multiple-Choice Questions
21. (a)   **23.** (e)   **25.** (e)   **27.** (d)   **29.** (b)   **31.** (b)   **33.** (a)   **35.** (b)
37. (b)   **39.** (c)

Practice Problems
41. $j2$; $j9.6$; $j7/2$; jA **43.** $-391 + j921$ **45.** $11 - j7$ **47.** $24.7\angle 55.3°$ **49.** $159\angle 105°$
51. $15\angle 36.9°$ **53.** $32.56\angle 132.5°$ **55.** $-1.231 + j0.8642 = 1.4936\angle 145.5°$
57. $-14.4 - j6.45 = 15.8\angle -156°$ **59.** $24 + j13 = 27.3\angle 28.4°$

Advanced Problems
61. $8.36 \angle 26.3°$ mA **63.** 45.3 mA; 21.1 mA **65.** $-174 - j195$
67. $35.1 - j17.0 = 39.0 \angle -25.8°$ **69.** $32,500 \angle -117°$ **71.** 15 mA
73. $21.4 \angle -16.2° = 20.6 - j5.98$ V **75.** 253 pF; $Z = 37.7 + j0$ kΩ
77. $-60.9 - j21.5 = 64.5 \angle -160.5°$ **79.** $2.27 - j0.175$ A; $4.48 - j8.57$ V

Chapter 17 Self-Review

True–False Questions
1. F **2.** F **3.** F **4.** T **5.** F **6.** T **7.** T **8.** F **9.** T **10.** T **11.** F
12. T **13.** F **14.** F **15.** T **16.** F **17.** T **18.** F **19.** F **20.** F

Multiple-Choice Questions
21. (d) **23.** (a) **25.** (a) **27.** (b) **29.** (d) **31.** (a) **33.** (d) **35.** (e)
37. (e) **39.** (e)

Practice Problems
41. $3 - j2$ mA **43.** $1 + j1$ mA **45.** $-2 + j3$ mA **47.** $\frac{2}{3} + j0$ kΩ **49.** $\frac{2}{3} + j0$ kΩ
51. $-2.46 + j1.47$ mA; $-1.35 + j3.78$ mA **53.** $-18.8 + j9.89$ V **55.** $1.11 + j2.30$ V

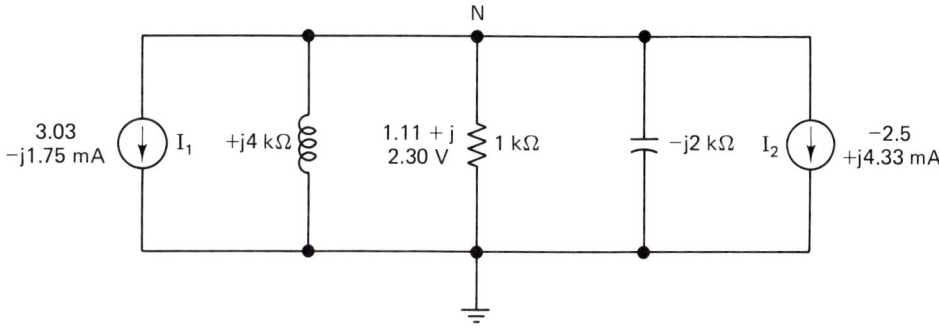

57. $0 + j1000$ Ω; $1000 + j0$ Ω; $0 - j1000$ Ω **59.** $4.15 + j0.939$ V

Advanced Problems
61. $1.26 - j0.523$ mA **63.** $1.09 + j0.0275$ mA **65.** $\frac{1}{2\pi\sqrt{6}\,RC}$ Hz; $-\frac{1}{29}$
67. $15.7 + j34.8$ mA **69.** $0.832 + j0.383$ kΩ **71.** $2.97 + j3.47$ mA
73. $-18.6 + j6.29$ V **75.** $-4.77 - j6.74$ mA; $5.16 + j7.25$ mA **77.** $1.68 + j0.723$ mA
79. $0.667 + j0$ kΩ; $0 - j0.667$ kΩ; $0.67 - j0.667$ kΩ

Chapter 18 Self-Review

True–False Questions
1. T **2.** T **3.** F **4.** F **5.** F **6.** F **7.** T **8.** T **9.** T **10.** F **11.** T
12. T **13.** F **14.** F **15.** T **16.** T **17.** F **18.** T **19.** F **20.** F

Multiple-Choice Question
21. (b) **23.** (d) **25.** (b) **27.** (d) **29.** (d) **31.** (a) **33.** (e) **35.** (c)
37. (e) **39.** (b)

Practice Problems
41. 2165 V **43.** 220 V **45.** 75 A **47.** delta-connected **49.** 0.693 A **51.** 60 Hz
53. 15 kW **55.** 23.2 A **57.** 2.61 A **59.** 110 V

Advanced Problems
61. 6600 rpm **63.** wye-connected alternator and delta-connected load
65. Apparent power = Real power = 5 kW; Reactive Power = 0 **67.** 16.4 A
69. 691 W **71.** 59 Hz **73.** 20.8 A; 29.5 A **75.** 14.7 A
77. Group 1: 4.4 for whole secondary winding; Group 2: 7.62 for whole secondary winding
79. 3.13 A

Chapter 19 Self-Review

True–False Questions
1. T 2. F 3. F 4. F 5. T 6. T 7. F 8. T 9. T 10. F 11. T
12. T 13. F 14. T 15. T 16. F 17. F 18. T 19. T 20. F

Multiple-Choice Questions
21. (b) 23. (a) 25. (b) 27. (b) 29. (d) 31. (e) 33. (e) 35. (d)
37. (e) 39. (b)

Practice Problems
41. 15.9 V 43. 28.4 V; 3.15 V; 1.13 V; 0.579 V 45. 4.27 V 47. 5.73 V
49. $-7.96 \cos 377t - 1.99 \cos 754t - 0.884 \cos 1131t$ mA; 5.83 mA 51. 5.066 V
53. 31.8 V 55. 0 57. 15.88 V 59. 0.3 A

Advanced Problems
61. 27.7 V; 0.383 W 63. 13.7 V; 13.7 V 65. 118 V 67. $16.05 \sin(628t + 51.5°) + 54.07 \sin(1256t + 68.3°) - 46.78 \sin(1884t + 75.1°)$ mA; 51.8 mA; 0.644 W
69. $45{,}200 \sin(628t - 63.73°) + 40{,}100 \sin(1256t - 4.06°) - 33{,}900 \sin(1884t + 45°)$ μA; 49.0 mA
71. 408 mV 73. $18.75 - 15.2 \cos(6.28 \times 10^4)t - 1.69 \cos(18.8 \times 10^4)t + 23.9 \sin(6.28 \times 10^4)t - 11.9 \sin(12.6 \times 10^4)t + 7.96 \sin(18.8 \times 10^4)t + \ldots$ V
75. Table with effective value and error for last harmonic:

Last Harmonic	Effective Value	Error (percent)
Fund.	0.5732 V	0.42
3rd	0.5767 V	0.07
5th	0.5771 V	0.02
7th	0.5773 V	0.01
9th	0.5773 V	0.005
11th	0.5773 V	0.003

77. Table of frequency for high-pass filter:

Frequency	V_0/V_{IN}
Fund.	0.9003 ∠45°
3rd	0.4026 ∠18.43°
5th	0.2497 ∠11.31°
7th	0.1801 ∠8.13°
9th	0.1406 ∠6.34°
11th	0.1153 ∠5.19°
13th	0.0977 ∠4.40°
15th	0.0847 ∠3.81°

79. $V_0 = -\left(\dfrac{4D}{\pi\omega}\right)\left(\cos\omega t + \dfrac{\cos 3\omega t}{3^2} + \ldots\right) = -\left(\dfrac{-\pi D}{2\omega}\right)\left(\dfrac{8D}{\pi^2}\right)\left(\cos\omega t + \dfrac{\cos 3\omega t}{3^2} + \ldots\right)$ V

The output is an inverted triangular wave whose peak value is $\dfrac{\pi D}{2\omega}$ V.

Chapter 20 Self-Review

True–False Questions
1. T 2. F 3. T 4. T 5. F 6. F 7. T 8. F 9. T 10. T 11. F
12. T 13. F 14. T 15. T 16. T 17. F 18. T 19. F 20. T

Multiple-Choice Questions
21. (a) 23. (c) 25. (a) 27. (d) 29. (c) 31. (e) 33. (b) 35. (e)
37. (a) 39. (b)

Practice Problems
41. 159 mA 43. -29% 45. 19.9999 47. 1.5 kW-hr. 49. 5 μs/div. 51. 103.5 V
53. 6 V 55. 0.175% 57. 2.83 V 59. 8333 Hz; "B" lags "A" by 90°

Advanced Problems
61. 444 mA 63. 10,000 Ω 65. 1.645 kW-hr. 67. 100 mV; 0.25 69. $X_C = 7.5$ kΩ
71. 5.55 V 73. 24 V 75. 5×10^{-6} 77. 0.1 kΩ; 3% 79. 3.10 V

Index

Absolute permeability, 297
Absolute permittivity, 365
Acceleration, 3–4
   unit (m/s^2), 3
Accuracy, meter, 420, 766
Ac current:
   effective, 459–60, 758–63
   peak, 457–60, 758
Ac electrical measurement, 756–76
Ac generator, 452–53, 704–15 (See also Alternator)
Algebraic sum, 229–31, 736
Alnico, 279–80
Alternating:
   electromotive force (EMF), 608
   magnetic flux, 608
   magnetic field, 762
Alternating currents (alternating current circuits), 22, 437–73
   in a capacitor, 469–72
   full-wave rectified, 462, 742
   fundamentals of, 444–52
   half-wave rectified, 462, 742
   in an inductor, 463–69
   power and effective value of, 457–63
   power circuits (See Power circuits)
   resonant circuits (See Resonant circuits)
   sine-wave, 452–57
   source, 703
   triangular waveform, 742
   waveforms of, 450–52, 735–36
Alternations, 457, 461, 758
   average value, 463
Alternator (Ac generator), 703–15
   with Δ-connected windings, 718–21
   revolving field, 703
   single phase, 703–7
   three-phase, 711–15
   two-phase, 703, 707–10
   with Y-connected windings, 715–17

American Wire Gage, 37–38
Ammeter, 28, 421–24, 757–69
   milliammeter, 421–22
   moving-iron, 432–33, 761
Ampacity, 35
Ampere-hour (Ah), 23, 198
Ampere, 23–24, 289
Amplitude of sine wave, 474, 734, 737
Angular velocity (frequency), 474, 709
Anode, 22–23
Apparent power, 534–35
Approximation, 737
Armature, 312–13, 703
Atomic numbers, 17
Atomic structure, 14–21
   electron, 15
   neutron, 17
   nucleus, 15
   proton, 15
Atomic weights, 17
Atoms, 15, 17–21
Attenuation, 579–80
Attraction moving-iron meter, 433, 761
Average:
   current value (I_{av}), 758
   responding meter, 759
   voltage value (V_{av}), 758
Ayrton shunt, 423–24

Back-off scale, 429
Balanced load(s), 707, 709, 711–19
Bandpass filter, 557
Bandwidth, 557
Bar magnets, 279–81
Batteries:
   capacity, 198
   lead-acid (secondary), 196–200
   primary dry, 194–96
Bleeder current, 80
Bode plot, 581–83
Branch currents, 114–15, 665
   current division rule, 134–36

Bridge circuits, 171–73, 248–49, 253–55, 686
Brush (carbon), 310, 703

Capacitance, 363–404
   definition of, 363
   stray inductance and, 388
Capacitive reactance, 470–71
   series ac circuit with resistance and, 488–93
Capacitance checker, 388
Capacitor(s), 366–68
   alternating current in, 469–72
   ceramic, 367
   charging of a, 379–82
   color code, 786–87
   compression, 369
   dielectric hysteresis, 388
   dielectric of, 363–66
   discharging of a, 382–88
   electrolytic, 367–68, 387
   energy stored in, 370–71
   internal inductance of, 388
   mica, 366–67
   multiplate, 374–75
   with an open circuit, 389
   paper, 366
   in parallel, 373–75
   partial short-circuit, 389–90
   physical factors, 364–65
   power factor of, 388
   in series, 371–73
   short-circuit, 389
   troubleshooting, 463–65
   types of, 439–43
   variable, 368–69
Capacity (See Capacitance)
Cascaded filter sections, 586
Cathode, 22, 769
Cathode ray oscilloscope, 759, 769–76
   controls, 769, 770–72
   deflection, 769
   electron gun, 769

Cathode ray oscilloscope *(cont.)*
  terminals, 771
Cathode ray tube, 769
  fluorescent screen, 769
Cells (*See also* Voltage sources)
  dry primary, 194–96
  in parallel, 138–39, 213–15
  secondary, 196–200
  in series, 88–89, 211–13
Centrifugal force, 15
Ceramic capacitors, 367
Charge density, 360–61
Charge(s), 23–24
  for capacitor with ac current, 470
  forces of attraction and repulsion, 357–60
  instantaneous value, 470
  quantity of electricity, 23
  sine wave, 470
  units of, 23
Chord distribution factor, 706
Circuit breaker, 52
Circuit efficiency, 205–7
Circuits (*See also* Networks; Parallel Circuits; Series circuits; Series-parallel circuits)
  clockwise, 234, 634, 638
  magnetic versus electrical, 298–99
  mutually coupled (*See* Mutually coupled circuits)
  open, 92–93, 97–100, 140–41, 179–80
  power (*See* Power circuits)
  resonant (*See* Resonant circuits)
  short, 99–100, 142–43, 145, 179–80
Circular mil, 37
Coefficient of coupling, 613–18
Coercive force, magnetic curves and, 301
Coil, 329–31 (*See also* Inductors)
  magnetic field, 333
  self-inductance of, 329–31
Color code:
  bar magnet poles, 280
  capacitors, 786–87
  low value inductors, 785
  resistors, 48–49
Commutator, 310
Complex number, 632
Composition resistors, 44–49
Compression capacitors, 369
Condensers (*See* Capacitors)
Conductance, 33–34, 652
  total equivalent, of parallel circuits, 127, 651
Conductivity (σ), 38–39
Conductors, 20–21
Constantan, 37, 49
Constant current:
  generator, 671
  source 212–13, 667
Continuity, 93
Conventional current, 22
Coordinate conversion, 632–37
  polar to rectangular, 635–37
  rectangular to polar, 632–34

Copper loss, 606
Coulomb, 23
Coulomb's law, 359
Counterclockwise direction, 627
Counter EMF, inductance and, 325
Cosine:
  and power factor, 535–36
  terms, 741
CR circuits, time constant of, 453–61
Cryogenics, 42
Current(s) (*See also* Kirchhoff's current law)
  actual or physical electron flow of, 22
  alternating current, 660–64
  branch, 114–15, 648
  carrying capacity, 35
  conventional or mathematical, 22
  effective, 459–60, 758
  instantaneous value, 453
  line, 718–27
  peak, 457–58, 758
  sine-wave, 457
  source, 212–13
  time rate of charge movement, 470
  units of, 23–24
Current division rule, 134–36
Current flow in a series circuit, 61–64, 481–93
Current sources:
  constant, 212–13
  Millman's theorem and, 239–40, 667–69
Current-squared meter, 434, 761
Cycle, 448
Cycle per second (hertz), 448–49

D'Arsonval (moving coil) meter movement, 417–21, 757–59
Dc components, 771
Dc generators (dynamos), 309–13
  output voltage of, 310–13
Decibel, 578–81
Degree, 453–57, 472–73
Delta(Δ)-connection, 718–21
Delta-wye transformation, 256–61, 686
Demagnetization, 303
Diamagnetic materials, 296–97
Dielectric, 363–67
Dielectric constant (*See* Relative permittivity)
Dielectric hysteresis, 388
Dielectric strength, 365–66
Difference of potential (*See* Potential difference)
Differentiation, 390–95
  square wave, of, 394–95
  sawtooth wave, of, 398–99
Differentiator circuits:
  application, 399–400
  LR, 397–98
  RC, 395–97
Digital meter, 765–69
Digital multimeter, 765–69
  accuracy, 766–67

  number of digits, 765
  overrange feature, 791
  resolution, 765
  sensitivity, 765
Direct current (dc), 22
Direct current (dc) circuits, inductance in (*See* Self-inductance in dc circuits)
Discontinuity, 93
Dissipation, 31–32
Division of numbers, 8–9
Domain theory of magnetism, 280–82
Dropping resistors, 94–96
Dry primary cell, 194–96
Dynamos (*See* Dc generators), 309–13

E (symbol), 25
Eddy currents, 606
Effective (RMS) value:
  alternating current, 457–63, 758–63
  nonsinusoidal wave, 743–44
Efficiency, 206–10, 696–707
Electrical filter, 567–88
Electric field intensity (electrostatic field strength), 360–62
Electric flux, 357–59
Electrically connected, 614–18
Electrochemical equivalents, 23–24
Electrodes, 22
Electrodynamometer meter movement, 435–36, 761–63
Electrolysis, 22–23
Electrolyte, 192, 198
Electrolytic capacitor, 367–68, 387
Electrolytic conduction, 22–23
Electromagnetic induction, 304–8
Electromagnetism, 288–93
Electromagnets, 293
Electromechanical meters, 757–63
Electrometer, 764
Electromotive force (EMF), 25–26
Electron current flow, 22, 660
Electronic:
  amplifier, 757, 763
  voltmeter, 763–64
Electron(s), 15
  charge, 23
  free, 20
Electron-volt (eV), 25
Electroplating, 22
Electrostatic induction, 358
Electrostatics, 356–62
Elements, 15–16
EMF (electromotive force), 25–26
Energy:
  unit (joule (J)), 54–55
  watthour, 7, 763
E_P, 597–606
E_S, 597–606
Exponential growth (or decay):
  in CR circuits, 380–83
  in LR circuits, 342–46
Exponents, 7–11

negative, 9
zero, 9

Faraday's laws
electrolysis, 22–23
electromagnetic induction, 304–8
Ferromagnetic core, 612
Field intensity, (See Magnetic field intensity or Electric field intensity)
Film resistors, 50
Filter:
bandpass, 557, 574–77
bandwidth, 557–88
cutoff frequency, 571–72
half-power points, 557–72
high-pass, 572–74
LC section, 583
low-pass, 567
L-sections, 584
π-section, 584
ratio, (V_{OUT}/V_{IN}), 567–83
T-section, 584
Flux density:
electric, 361
magnetic, 286–88
magnetization curve and, 296–97
relativity permeability, 296–97
relative permittivity, 365–66
Flux linkage, 330
Force:
electromotive, 25–26
unit [newton (N)], 3–4
Free electrons, 20
Free space:
permeability of, 295–96
permittivity of, 361–62, 364–65
Frequency of a waveform (f), 448–52
Frequency response, 580–83
Full-scale deflection current, 420
Fuses, 50–52

Generator effect, 304–5
Generators:
ac, 455–57 (See Alternator)
dc, (See Dc generators)
Giga-, 11
Graph, 571–72, 581–83
Gravity, 4
Ground:
in series circuits, 81–85
in series-parallel circuits, 168–71
as voltage reference level, 83

Heat, 31
Helium, 17
Henry, 327–28
Henry per ampere-turn, 298
Hertz (Hz), 448–52
Horizontal component, 637
Horizontal retrace (TV), 400
Horsepower, 6
Hydrogen, 15–16, 20
Hypotenuse, 482
Hysteresis:
dielectric, 388
loop, 301–3

loss, 606
magnetic, 300–303, 606

Imaginary number (j), 629
Impedance, 484–88
linear, 673
load, 677
Inductance:
alternating currents in, 463–69
definition of, 324
in parallel arrangement, 335–36
in series arrangement, 334–35
internal, of capacitors, 388
mutual, 604–14
resistance compared to, 324–28
skin effect, 564
stray capacitance and, 388
Induction:
electromagnetic, 304–8
electrostatic, 358
magnetic, 284–85
Inductive reactance, 466–69
series ac circuit with resistance and, 481–88
Inductor(s), 323 (See also Self-inductance in dc circuits)
alternating current in, 463–69
chokes, 331
color code, 785
decay of current in, 343–46
energy stored in magnetic field of, 333
growth of current in, 340–46
open circuits and short circuits in the coil winding of, 348–49
in parallel arrangement, 335–36
in series arrangement, 334–35
slugs of, 331
troubleshooting, 348–49
types of, 331
Inert gases, 18–19
Insulators, 21
Integration, 401
square wave, of, 401–2
Integrator circuits:
applications, 404–6
LR, 404
RC, 402–4
Internal resistance:
of constant current source, 212
of constant voltage source, 200
of dry primary cells, 196
of lead-acid battery, 199–200
Ions, 20
Iron, 280
Iron disk, 433
Iron vane, 432–33, 761
IR voltage, 31
Isotope, 18

J operator, 626–31
Joule, 4–5

Kelvin (K) temperature scale, 42
Kilo-, 11
kilogram, 2
kilowatthour, 7

Kirchhoff's current law (KCL), 115, 231, 229–33, 660
mesh analysis and, 234–36, 665
Kirchhoff's voltage law, 229, 660
mesh analysis and, 290–93
Kirchhoff's voltage law, 847

LC filter section, 583–88
L-section, 584
π-section, 584
T-section, 584
Lead-acid batteries, 196–200
Leakage flux, 294
Leakage resistance of capacitor, 387–88
Left-hand rule, 292
Left-hand generator rule, 306
Lenz's law, 307–9, 326, 328
Linear meter scale, 757
Linear mil, 37
Linear resistance, 229
Line currents, 718–27
Line neutral, 710–15
Line(s), 707–27
Line voltages, 715–27
Linkage flux, 611–12
Loaded voltage divider circuit, 174–75
Load power, maximum, 205–7, 536–41
Load resistance, 201–2
matching resistance, 205–7
Loads, 112
Logarithm:
common, 578
logarithmic scale, 581
LR circuits, time constant, 405–11

Magnetic circuits:
electrical circuits compared to, 298–99
reluctance of, 297–98
Magnetic field(s):
around a coil, 291–93
flux lines, 282–86, 612
energy stored in, 333
linkage, 612
patterns of, 282–84
Magnetic field intensity (magnetizing force), 294–96
magnetization curve and, 300–301
permeability and, 296–98
Magnetic flux, 282–86 (See also Flux density; Leakage flux)
Magnetic induction, 284–85
Magnetic material, 612
Magnetic poles, 280
Magnetic shielding, 298
Magnetism, 279–86 (See also Electromagnetism)
domain theory of, 280–81
residual (or remanent), 301
Magnetization curve, 301–2
Magnetomotive force (MMF), 293
Magnets:
artificial, 279
natural, 279

Magnets (cont.)
　permanent, 280–82
　temporary, 280
Mass unit (kilogram), 2
Matching, 205–7
Mathematical expressions, 741
Maximum, 453
Maximum load power, 205–7, 536–41
Maximum power transfer:
　dc, 205–7
　ac, 536–41
Mega-, 11
Mesh analysis, 234–36, 665–67
　equations, 665
Meter sensitivity, 419–20
Meter, unit of length, 2
Mica capacitors, 366–67
Micro-, 11
Microvoltmeter, 425
Milli-, 11
Millivoltmeter, 425
Millman's theorem, 239–40, 667–69
Minimum, 452
Modified sawtooth waveform, 741–42
Modulator unit (radar), 419–20
Molecules, 15, 18, 20
Motor effect, 290–91
Moving-iron instrument, 432–33, 761
Moving-coil (D'Arsonval) meter movement, 417–21, 757–64
Moving-iron ammeter and voltmeter, 432–33, 761–62
Multiplication of numbers, 8
Multiplier, 759
Multiplier resistor, 425–26
Mutual inductance, 604–14
　aiding or opposing, 614–18
Mutually coupled circuits, 595–618
　coefficient of coupling, 613
　mutual inductance in, 608–14
　transformers, 596–608

Nano-, 11
Nanometer, 764
Negative direction, 446
Negative ions, 20
Networks:
　delta-wye or wye-delta transformations, 256–61, 685–86
　Kirchhoff's laws, 229–33, 660
　mesh analysis, 234–36, 665–67
　Millman's theorem, 239–40, 667
　nodal analysis, 237–38, 670–72
　Norton's theorem, 252–55, 681–5
　superposition theorem of, 242–43, 673–77
　Thévenin's theorem, 245–50, 677–81
Neutral line, 708
Neutron, 17
Newton, 3–4
Newton-meters, 5–6
Nodal analysis, 237–38, 670–72
Node, 237

Nonlinear meter scale, 433–35, 761
Nonsinusoidal current:
　in series circuit, 744–46
　in series-parallel circuit, 746–49
North pole, 280
Norton:
　equivalent source, 252, 681
　impedance (Z_N), 681, 683
Norton's theorem, 252–55, 681–85
N_P, 597–606
N_S, 597–606
Nucleus, 15

Ohmmeter, 97–98, 429–31
　capacitor defects detected with, 389–90
　finding an open circuit with, in parallel circuits, 141
　finding an open circuit with, in series circuits, 97–98
Ohm's law, 27–33
　memory aid for, 29, 33
Open circuit(s):
　capacitors with, 389
　in the coil winding of inductors, 348–49
　in parallel circuits, 140–41
　in series circuits, 97–98
　in series-parallel circuits, 179–80
Open-circuit terminal voltage, 201
Opposition to ac current flow,
　in an inductor, 466
　in a capacitor, 471

Paper capacitors, 366
Parallel circuits:
　branch currents in, 114–15, 132, 495–501
　capacitors in, 373–75, 493–500, 512–23
　combining branch currents, 495–501
　current division rule and, 134–36
　identical cells connected in, 138–39, 213–15
　inductors in, 335–36, 493–96
　open circuit's effect on, 140–41
　potential difference, 112–14
　power relationships in, 128–29, 136
　practical parallel resonant circuit, 565
　practical voltage sources in, 213–15
　product-over-sum formula for total equivalent resistance of, 120–21, 133–34
　reciprocal formula for total equivalent resistance of, 118–19, 133
　resonant, 559–67
　RL and RC circuits, 481–501
　RLC circuits, 515–23
　R_P, 601–6
　R_S, 601–6
　short circuit's effect, 142–43
　step-by-step analysis, 132–34
　total current drawn from the voltage source in, 114–15, 132–33, 495
　total equivalent conductance of, 127
　total equivalent impedance, 493, 499, 501–23
　total equivalent resistance of, 117–21, 132–34
　troubleshooting, 143–45
　two resistors in, 120–21
Parallel resonance, 559–67
Paramagnetic materials, 296
Peak:
　current value (I_P), 758
　responding meter, 759
　voltage value (V_P), 758
Period (T), 447–52, 473
Periodic table of the elements, 15
Periodic waves, 447–52, 734
Permanent magnet field, 280–82, 757
Permeability:
　absolute, 297, 301
　of diamagnetic materials, 296–97
　of free space, 295–96
　of paramagnetic materials, 296
　relative, 296–97, 301–2
Permeance, 298
Permittivity:
　absolute, 365
　of free space, 361–62
　relative, 365
Phase angle(s), 474
　determination, 483
Phase currents, 709–26
Phase difference, 469
Phase voltage(s), 712–27
Phasors (phasor analysis), 469, 707–27, 626–53
　addition and subtraction of, 639–42
　diagram, 481–523
　j operator, 626–31
　multiplication and division of, 642–49
Pico-, 11
Pointer deflection, 417–21
Polar coordinates for impedance resistor and inductor in series, 485
Polarity, 230
Polarization in primary dry cells, 196
Polar notation, 485, 630
Poles of magnets, 280
Positive ions, 20, 22
Positive direction, 446
Potential difference, 30–31
　across parallel resistors, 112–14
Potentiometers, 49–50
　in series circuits, 85–86
Power, 6–7, 457
　apparent, 534–35
　average, 459
　dissipated in a resistor, 31–32, 457
　factor, 535–36
　half values of, 557
　instantaneous value, 457, 713
　peak, 458
　reactive, 532–33

transfer, maximum, 205–7, 536–41
unit [watt (W)], 6
Power circuits, 532–36
Δ-connection in, 718–21
Y-connection in, 715–17
Power dissipation in parallel circuit, 128–29
Power dissipation in series circuit, 70–72
Power factor, 535–36
Power losses of transformers, 606
Power supply, 463, 600
Precision resistors, 50
Prefixes in electronics, 11–12
Primary (winding of transformer)
power (P_P), 599
voltage (E_P), 597–606
winding reactance (X_P), 608–9
Primary dry cells, 194–96
Product-over-sum formula, 120–21, 133–34
Proton(s), 15

Radian, 473
Radio receivers, variable capacitors of, 368–69
Radius of circle, 472
RC circuits, parallel, 497–501
RC circuits, series, 488–93
Reactance:
capacitive (*See* Capacitive reactance)
inductive (*See* Inductive reactance)
Reactive power, 533–34
Real number, 632
Reciprocal formula, 118–19, 133
Reciprocals, 9, 499
Reciprocity theorem, 689–91
Rectangular conversion of phasors, 631–39
Rectangular coordinates:
for impedances of L and R in series, 485
for current, 632
Rectifier, 463, 757–58
full-wave, 459, 742, 758
half-wave, 459, 742, 758
Regulation, 203
Relative permeability, 296–97, 301
Reluctance of a magnetic circuit, 297–98
Repulsion moving-iron meter, 433
Residual (or remanent) magnetism, 301
Resistance:
for ac, 457–63
of a cylindrical conductor, 35
inductance compared to, 324–28
internal, of constant voltage source, 201–3
internal, of dry primary cells, 196
internal, of lead-acid battery, 199–200

load (*See* Load resistance)
in Ohm's law, 28–29
series ac circuit with capacitive reactance and, 488–93
series ac circuit with inductive reactance and, 481–88
temperature coefficient, 40–42
total equivalent, of a parallel circuit, 118–21, 133
total equivalent, of a series circuit, 67–68
Resistivity (ρ) (*See* Specific resistance)
Resistors:
color code, 48–49
composition, 44–49
delta-wye transformation of, 256–61
film, 50
fixed, 49
in parallel:
(*See* Parallel circuits)
potential difference across, 30–31
power dissipated in, 31–32
precision, 50
in series (*See* Series circuits)
series-parallel arrangement of (*See* Series-parallel circuits)
series voltage-dropping, 94–96
tapped, 49
thin-film, 50
tolerance of, 44–48
variable (*See* Potentiometers)
wirewound, 49–50
Resonant circuits, 538–41
maximum power transfer in, 538
parallel, 559–67
series, 548–59
Resonant frequency, 551, 559
Rheostat, 49–50
in series circuits, 49–50
Right-hand rule, 290, 306–7
Right triangle, 482, 534
Ripple, 713
RLC circuits:
parallel, 512–23, 559–67
series, 501–12, 549–58
RL circuits, parallel, 488–93
RL circuits, series, 481–88
Root-mean-square (rms) voltage or current, 560, 743–44
Rotor, 705
Rowland's law, 297–98

Saturation, 301
Sawtooth wave, 741–42
Scale, 581–83
Scientific notation, 7–11
coefficient, 7
division of numbers, 8–9
exponent, 7
multiplication of numbers, 8
Secondary cells, 196–200
Secondary winding:
load power (P_S), 599

reactance (X_S), 608–9
voltage (E_S), 597–606
Seebeck effect, 434
Self-inductance in dc circuits, 324–48
energy stored in magnetic field of inductor and, 333
open circuits and short circuits in the coil winding and, 348–49
physical factors that determine, 339–41
resistance and inductance, comparison between, 324–28
series and parallel arrangements of inductors and, 334–36
time constants of LR circuits and, 340–46
Semiconductor, 21
diode, 758
Sensitivity, meter, 419–20, 765–66
Series-aiding, voltage sources connected in, 88–89
Series:
Fourier, 734, 740–43
infinite, 741
Series circuits:
dc circuits with resistors, 74–135
ac circuit with resistance and capacitive reactance, 488–93
ac circuit with resistance and inductive reactance, 481–88
ac RLC circuits, 501–12
capacitors in, 371–73
cells in, 88–89, 211–13
current flow in, 61–64, 481–93
ground in, 81–85
inductors in, 334–35
open circuit's effect on, 97–98
potentiometer in, 85–86
powers dissipated by the resistors in, 70–82
resonant, 548–58
rheostat in, 86–87
RLC circuits, 501–12
series-aiding or series-opposing voltage sources in, 88–89
step-by-step analysis of, 73–76
total equivalent impedance, 484–88, 501–12
total equivalent resistance of, 67–68
troubleshooting, 97–100
voltage divider, 78–80, 567–74
voltage division rule and, 76–77
voltage drops in, 64–66
Series-opposing, voltage sources connected in, 89
Series-parallel circuits:
complex algebra for analysis of, 650–53
ground connections in, 81–85
loaded voltage divider circuit, 174–75
open circuit's effect on, 179–80
power dissipation in, 158–66
short circuit's effect on, 217–20
step-by-step analysis of, 163–66

Index 801